Experiments

Experiments

Planning, Analysis, and Parameter Design Optimization

C. F. Jeff Wu   Michael Hamada

A Wiley-Interscience Publication
JOHN WILEY & SONS, INC.
New York • Chichester • Weinheim • Brisbane • Singapore • Toronto

Copyright ©2000 by John Wiley & Sons, Inc. All rights reserved.

Published simultaneously in Canada.

Library of Congress Cataloging in Publication Data:

Wu, Chien-Fu Jeff.
   Experiments: planning, analysis, and parameter design optimization / C.F. Jeff Wu, Michael Hamada.
     p. cm.-- (Wiley series in probability and statistic)
   Includes bibliographical references and index.
   ISBN 0-471-25511-4 (alk. paper)
   1. Experimental design. I. Hamada, Michael, 1955- II. Title. III. Series

  QA279.W7 2000
  519.5--dc21   99-052840

To my parents and Jung Hee, Christina, and Alexandra
MH

To my mother and family
CFJW

Contents

Preface

Statistical experimental design and analysis is an indispensable tool for experimenters and one of the core topics in a statistics curriculum. Because of its importance in the development of modern statistics, many textbooks and several classics have been written on the subject, including the influential 1978 book *Statistic for Experimenters* by Box, Hunter, and Hunter. There have been many new methodological developments since 1978 and thus are not covered in standard texts. The writing of this book was motivated in part by the desire to make these modern ideas and methods accessible to a larger readership in a reader friendly fashion.

Among the new methodologies, robust parameter design stands out as an innovative statistical/engineering approach to off-line quality and productivity improvement. It attempts to improve a process or product by making it less sensitive to noise variation through statistically designed experiments. Another important development in theoretical experimental design is the widespread use of the minimum aberration criterion for optimal assignment of factors to columns of a design table. This criterion is more powerful than the maximum resolution criterion for choosing fractional factorial designs. The third development is the increasing use of designs with complex aliasing in conducting economical experiments. It turns out that many of these designs can be used for the estimation of interactions, which is contrary to the prevailing practice that they be used for estimating the main effects only. The fourth development is the widespread use of Generalized Linear Models (GLMs) and Bayesian methods for analyzing nonnormal data. Many experimental responses are nonnormally distributed, such as binomial and Poisson counts, ordinal frequencies or have lifetime distributions and are observed with censoring which arise in reliability and survival studies. With the advent of modern computing, these tools have been incorporated in texts on medical statistics and social statistics. They should also be made available to experimenters in science and engineering. There are also other experimental methodologies that originated more than 20 years ago but have received scant attention in standard application-oriented texts. These include mixed two- and four-level designs, the method of collapsing for generating orthogonal main-effect plans, Plackett–Burman designs, and mixed-level orthogonal arrays. The main goal of writing this book is to fill in these gaps and present a new and integrated system of experimental design and analysis, which may help in defining a new fashion of teaching and for conducting research on this subject.

The intended readership of this book includes general practitioners as well as specialists. As a textbook, it covers standard material like analysis of variance (ANOVA), two- and three-level factorial and fractional factorial designs and response surface methodologies. For reading most of the book, the only prerequisite is an undergraduate level course on statistical methods and a basic knowledge of regression analysis. Because of the multitude of topics covered in the book, it can be used for a variety of courses. The material contained here has been taught at the Department of Statistics and the Department of Industrial and Operations Engineering at the University of Michigan to undergraduate seniors, masters, and doctoral students. To help instructors choose which material to use from the book, a separate "Suggestions of Topics for Instructors" follows this preface.

Some highlights and new material in the book are outlined as follows. Chapters 1 and 2 contain standard material on analysis of variance, one-way and multi-way layout, randomized block designs, Latin squares, balanced incomplete block designs, and analysis of covariance. Chapter 3 addresses two-level factorial designs and provides new material in Sections 3.13–3.17 on the use of formal tests of effect significance in addition to the informal tests based on normal and half-normal plots. Chapter 4, on two-level fractional factorial designs, uses the minimum aberration criterion for selecting optimal fractions and emphasizes the use of follow-up experiments to resolve the ambiguities in aliased effects. In Chapter 5, which deals with three-level designs, the linear-quadratic system and the variable selection strategy for handling and analyzing interaction effects are new. A new strategy for handling multiple responses is also presented. Most of the material in Chapter 6 on mixed two- and four-level designs and the method of sliding levels is new. Chapter 7, on nonregular designs, is the only theoretical chapter in the book. It emphasizes statistical properties and applications of the designs rather than their construction and mathematical structure. For practitioners, only the collections of tables in its appendices and some discussions in the sections on their statistical properties may be of interest. Chapter 7 paves the way for the new material in Chapter 8. Both frequentist and Bayesian analysis strategies are presented. The latter employs Gibbs sampling for efficient model search. Supersaturated designs are also briefly discussed. Chapter 9 contains a standard treatment of response surface methodologies. Chapters 10 and 11 present robust parameter design. The former deals with problems with a simple response while the latter deals with those with a signal-response relationship. The three important aspects of parameter design are considered: choice of performance measures, planning techniques, and modeling and analysis strategies. Chapter 12 is concerned with experiments for reliability improvement. Both failure time data and degradation data are considered. Chapter 13 is concerned with experiments with nonnormal responses. Several approaches to analysis are considered, including generalized linear models and Bayesian methods.

The book has some interesting features not commonly found in experimental design texts. Each of Chapters 3 to 13 starts with one or more case studies, which include the goal of the investigation, the data, the experimental plan and the factors and their levels. It is then followed by sections devoted to the description of experimental plans (i.e., experimental designs). Required theory or methodology for the experimental designs are developed in these sections. They are followed by sections on modeling and analysis strategies. The chapter then returns to the original data, analyzes it using the strategies just outlined, and discusses the implications of the analysis results to the original case studies. The book contains more than 80 experiments, mostly based on actual case studies; of these, 30 sets are analyzed in the text and more than 50 are given in the exercises. Each chapter ends with a practical summary which provides an easy guide to the methods covered in that chapter and is particularly useful for readers who want to find a specific tool but do not have the patience to go through the whole chapter. The book takes a novel approach to design tables. Many tables are new and based on recent research in experimental design theory and algorithms. For regular designs, only the design generators are given. Full designs can be easily generated by the readers from these generators. The collections of clear effects are given in these tables, however, because it would require some effort, especially for the less mathematically oriented readers, to derive them. The complete layouts of the orthogonal arrays are given in Chapter 7 for the convenience of the readers. With our emphasis on methodologies and applications, mathematical derivations are given sparingly. Unless the derivation itself is crucial to the understanding of the methodology, we omit it and refer to the original source.

The majority of the writing of this book was done at the University of Michigan. Most of the authors' research that is cited in the book was done at the University of Michigan with support from the National Science Foundation (1994–1999) and at the University of Waterloo (1988-1995) with support from the Natural Sciences and Engineering Research Council of Canada and the GM/NSERC Chair in Quality and Productivity. We have benefitted from the comments and assistance of many colleagues and former students, including Julie Bérubé, Derek Bingham, Ching-Shui Cheng, Hugh Chipman, David Fenscik, Xiaoli Hou, Longcheen Huwang, Bill Meeker, Randy Sitter, Huaiqing Wu, Hongquan Xu, Qian Ye, Runchu Zhang, and Yu Zhu. Shao-Wei Cheng played a pivotal supporting role as the book was completed; Jock MacKay read the first draft of the entire book and made numerous penetrating and critical comments; Jung-Chao Wang provided invaluable assistance in the preparation of tables for Chapter 7. We are grateful to all of them. Without their efforts and interest, this book could not have been completed.

C. F. JEFF WU
Ann Arbor, Michigan

MICHAEL HAMADA
Los Alamos, New Mexico

Suggestions of Topics for Instructors

One term for senior and master students in Statistics, Engineering, Physical, Life and Social Sciences (with a background in introductory statistics including regression analysis):

> Chapters 1, 2, 3 (3.1–3.12), 4; optional material from Chapters 10 (10.1–10.5), 5 (5.1–5.6), 8 (8.1–8.4), 9 (9.1–9.3, 9.5, 9.7). For students without a background in regression analysis, a 1–2 week review of regression analysis should suffice.

One term for master and Ph.D. students in Statistics/Biostatistics:

> Chapters 1, 2, 3 (3.13–3.17 may be skipped), 4, 5 (5.7–5.8 may be skipped), 9 (9.4 and 9.8 may be skipped), 10 (10.1–10.5); optional material from Chapters 6 (6.1–6.5, 6.9), 7 (7.1–7.5), 8 (except 8.5). Coverage of Chapters 1 and 2 can be accelerated for those with a background in ANOVA.

Two-term sequence for master and Ph.D. students in Statistics/Biostatistics:

> First term: Chapters 1 and 2 (can be accelerated if ANOVA is a prerequisite), Chapters 3 (3.14–3.17 may be skipped), 4, 5, 6, 7 (the more theoretical material may be skipped).
>
> Second term: Chapters 8 (8.5 may be skipped), 9 (9.8 may be skipped), 10, 11 (11.6 may be skipped), 12 (12.5, 12.7–12.8 may be skipped), 13.

One-term advanced topics course for Ph.D. students with background in introductory graduate experimental design course:

> Selected topics from Chapters 6 to 13 depending on the interest and background of the students.

One-term course on theoretical experimental design for Ph.D. students in Statistics and Mathematics:

> Chapters 1 (1.3), 2 (2.6–2.9), 3 (3.3–3.5, 3.12), 4 (4.2–4.6), 5 (5.3–5.4, 5.8), 6 (6.2–6.3, 6.7–6.8), 7, 9 (9.4, 9.7–9.8), 10 (10.6–10.9), 11 (11.6).

List of Experiments and Data Sets

CHAPTER 3

CHAPTER 4

CHAPTER 5

CHAPTER 6

CHAPTER 7

CHAPTER 8

CHAPTER 11

CHAPTER 12

CHAPTER 13

Experiments

CHAPTER 1

Basic Principles and Experiments with a Single Factor

Some basic concepts and principles in experimental design are introduced in this chapter. The simplest class of experiments, that with a single factor, is then considered. Analysis techniques like regression analysis, variable selection, analysis of variance (ANOVA), multiple comparisons, and residual analysis are presented. These techniques are applicable to more complex experiments to be considered in later chapters.

1.1 INTRODUCTION AND HISTORICAL PERSPECTIVE

Experimentation is one of the most common activities that people engage in. It covers a wide range of applications from household work like food preparation to technological innovation in material science, semiconductors, robotics, life science, etc. It allows an investigator to find out what happens to the output or response when the settings of the input variables in a system are purposely changed. Statistical or common-sense analysis can then be used to study the relationship between the input and output values. A better understanding of how the input variables affect the performance of a system can thereby be achieved. This provides a basis for selecting optimum input settings. Experimental design is a body of knowledge and techniques that enables an investigator to conduct better experiments, analyze data efficiently, and make the connections between the conclusions from the analysis and the original objectives of the investigation.

Experimentation is used to understand and/or improve a system. A system can be a product or process. A product can be one developed in engineering, biology, or the physical sciences. A process can be a manufacturing process, a process that describes a physical phenomenon, or a non-physical process such as those found in service or administration. Although most examples in the book are from engineering or the physical and biological sciences, the methods can also be applied to other disciplines, such as business, medicine, and psychology. For example, in studying the efficiency and cost of a payroll operation, the entire payroll operation can be viewed as a process with key input variables such as the number of supervisors, the number of clerks, method of bank deposit, level of automation, administrative structure, etc. A computer simulation model can then be used to study the effects of changing these input variables on cost and efficiency.

Modern experimental design dates back to the pioneering work of the great statistician R. A. Fisher in the 1930s at the Rothamsted Agricultural Experimental Station in the United Kingdom. Fisher's work and the notable contributions by F. Yates and D. J. Finney were motivated by problems in agriculture and biology. Because of the nature of agricultural experiments, they tend to be large in scale, take a long time to complete, and must cope with variations in the field. Such considerations led to the development of blocking, randomization, replication, orthogonality, and the use of analysis of variance and fractional factorial designs. The theory of combinatorial designs, to which R. C. Bose has made fundamental contributions, was also stimulated by problems in block designs and fractional factorial designs. The work in this era also found applications in social science research and in the textile and woolen industries.

The next era of rapid development came soon after World War II. In attempting to apply previous techniques to solve problems in the chemical industries, G. E. P. Box and co-workers at Imperial Chemical Industries discovered that new techniques and concepts had to be developed to cope with the unique features in process industries. The new techniques focused on process modeling and optimization rather than on treatment comparisons, which was the primary objective in agricultural experiments. The experiments in process industries tend to take less time but put a premium on run size economy because of the cost of experimentation. These time and cost factors naturally favor sequential experimentation. The same considerations led to the development of new techniques for experimental planning, notably central composite designs and optimal designs. Their analysis relies more heavily on regression modeling and graphical analysis. Process optimization based on the fitted model is also emphasized. Because the choice of design is often linked to a particular model (e.g., a second-order central composite design for a second-order regression model) and the experimental region may be irregularly shaped, a flexible strategy for finding designs to suit a particular model and/or experimental region is called for. With the availability of fast computational algorithms, optimal designs (which was pioneered by J. Kiefer) have become an important part of this strategy.

The relatively recent emphasis on variation reduction has provided a new source of inspiration and techniques in experimental design. In manufacturing the ability to make many parts with few defects is a competitive advantage. Therefore variation reduction in the quality characteristics of these parts has become a major focus of quality and productivity improvement. G. Taguchi advocated the use of robust parameter design to improve a system (i.e., a product or process) by making it less sensitive to variation, which is hard to control during normal operating or use conditions of the product or process. The input variables of a system can be divided into two broad types: control factors, whose values remain fixed once they are chosen, and noise factors, which are hard to control during normal conditions. By exploiting the interactions between the control and noise factors, one can achieve robustness by choosing control factor settings that make the system less sensitive to noise variation. The new paradigm is variation modeling and reduction. Traditionally, when the mean and variance are both considered, variance is used to assess the variability of the sample mean as in the t test or of the treatment comparisons as in the analysis of variance. The focus on variation and the division of factors into two types led to the development of new concepts and techniques in the planning and analysis of robust parameter design experiments. The original problem formulation and some basic concepts were developed by G. Taguchi. Other basic concepts and many sound statistical techniques have been developed by statisticians since the mid-1980s.

Given this historical background, we now classify experimental problems into five broad categories according to their objectives.

1. *Treament Comparisons*. The main purpose is to compare several treatments and select the best ones. For example, in the comparison of six barley varieties, are they different in terms of yield and resistance to drought? If they are indeed different, how are they different and which are the best? Examples of treatments include varieties (rice, barley, corn, etc.) in agricultural trials, sitting positions in ergonomic studies, instructional methods, machine types, suppliers, etc.

2. *Variable Screening*. If there is a large number of variables in a system but only a relatively small number of them is important, a screening experiment can be conducted to identify the important variables. Such an experiment tends to be economical in that it has few degrees of freedom left for estimating error variance and higher order terms like quadratic effects or interactions. Once the important variables are identified, a follow-up experiment can be conducted to study their effects more thoroughly. This latter phase of the study falls into the category discussed next.

3. *Response Surface Exploration*. Once a smaller number of variables is identified as important, their effects on the response need to be explored. The relationship between the response and these variables is sometimes referred to as a response surface. Usually the experiment is based on a design that allows the linear and quadratic effects of the variables and some of the interactions between the variables to be estimated. This experiment tends to be larger (relative to the number of variables under study) than the screening experiment. Both parametric and semi-parametric models may be considered. The latter is more computer intensive but also more flexible in model fitting.

4. *System Optimization*. In many investigations, interest lies in the optimization of the system. For example, the throughput of an assembly plant or

the yield of a chemical process is to be maximized; the amount of scrap or number of reworked pieces in a stamping operation is to be minimized; the time required to process a travel claim reimbursement is to be reduced. If a response surface has been identified, it can be used for optimization. For the purpose of finding an optimum, it is, however, not necessary to map out the whole surface as in a response surface exploration. An intelligent sequential strategy can quickly move the experiment to a region containing the optimum settings of the variables. Only within this region is a thorough exploration of the response surface warranted.

5. *System Robustness*. Besides optimizing the response, it is important in quality improvement to make the system robust against noise (i.e., hard-to-control) variation. This is often achieved by choosing control factor settings at which the system is less sensitive to noise variation. Even though the noise variation is hard to control in normal conditions, it needs to be systematically varied during experimentation. The response in the statistical analysis is often the variance (or its transformation) among the noise replicates for a given control factor setting.

1.2   A SYSTEMATIC APPROACH TO THE PLANNING AND IMPLEMENTATION OF EXPERIMENTS

In this section, we provide some guidelines on the planning and implementation of experiments. The following seven-step procedure summarizes the important steps that the experimenter must address.

1. *State Objective*. The objective of the experiment needs to be clearly stated. All stakeholders should provide input. For example, for a manufactured product, the stakeholders may include design engineers who design the product, process engineers who design the manufacturing process, line engineers who run the manufacturing process, suppliers, lineworkers, customers, marketers, and managers.

2. *Choose Response*. The response is the experimental outcome or observation. There may be multiple responses in an experiment. Several issues arise in choosing a response. Responses may be *discrete* or *continuous*. Discrete responses can be counts or categories, e.g., binary (good, bad) or ordinal (easy, normal, hard). Continuous responses are generally preferable. For example, a continuous force measurement for opening a door is better than an ordinal (easy, normal, hard to open) judgment; the recording of a continuous characteristic is preferred to the recording of the percent that the characteristic is within its specifications. Trade-offs may need to be made. For example, an ordinal measurement of force to open a door may be preferable to delaying the experiment until a device to take continuous measurements can be developed. Most importantly, there should be a good

measurement system for measuring the response. In fact, an experiment called a *gauge repeatability and reproducibility* (R&R) *study* can be performed to assess a continuous measurement system (AIAG, 1990). When there is a single measuring device, the variation due to the measurement system can be divided into two types: variation between the operators and variation within the operators. Ideally, there should be no between-operator variation and small within-operator variation. The gauge R&R study provides estimates for these two components of measurement system variation. Finally, the response should be chosen to increase understanding of mechanisms and physical laws involved in the problem. For example, in a process that is producing underweight soap bars, soap bar weight is the obvious choice for the response in an experiment to improve the underweight problem. By examining the process more closely, there are two subprocesses that have a direct bearing on soap bar weight: the mixing process that affects the soap bar density and the forming process that impacts the dimensions of the soap bars. In order to better understand the mechanism that causes the underweight problem, soap bar density and soap bar dimensions are chosen as the responses. Even though soap bar weight is not used as a response, it can be easily determined from its density and dimensions. Therefore, no information is lost in studying the density and dimensions. Such a study may reveal new information about the mixing and forming subprocesses, which can in turn lead to a better understanding of the underweight problem. Further discussions on and other examples of the choice of responses can be found in Phadke (1989) and León, Shoemaker and Tsui (1993).

The chosen responses can be classified according to the stated objective. Three broad categories will be considered in this book: **nominal-the-best**, **larger-the-better**, and **smaller-the-better**. The first one will be addressed in Section 3.2 and the last two in Section 5.2.

3. *Choose Factors and Levels.* A **factor** is a variable that is studied in the experiment. In order to study the effect of a factor on the response, two or more values of the factor are used. These values are referred to as **levels** or settings. A **treatment** is a combination of factor levels. When there is a single factor, its levels are the treatments. For the success of the experiment, it is crucial that potentially important factors be identified at the planning stage. There are two graphical methods for identifying potential factors. First, a **flow chart** of the process or system is helpful to see where the factors arise in a multi-stage process. In Figure 1.1, a rough sketch of a paper pulp manufacturing process is given which involves raw materials from suppliers, a chemical process to make a slurry which is passed through a mechanical process to produce the pulp. Involving all the stakeholders is invaluable in capturing an accurate description of the process or system. Second, a **cause-and-effect diagram** can be used to list and organize the potential factors that may impact the response. In Figure 1.2, a cause-and-effect diagram is given which lists the factors thought to affect the product quality of an injection molding process. Traditionally, the factors are organized under the headings:

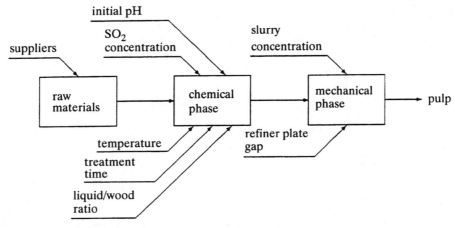

Figure 1.1. Flow Chart, Pulp Manufacturing Process.

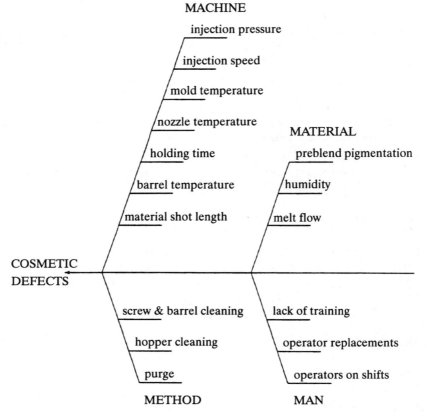

Figure 1.2. Cause-and-Effect Diagram, Injection Molding Experiment.

Man, Machine, Measurement, Material, Method, and Environment (Mother Nature for those who like M's). Because of their appearance, cause-and-effect diagrams are also called *fishbone diagrams*. Different characteristics of the factors need to be recognized because they can affect the choice of the experimental design. For example, a factor such as furnace temperature is *hard to change*. That is, after changing the temperature setting, it may take a considerable amount of time before the temperature stabilizes at the new setting. A factor may also be *hard to set* so that the actual level used in the experiment may be different than the intended level. For example, the actual impact force of a pellet projected at an automobile windshield can only be set within 3 psi of the intended impact force. Other factors that may be hard or impossible to control are referred to as *noise* factors. Examples of noise factors include environmental and customer use conditions. (An in-depth discussion of noise factors will be given in Section 10.3.)

Factors may be *quantitative* and *qualitative*. Quantitative factors like temperature, time, and pressure take values over a continuous range. Qualitative factors take on a discrete number of values. Examples of qualitative factors include operation mode, supplier, position, line, etc. Of the two types of factors, there is more freedom in choosing the levels of quantitative factors. For example, if temperature (in °C) is in the range 100–200°C, one could choose 130 and 160°C for two levels or 125, 150, and 175°C for three levels. If only a linear effect is expected, two levels should suffice. If curvature is expected, then three or more levels are required. In general, the levels of quantitative factors must be chosen far enough apart so that an effect can be detected but not too far so that different physical mechanisms are involved (which would make it difficult to do statistical modeling and prediction). There is less flexibility in choosing the levels of qualitative factors. Suppose there are three testing methods under comparison. All three must be included as three levels of the factor "testing method," unless the investigator is willing to postpone the study of one method so that only two methods are compared in a two-level experiment.

When there is flexibility in choosing the number of levels, the choice may depend on the availability of experimental plans for the given combination of factor levels. In choosing factors and levels, *cost* and *practical constraints* must be considered. If two levels of the factor "material" represent expensive and cheap materials, a negligible effect of material on the response will be welcomed because the cost can be drastically reduced by replacing the expensive material by the cheap alternative. Factor levels must be chosen to meet practical constraints. If a factor combination (e.g., high temperature and long time in an oven) can potentially lead to disastrous results (e.g., burned or overbaked), it should be avoided and a different plan should be chosen.

4. *Choose Experimental Plan.* Use the fundamental principles discussed in Section 1.3 as well as other principles presented throughout the book. The choice of the experimental plan is crucial. A poor design may capture little

information which no analysis can rescue. On the other hand, if the experiment is well planned, the results may be obvious so that no sophisticated analysis is needed.

5. *Perform the Experiment.* The use of a *planning matrix* is recommended. This matrix describes the experimental plan in terms of the actual values or settings of the factors. For example, it lists the actual levels such as 50 or 70 psi if the factor is pressure. To avoid confusion and eliminate potential problems of running the wrong combination of factor levels in a multifactor experiment, each of the treatments, such as temperature at 30°C and pressure at 70 psi, should be put on a separate piece of paper and given to the personnel performing the experiment. It is also worthwhile to perform a *trial run* to see if there will be difficulties in running the experiment, namely, if there are problems with setting the factors and measuring the responses. Any deviations from the planned experiment need to be recorded. For example, for hard-to-set factors, the actual values should be recorded.

6. *Analyze the Data.* An analysis appropriate for the design used to collect the data needs to be carried out. This includes model fitting and assessment of the model assumptions through an analysis of residuals. Many analysis methods will be presented throughout the book.

7. *Draw Conclusions and Make Recommendations.* Based on the data analysis, conclusions are presented which include the important factors and a model for the response in terms of the important factors. Recommended settings or levels for the important factors may also be given. The conclusions should refer back to the stated objectives of the experiment. A *confirmation experiment* is worthwhile, for example, to confirm the recommended settings. Recommendations for further experimentation in a *follow-up experiment* may also be given. For example, a follow-up experiment is needed if two models explain the experimental data equally well and one must be chosen for optimization.

For further discussion on the planning of experiments, see Coleman and Montgomery (1993), Knowlton and Keppinger (1993), and Barton (1997).

1.3 FUNDAMENTAL PRINCIPLES: REPLICATION, RANDOMIZATION, AND BLOCKING

There are three fundamental principles that need to be considered in the design of an experiment: **replication, randomization**, and **blocking**. Other principles will be introduced later in the book as they arise.

An *experimental unit* is a generic term that refers to a basic unit such as material, animal, person, machine, or time period, to which a treatment is applied. By *replication*, we mean that each treatment is applied to experimental units that are representative of the population of units to which the conclusions of the experiment will apply. It enables the estimation of the magnitude of experimental error (i.e., the error variance) against which the differences among treatments are judged. Increasing the number of replications, or *replicates*, decreases the variance of the treatment effect estimates and provides more power for detecting differences in treatments. A distinction needs to be made between replicates and *repetitions*. For example, three readings from the same experimental unit are repetitions while the readings from three separate experimental units are replicates. The error variance from the former is less than that from the latter because repeated readings only measure the variation due to errors in reading while the latter also measures the unit-to-unit variation. Underestimation of the true error variance can result in the false detection of effect significance.

The second principle is that of *randomization*. It should be applied to the allocation of units to treatments, the order in which the treatments are applied in performing the experiment, and the order in which the responses are measured. It provides protection against variables that are unknown to the experimenter but may impact the response. It reduces the unwanted influence of subjective judgment in treatment allocation such as in a physician's assignment of medical treatments to patients. Moreover, randomization ensures validity of the estimate of experimental error and provides a basis for inference in analyzing the experiments. For an in-depth discussion on randomization, see Hinkelmann and Kempthorne (1994).

A group of homogeneous units is referred to as a *block*. Examples of blocks include days, weeks, morning vs. afternoon, batches, lots, sets of twins, and pairs of kidneys. For *blocking* to be effective, the units should be arranged so that the within-block variation is much smaller than the between-block variation. By comparing the treatments within the same block, the block effects are eliminated in the comparison of the treatment effects, thereby making the experiment more efficient. For example, there may be a known day effect on the response so that if all the treatments can be applied within the same day, the day-to-day variation is eliminated.

If blocking is effective, it should be applied to remove the block-to-block variation. Randomization can then be applied to the assignments of treatments to units within the blocks to further reduce the influence of unknown variables. This strategy of **block what you can and randomize what you cannot** is used in randomized block designs, to be discussed in Section 2.2.

These three principles are generally applicable to physical experiments but not to computer experiments because the same input in a computer experiment gives rise to the same output. Computer experiments are not considered in the book.

A simple example will be used to explain these principles. Suppose two keyboards denoted by A and B are being compared in terms of typing efficiency. Six different manuscripts denoted by 1–6 are given to the same typist. First the test is arranged in the following sequence:

$$1.\ A, B, \quad 2.\ A, B, \quad 3.\ A, B, \quad 4.\ A, B, \quad 5.\ A, B, \quad 6.\ A, B.$$

Even though the experiment is replicated six times (with six manuscripts) and blocking is used to compare two keyboards on the same manuscript, the design has a serious flaw. After typing the manuscript on keyboard A, the typist will get more familiar with the content of the manuscript when he or she is typing the same manuscript on keyboard B. This "learning effect" will unfairly help the performance of keyboard B. The observed difference between A and B is the combination of the treatment effects (which measures the intrinsic difference between A and B) and the learning effect. For the given test sequence, it is impossible to disentangle the learning effect from the treatment effect. Randomization would help reduce the unwanted influence of the learning effect, which might not have been known to the investigator who planned the study. By randomizing the typing order for each manuscript, the test sequence may appear as follows:

$$1. \ A, B, \quad 2. \ B, A, \quad 3. \ A, B, \quad 4. \ B, A, \quad 5. \ A, B, \quad 6. \ A, B.$$

With four AB's and two BA's in the sequence, it is a better design than the first one. A further improvement can be made. The design is not balanced because B benefits from the learning effect in four trials while A only benefits from two trials. There is still a residual learning effect not completely eliminated by the second design. The learning effect can be completely eliminated by requiring that half of the trials have the order AB and the other half the order BA. The actual assignment of AB and BA to the six manuscripts should be done by randomization. This method is referred to as *balanced randomization*. Balance is a desirable design property, which will be discussed later.

For simplicity of discussion, we have assumed that only one typist was involved in the experiment. In a practical situation, such an experiment should involve several typists that are representative of the population of typists so that the conclusions made from the study would apply more generally. This and other aspects of the typing experiment will be addressed in the exercises.

With these principles in mind, a useful addition to the cause-and-effect diagram is to indicate how the proposed experimental design addresses each listed factor. The following designations are suggested: **E** for an experimental factor, **B** for a factor handled by blocking, **O** for a factor held constant at one value, and **R** for a factor handled by randomization. This designation clearly indicates how the proposed design deals with each of the potentially important factors. The designation **O**, for "one value," serves to remind the experimenter that the factor is held constant during the current experiment but may be varied in a future experiment. An illustration is given in Figure 1.3 from the injection molding experiment discussed in Section 1.2.

Other designations of factors can be considered. For example, experimental factors can be further divided into two types (control factors and noise factors), as in the discussion on the choice of factors in Section 1.2. For the

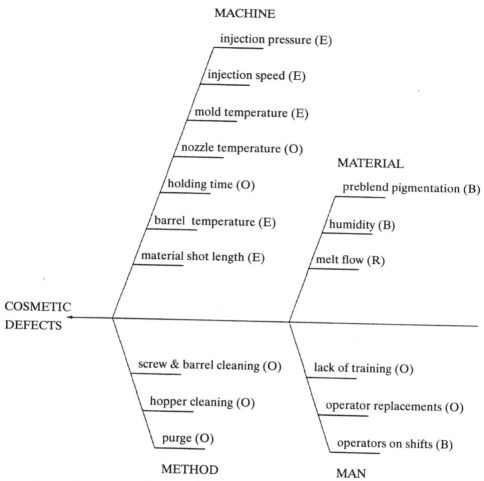

Figure 1.3. Revised Cause-and-Effect Diagram, Injection Molding Experiment.

implementation of experiments, we may also designate an experimental factor as "hard-to-change" or "easy-to-change." These designations will be considered later as they arise.

1.4   THE GENERAL LINEAR MODEL

Experimental data can often be modeled by the general linear model (also called the multiple regression model). Suppose that the response y is related to p covariates (also called explanatory variables, regressors, predictors) $x_1, x_2, \ldots, x_p$ as follows:

$$y = \beta_0 + \beta_1 x_1 + \cdots + \beta_p x_p + \epsilon, \tag{1.1}$$

where ϵ is the random part of the model which is assumed to be normally distributed with mean 0 and variance σ^2, i.e., $\epsilon \sim N(0, \sigma^2)$; because ϵ is normally distributed, so is y and $Var(y) = \sigma^2$. The structural part of the model is

$$E(y) = \beta_0 + \beta_1 x_1 + \cdots + \beta_p x_p + E(\epsilon)$$
$$= \beta_0 + \beta_1 x_1 + \cdots + \beta_p x_p.$$

Here, $E(y)$ is linear in the β's, the regression coefficients, which explains the term **linear model**.

If N observations are collected in an experiment, the model for them takes the form

$$y_i = \beta_0 + \beta_1 x_{i1} + \cdots + \beta_p x_{ip} + \epsilon_i, \qquad i = 1, \ldots, N, \qquad (1.2)$$

where y_i is the ith value of the response and $x_{i1}, \ldots, x_{ip}$ are the corresponding values of the p covariates.

These N equations can be written in matrix notation as:

$$\mathbf{y} = \mathbf{X}\boldsymbol{\beta} + \boldsymbol{\epsilon}, \qquad (1.3)$$

where $\mathbf{y} = (y_1, \ldots, y_N)^T$ is the $N \times 1$ vector of responses, $\boldsymbol{\beta} = (\beta_0, \beta_1, \ldots, \beta_p)^T$ is the $(p+1) \times 1$ vector of regression coefficients, $\boldsymbol{\epsilon} = (\epsilon_1, \ldots, \epsilon_N)^T$ is the $N \times 1$ vector of errors, and $\mathbf{X}$, the $N \times (p+1)$ **model matrix**, is given as

$$\mathbf{X} = \begin{pmatrix} 1 & x_{11} & \cdots & x_{1p} \\ \vdots & \vdots & \ddots & \vdots \\ 1 & x_{N1} & \cdots & x_{Np} \end{pmatrix}. \qquad (1.4)$$

The unknown parameters in the model are the regression coefficients $\boldsymbol{\beta}$ and the error variance σ^2. Thus, the purpose for collecting the data is to estimate and make inferences about these parameters. For estimating $\boldsymbol{\beta}$, the least squares criterion is used; i.e., the least squares estimators (LSEs), denoted by $\hat{\boldsymbol{\beta}}$, minimize the following quantity:

$$\sum_{i=1}^{N} \left(y_i - \left(\beta_0 + \beta_1 x_{i1} + \cdots + \beta_p x_{ip} \right) \right)^2 \qquad (1.5)$$

which in matrix notation is

$$(\mathbf{y} - \mathbf{X}\boldsymbol{\beta})^T (\mathbf{y} - \mathbf{X}\boldsymbol{\beta}). \qquad (1.6)$$

In other words, the squared distance between the response vector $\mathbf{y}$ and the vector of fitted values $\mathbf{X}\hat{\boldsymbol{\beta}}$ is minimized. In order to minimize the sum of squared residuals, the vector of **residuals**

$$\mathbf{r} = \mathbf{y} - \mathbf{X}\hat{\boldsymbol{\beta}} \tag{1.7}$$

needs to be perpendicular to the vector of *fitted values*

$$\hat{\mathbf{y}} = \mathbf{X}\hat{\boldsymbol{\beta}}, \tag{1.8}$$

that is, the cross product between these two vectors should be zero:

$$\mathbf{r}^T\hat{\mathbf{y}} = \mathbf{r}^T\mathbf{X}\hat{\boldsymbol{\beta}} = \mathbf{0}.$$

An equivalent way of stating this is that the columns of the model matrix $\mathbf{X}$ need to be perpendicular to $\mathbf{r}$, the vector of residuals, and thus satisfy

$$\mathbf{X}^T(\mathbf{y} - \mathbf{X}\hat{\boldsymbol{\beta}}) = \mathbf{X}^T\mathbf{y} - \mathbf{X}^T\mathbf{X}\hat{\boldsymbol{\beta}} = \mathbf{0}. \tag{1.9}$$

The solution to this equation is the **least squares estimate** which is

$$\hat{\boldsymbol{\beta}} = (\mathbf{X}^T\mathbf{X})^{-1}\mathbf{X}^T\mathbf{y}. \tag{1.10}$$

In fitting the model, one wants to know if any of the variables (regressors, predictors, covariates) has explanatory power. None of them has explanatory power if the null hypothesis

$$H_0 : \beta_1 = \cdots = \beta_p = 0 \tag{1.11}$$

holds. In order to test this null hypothesis, one needs to assess how much of the total variation in the response data can be explained by the model relative to the remaining variation after fitting the model, which is contained in the residuals.

Recall how the model was fitted: the residuals are perpendicular to the fitted values so that we have a right triangle. This brings to mind the Pythagorean theorem: the squared length of the hypotenuse is equal to the sum of the squared lengths of its opposite sides. In vector notation, the squared distance of a vector $\mathbf{a}$ is simply $\mathbf{a}^T\mathbf{a} = \Sigma a_i^2$. Thus, from the least squares fit, we obtain

$$\mathbf{y}^T\mathbf{y} = (\mathbf{X}\hat{\boldsymbol{\beta}})^T(\mathbf{X}\hat{\boldsymbol{\beta}}) + (\mathbf{y} - \mathbf{X}\hat{\boldsymbol{\beta}})^T(\mathbf{y} - \mathbf{X}\hat{\boldsymbol{\beta}})$$

$$= \hat{\boldsymbol{\beta}}^T\mathbf{X}^T\mathbf{X}\hat{\boldsymbol{\beta}} + (\mathbf{y} - \mathbf{X}\hat{\boldsymbol{\beta}})^T(\mathbf{y} - \mathbf{X}\hat{\boldsymbol{\beta}}),$$

where $\mathbf{y}^T\mathbf{y}$ is the *total sum of squares (uncorrected)*, $\hat{\boldsymbol{\beta}}^T\mathbf{X}^T\mathbf{X}\hat{\boldsymbol{\beta}}$ is the *regression sum of squares (uncorrected)*, and

$$RSS = (\mathbf{y} - \mathbf{X}\hat{\boldsymbol{\beta}})^T(\mathbf{y} - \mathbf{X}\hat{\boldsymbol{\beta}})$$

Table 1.1   ANOVA Table for General Linear Model

Source	Degrees of Freedom	Sum of Squares	Mean Squares
regression	p	$\hat{\boldsymbol{\beta}}^T \mathbf{X}^T \mathbf{X} \hat{\boldsymbol{\beta}} - N\bar{y}^2$	$(\hat{\boldsymbol{\beta}}^T \mathbf{X}^T \mathbf{X} \hat{\boldsymbol{\beta}} - N\bar{y}^2)/p$
residual	$N - p - 1$	$(\mathbf{y} - \mathbf{X}\hat{\boldsymbol{\beta}})^T(\mathbf{y} - \mathbf{X}\hat{\boldsymbol{\beta}})$	$(\mathbf{y} - \mathbf{X}\hat{\boldsymbol{\beta}})^T(\mathbf{y} - \mathbf{X}\hat{\boldsymbol{\beta}})/(N - p - 1)$
total (corrected)	$N - 1$	$\mathbf{y}^T\mathbf{y} - N\bar{y}^2$	

is the *residual* (or *error*) *sum of squares*. In order to test the null hypothesis (1.11), the contribution from estimating the intercept β_0 needs to be removed. Subtracting off its contribution $N\bar{y}^2$, where $\bar{y}$ is the average of the N observations, yields

$$CTSS = \mathbf{y}^T\mathbf{y} - N\bar{y}^2 = RegrSS + RSS$$

$$= \left(\hat{\boldsymbol{\beta}}^T \mathbf{X}^T \mathbf{X} \hat{\boldsymbol{\beta}} - N\bar{y}^2\right) + \left(\mathbf{y} - \mathbf{X}\hat{\boldsymbol{\beta}}\right)^T\left(\mathbf{y} - \mathbf{X}\hat{\boldsymbol{\beta}}\right), \qquad (1.12)$$

where $CTSS$ is called the *corrected total sum of squares* and is equal to $\sum_{i=1}^{N}(y_i - \bar{y})^2$, which measures the variation in the data, and $RegrSS$ is called the *corrected regression sum of squares*. In the remainder of this book, "corrected" will be dropped in reference to various sums of squares but will be implied. Thus, the variation in the data is split into the variation explained by the regression model plus the residual variation. This relationship is given in a table called the ANalysis Of VAriance or ANOVA table displayed in Table 1.1.

Based on (1.12), we can define

$$R^2 = \frac{RegrSS}{CTSS} = 1 - \frac{RSS}{CTSS}. \qquad (1.13)$$

Because the R^2 value measures the "proportion of total variation explained by the fitted regression model $\mathbf{X}\hat{\boldsymbol{\beta}}$," a higher R^2 value indicates a better fit of the regression model. It can be shown that R is the correlation between $\mathbf{y} = (y_i)_{i=1}^{N}$ and $\hat{\mathbf{y}} = (\hat{y}_i)_{i=1}^{N}$ and thus is called the *multiple correlation coefficient*.

The degrees of freedom are those associated with each sum of squares. The mean square is the sum of squares divided by the corresponding degrees of freedom. The residual mean square is commonly referred to as the *mean-squared error* (MSE) and is an estimate $\hat{\sigma}^2$ for σ^2, i.e.,

$$\hat{\sigma}^2 = \left(\mathbf{y} - \mathbf{X}\hat{\boldsymbol{\beta}}\right)^T\left(\mathbf{y} - \mathbf{X}\hat{\boldsymbol{\beta}}\right)/(N - p - 1). \qquad (1.14)$$

If the null hypothesis (1.11) holds, the F statistic

$$\frac{\left(\hat{\boldsymbol{\beta}}^T \mathbf{X}^T \mathbf{X} \hat{\boldsymbol{\beta}} - N\bar{\mathbf{y}}^2\right)/p}{\left(\mathbf{y} - \mathbf{X}\hat{\boldsymbol{\beta}}\right)^T \left(\mathbf{y} - \mathbf{X}\hat{\boldsymbol{\beta}}\right)/(N-p-1)} \qquad (1.15)$$

(the regression mean square divided by the residual mean square) has an F distribution with parameters p and $N-p-1$, which are the degrees of freedom of its numerator and denominator, respectively. The p value is calculated by evaluating

$$Prob\left(F_{p, N-p-1} > F_{obs}\right), \qquad (1.16)$$

where $Prob(\cdot)$ denotes the probability of an event, $F_{p, N-p-1}$ has an F distribution with parameters p and $N-p-1$, and F_{obs} is the observed value of the F statistic. The F critical values can be found in Appendix D. The p value in (1.16) can be obtained from certain pocket calculators or by interpolating the values given in Appendix D. An example of an F distribution is given in Figure 1.4 along with its critical values.

Note that the **p value** gives the probability under the null hypothesis that the F statistic value for an experiment conducted in comparable conditions will exceed the observed value F_{obs}. The smaller the p value, the stronger is the evidence that the null hypothesis does not hold. Therefore it provides a quantitative measure of the significance of effects in the experiment under study. The same interpretation can be applied when other test statistics and null hypotheses are considered.

It can be shown that the least squares estimate $\hat{\boldsymbol{\beta}}$ has a multivariate normal distribution with mean vector $\boldsymbol{\beta}$ and variance-covariance matrix

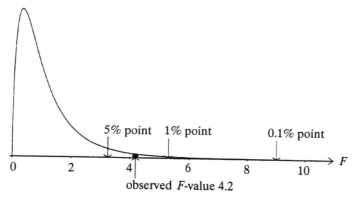

Figure 1.4. Observed F Value of 4.20 in Relation to an F Distribution With 3 and 16 Degrees of Freedom, Pulp Experiment.

$\sigma^2(\mathbf{X}^T\mathbf{X})^{-1}$, i.e.,

$$\hat{\boldsymbol{\beta}} \sim MN\left(\boldsymbol{\beta}, \sigma^2(\mathbf{X}^T\mathbf{X})^{-1}\right), \tag{1.17}$$

where MN stands for multivariate normal. The (i, j)th entry of the variance-covariance matrix is $Cov(\hat{\beta}_i, \hat{\beta}_j)$ and the jth diagonal element is $Cov(\hat{\beta}_j, \hat{\beta}_j)$ $= Var(\hat{\beta}_j)$. Therefore, the distribution for the individual $\hat{\beta}_j$ is $N(\beta_j, \sigma^2(\mathbf{X}^T\mathbf{X})_{jj}^{-1})$, which suggests that for testing the null hypothesis

$$H_0 : \beta_j = 0, \tag{1.18}$$

the following t statistic be used:

$$\frac{\hat{\beta}_j}{\sqrt{\hat{\sigma}^2(\mathbf{X}^T\mathbf{X})_{jj}^{-1}}}. \tag{1.19}$$

Under H_0, it has a t distribution with $N - p - 1$ degrees of freedom. This can also be used to construct confidence intervals since the denominator of the t statistic is the standard error of its numerator $\hat{\beta}_j$:

$$\hat{\beta}_j \pm t_{N-p-1, \alpha/2} \sqrt{\hat{\sigma}^2(\mathbf{X}^T\mathbf{X})_{jj}^{-1}}, \tag{1.20}$$

where $t_{N-p-1, \alpha/2}$ is the upper $\alpha/2$ quantile of the t distribution with $N - p - 1$ degrees of freedom. See Appendix C for t critical values.

Besides testing the individual β_j's, testing linear combinations of the β_j's can be useful. For testing $\mathbf{a}^T\boldsymbol{\beta} = \sum_{j=0}^{p} a_j \beta_j$, where $\mathbf{a}$ is a $(p+1) \times 1$ vector, it can be shown that

$$\mathbf{a}^T\hat{\boldsymbol{\beta}} \sim N\left(\mathbf{a}^T\boldsymbol{\beta}, \sigma^2\mathbf{a}^T(\mathbf{X}^T\mathbf{X})^{-1}\mathbf{a}\right). \tag{1.21}$$

This suggests using the test statistic

$$\frac{\mathbf{a}^T\hat{\boldsymbol{\beta}}}{\sqrt{\hat{\sigma}^2\mathbf{a}^T(\mathbf{X}^T\mathbf{X})^{-1}\mathbf{a}}}, \tag{1.22}$$

which has a t distribution with $N - p - 1$ degrees of freedom.

Extra Sum of Squares Principle

The *extra sum of squares principle* will be useful later for developing test statistics in a number of situations. Suppose that there are two models, say Model I and Model II. Model I is a special case of Model II, denoted by Model I $\subset$ Model II. Let

$$\text{Model I: } y_i = \beta_0 + \beta_1 x_{i1} + \cdots + \beta_q x_{iq} + \epsilon_i \tag{1.23}$$

and

Model II: $y_i = \beta_0 + \beta_1 x_{i1} + \cdots + \beta_q x_{iq} + \beta_{q+1} x_{i,q+1} + \cdots + \beta_p x_{ip} + \epsilon_i'.$

$$(1.24)$$

Model I $\subset$ Model II since $\beta_{q+1} = \cdots = \beta_p = 0$ in Model I. Then, for testing the null hypothesis that Model I is adequate, i.e.,

$$H_0 : \beta_{q+1} = \cdots = \beta_p = 0 \qquad (1.25)$$

holds, the extra sum of squares principle employs the F statistic:

$$\frac{(RSS(\text{Model I}) - RSS(\text{Model II}))/(p-q)}{RSS(\text{Model II})/(N-p-1)}, \qquad (1.26)$$

where RSS stands for the residual sum of squares. It follows that

$$RSS(\text{Model I}) - RSS(\text{Model II})$$
$$= RegrSS(\text{Model II}) - RegrSS(\text{Model I}), \qquad (1.27)$$

where $RegrSS$ denotes the regression sum of squares; thus, the numerator of the F statistic in (1.26) is the gain in the regression sum of squares for fitting the more general Model II relative to Model I, i.e., the *extra sum of squares*. When (1.25) holds, the F statistic has an F distribution with parameters $p - q$ (the difference in the number of estimated parameters between Models I and II) and $N - p - 1$. The extra sum of squares technique can be implemented by fitting Models I and II separately, obtaining their respective residual sums of squares, calculating the F statistic above, and then computing its p value.

1.5 VARIABLE SELECTION IN REGRESSION ANALYSIS

(The material in this section will not be used until Chapter 5.)

In the regression fitting of the linear model (1.2), those covariates whose regression coefficients are not significant may be removed from the full model. A more parsimonious model (i.e., one with fewer covariates) is preferred as long as it can explain the data well. It is also known that a model that fits the data too well may give poor predictions. The goal of variable selection in regression analysis is to identify the smallest subset of the covariates that explains the data well; one hopes to capture the true model or at least the covariates of the true model with the largest regression coefficients. One class of strategies is to use a model selection criterion to evaluate all possible subsets of the covariates and select the subset (which corresponds to a model) with the best value of the criterion. This is referred to as **best subset regression**. To maintain a balance between data fitting and prediction, a good model selection criterion should reward good model fitting as well as penalize model complexity. The R^2 in (1.13) is not a suitable criterion because it increases as the number of covariates increases. That is, it does not penalize excessively large models.

A commonly used criterion is the C_p **statistic** (Mallows, 1973). Suppose there are a total of q covariates. For a model that contains p regression coefficients corresponding to $p-1$ covariates and an intercept term β_0, define its C_p value as

$$C_p = \frac{RSS}{s^2} - (N - 2p), \tag{1.28}$$

where RSS is the residual sum of squares for the model, s^2 is the mean-squared error (see (1.14)) for the model containing all q covariates and β_0, and N is the total number of observations. As the model gets more complicated, the RSS term in (1.28) decreases while the p value in the second term increases. The counteracting effect of these two terms prevents the selection of extremely large or small models. If the model is true, $E(RSS) = (N-p)\sigma^2$. Assuming that $E(s^2) = \sigma^2$, it is then approximately true that

$$E(C_p) \approx \frac{(N-p)\sigma^2}{\sigma^2} - (N - 2p) = p.$$

Thus one should expect the best fitting models to be those with $C_p \approx p$. Further theoretical and empirical studies suggest that models whose C_p values are low and are close to p should be chosen.

For moderate to large q, fitting all subsets is computationally infeasible. An alternative strategy is based on adding or dropping one covariate at a time from a given model, which requires fewer model fittings but can still identify good fitting models. It need not identify the best fitting models as in any optimization that optimizes sequentially (and locally) rather than globally. The main idea is to compare the current model with a new model obtained by adding or deleting a covariate from the current model. Call the smaller and bigger models Model I and Model II, respectively. Based on the extra sum of squares principle in Section 1.4, one can compute the F statistic in (1.26), also known as a *partial F*, to determine if the covariate should be added or deleted. The partial F statistic takes the form

$$\frac{RSS(\text{Model I}) - RSS(\text{Model II})}{RSS(\text{Model II})/\nu}, \tag{1.29}$$

where ν is the degrees of freedom of the RSS (residual sum of squares) for Model II. Three versions of the strategy are considered next.

One version is known as **backward elimination**. It starts with the full model containing all q covariates and computes partial F's for all models with $q - 1$ covariates. At the kth step, Model II has $q - k + 1$ covariates and Model I has $q - k$ covariates, so that $v = N - (q - k + 1) - 1 = N - q + k - 2$ in the partial F in (1.29). At each step, compute the partial F value for each covariate being considered for removal. The one with the lowest partial F, provided it is smaller than a preselected value, is dropped. The procedure continues until no more covariates can be dropped. The preselected value is often chosen to be $F_{1, v, \alpha}$, the upper α critical value of the F distribution with 1 and v degrees of freedom. Choice of the α level determines the stringency level for eliminating covariates. Typical α's range from $\alpha = 0.1$ to 0.2. A conservative approach would be to choose a smaller F (i.e., a large α) value so that important covariates are not eliminated. Note that the statistic in (1.29) does not have a proper F distribution so that the F critical values serve only as guidelines. The literature often refers to them as *F-to-remove* values to make this distinction.

Another version is known as **forward selection**, which starts with the model containing an intercept and then adds one covariate at a time. The covariate with the largest partial F [as computed by (1.29)] is added, provided it is larger than a preselected F critical value, which is referred to as an *F-to-enter* value. The forward selection procedure is not recommended as it often misses important covariates. It is combined with backward elimination to form the following stepwise selection procedure.

The **stepwise selection** procedure starts with two steps of the forward selection and then alternates between one step of backward elimination and one step of forward selection. The F-to-remove and F-to-enter values should be chosen to be the same. A typical choice is $F_{1, v, \alpha}$ with $\alpha = 0.05, 0.1, 0.15$. The choice varies from data to data and can be changed as experience dictates. Among the three selection procedures, stepwise selection is known to be the most effective and is therefore recommended for general use.

For a comprehensive discussion on variable selection, see Draper and Smith (1998).

1.6   ONE-WAY LAYOUT

Consider the following experiment, reported by Sheldon (1960), which was performed at a pulp mill. Plant performance is based on pulp brightness as measured by a reflectance meter. Each of the four shift operators (denoted by A, B, C, and D) made five pulp handsheets from unbleached pulp. Reflectance was read for each of the handsheets using a brightness tester, as reported in Table 1.2. A goal of the experiment is to determine whether there are differences between the operators in making the handsheets and reading their brightness.

The experimental plan used in the pulp experiment was a one-way layout. The *one-way layout* is a single-factor experiment with k levels. Recall that a

Table 1.2   Reflectance Data, Pulp Experiment

	Operator		
A	*B*	*C*	*D*
59.8	59.8	60.7	61.0
60.0	60.2	60.7	60.8
60.8	60.4	60.5	60.6
60.8	59.9	60.9	60.5
59.8	60.0	60.3	60.5

treatment is a factor level combination applied to the experimental units. Since there is a single factor, the k treatments correspond to the k levels of the factor. Replication is used here with n_i observations taken for treatment i. For the pulp experiment, $k = 4$, $n_1 = n_2 = n_3 = n_4 = 5$, and $N = 20$.

Although Sheldon (1960) did not provide further details, randomization could have been applied in several ways in this experiment. First, 20 containers holding enough pulp to make a handsheet could have been set up and the 20 containers randomly distributed among the four shift operators. For randomization, there are

$$\binom{20}{5\ 5\ 5\ 5} = \frac{20!}{5!5!5!5!} \approx 11.7 \times 10^9$$

different allocations of the units and one such allocation can be randomly chosen by taking a random permutation of the numbers 1–20, then assigning the first five numbers to operator A, and so on. Second, the order of brightness measurements for the 20 handsheets could have been randomized.

The linear model for the one-way layout is

$$y_{ij} = \eta + \tau_i + \epsilon_{ij}, \quad i = 1,\ldots,k; \quad j = 1,\ldots,n_i, \tag{1.30}$$

where y_{ij} is the jth observation with treatment i, τ_i is the ith treatment effect, the errors ϵ_{ij} are independent $N(0, \sigma^2)$ with mean 0 and variance σ^2, k is the number of treatments, and n_i is the number of observations with treatment i.

In terms of the general linear model (1.3), for the pulp experiment $\boldsymbol{\beta} = (\eta, \tau_1, \tau_2, \tau_3, \tau_4)^T$, $\mathbf{y}$ is the column vector (59.8, 59.8, 60.7, 61.0, 60.0, 60.2, 60.7, 60.8, 60.4, 60.5, 60.6, 60.8, 59.9, 60.9, 60.5, 59.8, 60.0, 60.3, 60.5)^T; that is, operator A's first reading, operator B's first reading, and so on. The corresponding model matrix $\mathbf{X}$ is the 20×5 matrix

$$\mathbf{X} = \begin{pmatrix}
1 & 1 & 0 & 0 & 0 \\
1 & 0 & 1 & 0 & 0 \\
1 & 0 & 0 & 1 & 0 \\
1 & 0 & 0 & 0 & 1 \\
1 & 1 & 0 & 0 & 0 \\
1 & 0 & 1 & 0 & 0 \\
1 & 0 & 0 & 1 & 0 \\
1 & 0 & 0 & 0 & 1 \\
1 & 1 & 0 & 0 & 0 \\
1 & 0 & 1 & 0 & 0 \\
1 & 0 & 0 & 1 & 0 \\
1 & 0 & 0 & 0 & 1 \\
1 & 1 & 0 & 0 & 0 \\
1 & 0 & 1 & 0 & 0 \\
1 & 0 & 0 & 1 & 0 \\
1 & 0 & 0 & 0 & 1 \\
1 & 1 & 0 & 0 & 0 \\
1 & 0 & 1 & 0 & 0 \\
1 & 0 & 0 & 1 & 0 \\
1 & 0 & 0 & 0 & 1
\end{pmatrix}. \qquad (1.31)$$

The ANOVA table for the general linear model in Table 1.1 can be shown to reduce to Table 1.3 for the one-way layout, where $N = \sum_{i=1}^{k} n_k$, $p = k - 1$, and $\bar{y}_{..}$ denotes the average of all N observations. Generally, the dot subscript indicates the summation over the particular index, e.g., $\bar{y}_{i.}$ is the mean of observations for the ith treatment (i.e., averaged over the second index, $j = 1, \ldots, n_i$).

Instead of using the matrix algebra of Section 1.4, the ANOVA for the one-way layout can be derived directly as follows. Using the decomposition

$$y_{ij} = \hat{\eta} + \hat{\tau}_i + r_{ij}$$

$$= \bar{y}_{..} + (\bar{y}_{i.} - \bar{y}_{..}) + (y_{ij} - \bar{y}_{i.}), \qquad (1.32)$$

Table 1.3   ANOVA Table for One-Way Layout

Source	Degrees of Freedom	Sum of Squares
treatment	$k - 1$	$\sum_{i=1}^{k} n_i (\bar{y}_{i.} - \bar{y}_{..})^2$
residual	$N - k$	$\sum_{i=1}^{k} \sum_{j=1}^{n_i} (y_{ij} - \bar{y}_{i.})^2$
total	$N - 1$	$\sum_{i=1}^{k} \sum_{j=1}^{n_i} (y_{ij} - \bar{y}_{..})^2$

where

$$\hat{\eta} = \bar{y}_{..}, \quad \hat{\tau}_i = \bar{y}_{i.} - \bar{y}_{..}, \quad r_{ij} = y_{ij} - \bar{y}_{i.}, \tag{1.33}$$

then using

$$y_{ij} - \bar{y}_{..} = (\bar{y}_{i.} - \bar{y}_{..}) + (y_{ij} - \bar{y}_{i.}), \tag{1.34}$$

and squaring both sides and summing over i and j yield

$$\sum_{i=1}^{k} \sum_{j=1}^{n_i} (y_{ij} - \bar{y}_{..})^2 = \sum_{i=1}^{k} n_i (\bar{y}_{i.} - \bar{y}_{..})^2 + \sum_{i=1}^{k} \sum_{j=1}^{n_i} (y_{ij} - \bar{y}_{i.})^2. \tag{1.35}$$

For each i,

$$(\bar{y}_{i.} - \bar{y}_{..}) \sum_{j=1}^{n_i} (y_{ij} - \bar{y}_{i.}) = 0, \tag{1.36}$$

so that these cross-product terms do not appear in (1.35). The corrected total sum of squares on the left equals the treatment sum of squares plus the residual sum of squares. These three terms are given in the ANOVA table in Table 1.3. The treatment sum of squares is also called the *between-treatment sum of squares* and the residual sum of squares the *within-treatment sum of squares*.

Thus, the F statistic for the null hypothesis that there is no difference between the treatments, i.e.,

$$H_0 : \tau_1 = \cdots = \tau_k, \tag{1.37}$$

is

$$F = \frac{\sum_{i=1}^{k} n_i (\bar{y}_{i.} - \bar{y}_{..})^2 / (k-1)}{\sum_{i=1}^{k} \sum_{j=1}^{n_i} (y_{ij} - \bar{y}_{i.})^2 / (N-k)}, \tag{1.38}$$

which has an F distribution with parameters $k-1$ and $N-k$.

For the pulp experiment,

$$\bar{y}_{1.} = 60.24, \quad \bar{y}_{2.} = 60.06, \quad \bar{y}_{3.} = 60.62, \quad \bar{y}_{4.} = 60.68, \quad \bar{y}_{..} = 60.40,$$

$$n_1 = n_2 = n_3 = n_4 = 5.$$

Therefore the treatment (i.e., operator) sum of squares in Table 1.3 is

$$5(60.24 - 60.40)^2 + 5(60.06 - 60.40)^2$$
$$+ 5(60.62 - 60.40)^2 + 5(60.68 - 60.40)^2 = 1.34.$$

Table 1.4 ANOVA Table, Pulp Experiment

Source	Degrees of Freedom	Sum of Squares	Mean Squares	F
operator	3	1.34	0.447	4.20
residual	16	1.70	0.106	
total	19	3.04		

Because there are four operators, $k = 4$ and the degrees of freedom for the treatment sum of squares is $3 \,(= 4 - 1)$. Its mean square is then

$$\frac{1.34}{3} = 0.447.$$

Both 1.34 and 0.447 are given in the "operator" row of the ANOVA table in Table 1.4. Similarly, the residual sum of squares is

$$(59.8 - 60.24)^2 + (60 - 60.24)^2 + \cdots + (60.5 - 60.68)^2 = 1.70,$$

which has $16 \,(= N - k = 20 - 4)$ degrees of freedom. Then the mean-squared error is

$$\hat{\sigma}^2 = \frac{1.70}{16} = 0.106.$$

Both 1.70 and 0.106 are given in the "residual" row of the ANOVA table in Table 1.4. The F statistic in (1.38) has the value

$$\frac{0.447}{0.106} = 4.20,$$

which is given in Table 1.4 under the column F. Under the null hypothesis H_0, the F statistic has an F distribution with 3 and 16 degrees of freedom. The area under the curve (in Figure 1.4) to the right of the observed F value of 4.20 is the p value

$$Prob(F_{3, 16} > 4.20) = 0.02.$$

Recalling the interpretation of p values given after (1.16), the small value of 0.02 provides some evidence that there is an operator-to-operator difference. Another way to interpret the value 0.02 is that for the pulp experiment the F test rejects the null hypothesis H_0 at the 0.02 level.

Once H_0 is rejected, an immediate question is: what pairs of treatments are different? This question will be addressed by the method of multiple comparisons in the next section.

So far we have not considered the estimation of the treatment effects τ_i in (1.30). Because there are k types of observations but $k+1$ regression parameters in (1.30), the model (1.30) is *over-parameterized*. If one attempts to fit the model, i.e., to calculate the least squares estimate β in (1.10), the matrix $(\mathbf{X}^T\mathbf{X})^{-1}$ based on (1.31) does not exist. The matrix $\mathbf{X}$ is not of full rank since the sum of columns 2–5 equals column 1; that is, the five columns are not linearly independent so that $\mathbf{X}^T\mathbf{X}$ is singular. In order to make $\mathbf{X}^T\mathbf{X}$ a nonsingular matrix, one constraint needs to be put on the parameters. Two types of constraints will be considered in the remaining part of the section.

Constraint on the Parameters

The more commonly used constraint is

$$\sum_{i=1}^{k} \tau_i = 0, \tag{1.39}$$

which is called the **zero-sum constraint**. An example is the $\hat{\tau}_i$ in (1.33) for the ANOVA decomposition. It is readily verified that, for $n_i = n$,

$$\sum_{i=1}^{k} \hat{\tau}_i = \sum_{i=1}^{k} (\bar{y}_{i\cdot} - \bar{y}_{\cdot\cdot}) = 0.$$

Given τ_i, $i = 1, \ldots, k-1$, $\tau_k = -\sum_{i=1}^{k-1}\tau_i$. In substituting τ_k by $-\sum_{i=1}^{k-1}\tau_i$ in the model (1.30), the number of parameters is reduced by 1, i.e., $\beta = (\eta, \tau_1, \tau_2, \ldots, \tau_{k-1})^T$. The remaining parameters have a different meaning. For example,

$$\frac{1}{k} \sum_{i=1}^{k} E(y_{ij}) = \frac{1}{k} \sum_{i=1}^{k} (\eta + \tau_i) = \eta + 0 = \eta,$$

i.e., η represents the grand mean. Also,

$$E(y_{ij}) - \eta = \eta + \tau_i - \eta = \tau_i,$$

which is the *offset between the expected treatment i response and the average response*. Since treatment k is a linear combination of the remaining treatments, the treatment k offset can be represented by the linear combination $\mathbf{a}^T\beta = -\sum_{i=1}^{k-1}\tau_i$, where $\mathbf{a} = (0, -1, \ldots, -1)^T$.

With $\boldsymbol{\beta} = (\eta, \tau_1, \tau_2, \tau_3, \tau_4)^T$, (1.30) leads to

$$\mathbf{X}\boldsymbol{\beta} = \eta \begin{pmatrix} 1 \\ 1 \\ 1 \\ 1 \end{pmatrix} + \tau_1 \begin{pmatrix} 1 \\ 0 \\ 0 \\ 0 \end{pmatrix} + \tau_2 \begin{pmatrix} 0 \\ 1 \\ 0 \\ 0 \end{pmatrix} + \tau_3 \begin{pmatrix} 0 \\ 0 \\ 1 \\ 0 \end{pmatrix} + \tau_4 \begin{pmatrix} 0 \\ 0 \\ 0 \\ 1 \end{pmatrix} \tag{1.40}$$

for the first four rows of the $\mathbf{X}$ matrix. Substituting $\tau_4 = -\tau_1 - \tau_2 - \tau_3$ in (1.40) leads to

$$\mathbf{X}\boldsymbol{\beta} = \eta \begin{pmatrix} 1 \\ 1 \\ 1 \\ 1 \end{pmatrix} + \tau_1 \begin{pmatrix} 1 \\ 0 \\ 0 \\ 0 \end{pmatrix} + \tau_2 \begin{pmatrix} 0 \\ 1 \\ 0 \\ 0 \end{pmatrix} + \tau_3 \begin{pmatrix} 0 \\ 0 \\ 1 \\ 0 \end{pmatrix} + (-\tau_1 - \tau_2 - \tau_3) \begin{pmatrix} 0 \\ 0 \\ 0 \\ 1 \end{pmatrix}$$

$$= \eta \begin{pmatrix} 1 \\ 1 \\ 1 \\ 1 \end{pmatrix} + \tau_1 \begin{pmatrix} 1 \\ 0 \\ 0 \\ -1 \end{pmatrix} + \tau_2 \begin{pmatrix} 0 \\ 1 \\ 0 \\ -1 \end{pmatrix} + \tau_3 \begin{pmatrix} 0 \\ 0 \\ 1 \\ -1 \end{pmatrix},$$

which leads to the following model matrix $\mathbf{X}$ with $\boldsymbol{\beta} = (\eta, \tau_1, \tau_2, \tau_3)^T$:

$$\mathbf{X} = \begin{pmatrix} 1 & 1 & 0 & 0 \\ 1 & 0 & 1 & 0 \\ 1 & 0 & 0 & 1 \\ 1 & -1 & -1 & -1 \\ 1 & 1 & 0 & 0 \\ 1 & 0 & 1 & 0 \\ 1 & 0 & 0 & 1 \\ 1 & -1 & -1 & -1 \\ 1 & 1 & 0 & 0 \\ 1 & 0 & 1 & 0 \\ 1 & 0 & 0 & 1 \\ 1 & -1 & -1 & -1 \\ 1 & 1 & 0 & 0 \\ 1 & 0 & 1 & 0 \\ 1 & 0 & 0 & 1 \\ 1 & -1 & -1 & -1 \\ 1 & 1 & 0 & 0 \\ 1 & 0 & 1 & 0 \\ 1 & 0 & 0 & 1 \\ 1 & -1 & -1 & -1 \end{pmatrix}. \tag{1.41}$$

For the pulp experiment,

$$\hat{\boldsymbol{\beta}} = \left(\hat{\eta}, \hat{\tau}_1, \hat{\tau}_2, \hat{\tau}_3\right)^T = (60.40, -0.16, -0.34, 0.22)^T. \qquad (1.42)$$

Although $\boldsymbol{\beta}$ and $\hat{\boldsymbol{\beta}}$ depend on the choice of constraints, $\mathbf{X}\boldsymbol{\beta}$ and $\mathbf{X}\hat{\boldsymbol{\beta}}$ do not. That is, $\mathbf{X}\boldsymbol{\beta}$ in (1.31) and $\boldsymbol{\beta} = (\eta, \tau_1, \tau_2, \tau_3, \tau_4)^T$ is equal to $\mathbf{X}\boldsymbol{\beta}$ for $\mathbf{X}$ in (1.41) and $\boldsymbol{\beta} = (\eta, \tau_1, \tau_2, \tau_3)^T$.

The second type of constraint is

$$\tau_1 = 0, \qquad (1.43)$$

which amounts to dropping τ_1 from the vector of parameters $\boldsymbol{\beta}$ and dropping the corresponding column in the model matrix $\mathbf{X}$. Now $(\mathbf{X}^T\mathbf{X})^{-1}$ exists, so that the least squares estimate $\hat{\boldsymbol{\beta}}$ can be obtained. How should the remaining parameters be interpreted? It can be shown that $\eta = E(y_{1j})$, i.e., the expected response value from treatment 1;

$$E(y_{2j}) - E(y_{1j}) = \eta + \tau_2 - \eta = \tau_2, \qquad (1.44)$$

i.e., τ_2 represents the comparison of treatments 1 and 2 in terms of their expected responses; and the remaining τ parameters have a similar interpretation, e.g., $\tau_i = E(y_{ij}) - E(y_{1j})$. Thus, τ_i, $i \geq 2$, represent the comparisons between the first and the rest of the k treatments. Other treatment comparisons can be expressed as linear combinations of the τ_i, e.g., the comparison of treatments 2 and 3 is given by $\tau_3 - \tau_2$, which is a linear combination, $\mathbf{a}^T\boldsymbol{\beta}$, of $\boldsymbol{\beta}$, where $\mathbf{a}^T = (0, -1, 1, 0, \ldots, 0)$. Any linear combination $\mathbf{a}^T\boldsymbol{\beta}$ of $\boldsymbol{\beta}$ with $\Sigma a_j = 0$ is called a *contrast*.

The constraint in (1.43) is natural when treatment 1 is a standard or existing treatment and the other $k - 1$ treatments are new treatments. The performance of the new treatments is measured by their comparisons with the standard treatment. Treatment 1 can also be interpreted as the baseline for studies in medicine and social sciences. It is referred to as the **baseline constraint** in the book.

For the pulp experiment, $\boldsymbol{\beta} = (\eta, \tau_2, \tau_3, \tau_4)^T$. Then, $\mathbf{X}$ is obtained by dropping the second column of (1.31) and

$$\hat{\boldsymbol{\beta}} = (60.24, -0.18, 0.38, 0.44)^T. \qquad (1.45)$$

Again $\mathbf{X}\boldsymbol{\beta}$ (and respectively, $\mathbf{X}\hat{\boldsymbol{\beta}}$) under $\tau_1 = 0$ is the same as $\mathbf{X}\boldsymbol{\beta}$ (and respectively, $\mathbf{X}\hat{\boldsymbol{\beta}}$) under $\Sigma_{i=1}^{k} \tau_i = 0$.

1.7 MULTIPLE COMPARISONS

Because difference pairs (and sets) of treatments are being compared, this problem is referred to as **multiple comparisons**. To test for the difference between the ith and jth treatments in the one-way layout, it is common to use the t statistic:

$$t_{ij} = \frac{\bar{y}_{j\cdot} - \bar{y}_{i\cdot}}{\hat{\sigma}\sqrt{1/n_j + 1/n_i}}, \tag{1.46}$$

where $\bar{y}_{i\cdot}$ denotes the average of the n_i observations for treatment i and $\hat{\sigma}$ is the square root of the MSE from the ANOVA table (which is also called the root-mean-square error, or RMSE). Note that the denominator in (1.46) is the standard error of the numerator.

Suppose that the null hypothesis $H_0: \tau_1 = \cdots = \tau_k$ is rejected. An immediate question is to determine which pairs of treatments are significantly different. Using the *two-sample t test*, treatments i and j are declared significantly different at level α if

$$|t_{ij}| > t_{N-k, \alpha/2}, \tag{1.47}$$

where $t_{N-k, \alpha/2}$ is the upper $\alpha/2$ quantile of a t distribution with $N - k$ degrees of freedom. This test is valid for testing one pair of treatments.

Suppose that k' such tests are performed. It is easy to show that, under H_0, the probability of declaring at least one pair of treatments significantly different (which is called the *experimentwise error rate*) exceeds α for $k' > 1$. For larger k', the experimentwise error rate is higher. (Its proof is left as an exercise.) Therefore, the standard t test cannot be applied in the multiple comparisons of treatments.

To control the experimentwise error rate, two methods are available: the Bonferroni and the Tukey methods. They are convenient to use and have known theoretical properties.

The Bonferroni Method

The Bonferroni method for testing $\tau_i = \tau_j$ versus $\tau_i \neq \tau_j$ declares "τ_i different from τ_j at level α/k'" if

$$|t_{ij}| > t_{N-k, \alpha/2k'}, \tag{1.48}$$

where k' denotes the number of pairs being tested. In the case of the one-way layout with k treatments,

$$k' = \binom{k}{2} = \frac{1}{2}k(k-1).$$

Denote the set of observed data that satisfy (1.48) by A_{ij}. From the distribution of the t statistic,

$$Prob(A_{ij}|\tau_i = \tau_j) = \frac{\alpha}{k'}. \tag{1.49}$$

For any pair (i, j), declare "τ_i different from τ_j" if (1.48) for (i, j) is satisfied. Under $H_0 : \tau_1 = \cdots = \tau_k$,

$$Prob(\text{at least one pair is declared significantly different} | H_0)$$

$$= Prob\left(\bigcup_{i<j} A_{ij} | H_0 \right) \leq \sum_{i<j} Prob\{A_{ij} | \tau_i = \tau_j\} \tag{1.50}$$

$$= \sum_{i<j} \alpha/k' = \frac{k'\alpha}{k'} = \alpha.$$

Therefore, the probability of mistakenly declaring any pair of treatments significantly different when they are not (i.e., the experimentwise error rate) is at most α. This method is very easy to use. It is conservative but works for very general problems because the so-called *Bonferroni inequality* in (1.50) is general. For the one-way layout, the Tukey method (to be introduced next) is recommended. For multiple comparison problems for which the Tukey method is not applicable, the Bonferroni method is recommended.

The method can also be adapted to construct conservative *simultaneous confidence intervals* for the k' pairs of differences $\{\tau_i - \tau_j\}_{i<j}$. Solving for

$$\left| \bar{y}_{i\cdot} - \bar{y}_{j\cdot} - (\tau_i - \tau_j) \right| \leq t_{N-k, \alpha/2k'} \hat{\sigma} \sqrt{(1/n_i + 1/n_j)} \tag{1.51}$$

leads to the confidence interval for $\tau_i - \tau_j$ as

$$\bar{y}_{i\cdot} - \bar{y}_{j\cdot} \pm t_{N-k, \alpha/2k'} \hat{\sigma} \sqrt{1/n_i + 1/n_j} . \tag{1.52}$$

That is, after identifying which pairs are different, the confidence interval in (1.52) quantifies how different the two treatment effects are.

For the pulp experiment, the means for operators A–D are 60.24, 60.06, 60.62, and 60.68, respectively. The t statistics (1.46) are given in Table 1.5 for the six pairs of treatments. For example, the A-vs.-B t statistic is calculated as

$$\frac{60.06 - 60.24}{\sqrt{0.106} \sqrt{1/5 + 1/5}} = -0.87, \tag{1.53}$$

Table 1.5   Multiple Comparison t Statistics, Pulp Experiment

A vs. B	A vs. C	A vs. D	B vs. C	B vs. D	C vs. D
− 0.87	1.85	2.14	2.72	3.01	0.29

where $\hat{\sigma}^2 = 0.106$ is from Table 1.4 and $n_1 = n_2 = 5$. Notice that for the A-vs.-B comparison, the mean of A is subtracted from the mean of B in (1.53). *This convention is followed throughout the book when the t statistic for a pairwise comparison is presented in a table.* To apply the Bonferroni method in (1.48) at level $\alpha = 0.05$, first compute the $t_{N-k, \alpha/2k'}$ value, which is

$$t_{16, 0.05/2 \times 6} = 3.008, \tag{1.54}$$

because $N = 20$, $k = 4$, and $k' = 6$. By comparing the t values in Table 1.5 with 3.008, only the B-vs.-D comparison has a t value that exceeds 3.008. Therefore, only operators B and D are significantly different at the 0.05 level.

The Tukey Method

The only difference between the Tukey and Bonferroni methods is in the choice of the critical value. The Tukey method is described as follows: for any pair (i, j) with $1 \le i < j \le k$, declare "τ_i different from τ_j" if

$$|t_{ij}| > \frac{1}{\sqrt{2}} q_{k, N-k, \alpha}, \tag{1.55}$$

where t_{ij} is defined in (1.46) and $q_{k, N-k, \alpha}$ is the upper α quantile of the **Studentized range** distribution with parameter k and $N - k$ degrees of freedom. Recall that k is the number of treatments. See Appendix E for the Studentized range critical values. This method for equal sample sizes has been widely used for many years. A proof that the procedure (1.55) works for general n_i and n_j, i.e., the experimentwise error rate is at most α, can be found in Hochberg and Tamhane (1987). Details on the Studentized range distribution can be found in the same book.

This method is known to be generally the most effective among conservative methods for the one-way ANOVA, that is, its Type II error is generally the smallest (or equivalently, its confidence bound is the tightest). It is recommended unless the critical value $q_{k, N-k, \alpha}$ is not tabled.

For the balanced one-way layout (i.e., $n_i = n$), the experimentwise error rate for the Tukey method is exactly α. To prove this, note that

$$Prob(\text{at least one pair is declared significantly different} | H_0)$$

$$= Prob\left(\max_{i,j} \frac{|\bar{y}_{i\cdot} - \bar{y}_{j\cdot}|}{\hat{\sigma}\sqrt{(1/n + 1/n)}} > \frac{1}{\sqrt{2}} q_{k, N-k, \alpha} | H_0 \right)$$

$$= Prob\left(\frac{\max \bar{y}_{i\cdot} - \min \bar{y}_{i\cdot}}{\sigma\sqrt{1/n}} > q_{k, N-k, \alpha} | H_0 \right) = \alpha. \tag{1.56}$$

The last equality in (1.56) holds because under H_0

$$\sqrt{n}\,(\max \bar{y}_{i\cdot} - \min \bar{y}_{i\cdot})/\hat{\sigma}$$

is the *Studentized range* statistic with parameters k and $N-k$.

By solving

$$\frac{\left|(\bar{y}_{j\cdot} - \bar{y}_{i\cdot}) - (\tau_j - \tau_i)\right|}{\hat{\sigma}\sqrt{1/n_j + 1/n_i}} \le \frac{1}{\sqrt{2}}\, q_{k,\,N-k,\,\alpha}$$

for $\tau_j - \tau_i$, the simultaneous confidence intervals for $\tau_j - \tau_i$ are

$$\bar{y}_{j\cdot} - \bar{y}_{i\cdot} \pm \frac{1}{\sqrt{2}}\, q_{k,\,N-k,\,\alpha}\, \hat{\sigma}\sqrt{\frac{1}{n_j} + \frac{1}{n_i}}\,, \qquad (1.57)$$

for all i and j pairs. Since the Bonferroni method is conservative, the simultaneous confidence intervals based on the Tukey method are shorter.

For the pulp experiment, according to (1.55) at $\alpha = 0.05$,

$$\frac{1}{\sqrt{2}}\, q_{k,\,N-k,\,0.05} = \frac{1}{\sqrt{2}}\, q_{4,\,16,\,0.05} = \frac{4.05}{\sqrt{2}} = 2.86.$$

By comparing 2.86 with the t values in Table 1.5, the Tukey method also identifies that operators B and D are different. The 2.86 used here is smaller than the 3.008 of the Bonferroni method because the Bonferroni method is more conservative. For multiple comparisons at the 0.05 level, the two methods reach the same conclusion, but the simultaneous confidence intervals for the Tukey method are shorter.

1.8 QUANTITATIVE FACTORS AND ORTHOGONAL POLYNOMIALS

Mazumdar and Hoa (1995) performed an experiment which dealt with the laser-assisted manufacturing of a thermoplastic composite. The experimental factor is laser power at 40, 50, and 60 watts. The response is interply bond strength of the composite as measured by a short-beam-shear test. The strength data for the composite experiment are displayed in Table 1.6.

By treating the experimental design as a one-way layout, the ANOVA table for the experiment is computed and given in Table 1.7. The p value for the test of significance for the laser factor $Prob(F_{2,6} > 11.32)$ is 0.009, where $11.32 = 112.09/9.90$ is the observed F statistic value F_{obs}. Thus, the experiment provides strong evidence that laser power affects the strength of the composite. Because the laser power factor has three levels in the experiment,

Table 1.6   Strength Data, Composite Experiment

	Laser Power	
40 W	50 W	60 W
25.66	29.15	35.73
28.00	35.09	39.56
20.65	29.79	35.66

Table 1.7   ANOVA Table, Composite Experiment

Source	Degrees of Freedom	Sum of Squares	Mean Squares	F
laser	2	224.184	112.092	11.32
residual	6	59.422	9.904	
total	8	283.606		

its significance can be further studied by decomposing the sum of squares for the laser factor (with two degrees of freedom) into a linear component and a quadratic component.

Suppose that a factor is quantitative and has three levels with evenly spaced values. For example, laser power in the composite experiment has evenly spaced levels 40, 50, and 60. Then, one can investigate whether the relationship between the factor and response is linear or quadratic over the three levels. Denote the response value at the low, medium, and high levels by y_1, y_2, and y_3, respectively. Then the linear relationship can be evaluated using

$$y_3 - y_1 = -1y_1 + 0y_2 + 1y_3,$$

which is called the **linear contrast**. To define a quadratic effect, one can use the following argument. If the relationship is linear, then $y_3 - y_2$ and $y_2 - y_1$ should approximately be the same, i.e., $(y_3 - y_2) - (y_2 - y_1) = 1y_1 - 2y_2 + 1y_3$ should be close to zero. Otherwise, it should be large. Therefore, the **quadratic contrast**

$$y_1 - 2y_2 + y_3$$

can be used to investigate a quadratic relationship. The linear and quadratic contrasts can be written as $(-1, 0, 1)(y_1, y_2, y_3)^T$ and $(1, -2, 1)(y_1, y_2, y_3)^T$. The coefficient vectors $(-1, 0, 1)$ and $(1, -2, 1)$ are called the *linear contrast vector* and *the quadratic contrast vector*. Two vectors $\mathbf{u} = (u_i)_1^l$ and $\mathbf{v} = (v_i)_1^l$ are said to be **orthogonal** if their cross product $\mathbf{u}\mathbf{v}^T = \sum_{i=1}^l u_i v_i = 0$. It is easy to verify that the linear and quadratic contrast vectors are orthogonal, i.e., their cross product $(-1, 0, 1)(1, -2, 1)^T = (-1)(1) + (0)(-2) + (1)(1) = -1 + 0 + 1 = 0$. Orthogonality is a desirable property because it ensures that the contrasts and the tests based on them are statistically independent. To provide a consistent comparison of their regression coefficients, these vectors should be scaled by their lengths, i.e., $\sqrt{2}$ ($= [(-1)^2 + 0^2 + 1^2]^{1/2}$) and $\sqrt{6}$ ($= [(-1)^2 + 2^2 + (-1)^2]^{1/2}$), respectively. These scaled vectors become the covariates in the model matrix.

To see whether laser power has a linear and/or quadratic effect on strength, the linear model with linear and quadratic contrasts [i.e., $(-1, 0, 1)/\sqrt{2}$ for linear, $(1, -2, 1)/\sqrt{6}$ for quadratic] can be fitted and their effects tested for significance. Thus,

$$\mathbf{y} = (25.66, 29.15, 35.73, 28.00, 35.09, 39.56, 20.65, 29.79, 35.66)^T$$

and the model matrix $\mathbf{X}$ is

$$\mathbf{X} = \begin{pmatrix} 1 & -1 & 1 \\ 1 & 0 & -2 \\ 1 & 1 & 1 \\ 1 & -1 & 1 \\ 1 & 0 & -2 \\ 1 & 1 & 1 \\ 1 & -1 & 1 \\ 1 & 0 & -2 \\ 1 & 1 & 1 \end{pmatrix}, \tag{1.58}$$

whose second and third columns need to be divided by $\sqrt{2}$ and $\sqrt{6}$, respectively. The formulas in (1.10) and (1.17) are used to calculate the estimates and standard errors, which are given in Table 1.8 along with the corresponding t statistics. (The least squares estimate for the intercept is 31.0322.) The results in Table 1.8 indicate that laser power has a strong linear effect but no quadratic effect on composite strength. While the ANOVA in Table 1.7 indicates that laser power has a significant effect on composite strength, the additional analysis in Table 1.8 identifies the linear effect of laser power as the main contributor to the significance.

Suppose that the investigator of the composite experiment would like to predict the composite strength for other settings of the laser power, such as 55 or 62 watts. In order to answer this question, we need to extend the notion of orthogonal contrast vectors to **orthogonal polynomials**. Denote the three

Table 1.8 Tests for Polynomial Effects, Composite Experiment

Effect	Estimate	Standard Error	t	p value
linear	8.636	1.817	4.75	0.003
quadratic	-0.381	1.817	-0.21	0.841

evenly spaced levels by $m - \Delta$, m, and $m + \Delta$, where m denotes the middle level and Δ the distance between consecutive levels. Then define the first- and second-degree polynomials

$$P_1(x) = \frac{x - m}{\Delta}, \tag{1.59}$$

$$P_2(x) = 3\left[\left(\frac{x - m}{\Delta}\right)^2 - \frac{2}{3}\right]. \tag{1.60}$$

It is easy to verify that $P_1(x) = -1, 0, 1$ and $P_2(x) = 1, -2, 1$ for x equal to $m - \Delta$, m, and $m + \Delta$, respectively. These two polynomials are more general than the linear and quadratic contrast vectors because they are defined for a whole range of the quantitative factor and give the same values as the contrast vectors at the three levels of the factor where the experiment is conducted. Based on P_1 and P_2, we can use the following model for predicting the y value at *any* x value in the range,

$$y = \beta_0 + \beta_1 P_1(x)/\sqrt{2} + \beta_2 P_2(x)/\sqrt{6} + \epsilon, \tag{1.61}$$

where $\sqrt{2}$ and $\sqrt{6}$ are the scaling constants used for the linear and quadratic contrast vectors and ϵ are independent $N(0, \sigma^2)$. Because the y values are observed in the experiment only at $x = m - \Delta$, m and $m + \Delta$, the least squares estimates of β_0, β_1, and β_2 in (1.61) are the same as those using regression analysis with the **X** matrix in (1.58).

For the composite experiment, fitting model (1.61) including the quadratic effect (that was not significant) leads to the prediction model:

$$\text{Predicted strength} = 31.0322 + 8.636 P_1(\text{laser power})/\sqrt{2}$$

$$- 0.381 P_2(\text{laser power})/\sqrt{6}, \tag{1.62}$$

where the estimated coefficients 8.636 and -0.381 are the same as in Table 1.8. The model in (1.62) can be used to predict the strength for laser power settings in $[40, 60]$ and its immediate vicinity. For example, at a laser power of 55 watts,

$$P_1(55) = \frac{55 - 50}{10} = \frac{1}{2}, \qquad \frac{1}{\sqrt{2}} P_1(55) = \frac{1}{2\sqrt{2}} = 0.3536,$$

$$P_2(55) = 3\left[\left(\frac{55 - 50}{10}\right)^2 - \frac{2}{3}\right] = 3\left(\frac{1}{2}\right)^2 - 2 = -\frac{5}{4},$$

and

$$\frac{1}{\sqrt{6}} P_2(55) = \frac{-5}{4\sqrt{6}} = -0.5103,$$

where $m = 50$ and $\Delta = 10$. Therefore at 55 watts

$$\text{Predicted strength} = 31.0322 + 8.636(0.3536) - 0.381(-0.5103)$$
$$= 34.2803.$$

Using (1.62) to extrapolate far outside the experimental region $[40, 60]$, where the model may no longer hold, should be avoided.

For k evenly spaced levels, orthogonal polynomials of degree $1, \ldots, k - 1$ can be constructed. Orthogonal polynomials of degree 1–4 are given as follows:

$$P_1(x) = \lambda_1\left(\frac{x - m}{\Delta}\right),$$

$$P_2(x) = \lambda_2\left[\left(\frac{x - m}{\Delta}\right)^2 - \left(\frac{k^2 - 1}{12}\right)\right],$$

$$P_3(x) = \lambda_3\left[\left(\frac{x - m}{\Delta}\right)^3 - \left(\frac{x - m}{\Delta}\right)\left(\frac{3k^2 - 7}{20}\right)\right],$$

$$P_4(x) = \lambda_4\left[\left(\frac{x - m}{\Delta}\right)^4 - \left(\frac{x - m}{\Delta}\right)^2\left(\frac{3k^2 - 13}{14}\right) + \frac{3(k^2 - 1)(k^2 - 9)}{560}\right],$$

where Δ is the distance between the levels of x, k is the total number of levels, and $\{\lambda_i\}$ are constants such that the polynomials have integer values. Coefficients of the orthogonal polynomials and values of the λ_i for $k \leq 7$ are given in Appendix G. The value C in each column of the table in Appendix G represents the sum of squares of the coefficients. The contrast vector (whose entries are the coefficients in the column) can be scaled (i.e., divided) by the corresponding $\sqrt{C}$ value so that the regression coefficients of the polynomial effects can be consistently compared.

The model in (1.61) can be extended to a k-level factor by using higher degree polynomials. Polynomials with fourth and higher degrees, however, should not be used unless they can be justified by a physical model. Data can be well fitted by using a high-degree polynomial model but the resulting fitted model will lack predictive power. In regression analysis this phenomenon is known as *overfitting*. The undesirable consequence is that the average variance of the regression parameter estimates is proportional to the number of regression parameters in the model. Therefore, overfitting leads to a bigger variance in estimation and prediction. More on this can be found in Draper and Smith (1998). Another problem with the use of high-degree polynomials is that they are difficult to interpret. If a low-degree polynomial does not fit the data well, a semiparametric model such as those based on splines (Simonoff, 1996) can be employed. The main idea behind this class of models is to fit a low-degree polynomial over a small interval or region and patch these polynomials together over the entire region to make it into a smooth function or surface.

1.9 RESIDUAL ANALYSIS: ASSESSMENT OF MODEL ASSUMPTIONS

Before making inferences using hypothesis testing and confidence intervals, it is important to assess the model assumptions:

(i) Have all important effects been captured?
(ii) Are the errors independent and normally distributed?
(iii) Do the errors have constant (the same) variance?

We can assess these assumptions graphically by looking at the **residuals**

$$r_i = y_i - \hat{y}_i, \qquad i = 1, \dots, N,$$

where $\hat{y}_i = \mathbf{x}_i^T \hat{\boldsymbol{\beta}}$ is the fitted (or predicted) response at $\mathbf{x}_i$ and $\mathbf{x}_i$ is the ith row of the matrix $\mathbf{X}$ in (1.4). Writing $\mathbf{r} = (r_i)_{i=1}^N$, $\mathbf{y} = (y_i)_{i=1}^N$, $\hat{\mathbf{y}} = (\hat{y}_i)_{i=1}^N = \mathbf{X}\hat{\boldsymbol{\beta}}$, we have

$$\mathbf{r} = \mathbf{y} - \hat{\mathbf{y}} = \mathbf{y} - \mathbf{X}\hat{\boldsymbol{\beta}}. \tag{1.63}$$

In the decomposition, $\mathbf{y} = \hat{\mathbf{y}} + \mathbf{r}$, $\hat{\mathbf{y}}$ represents information about the assumed model, and $\mathbf{r}$ can *reveal* information about possible model violations.

Based on the model assumptions it can be shown (the proofs are left as two exercises) that the residuals have the following properties:

(a) $E(\mathbf{r}) = \mathbf{0}$,
(b) $\mathbf{r}$ and $\hat{\mathbf{y}}$ are independent, and
(c) $\mathbf{r} \sim MN(\mathbf{0}, \sigma^2(\mathbf{I} - \mathbf{H}))$, where $\mathbf{I}$ is the $N \times N$ identity matrix and

$$\mathbf{H} = \mathbf{X}(\mathbf{X}^T\mathbf{X})^{-1}\mathbf{X}^T$$

is the so-called *hat matrix* since $\hat{\mathbf{y}} = \mathbf{H}\mathbf{y}$, i.e., $\mathbf{H}$ puts the hat $\hat{}$ on $\mathbf{y}$. Thus,

$$Var(r_i) = \sigma^2(1 - h_{ii}),$$

where h_{ii} is the ith diagonal entry of $\mathbf{H}$. We might use $r_i' = r_i/\sqrt{1 - h_{ii}}$ instead, which have constant variance. Because h_{ii} do not vary much, especially for well-balanced experiments, the r_i are often used.

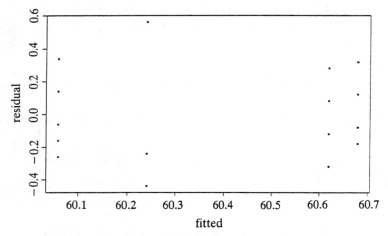

Figure 1.5. r_i vs. $\hat{y}_i$, Pulp Experiment.

Violation of the properties in (a)–(c) would suggest where the model assumptions may be going wrong and how to use the following plots to detect them:

1. **Plot** r_i **versus** $\hat{y}_i$—The plot should appear as a parallel band [from property (b)] centered about zero [from property (a)]. An example is given in Figure 1.5 for the pulp experiment. If the spread of r_i increases as $\hat{y}_i$ increases, it suggests that the error variance increases with the mean. An example is given in Figure 1.6, which is not related to the pulp experiment. Such a pattern would suggest that the response y needs to be transformed. Transformation of y will be considered in Section 2.5.

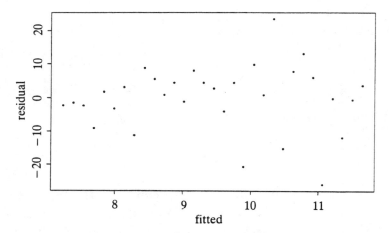

Figure 1.6. r_i vs. $\hat{y}_i$ Plot Showing a Pattern.

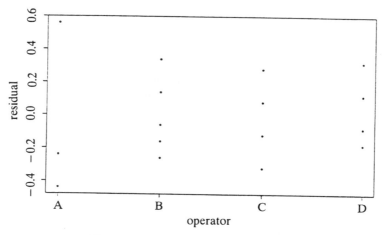

Figure 1.7. r_i vs. x_i, Pulp Experiment.

2. **Plot r_i versus x_i**—Property (c) suggests that the plot should appear as a parallel band centered about zero. An example is given in Figure 1.7 for the pulp experiment. If there is a systematic pattern, it would suggest that the relationship between x_i and the response has not been captured fully in the structural part of the model.

3. **Plot r_i versus time sequence i**, where i is the time sequence in which the observations were taken—The plot should be a parallel band centered about zero. If there is a systematic pattern, it would suggest that the observations are not independent and there is possibly correlation over time.

4. **Plot r_i from replicates grouped by treatment**—The spread of the residuals should be the same for all treatments. Unequal spreads would suggest that the error variance σ^2 also depends on the experimental factor(s). An example is displayed in Figure 1.7 for the pulp experiment; in this case, because only a single experimental factor was studied, this plot is the same as the r_i versus x_i plot.

If there is a large number of replicates per treatment, a **box-whisker** plot of the residuals for each treatment is recommended. A box-whisker plot given in Figure 1.8 displays the minimum, 25th percentile, median, 75th percentile, and maximum, where the box ends correspond to the 25th and 75th percentiles and the line inside the box is the median. Denote the 25th percentile by Q_1, the 75th percentile by Q_3, and the interquartile range $Q_3 - Q_1$ by IQR. Then the two whiskers denote the minimum and maximum values within the range $[Q_1 - 1.5\ IQR, Q_3 + 1.5\ IQR]$. Any values outside the whiskers are considered to be outliers. For example, the plot in Figure 1.8

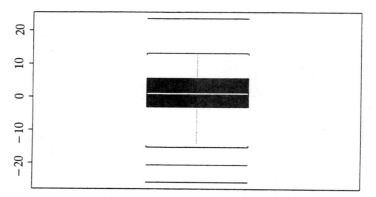

Figure 1.8. Box-Whisker Plot.

indicates one positive outlier and two negative outliers. The box-whisker plot enables the location, dispersion, skewness, and extreme values of the replicated observations to be displayed. Its use will be demonstrated later for the bolt experiment discussed in Section 2.3.

The normality assumption of the errors can be assessed by the following method. Let $r_{(1)} \leq \cdots \leq r_{(N)}$ denote the ordered residuals. If the errors were normally distributed, then the plot of the cumulative probabilities $p_i = (i - 0.5)/N$ versus the ordered residuals $r_{(i)}$ should ideally be S-shaped, which is the shape of the normal cumulative distribution function as depicted in Figure 1.9(a). The human eye has trouble recognizing departures from a curved line but can easily detect departures from a straight line. By stretching the scale at both ends, the ideal curve becomes a straight line on the

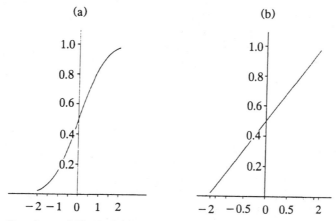

Figure 1.9. Regular and Normal Probability Plots of Normal Cumulative Distribution Function.

transformed scale, as shown in Figure 1.9(b). By plotting the ordered residuals on the transformed scale, any deviation of the plot from a straight line is an indication of the violation of normality. This method is developed and justified as follows. Suppose that the residuals r_i are normally distributed with the same variance (which is true for most balanced designs considered in the book.) Then, $\Phi(r_i)$ has a uniform distribution over $[0, 1]$. This implies that the expected values of $\Phi(r_{(i)})$, $i = 1, \ldots, N$, are spaced uniformly over $[0, 1]$, i.e., the N points $(p_i, \Phi(r_{(i)}))$, $p_i = (i - 0.5)/N$, should fall on a straight line. By applying the Φ^{-1} transformation to the horizontal and vertical scales, the N points

$$\left(\Phi^{-1}(p_i), r_{(i)}\right), \qquad i = 1, \ldots, N, \tag{1.64}$$

which form the **normal probability plot** of residuals, should plot roughly as a straight line. (Its rigorous justification can be found in Meeker and Escobar, 1998.) A marked deviation of the plot from a straight line would indicate that the normality or constant variance assumptions for the errors do not hold. The normal probability plot can also be used for quantities other than the residuals. A major application is in factorial designs, where the $r_{(i)}$ in (1.64) are replaced by ordered factorial effect estimates. (See Section 3.9.)

For the pulp experiment, the r_i vs. $\hat{y}_i$ and r_i vs. x_i plots are displayed in Figures 1.5 and 1.7. No patterns are evident in these plots. Moreover, the normal probability plot in Figure 1.10 appears close to a straight line. Thus, the residuals are consistent with the model assumptions. An unusually large residual would suggest that the associated observation may be an *outlier*. An outlier is an indication of model inadequacy or suggests that something peculiar happened to the experimental run for the associated observation.

For more details on residual analysis, see Draper and Smith (1998).

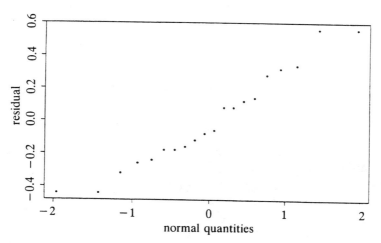

Figure 1.10. Normal Plot of r_i, Pulp Experiment.

1.10   PRACTICAL SUMMARY

1. Experimental problems can be divided into five broad categories:
   - (i)   treatment comparisons,
   - (ii)   variable screening,
   - (iii) response surface exploration,
   - (iv)   system optimization,
   - (v)   system robustness.

2. Statistical process control tools such as control charts are often used to monitor and improve a process. If a process is stable but needs to be further improved, more active intervention like experimentation should be employed.

3. There are seven steps in the planning and implementation of experiments:
   - (i)   state objective,
   - (ii)   choose response,
   - (iii) choose factors and levels,
   - (iv)   choose experimental plan,
   - (v)   perform the experiment,
   - (vi)   analyze the data,
   - (vii) draw conclusions and make recommendations.

4. Guidelines for choosing the response:
   - (i)   It should help understand the mechanisms and physical laws involved in the problem.
   - (ii)   A continuous response is preferred to a discrete response.
   - (iii) A good measurement system should be in place to measure the response.

5. For response optimization, there are three types of responses: nominal-the-best, larger-the-better, and smaller-the-better.

6. A cause-and-effect diagram or a flowchart should be used to facilitate the identification of potentially important factors and to provide a system view of the problem.

7. Three fundamental principles need to be considered in experimental design: replication, randomization, and blocking. Blocking is effective if the within-block variation is much smaller than the between-block variation.

8. Factors can be designated as **E** (experimental), **B** (blocking), **O** (constant level), and **R** (randomization).

9. A summary of linear model theory is given in Section 1.4 as the foundation for regression analysis used in the book. Variable selection procedures are given in Section 1.5.

10. One-way layout (comparison of treatments with no blocking):

    (i) Use model (1.30) with either of the constraints $\sum_1^k \tau_i = 0$ (zero sum) or $\tau_1 = 0$ (baseline). Interpretation of the τ_i for each constraint can be found in Section 1.6.

    (ii) Use the ANOVA table in Table 1.3 and the F test in (1.38) for testing the null hypothesis: $\tau_1 = \tau_2 = \cdots = \tau_k$.

    (iii) If the null hypothesis is rejected, multiple comparisons of the τ_i's should be considered. The Tukey method in (1.55) is recommended. The Bonferroni method in (1.48) can be used in very general situations. It is recommended in situations where the critical values for the Tukey method are not available.

11. For a quantitative factor, use orthogonal polynomials to further model the main effect of the factor. First- and second-degree polynomials are commonly used. Fourth- and higher degree polynomials are rarely used because of the problems associated with overfitting and interpretation.

12. For checking the model assumptions, use the following residual plots:

    (i) plot r_i versus $\hat{y}_i$,

    (ii) plot r_i versus x_i,

    (iii) plot r_i versus time sequence i,

    (iv) plot r_i from replicates grouped by treatment.

If any of these plots shows a systematic pattern, one or more of the model assumptions are violated. Countermeasures as described in Section 1.9 should be taken. If there is a large number of replicates per treatment, a box-whisker plot is recommended. It enables the location, dispersion, skewness, and extreme values of the replicated observations to be visually compared.

EXERCISES

1. Use a real example to illustrate the seven-step procedure in Section 1.2.

2. Use two examples, one from manufacturing and another from service, to illustrate the construction of the cause-and-effect diagram. Designate each factor on the diagram as **E**, **B**, **O**, or **R**.

3. Give examples of hard-to-change factors. How do you reconcile the hard-to-change nature of the factor with the need for randomization?

4. (a) For the typing experiment considered in Section 1.3, use a statistical model to quantify the gains from using randomization (as illustrated in the second sequence) and from using balance in addition to randomization.

(b) Suppose that the following sequence is obtained from using balanced randomization:

$$1\ A, B, \quad 2.\ A, B, \quad 3.\ A, B, \quad 4.\ B, A, \quad 5.\ B, A, \quad 6.\ B, A.$$

Would you use it for the study? If not, what would you do? What aspect of the sequence makes you uneasy? Can you relate it to the possibility that the advantage of the learning effect may diminish over time and express it in more rigorous terms? (Hint: The terms in the model should represent the effects you identified as potentially influencing the comparison.)

5. The typing experiment can be further improved by employing more typists that are representative of the population of typists. Suppose three typists are chosen for the study. Devise an experimental plan and discuss its pros and cons. (Some of the more elaborate plans may involve strategies that will be introduced in the next chapter.)

6. For the pulp experiment obtain the 95% simultaneous confidence intervals for the six pairs of treatment differences using the Bonferroni method and the Tukey method. Which gives shorter intervals?

7. (a) For the pulp experiment show that neither the Bonferroni nor the Tukey method declares any pair of treatments as different at the 0.01 level.

(b) How do you reconcile the finding in (a) with the result in Section 1.6 that the F test rejects the null hypothesis H_0 at the 0.05 level? After rejecting the null hypothesis, do you expect the multiple comparison method to identify at least one pair of treatments as different? (Hint: One is at the 0.01 level while the other is at the 0.05 level.)

(c) Recall that the p value for the observed F statistic value 4.20 is 0.02. How can you use this fact to reach the same conclusion in (a) without actually performing the multiple comparisons? (Hint: Use the relationship between the p value and the significance level of the F test.)

8. Make various residual plots for the composite experiment data to support the finding in Table 1.8 that the linear effect is significant while the quadratic effect is not.

9. Use the prediction model in (1.62) to predict the composite strength at 62 watts. If it is suggested to you that the model be used to predict the composite strength at 80 watts, what argument would you use to dissuade such a practice?

10. Prove that $E(\mathbf{r}) = 0$ and that the covariance matrix between $\mathbf{r}$ and $\hat{\mathbf{y}}$ is zero, i.e., $\mathbf{r}$ and $\hat{\mathbf{y}}$ are independent.

11. Show that $\mathbf{r} \sim MN(\mathbf{0}, \sigma^2(\mathbf{I} - \mathbf{H}))$, where $\mathbf{I}$ is the $N \times N$ identity matrix and $\mathbf{H} = \mathbf{X}(\mathbf{X}^T\mathbf{X})^{-1}\mathbf{X}^T$.

12. Prove that under H_0 in (1.37) the probability of declaring at least one pair of treatments significantly different, based on (1.47), exceeds α for $k' > 1$ and increases as k' increases. (Hint: Write the event in (1.47) as C_{ij} and express the rejection region as a union of the C_{ij}'s.)

13. If the plot of residuals against time exhibits a quadratic trend (going up and then going down), what does it suggest to you regarding the model currently entertained and what remedial measures would you take?

14. In order to analyze possible differences between five treatments, a one-way layout experiment was carried out. Each of the treatments was tested on three machines, resulting in a total of 15 experimental runs. After fitting the one-way model (1.30) (which has no block effect) to the data, the residuals were plotted against machine number, as shown in Figure 1.11. What do you learn from the plot? How would you modify your model and analysis?

15. The bioactivity of four different drugs A, B, C, D for treating a particular illness was compared in a study and the following ANOVA table was

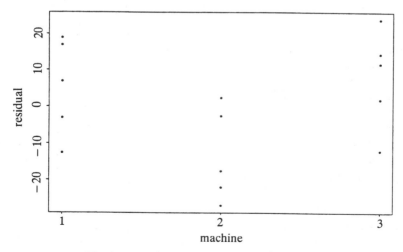

Figure 1.11. Residuals by Machine Number.

given for the data:

Source	Sum of Squares	Degrees of Freedom	Mean Square
between treatments	64.42	3	21.47
within treatments	62.12	26	2.39
total	126.54	29	

(a) Describe a proper design of the experiment to allow valid inferences to be made from the data.

(b) Use an F test to test at the 0.01 level the null hypothesis that the four treatments have the same bioactivity. Compute the p value of the observed F statistic.

(c) The treatment averages are as follows: $\bar{y}_A = 66.10$ (7 samples), $\bar{y}_B = 65.75$ (8 samples), $\bar{y}_C = 62.63$ (9 samples), $\bar{y}_D = 63.85$ (6 samples). Use the Tukey method to perform multiple comparisons of the four treatments at the 0.01 level.

(d) It turns out that A and B are brand-name drugs and C and D are generic drugs. To compare brand-name vs. generic drugs, the contrast $\frac{1}{2}(\bar{y}_A + \bar{y}_B) - \frac{1}{2}(\bar{y}_C + \bar{y}_D)$ is computed. Obtain the p value of the computed contrast and test its significance at the 0.01 level. Comment on the difference between brand-name and generic drugs.

16. In the winter, a plastic rain gauge cannot be used to collect precipitation data because it will freeze and crack. As a way to record snowfall, weather observers were instructed to collect the snow in a metal standard 2.5 can, allow the snow to melt indoors, pour it into a plastic rain gauge, and then record the measurement. An estimate of the snowfall is then obtained by multiplying the measurement by 0.44. (The factor 0.44 was theoretically derived as the ratio of the surface area of the rectangular opening of the rain gauge and of the circular metal can.) One observer questioned the validity of the 0.44 factor for estimating snowfall. Over one summer, the observer recorded the following rainfall data collected in the rain gauge and in the standard 2.5 can, both of which were mounted next to each other at the same height. The data (courtesy of Masaru Hamada) appear in Table 1.9, where the first column is the amount of rain collected in the standard 2.5 can (x) and the second column is the amount of rain collected in the rain gauge (y).

(a) Plot the residuals $y_i - 0.44 x_i$ for the data. Do you observe any systematic pattern to question the validity of the formula $y = 0.44x$?

(b) Use regression analysis to analyze the data in Table 1.9 by assuming a general β_0 (i.e., an intercept term) and $\beta_0 = 0$ (i.e., regression line through the origin). How well do the two models fit the data? Is the intercept term significant?

Table 1.9 Rainfall Data

x	y	x	y	x	y
0.11	0.05	2.15	0.96	1.25	0.62
1.08	0.50	0.53	0.32	0.46	0.23
1.16	0.54	5.20	2.25	0.31	0.17
2.75	1.31	0.00	0.06	0.75	0.33
0.12	0.07	1.17	0.60	2.55	1.17
0.60	0.28	6.67	3.10	1.00	0.43
1.55	0.73	0.04	0.04	3.98	1.77
1.00	0.46	2.22	1.00	1.26	0.58
0.61	0.35	0.05	0.05	5.40	2.34
3.18	1.40	0.15	0.09	1.02	0.50
2.16	0.91	0.41	0.25	3.75	1.62
1.82	0.86	1.45	0.70	3.70	1.70
4.75	2.05	0.22	0.12	0.30	0.14
1.05	0.58	2.22	1.00	0.07	0.06
0.92	0.41	0.70	0.38	0.58	0.31
0.86	0.40	2.73	1.63	0.72	0.35
0.24	0.14	0.02	0.02	0.63	0.29
0.01	0.03	0.18	0.09	1.55	0.73
0.51	0.25	0.27	0.14	2.47	1.23

Note: x = amount of rain collected in metal can, y = amount of rain collected in plastic gauge.

(c) Because of evaporation during the summer and the can being made of metal, the formula $y = 0.44x$ may not fit the rainfall data collected in the summer. An argument can be made that supports the model with an intercept. Is this supported by your analyses in (a) and (b)?

17. Analyze the mandrel portion of the torque data in Table 2.8 by treating it as a one-way layout. Your analysis should include ANOVA, residual analysis, and multiple comparisons of the three plating methods.

18. Data from a one-way layout are given in Table 1.10. The response is the muzzle velocity of mortar-like antipersonnel weapon. The quantitative factor is the discharge hole area (in inches), which has four levels in the experiment. An inverse relationship between muzzle velocity and discharge hole area was expected because a smaller hole would increase the pressure pulse of the propellant gases. Analyze the data in two ways: (i) by treating it as a one-way layout and using an F test and multiple comparisons, (ii) by using orthogonal polynomials to model the linear and quadratic effects. (Note: These data are obtained by collapsing and adapting three-way layout data given in an exercise in Chapter 2.)

Table 1.10   Adapted Muzzle Velocity Data

	Discharge Hole Area		
0.016	0.030	0.044	0.058
294.9	295.0	270.5	258.6
294.1	301.1	263.2	255.9
301.7	293.1	278.6	257.1
307.9	300.6	267.9	263.6
285.5	285.0	269.6	262.6
298.6	289.1	269.1	260.3
303.1	277.8	262.2	305.3
305.3	266.4	263.2	304.9
264.9	248.1	224.2	216.0
262.9	255.7	227.9	216.0
256.0	245.7	217.7	210.6
255.3	251.0	219.6	207.4
256.3	254.9	228.5	214.6
258.2	254.5	230.9	214.3
243.6	246.3	227.6	222.1
250.1	246.9	228.6	222.2

19. In tree crop spraying, an airblast sprayer was used with and without an air oscillator on grapefruit and orange trees in an experiment to evaluate the delivery of a solution. Data for the four treatments (grapefruit trees with oscillator, grapefruit trees without oscillator, orange trees with oscillator, orange trees without oscillator) consisted of 72 observations. The corresponding sample means and sample standard deviations of the solution deposited in nanograms per square centimeter (ng/cm^2) appear in Table 1.11.

   (a) Analyze the data as a one-way layout by constructing the corresponding ANOVA table. Are there significant differences between the treatments? (Hint: The mean-squared error can be calculated by pooling the sample variances and the treatment sum of squares can be determined from the sample means.)

   (b) The analysis in (a) assumes the error variance does not depend on the particular treatment. Are the data consistent with this assumption?

Table 1.11   Summary Data, Airsprayer Experiment

Treatment	Mean	Standard Deviation
grapefruit trees with oscillator	514	330
orange trees with oscillator	430	360
grapefruit trees without oscillator	590	460
orange trees without oscillator	450	325

REFERENCES

AIAG (Automotive Industry Action Group) (1990), *Measurement Systems Analysis Reference Manual*, AIAG. Troy, MI.

Barton, R.R. (1997), "Pre-Experiment Planning for Designed Experiments: Graphical Methods," *Journal of Quality Technology*, 29, 307–316.

Coleman, D.E. and Montgomery, D.C. (1993), "A Systematic Approach to Planning for a Designed Industrial Experiment" (with discussion), *Technometrics*, 35, 1–27.

Draper, N.R. and Smith, H. (1998), *Applied Regression Analysis*, 3rd ed., New York: John Wiley & Sons.

Hinkelmann, K. and Kempthorne, O. (1994), *Design and Analysis of Experiments*, Vol. 1, New York: John Wiley & Sons.

Hochberg, Y. and Tamhane, A.C. (1987), *Multiple Comparison Procedures*, New York: John Wiley & Sons.

Knowlton, J. and Keppinger, R. (1993), "The Experimentation Process," *Quality Progress*, February, 43–47.

León, R.V., Shoemaker, A.C., and Tsui, K.L. (1993), Discussion of "A Systematic Approach to Planning for a Designed Industrial Experiment" by Coleman, D.E. and Montgomery, D.C. , *Technometrics*, 35, 21–24.

Mallows, C.L. (1973), "Some Comments on C_p," *Technometrics*, 15, 661–676.

Mazumdar, S.K. and Hoa, S.V. (1995), "Application of Taguchi Method for Process Enhancement of On-line Consolidation Technique," *Composites*, 26, 669–673.

Meeker, W.Q. and Escobar, L.A. (1998), *Statistical Methods for Reliability Data*, New York: John Wiley & Sons.

Phadke, M.S. (1989), *Quality Engineering Using Robust Design*, Englewood Cliffs, NJ: Prentice-Hall.

Sheldon, F.R. (1960), "Statistical Techniques Applied to Production Situations," *Industrial and Engineering Chemistry*, 52, 507–509.

Simonoff, J.S. (1996), *Smoothing Methods in Statistics*, New York: Springer.

CHAPTER 2

Experiments with More Than One Factor

This chapter is a continuation of Chapter 1. Experiments with more than one treatment factor or with blocking are considered, which include paired comparison designs, randomized block designs, two-way and multi-way layouts, Latin square designs, Graeco-Latin square designs, and balanced incomplete block designs. Analysis techniques like data transformation and analysis of covariance are also introduced.

2.1 PAIRED COMPARISON DESIGN

Consider an experiment (Lin, Hullinger, and Evans, 1971) to compare two methods, denoted by MSI and SIB, that determine chlorine content in sewage effluents. The two methods were applied to each of the eight samples of Cl_2-demand-free water. These samples were collected at various doses and using various contact times. The residual chlorine readings in milligrams per liter are given in Table 2.1. The design employed in the sewage experiment is called a **paired comparison design** because a pair of treatments are compared in each of the eight samples.

A paired comparison design can be viewed as a randomized block design with blocks of size 2. Within each block (of two homogeneous units), two treatments are randomly assigned to the two units. The class of randomized block designs will be considered in the next section. Blocks of size 2 deserve special attention. They occur naturally in animal and medical studies. Examples include twins, pairs of eyes, pairs of kidneys, left and right feet, etc. If homogeneous units such as twins are not grouped into blocks, the two treatments would be compared using more heterogeneous units. This can result in unnecessarily large variation of the treatment comparison due to the unit-to-unit variation. The latter design is referred to as an **unpaired design**. Thus, if a paired comparison experiment is not recognized as such, its data may be incorrectly analyzed as an unpaired design.

Table 2.1   Residual Chlorine Readings, Sewage Experiment

Sample	Method MSI	SIB	d_i
1	0.39	0.36	-0.03
2	0.84	1.35	0.51
3	1.76	2.56	0.80
4	3.35	3.92	0.57
5	4.69	5.35	0.66
6	7.70	8.33	0.63
7	10.52	10.70	0.18
8	10.92	10.91	-0.01

The difference between paired and unpaired designs is illustrated with the sewage experiment. In the paired design, the two treatments are compared using a sample with the *same* dose and contact time. This eliminates a source of variation due to the difference in dose and contact time among the samples. Note that the sample must be large enough to accommodate the comparison of two treatments. If an unpaired design were used for the sewage experiment, 16 samples with possibly different doses and contact times would be prepared, out of which 8 would be selected randomly for MSI and the other 8 for SIB. Its precision in treatment comparison would be severely affected by the much larger variation among the 16 samples. If homogeneous samples with similar doses and contact times can be prepared, the unpaired design would have the advantage that it has more error degrees of freedom. For the sewage experiment it would have 14 degrees of freedom while the paired design in Table 2.1 has only 7. This point will become clearer after the analysis is presented.

The treatments in a paired comparison experiment can be compared by using the following paired t statistic:

$$t_{paired} = \sqrt{N}\,\bar{d}/s_d, \qquad (2.1)$$

where

$$d_i = y_{i2} - y_{i1}$$

is the difference between the two treatments for the ith block and

$$\bar{d} = \frac{1}{N}\sum_{i=1}^{N} d_i, \qquad s_d = \left\{ \frac{1}{N-1}\sum_{i=1}^{N}\left(d_i - \bar{d}\right)^2 \right\}^{1/2}$$

are the sample mean and sample standard deviation of the d_i, $i = 1,\ldots,N$. Under the null hypothesis of no difference between the two treatments, t_{paired} has a t distribution with $N-1$ degrees of freedom. The two treatments are declared to be significantly different at level α if

$$|t_{paired}| > t_{N-1,\,\alpha/2}. \tag{2.2}$$

This test is referred to as the **paired t test**. Because a paired comparison design is a special case of a randomized block design, it can also be analyzed by using the ANOVA for the randomized block design (to be given in the next section). It can be shown that the F statistic in the ANOVA for a paired comparison experiment is equal to the square of the paired t statistic. (Its proof is left as an exercise.) Inference using the paired t statistic or the ANOVA F statistic is equivalent since t^2 has an F distribution with parameters 1 and $N-1$. In an unpaired design, the $2N$ units are not grouped into N blocks of two units. Each treatment receives N out of the $2N$ units by random allocation. The corresponding t test is called the **unpaired t test**, or the two-sample t test,

$$t_{unpaired} = (\bar{y}_2 - \bar{y}_1)\big/\sqrt{(s_2^2/N) + (s_1^2/N)}, \tag{2.3}$$

where $\bar{y}_i$ and s_i^2 are the sample mean and sample variance for the ith treatment, $i = 1, 2$. Because it has $2N - 2$ degrees of freedom, the unpaired t test would have more power than the paired t test if the blocking of units is not effective. On the other hand, if the blocking is effective, the paired t test will be more powerful because it would have a smaller standard error [in the denominator of the t statistic in (2.1)] than the unpaired t test.

For the sewage experiment, the d_i values in the last column of Table 2.1 give $\bar{d} = 0.4138$, $s_d = 0.321$, and

$$t_{paired} = \frac{0.4138}{0.321/\sqrt{8}} = \frac{0.4138}{0.1135} = 3.645, \tag{2.4}$$

whose p value is $Prob(|t_7| > 3.645) = 0.008$. If the data were incorrectly analyzed as coming from an unpaired design, $\bar{y}_1 = 5.0212$, $\bar{y}_2 = 5.435$, $s_1^2 = 17.811$, $s_2^2 = 17.012$, and

$$t_{unpaired} = \frac{5.435 - 5.0212}{\sqrt{(17.811 + 17.012)/8}} = \frac{0.4138}{2.0863} = 0.198, \tag{2.5}$$

whose p value is $Prob(|t_{14}| > 0.198) = 0.848$, so the two treatments would not be declared significantly different. By comparing the denominators in the second ratio expressions of (2.4) and (2.5), it is easily seen that the standard error used in $t_{unpaired}$ is much larger because it includes the sample-to-sample variation component. Therefore, the unpaired t test has little power in detecting the difference between the two treatments.

Table 2.2   ANOVA Table, Sewage Experiment

Source	Degrees of Freedom	Sum of Squares	Mean Squares	F
sample	7	243.4042	34.77203	674.82
method	1	0.6848	0.68476	13.29
residual	7	0.3607	0.05153	

Table 2.3   ANOVA Table Ignoring Pairing, Sewage Experiment

Source	Degrees of Freedom	Sum of Squares	Mean Squares	F
method	1	0.6848	0.68476	0.04
residual	14	243.7649	17.41178	

The analysis can also be done by using ANOVA. From the results in Table 2.2, the value of the F statistic for the treatment method is 13.29, which is roughly equal to the square of the t_{paired} statistic 3.645. This confirms the previous statement that the F statistic is the square of the t_{paired} statistic.

The effectiveness of pairing (i.e., blocking) for the sewage experiment is evident from the ANOVA table in Table 2.2: the eight samples were significantly different because the p value of its F statistic $Prob(F_{7,7} > 674.82)$ is nearly 0. If the pairing had been ignored, it would have been treated as a one-way layout with $k = 2$ (see Section 1.6) and the ANOVA table in Table 2.3 would have been produced, which would lead to an incorrect conclusion. The two treatments would not have been identified as different because the p value $Prob(F_{1,14} > 0.04) = 0.85$. Here the F statistic 0.04 is roughly equal to the square of the $t_{unpaired}$ statistic 0.198.

2.2   RANDOMIZED BLOCK DESIGN

Consider an experiment (Narayanan and Adorisio, 1983) to compare four methods for predicting the shear strength for steel plate girders. Data for nine girders in the form of the ratio of predicted to observed load for these procedures are given in Table 2.4. Each of the four methods was used to predict the strength of each of the nine girders.

This experiment used a **randomized block design**, which has one experimental factor with k treatments and b blocks of size k. Since all the treatments are compared in each block, it is called a *complete* randomized block. The k treatments are randomly assigned to the k units in each block. Recall from the discussion in Section 1.3 that in order for blocking to be effective, the units within a block should be more homogeneous than units between blocks.

Table 2.4 Strength Data, Girder Experiment

(Block) Girder	Aarau	Method Karlsruhe	Lehigh	Cardiff
S1/1	0.772	1.186	1.061	1.025
S2/1	0.744	1.151	0.992	0.905
S3/1	0.767	1.322	1.063	0.930
S4/1	0.745	1.339	1.062	0.899
S5/1	0.725	1.200	1.065	0.871
S1/2	0.844	1.402	1.178	1.004
S2/2	0.831	1.365	1.037	0.853
S3/2	0.867	1.537	1.086	0.858
S4/2	0.859	1.559	1.052	0.805

The linear model for the randomized block design is

$$y_{ij} = \eta + \alpha_i + \tau_j + \epsilon_{ij}, \qquad i = 1,\ldots, b; \quad j = 1,\ldots, k, \qquad (2.6)$$

where y_{ij} represents the observation of the jth treatment in the ith block, α_i is the ith block effect, τ_j is the jth treatment effect, and ϵ_{ij} are errors which are independent $N(0, \sigma^2)$. For the girder experiment, there are nine blocks and four treatments (i.e., $b = 9$ and $k = 4$).

As with the one-way layout, constraints on the parameters need to be made since there are more parameters than the degrees of freedom in the observations. One can use the zero-sum constraints

$$\sum_{i=1}^{b} \alpha_i = \sum_{j=1}^{k} \tau_j = 0$$

or the baseline constraints

$$\alpha_1 = \tau_1 = 0.$$

The remaining parameters have similar interpretations as before and are left as an exercise.

The ANOVA for the randomized block design can be derived as follows. Using the zero-sum constraints, the following decomposition holds:

$$y_{ij} = \hat{\eta} + \hat{\alpha}_i + \hat{\tau}_j + r_{ij}$$
$$= \bar{y}.. + (\bar{y}_{i.} - \bar{y}..) + (\bar{y}_{.j} - \bar{y}..) + (y_{ij} - \bar{y}_{i.} - \bar{y}_{.j} + \bar{y}..), \qquad (2.7)$$

where

$$\hat{\eta} = \bar{y}.., \qquad \hat{\alpha}_i = \bar{y}_{i.} - \bar{y}.., \qquad \hat{\tau}_j = \bar{y}_{.j} - \bar{y}.., \qquad r_{ij} = y_{ij} - \bar{y}_{i.} - \bar{y}_{.j} + \bar{y}..,$$

$$\bar{y}_{i.} = k^{-1}\sum_{j=1}^{k} y_{ij}, \qquad \bar{y}_{.j} = b^{-1}\sum_{i=1}^{b} y_{ij}, \qquad \bar{y}.. = (bk)^{-1}\sum_{i=1}^{b}\sum_{j=1}^{k} y_{ij}. \qquad (2.8)$$

Table 2.5 ANOVA Table for Randomized Block Design

Source	Degrees of Freedom	Sum of Squares
block	$b - 1$	$\sum_{i=1}^{b} k(\bar{y}_i. - \bar{y}..)^2$
treatment	$k - 1$	$\sum_{j=1}^{k} b(\bar{y}._j - \bar{y}..)^2$
residual	$(b - 1)(k - 1)$	$\sum_{i=1}^{b} \sum_{j=1}^{k} (y_{ij} - \bar{y}_i. - \bar{y}._j + \bar{y}..)^2$
total	$bk - 1$	$\sum_{i=1}^{b} \sum_{j=1}^{k} (y_{ij} - \bar{y}..)^2$

Then, subtracting $\bar{y}..$, squaring both sides, and summing over i and j yield

$$\sum_{i=1}^{b} \sum_{j=1}^{k} (y_{ij} - \bar{y}..)^2 = \sum_{i=1}^{b} k(\bar{y}_i. - \bar{y}..)^2 + \sum_{j=1}^{k} b(\bar{y}._j - \bar{y}..)^2$$

$$+ \sum_{i=1}^{b} \sum_{j=1}^{k} (y_{ij} - \bar{y}_i. - \bar{y}._j + \bar{y}..)^2$$

$$= SS_b + SS_t + SS_r. \qquad (2.9)$$

That is, the corrected total sum of squares on the left equals the sum of the block sum of squares SS_b, the treatment sum of squares SS_t, and the residual sum of squares SS_r. The ANOVA table for the randomized block design appears in Table 2.5.

The null hypothesis of no treatment effect difference, $H_0 : \tau_1 = \cdots = \tau_k$, can be tested by using the F statistic

$$F = \frac{SS_t/(k - 1)}{SS_r/(b - 1)(k - 1)}, \qquad (2.10)$$

where SS_t and SS_r are the treatment and residual sums of squares in (2.9). The F test rejects H_0 at level α if the F value in (2.10) exceeds $F_{k-1, (b-1)(k-1), \alpha}$.

If H_0 is rejected, multiple comparisons of the τ_j should be performed. The t statistics for making multiple comparisons can be shown to have the form

$$t_{ij} = \frac{\bar{y}._j - \bar{y}._i}{\hat{\sigma}\sqrt{1/b + 1/b}}, \qquad (2.11)$$

where $\hat{\sigma}^2$ is the mean-squared error (i.e., the residual sum of squares divided by its degrees of freedom) in the ANOVA table. Note that the denominator in the t expression of (2.11) is the standard error of its numerator $\bar{y}._j - \bar{y}._i$. Under the null hypothesis $H_0 : \tau_1 = \cdots = \tau_k$, each t_{ij} has a t distribution with $(b - 1)(k - 1)$ degrees of freedom. At level α, the Tukey

Table 2.6   ANOVA Table, Girder Experiment

Source	Degrees of Freedom	Sum of Squares	Mean Squares	F
girder	8	0.089	0.011	1.62
method	3	1.514	0.505	73.03
residual	24	0.166	0.007	

multiple comparison method identifies "treatments i and j as different" if

$$|t_{ij}| > \frac{1}{\sqrt{2}} q_{k,(b-1)(k-1),\alpha}.$$

By solving

$$\frac{\left|(\bar{y}_{\cdot j} - \bar{y}_{\cdot i}) - (\tau_j - \tau_i)\right|}{\hat{\sigma}\sqrt{2/b}} \leq \frac{1}{\sqrt{2}} q_{k,(b-1)(k-1),\alpha}$$

for $\tau_j - \tau_i$, the simultaneous confidence intervals for $\tau_j - \tau_i$ are

$$\bar{y}_{\cdot j} - \bar{y}_{\cdot i} \pm q_{k,(b-1)(k-1),\alpha} \, \hat{\sigma}/\sqrt{b} \qquad (2.12)$$

for all i and j pairs.

We now consider the analysis of the girder experiment. Its ANOVA table appears in Table 2.6. The F statistic in (2.10) has the value

$$\frac{1.514/3}{0.166/24} = 73.03.$$

Therefore, the p value for testing the difference between methods is $Prob(F_{3,24} > 73.03) = 0.00$. The small p value suggests that the methods are different. Similarly we can compute the F statistic for testing the difference between girders (which are treated as blocks):

$$\frac{0.089/8}{0.166/24} = 1.62.$$

Its p value $Prob(F_{8,24} > 1.62) = 0.17$ is large, suggesting that there is no significant difference between the nine girders used in the experiment.

The means for the four methods denoted by A for Aarau, K for Karlsruhe, L for Lehigh, and C for Cardiff are 0.7949, 1.3401, 1.0662, and 0.9056, respectively. The multiple comparison t statistics based on (2.11) are

Table 2.7 Multiple Comparison t Statistics, Girder Experiment

A vs. K	A vs. L	A vs. C	K vs. L	K vs. C	L vs. C
13.91	6.92	2.82	-6.99	-11.09	-4.10

displayed in Table 2.7. For example, the A-vs.-K t statistic is

$$t_{12} = \frac{1.3401 - 0.7949}{\sqrt{0.007}\,\sqrt{2/9}} = 13.91.$$

With $\alpha = 0.05$,

$$t_{24,\,0.05/(6\times2)} = 2.875$$

for the Bonferroni method, since $k = 4$ and $\binom{k}{2} = 6$, and

$$\frac{1}{\sqrt{2}}q_{4,\,24,\,0.05} = \frac{3.90}{1.414} = 2.758$$

for the Tukey method. By comparing 2.875 with the t values in Table 2.7, at the 0.05 level the Bonferroni method identifies all pairs of methods as different except Aarau vs. Cardiff. Because the t statistic for A vs. C is 2.82, which exceeds 2.758 for the Tukey method, the Tukey method identifies all pairs of methods (including Aarau and Cardiff) as different. Analysis of the residuals $r_{ij} = y_{ij} - \bar{y}_i. - \bar{y}._j + \bar{y}..$ should be performed and is left as an exercise.

2.3 TWO-WAY LAYOUT

A manufacturer was finding unwanted differences in the torque values of a locknut that it made. Torque is the work (which is force $\times$ distance) required to tighten the nut. Consequently, the manufacturer conducted an experiment to determine what factors affected the torque values (Meek and Ozgur, 1991). The type of plating process was isolated as the most probable factor to impact torque, especially using no plating versus using plating. Another factor is the test medium, that is, whether the locknut is threaded onto a bolt or a mandrel. A mandrel is like a bolt but harder. Thus the two experimental factors were type of plating, whose three levels were no plating, cadmium plated, and phosphate plated, and test medium, whose levels were mandrel and bolt. The torque values for 10 locknuts for each of the six treatments are given in Table 2.8, where a different set of locknuts is used for each treatment. The terms C&W, HT, and P&O are for the three types of plating,

Table 2.8   Torque Data, Bolt Experiment

	C&W	HT	P&O
bolt	20, 16, 17, 18, 15, 16, 19, 14, 15, 24	26, 40, 28, 38, 38, 30, 26, 38, 45, 38	25, 40, 30, 17, 16, 45, 49, 33, 30, 20
mandrel	24, 18, 17, 17, 15, 23, 14, 18, 12, 11	32, 22, 30, 35, 32, 28, 27, 28, 30, 30	10, 13, 17, 16, 15, 14, 11, 14, 15, 16

which are cadmium and wax, heat treated only (i.e., no plating), and phosphate and oil. The industry standard is 45 foot-pound maximum when the locknut is first threaded onto its mating partner (i.e., bolt or mandrel) as measured by a manual torque wrench. The Industrial Fastener Institute torque specifications are given for locknuts prior to plating. One of the experimental goals was to assess whether these specifications should apply to plated nuts.

The bolt experiment employed a *two-way layout*, which involves two treatment factors. Suppose that the number of replicates is n and for each replicate IJ units are randomly assigned to the IJ treatments, the number of factor level combinations in the experiment. Then, the linear model for the two-way layout is

$$y_{ijl} = \eta + \alpha_i + \beta_j + \omega_{ij} + \epsilon_{ijl},$$

$$i = 1, \ldots, I; \quad j = 1, \ldots, J; \quad l = 1, \ldots, n, \qquad (2.13)$$

where y_{ijl} is the observation for the lth replicate of the ith level of factor A and the jth level of factor B, α_i is the ith main effect for A, β_j is the jth main effect for B, ω_{ij} is the (i, j)th interaction effect between A and B, and ϵ_{ijl} are independent errors distributed as $N(0, \sigma^2)$. Interpretation of main effects and interaction effects will be considered later in this section. For the bolt experiment, $I = 2$, $J = 3$, $n = 10$, $i = 1$ for bolt, $i = 2$ for mandrel, $j = 1$ for C&W, $j = 2$ for HT, and $j = 3$ for P&O.

For the two-way layout, more constraints are needed than for the randomized block design because of the presence of interaction effects. As in the one-way layout and the randomized block design, there are two types of constraints:

(a)   Baseline constraints: $\alpha_1 = \beta_1 = 0$ and $\omega_{1j} = \omega_{i1} = 0$ for $i = 1, \ldots, I$; $j = 1, \ldots, J$;

$$(2.14)$$

(b)   Zero-sum constraints: $\sum_{i=1}^{I} \alpha_i = \sum_{j=1}^{J} \beta_j = 0$, $\sum_{i=1}^{I} \omega_{ij} = 0$ for $j = 1, \ldots, J$ and $\sum_{j=1}^{J} \omega_{ij} = 0$ for $i = 1, \ldots, I$.

The ANOVA for the two-way layout can be derived as follows. Using the zero-sum constraints in (b) above, the following holds:

$$y_{ijl} = \hat{\eta} + \hat{\alpha}_i + \hat{\beta}_j + \hat{\omega}_{ij} + r_{ijl}$$

$$= \bar{y}_{...} + (\bar{y}_{i..} - \bar{y}_{...}) + (\bar{y}_{.j.} - \bar{y}_{...}) + (\bar{y}_{ij.} - \bar{y}_{i..} - \bar{y}_{.j.} + \bar{y}_{...})$$

$$+ (y_{ijl} - \bar{y}_{ij.}), \qquad (2.15)$$

where

$$\hat{\eta} = \bar{y}_{...}, \qquad \hat{\alpha}_i = \bar{y}_{i..} - \bar{y}_{...}, \qquad \hat{\beta}_j = \bar{y}_{.j.} - \bar{y}_{...},$$

$$\hat{\omega}_{ij} = \bar{y}_{ij.} - \bar{y}_{i..} - \bar{y}_{.j.} + \bar{y}_{...}, \qquad r_{ijl} = y_{ijl} - \bar{y}_{ij.}, \qquad (2.16)$$

and the definitions of $\bar{y}_{i..}$, $\bar{y}_{.j.}$, and $\bar{y}_{ij.}$ are similar to those in (2.8). (Note that the "$\cdot$" in the subscript indicates averaging over the levels of the particular variable or factor.) Then, subtracting $\bar{y}_{...}$, squaring both sides, and summing over i, j, and l yield

$$\sum_{i=1}^{I} \sum_{j=1}^{J} \sum_{l=1}^{n} (y_{ijl} - \bar{y}_{...})^2 = \sum_{i=1}^{I} nJ(\bar{y}_{i..} - \bar{y}_{...})^2 + \sum_{j=1}^{J} nI(\bar{y}_{.j.} - \bar{y}_{...})^2$$

$$+ \sum_{i=1}^{I} \sum_{j=1}^{J} n(\bar{y}_{ij.} - \bar{y}_{i..} - \bar{y}_{.j.} + \bar{y}_{...})^2$$

$$+ \sum_{i=1}^{I} \sum_{j=1}^{J} \sum_{l=1}^{n} (y_{ijl} - \bar{y}_{ij.})^2$$

$$= SS_A + SS_B + SS_{AB} + SS_r. \qquad (2.17)$$

That is, the corrected total sum of squares on the left equals the sum of the factor A sum of squares SS_A, the factor B sum of squares SS_B, the A-by-B interaction (denoted by $A \times B$) sum of squares SS_{AB}, and the residual sum of squares SS_r. The ANOVA table for the two-way layout is displayed in Table 2.9.

The ANOVA table can be used to test the hypothesis on the various effects in (2.13). Consider, for example, the null hypothesis $H_0: \alpha_1 = \cdots = \alpha_I$ (i.e., no difference between the factor A main effects). The F test rejects H_0 at level α if the F statistic

$$F = \frac{SS_A/(I-1)}{SS_r/IJ(n-1)} > F_{I-1, IJ(n-1), \alpha}, \qquad (2.18)$$

Table 2.9   ANOVA Table for Two-Way Layout

Source	Degrees of Freedom	Sum of Squares
A	$I-1$	$nJ\sum_{i=1}^{I}(\bar{y}_{i..}-\bar{y}...)^2$
B	$J-1$	$nI\sum_{j=1}^{J}(\bar{y}_{.j.}-\bar{y}...)^2$
$A\times B$	$(I-1)(J-1)$	$n\sum_{i=1}^{I}\sum_{j=1}^{J}(\bar{y}_{ij.}-\bar{y}_{i..}-\bar{y}_{.j.}+\bar{y}...)^2$
residual	$IJ(n-1)$	$\sum_{i=1}^{I}\sum_{j=1}^{J}\sum_{l=1}^{n}(y_{ijl}-\bar{y}_{ij.})^2$
total	$IJn-1$	$\sum_{i=1}^{I}\sum_{j=1}^{J}\sum_{l=1}^{n}(y_{ijl}-\bar{y}...)^2$

where SS_A and SS_r are given in (2.17). Tests of the factor B main effects and of the $A\times B$ interaction effects can be similarly developed.

So far we have assumed that the experiment is replicated with $n>1$. For an unreplicated experiment, $n=1$ and there is not enough degrees of freedom for estimating the interaction effects. The SS_{AB} in (2.17) should be used as the residual sum of squares. Some aspect of the interaction effects can still be tested by using Tukey's "one-degree of freedom for non-additivity" (see Neter et al., 1996).

For the bolt experiment, the F statistics from the ANOVA table given in Table 2.10 suggest that test, plating, and test-by-plating interaction are all significant. For example, the p value for the test main effect equals 0.00001 $[=Prob(F_{1,54}>22.46)]$.

While the randomized block design involves two factors, one blocking and one experimental, it is fundamentally different from a two-way layout which involves two experimental factors. In the two-way layout, the IJ units of each replicate are randomly assigned to the IJ level combinations of the two factors. Also, in the ANOVA for randomized block design, there is no *block × treatment* interaction; in fact, because there is only a single replicate, the *block × treatment* interaction sum of squares and the residual sum of squares are identical.

The main effects and interactions can be modeled and estimated. The development is given in the next three subsections according to the nature of the factors.

Table 2.10   ANOVA Table, Bolt Experiment

Source	Degrees of Freedom	Sum of Squares	Mean Squares	F
test	1	821.400	821.400	22.46
plating	2	2290.633	1145.317	31.31
test × plating	2	665.100	332.550	9.09
residual	54	1975.200	36.578	

2.3.1 Two Qualitative Factors

Both factors in the bolt experiment are qualitative. For model (2.13) with qualitative factors, which type of constraints on its parameters is more appropriate? Because pairwise comparisons of the levels of a qualitative factor are more natural and interpretable than comparisons of the levels to their "average," the baseline constraints in (2.14) are preferred to the zero-sum constraints. For the bolt experiment, the baseline constraints are $\alpha_1 = \beta_1 = 0$ and $\omega_{1j} = \omega_{i1} = 0$ for $i = 1, 2$, $j = 1, 2, 3$. Under these constraints, model (2.13) for the bolt experiment simplifies to the following:

$$E(y_{11}) = \eta, \qquad E(y_{12}) = \eta + \beta_2, \qquad E(y_{13}) = \eta + \beta_3,$$
$$E(y_{21}) = \eta + \alpha_2, \qquad E(y_{22}) = \eta + \alpha_2 + \beta_2 + \omega_{22},$$
$$E(y_{23}) = \eta + \alpha_2 + \beta_3 + \omega_{23}. \qquad (2.19)$$

The corresponding model matrix (for one replicate) is

$$\mathbf{X} = \begin{pmatrix} 1 & 0 & 0 & 0 & 0 & 0 \\ 1 & 0 & 1 & 0 & 0 & 0 \\ 1 & 0 & 0 & 1 & 0 & 0 \\ 1 & 1 & 0 & 0 & 0 & 0 \\ 1 & 1 & 1 & 0 & 1 & 0 \\ 1 & 1 & 0 & 1 & 0 & 1 \end{pmatrix}. \qquad (2.20)$$

From (2.19) it can be shown that η, α_2, β_2, β_3, ω_{22}, and ω_{23} have the following interpretation:

$$\eta = E(y_{11}),$$
$$\alpha_2 = E(y_{21}) - E(y_{11}),$$
$$\beta_2 = E(y_{12}) - E(y_{11}),$$
$$\beta_3 = E(y_{13}) - E(y_{11}), \qquad (2.21)$$
$$\omega_{22} = (E(y_{22}) - E(y_{21})) - (E(y_{12}) - E(y_{11})),$$
$$\omega_{23} = (E(y_{23}) - E(y_{21})) - (E(y_{13}) - E(y_{11})).$$

Note that α_2 measures the difference between levels 2 and 1 of factor A, β_2 measures the difference between levels 2 and 1 of factor B, and ω_{22} compares whether the difference between levels 1 and 2 of B is the same at both levels of A. Similarly, ω_{23} compares whether the difference between levels 1 and 3 of B is the same at both levels of A.

The least squares estimates of the parameters in (2.21) are obtained by regression analysis with $\mathbf{X}$ in (2.20) as the model matrix (for each of the 10 replicates). For the bolt experiment, the estimates, standard errors, t statistics, and p values are given in Table 2.11. Both $\hat{\beta}_2$ and $\hat{\beta}_3$ are significant,

Table 2.11 Tests, Bolt Experiment

Effect	Estimate	Standard Error	t	p Value
η	17.4000	1.9125	9.10	0.000
α_2	-0.5000	2.7047	-0.18	0.854
β_2	17.3000	2.7047	6.40	0.000
β_3	13.1000	2.7047	4.84	0.000
ω_{22}	-4.8000	3.8251	-1.25	0.215
ω_{23}	-15.9000	3.8251	-4.16	0.000

suggesting that at $i = 1$ (bolt) there are differences between the levels of the plate factor (C&W vs. HT, C&W vs. P&O). The fact that $\hat{\alpha}_2$ is not significant suggests that there is no difference between bolt and mandrel at $j = 1$ (C&W) but the significant $\hat{\omega}_{23}$ suggests that the difference between C&W and P&O varies from bolt to mandrel.

Because five effects (excluding η) are being tested in Table 2.11, strictly speaking, the critical values for the t tests should be adjusted for simultaneous testing. The Bonferonni method of Section 1.7 can be shown to be equivalent to taking the individual p values and multiplying them by the number of tests to obtain **adjusted p values**. For example, the adjusted p value for ω_{23} is 0.0005 ($= 0.0001 \times 5$), which is still very small and suggests that ω_{23} is significant. Details are left as an exercise.

Box-whisker plots of the residuals for each of the six treatments for the bolt experiment appear in Figure 2.1. The bolt and P&O treatment has the largest variability and the bolt and HT treatment has the second largest variability among the treatments. The plot may suggest that the constant variance assumption in (2.13) does not hold and that the variance of y for bolt is larger than that for mandrel. Further analysis of this aspect of the bolt experiment is left as an exercise.

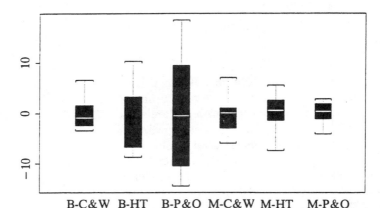

Figure 2.1. Box-Whisker Plots of Residuals, Bolt Experiment.

Table 2.12   Yield Data, Tomato Experiment

Variety	Plant Density 10,000	20,000	30,000
Ife #1	16.1, 15.3, 17.5	16.6, 19.2, 18.5	20.8, 18.0, 21.0
Pusa Early Dwarf	8.1, 8.6, 10.1	12.7, 13.7, 11.5	14.4, 15.4, 13.7

2.3.2   One Qualitative Factor and One Quantitative Factor

Consider an experiment (Adelana, 1976) to study the effects of variety and planting density on tomato yield. Two varieties of tomato, Ife # 1 and Pusa Early Dwarf, and three plant densities were investigated. The plant density is the number of plants per hectare. The experiment was replicated three times. The yield data (in tons per hectare) for the tomato experiment appear in Table 2.12.

When one factor is qualitative and one is quantitative, performing an ANOVA will indicate whether there is interaction between the two factors. If there is no interaction, then comparisons between the levels of the qualitative factor are of interest as well as the nature of the quantitative factor's impact on the response in terms of polynomial effects.

If there is an interaction between the qualitative and quantitative factors, say A and B, for ease of interpretation, it is best to handle the interaction by looking at the polynomial effects (linear, quadratic, etc.) of the quantitative factor at each level of the qualitative factor. This approach is referred to as a *nested analysis*. There are comparison effects between the different levels of the qualitative factor as well as the polynomial effects within each level of the qualitative factor. The polynomial effects can be studied by using the orthogonal polynomials in Section 1.8.

The corresponding linear model is

$$y = \beta_0 + \beta_1 x_A + \beta_2 x_{B_l|A_1} + \beta_3 x_{B_l|A_2} + \beta_4 x_{B_q|A_1}$$
$$+ \beta_5 x_{B_q|A_2} + \epsilon, \tag{2.22}$$

where A represents the comparison between the two levels of factor A and $B_l|A_i$ and $B_q|A_i$ denote the linear and quadratic effects of B at level i of A. Recall from Section 1.8 that the linear and quadratic effects are represented by the contrast vectors $(1/\sqrt{2})(-1, 0, 1)$ and $(1/\sqrt{6})(1, -2, 1)$ respectively. Therefore the model matrix (for one replicate) is

$$\mathbf{X} = \begin{pmatrix} 1 & 0 & -1 & 0 & 1 & 0 \\ 1 & 0 & 0 & 0 & -2 & 0 \\ 1 & 0 & 1 & 0 & 1 & 0 \\ 1 & 1 & 0 & -1 & 0 & 1 \\ 1 & 1 & 0 & 0 & 0 & -2 \\ 1 & 1 & 0 & 1 & 0 & 1 \end{pmatrix}, \tag{2.23}$$

Table 2.13   Tests, Tomato Experiment

Effect	Estimate	Standard Error	t	p Value
intercept	15.0667	0.2869	52.51	0.000
tomato	-6.0889	0.5738	-10.61	0.000
density$_l$|tomato$_1$	2.5692	0.7028	3.66	0.003
density$_l$|tomato$_2$	3.9362	0.7028	5.60	0.000
density$_q$|tomato$_1$	0.0136	0.7028	0.02	0.985
density$_q$|tomato$_2$	-0.7485	0.7028	-1.07	0.308

where columns 3 and 4 need to be divided by the scaling constant $\sqrt{2}$ and columns 5 and 6 by the scaling constant $\sqrt{6}$. Thus, the parameters in (2.22) can be estimated by regression analysis with $\mathbf{X}$ in (2.23) as its model matrix.

For the tomato experiment, Table 2.13 displays the least squares estimates of the β_i in (2.22). It shows that there is a difference between the two varieties and that yield is linearly increasing in density for both varieties. The question remaining is whether the two linear coefficients are the same. The test for this comparison is based on the linear combination $\mathbf{a}^T\boldsymbol{\beta} = \beta_3 - \beta_2$, where $\mathbf{a} = (0, 0, -1, 1, 0, 0)^T$. Here since the columns corresponding to the two linear effects are orthogonal, the two estimates given in Table 2.13 are independent so that the t statistic for testing $\beta_3 - \beta_2 = 0$ is

$$(-2.56972 + 3.9362)/(\sqrt{2}\,(0.7028)) = 1.37. \qquad (2.24)$$

The p value for a t distribution with 12 degrees of freedom is $Prob(|t_{12}| > 1.37) = 0.196$; thus, there is no difference between the linear effects. This suggests the simplified model with A and B_l:

$$y = \beta_0 + \beta_1 x_A + \beta_2 x_{B_l} + \epsilon.$$

2.3.3   Two Quantitative Factors

Suppose that the two factors A and B are quantitative, both at three levels. Then A and B can both be modeled by linear and quadratic main effects. The $A \times B$ interaction can be broken up into linear-by-linear, linear-by-quadratic, quadratic-by-linear, and quadratic-by-quadratic components. The corresponding linear model is

$$y = \beta_0 + \beta_1 x_{A_l} + \beta_2 x_{A_q} + \beta_3 x_{B_l} + \beta_4 x_{B_q}$$
$$+ \beta_5 x_{A_l B_l} + \beta_6 x_{A_l B_q} + \beta_7 x_{A_q B_l}$$
$$+ \beta_8 x_{A_q B_q} + \epsilon, \qquad (2.25)$$

and the model matrix is

$$\mathbf{X} = \begin{pmatrix} 1 & -1 & 1 & -1 & 1 & 1 & -1 & -1 & 1 \\ 1 & -1 & 1 & 0 & -2 & 0 & 2 & 0 & -2 \\ 1 & -1 & 1 & 1 & 1 & -1 & -1 & 1 & 1 \\ 1 & 0 & -2 & -1 & 1 & 0 & 0 & 2 & -2 \\ 1 & 0 & -2 & 0 & -2 & 0 & 0 & 0 & 4 \\ 1 & 0 & -2 & 1 & 1 & 0 & 0 & -2 & -2 \\ 1 & 1 & 1 & -1 & 1 & -1 & 1 & -1 & 1 \\ 1 & 1 & 1 & 0 & -2 & 0 & -2 & 0 & -2 \\ 1 & 1 & 1 & 1 & 1 & 1 & 1 & 1 & 1 \end{pmatrix}. \quad (2.26)$$

Columns 2 and 4 for the linear main effects need to be divided by $\sqrt{2}$. They correspond to the linear contrast vectors for factors A and B, respectively. Columns 3 and 5 for the quadratic main effects need to be divided by $\sqrt{6}$. They correspond to the quadratic contrast vectors for factors A and B, respectively. Because column 6, the linear-by-linear component, is the product of columns 2 and 4, the two linear main effects, column 6 needs to be divided by 2 ($= \sqrt{2}\sqrt{2}$). Similarly, columns 7–9 need to be divided by $\sqrt{2}\sqrt{6}$, $\sqrt{6}\sqrt{2}$, and $\sqrt{6}\sqrt{6}$, respectively. Thus, a two-way layout with two quantitative factors can be analyzed by fitting (2.25) and identifying significant effects.

The model in (2.25) applies to replicated experiments because there are enough degrees of freedom for error variance estimation. For unreplicated experiments, some higher order interaction components need to be sacrificed and used for the residual sum of squares. For example, in an unreplicated experiment with two three-level quantitative factors, only the linear-by-linear interaction might be entertained. That is, the model

$$y = \beta_0 + \beta_1 x_{A_l} + \beta_2 x_{A_q} + \beta_3 x_{B_l} + \beta_4 x_{B_q}$$
$$+ \beta_5 x_{A_l B_l} + \epsilon \quad (2.27)$$

might be fitted. Then, the residual sum of squares would consist of the sums of squares corresponding to the linear-by-quadratic, quadratic-by-linear, and quadratic-by-quadratic interaction components.

2.4 MULTI-WAY LAYOUT

The two-way layout studied in the last section generalizes to more factors and is referred to as a multi-way layout. Consider a multi-way layout with three factors A, B, and C whose number of levels are I, J, and K, respectively. Suppose that the number of replicates is n. Then for each replicate, IJK units are randomly assigned to the IJK treatments, the number of factor level combinations in the experiment. The linear model for the three-way layout is

$$y_{ijkl} = \eta + \alpha_i + \beta_j + \delta_k + (\alpha\beta)_{ij} + (\alpha\delta)_{ik} + (\beta\delta)_{jk} + \gamma_{ijk} + \epsilon_{ijkl},$$

$$i = 1, \ldots, I; \quad j = 1, \ldots, J; \quad k = 1, \ldots, K; \quad l = 1, \ldots, n, \quad (2.28)$$

where α_i, β_j, and δ_k are the A, B, and C main effects; $(\alpha\beta)_{ij}$, $(\alpha\delta)_{ik}$, and $(\beta\delta)_{jk}$ are the $A \times B$, $A \times C$, and $B \times C$ two-factor interaction effects; and γ_{ijk} is the $A \times B \times C$ three-factor interaction effect. The errors ϵ_{ijkl} are independent $N(0, \sigma^2)$.

The ANOVA for the three-way layout can be derived as follows. The zero-sum constraints are

$$\sum_{i=1}^{I} \alpha_i = \sum_{j=1}^{J} \beta_j = \sum_{k=1}^{K} \delta_k = 0,$$

$$\sum_{i=1}^{I} (\alpha\beta)_{ij} = \sum_{i=1}^{I} (\alpha\delta)_{ik} = \sum_{j=1}^{J} (\alpha\beta)_{ij} = \sum_{j=1}^{J} (\beta\delta)_{jk}$$

$$= \sum_{k=1}^{K} (\alpha\delta)_{ik} = \sum_{k=1}^{K} (\beta\delta)_{jk} = 0,$$

$$\sum_{i=1}^{I} \gamma_{ijk} = \sum_{j=1}^{J} \gamma_{ijk} = \sum_{k=1}^{K} \gamma_{ijk} = 0$$

$$\text{for } i = 1, \ldots, I, \, j = 1, \ldots, J, \text{ and } k = 1, \ldots, K.$$

Using these zero-sum constraints, the following holds:

$$y_{ijkl} = \hat{\eta} + \hat{\alpha}_i + \hat{\beta}_j + \hat{\delta}_k + \widehat{(\alpha\beta)}_{ij} + \widehat{(\alpha\delta)}_{ik} + \widehat{(\beta\delta)}_{jk} + \hat{\gamma}_{ijk} + r_{ijkl}, \quad (2.29)$$

where

$$\hat{\eta} = \bar{y}....,$$

$$\hat{\alpha}_i = \bar{y}_i... - \bar{y}....,$$

$$\hat{\beta}_j = \bar{y}._{j}.. - \bar{y}....,$$

$$\hat{\delta}_k = \bar{y}.._{k}. - \bar{y}....,$$

$$\widehat{(\alpha\beta)}_{ij} = \bar{y}_{ij}.. - \bar{y}_i... - \bar{y}._{j}.. + \bar{y}...., \quad\quad\quad (2.30)$$

$$\widehat{(\alpha\delta)}_{ik} = \bar{y}_i._{k}. - \bar{y}_i... - \bar{y}.._{k}. + \bar{y}....,$$

$$\widehat{(\beta\delta)}z_{jk} = \bar{y}._{jk}. - \bar{y}._{j}.. - \bar{y}.._{k}. + \bar{y}....,$$

$$\hat{\gamma}_{ijk} = \bar{y}_{ijk}. - \bar{y}_{ij}.. - \bar{y}_i._{k}. - \bar{y}._{jk}. + \bar{y}_i... + \bar{y}._{j}.. + \bar{y}.._{k}. - \bar{y}....,$$

$$r_{ijkl} = y_{ijkl} - \bar{y}_{ijk}..$$

Table 2.14   ANOVA Table for Three-Way Layout

Source	Degrees of Freedom	Sum of Squares
A	$I-1$	$\sum_{i=1}^{I} nJK(\hat{\alpha}_i)^2$
B	$J-1$	$\sum_{j=1}^{J} nIK(\hat{\beta}_j)^2$
C	$K-1$	$\sum_{k=1}^{K} nIJ(\hat{\delta}_k)^2$
$A \times B$	$(I-1)(J-1)$	$\sum_{i=1}^{I} \sum_{j=1}^{J} nK\,((\widehat{\alpha\beta})_{ij})^2$
$A \times C$	$(I-1)(K-1)$	$\sum_{i=1}^{I} \sum_{k=1}^{K} nJ((\widehat{\alpha\delta})_{ik})^2$
$B \times C$	$(J-1)(K-1)$	$\sum_{j=1}^{J} \sum_{k=1}^{K} nI((\widehat{\beta\delta})_{jk})^2$
$A \times B \times C$	$(I-1)(J-1)(K-1)$	$\sum_{i=1}^{I} \sum_{j=1}^{J} \sum_{k=1}^{K} n(\hat{\gamma}_{ijk})^2$
residual	$IJK(n-1)$	$\sum_{i=1}^{I} \sum_{j=1}^{J} \sum_{k=1}^{K} \sum_{l=1}^{n}(y_{ijkl} - \bar{y}_{ijk\cdot})^2$
total	$IJKn-1$	$\sum_{i=1}^{I} \sum_{j=1}^{J} \sum_{k=1}^{K} \sum_{l=1}^{n}(y_{ijkl} - \bar{y}....)^2$

As before, the "·" in the subscript indicates averaging over the levels of the corresponding variable or factor.

Then, subtracting $\bar{y}$.... from both sides of (2.29) and summing over i, j, k, and l yield

$$\sum_{i=1}^{I} \sum_{j=1}^{J} \sum_{k=1}^{K} \sum_{l=1}^{n} (y_{ijkl} - \bar{y}....)^2$$

$$= \sum_{i=1}^{I} nJK(\hat{\alpha}_i)^2 + \sum_{j=1}^{J} nIK(\hat{\beta}_j)^2 + \sum_{k=1}^{K} nIJ(\hat{\delta}_k)^2$$

$$+ \sum_{i=1}^{I} \sum_{j=1}^{J} nK((\widehat{\alpha\beta})_{ij})^2 + \sum_{i=1}^{I} \sum_{k=1}^{K} nJ((\widehat{\alpha\delta})_{ik})^2 + \sum_{j=1}^{J} \sum_{k=1}^{K} nI((\widehat{\beta\delta})_{jk})^2$$

$$+ \sum_{i=1}^{I} \sum_{j=1}^{J} \sum_{k=1}^{K} n(\hat{\gamma}_{ijk})^2 + \sum_{i=1}^{I} \sum_{j=1}^{J} \sum_{k=1}^{K} \sum_{l=1}^{n} r_{ijkl}^2. \qquad (2.31)$$

That is, the corrected total sum of squares on the left equals the sum of the A, B, and C sums of squares, the $A \times B$, $A \times C$, and $B \times C$ sums of squares, the $A \times B \times C$ sum of squares, and the residual sum of squares.

The ANOVA table for the three-way layout is displayed in Table 2.14. To complete the ANOVA table, the mean squares are computed by dividing the sums of squares by their corresponding degrees of freedom. The residual mean square (i.e., the mean-squared error) is an estimate of σ^2. To test the significance of a factor main effect or interaction effect, the F statistics are computed by dividing the corresponding mean squares by the residual mean square. Under the null hypothesis that a particular effect equals zero, the corresponding F statistic has an F distribution whose parameters are the degrees of freedom for the particular effect and for the residual mean square.

For more factors, the ANOVA for the multi-way layout generalizes accordingly.

2.5   TRANSFORMATION OF THE RESPONSE

Thus far, we have considered linear models for the response y. A more general model is for the linear model to hold for some transformation of y. Typically, the transformed scale should have some meaning. For example, Box and Cox (1964) presented an example of a two-way layout which measured survival time. They showed that analyzing reciprocal survival times (i.e., y^{-1}) led to a simpler model which did not involve the two-factor interaction while the analysis of the original survival times did. Moreover, the reciprocal survival time has the simple interpretation of being the rate of dying.

One motivation for transformation is for variance stabilization, which is explained as follows. Suppose that

$$y = \mu + \epsilon, \tag{2.32}$$

where the standard deviation of y, $\sigma_y = (Var(y))^{1/2} = (Var(\epsilon))^{1/2}$, is proportional to some power α of the mean μ of y, i.e.,

$$\sigma_y \propto \mu^{\alpha}. \tag{2.33}$$

This relationship may be detected via a residual plot. For a replicated experiment, we use y_{ij} to denote the observation for the jth replicate of the ith condition, $\bar{y}_i$. the average over the replicated observations for the ith condition, and $r_{ij} = y_{ij} - \bar{y}_i$. its residual. Then the residual plot r_{ij} vs. $\bar{y}_i$. can be used to detect the relationship in (2.33). For an unreplicated experiment, we use y_i to denote the observation for the ith condition, $\hat{y}_i$ its fitted value based on a model, and $r_i = y_i - \hat{y}_i$ its residual. The residual plot r_i vs. $\hat{y}_i$ is used to detect the relationship in (2.33).

Consider the family of **power transformations**

$$z = f(y) = \begin{cases} \dfrac{y^{\lambda} - 1}{\lambda}, & \lambda \neq 0, \\ \ln y, & \lambda = 0. \end{cases} \tag{2.34}$$

Notice that the family is continuous in λ because $\lim_{\lambda \to 0}(y^{\lambda} - 1)/\lambda = \ln y$. The standard deviation of z can be derived from that of y as follows.

Suppose $z = f(y)$ is a random variable defined as a smooth function of another random variable y. Denote the mean and variance of y by μ and σ_y^2. Using a Taylor series expansion of $f(y)$ around μ,

$$z = f(y) \approx f(\mu) + f'(\mu)(y - \mu).$$

Then

$$\sigma_z^2 = Var(z) \approx (f'(\mu))^2 Var(y) = (f'(\mu))^2 \sigma_y^2. \qquad (2.35)$$

For the power transformation in (2.34), $f'(\mu) = \mu^{\lambda-1}$ and (2.35) becomes

$$\sigma_z \approx |f'(\mu)| \sigma_y = \mu^{\lambda-1} \sigma_y \propto \mu^{\lambda-1} \mu^\alpha = \mu^{\lambda+\alpha-1}. \qquad (2.36)$$

By setting the exponent of μ in (2.36) to be zero, i.e., $\lambda + \alpha - 1 = 0$, σ_z can be made constant by choosing

$$\lambda = 1 - \alpha \qquad (2.37)$$

as the power of the transformation.

Since α is unknown, λ needs to be determined. A method based on the likelihood function to determine λ was proposed by Box and Cox (1964) the details of which can be found in Box, Hunter, and Hunter (1978). For a general discussion of data transformation, see Carroll and Ruppert (1988).

A much simpler and equally effective approach is to try a few selective transformations, e.g., the eight transformations in Table 2.15. For each transformation, analyze the z data and choose a transformation according to the following criteria:

(i) It gives the most parsimonious model, i.e., with the fewest terms, particularly the omission of higher order terms like cubic effects and interactions.

Table 2.15 Variance Stabilizing Transformations

$\sigma_y \propto \mu^\alpha$	α	$\lambda = 1 - \alpha$	Transformation
$\sigma_y \propto \mu^3$	3	-2	reciprocal squared
$\sigma_y \propto \mu^2$	2	-1	reciprocal
$\sigma_y \propto \mu^{3/2}$	$\frac{3}{2}$	$-\frac{1}{2}$	reciprocal square root
$\sigma_y \propto \mu$	1	0	log
$\sigma_y \propto \mu^{1/2}$	$\frac{1}{2}$	$\frac{1}{2}$	square root
$\sigma_y \propto$ constant	0	1	original scale
$\sigma_y \propto \mu^{-1/2}$	$-\frac{1}{2}$	$\frac{3}{2}$	$\frac{3}{2}$ power
$\sigma_y \propto \mu^{-1}$	-1	2	square

Table 2.16   Design Matrix and Response Data, Drill Experiment

Run	A	B	C	D	Advance Rate
			Factor		
1	−	−	−	−	1.68
2	+	−	−	−	1.98
3	−	+	−	−	3.28
4	+	+	−	−	3.44
5	−	−	+	−	4.98
6	+	−	+	−	5.70
7	−	+	+	−	9.97
8	+	+	+	−	9.07
9	−	−	−	+	2.07
10	+	−	−	+	2.44
11	−	+	−	+	4.09
12	+	+	−	+	4.53
13	−	−	+	+	7.77
14	+	−	+	+	9.43
15	−	+	+	+	11.75
16	+	+	+	+	16.30

(ii) There are no unusual patterns in the residual plots.

(iii) The interpretability of the transformation is good.

Typically, if σ_y increases as μ increases, try the first five transformations in Table 2.15. In order to achieve "variance stabilization," these five transformations should shrink larger y values more severely. This is obvious for the log transformation, the reciprocal transformation, and the square root transformation. Only when σ_y increases as μ decreases, try the last two transformations in Table 2.15.

We illustrate the use of transformation by an experiment presented in Daniel (1976). The experiment studied the effect of four factors on the rate of advance of a drill. Each of the four factors (denoted by A, B, C, and D) were studied at two levels. The design and data appear in Table 2.16. The "$-$" and "$+$" denote the two levels used for each factor. The design is a two-level full factorial which will be studied in detail in the next chapter. For purposes of illustration, we only consider models with main effects and two-factor interactions, which leaves five degrees of freedom for estimating error variance. The definitions of main effects and two-factor interactions are the same as those given in (2.13) and (2.16). Because the experiment is unreplicated, the r_i vs. $\hat{y}_i$ plot in Figure 2.2 is used to detect any relationship between its standard deviation and mean. The plot shows an increasing pattern in the range of the residuals as the fitted value increases. For the

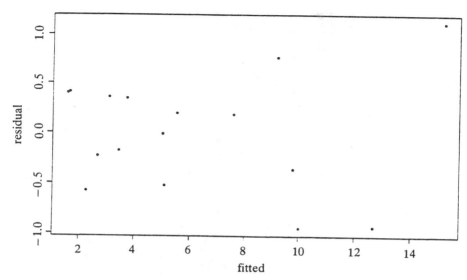

Figure 2.2. r_i vs. $\hat{y}_i$, Drill Experiment.

larger $\hat{y}_i$ values the corresponding r_i values have a range about twice that of the r_i values for the smaller $\hat{y}_i$ values. This increasing pattern suggests that one of the first five transformations in Table 2.15 should achieve variance stabilization and possibly other benefits as described above.

For each of the eight transformations in Table 2.15, a model was fitted and the t statistics corresponding to the four main effects and six two-factor interactions were calculated. These t statistics are displayed in the scaled lambda plot of Figure 2.3. Notice that for the log transformation ($\lambda = 0$), the largest t statistics (C, B, D, and possibly A) stand apart from those of the remaining effects; this illustrates how transformation can make the analysis more sensitive. Therefore C, B, D, and A are included in the model for $\ln y$. The next best is the reciprocal square root ($\lambda = -\frac{1}{2}$) transformation which identifies, C, B, D, BC, and possibly A. The identified model for $y^{-0.5}$ is less parsimonious than the main effects model for the log transformation, because the former model has one more term than the latter model and the additional term happens to be an interaction effect. On the original scale ($\lambda = 1$), the identified model also has the four main effects, but these effects do not stand apart from the remaining effects. This last point can be supported quantitatively by computing the normalized likelihood values for the respective transformations. Details are not given here, however.

An alternative to transformation of the response, generalized linear models, or GLMs, will be presented in Chapter 13.

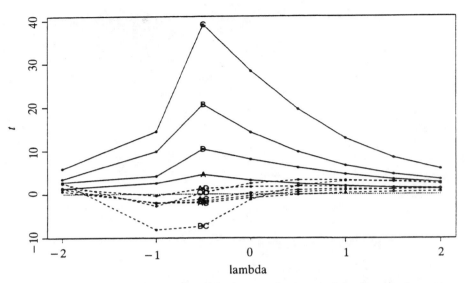

Figure 2.3. Scaled Lambda Plot [lambda denotes the power λ in the transformation (2.34)].

2.6 LATIN SQUARE DESIGN: TWO BLOCKING VARIABLES

Consider an experiment (Davies, 1954, p. 163) for testing the abrasion resistance of rubber-covered fabric in a Martindale wear tester. There are four types of material denoted by $A-D$. The response is the loss in weight in 0.1 milligrams (mgm) over a standard period of time. There are four positions on the tester so that four samples can be tested at a time, where past experience indicates that there are slight differences between the four positions. Each time the tester was used, the setup could be a bit different, i.e., there might be a systematic difference from application to application. Therefore "application" should be treated as a blocking variable. The experiment was designed to remove the variation due to position and application. The design used in this experiment is given in Table 2.17 and the weight loss data is given in Table 2.18.

The wear experiment used a **Latin square** design which has two blocking factors (represented by the rows and columns of the square) and one experimental factor (represented by the Latin letters in the square) with all factors having k levels. Formally a *Latin square* of order k is a $k \times k$ square of k Latin letters so that each Latin letter appears only once in each row and only once in each column. A 4×4 Latin square design is given in Table 2.17. The levels of the two blocking factors are randomly assigned to the rows and columns; the treatments of the experimental factor are randomly assigned to the Latin letters $A, B, C, \ldots$. A collection of Latin squares is given in Appendix 2A.

Table 2.17   Latin Square Design for Materials A–D, Wear Experiment

Application	Position			
	1	2	3	4
1	C	D	B	A
2	A	B	D	C
3	D	C	A	B
4	B	A	C	D

Table 2.18   Weight Loss Data, Wear Experiment

Application	Position			
	1	2	3	4
1	235	236	218	268
2	251	241	227	229
3	234	273	274	226
4	195	270	230	225

The linear model for the Latin square design is

$$y_{ijl} = \eta + \alpha_i + \beta_j + \tau_l + \epsilon_{ijl}, \tag{2.38}$$

where $i, j = 1, \ldots, k$ and l is the Latin letter in the (i, j) cell of the Latin square, α_i is the ith row effect, β_j is the jth column effect, τ_l is the lth treatment effect, and the errors ϵ_{ijl} are independent $N(0, \sigma^2)$. The triplet (i, j, l) takes on only the k^2 values dictated by the particular Latin square chosen for the experiment. The set of k^2 values of (i, j, l) is denoted by S. There are no interaction terms in (2.38) because they cannot be estimated in an unreplicated experiment. Constraints similar to those used for the randomized block design apply here. For the 16 runs of the wear experiment, the set S consists of the following (i, j, l) values: $(1, 1, 3)$, $(1, 2, 4)$, $(1, 3, 2)$, $(1, 4, 1)$, $(2, 1, 1), \ldots, (4, 4, 4)$.

The ANOVA for the Latin square design can be derived as follows. Using the zero-sum constraints, the following holds:

$$
\begin{aligned}
y_{ijl} &= \hat{\eta} + \hat{\alpha}_i + \hat{\beta}_j + \hat{\tau}_l + r_{ijl} \\
&= \bar{y}_{\ldots} + (\bar{y}_{i\cdot\cdot} - \bar{y}_{\ldots}) + (\bar{y}_{\cdot j\cdot} - \bar{y}_{\ldots}) + (\bar{y}_{\cdot\cdot l} - \bar{y}_{\ldots}) \\
&\quad + (y_{ijl} - \bar{y}_{i\cdot\cdot} - \bar{y}_{\cdot j\cdot} - \bar{y}_{\cdot\cdot l} + 2\bar{y}_{\ldots}),
\end{aligned} \tag{2.39}
$$

where

$$\hat{\eta} = \bar{y}..., \qquad \hat{\alpha}_i = \bar{y}_{i}.. - \bar{y}..., \qquad \hat{\beta}_j = \bar{y}._{j}. - \bar{y}...,$$

$$\hat{\tau}_l = \bar{y}.._{l} - \bar{y}..., \qquad r_{ijl} = y_{ijl} - \bar{y}_{i}.. - \bar{y}._{j}. - \bar{y}.._{l} + 2\bar{y}... \qquad (2.40)$$

Then, subtracting $\bar{y}...$, squaring both sides, and summing over all the observations,

$$\sum_{(i,j,l)\in S} \left(y_{ijl} - \bar{y}...\right)^2$$

$$= \sum_{i=1}^{k} k\left(\bar{y}_{i}.. - \bar{y}...\right)^2 + \sum_{j=1}^{k} k\left(\bar{y}._{j}. - \bar{y}...\right)^2 + \sum_{l=1}^{k} k\left(\bar{y}.._{l} - \bar{y}...\right)^2$$

$$+ \sum_{(i,j,l)\in S} \left(y_{ijl} - \bar{y}_{i}.. - \bar{y}._{j}. - \bar{y}.._{l} + 2\bar{y}...\right)^2. \qquad (2.41)$$

That is, the corrected total sum of squares on the left equals the sum of the row sum of squares, the column sum of squares, the treatment sum of squares, and the residual sum of squares. The ANOVA table for the Latin square design appears in Table 2.19. The mean squares are computed by dividing the sums of squares by their corresponding degrees of freedom. To test the null hypothesis of no treatment effect difference, $H_0 : \tau_1 = \cdots = \tau_k$, the F statistic is computed by dividing the treatment mean square by the residual mean square. If this value exceeds $F_{k-1, (k-1)(k-2), \alpha}$, H_0 is rejected at level α. A similar F test can be developed for testing the significance of row effects or column effects. If there are several replicates of the $k \times k$ Latin square, the residual sum of squares can be further decomposed into two terms: the within-replicate sum of squares and the between-replicate sum of squares. The derivation is left as an exercise.

Table 2.19   ANOVA Table for Latin Square Design

Source	Degrees of Freedom	Sum of Squares
row	$k - 1$	$k\sum_{i=1}^{k}(\bar{y}_{i}.. - \bar{y}...)^2$
column	$k - 1$	$k\sum_{j=1}^{k}(\bar{y}._{j}. - \bar{y}...)^2$
treatment	$k - 1$	$k\sum_{l=1}^{k}(\bar{y}.._{l} - \bar{y}...)^2$
residual	$(k-1)(k-2)$	$\sum_{(i,j,l)\in S}(y_{ijl} - \bar{y}_{i}.. - \bar{y}._{j}. - \bar{y}.._{l} + 2\bar{y}...)^2$
total	$k^2 - 1$	$\sum_{(i,j,l)\in S}(y_{ijl} - \bar{y}...)^2$

If H_0 is rejected, multiple comparisons of the k levels of treatment should be performed. The multiple comparison t statistics have the form

$$t_{ij} = \frac{\bar{y}_{..j} - \bar{y}_{..i}}{\hat{\sigma}\sqrt{1/k + 1/k}}, \qquad (2.42)$$

where $\hat{\sigma}^2$ is the residual mean square in the ANOVA table, i and j denote treatments i and j, and k is the number of treatments. Under the null hypothesis $H_0 : \tau_1 = \cdots = \tau_k$, each t_{ij} has a t distribution with $(k-1)(k-2)$ degrees of freedom. The Tukey multiple comparison method identifies treatments i and j as different if

$$|t_{ij}| > \frac{1}{\sqrt{2}} q_{k,(k-1)(k-2),\alpha}.$$

Based on the Tukey multiple comparison method, the simultaneous confidence intervals for $\tau_j - \tau_i$ are

$$\bar{y}_{..j} - \bar{y}_{..i} \pm \frac{1}{\sqrt{k}} q_{k,(k-1)(k-2),\alpha} \hat{\sigma} \qquad (2.43)$$

for all i and j pairs.

Returning to the wear experiment, the ANOVA table is displayed in Table 2.20. It indicates that blocking in the wear experiment is indeed important. The p values for application and position are $0.039 \ [= Prob(F_{3,6} > 5.37)]$ and $0.016 \ [= Prob(F_{3,6} > 7.99)]$, respectively. The treatment factor (material) has the most significance as indicated by a p value of $0.0008 \ [= Prob(F_{3,6} > 25.15)]$. With $k = 4$ and $(k-1)(k-2) = 6$, the critical value for the Tukey multiple comparison method is

$$\frac{1}{\sqrt{2}} q_{4,6,0.05} = \frac{4.90}{\sqrt{2}} = 3.46$$

at the 0.05 level. The mean weight losses are 265.75, 220, 241.75, and 230.5

Table 2.20   ANOVA Table, Wear Experiment

Source	Degrees of Freedom	Sum of Squares	Mean Squares	F
application	3	986.5	328.833	5.37
position	3	1468.5	489.500	7.99
material	3	4621.5	1540.500	25.15
residual	6	367.5	61.250	

Table 2.21 Multiple Comparison t Statistics, Wear Experiment

A vs. B	A vs. C	A vs. D	B vs. C	B vs. D	C vs. D
-8.27	-4.34	-6.37	3.93	1.90	-2.03

Table 2.22 ANOVA Table (Ignoring Blocking), Wear Experiment

Source	Degrees of Freedom	Sum of Squares	Mean Squares	F
material	3	4621.5	1540.50	6.55
residual	12	2822.5	235.21	

for materials $A-D$, respectively, and the corresponding multiple comparison t statistics are given in Table 2.21. By comparing these values with 3.46, materials A and B, A and C, A and D, and B and C are identified as different, with A wearing more than B, C, and D and C wearing more than B.

To appreciate the usefulness of blocking, consider the ANOVA table in Table 2.22 by ignoring the blocking of the applications and positions. The residual sum of squares in Table 2.22 is obtained by merging the application, position, and residual sums of squares from Table 2.20. While the test for material is still significant, the p value $[= Prob(F_{3, 12} > 6.55)]$ is an order of magnitude larger (0.007 rather than 0.0008). This suggests the possibility that, if the differences between the treatments were smaller (than those observed in this experiment), they could be missed if blocking is not used.

2.7 GRAECO-LATIN SQUARE DESIGN

Two Latin squares are said to be *orthogonal* if the two squares when superimposed have the property that each pair of letters appears once. For example, by superimposing the two 3×3 squares in Appendix 2A, each of the nine combinations of A, B, and C appears once. To distinguish the variables represented by the letters in the two squares, it is customary to replace the Latin letters A, B, and C in the second square by the Greek letters α, β, and γ. The superimposed square as shown below is called a **Graeco-Latin square**.

$$
\begin{array}{ccc}
A\alpha & B\beta & C\gamma \\
B\gamma & C\alpha & A\beta \\
C\beta & A\gamma & B\alpha
\end{array}
$$

It is known that Graeco-Latin squares of order k, with prime power k, are equivalent to k-level fractional factorial designs. Therefore the construction of the former can be obtained from that of the latter, which will be discussed in Section 6.8. Graeco-Latin squares of order 3, 4, 5, 7, and 8 are given in Appendix 2A. Note that Latin and Graeco-Latin square designs of order higher than 8 are seldom used because blocking based on larger squares is less effective as it becomes more difficult to control the variation within larger blocks.

The Graeco-Latin square design can be used for treatment comparisons, with Latin letters representing treatments and rows, columns, and Greek letters representing three blocking variables. Consider, for example, an experiment to compare three gasoline additives by testing them on three cars with three drivers over three days. The additives are the treatment factor assigned to the three Latin letters; the cars, drivers, and days are blocking variables assigned respectively to the three rows, columns, and Greek letters of the 3×3 Graeco-Latin square given above. Alternatively, it can be used to study two treatment factors represented by Latin and Greek letters in the presence of two blocking variables (represented by rows and columns).

Because the analysis is similar to that for a Latin square design, the following discussion will be brief. The linear model for a Graeco-Latin square design is

$$y_{ijlm} = \eta + \alpha_i + \beta_j + \tau_l + \zeta_m + \epsilon_{ijlm}, \qquad (2.44)$$

where i and j represent row and column, $i, j = 1, \ldots, k$, l and m are varied according to the settings of a $k \times k$ Graeco-Latin square, α_i is the ith row effect, β_j is the jth column effect, τ_l is the lth treatment effect of Latin letters, ζ_m is the mth effect of Greek letters, and ϵ_{ijlm} are independent $N(0, \sigma^2)$. The k^2 values of (i, j, l, m) are denoted by the set S.

The ANOVA for a Graeco-Latin square design can be derived in the same way as in Section 2.6 and is summarized in Table 2.23. Note that the residual sum of squares in Table 2.23 is obtained by subtracting the sums of squares in the first four rows from the (corrected) total sum of squares. Multiple comparisons of treatment effects are similar to those in Section 2.6 and the details are left as an exercise.

Table 2.23   ANOVA Table for Graeco-Latin Square Design

Source	Degrees of Freedom	Sum of Squares
row	$k - 1$	$k\sum_{i=1}^{k} (\bar{y}_{i\ldots} - \bar{y}_{\ldots})^2$
column	$k - 1$	$k\sum_{j=1}^{k} (\bar{y}_{.j..} - \bar{y}_{\ldots})^2$
Latin letter	$k - 1$	$k\sum_{l=1}^{k} (\bar{y}_{..l.} - \bar{y}_{\ldots})^2$
Greek letter	$k - 1$	$k\sum_{m=1}^{k} (\bar{y}_{\ldots m} - \bar{y}_{\ldots})^2$
residual	$(k - 3)(k - 1)$	by subtraction
total	$k^2 - 1$	$\sum_{(i,j,l,m) \in S} (y_{ijlm} - \bar{y}_{\ldots})^2$

Hyper-Graeco-Latin squares can be obtained as extensions of Graeco-Latin squares by superimposing three or more mutually orthogonal Latin squares. A small collection of hyper-Graeco-Latin squares is given in Appendix 2A.

2.8 BALANCED INCOMPLETE BLOCK DESIGN

Consider an experiment (Davies, 1954, p. 200) to assess the effect of tires made out of different compounds on the lifetime of the tires. The design used in the tire experiment as well as the tire wear data are given in Table 2.24. Because of a manufacturing limitation, a tire can be divided into at most three sections with each section being made out of a different compound. Consequently, only three compounds can be compared on the same unit (i.e., tire). As a particular tire is tested for resistance to wear, the three compounds on the tire are subjected to the same road conditions. Therefore, each tire is treated as a block. The wear data are reported as relative wear in arbitrary units.

For the tire experiment, the block size of 3 is smaller than the number of treatments, which is 4. Since not all of the treatments can be compared within a block, as is the case for the (complete) randomized block design, this experiment used what is known as a **balanced incomplete block design**, or BIBD.

In a BIBD, the block size k is less than the number of treatments t, which explains the term *incomplete*. There are b blocks and each treatment is replicated r times. (Previously k was used for the number of treatments and n for the number of replicates, but the notation is changed to be consistent with the BIBD literature.) The design is called *balanced* because each pair of treatments is compared in the *same* number of blocks. Denote this number by λ. (For the tire experiment, $k = 3$, $t = 4$, $b = 4$, $r = 3$, and $\lambda = 2$.) Randomization is employed by randomizing the treatments to the units within a block as well as randomizing the block labels to the blocks. Tables of BIBDs can be found in Cochran and Cox (1957).

Among the five parameters in a BIBD, there are two basic relations:

$$bk = rt, \tag{2.45}$$

$$r(k - 1) = \lambda(t - 1). \tag{2.46}$$

Table 2.24 Wear Data, Tire Experiment

Tire	Compound			
	A	*B*	*C*	*D*
1	238	238	279	
2	196	213		308
3	254		334	367
4		312	421	412

The first identity is easily proved by noting that bk and rt are equivalent ways of counting the total number of units in the experiment. To prove (2.46), consider the total number of units (other than treatment 1) that appear with treatment 1 in the same blocks. This number equals $r(k-1)$ because treatment 1 appears in r blocks and within each of the r blocks there are $k-1$ remaining units. Another way to enumerate this number is to consider the $t-1$ treatments that are different from treatment 1. Each of the $t-1$ treatments appears with treatment 1 in λ blocks, giving $\lambda(t-1)$ as the desired number and proving (2.46). The following four inequalities are also of interest:

$$t > k, \tag{2.47}$$

$$r > \lambda, \tag{2.48}$$

$$b > r, \tag{2.49}$$

$$rk > \lambda t . \tag{2.50}$$

The first two inequalities follow easily from the definition of "incompleteness"; (2.49) follows from (2.45) and (2.47); and (2.50) follows from (2.46) and (2.48).

The linear model for a BIBD experiment is

$$y_{ij} = \eta + \alpha_i + \tau_j + \epsilon_{ij}, \qquad i = 1, \ldots, b; \quad j = 1, \ldots, t, \tag{2.51}$$

where α_i is the ith block effect, τ_j is the jth treatment effect, and ϵ_{ij} are independent $N(0, \sigma^2)$. Because not all the treatments are compared within a block, data are not observed for all (i, j) pairs. This complicates the analysis but can still be handled by using the extra sum of squares principle as follows.

To test no difference between the treatments, let Model II be (2.51) and Model I be

$$y_{ij} = \eta + \alpha_i + \epsilon_{ij}, \qquad i = 1, \ldots, b; \quad j = 1, \ldots, t. \tag{2.52}$$

The ANOVA table for the tire experiment appears in Table 2.25. The extra sum of squares principle given in Section 1.4 is used to obtain the

Table 2.25 ANOVA Table, Tire Experiment

Source	Degrees of Freedom	Sum of Squares	Mean Squares	F
tire (block)	3	39,122.67	13,040.89	37.24
compound	3	20,729.08	6,909.69	19.73
residual	5	1,750.92	350.10	

ANOVA table as follows. The *RSS* for Model II (tire and compound) is 1750.92, which is the residual sum of squares in Table 2.25. The *RegrSS* for Model I (tire factor alone) is 39,122.67, i.e., the tire sum of squares in Table 2.25. The *RSS* for Model I (tire factor alone) is 22,480.00, so that the sum of squares for the compound after the tire has been accounted for is

$$22,480.00 - 1750.92 = 20,729.08,$$

which is the compound sum of squares in Table 2.25. For the tire experiment, the ANOVA table shows that blocking was important [with a p value of $0.0008 = Prob(F_{3,5} > 37.24)$]. There are also differences between the compounds but with a larger p value of 0.0034 [$= Prob(F_{3,5} > 19.73)$]. If blocking is ignored in the analysis, the treatment factor compound would not be as significant. Thus, blocking is effective here because it removes the significant tire-to-tire variation.

If the hypothesis of no treatment differences is rejected, then the multiple-comparison procedures can be used to identify which treatments are different. Under model (2.51) with the zero-sum constraints $\sum_{j=1}^{t} \tau_j = 0$ and $\sum_{i=1}^{b} \alpha_i = 0$, it can be shown that the least squares estimate $\hat{\tau}_j$ of τ_j is

$$\hat{\tau}_j = \frac{k}{\lambda t} Q_j, \qquad (2.53)$$

where

$$Q_j = y_{\cdot j} - \sum_{i=1}^{b} n_{ij} \bar{y}_i.$$

is the *adjusted treatment total* for the jth treatment, $y_{\cdot j}$ is the sum of y values for the jth treatment, $\bar{y}_i$ is the average of y values in the ith block, and $n_{ij} = 1$ if treatment j appears in block i and $n_{ij} = 0$ otherwise. It can also be shown that

$$Var(\hat{\tau}_j - \hat{\tau}_i) = \frac{2k}{\lambda t} \sigma^2. \qquad (2.54)$$

The proofs of (2.53) and (2.54) can be found in John (1971). Then the t statistics for testing $\tau_i = \tau_j$, $i, j = 1, \ldots, t$, have the form

$$t_{ij} = \frac{\hat{\tau}_j - \hat{\tau}_i}{\hat{\sigma}\sqrt{2k/\lambda t}}, \qquad (2.55)$$

where $\hat{\sigma}^2$ is the mean-squared error in the ANOVA table. Under the null hypothesis $H_0: \tau_1 = \cdots = \tau_t$, each t_{ij} has a t distribution with $tr - b - t + 1$ degrees of freedom. Because the $\hat{\tau}_i$'s have equal correlation, it can be shown that the Tukey multiple comparison method is still valid (Miller, 1981), which identifies treatments i and j as different if

$$|t_{ij}| > \frac{1}{\sqrt{2}} q_{t,tr-b-t+1,\alpha}.$$

Based on the Tukey method, the simultaneous confidence intervals for $\tau_j - \tau_i$ are

$$\hat{\tau}_j - \hat{\tau}_i \pm q_{t,tr-b-t+1,\alpha} \hat{\sigma} \sqrt{\frac{k}{\lambda t}} \tag{2.56}$$

for all i and j pairs.

Using the baseline constraints $\tau_1 = 0$ and $\alpha_1 = 0$ in (2.51) (so that τ_j measures the difference between compounds j and 1, $j > 1$), regression analysis leads to the estimates

$$\hat{\tau}_2 = 4.375, \qquad \hat{\tau}_3 = 76.250, \qquad \hat{\tau}_4 = 100.875. \tag{2.57}$$

Then, the remaining comparisons between the compounds can be computed as

$$\hat{\tau}_3 - \hat{\tau}_2 = 71.875, \qquad \hat{\tau}_4 - \hat{\tau}_2 = 96.500, \qquad \hat{\tau}_4 - \hat{\tau}_3 = 24.625, \tag{2.58}$$

where $\hat{\tau}_3 - \hat{\tau}_2$ estimates the difference between compounds 3 and 2, etc. The standard error for each of these comparisons is

$$\hat{\sigma} \sqrt{\frac{2k}{\lambda t}} = \sqrt{350} \sqrt{\frac{2(3)}{2(4)}} = 16.206,$$

where $k = 3$, $t = 4$, and $\lambda = 2$. The multiple comparison t statistics according to (2.55) are given in Table 2.26. With $t = 4$ and $tr - b - t + 1 = 4 \cdot 3 - 4 - 4 + 1 = 5$, the 0.05 critical value for the Tukey method is

$$\frac{1}{\sqrt{2}} q_{4,5,0.05} = \frac{5.22}{1.414} = 3.69.$$

By comparing 3.69 with the t values in Table 2.26, we conclude that at the 0.05 level compounds A and B are different from C and D, with C and D wearing more than A and B.

Table 2.26 Multiple Comparison t Statistics, Tire Experiment

A vs. B	A vs. C	A vs. D	B vs. C	B vs. D	C vs. D
0.27	4.71	6.22	4.44	5.95	1.52

Table 2.27   Data (x = film thickness, y = strength), Starch Experiment

Canna		Starch Corn		Potato	
y	x	y	x	y	x
791.7	7.7	731.0	8.0	983.3	13.0
610.0	6.3	710.0	7.3	958.8	13.3
710.0	8.6	604.7	7.2	747.8	10.7
940.7	11.8	508.8	6.1	866.0	12.2
990.0	12.4	393.0	6.4	810.8	11.6
916.2	12.0	416.0	6.4	950.0	9.7
835.0	11.4	400.0	6.9	1282.0	10.8
724.3	10.4	335.6	5.8	1233.8	10.1
611.1	9.2	306.4	5.3	1660.0	12.7
621.7	9.0	426.0	6.7	746.0	9.8
735.4	9.5	382.5	5.8	650.0	10.0
990.0	12.5	340.8	5.7	992.5	13.8
862.7	11.7	436.7	6.1	896.7	13.3
		333.3	6.2	873.9	12.4
		382.3	6.3	924.4	12.2
		397.7	6.0	1050.0	14.1
		619.1	6.8	973.3	13.7
		857.3	7.9		
		592.5	7.2		

2.9   ANALYSIS OF COVARIANCE: INCORPORATING AUXILIARY INFORMATION

Consider an experiment (Flurry, 1939) to study the breaking strength (y) in grams of three types of starch film. The breaking strength is also known to depend on the thickness of the film (x) as measured in 10^{-4} inches. Because film thickness varies from run to run and its values cannot be controlled or chosen prior to the experiment, it should be treated as a covariate whose effect on strength needs to be accounted for before comparing the three types of starch. See Table 2.27 for the experimental data.

The starch experiment is an example of treatment comparison experiments, in which the response depends on auxiliary covariates whose information is available to the investigator. By adjusting for the effects of these covariates on the response, the comparison of treatments (which is the primary goal of the study) can be done more efficiently. The following linear model is assumed to hold:

$$y_{ij} = \eta + \tau_i + \gamma x_{ij} + \epsilon_{ij}, \qquad i = 1,\ldots,k; \quad j = 1,\ldots,n_i, \qquad (2.59)$$

where y_{ij} is the response value for the jth observation receiving the ith treatment, x_{ij} is the corresponding covariate value, γ is the regression coefficient for the covariate, τ_i is the ith treatment effect, and the errors ϵ_{ij} are independent $N(0, \sigma^2)$. As before, constraints on the treatment effects are needed—either $\tau_1 = 0$ or $\sum_{i=1}^{k} \tau_i = 0$. The ANOVA associated with this linear model is referred to as ANalysis of COVAriance or ANCOVA. The inclusion of a covariate is effective if it can explain a substantial portion of the variation in the response, i.e., γ is significant.

For the starch experiment, we should use the baseline constraint $\tau_1 = 0$ because it can facilitate pairwise comparisons of the three types of starch film. Then the model in (2.59) takes the following form:

$$
\begin{aligned}
y_{1j} &= \eta + \gamma x_{1j} + \epsilon_{1j} & j &= 1, \cdots, 13, \quad i = 1 \quad \text{(canna)}; \\
y_{2j} &= \eta + \tau_2 + \gamma x_{2j} + \epsilon_{2j}, & j &= 1, \cdots, 19, \quad i = 2 \quad \text{(corn)}; \quad (2.60) \\
y_{3j} &= \eta + \tau_3 + \gamma x_{3j} + \epsilon_{3j}, & j &= 1, \cdots, 17, \quad i = 3 \quad \text{(potato)},
\end{aligned}
$$

where η is the intercept term, γ the regression coefficient for thickness, τ_2 the comparison between canna and corn, and τ_3 the comparison between canna and potato. Fitting the model (2.60) leads to the estimates given in Table 2.28 above the dashed line. The comparison between τ_2 and τ_3 (corn vs. potato) is obtained by using the estimate

$$
\hat{\tau}_3 - \hat{\tau}_2 = 70.360 - (-83.666) = 154.026.
$$

Its standard error is 107.76 (method for computing this standard error will be discussed toward the end of the section) and its corresponding t statistic equals $154.02/107.76 = 1.43$. These values are reported in the last row of Table 2.28 for "corn vs. potato." Because none of the last three t values in the table are significant, no pair of film types is declared to be significantly different. (A practical implication of this finding is that other considerations, such as cost of starch film types, should be used in the optimal selection of starch film.) The same conclusion is reached by using the following ANCOVA procedure.

To test the null hypothesis of no difference between the treatments, the extra sum of squares principle (see Section 1.4) can be used. Model II is

Table 2.28   Tests, Starch Experiment

Effect	Estimate	Standard Error	t	p Value
intercept	158.261	179.775	0.88	0.38
thickness	62.501	17.060	3.66	0.00
canna vs. corn	−83.666	86.095	−0.97	0.34
canna vs. potato	70.360	67.781	1.04	0.30
corn vs. potato	154.026	107.762	1.43	0.16

Table 2.29 ANCOVA Table, Starch Experiment

Source	Degrees of Freedom	Sum of Squares	Mean Squares	F
thickness	1	2,553,357	2,553,357	94.19
starch	2	56,725	28,362	1.05
residual	45	1,219,940	27,110	

(2.59) and Model I is

$$y_{ij} = \eta + \gamma x_{ij} + \epsilon_{ij}, \qquad i = 1, \ldots, k, \qquad j = 1, \ldots, n_i. \qquad (2.61)$$

For the starch experiment, the regression and residual sum of squares for Model I (thickness alone) are $RegrSS(I) = 2,553,357$ and $RSS(I) = 1,276,665$, i.e., the sum of squares for thickness is 2,553,357. For both thickness and starch (Model II), the residual sum of squares $RSS(II)$ is 1,219,940. Thus, the sum of squares due to starch (after thickness has been accounted for) is

$$1,276,665 - 1,219,940 = 56,725.$$

The ANCOVA table is constructed and displayed in Table 2.29. It indicates that there is no difference between the three starch types with a p value of 0.3597 [$= Prob(F_{2,45} > 1.05)$]. Table 2.30 presents the ANOVA table obtained from incorrectly ignoring the covariate, film thickness, i.e., by treating it as a one-way layout. The results incorrectly indicate a significant difference between the starch types.

In order to perform multiple comparisons, estimates for $\boldsymbol{\beta}$ in the general linear model framework (see Section 1.4) need to be obtained. Suppose that the constraint $\tau_1 = 0$ is used. Then the last $k - 1$ entries of $\hat{\boldsymbol{\beta}}$ are $\hat{\tau}_2, \ldots, \hat{\tau}_k$, which are the comparisons between level 1 and level 2 through level k, respectively. The variance estimates for these comparisons are the appropriate entries of $\hat{\sigma}^2 (\mathbf{X}^T \mathbf{X})^{-1}$. Recall that $\hat{\sigma}^2$ is the MSE from Model II which has $N - 1 - (k - 1) - 1 = N - k - 1$ degrees of freedom, where N is the total number of observations. The other comparisons, e.g., between levels 2 and 3, can be expressed as a linear combination of the model parameters, $\mathbf{a}^T \boldsymbol{\beta}$, where $\mathbf{a}^T = (0, 0, -1, 1, 0, \ldots, 0)$ and $\boldsymbol{\beta} = (\eta, \gamma, \tau_2, \tau_3, \ldots, \tau_k)$. Its estimate is $\mathbf{a}^T \hat{\boldsymbol{\beta}}$, whose variance estimate is $\hat{\sigma}^2 \mathbf{a}^T (\mathbf{X}^T \mathbf{X})^{-1} \mathbf{a}$ [see (1.22)]. Finally,

Table 2.30 ANOVA Table Ignoring Thickness, Starch Experiment

Source	Degrees of Freedom	Sum of Squares	Mean Squares	F
starch	2	2,246,204	1,123,102	32.62
residual	46	1,583,818	34,431	

the t statistics can be computed by dividing each estimated comparison by its standard error (square root of its estimated variance). These t statistics can be simultaneously tested by using the adjusted p values as discussed in Section 2.3.1.

In (2.59), the same regression coefficient γ is used for all the treatment groups. If a separate regression coefficient is used for each group, fitting of the parameters can be done by performing regression analysis separately for each group. Only the variance σ^2 is estimated by pooling data from all groups. The ANCOVA procedure as described above cannot be used with this approach. A reanalysis of the starch experiment using this approach is left as an exercise.

In this section the ANCOVA method is developed for experiments with a single treatment factor. The information in the auxiliary covariates can be similarly incorporated in the analysis of more general experiments that involve blocking variables and more than one treatment factor, such as randomized block designs, Latin square designs, etc. See Neter et al. (1996) for details.

2.10 PRACTICAL SUMMARY

1. Use a randomized block design for treatment comparisons if blocking is effective. The corresponding linear model, ANOVA table, and multiple comparison procedure can be found in Section 2.2.

2. Paired comparison designs are a special case of randomized block designs with block size 2. Pairing is effective if blocks of size 2 occur naturally such as in animal and medical studies with twins, pairs of eyes, etc. Paired designs are better than unpaired designs when pairing is effective. Use the paired t test statistic in (2.1) for testing the difference between two treatments. Incorrect use of the unpaired t test (also called the two-sample t test) will result in a severe loss of efficiency. The square of the paired t test is equivalent to the F test in the ANOVA for randomized block designs.

3. Use a two-way layout for experiments with two treatment factors (and no blocking variable). Different models are available in Section 2.3 for three cases: two qualitative factors, one qualitative factor and one quantitative factor, and two quantitative factors.

4. Box-whisker plots of residuals are an effective graphical tool for experiments with at least 10 replicates per treatment combination. They facilitate an easy graphical comparison of the locations and dispersions of the various treatment combinations. An example is given in Figure 2.1 for the bolt experiment.

5. Use a multi-way layout for experiments with more than two treatment factors (and no blocking variable).

6. Using a power transformation in (2.34) on the response may lead to a more parsimonious model, including possibly the elimination of spurious interactions or of variance heterogeneity. A simple and easy-to-follow procedure to select an appropriate power is given in Section 2.5.

7. Use a Latin square design for experiments with one treatment factor and two blocking variables. Use a Graeco-Latin square design for experiments with one treatment factor and three blocking variables or two treatment factors and two blocking variables.

8. Use a balanced incomplete block design for treatment comparisons when the block size is smaller than the total number of treatments.

9. When auxiliary covariates are available, use analysis of covariance and regression analysis to incorporate such information in the comparison of treatments.

EXERCISES

1. For the two alternative constraints given after (2.6), interpret the remaining parameters for the randomized block design.

2. Prove the equivalence of the F test statistic in the ANOVA for the paired comparison experiment and the square of the paired t statistic given in (2.1).

3. The following are the weights of six rock samples (from different lakes) measured on two different scales. The purpose is to test whether the two scales are different.

	Weight in Grams by Sample Number					
Scale	1	2	3	4	5	6
I	8	14	16	19	18	12
II	11	16	20	18	20	15

(a) Perform an appropriate test at the 0.05 level to test the null hypothesis that scale I and scale II give the same measurement. Formulate this hypothesis and give details of your computations.

(b) Compute the p value of the test statistic you obtained in (a).

4. Perform a residual analysis of the girder experiment data. Does it lead to any new insights on the model assumptions?

5. Perform the Tukey multiple comparisons at the 0.01 level for the girder experiment in Section 2.2. Are the conclusions different from those in Section 2.2 at the 0.05 level?

Table 2.31 Strength Data, Original Girder Experiment

	Method									
Girder	Aarau	Karlsruhe	Lehigh	Cardiff	Gothenburg	Osaka	Prague-Cardiff	Stockholm	Tokyo	Zurich
S1/1	0.772	1.186	1.061	1.025	0.951	1.136	1.053	1.076	1.033	1.181
S2/1	0.744	1.151	0.992	0.905	0.878	0.976	0.081	0.907	0.932	0.949
S3/1	0.767	1.322	1.063	0.930	0.930	0.958	0.739	0.918	0.910	0.916
S4/1	0.745	1.339	1.062	0.899	0.916	0.907	0.650	0.861	0.870	0.821
S5/1	0.725	1.200	1.065	0.871	0.963	0.872	0.560	0.786	0.838	0.737
S1/2	0.844	1.402	1.178	1.004	0.960	1.178	1.120	1.100	1.180	1.234
S2/2	0.831	1.365	1.037	0.853	0.929	0.950	1.110	0.962	0.973	0.975
S3/2	0.867	1.537	1.086	0.858	0.990	0.925	0.790	0.969	0.918	0.927
S4/2	0.859	1.559	1.052	0.805	0.966	0.844	0.670	0.867	0.828	0.811

Table 2.32 ANOVA Table, Randomized Block Experiment Exercise

Source	Degrees of Freedom	Sum of Squares	Mean Squares
block		520	
treatment		498	
residual		40	
total	14		

6. The original girder experiment studied 10 methods (instead of the four methods reported in Section 2.2). The strength data appear in Table 2.31. Analyze the full experiment, including ANOVA, multiple comparisons, and residual analysis.

7. The entries in Table 2.32 were determined from some data collected according to a randomized block design with three blocks. The treatment averages were 45, 58, 46, 45, and 56 for the five groups.

(a) Complete the ANOVA table in Table 2.32. (Are any entries impossible to determine with the information given? Explain.)

(b) Use the Tukey method to compare the five treatment means simultaneously at level 0.01. Report only those pairs that are found to be significantly different.

(c) Without doing the actual computations, explain why, from the findings in (b), the usual F test at level 0.01 will reject the null hypothesis that the treatment effects are the same.

8. A chemical reaction experiment was carried out with the objective of comparing if a new catalyst B would give higher yields than the old catalyst A. The experiment was run on six different batches of raw material which were known to be quite different from one another. Each

Table 2.33   Yield Data, Chemical Reaction Experiment

Catalyst	Batch					
	1	2	3	4	5	6
A	9	19	28	22	18	8
B	10	22	30	21	23	12

batch was divided into two portions to which A or B was applied at random. The data collected are given in Table 2.33.

(a) Explain the experimental design.

(b) Carry out the appropriate t test.

(c) Construct a 95% confidence interval for the difference between B and A.

9. An experimenter believes that drug A lowers the diastolic blood pressure in people with high blood pressure and that it takes effect in two hours. Design a simple comparative experiment with a control group to assess the effect of the drug A. Tell the experimenter exactly what steps to take to carry out the experiment.

10. Assume that the true model is

$$y_i = f(\mathbf{x}_i^T \boldsymbol{\beta}) + \psi. \tag{2.62}$$

Transforming y to $f^{-1}(y)$ would allow the transformed response to be represented approximately by a linear model. The nonlinearity in (2.62) is referred to as *transformable non-additivity*. Suppose that the usual linear model

$$y_i = \mathbf{x}_i^T \boldsymbol{\beta} + \epsilon \tag{2.63}$$

is fitted to the data. Letting $\mu = \mathbf{x}^T \boldsymbol{\beta}$, a Taylor series approximation yields

$$f(\mu) \sim \gamma_1 \mu + \gamma_2 \mu^2, \tag{2.64}$$

where γ_1 and γ_2 are constants. Does (2.64) suggest that μ^2 is an important covariate of y? What would a plot of $\hat{y}_i$ versus r_i look like? [Hint: compare (2.63) and (2.64).]

11. Compute the p values of the F tests in Table 2.10 for test, plating, and test-by-plating. Comment on the significance of the three effects.

12. Develop the Tukey multiple comparison procedure for the row effects in a two-way layout.

Table 2.34 Strength Data, Revised Composite Experiment

Tape Speed	Laser Power		
	40 W	50 W	60 W
6.42	25.66	29.15	35.73
13.00	28.00	35.09	39.56
27.00	20.65	29.79	35.66

13. Use the Bonferonni inequality in (1.50) to devise a conservative test of the five effects in Table 2.11. At which significance levels are β_2, β_3, and ω_{23} still significant?

14. In the two-way layout, let s_{ij}^2 be the sample variance of the n_{ij} observations for the ith level of factor A and the jth level of factor B. By treating $z_{ij} = \ln s_{ij}^2$ as the response, analyze z_{ij} for the bolt experiment and compare the results with the findings from the box-whisker plot in Figure 2.1. (Note: The use of the log transformation of s_{ij}^2 will be explained in Section 3.8.)

15. For the composite experiment of Section 1.8, the original paper by Mazumdar and Hoa (1995) reported a second factor, tape speed. Table 2.34 shows the three replicates for each level of laser power corresponding to tape speed of 6.42, 13.0, and 27.0 m/s, respectively. The levels of tape speed are roughly evenly spaced on the log scale, so that linear and quadratic effects can be entertained for the second quantitative factor. Analyze the experiment as a two-way layout with a single replicate, including model building, parameter estimation, ANOVA and residual analysis.

16. Perform a residual analysis of the bolt experiment from Table 2.8.

17. Reanalyze the bolt experiment by using the zero-sum constraints. Compare the results with those in Section 2.3.1. Which approach is preferred?

18. (a) For the tomato experiment, a measure of productivity can be defined as the ratio of yield to plant density, which equals the ratio of tons of tomatoes to the number of plants per hectare. Discuss the productivity of the three plant densities using this measure and the fitted model in Section 2.3.2.

(b) Another measure in terms of profit is defined as $c_1 \times$ tons (per hectare) $- c_2 \times$ no. of plants (per hectare), where c_1 is the wholesale price of tomato per ton and c_2 is the cost of each plant. Discuss the profit of the three plant densities using this measure (with $c_1 = \$2500$ and $c_2 = \$1.00$) and the fitted model in Section 2.3.2.

(c) For each variety, choose the best plant density in terms of productivity and profit.

Table 2.35   Worsted Yarn Data

A	B	C	y	A	B	C	y	A	B	C	y
0	0	0	674	1	0	0	1414	2	0	0	3636
0	0	1	370	1	0	1	1198	2	0	1	3184
0	0	2	292	1	0	2	634	2	0	2	2000
0	1	0	338	1	1	0	1022	2	1	0	1568
0	1	1	266	1	1	1	620	2	1	1	1070
0	1	2	210	1	1	2	438	2	1	2	566
0	2	0	170	1	2	0	442	2	2	0	1140
0	2	1	118	1	2	1	332	2	2	1	884
0	2	2	90	1	2	2	220	2	2	2	360

19. For the drill experiment, the scaled lambda plot in Figure 2.3 suggests that the best three transformations are $\log (\lambda = 0)$, reciprocal square root $(\lambda = -\frac{1}{2})$, and square root $(\lambda = \frac{1}{2})$. Compare the three transformations in terms of some model adequacy measures for the chosen models for the three transformed responses. (Hint: Use residual plots and a model selection criterion such as the C_p criterion given in (1.28).)

20. In a wool textile experiment, an unreplicated three-way layout was employed to study the effect on y, the number of cycles to failure, of the lengths of worsted yarn under cycles of repeated loading. Three factors are A, length of test specimen (250, 300, 350 mm); B, amplitude of loading cycle (8, 9, 10 mm); and C, load (40, 45, 50 g). The data (Cox and Snell, 1981) are given in Table 2.35. Analyze the data by transforming the y values using the family of power transformations. Choose the best transformation based on the three criteria given in Section 2.5.

21. For the Latin square design, find two sets of constraints similar to those used for the randomized block design. Interpret the remaining parameters for both sets of constraints.

21. Develop the Bonferroni multiple comparison method for the Latin square design and apply it to the wear experiment data. Compare the results with those in Section 2.6 based on the Tukey method. For Latin square designs, which method is preferred?

23. Develop the ANOVA table for a replicated $k \times k$ Latin square with r replicates. Assume that the levels of the two blocking factors remain the same for each of the replicates.

24. Derive the ANOVA table in Table 2.23 for the Graeco-Latin square design.

Table 2.36 Design, Gasoline Additive Experiment

Day	Time of Day			
	1	2	3	4
I	A	B	D	C
II	D	C	A	B
III	B	D	C	A
IV	C	A	B	D

25. Develop the Tukey multiple comparison procedure for the Graeco-Latin square design, where the treatment variable is represented by the Latin letters.

26. To compare the effects of four gasoline additives A, B, C, and D, 16 tests are made on one car in 4 days, I, II, III, IV, which are representative of four weather conditions, and at four times of day: 1:30 p.m., 2:30 p.m., 3:30 p.m., and 4:30 p.m. in the order 1, 2, 3, 4. Comment on the pros and cons of the design given in Table 2.36.

27. An experiment is run to compare four different levels of a treatment. Forty runs are to be made, with at most eight per day. However, the experimenter suspects that some day-to-day variations may exist.
    (a) Should the experiment be blocked? What factors would be relevant in making that decision?
    (b) Whatever your decision (blocking or not), briefly say how you would set up the experiment.

28. Suppose your state governor asks you to recommend an experimental design to test the claim that a certain chemical, W0W3, added to gasoline will give better mileage for government cars. Available for this experiment are 50 cars, which will be driven in their usual manner during the month of June. These cars are not all the same make, model, and age. All cars will be tuned at the beginning of this experiment. During June, each time a car needs fuel, gasoline with or without W0W3 will be used in accordance with your instructions. Each time the car gets fuel, data will be recorded on mileage driven and gallons of fuel consumed. What design would you recommend? Provide support for your recommendation.

29. One of two shields (either I or II) is set up on an experimental apparatus. The sun shines through it and a response (either y_I or y_{II}) is measured on a certain piece of equipment. Each such measurement requires about 15 minutes. The equipment has three adjustable knobs (a, b, c), each at two levels. Given that you have 4 hours of suitable sunshine available on

Table 2.37   Data, Resistor Experiment

Plate	Shape			
	A	B	C	D
1	1.11		0.95	0.82
2	1.70	1.22		0.97
3	1.60	1.11	1.52	
4		1.22	1.54	1.18

each of four successive days, how would you design an experiment to find the best settings of the three knobs if the object is to find that single setting of the knobs that will produce the largest difference in the measurements made with the two shields?

30. Prove that $Var(\hat{\tau}_{.j} - \hat{\tau}_{.i}) = (2k/\lambda t)\sigma^2$ in a BIBD. If the experiment were run as a randomized complete block, $Var(\hat{\tau}_{.j} - \hat{\tau}_{.i})$ would equal $Var(\bar{y}_{.j} - \bar{y}_{.i}) = 2\sigma^2/b$. Prove that $2k/\lambda t > 2/b$ and give it a statistical interpretation.

31. Natrella (1963, pp. 13–14) described an experiment on a resistor mounted on a ceramic plate in which the impact of four geometrical shapes of the resistors on the current noise of the resistor is studied. Only three resistors can be mounted on one plate. The design and data for the resistor experiment are given in Table 2.37. Describe the design and analyze the experiment.

32. (a) What experimental design would be the best to use in the following situation? Be specific. Explain the good features of your proposed design.

A taste panel will convene this afternoon to compare six different brands of ice cream. The panel is comprised of 10 persons who are expert tasters. The maximum number of different brands that an individual taster will taste is 3.

(b) What would you do if three of the expert tasters failed to come this afternoon because of illness, so that you could not run the design recommended in (a) above?

33. Kodlin (1951) reported an experiment that compared two substances for lowering blood pressure, denoted as substances A and B. Two groups of animals are randomized to the two substances and a decrease in pressure is recorded. The initial pressure is also recorded. The data for the blood pressure experiment appear in Table 2.38. Compare the two substances by accounting for possible effect of the initial pressure on the decrease in pressure. Discuss the results of the ANCOVA analysis and compare

Table 2.38   Data, Blood Pressure Experiment

	Substance A			Substance B	
Animal	Initial Pressure	Decrease in Pressure	Animal	Initial Pressure	Decrease in Pressure
1	135	45	11	90	34
2	125	45	12	135	55
3	125	20	13	130	50
4	130	50	14	115	45
5	105	25	15	110	30
6	130	37	16	140	45
7	140	50	17	130	45
8	93	20	18	95	23
9	110	25	19	90	40
10	100	15	20	105	35

them with those obtained by ignoring the potential effect of initial pressure.

34. In Section 2.9, the same regression slope γ was used for each of the three types of starch film. If a separate regression slope is used for each type, reanalyze the data by performing a regression analysis with three intercept and slope terms. Compare the results with those obtained in Section 2.9.

35. The variance σ^2 of an individual observation can be estimated in a number of ways, depending on the situation. How many degrees of freedom are associated with the estimate s^2 (of σ^2) in each of the following circumstances?

(a) A randomized paired comparison of two treatments A and B, where there are $n_A = 22$ observations with A and $n_B = 22$ observations with B.

(b) A randomized unpaired comparison of two treatments A and B, where there are $n_A = 21$ observations with A and $n_B = 14$ observations with B.

(c) An 8×8 Latin square design.

(d) A 9×9 Graeco-Latin square design.

(e) A completely randomized design with five treatments, where $n_A = 24$, $n_B = 15$, $n_C = 12$, $n_D = 20$, $n_E = 17$.

(f) A randomized block design with 6 blocks and 14 treatments.

(g) An experiment on 24 mice to compare four diet types A, B, C, D, with $n_A = 6$, $n_B = 8$, $n_C = 5$, $n_D = 5$. The initial body weight of each mouse is recorded as the covariate value. Analysis of covariance is used to compare the effects of the four diet types.

Table 2.39   Throughput Data

Day	\multicolumn{10}{c}{Operator}
	\multicolumn{2}{c}{1}	\multicolumn{2}{c}{2}	\multicolumn{2}{c}{3}	\multicolumn{2}{c}{4}	\multicolumn{2}{c}{5}
1	$A\alpha$	(102)	$B\beta$	(105)	$C\gamma$	(82)	$D\delta$	(141)	$E\epsilon$	(132)
2	$B\gamma$	(92)	$C\delta$	(112)	$D\epsilon$	(131)	$E\alpha$	(112)	$A\beta$	(99)
3	$C\epsilon$	(96)	$D\alpha$	(130)	$E\beta$	(108)	$A\gamma$	(73)	$B\delta$	(129)
4	$D\beta$	(120)	$E\gamma$	(100)	$A\delta$	(111)	$B\epsilon$	(116)	$C\alpha$	(100)
5	$E\delta$	(123)	$A\epsilon$	(110)	$B\alpha$	(111)	$C\beta$	(85)	$D\gamma$	(100)

36. To compare the effects of five different assembly methods (denoted by the Latin letters A, B, C, D, E) on the throughput, an experiment based on a Graeco-Latin square was conducted which involved three blocking variables: day, operator, and machine type. The data are given in Table 2.39, where the machine type is denoted by the five Greek letters. The response, throughput, is the number of completed pieces per day and is given in the parentheses in the table. Analyze the data and compare the five assembly methods. Which of the methods are significantly better than the others?

37. A study was conducted with the objective of increasing muzzle velocity of mortar-like antipersonnel weapon with grenade-type golf-ball-size ammunition. Three factors were considered in a three-way layout experiment (with two replicates): A (addition or not of an O-ring, which was thought to reduce propellant gas escape in the muzzle and increase muzzle velocity), B (discharge hole area in inches, which might affect the pressure pulse of the propellant gases), and C (vent hole volume in cubic inches). The response is muzzle velocity. The data (King, 1992) are given in Table 2.40. Analyze the data by using the methods you have learned in

Table 2.40   Muzzle Velocity Data

Vent Volume	\multicolumn{8}{c}{Discharge Hole Area}
	\multicolumn{4}{c}{With O-ring}	\multicolumn{4}{c}{Without O-ring}
	0.016	0.030	0.048	0.062	0.016	0.030	0.048	0.062
0.29	294.9	295.0	270.5	258.6	264.9	248.1	224.2	216.0
	294.1	301.1	263.2	255.9	262.9	255.7	227.9	216.0
0.40	301.7	293.1	278.6	257.1	256.0	245.7	217.7	210.6
	307.9	300.6	267.9	263.6	255.3	251.0	219.6	207.4
0.59	285.5	285.0	269.6	262.6	256.3	254.9	228.5	214.6
	298.6	289.1	269.1	260.3	258.2	254.5	230.9	214.3
0.91	303.1	277.8	262.2	305.3	243.6	246.3	227.6	222.1
	305.3	266.4	263.2	304.9	250.1	246.9	228.6	222.2

the chapter. (Hint: Because the levels of B and C are not evenly spaced, their linear, quadratic, and cubic effects can be modeled by using the x, x^2, and x^3 terms directly, where x denotes the original value of the level. Alternatively, the orthogonal polynomials for four-level factors in Appendix G can still be used.)

APPENDIX 2A: TABLE OF LATIN SQUARES, GRAECO-LATIN SQUARES, AND HYPER-GRAECO-LATIN SQUARES

(Note: Randomization should be performed by randomly permuting rows and columns and assigning treatments to letters.)

3 × 3

```
A B C     A B C
B C A     C A B
C A B     B C A
```

These two 3×3 squares may be superimposed to form a Graeco-Latin square.

4 × 4

```
A B C D     A B C D     A C B D
B A D C     C D A B     D B C A
C D A B     D C B A     C A D B
D C B A     B A D C     B D A C
```

Two 4×4 squares may be superimposed to form a Graeco-Latin square; three 4×4 squares may be superimposed to form a hyper-Graeco-Latin square.

5 × 5

```
A B C D E     A B C D E     A B C D E     A B C D E
B C D E A     C D E A B     E A B C D     D E A B C
C D E A B     E A B C D     D E A B C     B C D E A
E A B C D     D E A B C     B C D E A     C D E A B
D E A B C     B C D E A     C D E A B     E A B C D
```

Two 5×5 squares may be superimposed to form a Graeco-Latin square; three or four 5×5 squares may be superimposed to form hyper-Graeco-Latin squares.

6 × 6

$$
\begin{array}{cccccc}
A & B & C & D & E & F \\
B & C & F & A & D & E \\
C & F & B & E & A & D \\
D & E & A & B & F & C \\
E & A & D & F & C & B \\
F & D & E & C & B & A
\end{array}
$$

No 6 × 6 Graeco-Latin square exists.

7 × 7

$$
\begin{array}{ccccccc}
A & B & C & D & E & F & G \\
B & C & D & E & F & G & A \\
C & D & E & F & G & A & B \\
D & E & F & G & A & B & C \\
E & F & G & A & B & C & D \\
F & G & A & B & C & D & E \\
G & A & B & C & D & E & F
\end{array}
\qquad
\begin{array}{ccccccc}
A & B & C & D & E & F & G \\
D & E & F & G & A & B & C \\
G & A & B & C & D & E & F \\
C & D & E & F & G & A & B \\
F & G & A & B & C & D & E \\
B & C & D & E & F & G & A \\
E & F & G & A & B & C & D
\end{array}
$$

These two 7 × 7 Latin squares may be superimposed to form a Graeco-Latin square.

8 × 8

$$
\begin{array}{cccccccc}
A & B & C & D & E & F & G & H \\
B & A & D & C & F & E & H & G \\
C & D & A & B & G & H & E & F \\
D & C & B & A & H & G & F & E \\
E & F & G & H & A & B & C & D \\
F & E & H & G & B & A & D & C \\
G & H & E & F & C & D & A & B \\
H & G & F & E & D & C & B & A
\end{array}
\qquad
\begin{array}{cccccccc}
A & B & C & D & E & F & G & H \\
G & H & E & F & C & D & A & B \\
E & F & G & H & A & B & C & D \\
C & D & A & B & G & H & E & F \\
H & G & F & E & D & C & B & A \\
B & A & D & C & F & E & H & G \\
D & C & B & A & H & G & F & E \\
F & E & H & G & B & A & D & C
\end{array}
$$

These two 8 × 8 Latin squares may be superimposed to form a Graeco-Latin square.

REFERENCES

Adelana, B.O. (1976), "Effects of Plant Density on Tomato Yields in Western Nigeria," *Experimental Agriculture*, 12, 43–47.

Box, G.E.P. and Cox, D.R. (1964), "An Analysis of Transformations" (with discussion), *Journal of the Royal Statistical Society, Series B*, 26, 211–252.

Box, G.E.P., Hunter, W.G., and Hunter, J.S. (1978), *Statistics for Experimenters*, New York: John Wiley & Sons.

Carroll, R. and Ruppert, D. (1988), *Transformation and Weighting in Regression*, New York: Chapman & Hall.

Cochran, W.G. and Cox, G.M. (1957), *Experimental Designs*, 2nd ed., New York: John Wiley & Sons.

Cox, D.R. and Snell, L. (1981), *Applied Statistics: Principles and Examples*, New York: Chapman & Hall.

Daniel, C. (1976), *Application of Statistics to Industrial Experimentation*, New York: John Wiley & Sons.

Davies, O.L., Ed. (1954), *The Design and Analysis of Industrial Experiments*, London: Oliver and Boyd.

Flurry, M.S. (1939), "Breaking Strength, Elongation and Folding Endurance of Films of Starches and Gelatin Used in Sizing," U.S. Department of Agriculture, Technical Bulletin 674.

John, P.W.M. (1971), *Statistical Design and Analysis of Experiments*, New York: MacMillan.

King, J.R. (1992), "Presenting Experimental Data Effectively," *Quality Engineering*, 4, 399–412.

Kodlin, D. (1951), "An Application of the Analysis of Covariance in Pharmacology," *Arch. Int. Pharmacodyn.*, 88, 207–211.

Lin, S., Hullinger, D.L., and Evans, R.L. (1971), "Selection of a Method to Determine Residual Chlorine in Sewage Effluents," *Water and Sewage Works*, 118, 360–364.

Mazumdar, S.K. and Hoa, S.V. (1995), "Application of Taguchi Method for Process Enhancement of On-line Consolidation Technique," *Composites*, 26, 669–673.

Meek, G.E. and Ozgur, C.O. (1991), "Torque Variation Analysis," *Journal of the Industrial Mathematics Society*, 41, 1–16.

Miller, R.G., Jr. (1981), *Simultaneous Statistical Inference*, 2nd ed., New York: Springer-Verlag.

Narayanan, R. and Adorisio, D. (1983), "Model Studies on Plate Girders," *Journal of Strain Analysis*, 18, 111–117.

Natrella, M.G. (1963), *Experimental Statistics*, National Bureau of Standards Handbook 91, Gaithersburg, MD: National Bureau of Standards.

Neter, J., Kutner, M.H., Nachtsheim, C.J., and Wasserman, W. (1996), *Applied Linear Statistical Models*, 4th ed., Chicago: Irwin.

CHAPTER 3

Full Factorial Experiments at Two Levels

In many scientific investigations the interest lies in the study of effects of two or more factors simultaneously. Factorial designs are most commonly used for this type of investigation. The special case of two-factor experiments was discussed in Section 2.3. This chapter considers the important class of factorial designs for k factors each at two levels. Because this class of designs requires $2 \times 2 \times \cdots \times 2 = 2^k$ observations, it is referred to as 2^k *factorial designs*. To distinguish it from the class of *fractional* factorial designs (which will be considered in the next few chapters), it is also called the class of 2^k *full* factorial designs. Since each factor only has two levels, they are used to study the linear effect of the response over the range of the factor levels chosen. Factorial effects, which include main effects and interaction effects of different order, are defined. Estimation and testing of factorial effects for location and dispersion models are considered for replicated and unreplicated experiments. (First-time readers of the chapter may skip Sections 3.14–3.17 and possibly Section 3.13.)

3.1 AN EPITAXIAL LAYER GROWTH EXPERIMENT

One of the initial steps in fabricating integrated circuit (IC) devices is to grow an epitaxial layer on polished silicon wafers. The wafers are mounted on a six-faceted cylinder (two wafers per facet), called a susceptor, which is spun inside a metal bell jar. The jar is injected with chemical vapors through nozzles at the top of the jar and heated. The process continues until the epitaxial layer grows to a desired thickness. The nominal value for thickness is 14.5 μm with specification limits of 14.5 $\pm$ 0.5 μm. In other words, it is desirable that the actual thickness be as close to 14.5 as possible and within the limits [14,15]. The current settings caused variation to exceed the specification of 1.0 μm. Thus, the experimenters needed to find process

Table 3.1   Design Matrix and Thickness Data, Adapted Epitaxial Layer Growth Experiment

Run	Factor A	B	C	D	Thickness						$\bar{y}$	s^2	$\ln s^2$
1	−	−	−	+	14.506	14.153	14.134	14.339	14.953	15.455	14.59	0.270	−1.309
2	−	−	−	−	12.886	12.963	13.669	13.869	14.145	14.007	13.59	0.291	−1.234
3	−	−	+	+	13.926	14.052	14.392	14.428	13.568	15.074	14.24	0.268	−1.317
4	−	−	+	−	13.758	13.992	14.808	13.554	14.283	13.904	14.05	0.197	−1.625
5	−	+	−	+	14.629	13.940	14.466	14.538	15.281	15.046	14.65	0.221	−1.510
6	−	+	−	−	14.059	13.989	13.666	14.706	13.863	13.357	13.94	0.205	−1.585
7	−	+	+	+	13.800	13.896	14.887	14.902	14.461	14.454	14.40	0.222	−1.505
8	−	+	+	−	13.707	13.623	14.210	14.042	14.881	14.378	14.14	0.215	−1.537
9	+	−	−	+	15.050	14.361	13.916	14.431	14.968	15.294	14.67	0.269	−1.313
10	+	−	−	−	14.249	13.900	13.065	13.143	13.708	14.255	13.72	0.272	−1.302
11	+	−	+	+	13.327	13.457	14.368	14.405	13.932	13.552	13.84	0.220	−1.514
12	+	−	+	−	13.605	13.190	13.695	14.259	14.428	14.223	13.90	0.229	−1.474
13	+	+	−	+	14.274	13.904	14.317	14.754	15.188	14.923	14.56	0.227	−1.483
14	+	+	−	−	13.775	14.586	14.379	13.775	13.382	13.382	13.88	0.253	−1.374
15	+	+	+	+	13.723	13.914	14.913	14.808	14.469	13.973	14.30	0.250	−1.386
16	+	+	+	−	14.031	14.467	14.675	14.252	13.658	13.578	14.11	0.192	−1.650

factors that could be set to minimize the epitaxial layer nonuniformity while maintaining average thickness as close to the nominal value as possible. In this section, a simplified version of this experiment is discussed. The adapted epitaxial layer growth experiment uses a 16-run full factorial design (with six replicates per run) as given in Table 3.1. The original experiment will be considered later in Section 3.11.

In the epitaxial layer growth process, suppose that the four experimental factors, susceptor rotation method, nozzle position, deposition temperature, and deposition time (labeled A, B, C and D) are to be investigated at the two levels given in Table 3.2. The purpose of this experiment is to find process conditions, i.e., combinations of factor levels for A, B, C, and D, under which the average thickness is close to the target 14.5 μm with variation as small as possible. The most basic experimental design or plan is the full factorial design, which studies all possible combinations of factors at two levels. Because of the dual requirements on the average and variation of thickness, analysis of the thickness data should focus on both aspects. More discussion is given in the next section.

Table 3.2   Factors and Levels, Adapted Epitaxial Layer Growth Experiment

Factor	Level −	+
A. susceptor-rotation method	continuous	oscillating
B. nozzle position	2	6
C. deposition temperature (°C)	1210	1220
D. deposition time	low	high

3.2 NOMINAL-THE-BEST PROBLEM AND QUADRATIC LOSS FUNCTION

In the epitaxial layer growth experiment, it is desirable to keep the thickness deviation from the target 14.5 μm as small as possible. In quality engineering this is referred to as the **nominal-the-best** problem because there is a nominal or target value based on engineering design requirements. First, a quantitative measure of "loss" due to the deviation from the nominal value needs to be defined. Suppose that y is a random variable representing the response or quality characteristic of the experiment, and t denotes the value of the desired target. For this particular experiment, y represents layer thickness and $t = 14.5$. It is convenient to measure the quality loss of the deviation of y from t by using the quadratic loss function

$$L(y,t) = c(y-t)^2, \tag{3.1}$$

where the constant c is determined by other considerations such as cost or financial loss. The expected loss $E(L(y,t))$ can be expressed as

$$\begin{aligned} E(L(y,t)) &= E\left(c(y-t)^2\right) \\ &= cE\left[(y-E(y)) + (E(y)-t)\right]^2 \\ &= c\,Var(y) + c(E(y)-t)^2, \end{aligned} \tag{3.2}$$

where $E(y)$ and $Var(y)$ are the mean and variance of y. From (3.2), the expected loss can be minimized or reduced by selecting some factor levels (i) that minimize the variance $Var(y)$ and (ii) that move the mean $E(y)$ closer to the target t. Often the two groups of factors associated with (i) and (ii) are mutually exclusive. In this situation the selection of levels can be done using the following procedure:

Two-Step Procedure for Nominal-the-Best Problem

> (*i*) *Select levels of some factors to minimize Var(y).*
> (*ii*) *Select the level of a factor not in* (*i*) *to move E(y) closer to t.* (3.3)

A factor in step (ii) is called an **adjustment factor** if it has a significant effect on the mean $E(y)$ but not on the variance $Var(y)$. The existence of an adjustment factor makes it possible to use the two-step procedure to minimize the quadratic loss. Otherwise, both the mean and the variance will change when adjusting the factor level in step (ii), which in turn may require the factor level identified in step (i) to be changed again in order to reduce the variance; several iterations between steps (i) and (ii) may then be needed. Therefore, it is clear that without the existence of an adjustment factor, the two-step procedure would not be effective.

This procedure will be applied to the data from the original epitaxial layer growth experiment in Section 3.11. The main technical advantage for employing the quadratic loss function is the decomposition (3.2), which makes it easy to perform the two-step design optimization just given. Quadratic loss often provides a reasonable approximation to the true quality loss. It is, however, not appropriate if the quality loss is highly skewed (i.e., loss for above target and loss for below target are very different) or if the quality loss remains a constant for any y with $|y - t| \geq c_0$, where c_0 is a predetermined value and the probability of $|y - t| \geq c_0$ is not negligible.

3.3 FULL FACTORIAL DESIGNS AT TWO LEVELS: A GENERAL DISCUSSION

Some basic design principles, including the choice of factors and factor levels, were given in Sections 1.2 and 1.3. In this section, we discuss some fundamental aspects of factorial designs that are applicable to full factorial and fractional factorial designs.

The 2^k full factorial design consists of all 2^k combinations of the k factors taking on two levels. A combination of factor levels is called a *treatment*. It is also referred to as a *run*. Hence, the 2^k full factorial design consists of 2^k runs, i.e., it has *run size* of 2^k. The experiment in Table 3.1 employs a 2^4 full factorial design with six replicates for each run. The notation $(-, +)$ in Table 3.1 is used to represent the two factor levels. For consistency, we suggest the use of $-$ for the lower level and $+$ for the higher level of quantitative factors, e.g., for factor C, 1210°F for the $-$ level and 1220°F for the $+$ level. For qualitative factors like A and B, there is no need to order the levels.

For carrying out the experiment, a **planning matrix** should be used to display the experimental design in terms of the *actual* factor levels. See the planning matrix for the adapted epitaxial layer growth experiment in Table 3.3. Instead of using $-$ and $+$ for levels, the matrix uses actual levels like continuous, oscillating, 2, 6, etc. The planning matrix avoids any confusion or misunderstanding of what the experimental factors are and what levels they should be set at for each experimental run.

Note that the order of the runs has been randomized so that they are no longer in the order given in Table 3.1. That is, the order given in Table 3.1 is not necessarily the actual order, or *run order*, in which the experiment is carried out. A general discussion of randomization was given in Section 1.3. Recall that randomization reduces the unwanted effects of other variables not in the planning matrix. A variable is called a *lurking variable* if its value changes during the course of the experiment and has a significant effect on the response but is not identified by the experimenter as being potentially important (and therefore is not included in the planning matrix). An example of a lurking variable is room humidity which may increase during the day.

Table 3.3   Planning Matrix

Run Order	Susceptor Rotation	Nozzle Position	Deposition Temperature	Deposition Time	Order from Table 3.1
			Factor		Order from Table 3.1
1	oscillating	6	1210	low	14
2	oscillating	2	1210	low	10
3	oscillating	2	1220	high	11
4	oscillating	6	1220	low	16
5	continuous	2	1220	high	3
6	continuous	2	1210	low	2
7	oscillating	2	1210	high	9
8	continuous	6	1220	low	8
9	continuous	2	1210	high	1
10	continuous	6	1220	high	7
11	oscillating	2	1220	low	12
12	continuous	2	1220	low	4
13	oscillating	6	1210	high	13
14	oscillating	6	1220	high	15
15	continuous	6	1210	high	5
16	continuous	6	1210	low	6

Note: Last column is for reference purposes and is not part of the planning matrix.

Suppose that the humidity control in the room works slowly so that it does not reach an equilibrium until noon. Then the room humidity over the day may look like that displayed in Figure 3.1. If the experiment is carried out over this time period and in the order given in Table 3.1, the continuous-versus-oscillating rotation method effect would be confounded with the humidity effect. That is, the observed effect of factor A based on column A of Table 3.1 is due to the difference between continuous and oscillating rotation as well as the difference between low and high humidity; these two effects cannot be disentangled. Randomizing the experiment would eliminate this confounding by canceling out the humidity effect.

Depending on the experiment, some factors may be *hard to change*. For example, deposition temperature in the epitaxial layer growth experiment may be hard to change more than once a day because the temperature takes a long time to stabilize. One can still do randomization of the other factor levels for each level of the hard-to-change factor, as shown in Table 3.4. This is referred to as *restricted randomization*, i.e., factors A, B, and D are randomized subject to the practical constraint that C can only change once. Examples of hard-to-change factors include furnace temperature and the alignment of dies in a press. It usually takes a long time for the temperature in a heat furnace to stabilize. Similarly, creating degrees of misalignment between the upper and lower dies is a time-consuming task.

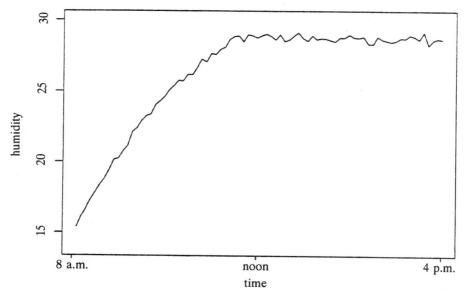

Figure 3.1. Room Humidity as a Lurking Variable.

Table 3.4   Planning Matrix Using Restricted Randomization

Run Order	Susceptor Rotation	Nozzle Position	Deposition Temperature	Deposition Time	Order from Table 3.1
			Factor		Order
1	continuous	6	1210	low	5
2	oscillating	6	1210	high	14
3	continuous	6	1210	high	6
4	oscillating	6	1210	low	13
5	continuous	2	1210	high	2
6	continuous	2	1210	low	1
7	oscillating	2	1210	high	10
8	oscillating	2	1210	low	9
9	oscillating	2	1220	high	12
10	oscillating	6	1220	low	15
11	continuous	6	1220	high	8
12	oscillating	6	1220	high	16
13	continuous	2	1220	low	3
14	continuous	6	1220	low	7
15	continuous	2	1220	high	4
16	oscillating	2	1220	low	11

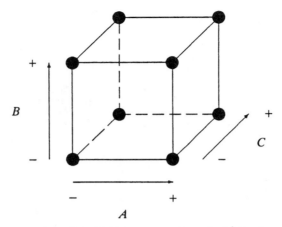

Figure 3.2. Cuboidal Representation of a 2^3 Design.

Table 3.5   Model Matrix and Data for 2^3 Design

1	2	3	12	13	23	123	Data
A	B	C	AB	AC	BC	ABC	z
−	−	−	+	+	+	−	14.090
−	−	+	+	−	−	+	14.145
−	+	−	−	+	−	+	14.295
−	+	+	−	−	+	−	14.270
+	−	−	−	−	+	+	14.195
+	−	+	−	+	−	−	13.870
+	+	−	+	−	−	−	14.220
+	+	+	+	+	+	+	14.205

An alternative to design tables is to represent the 2^3 and 2^4 factorial designs graphically. See Figure 3.2 for the 2^3 design, which is given in the first three columns of Table 3.5. Also, Figure 3.3 depicts the 2^4 design, which is listed in the first four columns of Table 3.1. Both figures show the design at the corners of the experimental region. Moreover, the symmetry and balance of the design is apparent, e.g., both factor levels appear the same number of times for each factor.

Corresponding to the 2^3 design in the first three columns of Table 3.5 is the **model matrix**, shown in the first seven columns of the same table. It facilitates the computation of main effects and interaction effects, which are discussed in the next section. The computation of the effects using the model matrix will be described later in Section 3.6.

Two key properties of the 2^k designs are *balance* and *orthogonality*. Balance means that each factor level appears in the same number of runs.

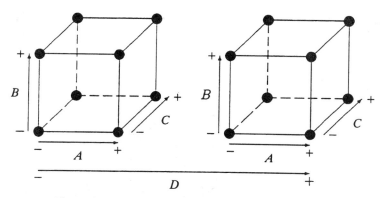

Figure 3.3. Cuboidal Representation of a 2^4 Design.

Two factors are said to be *orthogonal* if all their level combinations appear in the same number of runs. A design is called **orthogonal** if all pairs of its factors are orthogonal. Consider the 2^4 design in Table 3.1. The 2^4 design is balanced because for each factor, each factor level appears in eight runs. The 2^4 design is orthogonal because each of the four combinations $(-, -)$, $(-, +)$, $(+, -)$, and $(+, +)$ appears in four runs for each pair of factors. A more general discussion of orthogonality will be given later in terms of the concept of orthogonal arrays.

The experiment can be **replicated**, which means that the experimental plan is carried out two or more times. Each time the experiment is carried out, the order of the runs should be randomized. An **unreplicated** experiment is also said to have a single replicate. Unreplicated experiments are often performed when runs are expensive. For example, if a run consists of an engine prototype, it would be uneconomical to replicate runs (e.g., make more than one prototype) for each setting of the factors. Replication is also unnecessary for computer experiments because repeated computer runs with the same input give the same output. The need for replication and the distinction between replicates and repetitions were discussed in Section 1.3. Recall that mistaking repetitions for genuine replicates can lead to underestimation of the true error variance and false detection of effect significance.

3.4 FACTORIAL EFFECTS AND PLOTS

To achieve the two experimental objectives stated in Section 3.1, the mean (for location) and variance (for dispersion) need to be modeled separately as functions of the experimental factors.

Let y_{ij} denote the response of the jth replicate for the ith run. Using the experimental data, $\bar{y}_i$ and s_i^2, the sample mean and sample variance for the

ith run using the six replicates of the ith run are computed first, where

$$\bar{y}_i = \frac{1}{n_i} \sum_{j=1}^{n_i} y_{ij}, \quad s_i^2 = \frac{1}{n_i - 1} \sum_{j=1}^{n_i} \left(y_{ij} - \bar{y}_i\right)^2, \quad \text{and} \quad n_i = 6. \qquad (3.4)$$

Here the sample mean $\bar{y}_i$ and the sample variance s_i^2 are used to measure respectively the **location** and the **dispersion** of the data. They are the most commonly used measures when the data can be adequately modeled by a normal distribution and the underlying loss function is not far from quadratic. When these two conditions are not satisfied, other measures of location and dispersion should be used, which will be introduced in the book as they arise.

For analyzing unreplicated experiments, where no reliable estimate for σ^2 is available, the analysis strategy in this chapter is applicable except for analyzing sample variances s_i^2 or $\ln s_i^2$.

3.4.1 Main Effects

To avoid confusion in the definitions that follow, we use the generic notation z_i to denote the observation for the ith run. For the epitaxial layer growth experiment, z_i can be $\bar{y}_i$ if location is of interest or $\ln s_i^2$ if dispersion is of interest. (The reason for taking ln, the natural log, will be explained in Section 3.8.) To measure the average effect of a factor, say A, compute the difference between the average z_i value of all observations in the experiment at the high $(+)$ level of A and the average z_i value of all observations in the experiment at the low $(-)$ level of A. This difference is called the **main effect** of A. Notationally, $A+$ and $A-$ are used to represent the high and low levels of A, respectively, and

$$ME(A) = \bar{z}(A+) - \bar{z}(A-) \qquad (3.5)$$

to represent the main effect of A, where $\bar{z}(A+)$ is the average of the z_i values observed at $A+$ and $\bar{z}(A-)$ is similarly defined.

Because of the symmetry of full factorial designs (see Figure 3.2 or Table 3.1), $\bar{z}(A+)$ and $\bar{z}(A-)$ are computed over all level combinations of the other factors. For example, in Figure 3.2, the runs on the left (right) face of the cube correspond to the $A-$ $(A+)$ observations, where each face contains all four level combinations of the B and C factors. This property allows the main effect to have a relevant scientific interpretation beyond the current experiment. That is, a significant factor A effect in the current experiment is likely to hold true in other experiments or investigations involving the same factors and ranges of levels. Fisher (1971, Section 39) referred to this property of *reproducibility* as having a *wider inductive basis* for conclusions made from factorial experiments. It is a key reason for many successful applications of factorial experiments. Non-symmetric designs such as the "one-factor-at-a-time" plans to be discussed in Section 3.7 do not have this reproducibility property.

For the remainder of this section, consider modeling the $\bar{y}_i$ to identify factors that affect the mean thickness. By applying the previous definition to the $\bar{y}_i$ values (i.e., $z_i = \bar{y}_i$) of Table 3.1, the calculation for the A main effect for the adapted epitaxial layer growth experiment is

$$ME(A) = \bar{z}(A+) - \bar{z}(A-)$$
$$= \tfrac{1}{8}(14.67 + 13.72 + 13.84 + 13.90 + 14.56 + 13.88 + 14.30 + 14.11)$$
$$- \tfrac{1}{8}(14.59 + 13.59 + 14.24 + 14.05 + 14.65 + 13.94 + 14.40 + 14.14)$$
$$= 14.123 - 14.200$$
$$= -0.077 .$$

Similarly, the other main effects are computed as

$$ME(B) = \bar{z}(B+) - \bar{z}(B-) = 14.248 - 14.075 = +0.173,$$
$$ME(C) = \bar{z}(C+) - \bar{z}(C-) = 14.123 - 14.200 = -0.077,$$
$$ME(D) = \bar{z}(D+) - \bar{z}(D-) = 14.406 - 13.916 = +0.490.$$

These main effect calculations can be displayed graphically, as shown in Figure 3.4, which is referred to as a **main effects plot**. The main effects plot graphs the averages of all the observations at each level of the factor and connects them by a line. For two-level factors, the vertical height of the line is the difference between the two averages, which is the main effect. For example, the line in Figure 3.4 for factor A connects the two values $\bar{z}(A-) = 14.200$ and $\bar{z}(A+) = 14.123$.

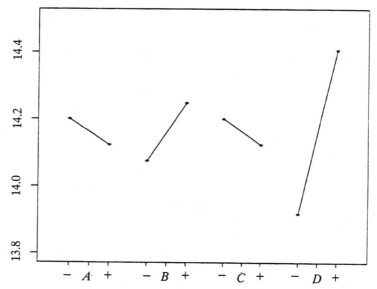

Figure 3.4. Main Effects Plot, Adapted Epitaxial Layer Growth Experiment.

3.4.2 Interaction Effects

To measure the joint effect of two factors, say A and B, the $A \times B$ **interaction effect** can be defined in three equivalent ways:

$$INT(A, B) = \tfrac{1}{2}\{\bar{z}(B + |A +) - \bar{z}(B - |A +)\}$$
$$- \tfrac{1}{2}\{\bar{z}(B + |A -) - \bar{z}(B - |A -)\} \qquad (3.6)$$
$$= \tfrac{1}{2}\{\bar{z}(A + |B +) - \bar{z}(A - |B +)\}$$
$$- \tfrac{1}{2}\{\bar{z}(A + |B -) - \bar{z}(A - |B -)\} \qquad (3.7)$$
$$= \tfrac{1}{2}\{\bar{z}(A + |B +) + \bar{z}(A - |B -)\}$$
$$- \tfrac{1}{2}\{\bar{z}(A + |B -) + \bar{z}(A - |B +)\}, \qquad (3.8)$$

where $\bar{z}(B + |A +)$ denotes the average of the z_i values with both A and B at the $+$ level with others being similarly defined. The expression in the first curly bracket of (3.6) measures the main effect of B given that A is at the $+$ level and is called the **conditional main effect** of B at the $+$ level of A. Notationally this difference is denoted by

$$ME(B|A +) = \bar{z}(B + |A +) - \bar{z}(B - |A +). \qquad (3.9)$$

Similarly, the expression in the second curly bracket of (3.6) is called the conditional main effect of B at the $-$ level of A and is denoted by $ME(B|A -)$. Then the $A \times B$ interaction effect is half of the difference between the conditional main effect of B at $A +$ and the conditional main effect of B at $A -$. This definition is equivalent to the definition in (3.7) which reverses the roles of A and B, that is, half of the difference between the conditional main effect of A at $B +$ and the conditional main effect of A at $B -$. Notationally,

$$INT(A, B) = \tfrac{1}{2}\{ME(B|A +) - ME(B|A -)\}$$
$$= \tfrac{1}{2}\{ME(A|B +) - ME(A|B -)\}. \qquad (3.10)$$

A large interaction indicates a large difference between $ME(B|A +)$ and $ME(B|A -)$ or between $ME(A|B +)$ and $ME(A|B -)$; that is, the effect of one factor, say B, depends on the level of the other factor, say A, which explains the use of the term "interaction."

The advantage of these two definitions is that the interaction (which is a second-order effect) is defined in terms of the difference of two conditional main effects (which are both first-order effects). The third definition in (3.8) computes the difference between two averages over the group with $(A, B) = [(+, +); (-, -)]$ and the group with $(A, B) = [(+, -); (-, +)]$. Although the third definition is algebraically simpler, it is not related to the main effects of A or B and the two groupings are not easily interpretable. Other advantages of the first two definitions will be discussed in the context of robust parameter design in Chapter 10.

Because of the symmetry of the full factorial design, the two terms in each of the three definitions (3.6)–(3.8) of $INT(A, B)$ are computed over all level combinations of the other factors. This fact also holds for higher order interactions, which will be defined later in (3.15). As in the discussion in Section 3.4.1 on the reproducibility of significant main effects, the factorial arrangement makes it more likely for significant interaction effects in the current experiment to be observed in similar studies.

Using the $\bar{y}_i$ data from the adapted epitaxial layer growth experiment, the estimate for the $A \times B$ interaction is obtained by first computing

$$\bar{z}(B + |A +) = \tfrac{1}{4}(14.56 + 13.88 + 14.30 + 14.11) = 14.213$$

and, similarly, computing

$$\bar{z}(B - |A +) = 14.033,$$
$$\bar{z}(B + |A -) = 14.283,$$
$$\bar{z}(B - |A -) = 14.118.$$

Then using (3.6), the estimate for the $A \times B$ interaction is

$$INT(A, B) = \tfrac{1}{2}[14.213 - 14.033] - \tfrac{1}{2}[14.283 - 14.118]$$
$$= +0.008.$$

Estimates for the other two-factor interactions are:

$$INT(A, C) = -0.093, \qquad INT(A, D) = -0.050,$$
$$INT(B, C) = +0.058, \qquad INT(B, D) = -0.030, \qquad INT(C, D) = -0.346.$$

Similar to the main effects plot, the interaction calculations can be displayed graphically as shown in Figure 3.5 which are referred to as **interaction plots**. They display the average at each of the combinations of the two factors A and B $[(-, -); (+, -); (-, +); (+, +)]$ using the B factor level as the horizontal axis and the average as the vertical axis, the averages having the same level of A, i.e., $[(-, -); (-, +)]$ and $[(+, -); (+, +)]$ are joined by a line. This particular plot is called the *A-against-B plot*. For the $+$ level of A, the vertical distance of the joined line in the plot is equal to $ME(B|A +)$, the conditional main effect of B at $A +$; similarly, for the $-$ level of A, the vertical distance of the joined line is equal to $ME(B|A -)$. Thus, the plot graphically represents the values $ME(B|A +)$ and $ME(B|A -)$. Unlike the interaction effect, which only gives the difference of the two values, the interaction effect plot shows both values. Judging the

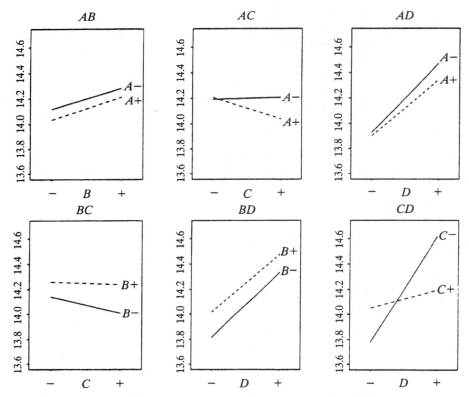

Figure 3.5. Interaction Plots, Adapted Epitaxial Layer Growth Experiment.

relative relationship between the two lines (or equivalently the four values) in the plot often gives more insight into the problem than using the single value of the interaction effect.

An A-against-B plot is called **synergistic** (and respectively, **antagonistic**) if the two conditional main effects of B at $A+$ and $A-$ have the same sign (and respectively, opposite signs), that is,

$$ME(B|A+)ME(B|A-) > 0 \qquad (3.11)$$

and respectively,

$$ME(B|A+)ME(B|A-) < 0. \qquad (3.12).$$

Switching the roles of the two factors in an interaction plot may reveal very different and valuable information. For example, in Figure 3.6 the C-against-D plot is synergistic, i.e., at each of the two temperature settings the average thickness increases as the deposition time increases. This makes obvious

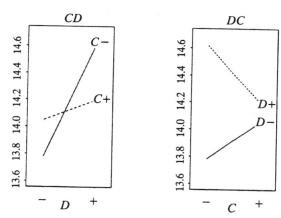

Figure 3.6. C-against-D and D-against-C Plots, Adapted Epitaxial Layer Growth Experiment.

physical sense because the layer grows thicker as more material is being deposited. By plotting D against C (as in Figure 3.6), an antagonistic pattern emerges. At $D-$ (low deposition time), the thickness increases as temperature increases; meanwhile at $D+$ (high deposition time), the thickness decreases as temperature increases. This relation reversal as deposition time changes may suggest that the underlying response surface is more complicated and calls for further investigation of the process.

Although the algebraic definitions in (3.6)–(3.7) are symmetric in the two factors A and B, the synergistic property may change when the roles of factors A and B are reversed. When this happens, particular attention should be paid to the interpretation of the physical meaning of significant interaction effects. An antagonistic interaction plot, say an A-against-B plot, may reveal valuable information about the relationship between the response and the factors A and B. If A and B are both quantitative, the change of signs in the two lines of the plot may suggest that, at an intermediate level between $A+$ and $A-$, the response is a flat function of the factor B, which will have a significant consequence for certain problems like robust parameter design.

In order to define the interaction between three factors A, B, and C, note that the conditional $A \times B$ interaction at the $+$ level of C is defined as the $A \times B$ interaction given in (3.6)–(3.8) for the observations at the $+$ level of C. Notationally, this is represented by $INT(A, B|C+)$. Then, the three-factor interaction $INT(A, B, C)$ can be defined in terms of two-factor interactions in three equivalent ways,

$$INT(A, B, C) = \tfrac{1}{2}INT(A, B|C+) - \tfrac{1}{2}INT(A, B|C-)$$
$$= \tfrac{1}{2}INT(A, C|B+) - \tfrac{1}{2}INT(A, C|B-)$$
$$= \tfrac{1}{2}INT(B, C|A+) - \tfrac{1}{2}INT(B, C|A-). \quad (3.13)$$

It is also easy to show that $INT(A, B, C)$ equals

$$\bar{z}[(A, B, C) = (+, +, +); (+, -, -); (-, +, -); (-, -, +)]$$
$$- \bar{z}[(A, B, C) = (-, -, -); (+, +, -); (+, -, +); (-, +, +)],$$
$$(3.14)$$

which measures the difference between the averages of two groups, one whose product of the levels for A, B, and C is $+$ and the other whose product of the levels is $-$.

The estimate of the $A \times B \times C$ interaction for the adapted epitaxial layer growth experiment, computed four ways using (3.13) and (3.14) in the order given above, is

$$INT(A, B, C) = \tfrac{1}{2}[0.1050 - (-0.0900)]$$
$$= \tfrac{1}{2}[0.0050 - (-0.1900)]$$
$$= \tfrac{1}{2}[0.1550 - (-0.0400)]$$
$$= (14.2100 - 14.1125)$$
$$= 0.0975 .$$

The $A \times B \times C$ interaction can be plotted in three different ways: (A, B) against C, (A, C) against B, and (B, C) against A. For the (A, B)-against-C plot, draw a line that connects the values at the $C+$ and $C-$ levels for each of the four-level combinations $(-, -)$, $(-, +)$, $(+, -)$, and $(+, +)$ of A and B. Examples will be given in Chapters 5 and 10.

Higher order interactions are similarly defined. For k factors, define their interaction in terms of the $(k - 1)$-factor interactions as follows:

$$INT(A_1, A_2, \dots, A_k) = \tfrac{1}{2}INT(A_1, A_2, \dots, A_{k-1}|A_k +)$$
$$- \tfrac{1}{2}INT(A_1, A_2, \dots, A_{k-1}|A_k -). \quad (3.15)$$

This definition is independent of the choice of conditioning factor; that is, the value in (3.15) remains the same when A_k is replaced by any A_i. The main effects and interaction effects of all orders are collectively called the *factorial effects*. It is easy to show (and left as an exercise) from (3.15) that any factorial effect can be represented as

$$\bar{z}_+ - \bar{z}_-, \quad (3.16)$$

where $\bar{z}_+$ is the average of the z values at the combinations for which the product of the levels of the factors that appear in the definition of the factorial effect is $+$; $\bar{z}_-$ is similarly defined.

Suppose that the observations for each experimental run are independent and normally distributed with variance σ^2. Denote the factorial effect in

Table 3.6 Factorial Effects, Adapted Epitaxial Layer
Growth Experiment

Effect	$\bar{y}$	$\ln s^2$
A	-0.078	0.016
B	0.173	-0.118
C	-0.078	-0.112
D	0.490	0.056
AB	0.008	0.045
AC	-0.093	-0.026
AD	-0.050	-0.029
BC	0.058	0.080
BD	-0.030	0.010
CD	-0.345	0.085
ABC	0.098	-0.032
ABD	0.025	0.042
ACD	-0.030	0.000
BCD	0.110	-0.003
$ABCD$	0.020	0.103

(3.16) by $\hat{\theta}$, where $\bar{z}_+$ is the average of $N/2$ observations at its $+$ level, $\bar{z}_-$ the average of $N/2$ observations at its $-$ level, and N the total number of observations. For an unreplicated experiment, $N = 2^k$; for a replicated experiment with m replicates per run, $N = m2^k$. Then

$$Var(\hat{\theta}) = \frac{\sigma^2}{N/2} + \frac{\sigma^2}{N/2} = \frac{4}{N}\sigma^2. \tag{3.17}$$

The variance σ^2 can be estimated from other sources or by the estimate in (3.41).

For simplicity, we may drop the "$\times$" in $A \times B$ and $A \times B \times C$ and refer to these interactions as AB or ABC. Both the $A \times B$ and AB notation will be used throughout the book.

For the $\bar{y}_i$ and $\ln s_i^2$ data of the adapted epitaxial layer growth experiment, all the factorial effects consisting of the main effects, two-factor, three-factor, and four-factor interactions, are presented in Table 3.6. Analysis of these effects will be considered later.

3.5 FUNDAMENTAL PRINCIPLES FOR FACTORIAL EFFECTS: HIERARCHICAL ORDERING, EFFECT SPARSITY, AND EFFECT HEREDITY

After defining main effects and interactions of different orders, an immediate question to be addressed is the relative importance and relationship among these effects. Three fundamental principles are considered in the book. They are often used to justify the development of factorial design theory and data analysis strategies. The first is the *hierarchical ordering principle*.

Hierarchical Ordering Principle

(i)   Lower order effects are more likely to be important than higher order effects.

(ii) Effects of the same order are equally likely to be important.

This principle suggests that when resources are scarce, priority should be given to the estimation of lower order effects. Its application is particularly effective when the number of factorial effects is large and cannot all be estimated. It is an *empirical* principle whose validity has been confirmed in many real experiments. Another reason for its wide acceptance is that higher order interactions are more difficult to interpret or justify physically. As a result, the investigators are less interested in estimating the magnitudes of these effects even when they are statistically significant. Unless substantive knowledge suggests a different ordering of factorial effects, this principle can be safely used. A modification of this principle for robust parameter design will be considered in Chapter 10.

The second principle is concerned with the number of relatively important effects in a factorial experiment. Based on empirical evidence observed in many investigations, the variation exhibited in the data is primarily attributed to a few effects. This useful rule is extensively used throughout the book.

Effect Sparsity Principle

The number of relatively important effects in a factorial experiment is small.

The term "effect sparsity" is due to Box and Meyer (1986). The principle may also be called the *Pareto Principle in Experimental Design* because of its focus on the "vital few" and not the "trivial many." For the use of the Pareto Principle in general quality investigations, see Juran (1993).

The third principle governs the relationship between an interaction and its parent factors.

Effect Heredity Principle

In order for an interaction to be significant, at least one of its parent factors should be significant.

This principle (Hamada and Wu, 1992) and a refined version (given in Chapter 8) will find many applications throughout the book. In particular, it can be used to weed out incompatible models in searching for a good model such as those in Chapters 5 and 8.

3.6 USING REGRESSION AND THE MODEL MATRIX TO COMPUTE FACTORIAL EFFECTS

This section discusses how the factorial effects can be computed using regression analysis, which was discussed in Chapter 1.

Consider the 2^3 design for factors A, B, and C and the corresponding model matrix given in Table 3.5. The data z are the means of the 12 observations from the corresponding two runs of the original experiment given in Table 3.1. The columns in Table 3.5 labeled A, B, and C form the 2^3 design. The other columns are computed by multiplying the appropriate A, B, or C columns componentwise. For example, column AB is obtained by multiplying column A with column B, where the product of two like signs is a $+$ and the product of two different signs is a $-$, i.e., $(++) = (--) = +$ and $(+-) = (-+) = -$. It is important to note the difference between the planning matrix and the model matrix. The planning matrix lists the factor level combinations in terms of the actual levels. It tells the experimenter exactly what levels to set the factors at for each experimental run. On the other hand, the model matrix uses the $(+, -)$ notation and has columns for each of the factorial effects. Its main purpose is to facilitate the computation of the various effects.

The estimated effect corresponding to a particular column of the model matrix is obtained by applying the signs in that column to the z data, then summing up and dividing by 4 (half the number of runs).

Take, for example, the $A \times B \times C$ interaction. Applying the signs in the column labeled ABC to the z data, then summing them up and dividing by 4 gives

$$INT(A, B, C) = \tfrac{1}{4}(-14.090 + 14.145 + 14.295 - 14.270$$
$$+ 14.195 - 13.870 - 14.220 + 14.205)$$
$$= 0.0975.$$

The factorial effects in the 2^k design can be estimated in the linear model framework discussed in Section 1.4. Take the 2^3 design in Table 3.5 to illustrate this approach. Denote the A, B, C columns by $\mathbf{x}_1 = (x_{i1})$, $\mathbf{x}_2 = (x_{i2})$, $\mathbf{x}_3 = (x_{i3})$, where $i = 1, \ldots, 8$. Then the AB, AC, BC, and ABC columns can be represented by $\mathbf{x}_4 = \mathbf{x}_1\mathbf{x}_2 = (x_{i1}x_{i2})$, $\mathbf{x}_5 = \mathbf{x}_1\mathbf{x}_3 = (x_{i1}x_{i3})$, $\mathbf{x}_6 = \mathbf{x}_2\mathbf{x}_3 = (x_{i2}x_{i3})$, and $\mathbf{x}_7 = \mathbf{x}_1\mathbf{x}_2\mathbf{x}_3 = (x_{i1}x_{i2}x_{i3})$ as previously explained. Consider the linear model

$$z_i = \beta_0 + \sum_{j=1}^{7} \beta_j\, x_{ij} + \epsilon_i \qquad (3.18)$$

for the ith run, $i = 1, \ldots, 8$. It is easy to show that for $j = 1, \ldots, 7$, the least squares estimate (LSE) of β_j is

$$\hat{\beta}_j = \frac{1}{1 - (-1)}\left(\bar{z}(x_{ij} = +1) - \bar{z}(x_{ij} = -1)\right), \qquad (3.19)$$

which is *half* the factorial effect given in Section 3.4. Therefore, each factorial effect can be obtained by doubling the corresponding regression estimate $\hat{\beta}_j$. The β_0 term in (3.18) is the grand mean or equivalently the intercept. The extension of (3.18) to the general 2^k design can be obtained in a similar manner.

Before computing was readily available, analysis of experiments was done by hand. Yates proposed an efficient algorithm which computed the factorial effects only by summing and subtracting pairs of numbers. In today's computing environment, Yates's algorithm is no longer needed. The procedure can be found, for example, in Section 10.7 of Box, Hunter, and Hunter (1978).

3.7 COMPARISONS WITH THE "ONE-FACTOR-AT-A-TIME" APPROACH

A commonly used approach in scientific and engineering investigations is to (i) identify the most important factor, (ii) investigate this factor by itself, ignoring the other factors, (iii) make recommendations on changes (or no change) to this factor, and (iv) move onto the next most important factor and repeat steps (ii) and (iii). The iterations end when a satisfactory solution is found. In the most fortunate situation, it may terminate after studying only one factor or it may study several factors one at a time. This mode of investigation is referred to as the **one-factor-at-a-time** approach. It is used explicitly or implicitly in many investigations, especially by those who have not been exposed to factorial experiments. By comparison with the factorial design method, it has the following disadvantages:

1. It requires more runs for the same precision in effect estimation.
2. It cannot estimate some interactions.
3. The conclusions from its analysis are not general.
4. It can miss optimal settings of factors.

We will explain these four points in the context of a hypothetical experiment on an injection molding process that is producing an unacceptable percentage of burned parts. Suppose a 2^3 design is used in an experiment that studies three factors. The factors and their two levels are injection pressure (1200–1400 psi), screw rpm control (0.3–0.6 turns counterclockwise), and injection speed (slow–fast). Table 3.7 shows the planning matrix and the percent burned response for the eight runs, where P, R, and S are used to denote the three factors injection pressure, screw rpm control, and injection speed, respectively.

In order to explain the four points, we need to use the following well-defined version of the "one-factor-at-a-time" approach for purposes of comparison.

Table 3.7   Planning Matrix for 2^3 Design and Response Data for Comparison with One-Factor-at-a-Time Approach

	Factor		Percent
P	R	S	Burned
1200	0.3	slow	11
1200	0.3	fast	17
1200	0.6	slow	25
1200	0.6	fast	29
1400	0.3	slow	02
1400	0.3	fast	09
1400	0.6	slow	37
1400	0.6	fast	40

Step 1. Factor P is thought to be the most important. By fixing the other factors at standard conditions ($R = 0.6$, $S =$ fast), two levels of P at 1200 and 1400 are compared. Here, $P = 1200$ is chosen as it gives a smaller percent burned than at $P = 1400$. See Table 3.7.

Step 2. The next most important factor is thought to be R. By fixing $P = 1200$ from step 1 and $S =$ fast (standard condition), the two levels of R at 0.3 and 0.6 are compared. Here, $R = 0.3$ is chosen as it gives a smaller percent burned than at $R = 0.6$.

Step 3. It is felt that further gains may be made by varying the remaining factor S. Two levels of S are compared with $P = 1200$ and $R = 0.3$ based on the previous studies in steps 1 and 2. The level $S =$ slow is chosen as it gives a smaller percent burned than at $S =$ fast.

This version of the "one-factor-at-a-time" plan is graphically depicted in Figure 3.7. Instead of conducting an experiment at each corner of the cube, it visits only four corners of the cube in the indicated direction. It will be used to illustrate the disadvantages mentioned above. To explain the first disadvantage, note that it would require taking four observations at the low and high levels of factor P (with $R = 0.6$, $S =$ fast) in order for the factor P effect (which is a conditional main effect at $R = 0.6$ and $S =$ fast) to have the same precision (i.e., variance) as the factor P effect from the 2^3 design. Recall that in the 2^3 design, each factorial effect is of the form $\bar{z}_+ - \bar{z}_-$, where $\bar{z}_+$ and $\bar{z}_-$ are the averages of four observations. Similarly, in order to get the same precision for the estimated effects for factors R and S, it would require taking four observations at the high and low levels of factor R (with $P = 1200$, $S =$ fast) and also four observations at the high and low levels of factor S (with $P = 1200$, $R = 0.3$). This would require at least 16 observations (i.e., four observations at each of the four corners indicated in Figure 3.7) as opposed to eight observations in the 2^3 design.

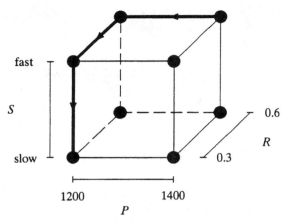

Figure 3.7. The Path of a One-Factor-at-a-Time Plan.

More generally, to achieve the same precision as a 2^k design for studying k factors, the "one-factor-at-a-time" plan would require at least $(k+1)2^{k-1}$ runs. This is because it requires taking 2^{k-1} observations at each level combination being studied and at least $k+1$ level combinations (i.e., $k+1$ corners of the 2^k design) need to be studied. Thus, *the runs required by the "one-factor-at-a-time" plan are $(k+1)/2$ times those for the 2^k design.*

The second disadvantage can be seen from Figure 3.7 where none of the interactions can be estimated from the observations in the "one-factor-at-a-time plan. For example, to estimate the $P \times R$ interaction, it would require having observations at the four level combinations of P and R: $(-,-)$, $(-,+)$, $(+,-)$, and $(+,+)$. Because the "one-factor-at-a-time" plan does not have any observations at $(+,-)$, it cannot estimate the $P \times R$ interaction. While subject-matter experts are good at identifying the important factors that have sizable main effects before the experiment, they are usually not as successful at identifying the important interactions. Consequently, running a 2^k design ensures that the experimenter can investigate all the interactions between pairs of factors, as well as all the higher order interactions.

Unlike factorial designs, the "one-factor-at-a-time" plan does not conduct a systematic or comprehensive search over the experimental region. This can be seen from the geometric representation in Figure 3.7 or from the fact that when a factor is being investigated, the levels of the other factors are held constant. This also explains the third disadvantage regarding the generality of the conclusions from "one-factor-at-a-time" experiments. Recall that in factorial designs, each factor effect is computed over *all* possible level combinations of the other factors. The same is not true for the factor effects computed from "one-factor-at-a-time" plans. Since in these plans some level combinations of the other factors are not included in the factor effect computation, a significant factor effect in the current experiment may not be reproduced in another study if the latter involves some level combinations not included in the current experiment.

The data in Table 3.7 can be used to illustrate the fourth disadvantage of potentially missing optimal factor settings. That is, using the "one-factor-at-a-time" plan as described in steps 1–3 would miss two better factor settings at $P = 1400$, $R = 0.3$, and $S =$ slow or fast. This problem can be explained by the large $P \times R$ interaction (which equals 10.0 from the data in Table 3.7). A large $P \times R$ interaction implies that, among the four settings for factors P and R, one of them has a much larger value than the corresponding value at the setting in the opposite corner of the P-by-R square. So if the search path in the "one-factor-at-a-time" plan is not chosen properly, it can miss the optimal setting. In some sense, the last three disadvantages are all rooted in the fact that the "one-factor-at-a-time" plan does not conduct a systematic search over the experimental region.

In spite of the criticisms made about the "one-factor-at-a-time" approach, it has sometimes been used effectively in solving problems, especially when the investigator is able to identify the few important factors and the levels of the other factors are also chosen properly. With the presence of these favorable conditions for the success of the "one-factor-at-a-time" approach, one can easily argue that using the more systematic method of factorial designs would lead to even bigger gains in the identification of optimal factor settings and the understanding of the underlying response surface relationship.

3.8 LOG TRANSFORMATION OF THE SAMPLE VARIANCES

To measure the dispersion effects of different factors, should the factorial effect formulas in Section 3.4 be applied directly to the sample variances s_i^2? The methods described in this chapter assume that the analyzed quantities are nearly normal so that a direct application is not appropriate because $(n_i - 1) s_i^2$ has a chi-squared distribution. A commonly used approach is to take the log transformation of s_i^2 and then analyze the $\ln s_i^2$ using standard methods, where ln denotes *natural logarithm*. A primary reason for the log transformation is that it maps positive values to real (both positive and negative) values, and by taking its inverse transformation, any predicted value on the log scale will be transformed back to a positive value on the original scale. Suppose $z_i = \ln s_i^2$ and let $\hat{z}_i$ denote the predicted value of $\ln \sigma_i^2$ for a new factor setting based on a regression model applied to the $\ln s_i^2$ data. By taking the inverse transformation, $\exp(\hat{z}_i)$ is obtained as the predicted value of σ_i^2 for the new factor setting. Many other transformations like the power transformation from Section 2.5, $z \rightarrow z^\lambda$, $\lambda \neq 0$, do not guarantee that the predicted value of σ_i^2 is always positive, however. Another justification stems from the nature of physical laws. Many physical laws have multiplicative relationships. Log transformation converts a multiplicative relationship into an additive relationship, which is much easier to model statistically.

The appropriateness of the log transformation can be assessed by the transformed response's proximity to the normal distribution. Noting that

$$(n_i - 1) \, s_i^2 = \sum_{j=1}^{n_i} (y_{ij} - \bar{y}_i)^2 \sim \sigma_i^2 \chi_{n_i-1}^2,$$

where χ_f^2 is the chi-squared distribution with f degrees of freedom, then taking logs yields

$$\ln s_i^2 = \ln \sigma_i^2 + \ln \frac{1}{n_i - 1} \chi_{n_i-1}^2. \qquad (3.20)$$

Letting X denote the random variable $f^{-1}\chi_f^2$, $\ln(X)$ has approximately mean zero since $E(X) = 1$ and $Var(X) = 2f^{-1}$. Its variance can be approximated by the following general formula:

$$Var(h(X)) \approx [h'(E(X))]^2 \, Var(X), \qquad (3.21)$$

where h is a smooth function of a random variable X. For $h(X) = \ln X$,

$$Var(\ln(X)) \approx (h'(1))^2 2f^{-1} = 2f^{-1},$$

which is not a constant but has a narrow range of values for $f = n_i - 1 \geq 9$. (See Bartlett and Kendall, 1946.) From the previous formula and (3.20), the distribution of $\ln s_i^2$ can be approximated by

$$N\left(\ln \sigma_i^2, 2(n_i - 1)^{-1}\right). \qquad (3.22)$$

From (3.22), it can be seen that the log transformation has the unique property that allows σ_i^2 to appear only in the mean of the normal approximation, thus leaving the variance of the normal approximation independent of σ_i^2. In this sense, $\ln s_i^2$ is a *variance-stabilizing* transformation.

For the s_i^2 data in the adapted epitaxial layer growth experiment, the factorial effect calculations were applied to the $\ln s_i^2$ and are reported in Table 3.6. Analysis of these effects will be considered later.

3.9   EFFECT OF SIMULTANEOUS TESTING

How does one judge the significance of the main effects and interaction effects computed from a factorial experiment? Denote the ith factorial effect estimate by $\hat{\theta}_i$. From (3.17) the variance of $\hat{\theta}_i$ is $4\sigma^2/N$, where N is the total number of observations. If the value of σ^2 is known (e.g., available from other sources), a $100(1 - \alpha)\%$ confidence interval for θ_i, the unknown ith factorial effect, is

$$\left(\hat{\theta}_i - z_{\alpha/2} \frac{2}{\sqrt{N}} \sigma, \; \hat{\theta}_i + z_{\alpha/2} \frac{2}{\sqrt{N}} \sigma \right), \tag{3.23}$$

where $z_{\alpha/2}$ is the upper $\alpha/2$ point of the standard normal distribution.

This approach has two limitations, however: (i) it cannot be used for *simultaneously* testing the significance of *all* the factorial effects and (ii) it assumes that σ is known. The problem with simultaneous testing can be illustrated as follows. For full factorial designs and other orthogonal designs (whose column vectors are orthogonal), the estimated factorial effects $\hat{\theta}_i$ are independent with the distribution $N(\theta_i, \tau^2)$, where $\tau^2 = 4\sigma^2/N$. Assuming τ is known, using the standard z test (based on the standard normal distribution) for testing $\theta_i = 0$, $i = 1, \ldots, I = 2^k - 1$, would declare θ_i significant for any i if

$$\frac{|\hat{\theta}_i|}{\tau} \geq z_{\alpha/2}. \tag{3.24}$$

Its experimentwise error, which is the probability of mistakenly declaring at least one θ_i as being nonzero when $\theta_1 = \cdots = \theta_I = 0$, is

$$Prob\left(\max_{1 \leq i \leq I} \frac{|\hat{\theta}_i|}{\tau} \geq z_{\alpha/2} | \theta_1 = \cdots = \theta_I = 0 \right)$$

$$= 1 - \left\{ P\left(|Z| \leq z_{\alpha/2} \right) \right\}^I = 1 - (1 - \alpha)^I, \tag{3.25}$$

which exceeds α for $I > 1$ and increases as I increases. In (3.25), Z denotes the standard normal random variable. Without adjusting the $z_{\alpha/2}$ value in (3.24), the problem with simultaneous testing becomes serious when I, the number of effects being tested, is large.

For a replicated experiment, σ can be replaced by a pooled estimate s [see (3.41)] and $z_{\alpha/2}$ by $t_{\alpha/2}$ (the upper $\alpha/2$ point of the t distribution with appropriate degrees of freedom) in the interval estimate given in (3.23). This t interval still suffers the same problem as described above. A method that takes into account the effect of simultaneous testing for replicated experiments will be described in Section 3.15.

For an unreplicated experiment, an objective estimate of σ is usually not available. Instead, a simple and effective graphical method for *informally* detecting effect significance can be used, which will be considered next.

3.10   NORMAL AND HALF-NORMAL PLOTS

Using a graphical method to judge effect significance is often preferred to using an extension of the t test such as the studentized maximum modulus test to be given in Section 3.15, because the latter depends on the estimate s^2 of σ^2, which may be unreliable, whereas the former depends on seeing a deviation from linearity, which is easier to judge. The graphical method uses the **normal probability plot** introduced in Section 1.9. Let $\hat{\theta}_{(1)} \leq \cdots \leq \hat{\theta}_{(I)}$ denote the ordered values of the factorial effect estimates $\hat{\theta}_i$. Plot them against their corresponding coordinates on the normal probability scale:

$$\hat{\theta}_{(i)} \text{ versus } \Phi^{-1}([i - 0.5]/I) \quad \text{for } i = 1, \ldots, I, \qquad (3.26)$$

where Φ is the cumulative distribution function (cdf) of a standard normal random variable. By fitting a straight line to the middle group of points in (3.26) whose $\hat{\theta}_{(i)}$ are near zero, *any effect whose corresponding point falls off the line is declared significant.*

   The rationale behind this graphical method is as follows. Assume that the estimated effects are normally distributed with means equal to the effects. (This assumption is usually not restrictive because the estimated effects involve the averaging of N observations and the central limit theorem ensures that averages are nearly normal for N as small as 8.) Then, under the null hypothesis that all the effects are zero, the means of all the estimated effects are zero. The resulting normal probability plot of the estimated effects will be a straight line. Therefore, the normal probability plot is testing whether all of the estimated effects have the same distribution (i.e., the same means), so that when some of the effects are non-zero, the corresponding estimated effects will tend to be larger and fall off the straight line. For positive effects, the estimated effects fall above the line while those for negative effects fall below the line.

   This ingenious method, first proposed by Daniel (1959), *does not require σ to be estimated.* In the t or F test for significance testing, a standard error s is used as a reference quantity for judging the significance of the effect being tested. Since such a quantity is not available for unreplicated experiments, Daniel's idea was to use the normal curve as the *reference distribution* against which the significance of effects is judged. Large deviations from the normal curve (which is a straight line on the normal probability scale) cannot be explained by chance occurrence. Therefore, the corresponding effects are judged to be significant in this procedure. It is important to note that use of the normal probability plot does not suffer from the effect of simultaneous testing, as was illustrated in (3.25). An explanation is given in an exercise.

   Figure 3.8 displays the normal probability plot of effects for the $\bar{y}_i$ data from the adapted epitaxial layer growth experiment. It suggests that only D, CD (and possibly B) are significant.

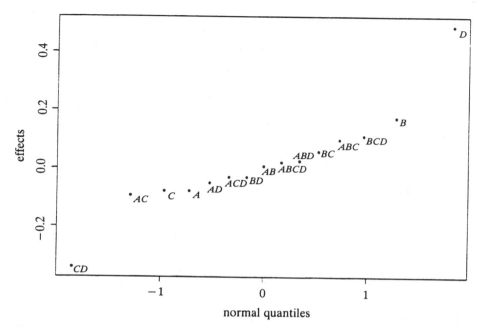

Figure 3.8. Normal Plot of Location Effects, Adapted Epitaxial Layer Growth Experiment.

A related graphical method is the half-normal probability plot. Let $|\hat{\theta}|_{(1)}$ $\leq \cdots \leq |\hat{\theta}|_{(I)}$ denote the ordered values of the unsigned effect estimates $|\hat{\theta}_{(i)}|$. Plot them against coordinates based on a half-normal distribution; the absolute value of a normal random variable is half-normal. The **half-normal probability plot** consists of the points

$$\left(\Phi^{-1}(0.5 + 0.5[i - 0.5]/I), |\hat{\theta}|_{(i)}\right) \quad \text{for } i = 1, \ldots, I. \qquad (3.27)$$

An advantage of the half-normal plot is that all the large estimated effects appear in the upper right corner and fall above the line through the small estimated effects. This advantage is demonstrated in Figure 3.9, which is based on hypothetical data. The normal plot (top plot in Figure 3.9) is visually misleading. If the user were looking only for departures from the line and ignored the magnitude of the estimated effects, he or she might in addition to C identify K and I as important. This certainly makes no sense because K and I are smaller than G and O in magnitude. By plotting all the absolute estimated effects in the bottom plot of Figure 3.9, the half-normal plot avoids this visual pitfall. In fact, it shows that C is the only important effect.

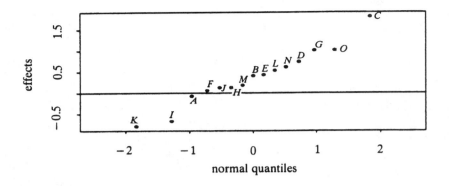

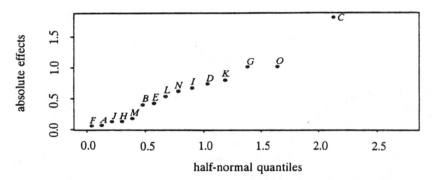

Figure 3.9. Comparison of Normal and Half-Normal Plots.

In this book half-normal plots are used for testing effect significance. For other purposes such as outlier detection, normal plots may be preferred. See Daniel (1959, 1971) for details.

Figure 3.10 displays the half-normal plot of the effects for the $\ln s_i^2$ data from the adapted epitaxial layer growth experiment. It suggests that no effects are significant. There appears to be no opportunity for variation reduction by changing the levels of the factors. This conclusion is also confirmed by a formal test which will be introduced later.

For those readers who will not study or use the formal test methods to be introduced in Sections 3.13–3.16, we recommend that they read the summary of analysis strategies in Section 3.17. These strategies involve the use of half-normal plots or other formal test methods. It is interesting to note, however, that each task in the strategies can be performed by using half-normal plots. Even without the mastery of more elaborate methods in later sections, half-normal plots provide a simple and effective tool for testing

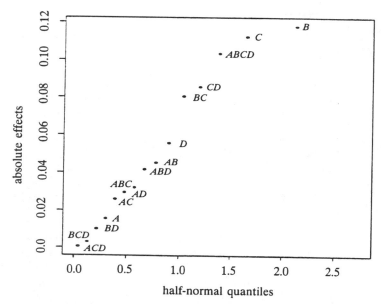

Figure 3.10. Half-Normal Plot of Dispersion Effects, Adapted Epitaxial Layer Growth Experiment.

effect significance in location and dispersion models for replicated and unreplicated experiments.

3.11 ANALYSIS OF LOCATION AND DISPERSION: REVISITING THE EPITAXIAL LAYER GROWTH EXPERIMENT

The half-normal plots of factorial effects estimated from $\bar{y}_i$ and $\ln s_i^2$ can be used to identify what factors influence location as measured by $\bar{y}_i$ and what factors influence dispersion as measured by s_i^2. Consider now the data in Table 3.8 from the original epitaxial layer growth experiment of Kackar and Shoemaker (1986) as reported in Shoemaker, Tsui, and Wu (1991). The factors and levels are the same as those given in Table 3.2 (but the data in Table 3.8 are different from those in Table 3.1). The corresponding estimates of the factorial effects are given in Table 3.9.

The half-normal plots for $\bar{y}_i$ and $\ln s_i^2$ are displayed in Figures 3.11 and 3.12, respectively. It is clear from Figure 3.12 that factor A (rotation method) is the single most influential factor for $\ln s_i^2$ and from Figure 3.11 that factor D (deposition time) is the single most influential factor for $\bar{y}_i$. (A formal test to be introduced in Section 3.13 will also confirm A and D as significant.) Based on the sign of the A effect, the rotation method needs to be changed

Table 3.8   Design Matrix and Thickness Data, Original Epitaxial Layer Growth Experiment

A	Design B	C	D	Thickness						$\bar{y}$	s^2	$\ln s^2$
−	−	−	+	14.812	14.774	14.772	14.794	14.860	14.914	14.821	0.003	−5.771
−	−	−	−	13.768	13.778	13.870	13.896	13.932	13.914	13.860	0.005	−5.311
−	−	+	+	14.722	14.736	14.774	14.778	14.682	14.850	14.757	0.003	−5.704
−	−	+	−	13.860	13.876	13.932	13.846	13.896	13.870	13.880	0.001	−6.984
−	+	−	+	14.886	14.810	14.868	14.876	14.958	14.932	14.888	0.003	−5.917
−	+	−	−	14.182	14.172	14.126	14.274	14.154	14.082	14.165	0.004	−5.485
−	+	+	+	14.758	14.784	15.054	15.058	14.938	14.936	14.921	0.016	−4.107
−	+	+	−	13.996	13.988	14.044	14.028	14.108	14.060	14.037	0.002	−6.237
+	−	−	+	15.272	14.656	14.258	14.718	15.198	15.490	14.932	0.215	−1.538
+	−	−	−	14.324	14.092	13.536	13.588	13.964	14.328	13.972	0.121	−2.116
+	−	+	+	13.918	14.044	14.926	14.962	14.504	14.136	14.415	0.206	−1.579
+	−	+	−	13.614	13.202	13.704	14.264	14.432	14.228	13.907	0.226	−1.487
+	+	−	+	14.648	14.350	14.682	15.034	15.384	15.170	14.878	0.147	−1.916
+	+	−	−	13.970	14.448	14.326	13.970	13.738	13.738	14.032	0.088	−2.430
+	+	+	+	14.184	14.402	15.544	15.424	15.036	14.470	14.843	0.327	−1.118
+	+	+	−	13.866	14.130	14.256	14.000	13.640	13.592	13.914	0.070	−2.653

Table 3.9   Factorial Effects, Original Epitaxial Layer Growth Experiment

Effect	$\bar{y}$	$\ln s^2$
A	−0.055	3.834
B	0.142	0.078
C	−0.109	0.077
D	0.836	0.632
AB	−0.032	−0.428
AC	−0.074	0.214
AD	−0.025	0.002
BC	0.047	0.331
BD	0.010	0.305
CD	−0.037	0.582
ABC	0.060	−0.335
ABD	0.067	0.086
ACD	−0.056	−0.494
BCD	0.098	0.314
ABCD	0.036	0.109

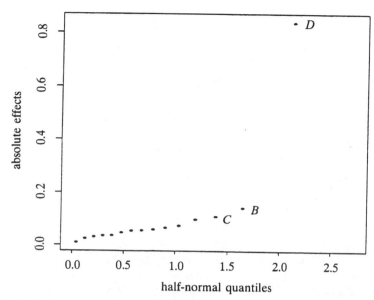

Figure 3.11. Half-Normal Plot of Location Effects, Original Epitaxial Layer Growth Experiment.

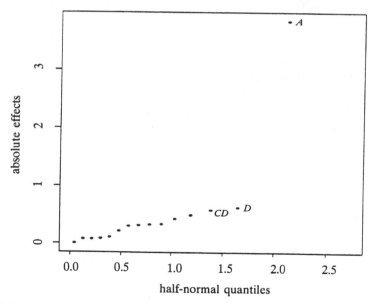

Figure 3.12. Half-Normal Plot of Dispersion Effects, Original Epitaxial Layer Growth Experiment.

from oscillating (level +) to continuous (level −) to reduce dispersion. (The positive A effect means that the dispersion is higher at the + level.) To utilize the information in $\bar{y}_i$ for adjusting the average thickness closer to the target 14.5, a more precise measure obtained through a regression equation is needed. Regression equations for both responses are

$$\hat{y} = \hat{\alpha} + \hat{\beta}_D x_D = 14.389 + 0.418 \, x_D,$$

$$\hat{z} = \hat{\gamma}_0 + \hat{\gamma}_A x_A = -3.772 + 1.917 x_A,$$

where $z = \ln s^2$. [Recall from (3.19) that the coefficients 0.418 and 1.917 for x_D and x_A are half the main effect estimates of D and A, respectively.] By equating

$$14.5 = \hat{y} = 14.389 + 0.418 x_D, \tag{3.28}$$

the setting for D (deposition time) should be changed to $x_D = 0.266$ so that the expected thickness is on target. Since the actual values for $D-$ and $D+$ were not reported in the original experiment, we assume that the − and + levels correspond to 30 and 40 seconds, respectively. Then the deposition time should be chosen to be

$$35 + 0.266(5) = 36.33$$

seconds. Based on the fitted equation, dispersion is reduced by changing x_A to -1 (i.e., continuous), resulting in the predicted variance

$$\hat{\sigma}^2 = \exp(-3.772 + 1.917(-1)) = (0.058)^2. \tag{3.29}$$

Because x_D does not appear in the model for $\hat{z}$, it is an adjustment factor and the two-step procedure in Section 3.2 can be applied. To summarize, the procedure first uses the continuous rotation to reduce the variation and then sets deposition time to 36.33 seconds to bring the process on target. We should, however, point out that the small $\hat{\sigma}^2$ value in (3.29) may be too optimistic. In practice, any recommended settings from a statistical model fitting and their predicted performance should be verified by performing confirmation experiments.

3.12 BLOCKING AND OPTIMAL ARRANGEMENT OF 2^k FACTORIAL DESIGNS IN 2^q BLOCKS

The need for blocking and examples of blocking variables were discussed in Section 1.3. We consider how to block full factorial experiments in this section.

First, consider the simple case of a 2^3 design arranged in two blocks. Recall that units within the same block are homogeneous and units between blocks are heterogeneous. Define the block effect estimate to be

$$\bar{y}(II) - \bar{y}(I), \tag{3.30}$$

where $\bar{y}(I)$ and $\bar{y}(II)$ are the averages of observations in blocks I and II, respectively. If the blocking is effective, the block effects should be large. Because the 2^3 design has seven degrees of freedom, it is impossible to estimate all seven factorial effects as well as the block effect. Consequently, one of the factorial effects needs to be sacrificed. Unless there is *a priori* knowledge about the relative importance of the seven effects, based on the hierarchical ordering principle (see Section 3.5), it is reasonable to assume that the three-factor interaction is the least important and should be sacrificed. We can use the **123** column of Table 3.10 to define the following blocking scheme: a unit belongs to block I if its entry in the **123** column is $-$, and to block II if its entry in the **123** column is $+$. Therefore, among the eight runs in Table 3.10, block I consists of runs 1, 4, 6, 7, and block II consists of runs 2, 3, 5, 8. By comparing the **123** column and the blocking column (i.e., the last column) of Table 3.10, it is clear that they have the same numerical estimate, i.e., the block effect estimate in (3.30) is identical to the three-factor interaction effect estimate $\bar{y}(\mathbf{123} = +) - \bar{y}(\mathbf{123} = -)$. Because these two effects have the same estimate, the block effect is said to be **confounded** (i.e., confused) with the three-factor interaction effect. Notationally this confounding relation is represented by

$$\mathbf{B} = \mathbf{123},$$

where **B** denotes the block effect in (3.30). By giving up our ability to estimate one factorial effect, typically an unimportant one, we gain in return a higher precision in estimating the main effects and two-factor interactions by arranging the eight runs in two homogeneous blocks in such a way that each block contains two runs at the $-$ level and two runs at the $+$ level for each of the main effects and two-factor interactions. Using the same principle, any 2^k design can be arranged in two blocks by confounding the block effect with the k-factor interaction.

Table 3.10   Arranging a 2^3 Design in Two Blocks of Size 4

Run	1	2	3	12	13	23	123	Block
1	$-$	$-$	$-$	$+$	$+$	$+$	$-$	I
2	$-$	$-$	$+$	$+$	$-$	$-$	$+$	II
3	$-$	$+$	$-$	$-$	$+$	$-$	$+$	II
4	$-$	$+$	$+$	$-$	$-$	$+$	$-$	I
5	$+$	$-$	$-$	$-$	$-$	$+$	$+$	II
6	$+$	$-$	$+$	$-$	$+$	$-$	$-$	I
7	$+$	$+$	$-$	$+$	$-$	$-$	$-$	I
8	$+$	$+$	$+$	$+$	$+$	$+$	$+$	II

Table 3.11   Arranging a 2^3 Design in Four Blocks of Size Two

Run	1	2	3	$B_1(=12)$	$B_2(=13)$	23	123	Block
1	−	−	−	+	+	+	−	IV
2	−	−	+	+	−	−	+	III
3	−	+	−	−	+	−	+	II
4	−	+	+	−	−	+	−	I
5	+	−	−	−	−	+	+	I
6	+	−	+	−	+	−	−	II
7	+	+	−	+	−	−	−	III
8	+	+	+	+	+	+	+	IV

$$
\begin{array}{c c c c}
 & & & \mathbf{B_1} \\
\mathbf{B_2} & & - & + \\
 & - & \mathrm{I} & \mathrm{III} \\
 & + & \mathrm{II} & \mathrm{IV}
\end{array}
$$

To illustrate a general approach to be taken later, let us consider the problem of arranging a 2^3 design in four blocks. Define two blocking variables $\mathbf{B_1}$ and $\mathbf{B_2}$ as in Table 3.11, which divides the eight runs into four blocks I, II, III, IV, according to $(\mathbf{B_1}, \mathbf{B_2}) = (-,-), (-,+), (+,-), (+,+)$, respectively. For example, $\mathbf{B_1}$ could be the supplier of raw material (suppliers A and B) and $\mathbf{B_2}$ could be the workers on a given shift (day and night). Since there are three block effects among four blocks, we must know which three factorial effects are confounded with the block effects. By following the previous argument, the **12** interaction is confounded with the block effect, which is defined to be the difference between blocks I, II and blocks III, IV. This can be easily seen by comparing the $\mathbf{B_1}$ column and the blocking column of Table 3.11, i.e., the + entries (and, respectively, the − entries) in the $\mathbf{B_1}$ column are the same as blocks III, IV (and, respectively, blocks I, II) in the blocking column. We say $\mathbf{B_1}$ is confounded with the **12** interaction or notationally $\mathbf{B_1} = \mathbf{12}$. Similarly, we have $\mathbf{B_2} = \mathbf{13}$. The third block effect is the difference between blocks I, IV, and blocks II, III. From the lower part of Table 3.11, it is easy to see that this difference is identical to the difference between $\mathbf{B_1}\mathbf{B_2} = +$ and $\mathbf{B_1}\mathbf{B_2} = -$. By comparing the **23** column and the $\mathbf{B_1}$, $\mathbf{B_2}$ columns of Table 3.11, **23** is identical to the product of $\mathbf{B_1}$ and $\mathbf{B_2}$. Notationally, we can derive this relation as

$$\mathbf{B_1}\mathbf{B_2} = \mathbf{12} \times \mathbf{13} = \mathbf{1} \times \mathbf{1} \times \mathbf{2} \times \mathbf{3} = \mathbf{I} \times \mathbf{2} \times \mathbf{3} = \mathbf{23}. \tag{3.31}$$

In deriving (3.31), note that (i) **12** is the product of columns **1** and **2** and (ii) $\mathbf{1} \times \mathbf{1}$ is the componentwise product of column **1** and itself, which equals the column of $+$'s and is denoted by **I**. To summarize, the blocking scheme in Table 3.11 confounds the effects **12**, **13**, and **23** with the three block effects.

An alternative scheme would be to choose $\mathbf{B}_1 = 123$ and $\mathbf{B}_2 = 13$. At first glance, this appears to be better because $\mathbf{B}_1$ is confounded with the three-factor interactions 123 (instead of the 12 interaction in the first scheme). By multiplying $\mathbf{B}_1$ and $\mathbf{B}_2$ to find the third block effect, we have $\mathbf{B}_1\mathbf{B}_2 = 123 \times 13 = 2$. Thus, the main effect 2 is confounded with a block effect. Since it is essential that none of the main effects be confounded with a block effect (otherwise the factors should not be included in the study), the alternative blocking scheme is inferior and should not be used.

The previous derivations and discussion are based on the following assumption:

Assumption *The block-by-treatment interactions are negligible.*

This assumption, which is made throughout the book, states that the treatment effects do not vary from block to block. Without it, the estimability of the factorial effects will be very complicated. Take, for example, the relation $\mathbf{B}_1 = 12$ in the first blocking scheme. It implies two other relations: $\mathbf{B}_1 1 = 2$ and $\mathbf{B}_1 2 = 1$. If there is a significant interaction between the block effect $\mathbf{B}_1$ and the main effect 1, then the main effect 2 is confounded with $\mathbf{B}_1 1$ and therefore is not estimable. A similar argument can be made for $\mathbf{B}_1 2 = 1$. The validity of the assumption has been confirmed in many empirical studies. It can be checked by plotting the residuals for all the treatments within each block. If the pattern varies from block to block, the assumption may be violated. A block-by-treatment interaction often suggests interesting information about the treatment and blocking variables. Suppose that the blocks consist of blends of raw material and one blend is contaminated with impurities. If the impurities compromise the effect of one of the treatments, then the particular treatment will be ineffective in the contaminated blend but not in the other blends. Detection of a block-by-treatment interaction may suggest that the particular blend be investigated further, which may lead to the discovery of the relationship between the impurities and the particular treatment.

We are now ready to outline a general approach for arranging a 2^k design in 2^q blocks of size 2^{k-q}. Denote the q independent variables for defining the 2^q blocks by $\mathbf{B}_1, \mathbf{B}_2, \ldots, \mathbf{B}_q$. Suppose that the factorial effect $\mathbf{v}_i$ is confounded with $\mathbf{B}_i$, i.e.,

$$\mathbf{B}_1 = \mathbf{v}_1, \mathbf{B}_2 = \mathbf{v}_2, \ldots, \mathbf{B}_q = \mathbf{v}_q.$$

The remaining block effects can be obtained by multiplying the $\mathbf{B}_i$'s, e.g.,

$$\mathbf{B}_1\mathbf{B}_2 = \mathbf{v}_1\mathbf{v}_2, \quad \mathbf{B}_1\mathbf{B}_3 = \mathbf{v}_1\mathbf{v}_3, \ldots, \quad \mathbf{B}_1\mathbf{B}_2 \cdots \mathbf{B}_q = \mathbf{v}_1\mathbf{v}_2 \cdots \mathbf{v}_q.$$

The $2^q - 1$ possible products of the $\mathbf{B}_i$'s and the column $\mathbf{I}$ (whose components are $+$) form a group with $\mathbf{I}$ being its identity element. This group is called the **block defining contrast subgroup**. For example, a 2^5 design can be arranged in eight blocks using $\mathbf{B}_1 = 135$, $\mathbf{B}_2 = 235$, and $\mathbf{B}_3 = 1234$. The remaining block effects are confounded with the following interactions: $\mathbf{B}_1\mathbf{B}_2 = 12$, $\mathbf{B}_1\mathbf{B}_3 = 245$, $\mathbf{B}_2\mathbf{B}_3 = 145$, $\mathbf{B}_1\mathbf{B}_2\mathbf{B}_3 = 34$. In this blocking scheme, the seven block effects are confounded with the seven interactions

$$12, 34, 135, 145, 235, 245, 1234. \tag{3.32}$$

Is this an optimal blocking scheme? Based on the hierarchical ordering principle, a good blocking scheme should confound as few low-order interactions as possible. As a comparison, consider an alternative scheme with $\mathbf{B}_1 = 12$, $\mathbf{B}_2 = 13$, and $\mathbf{B}_3 = 45$. It is easy to show that it confounds the following interactions:

$$12, 13, 23, 45, 1245, 1345, 2345. \tag{3.33}$$

By comparing (3.32) and (3.33), the second scheme confounds four two-factor interactions, while the first scheme confounds only two two-factor interactions. Since two-factor interactions are likely to be more important than three- or four-factor interactions, the first scheme is superior.

In general, we use the following criterion to compare any two blocking schemes. For any blocking scheme b, denote by $g_i(b)$ the number of i-factor interactions that are confounded with block effects. Because no main effect should be confounded with any block effect, $g_1(b) = 0$ and the definition of $g_i(b)$ starts with $i = 2$. Also the sum of the $g_i(b)$'s is $2^q - 1$, the total number of block effects. For any two blocking schemes, b_1 and b_2, let r be the smallest value such that $g_r(b_1) \neq g_r(b_2)$. The hierarchical ordering principle suggests that fewer lower order effects should be confounded with block effects. Therefore, we say that scheme b_1 has *less aberration* than scheme b_2 if $g_r(b_1) < g_r(b_2)$.

Minimum Aberration Blocking Scheme

A blocking scheme is said to have minimum aberration if there is no blocking scheme with less aberration.

In the example in the preceding paragraph, it can be verified that the first scheme has minimum aberration. From (3.32) and (3.33), both schemes have the property that all the main effects are not confounded with block effects and therefore are estimable. This is called estimability of order 1. In general, a blocking scheme is said to have **estimability of order** e if the lowest order interactions confounded with block effects are of order $e + 1$. In other words, estimability of order e ensures that all factorial effects of order e are estimable in the blocking scheme. Here the main effects are of order 1, and the i-factor interactions are of order i, $i \geq 2$. Details on estimability of order e, the minimum aberration criterion, and its relation to the same criterion for fractional factorial designs (to be discussed in Section 4.2) can be found in Sun, Wu, and Chen (1997).

Table 3A.1 gives minimum aberration blocking schemes for 2^k designs in 2^q blocks, $1 \leq q \leq k - 1$, $3 \leq k \leq 8$. For each scheme, the order of estimability (column denoted by "e"), the independent generators that define the blocking scheme, and the list of confounded interactions are given. Knowing what interactions are being confounded can help the experimenter to plan the experiment better. This point is illustrated with the case $k = 4$ and $q = 2$. From Table 3A.1, the **12**, **134**, and **234** interactions are confounded. Care should then be taken in assigning the four factors to the columns **1**, **2**, **3**, and **4** so that the interaction between the factors assigned to columns **1** and **2** is not likely to be important.

The generators given in Table 3A.1 are the longest among different sets of generators that generate the same blocking scheme. These generators were found by using an algorithm due to Cheng, Wu, and Wu (1999). To explain what is meant by "longest," consider, for example, the case of $k = 4$ and $q = 2$ in the table. There are three equivalent sets of generators: (**134**, **234**), (**12**, **134**), and (**12**, **234**). It is obvious that the first set has longer "words" than the other two sets. In general, the lengths of the generators in any two sets can be compared by applying the aberration criterion given above. The set with the highest ranking (i.e., the maximum aberration) is given in the table. Why does the choice of generators matter if they generate the same blocking scheme? It depends on the practical situation. If all the $2^q - 1$ block effects in a blocking scheme are thought to be equally important, then the choice does not matter. For example, if the 2^q blocks correspond to 2^q suppliers or 2^q days, then the difference between any two blocks may be potentially significant. In other situations, the generalized block effects $\mathbf{B}_{i_1} \cdots \mathbf{B}_{i_j}$, which are defined to be the interactions between the blocking variables $\mathbf{B}_{i_1}, \ldots, \mathbf{B}_{i_j}$, do not have a natural interpretation and are therefore likely to be less significant than the block effects $\mathbf{B}_1, \ldots, \mathbf{B}_q$. Suppose that $\mathbf{B}_1$ represents two suppliers and $\mathbf{B}_2$ represents the day or night shift. It would be hard to give a meaning for a $\mathbf{B}_1 \times \mathbf{B}_2$ interaction. Therefore, it is reasonable to assume that $\mathbf{B}_1$ and $\mathbf{B}_2$ are more significant than $\mathbf{B}_1 \times \mathbf{B}_2$. If the block effects $\mathbf{B}_1, \ldots, \mathbf{B}_q$ are more significant than their interactions, longer words should be assigned to $\mathbf{B}_1, \ldots, \mathbf{B}_q$, which are the block generators.

3.13 A FORMAL TEST OF EFFECT SIGNIFICANCE FOR UNREPLICATED EXPERIMENTS: WITHOUT s^2

The half-normal and normal plots are informal graphical methods involving visual judgment. It would be desirable to judge the deviation from a straight line quantitatively based on a formal test of significance. Many such methods have been proposed in the literature. A comprehensive review of these methods and their performance can be found in Hamada and Balakrishnan (1998). In this section we present one method that is simple to compute and generally performs well according to the study of Hamada and Balakrishnan.

Let $\hat{\theta}_1, \ldots, \hat{\theta}_I$ denote the estimated factorial effects $\theta_1, \ldots, \theta_I$, where I denotes the number of effects being tested. For the 2^k design, $I = 2^k - 1$; for the 2^{k-p} designs considered in the next chapter, $I = 2^{k-p} - 1$. It is important that the $\hat{\theta}_i$ have the same standard deviation τ. Lenth (1989) considered a *robust* estimator of the standard deviation of $\hat{\theta}_i$, which he called the pseudo standard error, or *PSE*:

$$PSE = 1.5 \cdot \text{median}_{\{|\hat{\theta}_i| < 2.5 \, s_0\}} |\hat{\theta}_i|, \qquad (3.34)$$

where the median is computed among the $|\hat{\theta}_i|$ with $|\hat{\theta}_i| < 2.5 s_0$ and

$$s_0 = 1.5 \cdot \text{median} |\hat{\theta}_i|. \qquad (3.35)$$

The initial standard error s_0 in (3.35), which uses the scaling factor 1.5, is a consistent estimator of the standard deviation of $\hat{\theta}$ when the θ_i's are zero and the underlying error distribution is normal. Its proof is left as an exercise. (If $\hat{\tau}_n$ is an estimate of τ based on a sample of size n, $\hat{\tau}_n$ is *consistent* if $\hat{\tau}_n \to \tau$ in probability as $n \to \infty$.)

Because $Prob(|N(0, 1)| \geq 2.57) = 0.01$, $\{|\hat{\theta}_i| < 2.5 s_0\}$ in (3.34) trims about 1% of the $\hat{\theta}_i$'s if all the θ_i's are zero. Therefore, *PSE* is still approximately a consistent estimator. More importantly, the trimming attempts to remove the $\hat{\theta}_i$'s associated with *active* or non-zero effects. By using the median in combination with the trimming, *PSE* is then a robust estimator of the standard deviation of the $\hat{\theta}_i$'s. Here, a *robust* estimator means that its performance is not sensitive to the $\hat{\theta}_i$'s which are associated with active effects.

By dividing the $\hat{\theta}_i$'s by the *PSE*, *t*-like statistics are obtained:

$$t_{PSE,i} = \frac{\hat{\theta}_i}{PSE}. \qquad (3.36)$$

An effect $\hat{\theta}_i$ is declared significant if the $|t_{PSE,i}|$ value exceeds the critical value given in Appendix H. This method of testing effect significance is referred to as **Lenth's method** in the book. Because it does not require an unbiased estimate s^2 of σ^2, it is particularly useful for unreplicated experiments. It can still be used for replicated experiments if the investigator cannot obtain or chooses not to use an unbiased estimate s^2.

There are two types of critical values given in Appendix H, those that control the *individual error rate* (IER) and those that control the *experiment-wise error rate* (EER). The IER and EER are best explained through an example. From Appendix H, the IER-based critical value for $\alpha = 0.01$ and

$I = 7$ is 5.07. This means that on average 1% of the $\hat{\theta}_i$'s would be declared significant if all the θ_i's were zero. Using the EER-based critical value 9.72 for $\alpha = 0.01$ and $I = 7$ means that on average in 1% of the experiments (in which seven $\hat{\theta}_i$'s are tested for each experiment) one or more of the $\hat{\theta}_i$'s would be declared significant (i.e., those with $|t_{PSE,i}| > 9.72$) if all the θ_i's were zero.

Generally, we can define the two error rates as follows. Let IER_α and EER_α denote the IER- and EER-based critical values of t_{PSE} at level α and H_0 denote the null hypothesis that all the θ_i's are zero. Then, under the normal error assumption

$$Prob(|t_{PSE,i}| > \text{IER}_\alpha | H_0) = \alpha, \tag{3.37}$$

where $i = 1, \ldots, I$ because under H_0, $t_{PSE,i}$ has the same distribution for all i, and

$$Prob(|t_{PSE,i}| > \text{EER}_\alpha \text{ for at least one } i, i = 1, \ldots, I | H_0)$$
$$= Prob\left(\max_{1 \leq i \leq I} |t_{PSE,i}| > \text{EER}_\alpha | H_0 \right) = \alpha. \tag{3.38}$$

The EER-based critical values account for the number of tests done in an experiment in the spirit of multiple comparison procedures. For factorial experiments, however, the IER version of Lenth's method is preferable, because it has more power to detect significant effects. Especially in a screening context, misidentifying an inert factor as active is less serious than missing an important (active) factor. For large I, the EER version is too conservative because many of the θ_i's which are likely to be negligible need not be tested. Thus, the EER critical values are inflated by considering many inert θ_i's. Throughout the book, we refer to the IER and EER versions of Lenth's method as those using the IER and EER critical values of t_{PSE}, respectively.

The critical values in Appendix H are taken from Ye and Hamada (2000). Because of the complexity of the definitions, these critical values behave somewhat erratically. For fixed α, they can increase or decrease as I increases. The IER values decrease for $\alpha \leq 0.04$ but increase for $\alpha \geq 0.2$ and go down, up, and then down for $0.05 \leq \alpha \leq 0.1$. The EER values decrease for $\alpha \leq 0.05$ but increase for $\alpha \geq 0.3$. There is no discernible pattern for α between 0.06 and 0.2.

Returning to the adapted epitaxial layer growth experiment, for the location effects in Table 3.6, median$|\hat{\theta}_i| = 0.078$ so that

$$s_0 = 1.5 \times 0.078 = 0.117.$$

The trimming constant is

$$2.5s_0 = 2.5 \times 0.117 = 0.292,$$

Table 3.12 $|t_{PSE}|$ **Values, Adapted Epitaxial Layer Growth Experiment**

Effect	$\bar{y}$	$\ln s^2$
A	0.90	0.25
B	1.99	1.87
C	0.90	1.78
D	5.63	0.89
AB	0.09	0.71
AC	1.07	0.41
AD	0.57	0.46
BC	0.67	1.27
BD	0.34	0.16
CD	3.97	1.35
ABC	1.13	0.51
ABD	0.29	0.67
ACD	0.34	0.00
BCD	1.26	0.05
ABCD	0.23	1.63

which according to (3.34) eliminates 0.490 and 0.345 associated with D and CD, so that median$_{\{|\hat{\theta}_i| < 2.5 s_0\}} |\hat{\theta}_i| = 0.058$, yielding

$$PSE = 1.5 \times 0.058 = 0.087.$$

The corresponding $|t_{PSE}|$ values appear in Table 3.12 under $\bar{y}$. At the 0.01 level the IER critical value is 3.63 for $I = 15$. By comparing the $|t_{PSE}|$ values in Table 3.12 against 3.63, D and CD are found to be significant. Note that none of the effects is significant if the EER version is applied at the 0.01 level. (The EER critical value is 6.45 for $I = 15$ and $\alpha = 0.01$.) For the dispersion effects in Table 3.6, it can be verified that $s_0 = PSE = 0.063$. The corresponding $|t_{PSE}|$ values are given in Table 3.12 under $\ln s^2$. None of them exceed 2.16, the critical value in Appendix H for $\alpha = 0.05$ and $I = 15$, which indicates that none of the effects are significant even at the 0.05 level. Use of Lenth's method confirms and provides a quantitative basis for the informal conclusions in Section 3.11 based on the half-normal plots.

    The p value of an effect estimate $\hat{\theta}_i$ can also be determined by interpolating the critical values in the tables in Appendix H. For example, the $|t_{PSE}|$ value for CD is 3.97. For the IER version of Lenth's method with $I = 15$, the closest values are 3.86 for $\alpha = 0.008$ and 4.00 for $\alpha = 0.007$. The p value of the CD effect is the α value that corresponds to 3.97, which can be determined by the following linear interpolation:

$$0.008 - \frac{3.97 - 3.86}{4.0 - 3.86}(0.008 - 0.007) = 0.008 - \left(\frac{11}{14}\right)(0.001)$$

$$= 0.008 - 0.0008 = 0.0072.$$

Therefore the p value of CD is 0.0072, suggesting that CD is significant at level α for any $\alpha \geq 0.0072$.

3.14 TESTING VARIANCE HOMOGENEITY

A more formal method than the normal and half-normal plots for testing the homogeneity (i.e., equality) of the σ_i^2 is considered in this section. Consider the null hypothesis of variance homogeneity

$$H_0 : \sigma_1^2 = \cdots = \sigma_m^2$$

against the alternative

$$H_1 : \text{some of the } \sigma_i^2 \text{ are not equal.}$$

Based on the approximation $\ln s_i^2 \sim N(\ln \sigma_i^2, 2(n_i - 1)^{-1})$ given in (3.22), the equality of $\ln \sigma_i^2$ can be tested by using an analysis-of-variance procedure with known variance $[= 2(n_i - 1)^{-1}]$. For equal $n_i = n$, the test rejects H_0 at level α if

$$\frac{n-1}{2} \sum_{i=1}^{m} (z_i - \bar{z})^2 > \chi_{m-1,\alpha}^2, \tag{3.39}$$

where

$$z_i = \ln s_i^2, \qquad \bar{z} = \frac{1}{m}(z_1 + \cdots + z_m),$$

$\chi_{m-1,\alpha}^2$ is the upper α point of the χ_{m-1}^2 random variable. For unequal n_i's, the test rejects H_0 if

$$\sum_{i=1}^{m} \frac{n_i - 1}{2}(z_i - \bar{z}_w)^2 > \chi_{m-1,\alpha}^2, \tag{3.40}$$

where

$$\bar{z}_w = \left(\sum_{i=1}^{m} \frac{n_i - 1}{2} z_i \right) \Big/ \left(\sum_{i=1}^{m} \frac{n_i - 1}{2} \right).$$

A theoretical justification of (3.40) can be found in Lehmann (1986, p. 377).

Computing (3.39) for the $\ln s_i^2$ data of the adapted epitaxial layer growth experiment in Table 3.1 gives a value of 0.608 whose p value is 0.999 based on the χ_{15}^2 distribution. That is, $\bar{z} = -1.449$ and

$$\frac{n-1}{2} \sum_{i=1}^{m} (z_i - \bar{z})^2$$

$$= \frac{6-1}{2} \{ (-1.309 - (-1.449))^2 + \cdots + (-1.650 - (-1.449))^2 \}$$

$$= 2.5 \, (0.2432) = 0.608.$$

The p value is calculated as

$$Prob\left(\chi_{15}^2 \geq 0.608 \right) = 0.999,$$

which is nearly 1.000. Consequently, the data suggest no difference in the variances.

The tests in (3.39) and (3.40) which involve the constant $(n_i - 1)/2$ are highly sensitive to the normality assumption (Lehmann, 1986, p. 378). An alternative is to use Bartlett's test based on the likelihood ratio statistic. Although it is more efficient, it too is sensitive to the normality assumption. Consequently, (3.39) and (3.40) are preferable because of their simplicity. In most practical situations, testing variance homogeneity is an intermediate step of the analysis and therefore should not be taken too formally. For this purpose, the non-robustness to normality of (3.39) and (3.40) should not be a serious problem. More robust tests can be found in Conover, Johnson, and Johnson (1981).

If H_0 is not rejected, or alternatively, if a half-normal plot of the $\ln s_i^2$ shows no significant effects, the s_i^2 can be pooled to get an estimate s^2 of the common variance σ^2 for the experimental runs,

$$s^2 = \frac{\sum_{i=1}^m (n_i - 1) s_i^2}{\sum_{i=1}^m (n_i - 1)}. \tag{3.41}$$

Using the s^2 data from Table 3.1,

$$s^2 = \frac{(6-1)}{16(6-1)}(0.270 + 0.291 + \ldots + 0.192) = 0.2376. \tag{3.42}$$

This variance estimate will be used in the test of effect significance of $\bar{y}_i$ in the next section. If the σ_i^2 are not homogeneous, one should try to explain or model their heterogeneity in terms of the factor levels as was done in Section 3.11.

3.15 A FORMAL TEST OF EFFECT SIGNIFICANCE FOR $\bar{y}_i$: WITH s^2

From (3.16), the estimated ith factorial effect $\hat{\theta}_i$ based on the sample means $\bar{y}_i$ can be represented as $\bar{y}_+ - \bar{y}_-$, where N is the total number of observations. It is further assumed that $n_i = n$ in (3.41) and therefore $N = nm$. For the 2^k full factorial designs, $m = 2^k$; for the 2^{k-p} fractional factorial designs (presented in the next chapter), $m = 2^{k-p}$. By replacing σ in the confidence interval in (3.23) by an estimate s from an independent source (if available) or the pooled estimate in (3.41), one obtains a $100(1 - \alpha)\%$ confidence interval

$$\left(\hat{\theta}_i - t_{\alpha/2} \frac{2s}{\sqrt{N}}, \ \hat{\theta}_i + t_{\alpha/2} \frac{2s}{\sqrt{N}} \right) \tag{3.43}$$

for testing the significance of θ_i, where $t_{\alpha/2}$ is the upper $\alpha/2$ point of the t distribution with $\sum_{i=1}^{m} (n-1) = N - m$ degrees of freedom. Therefore, $\hat{\theta}_i$ is declared significant at level α if the interval in (3.43) does not contain zero.

To simultaneously test the significance of several $\hat{\theta}_i$'s, say $i = 1, \ldots, I$, a different procedure that controls the experimentwise error is needed. Let

$$R = \max_{1 \le i \le I} \frac{|\hat{\theta}_i|}{sd(\hat{\theta}_i)} \tag{3.44}$$

be the **studentized maximum modulus** statistic (Tukey, 1953, 1994), where $sd(\hat{\theta}_i)$ is the standard error for $\hat{\theta}_i$ and equals $2s/\sqrt{N}$ in this case. Under the null hypothesis of $\theta_1 = \cdots = \theta_I = 0$, i.e., no effect is significant, R is the maximum of I absolute values of the t statistics. Since the $\hat{\theta}_i$ are independent as noted in Section 3.9, R has the *studentized maximum modulus* distribution with parameters I and $\upsilon \ (= N - m)$. [Further details on this distribution and its use in multiple comparisons can be found in Hochberg and Tamhane (1987).] Denote the upper α point of this distribution by $|M|_{I,N-m,\alpha}$, whose values are given in Appendix F. Then the *studentized maximum modulus test* declares any $\hat{\theta}_i$ significant if

$$\frac{|\hat{\theta}_i|}{sd(\hat{\theta}_i)} > |M|_{I,N-m,\alpha}, \qquad i = 1, \ldots, I. \tag{3.45}$$

It is obvious from the definition of $|M|_{I,N-m,\alpha}$ that under the normal assumption the test in (3.45) has an experimentwise error α, i.e.,

$$Prob\left(\text{at least one } \hat{\theta}_i \text{ is declared significant} \mid \theta_1 = \cdots = \theta_I = 0\right) = \alpha.$$

If n_i, the number of replicates for the ith condition, is not a constant, the simultaneous test with the following modification is still valid. For general n_i's, an extension of (3.17) gives

$$Var\left(\hat{\theta}_i\right) = \frac{\sigma^2}{N_{i+}} + \frac{\sigma^2}{N_{i-}} = \frac{N\sigma^2}{N_{i+}N_{i-}}, \tag{3.46}$$

where N_{i+} (and respectively N_{i-}) is the number of observations at the $+$ level (and respectively at the $-$ level) of the effect $\hat{\theta}_i$, and $N = N_{i+} + N_{i-}$. By replacing $sd(\hat{\theta}_i) = 2s/\sqrt{N}$ in (3.44) by

$$sd(\hat{\theta}_i) = s\sqrt{\frac{N}{N_{i+}N_{i-}}}, \tag{3.47}$$

the test in (3.45) still has experimentwise error α. The proof is left as an exercise.

For the adapted epitaxial layer growth experiment, $s^2 = 0.2376$ from (3.42) and

$$sd(\hat{\theta}_i) = 2s/\sqrt{N} = 2\sqrt{0.2376}/\sqrt{96} = 0.0995.$$

At level $\alpha = 0.05$, $I = 15$, $N = 96$, $m = 16$, and $|M|_{I, N-m, \alpha} = |M|_{15, 80, 0.05} = 3.01$. From (3.45), the studentized maximum modulus test declares any $\hat{\theta}_i$ significant if

$$|\hat{\theta}_i| > sd(\hat{\theta}_i)|M|_{15, 80, 0.05} = (0.0995)\,(3.01) = 0.30.$$

Since only the values for D and CD in Table 3.6 exceed 0.30, the studentized maximum modulus test declares D and CD significant at the 0.05 level. This finding is consistent with the one based on the normal probability plot in Figure 3.8. Because

$$sd(\hat{\theta}_i)|M|_{15, 80, 0.01} = (0.0995)(3.54) = 0.35,$$

only D is significant at the 0.01 level.

If the investigator is satisfied with controlling only the individual error rate, the IER version of Lenth's method can be used for significance testing. As shown in Section 3.13, both D and CD were declared significant at the 0.01 level. This does not mean that Lenth's method is more powerful than the studentized maximum modulus test because the former only controls the IER while the latter controls the EER. If the EER version of Lenth's method is used, the corresponding critical value for $I = 15$ and $\alpha = 0.05$ is 4.24 (from Appendix H). By comparing the t_{PSE} values in Table 3.12 with 4.24, only D is declared significant at the 0.05 level. This suggests that the studentized maximum modulus test may be more powerful than the EER version of Lenth's method since it declares both D and CD as significant at the 0.05 level. In making this comparison, one should, however, keep in mind that Lenth's method works for unreplicated experiments, while the studentized maximum modulus test does not.

Note that (3.45) is not as powerful for detecting important effects smaller than the largest effect since the same critical value is used. Some improvement is provided by using *step-up* or *step-down* procedures, as discussed in Hochberg and Tamhane (1987).

Another reason for the lack of power of the studentized maximum modulus test is that for experimental design applications I tends to be large.

For example, for a 2^k design, $I = 2^k - 1$. In many experimental situations the majority of the I effects are known *a priori* to be negligible (which is called the effect sparsity principle in Section 3.5). The test can then be applied to a smaller subset of effects to increase its power, assuming that no data snooping has been done, i.e., without first looking at the data to select the subset of effects to be tested. For example, if only the main effects and two-factor interactions are tested, $I = k + k(k - 1)/2 = k(k + 1)/2$. By lowering the value of I from $2^k - 1$ to $k(k + 1)/2$, the EER of the test is controlled over a smaller set of effects, which explains the increase in the power of the test. If it is thought that there is a substantial number of significant three-factor interactions, only four-factor and higher order interactions should be discarded in the test. The modification is also justifiable in view of the hierarchical ordering principle (see Section 3.5).

Using the adapted epitaxial layer growth experiment to illustrate the modified procedure, there are four main effects and six two-factor interactions; therefore, $I = 10$ and

$$sd\left(\hat{\theta}_i\right)|M|_{10,80,0.05} = (0.0995)(2.87) = 0.285.$$

Based on the values in Table 3.6, no additional effects are detected.

The pooled standard error s used in (3.43) was based on the assumption that the test of variance homogeneity in (3.39) does not reject the null hypothesis. Even if the null hypothesis is rejected, one can still use it for the studentized maximum modulus statistic in (3.44). If the s_i^2 in (3.41) are not homogeneous, the s^2 in (3.41) will be an overestimate, which makes the studentized maximum modulus test a conservative test. As long as it is acceptable to use a conservative test, this modification of the studentized maximum modulus test can be employed.

3.16   A FORMAL TEST OF EFFECT SIGNIFICANCE FOR $\ln s_i^2$

For testing the significance of factorial effects based on the log variances $\ln s_i^2$, the test in (3.45) needs a simple modification. To avoid confusion, denote $z_j = \ln s_j^2$. As in (3.16), the ith factorial effect $\hat{\theta}_i$ can be represented as $\bar{z}_+ - \bar{z}_-$, where $\bar{z}_+$ and $\bar{z}_-$ are the average z values corresponding to the $+$ and $-$ levels of $\hat{\theta}_i$. Recall from (3.22) that $\ln s_i^2$ is approximately normal with mean $\ln \sigma_i^2$ and variance $2(n_i - 1)^{-1}$. Using this approximation, it is easy to show that $\hat{\theta}_i$ has a known variance,

$$Var\left(\hat{\theta}_i\right) = Var(\bar{z}_+) + Var(\bar{z}_-)$$

$$= \left(\frac{2}{m}\right)^2 \sum_+ \frac{2}{n_j - 1} + \left(\frac{2}{m}\right)^2 \sum_- \frac{2}{n_j - 1},$$

$$= h_i^2, \tag{3.48}$$

where m is the total number of distinct runs, Σ_+ denotes the summation over all the runs at the $+$ level of $\hat{\theta}_i$, and Σ_- denotes the summation over all the runs at the $-$ level of $\hat{\theta}_i$. In the case of equal replicates $n_j = n$,

$$h_i^2 = \frac{8}{m(n-1)} . \tag{3.49}$$

As in Section 3.15, m equals 2^k for 2^k full factorial designs and 2^{k-p} for 2^{k-p} fractional factorial designs. Under the null hypothesis that $\sigma_1^2 = \cdots = \sigma_m^2$, $\hat{\theta}_i$ has mean zero and variance h_i^2. Orthogonality of the factorial or fractional factorial designs guarantees that the $\hat{\theta}_i$ are independent. Therefore under the null hypothesis, the statistic

$$R = \max_{1 \le i \le I} \frac{|\hat{\theta}_i|}{h_i} \tag{3.50}$$

has the *maximum modulus* distribution with parameter I. This is a special case of the studentized maximum modulus distribution with parameters I and $\nu = \infty$. Thus the upper α point of the maximum modulus distribution can be denoted by $|M|_{I, \infty, \alpha}$, which is tabled in Appendix F. The *maximum modulus test* declares any $\hat{\theta}_i$ significant, $i = 1, \ldots, I$, if

$$|\hat{\theta}_i|/h_i > |M|_{I, \infty, \alpha}, \qquad i = 1, \ldots, I. \tag{3.51}$$

It is obvious from the definition that under the normality assumption for $\ln s_i^2$, this test has an experimentwise error of α, i.e.,

$$P\left(\text{any } |\hat{\theta}_i|/h_i > |M|_{I, \infty, \alpha}, \quad i = 1, \ldots, I | \sigma_1^2 = \cdots = \sigma_m^2\right) = \alpha. \tag{3.52}$$

The validity of the test does not require the n_i to be equal. An illustration of the test will be given later in Section 4.3.

3.17 A SUMMARY OF ANALYSIS STRATEGIES

In this section we provide a summary of the analysis methods that are covered in various parts of the chapter. It is given separately for unreplicated and replicated experiments.

Unreplicated Experiments

Because there are no replicated observations, no dispersion modeling is considered. *Location modeling* is done by modeling y_i in terms of the various factorial effects. Effect significance can be tested by using *half-normal plots* or *Lenth's method*. The former is an informal and simpler method, while the latter is a formal test that provides quantitative evidence of effect significance.

Replicated experiments

Both *location and dispersion modeling* can be considered by modeling $\bar{y}_i$ and $\ln s_i^2$ in terms of the various factorial effects. There are three main steps:

1. Test the *variance homogeneity* by using the *test statistic* in Section 3.14 or the *half-normal plot* of factorial effects for $\ln s_i^2$. If variance homogeneity is accepted, go to step 2; otherwise, go to step 3.
2. Because the variances s_i^2 are homogeneous, there is no need to model dispersion. Use (3.41) to obtain a *pooled variance estimate s^2* and apply the *studentized maximum modulus test* in Section 3.15 to test the significance of the factorial effects in the location model. Alternatively *half-normal plots* or *Lenth's method* can be used for testing effect significance. Neither requires a variance estimate s^2.
3. *Half-normal plots* or *Lenth's method* can be used for testing the significance of factorial effects in the location model and the dispersion model. Alternatively, the *maximum modulus test* in Section 3.16 can be used for testing effect significance in the dispersion model.

When different methods are available for performing the same task, their pros and cons can be found in the discussions in various sections.

3.18 PRACTICAL SUMMARY

1. For the nominal-the-best problem, use the quadratic loss function $L(y, t) = c(y - t)^2$ to measure the quality loss of deviation of the response y from the target t. If there is an adjustment factor, use the two-step procedure to choose optimal factor settings:
   (i) Select levels of some factors to minimize $Var(y)$.
   (ii) Select the level of a factor not used in (i) to move $E(y)$ closer to t.

   To carry out this procedure in the data analysis, use linear regression to model the sample average $\bar{y}$ and the log sample variance $\ln s^2$ in terms of the factor effects. See the analysis in Section 3.11 for an illustration. An adjustment factor is a factor that has a significant effect on the mean but not on the variance.
2. If some of the factors are hard to change, randomization can only be applied to the remaining factors subject to the practical constraints on the hard-to-change factors. This is referred to as restricted randomization.
3. Use a planning matrix for the planning and implementation of the experiment and a model matrix for estimation of effects in a linear model. The former describes the experimental plan in terms of the

actual levels of the factors; the latter is the **X** matrix in the linear model $E(\mathbf{y}) = \mathbf{X}\boldsymbol{\beta}$.

4. Key properties of 2^k full factorial designs:

   (i) balance,

   (ii) orthogonality,

   (iii) reproducibility of conclusions in similar experiments.

5. Disadvantages of the "one-factor-at-a-time" approach relative to the factorial design method:

   (i) It requires more runs for the same precision in effect estimation.

   (ii) It cannot estimate some interactions.

   (iii) Conclusions from its analysis are not general.

   (iv) It can miss optimal factor settings.

6. Main effect of factor A: $ME(A) = \bar{z}(A+) - \bar{z}(A-)$, where $\bar{z}(A+)$ is the average of the z_i values observed at $A+$ and $\bar{z}(A-)$ is similarly defined. Interaction effect between factors A and B:

$$INT(A, B) = \tfrac{1}{2}[ME(B|A+) - ME(B|A-)]$$
$$= \tfrac{1}{2}[ME(A|B+) - ME(A|B-)]$$
$$= \tfrac{1}{2}\{\bar{z}(A+|B+) + \bar{z}(A-|B-)\}$$
$$- \tfrac{1}{2}\{\bar{z}(A+|B-) + \bar{z}(A-|B+)\},$$

where $ME(B|A+) = \bar{z}(B+|A+) - \bar{z}(B-|A+)$ is the conditional main effect of factor B at the $+$ level of factor A, and other ME terms can be similarly defined. A significant interaction between A and B suggests that the effect of one factor depends on the level of the other factor. For k factors define their interaction in terms of the $(k-1)$-factor interactions:

$$INT(A_1, A_2, \ldots, A_k) = \tfrac{1}{2}INT(A_1, A_2, \ldots, A_{k-1}|A_k+)$$
$$- \tfrac{1}{2}INT(A_1, A_2, \ldots, A_{k-1}|A_k-),$$

which can be written as $\bar{z}_+ - \bar{z}_-$, where $\bar{z}_+$ is the average of the z values at the factorial combinations whose product of its levels is $+$; $\bar{z}_-$ is similarly defined. These factorial effects can be estimated via regression analysis by using a model like (3.18).

7. The main effects plot and interaction plots are effective graphical tools. They often convey more information than the numerical values of main effect and interaction effect estimates. For factors A and B, there are the A-against-B plot and the B-against-A plot. An A-against-B plot is called synergistic if $ME(B|A+)ME(B|A-) > 0$ and antagonistic if $ME(B|A+)ME(B|A-) < 0$. An antagonistic A-against-B plot suggests that the underlying response relationship may be more complicated and that, at an intermediate setting between $A+$ and $A-$, the relationship between the response and B is flatter.

Similar definitions of antagonism and synergism hold for the *B*-against-*A* plot.

8. Three fundamental principles for factorial effects.

   Hierarchical ordering principle: (i) Lower order effects are more likely to be important than higher order effects; (ii) effects of the same order are equally likely to be important.

   Effect sparsity principle: The number of relatively important effects in a factorial experiment is small.

   Effect heredity principle: In order for an interaction to be significant, at least one of its parent factors should be significant.

9. The use of $\ln s^2$ has two advantages:

   (i) it maps positive values into real values; exponentiating a fitted model to $\ln s^2$ always results in a positive estimate of the variance;

   (ii) $\ln s^2 = \ln \sigma^2 + \ln[\chi^2_{n-1}/(n-1)]$ allows $\ln \sigma^2$ to be modeled separately.

10. The normal and half-normal plots can be used to judge effect significance for both unreplicated and replicated experiments and are particularly useful for unreplicated experiments. To avoid visual misjudgment as demonstrated in Figure 3.9, half-normal plots are preferred. Both plots are informal methods based on visual judgment. A more formal and quantitative method is based on Lenth's method of Section 3.13, which is particularly useful when the experiment is unreplicated and the control of the individual error rate is of primary concern.

11. Use the maximum modulus test in (3.51) to test effect significance for $\ln s^2$. If a reliable estimate of the variance is available either from other sources or from pooling the sample variances s_i^2 as in (3.41) [assuming they pass the variance homogeneity test in (3.39) or (3.40)], use the studentized maximum modulus test in (3.45) to test effect significance for $\bar{y}$. The experimentwise error of both tests is equal to the nominal level α.

12. A summary of data analysis strategies involving the tools mentioned in points 10 and 11 above is given in Section 3.17 for replicated and unreplicated experiments.

13. Use the minimum aberration criterion to select an optimal blocking scheme for arranging a 2^k design in 2^q blocks. These optimal blocking schemes can be found in Table 3A.1.

EXERCISES

1. Two possible violations of the quadratic loss assumption were discussed at the end of Section 3.2. Give real examples to demonstrate the possibility of (i) a highly skewed quality loss and (ii) a flat quality loss for $|y - t| \geq c_0$.

2. Give examples of hard-to-change factors and describe in realistic terms how they change the way in which randomization is performed.

3. Derive a simple necessary and sufficient condition for an interaction effect plot to be synergistic.

4. Show the equivalence of (3.13) and (3.14).

5. Prove the equivalence of (3.15) and (3.16). As a result, the definition in (3.15) is independent of the conditioning variable A_k.

6. Show the relationship given in (3.19) between the LSE $\hat{\beta}_j$ and the corresponding factorial effect. [Hint: Use $\hat{\beta} = (\mathbf{X}^T\mathbf{X})^{-1}\mathbf{X}^T\mathbf{y}$.]

7. Box and Bisgaard (1987) described an experiment to reduce the percentage of cracked springs produced by a particular manufacturing process. Three factors are investigated: temperature of the quenching oil, carbon content of the steel, and temperature of the steel before quenching. A 2^3 design was used; the design and response data, percentage of non-cracked springs, are given in Table 3.13. Show that the "one-factor-at-a-time" approach can miss the optimal settings obtained from analyzing the factorial effects. Assume that any non-zero estimated effects are significant.

8. (a) Explain why using the normal probability plot (or half-normal probability plot) takes into account the effect of simultaneous testing. Recall that in (3.25) the experimentwise error increases rapidly as I, the number of effects being tested, increases. (Hint: As I increases, while there are more large values of $\hat{\theta}_i$, they are counterbalanced by a more stretched normal probability scale as measured by $\Phi^{-1}([i - \frac{1}{2}]/I)$, $i = 1, \ldots, I$.)

   (b) Repeat (a) for Lenth's method.

Table 3.13   Planning Matrix and Response Data, Spring Experiment

Oil Temperature (°F)	Percentage of Carbon	Steel Temperature (°F)	Percentage of Non-Cracked Springs
70	0.50	1450	67
70	0.50	1600	79
70	0.70	1450	61
70	0.70	1600	75
120	0.50	1450	59
120	0.50	1600	90
120	0.70	1450	52
120	0.70	1600	87

9. Suppose that $\hat{\theta}_1, \ldots, \hat{\theta}_I$ are a random sample from $N(0, \sigma^2)$. It is known that the sample median of the $|\hat{\theta}_i|$ converges in probability to the population median of the $|\hat{\theta}_i|$ as $n \to \infty$.

   (a) Show that the median of the positive part of a $N(0, 1)$ random variable is 0.6745.

   (b) From (a), show that median$|\hat{\theta}_i|/\sigma$ converges to 0.6745 as $n \to \infty$, which is equivalent to $(1/0.6745)$ median$|\hat{\theta}_i| = 1.4826$ median$|\hat{\theta}_i| \to \sigma$ as $n \to \infty$. [In (3.35), the scaling factor 1.4826 is rounded off to 1.5.]

10. Complete the analysis of dispersion effects at the end of Section 3.13. In particular, find the smallest α values at which the IER version of Lenth's method will declare at least one effect significant.

11. For the $\bar{y}$ analysis of the adapted epitaxial layer growth experiment, three formal tests of significance are available: the IER and EER versions of Lenth's method and the studentized maximum modulus test. For each test, find the p values of the effects D and CD. Compare the p values for the three tests and comment.

12. For general n_i's, define R as in (3.44), but with $sd(\hat{\theta}_i) = s\sqrt{N/(N_{i+}N_{i-})}$, where s^2 is given in (3.41). Prove that R still has the studentized maximum modulus distribution with parameters I and $N - m$. [Hint: Under H_0, $\hat{\theta}_i/\sqrt{N/(N_{i+}N_{i-})}$ are independent normal with mean zero and variance σ^2.]

13. A food scientist has collected the data reported in Table 3.14.

   (a) Calculate the two-factor interaction between mixing speed and time. Does the interaction effect estimate use the information in runs 9–11? Explain why.

   (b) What observations among the 11 runs can be used to estimate the underlying process variance σ^2?

   (c) Assume each run follows a normal distribution with the same variance. Estimate the standard error of the interaction effect computed in (a).

   (d) Suppose s^2 is the estimate of σ^2 obtained in (b). Use the studentized maximum modulus test to test the significance of the seven factorial effects at $\alpha = 0.01$, 0.05. Determine the p value of the largest effect.

Table 3.14   Food Experiment

Factor	Level −	Level 0	Level +
A. temperature	25	30	35
B. mixing speed	low	medium	high
C. time (sec.)	60	75	90

Run	Factor *A*	Factor *B*	Factor *C*	Viscosity
1	−	−	−	60
2	+	−	−	62
3	−	+	−	82
4	+	+	−	86
5	−	−	+	85
6	+	−	+	85
7	−	+	+	61
8	+	+	+	61
9	0	0	0	84
10	0	0	0	87
11	0	0	0	81

(e) Repeat the task in (d) by using the IER version of Lenth's method at $\alpha = 0.01, 0.05$. Compare the results with those in (d).

(f) Repeat (e) by using the EER version of Lenth's method. Is it too conservative to advocate the control of the EER for this problem?

14. A 2^3 experiment was performed by an industrial engineer and the results are given in Table 3.15. Half of the experiment was done on one day and the remaining half on the next day.

(a) Compute the main effect of task.

(b) Use a half-normal plot to detect significant factorial effects. Compare your finding with those based on the IER version of Lenth's method at an appropriate α level. Would you recommend using the EER version of Lenth's method?

Table 3.15 Task Efficiency Experiment

Setup	Flasher Type	Inertia of Lever	Task	Average Response Time (tenths of a second)	Randomized Order in Which Experiments Were Performed
1	A	low	Y	11	1
2	B	low	Y	12	4
3	A	high	Y	10	5
4	B	high	Y	11	3
5	A	low	Z	16	2
6	B	low	Z	14	6
7	A	high	Z	15	7
8	B	high	Z	19	8

(c) With the timing mechanism operating properly, the standard deviation of the readings of average response time is known to be about 1 ($\sigma = 1$). Subsequent to carrying out this experiment, however, it was discovered that the timing device, although it had been operating properly on the first day, had not been operating properly on the second day. On that second day, the standard deviation was $\sigma = 4$. Given this information, what is the variance of the main effect of task computed in part (a) above?

15. Use the IER version of Lenth's method to test the significance of effects in Table 3.9 for $\bar{y}$ and $\ln s^2$ at $\alpha = 0.01, 0.05$. What is the p value of the A effect for the $\ln s^2$ analysis and of the D effect for the $\bar{y}$ analysis?

16. Four runs were planned for an experiment to study the effects of part width (in millimeters) and filler content (in percent) at two levels on the design of a new fastener device for a plastics application. The response y is the pull-off force (in pounds). After the first two runs, the experiment was put on the back burner because the engineers needed to "put out a fire." When the engineers returned to finish up the experiment, they discovered that they had lost the original experimental plan.

Run	Part Width	Filler	Temperature	Pull-off Force
1	36	40		80
2	50	20		45
3				
4				

(a) From the first two runs, can the engineers distinguish between the effects of width and filler content?

(b) The engineers want to use a factorial design. Can they? If so, fill in the factor levels for the last two runs.

(c) The engineers receive word that fastener joints are melting in the field. Both of the existing pull-off tests were performed at room temperature (20°C). If they decide that they would like to also include temperature as a two-level variable in their experiment (with 40°C as the other level), do they need to perform more than two additional runs to estimate all main effects? What assumption would they need to make? Recommend and clearly spell out an experimental plan for these engineers.

17. An experiment on a surface-finishing operation of an overhead cam block auxiliary drive shaft was reported by Sirvanci and Durmaz (1993). The following is an adaptation of the original experiment. The part had a target value of 75 μm. The three experimental factors were type of insert (A), speed in rpm (B), and feed rate in millimeters per minute (C). A 2^3 design was used; the $(-, +)$ levels for the factors are (#5023, #5074) for A, $(800, 1000)$ for B, and $(50, 80)$ for C. The surface roughness of each part was measured by a Surtronic-3 device. The design and data for the drive shaft experiment are displayed in Table 3.16. The current setting for the surface-finishing process is insert type #5023, 780 rpm, and 60 mm/min.

(a) Analyze the experiment for both location and dispersion effects by using appropriate methods presented in this chapter.

(b) Make recommendations on the optimal factor settings for location and dispersion effects. Is it appropriate to use the two-step procedure?

Table 3.16 Design Matrix and Roughness Data, Drive Shaft Experiment

Run	A	B	C	Roughness				
1	−	−	−	54.6	73.0	139.2	55.4	52.6
2	−	−	+	86.2	66.2	79.2	86.0	82.6
3	−	+	−	41.4	51.2	42.6	58.6	58.4
4	−	+	+	62.8	64.8	74.6	74.6	64.6
5	+	−	−	59.6	52.8	55.2	61.0	61.0
6	+	−	+	82.0	72.8	76.6	73.4	75.0
7	+	+	−	43.4	49.0	48.6	49.6	55.2
8	+	+	+	65.6	65.0	64.2	60.8	77.4

18. A biologist performed a 2^5 factorial design in four blocks of equal size by making use of the blocking variables: $B_1 = 1234$, $B_2 = 2345$. If block effects are present when the experiment is performed, how will the main effect of variable **5** be affected? How will the interaction **15** be affected?

19. Suppose a 2^7 design is arranged in 8 blocks with the generators $B_1 = 123$, $B_2 = 456$, $B_3 = 167$.

   (a) Find all the interactions that are confounded with block effects and use these to compute the $g_i(b)$ for $i \geq 2$ and to determine the order of estimability.

   (b) An alternative scheme is the one given in Table 3A.1 with $k = 7$ and $q = 3$. Compute the $g_i(b)$ for $i \geq 2$ and determine the order of estimability for the second scheme.

   (c) Compare the two schemes and show the clear advantages of the second scheme.

20. To arrange a 2^6 design in 16 blocks, two blocking schemes are being considered. The first has the generators $B_1 = 126$, $B_2 = 136$, $B_3 = 346$, $B_4 = 456$ and the second is given in Table 3A.1 with $k = 6$ and $q = 4$. Compare the two schemes in terms of their respective $g_i(b)$ and comment on the advantage of the second scheme.

21. *Difficulties with Non-Orthogonal Blocking*

   (a) Suppose that the 2^3 design in Table 3.10 is arranged in two blocks with runs 1, 2, 5, and 7 in block I and runs 3, 4, 6, and 8 in block II. Show that the corresponding blocking variable is not identical to any of the first seven columns of the table and, therefore, it is not confounded with any of the factorial effects.

   (b) Show that the block effect is not orthogonal to some of the factorial effects. Identify these effects. How does it affect the estimation of these effects? Discuss why it is undesirable to use this blocking scheme.

   (c) In general, is it advisable to choose blocking variables that are not confounded with any of the factorial effects?

APPENDIX 3A: TABLE OF 2^k FACTORIAL DESIGNS IN 2^q BLOCKS

Table 3A.1 2^k **Full Factorial Designs in 2^q Blocks (k = number of factors, 2^k = number of runs, 2^q = number of blocks, e = order of estimability)**

k	q	e	Block Generators	Confounded Interactions
3	1	2	$B_1 = 123$	123
3	2	1	$B_1 = 12$, $B_2 = 13$	12, 13, 23
4	1	3	$B_1 = 1234$	1234
4	2	1	$B_1 = 134$, $B_2 = 234$	12, 134, 234
4	3	1	$B_1 = 12$, $B_2 = 13$, $B_3 = 1234$	12, 13, 14, 23, 24, 34, 1234
5	1	4	$B_1 = 12345$	12345
5	2	2	$B_1 = 123$, $B_2 = 2345$	123, 145, 2345
5	3	1	$B_1 = 135$, $B_2 = 235$, $B_3 = 1234$	12, 34, 135, 145, 235, 245, 1234
5	4	1	$B_1 = 1234$, $B_2 = 1235$, $B_3 = 1245$, $B_4 = 1345$	all interactions of even order
6	1	5	$B_1 = 123456$	123456
6	2	3	$B_1 = 1234$, $B_2 = 1256$	1234, 1256, 3456
6	3	2	$B_1 = 123$, $B_2 = 2345$, $B_3 = 1346$	123, 145, 246, 356, 1256, 1346, 2345
6	4	1	$B_1 = 136$, $B_2 = 1234$, $B_3 = 3456$, $B_4 = 123456$	12, 34, 56, 135, 136, 145, 146, 235, 236, 245, 246, 1234, 1256, 3456, 123456
6	5	1	$B_1 = 1234$, $B_2 = 1235$, $B_3 = 1245$, $B_4 = 1345$, $B_5 = 123456$	all interactions of even order
7	1	6	$B_1 = 1234567$	1234567
7	2	3	$B_1 = 12567$, $B_2 = 34567$	1234, 12567, 34567
7	3	3	$B_1 = 1234$, $B_2 = 1256$, $B_3 = 1357$	1234, 1256, 1357, 1467, 2367, 2457, 3456
7	4	2	$B_1 = 1247$, $B_2 = 2345$, $B_3 = 4567$, $B_4 = 1234567$	123, 145, 167, 246, 257, 347, 356, 1256, 1247, 1346, 1357, 2345, 2367, 4567, 1234567
7	5	1	$B_1 = 12346$, $B_2 = 12347$, $B_3 = 124567$, $B_4 = 134567$, $B_5 = 234567$	12, 13, 23, 45, 67, 146, 147, 156, 157, 246, 247, 256, 257, 346, 347, 356, 357, 1245, 1267, 1345, 1367, 2345, 2367, 4567, 12346, 12347, 12356, 12357, 124567, 134567, 234567
7	6	1	$B_1 = 123456$, $B_2 = 123457$, $B_3 = 123467$, $B_4 = 123567$, $B_5 = 124567$, $B_6 = 134567$	all interactions of even order

Table 3A.1 (*Continued*)

k	q	e	Block Generators	Confounded Interactions
8	1	7	$B_1 = 12345678$	12345678
8	2	4	$B_1 = 12345$, $B_2 = 345678$	12345, 12678, 345678
8	3	3	$B_1 = 13578$, $B_2 = 23678$, $B_3 = 24578$	1234, 1256, 3456, 13578, 14678, 23678, 24578
8	4	3	$B_1 = 1234$, $B_2 = 1256$, $B_3 = 1357$, $B_4 = 12345678$	1234, 1256, 1278, 1357, 1368, 1458, 1467, 2358, 2367, 2457, 2468, 3456, 3478, 5678, 12345678
8	5	1	$B_1 = 12357$, $B_2 = 123456$, $B_3 = 123478$, $B_4 = 1345678$, $B_5 = 2345678$	12, 134, 156, 178, 234, 256, 278, 357, 368, 458, 467, 1358, 1367, 1457, 1468, 2358, 2367, 2457, 2468, 3456, 3478, 5678, 12357, 12368, 12458, 12467, 123456, 123478, 125678, 1345678, 2345678
8	6	1	$B_1 = 124578$, $B_2 = 125678$, $B_3 = 134578$, $B_4 = 134678$, $B_5 = 1234567$, $B_6 = 1234568$	12, 13, 23, 45, 46, 56, 78, 147, 148, 157, 158, 167, 168, 247, 248, 257, 258, 267, 268, 347, 348, 357, 358, 367, 368, (remaining 38 effects are higher order interactions)
8	7	1	$B_1 = 123456$, $B_2 = 123457$, $B_3 = 123467$, $B_4 = 123567$, $B_5 = 124567$, $B_6 = 134567$, $B_7 = 12345678$	all interactions of even order

REFERENCES

Bartlett, M.S. and Kendall, D.G. (1946), "The Statistical Analysis of Variance-Heterogeneity and the Logarithmic Transformation," *Journal of the Royal Statistical Society, Series B*, 8, 128–138.

Box, G.E.P. and Bisgaard, S. (1987), "The Scientific Context of Quality Improvement," *Quality Progress*, 20, June, 54–61.

Box, G.E.P and Meyer, R.D. (1986), "An Analysis for Unreplicated Fractional Factorials," *Technometrics*, 28, 11–18.

Box, G.E.P., Hunter, W.G., and Hunter, J.S. (1978), *Statistics for Experimenters*, New York: John Wiley & Sons.

Cheng, S.W., Wu, C.F.J., and Wu, H. (1999), "Finding Defining Generators with Extreme Lengths," Technical Report No. 332, Department of Statistics, University of Michigan.

Conover, W.J., Johnson, M.E., and Johnson, M.M. (1981), "A Comparative Study of Tests for Homogeneity of Variances, with Application to the Outer Continental Shelf Bidding Data," *Technometrics*, 23, 351–362.

Daniel, C. (1959), "Use of Half-Normal Plots in Interpreting Factorial Two-Level Experiments," *Technometrics*, 1, 311–341.

Daniel, C. (1971), *Applications of Statistics to Industrial Experimentation*, New York: John Wiley & Sons.

Fisher, R.A. (1971), *The Design of Experiments*, New York: Hafner.

Hamada, M. and Balakrishnan, N. (1998), "Analyzing Unreplicated Factorial Experiments: A Review with Some New Proposals (with discussion)," *Statistica Sinica*, 8, 1–41.

Hamada, M. and Wu, C.F.J. (1992), "Analysis of Designed Experiments with Complex Aliasing," *Journal of Quality Technology*, 24, 130–137.

Hochberg, Y. and Tamhane, A.C. (1987), *Multiple Comparison Procedures*, New York: John Wiley & Sons.

Juran, J.M. (1993), *Quality Planning and Analysis: From Product Development Through Use*, 3rd ed., New York: McGraw-Hill.

Kackar, R.N. and Shoemaker, A.C. (1986), "Robust Design: A Cost-Effective Method for Improving Manufacturing Processes," *AT&T Technical Journal*, 65, 39–50.

Lehmann, E.L. (1986), *Testing Statistical Hypotheses*, 2nd ed., New York: John Wiley & Sons.

Lenth, R.V. (1989), "Quick and Easy Analysis of Unreplicated Factorials," *Technometrics*, 31, 469–473.

Shoemaker, A.C., Tsui, K.L., and Wu, C.F.J. (1991), "Economical Experimentation Methods for Robust Design," *Technometrics*, 33, 415–427.

Sirvanci, M.B. and Durmaz, M. (1993), "Variation Reduction by the Use of Designed Experiments, *Quality Engineering*, 5, 611–618.

Sun, D.X., Wu, C.F.J., and Chen, Y. (1997), "Optimal Blocking Schemes for 2^n and 2^{n-p} Designs," *Technometrics*, 39, 298–307.

Tukey, J. (1953), "The Problem of Multiple Comparisons," unpublished manuscript.

Tukey, J. (1994), *The Collected Works of John W. Tukey*, Vol. VIII, *Multiple Comparisons: 1948–1983*, Ed. H.I. Braun, New York: Chapman & Hall.

Ye, K. and Hamada, M. S. (2000), "Critical Values of the Lenth Method for Unreplicated Factorial Designs," *Journal of Quality Technology*, 32, 57–66.

CHAPTER 4

Fractional Factorial Experiments at Two Levels

Because 2^k full factorial designs require that 2^k runs be performed, full factorial designs are rarely used in practice for large k (say, $k \geq 7$). For economic reasons, *fractional factorial designs*, which consist of a subset or *fraction* of full factorial designs, are commonly used. "Optimal" fractions are chosen according to the *resolution* and *minimum aberration* criteria. Aliasing of factorial effects is a consequence of using fractional factorial designs. Implications of effect aliasing are discussed. Data analysis strategies to resolve the ambiguities with aliased effects are considered.

4.1 A LEAF SPRING EXPERIMENT

Consider an experiment to improve a heat treatment process on truck leaf springs reported in Pignatiello and Ramberg (1985). The heat treatment which forms the camber (or curvature) in leaf springs consists of heating in a high temperature furnace, processing by a forming machine, and quenching in an oil bath. The height of an unloaded spring, known as free height, is an important quality characteristic whose target value is 8 inches. An experimental goal is to make the variation about the target 8.0 as small as possible. The five factors listed in Table 4.1 were chosen across the various stages of the process: furnace temperature (B) and heating time (C) are from the heating stage, transfer time (D) is the time it takes the conveyer to transport the springs from the furnace to the forming machine, hold-down time (E) is the time that the springs are held in a high pressure press to form the camber, and quench oil temperature (Q) is from the quenching stage. These factors are all quantitative but the levels of Q do not have fixed values. Because the quench oil temperature could not be controlled precisely, the experimenters could only give a range of values for each level. This aspect of the experiment will be re-examined later using parameter design methodology in Chapter 10.

153

Table 4.1 Factors and Levels, Leaf Spring Experiment

		Level	
	Factor	−	+
B.	high heat temperature (°F)	1840	1880
C.	heating time (seconds)	23	25
D.	transfer time (seconds)	10	12
E.	hold down time (seconds)	2	3
Q.	quench oil temperature (°F)	130–150	150–170

Table 4.2 Design Matrix and Free Height Data, Leaf Spring Experiment

B	C	D	E	Q	Free Height			$\bar{y}_i$	s_i^2	$\ln s_i^2$
−	+	+	−	−	7.78	7.78	7.81	7.7900	0.0003	− 8.1117
+	+	+	+	−	8.15	8.18	7.88	8.0700	0.0273	− 3.6009
−	−	+	+	−	7.50	7.56	7.50	7.5200	0.0012	− 6.7254
+	−	+	−	−	7.59	7.56	7.75	7.6333	0.0104	− 4.5627
−	+	−	+	−	7.94	8.00	7.88	7.9400	0.0036	− 5.6268
+	+	−	−	−	7.69	8.09	8.06	7.9467	0.0496	− 3.0031
−	−	−	−	−	7.56	7.62	7.44	7.5400	0.0084	− 4.7795
+	−	−	+	−	7.56	7.81	7.69	7.6867	0.0156	− 4.1583
−	+	+	−	+	7.50	7.25	7.12	7.2900	0.0373	− 3.2888
+	+	+	+	+	7.88	7.88	7.44	7.7333	0.0645	− 2.7406
−	−	+	+	+	7.50	7.56	7.50	7.5200	0.0012	− 6.7254
+	−	+	−	+	7.63	7.75	7.56	7.6467	0.0092	− 4.6849
−	+	−	+	+	7.32	7.44	7.44	7.4000	0.0048	− 5.3391
+	+	−	−	+	7.56	7.69	7.62	7.6233	0.0042	− 5.4648
−	−	−	−	+	7.18	7.18	7.25	7.2033	0.0016	− 6.4171
+	−	−	+	+	7.81	7.50	7.59	7.6333	0.0254	− 3.6717

The analysis in this chapter will treat the ranges 130–150° F and 150–170° F as the low and high levels of factor Q. The experimental design that was used was a half fraction of a 2^5 full factorial design with each run being replicated three times. The design matrix and free height data are presented in Table 4.2.

4.2 FRACTIONAL FACTORIAL DESIGNS: EFFECT ALIASING AND THE CRITERIA OF RESOLUTION AND MINIMUM ABERRATION

For the leaf spring experiment, what information could have been obtained if a full factorial 2^5 design had been used? The numbers of effects that could have been estimated are given in Table 4.3. Out of the 31 degrees of freedom

Table 4.3   Number of Effects in a 2^5 Design

Main Effects	Interaction			
	2-Factor	3-Factor	4-Factor	5-Factor
5	10	10	5	1

in a 32-run design, 16 are used for estimating three-factor or higher interactions. Is it practical to commit half of the degrees of freedom to estimate such effects? According to the hierarchical ordering principle (see Section 3.5), three-factor and higher order interactions are usually not significant (one exception being the control × control × noise interactions to be discussed in Chapter 10). Therefore, using a full factorial design is quite wasteful. A more economical approach is to use a fraction of the full factorial design that allows the lower order effects to be estimated.

Consider the design used in the leaf spring experiment for example. It studies five factors in 16 ($= 2^4$) runs, which is a $\frac{1}{2}$ fraction of a 2^5 design. The design is referred to as a 2^{5-1} design, where 5 denotes the number of factors and $2^{-1} = \frac{1}{2}$ denotes the fraction. A key issue is how the fraction should be chosen. For this experiment, factor E is assigned to the column that equals the product of the columns for factors B, C, and D; that is, each entry in the column for E is the same as the product of the corresponding entries in the columns for B, C, and D. Because the column for E is used for estimating the main effect of E and also for the interaction effect between B, C, and D, i.e.,

$$\bar{y}(E+) - \bar{y}(E-) = \bar{y}(BCD+) - \bar{y}(BCD-),$$

the data from such a design is not capable of distinguishing the estimate of E from the estimate of BCD. The factor main effect E is said to be **aliased** with the $B \times C \times D$ interaction. Notationally, this aliasing relation is denoted by

$$E = BCD \quad \text{or} \quad I = BCDE,$$

where I denotes the column of all $+$'s; that is, the product of the four columns B, C, D, E is all $+$'s. Note that the aliasing of effects uses the same mathematical definition as the confounding of a block effect with a factorial effect, which was discussed in Section 3.12. Aliasing of effects is a price one must pay for choosing a smaller design. Because the 2^{5-1} design has only 15 degrees of freedom for estimating factorial effects, it cannot estimate all 31 factorial effects among the factors B, C, D, E, and Q. It will be shown below that out of the 31 effects, 30 can be estimated in 15 pairs of aliased effects. (Because the experiment is replicated three times, it has another 32 degrees of freedom which can be used for estimating error or dispersion effects.)

The equation $I = BCDE$ is called the **defining relation** of the 2^{5-1} design. The design is said to have resolution IV because the defining relation consists of the "word" $BCDE$, which has "length" 4. By multiplying both sides of $\mathbf{I} = BCDE$ by column B, the relation $B = CDE$ is obtained; that is, column B is identical to the product of columns C, D, and E so that the B main effect is aliased with the $C \times D \times E$ interaction. By following the same method, all 15 aliasing relations can be inferred:

$$
\begin{array}{llll}
B = CDE, & C = BDE, & D = BCE, & E = BCD, \\
BC = DE, & BD = CE, & BE = CD, & \\
Q = BCDEQ, & BQ = CDEQ, & CQ = BDEQ, & DQ = BCEQ, \\
EQ = BCDQ, & BCQ = DEQ, & BDQ = CEQ, & BEQ = CDQ.
\end{array}
\tag{4.1}
$$

The relations in (4.1) and $\mathbf{I} = BCDE$ (meaning that the intercept and the $B \times C \times D \times E$ interaction are aliased) contain all the information on conditions for estimability of the 31 factorial effects in the leaf spring experiment. Each of the four main effects, B, C, D, and E, is estimable if the respective three-factor interaction alias in (4.1) is negligible, which is usually a reasonable assumption. The six two-factor interactions not involving Q are aliased in three pairs. For example, the $B \times C$ interaction is estimable only if prior knowledge would suggest that the $D \times E$ interaction is negligible. The Q main effect is estimable under the weak assumption that the $B \times C \times D \times E \times Q$ interaction is negligible. Similarly, the $B \times Q$ interaction is estimable under the weak assumption that the $C \times D \times E \times Q$ interaction is negligible. As in Wu and Chen (1992), we shall call a main effect or two-factor interaction **clear** if none of its aliases are main effects or two-factor interactions and **strongly clear** if none of its aliases are main effects, two-factor, or three-factor interactions. For the leaf spring experiment, the main effects B, C, D, and E are clear, and the main effect Q and the two-factor interactions $B \times Q$, $C \times Q$, $D \times Q$, and $E \times Q$ are strongly clear. The three-factor interactions involving Q cannot be estimated so easily. For example, based on the relation $BCQ = DEQ$, it would be difficult to claim that one of them can be estimated while assuming the other is negligible.

An alternate fractional plan is to define column Q as the product of columns B, C, D, E, i.e., $Q = BCDE$ or, equivalently, $\mathbf{I} = BCDEQ$. See the layout in Table 4.4. Because the word $BCDEQ$ in this defining relation has length 5, it is called a resolution V design. Traditionally, a design with higher resolution is favored. From the defining relation $\mathbf{I} = BCDEQ$, each main effect is aliased with a four-factor interaction (e.g., $B = CDEQ$). Therefore, all five main effects are strongly clear. Similarly, the ten two-factor interactions are clear because each of them is aliased with a three-factor interaction (e.g., $BC = DEQ$).

Which of the two fractional plans is better? The answer depends on what interactions are considered to be more important. The strength of the first plan is that any two-factor interaction involving Q is strongly clear, and the weakness is that the other two-factor interactions are not clear because they

Table 4.4 Alternate Design Matrix, Leaf Spring Experiment Using I = BCDEQ

B	C	D	E	Q
−	+	+	+	−
+	+	+	−	−
−	−	+	−	−
+	−	+	+	−
−	+	−	−	−
+	+	−	+	−
−	−	−	+	−
+	−	−	−	−
−	+	+	−	+
+	+	+	+	+
−	−	+	+	+
+	−	+	−	+
−	+	−	+	+
+	+	−	−	+
−	−	−	−	+
+	−	−	+	+

are aliased with other two-factor interactions. The second plan treats all factors equally, i.e., all two-factor interactions are clear, but not strongly clear. Thus, if prior knowledge strongly suggests that two-factor interactions involving Q are likely to be more important, the first plan is preferred. Otherwise the second plan would be the more common choice. Reasons for adopting the first plan in the original experiment will be given in Chapter 10 in the context of robust parameter design.

More generally, we define a 2^{k-p} design to be a fractional factorial design with k factors, each at two levels, consisting of 2^{k-p} runs. Therefore, it is a 2^{-p}-th fraction of the 2^k full factorial design in which the fraction is determined by p defining words, where a "word" consists of letters which are the names of the factors denoted by $1, 2, \ldots, k$ or $A, B, \ldots$. The number of letters in a word is its *wordlength* and the group formed by the p defining words is called the **defining contrast subgroup**. The group consists of $2^p - 1$ words plus the identity element **I**.

For example, consider the 2^{6-2} design d with **5 = 12** and **6 = 134**. That is, for design d, **1–4**, each consisting of entries of +'s and −'s, generate a 16-run full factorial design. By introducing **5** as the interaction **12** (i.e., **5 = 12**) and **6** as the interaction **134** (i.e., **6 = 134**), two defining words **I = 125** and **I = 1346** result. The first defining word is obtained by multiplying both sides of **5 = 12** by **5**, which gives **I = 125**. Similarly, multiplying both sides of **6 = 134** by **6** gives **I = 1346**. Then, taking the product of these two defining words gives **125 × 1346 = 23456** (using the rule that **11 = 22 = ··· = I**), another defining word. Thus, the defining contrast subgroup for design d is

$$I = 125 = 1346 = 23456. \tag{4.2}$$

The defining contrast subgroup can be used to study all the aliasing relations among effects and to see how each degree of freedom is used in estimating the aliased effects. For the design given in (4.2) and using the rules

$$11 = 22 = \cdots = I \quad \text{and} \quad \text{I}i = i,$$

we can easily obtain all the aliased effects of the main effect **1** as follows. By multiplying each term in (4.2) by **1**, we have $1 = 1\text{I} = 1(125) = 1(1346) = 1(23456)$, or equivalently, $1 = 25 = 346 = 123456$. Thus, the effects **1, 25, 346**, and **123456** are aliases of each other. Similarly, we can obtain the other aliasing relations for the remaining 14 degrees of freedom. All the aliasing relations for this design are

$$
\begin{aligned}
\mathbf{I} &= \mathbf{125} &&= \mathbf{1346} &&= \mathbf{23456}, \\
\mathbf{1} &= \mathbf{25} &&= \mathbf{346} &&= \mathbf{123456}, \\
\mathbf{2} &= \mathbf{15} &&= \mathbf{12346} &&= \mathbf{3456}, \\
\mathbf{3} &= \mathbf{1235} &&= \mathbf{146} &&= \mathbf{2456}, \\
\mathbf{4} &= \mathbf{1245} &&= \mathbf{136} &&= \mathbf{2356}, \\
\mathbf{5} &= \mathbf{12} &&= \mathbf{13456} &&= \mathbf{2346}, \\
\mathbf{6} &= \mathbf{1256} &&= \mathbf{134} &&= \mathbf{2345}, \\
\mathbf{13} &= \mathbf{235} &&= \mathbf{46} &&= \mathbf{12456}, \\
\mathbf{14} &= \mathbf{245} &&= \mathbf{36} &&= \mathbf{12356}, \\
\mathbf{16} &= \mathbf{256} &&= \mathbf{34} &&= \mathbf{12345}, \\
\mathbf{23} &= \mathbf{135} &&= \mathbf{1246} &&= \mathbf{456}, \\
\mathbf{24} &= \mathbf{145} &&= \mathbf{1236} &&= \mathbf{356}, \\
\mathbf{26} &= \mathbf{156} &&= \mathbf{1234} &&= \mathbf{345}, \\
\mathbf{35} &= \mathbf{123} &&= \mathbf{1456} &&= \mathbf{246}, \\
\mathbf{45} &= \mathbf{124} &&= \mathbf{1356} &&= \mathbf{236}, \\
\mathbf{56} &= \mathbf{126} &&= \mathbf{1345} &&= \mathbf{234}.
\end{aligned}
\tag{4.3}
$$

From these relations, one can infer that the main effects **3, 4, 6** and the interactions **23, 24, 26, 35, 45, 56** are clear while the other effects are aliased with at least one other two-factor interaction.

The previous example provides a good illustration of how the k columns in a 2^{k-p} design can be generated. First, let $1, \ldots, k-p$ denote the $k-p$ independent columns of the $+1$'s and -1's that generate the 2^{k-p} runs in the design. The remaining p columns, $k-p+1, \ldots, k$, can be generated as interactions of the first $k-p$ columns. Choice of these p columns determines the defining contrast subgroup of the design.

For a 2^{k-p} design, let A_i denote the number of words of length i in its defining contrast subgroup. The vector

$$W=(A_3,\ldots,A_k) \qquad (4.4)$$

is called the *wordlength pattern* of the design. The **resolution** of a 2^{k-p} design is defined to be the smallest r such that $A_r \geq 1$, that is, the length of the shortest word in the defining contrast subgroup. The **maximum resolution** criterion, proposed by Box and Hunter (1961), chooses the 2^{k-p} design with maximum resolution. This criterion can be justified by the hierarchical ordering principle (see Section 3.5). A lower-resolution design has defining words with shorter length, which imply the aliasing of lower order effects. According to the principle, lower order effects are more likely to be significant. It is, therefore, less desirable to choose a design which aliases lower order effects. If R is the resolution of the design, the design is sometimes denoted by 2_R^{k-p}. Note that the definition of W in (4.4) starts with A_3 because a design with a positive A_1 or A_2 would be useless.

Resolution R implies that no effect involving i factors is aliased with effects involving less than $R-i$ factors. For example, in a resolution III design, some main effects are aliased with two-factor interactions, but not with other main effects. Similarly, in a resolution IV design, some main effects are aliased with three-factor interactions and some two-factor interactions are aliased with other two-factor interactions. Therefore, in a resolution IV design, all the main effects are clear and some two-factor interactions are not clear. One can then use the number of clear two-factor interactions to compare any two resolution IV designs. In a resolution V design, some main effects are aliased with four-factor interactions and some two-factor interactions are aliased with three-factor interactions. From these properties, we derive the following important and useful rules:

Rules for Resolution IV and V Designs:

    (i) *In any resolution IV design, the main effects are clear.*

    (ii) *In any resolution V design, the main effects are strongly clear and the two − factor interactions are clear.*   (4.5)

    (iii) *Among the resolution IV designs with given k and p, those with the largest number of clear two-factor interactions are the best.*

Another implication of resolution is that a projection of the design in any $R-1$ factors is a full factorial in $R-1$ factors. For example, in Figure 4.1,

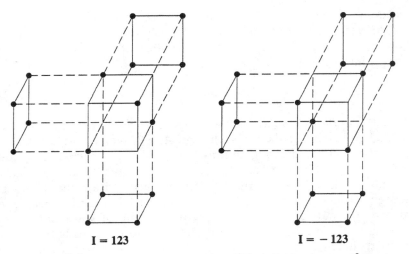

$$I = 123 \qquad\qquad I = -123$$

Figure 4.1. 2^{3-1} Designs Using $I = \pm 123$ and Their Projections to 2^2 Designs.

which displays the two 2^{3-1} designs $I = -123$ (i.e., $3 = -12$) and $I = 123$, respectively, any projection of two factors gives a 2^2 design in those two factors. This **projective** property (Box and Hunter, 1961) provides another rationale for using fractional factorial designs. If there are at most $R - 1$ important factors out of the k experimental factors, then the fractional factorial design yields a full factorial in the $R - 1$ factors, whatever the $R - 1$ factors are. (Its proof is left as an exercise.) This property can be exploited in data analysis. After analyzing the main effects, if only $R - 1$ of them are significant, then all the interactions among the $R - 1$ factors can also be estimated because the collapsed design on the $R - 1$ factors is a full factorial.

A design cannot be judged by its resolution alone. Consider the following two 2^{7-2} designs:

$$d_1 : \mathbf{I} = \mathbf{4567} = \mathbf{12346} = \mathbf{12357},$$
$$d_2 : \mathbf{I} = \mathbf{1236} = \mathbf{1457} = \mathbf{234567}. \tag{4.6}$$

While both designs are of resolution IV, they have different wordlength patterns:

$$W(d_1) = (0, 1, 2, 0, 0) \quad \text{and} \quad W(d_2) = (0, 2, 0, 1, 0).$$

The design d_1 has three pairs of aliased two-factor interactions, e.g., $(\mathbf{45}, \mathbf{67})$, $(\mathbf{46}, \mathbf{57})$, and $(\mathbf{47}, \mathbf{56})$, while d_2 has six pairs. This is because d_1 has one four-letter word while d_2 has two. To further characterize or discriminate between fractional factorial designs, Fries and Hunter (1980) proposed the following criterion.

Minimum Aberration Criterion

For any two 2^{k-p} designs d_1 and d_2, let r be the smallest integer such that $A_r(d_1) \neq A_r(d_2)$. Then d_1 is said to have *less aberration* than d_2 if $A_r(d_1) < A_r(d_2)$. If there is no design with less aberration than d_1, then d_1 has *minimum aberration*.

For given k and p, a minimum aberration design always exists. The minimum aberration criterion also allows any two designs to be rank ordered. It can be justified by the hierarchical ordering principle following an argument similar to the one used for resolution. Further discussion on the minimum aberration criterion and a collection of efficient designs according to this criterion will be given in Section 4.5.

4.3 ANALYSIS

Now we turn to the analysis of the leaf spring experiment. Because there are a total of 16 distinct combinations of factors B, C, D, E, and Q, there are 15 degrees of freedom for estimating the factorial effects. How are these 15 degrees of freedom allocated to the 31 ($= 2^5 - 1$) possible effects? From the analysis of the defining contrast subgroup $\mathbf{I} = BCDE$ in Section 4.2, the nine effects B, C, D, E, Q, BQ, CQ, DQ, and EQ are clear or strongly clear. For the remaining six pairs of aliased effects, $BC = DE$, $BD = CE$, $BE = CD$, $BCQ = DEQ$, $BDQ = CEQ$, $BEQ = CDQ$, an effect is estimable if its alias is negligible. Apart from the interpretation of significant effects that have aliases, the analysis methods given in Chapter 3 are applicable here.

First, consider the location effects based on the $\bar{y}_i$ values in Table 4.2. The estimated factorial effects are displayed in Table 4.5. A half-normal plot in Figure 4.2 indicates that Q, B, C, CQ, E, and perhaps BQ are significant. Note that none of the six pairs of aliased effects are significant. The $B \times Q$ and $C \times Q$ interaction plots in Figure 4.3 show that B and Q and C and Q are synergistic.

To what extent can this conclusion based on an informal graphical method be confirmed by using the studentized maximum modulus test? Pooling all the s_i^2 in Table 4.2 gives the estimate $s^2 = 0.017$, from which we can compute the standard error of each factorial effect,

$$se(\hat{\theta}_i) = 2s/\sqrt{N} = 2(0.017)^{1/2}/\sqrt{48} = 0.0376.$$

With $I = 15$, $N = 48$, $m = 16$, $|M|_{15, 32, 0.05} = 3.14$ or with $I = 12$, if three-factor interactions are not tested, $|M|_{12, 32, 0.05} = 3.06$. At the 0.05 level, the test declares θ_i significant if

$$|\hat{\theta}_i| \geq se(\hat{\theta}_i)|M|_{I, 32, 0.05} = \begin{cases} 0.1181 & \text{for } I = 15, \\ 0.1151 & \text{for } I = 12. \end{cases} \quad (4.7)$$

Table 4.5 Factorial Effects, Leaf Spring Experiment

Effect	$\bar{y}$	$\ln s^2$
B	0.221	1.891
C	0.176	0.569
D	0.029	−0.247
E	0.104	0.216
Q	−0.260	0.280
BQ	0.085	−0.589
CQ	−0.165	0.598
DQ	0.054	1.111
EQ	0.027	0.129
BC	0.017	−0.002
BD	0.020	0.425
CD	−0.035	0.670
BCQ	0.010	−1.089
BDQ	−0.040	−0.432
BEQ	−0.047	0.854

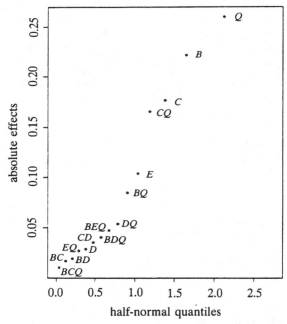

Figure 4.2. Half-Normal Plot of Location Effects, Leaf Spring Experiment.

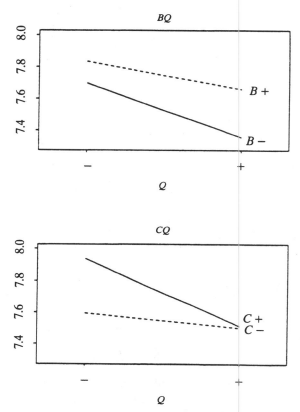

Figure 4.3. $B \times Q$ and $C \times Q$ Interaction Plots for Location, Leaf Spring Experiment.

By comparing the $\hat{\theta}_i$ values in Table 4.5 with (4.7), the test confirms the significance at $\alpha = 0.05$ of all the effects identified by the half-normal plot except BQ and E. The estimate $s^2 = 0.017$ may be conservative because the analysis of $\ln s_i^2$ that follows shows that the s_i^2 are not homogeneous. Consequently, the test as applied is conservative, which means that the actual level may be less than the nominal level 0.05. For purpose of illustration, the factor effect E is included in the following model in (4.8).

Next, we consider the $\ln s_i^2$ values given in Table 4.2. The estimated factorial effects are displayed in Table 4.5 with the corresponding half-normal plot appearing in Figure 4.4. The half-normal plot identifies B and possibly DQ, BCQ, BEQ as being significant. Their significance can be tested formally by using the maximum modulus test which was discussed in Section 3.16. Recall that the $\ln s_i^2$ are approximately $N(\ln \sigma_i^2, 2(n_i - 1)^{-1}) = N(\ln \sigma_i^2, 1)$, because $n_i = 3$. The factorial effects $\hat{\theta}_i$ have variance $4/N = \frac{4}{16} = \frac{1}{4}$. (Note that N equals 16, not 48, since there are only 16 $\ln s_i^2$ values.) Using the h_i notation in (3.48), $h_i^2 = \frac{1}{4}$. According to (3.51), the test declares

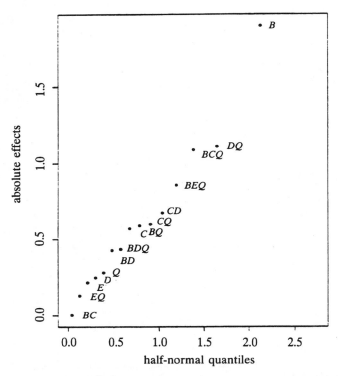

Figure 4.4. Half-Normal Plot of Dispersion Effects, Leaf Spring Experiment.

any $\hat{\theta}_i$ significant at level α if

$$\left| \hat{\theta}_i \right| > h_i |M|_{I,\infty,\alpha} = \tfrac{1}{2} |M|_{I,\infty,\alpha}.$$

At $\alpha = 0.05$,

$$\tfrac{1}{2} |M|_{15,\infty,0.05} = \tfrac{1}{2}(2.93) = 1.465.$$

By comparing the $\hat{\theta}_i$ values in Table 4.5 with 1.465, only B is significant. By increasing the α value to 0.1, $\tfrac{1}{2}|M|_{15,\infty,0.1} = \tfrac{1}{2}(2.70) = 1.35$. By comparing 1.35 with the $\hat{\theta}_i$ values in Table 4.5, no additional effects are declared significant. Even by restricting attention to the main effects and two-factor interactions so that $I = 12$ and $\tfrac{1}{2}|M|_{12,\infty,0.1} = \tfrac{1}{2}(2.62) = 1.31$, the conclusion remains unchanged. For illustrative purposes, we shall assume that B, DQ, and BCQ are significant for $\ln s^2$ and B, C, E, Q, BQ, and CQ for $\bar{y}$.

The regression equations for location and dispersion, respectively, are

$$\hat{y} = 7.6360 + 0.1106x_B + 0.0519x_E + 0.0881x_C - 0.1298x_Q$$

$$+ 0.0423x_B x_Q - 0.0827x_C x_Q \tag{4.8}$$

and

$$\hat{z} = \ln \hat{\sigma}^2$$
$$= -4.9313 + 0.9455x_B + 0.5556x_D x_Q - 0.5445x_B x_C x_Q, \qquad (4.9)$$

where $z = \ln s^2$. As explained in Section 3.6, the regression coefficients in (4.8) and (4.9) are half of the corresponding factorial effects in Table 4.5.

We can use the signs of the effects in (4.9) and the corresponding interaction plots to determine the factor settings that minimize dispersion. Starting with the largest effect B, based on its sign, the level of B should be $-$. Based on the $D \times Q$ interaction plot in Figure 4.5, D and Q should be set to $(+, -)$. Given that $(B, Q) = (-, -)$, $C = +$ should be chosen based on the negative sign of the coefficient for $x_B x_C x_Q$ in (4.9) or the $B \times C \times Q$ interaction plot in Figure 4.6. Therefore, we can choose $BCDQ = (-, +, +, -)$ to reduce the $\hat{\sigma}^2$ value. To complete the two-step procedure (see Section 3.2), we need to find an adjustment factor to move $\hat{y}$ closer to the target value of 8.0. Because factor E appears in (4.8) but not in (4.9),

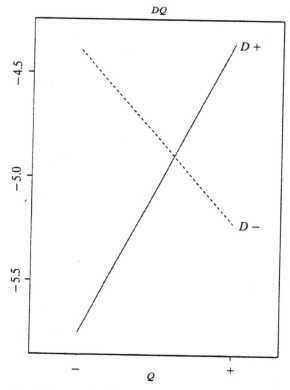

Figure 4.5. $D \times Q$ Interaction Plot for Dispersion, Leaf Spring Experiment.

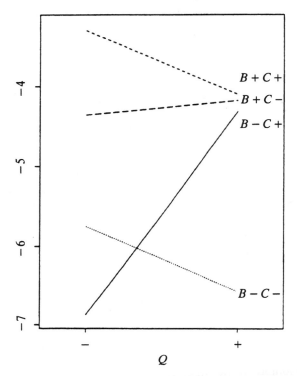

Figure 4.6. $B \times C \times Q$ Interaction Plot for Dispersion, Leaf Spring Experiment.

it can be used as an adjustment factor. To implement the second step of the two-step procedure, at $BCDQ = (-, +, +, -)$, the equation in (4.8) reduces to

$$\hat{y} = 7.6360 + 0.1106(-1) + 0.0519x_E + 0.0881(+1) - 0.1298(-1)$$
$$+ 0.0423(-1)(-1) - 0.0827(+1)(-1)$$
$$= 7.8683 + 0.0519x_E. \tag{4.10}$$

By solving $\hat{y} = 8.0$, x_E should be set at

$$\frac{8.0 - 7.8683}{0.0519} = 2.54.$$

The uncoded value for factor E is $2.5 + 2.54(0.5) = 3.77$ seconds. Because this value is much larger than the high level of the factor, there are no experimental data around 3.77 seconds to validate the fitted model $\hat{y}$ in (4.10). In general, any recommendation involving settings far outside the experimental region should be treated with caution.

An alternative to the two-step procedure is to minimize the estimated mean-squared error

$$\widehat{\text{MSE}} = (\hat{y} - 8)^2 + \hat{\sigma}^2 = (\hat{y} - 8)^2 + \exp(\ln \hat{\sigma}^2)$$

in a single step. By substituting the fitted models for $\hat{y}$ and $\ln \hat{\sigma}^2$ in (4.8) and (4.9) into the $\widehat{\text{MSE}}$ formula, $\widehat{\text{MSE}}$ is a function of the factors. A complete search over the range of the factor levels will find optimal factor settings that minimize the $\widehat{\text{MSE}}$. The computational details are left as an exercise. This approach is particularly attractive when no adjustment factor can be identified.

4.4 TECHNIQUES FOR RESOLVING THE AMBIGUITIES IN ALIASED EFFECTS

In model (4.9) for $\ln \hat{\sigma}^2$, the $B \times C \times Q$ interaction is aliased with the $D \times E \times Q$ interaction. Because the experimental data cannot distinguish BCQ from DEQ, one must find a way to resolve this ambiguity if BCQ (or DEQ) is significant. There are two general approaches for resolving the ambiguity of aliased effects.

First, prior knowledge may suggest that some of the aliased effects, say BCQ, are not likely to be important. Then the observed effect significance should be attributed to the other effect, say DEQ. If a higher order effect is aliased with a lower order effect, the former may be dismissed based on the hierarchical ordering principle. Consider an experiment that studies a whole process or whole system where factors can be grouped by subprocesses or subsystems. Then factors within the same subprocess may be more likely to have interactions than factors between subprocesses. Take, for example, wood pulp manufacturing in which a chemical subprocess is followed by a mechanical subprocess. Interactions between factors in the chemical subprocess and factors in the mechanical subprocess may be hard to justify or interpret. Therefore, such interactions may be dismissed.

If this approach is not successful in resolving the ambiguity, one then needs to conduct *follow-up* experiments at different factor settings than the original experiment. This provides additional information that can be used to de-alias the aliased effects. Three methods for conducting follow-up experiments are considered in this section.

4.4.1 Method of Adding Orthogonal Runs

The first method, referred to as the *method of adding orthogonal runs*, chooses additional factor settings for which the aliased effects in the original experiment are orthogonal. Orthogonality provides maximum separation between the aliased effects. Because the orthogonality constraint does not

Table 4.6 Augmented Design Matrix and Model Matrix, Leaf Spring Experiment

Run	B	C	D	E	Q	Block	BCQ	DEQ	DQ
1	−	+	+	−	−	−	+	+	−
2	+	+	+	+	−	−	−	−	−
3	−	−	+	+	−	−	−	−	−
4	+	−	+	−	−	−	+	+	−
5	−	+	−	+	−	−	+	+	+
6	+	+	−	−	−	−	−	−	+
7	−	−	−	−	−	−	−	−	+
8	+	−	−	+	−	−	+	+	+
9	−	+	+	−	+	−	−	−	+
10	+	+	+	+	+	−	+	+	+
11	−	−	+	+	+	−	+	+	+
12	+	−	+	−	+	−	−	−	+
13	−	+	−	+	+	−	−	−	−
14	+	+	−	−	+	−	+	+	−
15	−	−	−	−	+	−	+	+	−
16	+	−	−	+	+	−	−	−	−
17	+	+	−	−	+	+	+	+	−
18	−	+	−	−	−	+	+	−	+
19	−	+	+	+	+	+	−	+	+
20	+	+	+	+	−	+	−	−	−

uniquely determine the factor settings in the follow-up experiment, it takes additional work, often on an *ad hoc* basis, to find them. The method will be illustrated with the model (4.9) for ln $\hat{\sigma}^2$, which has the effects B, DQ, and BCQ. From the original experiment, the column for BCQ is identical to the column for DEQ (i.e., they are aliased), as shown in Table 4.6. Suppose four additional runs can be taken; these are runs 17–20 in Table 4.6. What factor settings should be chosen for these four runs? In order to de-alias BCQ and DEQ, we should choose two orthogonal vectors $(+, +, −, −)$ and $(+, −, +, −)$ as the values of BCQ and DEQ, respectively, for runs 17–20. We should then determine the settings for the largest effect B. Because $(+, −, −, +)$ is orthogonal to the two previous vectors, it is chosen as the settings for B. Noting that no other contrast vector (i.e., a column consisting of half +'s and half −'s) is orthogonal to these three vectors, the choice of settings for DQ is arbitrary. Choosing $(−, +, +, −)$ for DQ uniquely determines the settings for E. Because $E = DQ \times DEQ$, we choose $E = (−, −, +, +)$. Furthermore, if $(+, −, +, −)$ is chosen for Q, the settings for C and D are uniquely determined. See the resulting augmented design matrix and model matrix in Table 4.6. Because the additional runs are performed at a different time, a potential block effect should be accounted for by including an explanatory variable whose levels appear in the column labeled "block." The level − represents the original experiment and the level + represents the follow-up experiment.

Because the 20×9 model matrix in Table 4.6 for the augmented design is no longer orthogonal, analysis for an augmented experiment should apply regression analysis to the given model matrix. Standard factorial effect estimates will not be valid and should not be used.

This method is rather *ad hoc* and requires some ingenuity on the part of the experimenter. It may not be generally applicable. A general method will be considered in the next subsection.

4.4.2 Optimal Design Approach for Follow-up Experiments

The method starts with a model consisting of the significant effects identified by the original experiment. Because some of these effects are aliased with other effects, all the aliased effects (except those judged insignificant or unimportant *a priori*) should be included in the model. The model should include a block effect term to account for the systematic difference between runs in the original experiment and the runs in the follow-up experiment *if* this difference is judged to be potentially important. Once a model is specified, the factor settings in the follow-up experiment can be chosen to optimize some design criterion. A design criterion provides a performance measure of the combined factor settings (in the original experiment and the follow-up experiment) under the given model. Therefore *the approach is driven by the best model identified by the original experiment as well as a particular optimal design criterion*. In this subsection, only the D and D_s criteria will be considered. The approach is equally applicable to other design criteria.

We now return to the leaf spring experiment to illustrate this approach. To simplify the presentation, consider adding only two more runs, 17 and 18. For each of the five factors, there are four choices for the factor settings of runs 17 and 18, $(+, +)$, $(-, -)$, $(+, -)$, and $(-, +)$. This gives altogether $4^5 = 1024$ choices. Each choice corresponds to an 18×6 matrix, whose 18 rows are the 18 runs and 6 columns represent the variables B, DQ, BCQ, DEQ, the block variable, and the grand mean (column of all $+$'s). Formally, the model can be written as

$$E(z) = \beta_0 + \beta_{bl} x_{bl} + \beta_B x_B + \beta_{DQ} x_D x_Q$$

$$+ \beta_{BCQ} x_B x_C x_Q + \beta_{DEQ} x_D x_E x_Q, \qquad (4.11)$$

where $z = \ln s^2$, β_0 is the grand mean, β_{bl} is the block effect, $x_B = -1, 1$ according to the level of factor B (with definitions for the other x's being similar), β_B is half the main effect of B, β_{DQ} is half the $D \times Q$ interaction, etc. The effect β_{bl} is used to measure the systematic difference between the first 16 runs (with $x_{bl} = -1$) and the additional two runs (with $x_{bl} = +1$) because the two sets (i.e., blocks) of runs are conducted at different times. The block effect takes into account this difference so that it is not inadvertently included in the error term. For each of the 1024 choices, its corresponding model matrix is denoted by $\mathbf{X}_d, d = 1, \ldots, 1024$. Then the design d^* is chosen so that the determinant of the $\mathbf{X}_d^T \mathbf{X}_d$ matrix is maximized, i.e.,

$$\max_d |\mathbf{X}_d^T \mathbf{X}_d| = |\mathbf{X}_{d^*}^T \mathbf{X}_{d^*}|. \tag{4.12}$$

In optimal design theory, (4.12) is called the *D*-**optimal criterion** and d^* is called a *D*-**optimal design**, where D stands for determinant.

The D criterion has a natural interpretation. In the general linear model $\mathbf{y} = \mathbf{X}\boldsymbol{\beta} + \boldsymbol{\epsilon}$ of (1.3), the least squares estimator $\hat{\boldsymbol{\beta}} = (\mathbf{X}^T\mathbf{X})^{-1}\mathbf{X}^T\mathbf{y}$ [see (1.10)] has the variance-covariance matrix $Var(\hat{\boldsymbol{\beta}}) = \sigma^2(\mathbf{X}^T\mathbf{X})^{-1}$. It is known that the determinant $|\mathbf{X}^T\mathbf{X}|$ is proportional to the reciprocal of the volume of the confidence ellipsoid for $\boldsymbol{\beta}$ so that maximizing the D criterion is equivalent to minimizing the volume of the confidence ellipsoid.

For the leaf spring experiment, there are several choices that attain the maximum D value of 4,194,304. One of them is $(B, C, D, E, Q) = (+ + + - +)$ and $(+ + - + -)$ for runs 17 and 18, respectively. Note that, because C, D, E, and Q influence the linear model (4.11) through DQ, BCQ, and DEQ, there are several factor settings with the same values of DQ, BCQ, and DEQ.

As an alternative, one may primarily be interested in the estimation of BCQ and DEQ because these two effects could not be estimated separately in the original experiment. The variance-covariance matrix for estimating them is proportional to the 2×2 submatrix of $(\mathbf{X}_d^T\mathbf{X}_d)^{-1}$ corresponding to these two effects. If $\mathbf{X}_d^T\mathbf{X}_d$ is partitioned as

$$\begin{bmatrix} \mathbf{X}_1^T\mathbf{X}_1 & \mathbf{X}_1^T\mathbf{X}_2 \\ \mathbf{X}_2^T\mathbf{X}_1 & \mathbf{X}_2^T\mathbf{X}_2 \end{bmatrix},$$

where

$$\mathbf{X}_d = [\mathbf{X}_1, \mathbf{X}_2],$$

with $\mathbf{X}_1$ corresponding to the first four variables in (4.11) and $\mathbf{X}_2$ to the last two variables BCQ and DEQ, then the lower right 2×2 submatrix of $(\mathbf{X}_d^T\mathbf{X}_d)^{-1}$ can be shown to be

$$\left(\mathbf{X}_2^T\mathbf{X}_2 - \mathbf{X}_2^T\mathbf{X}_1(\mathbf{X}_1^T\mathbf{X}_1)^{-1}\mathbf{X}_1^T\mathbf{X}_2 \right)^{-1}$$

and the optimal choice of factor settings is one that satisfies

$$\max_d \left| \mathbf{X}_2^T\mathbf{X}_2 - \mathbf{X}_2^T\mathbf{X}_1(\mathbf{X}_1^T\mathbf{X}_1)^{-1}\mathbf{X}_1^T\mathbf{X}_2 \right|. \tag{4.13}$$

In optimal design theory, (4.13) is called the D_s-**optimal criterion**, where s stands for "subset" of regression parameters. For the leaf spring experiment, there are several choices that attain the maximum D_s value of 128. One of them is $(B, C, D, E, Q) = (+ + + - +)$ and $(+ + - + -)$ for runs 17 and 18, respectively, which also maximizes the D criterion. The set of D_s-optimal factor settings is, however, not identical to the set of D-optimal settings.

Because the model matrix for the augmented design is generally non-orthogonal, analysis of the augmented experiment should apply regression analysis to the given model matrix.

Formally, the optimal design approach can be described as follows. Suppose there are N runs in the original experiment and n runs are to be added. The working model for design optimization consists of the grand mean, the block variable, and the variables identified to be significant from the original experiment (which should include all the aliased effects unless they have been ruled out *a priori*). Assume that this model has p terms. Let d be the factor settings in the n runs and $\mathbf{X}_d$ the $(N + n) \times p$ matrix whose (i, j)th entry represents the setting of variable j for run i. Partition $\mathbf{X}_d = [\mathbf{X}_1, \mathbf{X}_2]$ with $\mathbf{X}_2$ denoting the $(N + n) \times q$ submatrix of $\mathbf{X}_d$, whose q columns represent the q aliased effects. Choice of design amounts to selecting the bottom $n \times p$ submatrix of $\mathbf{X}_d$ according to the D criterion,

$$\max_{d \in \mathscr{D}} \left| \mathbf{X}_d^T \mathbf{X}_d \right|,$$

or the D_s criterion,

$$\max_{d \in \mathscr{D}} \left| \mathbf{X}_2^T \mathbf{X}_2 - \mathbf{X}_2^T \mathbf{X}_1 \left(\mathbf{X}_1^T \mathbf{X}_1 \right)^{-1} \mathbf{X}_1^T \mathbf{X}_2 \right|,$$

where $\mathscr{D}$ represents a set of candidate $n \times p$ submatrices. The D_s criterion can be justified as follows. Write $E(\mathbf{y}) = \mathbf{X}_1 \boldsymbol{\beta}_1 + \mathbf{X}_2 \boldsymbol{\beta}_2$. Then the least squares estimate $\hat{\boldsymbol{\beta}}_2$ has the variance-covariance matrix

$$\sigma^2 \left(\mathbf{X}_2^T \mathbf{X}_2 - \mathbf{X}_2^T \mathbf{X}_1 \left(\mathbf{X}_1^T \mathbf{X}_1 \right)^{-1} \mathbf{X}_1^T \mathbf{X}_2 \right)^{-1},$$

whose determinant is minimized by maximizing the D_s criterion.

Further considerations like orthogonality between certain variables or balance (i.e., equal number of $+$ and $-$ levels) for certain variables will reduce the number of candidate designs in $\mathscr{D}$. Orthogonality between the aliased effects is a particularly sensible constraint. Because the original experiment cannot distinguish between the aliased effects, one may insist that, at the new settings, these effects should be orthogonal to each other. This modification of the optimal design approach is very similar to the method of Section 4.4.1. Both require orthogonality between aliased effects. The only difference is in the choice of settings for the other effects in the model. The optimal design approach provides a systematic search over all the settings while the method of adding orthogonal runs is vague about the determination of settings not specified by the aliased effects (as can be seen from its illustrative example). The latter can be used for smaller problems which typically require only a limited manual search. A manual search may also encourage the investigator to be more involved in the deliberation on the selection of factor settings. It cannot be used for large problems because a cumbersome manual search will be required over many factor settings that satisfy the orthogonality constraint.

The optimal design approach is very flexible. It works for any model and shape of experimental region for the design variables. Because of the availability of fast optimal design algorithms in standard software [see, for example, the OPTEX program in SAS/QC (SAS, 1989) or ECHIP (1998)], the approach can be implemented for very large problems. Take the example just considered. If four additional runs are required, there are $16 (= 2^4)$ combinations of factor settings for each factor and $(16)^5 = 1,048,576$ combinations of factor settings for five factors. An exhaustive search over a million combinations is not advisable. An intelligent and efficient search can be done by using one of the optimal design algorithms. If the orthogonality constraint is imposed, it can substantially reduce the number of candidate designs, but there is still a limit to what can be done by a manual or exhaustive search. The same example also illustrates the advantage of the method of adding orthogonal runs for small problems like adding four runs. As can be seen from the previous illustration, it is very easy to implement the method manually.

The readers must, however, be aware that the performance of optimal designs often depends critically on the validity of the model that drives the search. It can be fruitfully applied in the planning of follow-up experiments because the model is based on the objective information obtained from the original experiment. Its use in determining the design for the original experiment is not recommended as a routine practice because it would require some subjective assumptions about the model chosen which cannot be validated in many cases.

For more information on optimal design theory including the various optimality criteria, see Atkinson and Donev (1992) and Pukelsheim (1993). A Bayesian approach to design a follow-up experiment can be found in Meyer, Steinberg, and Box (1996).

4.4.3   Fold-Over Technique for Follow-up Experiments

If the original experiment is based on a resolution III design and either (i) all the main effects or (ii) one main effect and all the two-factor interactions involving this main effect need to be de-aliased, the fold-over technique can be used to find a follow-up experiment that meets this purpose.

For illustration, suppose the original experiment is based on a 2_{III}^{7-4} design with generators,

$$d_1: \ 4 = 12, \ 5 = 13, \ 6 = 23, \ 7 = 123. \tag{4.14}$$

Its defining contrast subgroup has eight words of odd length and seven words of even length:

$$I = 124 = 135 = 167 = 236 = 257 = 347 = 456 = 1234567 \tag{4.15}$$

$$= 1237 = 1256 = 1346 = 1457 = 2345 = 2467 = 3567. \tag{4.16}$$

From (4.15) and (4.16), it is easily seen that none of the main effects or two-factor interactions in d_1 are clear. To de-alias the seven main effects, consider adding another eight runs by switching the signs in all the seven columns of d_1. The augmented design of 16 runs is displayed in Table 4.7. The second design (which is used for the follow-up experiment) is obtained from the first design by "folding over" the first design and is known as **fold-over** design (Box and Hunter, 1961). Let the seven columns in d_1 be denoted by **1, 2, 3, 4, 5, 6,** and **7**. Then the seven columns in the second design d_2 are denoted by $-1, -2, -3, -4, -5, -6,$ and -7. It can then be

Table 4.7 Augmented Design Matrix Using Fold-Over Technique

Run	1	2	3	4 = 12	5 = 13	6 = 23	7 = 123	8
				d_1				
1	−	−	−	+	+	+	−	+
2	−	−	+	+	−	−	+	+
3	−	+	−	−	+	−	+	+
4	−	+	+	−	−	+	−	+
5	+	−	−	−	−	+	+	+
6	+	−	+	−	+	−	−	+
7	+	+	−	+	−	−	−	+
8	+	+	+	+	+	+	+	+

Run	−1	−2	−3	−4	−5	−6	−7	−8
				d_2				
9	+	+	+	−	−	−	+	−
10	+	+	−	−	+	+	−	−
11	+	−	+	+	−	+	−	−
12	+	−	−	+	+	−	+	−
13	−	+	+	+	+	−	−	−
14	−	+	−	+	−	+	+	−
15	−	−	+	−	+	+	+	−
16	−	−	−	−	−	−	−	−

shown that d_2 has generators:

$$d_2: 4 = -12, 5 = -13, 6 = -23, 7 = 123, \qquad (4.17)$$

and its defining contrast subgroup has eight words of odd length, which can be obtained by changing the sign of each word in (4.15), and seven words of even length, which are the same as in (4.16). Its proof is left as an exercise. Denote by d the augmented design consisting of d_1 and d_2. Because all the defining words of odd length have both $+$ and $-$ signs in the augmented design, they do not impose any constraint on the factors and therefore do not appear in the defining contrast subgroup of d. The defining words of even length in (4.16) are the same for both d_1 and d_2 and therefore comprise the defining contrast subgroup of d. Because these seven words all have length 4, d is a 2_{IV}^{7-3} design.

What has been achieved by adding the second design? Because the augmented design has resolution IV, all the seven main effects are clear. Out of the eight degrees of freedom in d_2, seven are used to de-alias the seven main effects (i.e., making them clear of other main effects and two-factor interactions). The eighth degree of freedom can be used to measure the difference between the first eight runs and the last eight runs, which is given in the last column of Table 4.7. It can be used either as a block effect as previously explained in the section or as the main effect of a new factor. In the latter case, the augmented design is a 2_{IV}^{8-4} design with the defining contrast subgroup:

$$\mathbf{I = 1248 = 1358 = 1678 = 2368 = 2578 = 3478 = 4568 = 12345678}$$
$$\mathbf{= 1237 = 1256 = 1346 = 1457 = 2345 = 2467 = 3567,}$$

where the words in the first line are obtained from those in (4.15) by adding the new factor 8.

Another version of the fold-over technique is useful for de-aliasing a main effect and all the two-factor interactions involving the main effect. Using the same d_1 in (4.14) as the first design, obtain the second design, denoted by d_3, by changing the sign of only one column, say column 5, in d_1. Thus,

$$d_3: 4 = 12, 5 = -13, 6 = 23, 7 = 123. \qquad (4.18)$$

Because both $\mathbf{5 = 13}$ and $\mathbf{5 = -13}$ appear in the augmented design, each of the eight combinations of factors 1, 3, and 5 appears in the augmented design, i.e., there is no constraint on the factors 1, 3 and 5. Therefore, the augmented design d', which consists of d_1 and d_3, is a 2_{III}^{7-3} design with three generators:

$$d': 4 = 12, 6 = 23, 7 = 123. \qquad (4.19)$$

Because $\mathbf{5}$ does not appear in any of the defining words for d', it is easy to verify that the main effect $\mathbf{5}$ is strongly clear and all the six two-factor interactions involving $\mathbf{5}$ are clear. Out of the eight degrees of freedom in d_3, seven are used to de-alias the main effect 5 and the six two-factor interactions involving 5. As in the previous case, the eighth degree of freedom can be used for a block effect or a new factor. Noting that the three defining words for d' are the same as in d_1, the estimability of the other main effects and two-factor interactions remains the same and is not helped by the addition of d_3.

If m, the number of factors in the original experiment, is less than 7, the same fold-over technique can be applied to m columns in Table 4.7. Similarly, both versions of the fold-over technique can be applied to other resolution III designs with more than eight runs (e.g., 16, 32 runs, etc.) and the estimability property of the augmented design is qualitatively the same as that discussed before. Details will not be repeated here. One example with 16 runs will be given in an exercise. Further discussion of design augmentation can be found in Daniel (1962, 1976).

Unlike the two previous methods, the augmented design after fold-over is still orthogonal. Its analysis should follow standard methods for fractional factorial designs.

How useful is fold-over as a technique for choosing follow-up experiments? As shown in the previous discussion, it is effective for de-aliasing (i) all the main effects or (ii) one main effect and all the two-factor interactions involving this main effect in a resolution III design for the original experiment. These two objectives are quite narrow because results from analyzing the original experiment may suggest a very different set of effects to be de-aliased. One such example is given by the leaf spring experiment previously analyzed. The fold-over technique also requires that the follow-up experiment have the *same* run size as the original experiment. This can be quite wasteful if the number of effects to be de-aliased is much smaller than the run size. This point is again illustrated nicely by the leaf spring experiment. By comparison, the optimal design approach is more flexible and economical. It can handle the de-aliasing of any set of effects identified from analyzing the original experiment. The required run size for the follow-up experiment depends on the number of effects to be de-aliased and can therefore be much smaller than the run size of the original experiment. Because the number of effects to be de-aliased does not grow in proportion to the size of the original experiment, the disadvantage of the fold-over approach relative to the optimal design approach becomes even more pronounced when the original experiment has 16 or more runs.

4.5 SELECTION OF 2^{k-p} DESIGNS USING MINIMUM ABERRATION AND RELATED CRITERIA

An important question that arises in fractional factorial experiments is how to choose a good 2^{k-p} design and what criteria should be used to judge "goodness." The most commonly used criterion is the minimum aberration criterion introduced in Section 4.2. There are, however, situations in which minimum aberration designs are not the best according to other sensible criteria. To illustrate this point, consider the two 2_{IV}^{9-4} designs given in Table 4A.3. Both are of resolution IV. The first design, denoted by d_1, has minimum aberration and has the following six defining words of length 4:

$$I = 1236 = 1247 = 1258 = 3467 = 3568 = 4578. \qquad (4.20)$$

Because **9** does not appear in any of the four-letter words in (4.20), it is easily shown that the eight two-factor interactions involving factor 9 are clear. The second design, denoted by d_2, has the following seven defining words of length 4:

$$I = 1236 = 1247 = 1348 = 3467 = 2468 = 2378 = 1678. \qquad (4.21)$$

By observing that neither **5** nor **9** appears in (4.21), it is easily shown that the 15 two-factor interactions involving factors 5 or 9 are clear. So, even though $A_4(d_1) = 6 < A_4(d_2) = 7$, d_2 has 15 clear two-factor interactions while d_1 has only 8. Because the concept of clear estimation of effects has a more direct and interpretable meaning than the length of defining words, from an operational point of view, one would judge d_2 to be superior to d_1. This comparison of resolution IV designs in terms of the number of clear two-factor interactions was stated as Rule (iii) in (4.5) of Section 4.2. By examining the designs in Tables 4A.1–4A.7, we find five other minimum aberration resolution IV designs 2^{13-7}, 2^{14-8}, 2^{15-9}, 2^{16-10}, 2^{17-11} (each having 64 runs) which are outperformed by non-minimum aberration designs according to this rule. The latter designs are given right after the minimum aberration designs in the tables.

The **number of clear effects** was proposed by Chen, Sun, and Wu (1993) as an additional criterion for design selection. To further exemplify its use, let us consider the two 2^{6-2} designs in Table 4A.2. The first design d_1 with

$$I = 1235 = 1246 = 3456$$

has resolution IV and six clear main effects but no clear two-factor interactions. The second design d_2 with

$$I = 125 = 1346 = 23456$$

has resolution III and only three clear main effects **3, 4, 6** but six clear two-factor interactions **23, 24, 26, 35, 45, 56**. A detailed account of effect aliasing and clear effects for this design was given in Section 4.2. Because d_2 has only three clear main effects and main effects are usually more important than two-factor interactions, one may argue that it is inferior to d_1. On the other hand, d_2 has altogether nine clear effects while d_1 has only six. If in an investigation, only three factors (out of six) and some of their two-factor interactions are believed to be important *a priori*, design d_2 may be a preferred choice. Other examples can be found in the exercises.

If it is important to estimate clearly one particular factor main effect and all the two-factor interactions involving this factor, one can choose a 2^{k-p} design whose defining contrast subgroup does not involve this particular factor. One such example is the 2_{III}^{7-3} design d' given in (4.19), where all the factors except **5** appear in its defining contrast subgroup. It was shown there that main effect **5** is strongly clear and all the two-factor interactions involving **5** are clear. The best way to construct a 2^{k-p} design whose defining contrast subgroup only involves $k-1$ factors is to choose a minimum aberration $2^{(k-1)-p}$ design for $k-1$ factors and then use its defining contrast subgroup for a 2^{k-p} design. The design matrix is obtained by writing down two copies of $2^{(k-1)-p}$ design. For the first copy, the kth factor is assigned level $-$; for the second copy, the kth factor is assigned level $+$. Because one factor does not appear in the defining contrast subgroup of the constructed 2^{k-p} design, the main effect for this factor is strongly clear and all the two-factor interactions involving this factor are clear. A more general version of this result and its proof are given in an exercise.

In Tables 4A.1–4A.7, we give selected 2^{k-p} designs for 8–128 runs. For 8, 16, and 32 runs, the collection of designs covers the whole range of factors. For 64 runs, the table covers $k = 7, \ldots, 32$, and for 128 runs, the table covers $k = 8, \ldots, 13$. For each k and p, if there is only one design in the table, it is the minimum aberration design. If there are two designs, the first has minimum aberration and the second, though slightly worse according to the aberration criterion, has more clear two-factor interactions than the first. If both designs have resolution IV, the second design is recommended based on Rule (iii) in (4.5). If at least one of the two designs has resolution III, the choice depends on which main effects and/or two-factor interactions are judged *a priori* to be important in a given situation. The designs in these tables are based on the computer search results of Chen, Sun, and Wu (1993) or the theoretical results of Chen (1992, 1998), Chen and Wu (1991), and Tang and Wu (1996).

In order to save space in the tables of Appendix 4A, only the generators of the designs are given. The other defining words of the designs can be obtained by multiplying the generators. Several examples were given in Section 4.2 to show how the complete set of defining words can be obtained from the generators. In the last column of the tables, the collection of clear main effects and two-factor interactions is given. If none or all of the main effects and two-factor interactions are clear for all the designs in a given table, this column will be omitted and this fact will be stated in boldface at the beginning of the table.

Sometimes a graphical representation of the set of clear two-factor interactions can be helpful in deciding how to assign experimental factors to columns. In this graph, which is referred to as an **interaction graph**, a node represents a factor and a line represents the interaction between the two factors on its two nodes. Consider again the 2_{III}^{6-2} design in (4.3) with **I = 125 = 1346 = 23456**. Its three clear main effects **3, 4, 6** are represented by

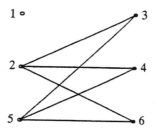

Figure 4.7. Interaction Graph for the 2_{III}^{6-2} Design with $5 = 12$, $6 = 134$.

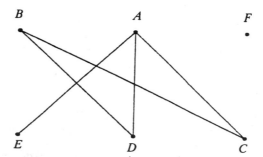

Figure 4.8. Graph That Represents Five Designated Interactions.

solid circles, the other main effects by hollow circles, and the six clear two-factor interactions by lines as displayed in Figure 4.7.

Suppose that in an investigation involving six factors A, B, C, D, E, F, it is believed that the interactions AC, AD, AE, BC, BD are potentially important and should therefore be estimated clearly of the main effects and the other two-factor interactions. These five interactions can be represented by the graph in Figure 4.8. By comparing the two graphs, one can easily determine that factors A, B, C, D, E should be assigned to columns 2, 5, 3, 4, 6, respectively. A general discussion on the use of interaction graphs for factor assignment (to columns) can be found in Kacker and Tsui (1990) and Wu and Chen (1992). Two examples of interaction graphs are given in the exercises.

In the tables of Appendix 4A, for each k and p, only one fraction of the full 2^k design is given. All 2^p fractions can be easily generated from the one given in the tables. Consider the 2_{III}^{7-4} design with $4 = 12$, $5 = 13$, $6 = 23$, and $7 = 123$. It is a sixteenth fraction of the 2^7 design. All 16 fractions can be obtained by using $+$ or $-$ signs in the four generators $4 = \pm 12$, $5 = \pm 13$, $6 = \pm 23$, $7 = \pm 123$. By randomly choosing the signs, a *randomized* choice of the fraction is obtained. This randomization may protect the experiment from the influence of lurking variables, as discussed in Sections 1.3 and 3.3. In some practical situations, the investigator may be able to identify certain combinations of factor levels as being undesirable. For example, high temperature and a long time in the furnace may lead to overheating and result in

Table 4.8   Eight-Run Design

1	2	3	12	13	23	123
−	−	−	+	+	+	−
−	−	+	+	−	−	+
−	+	−	−	+	−	+
−	+	+	−	−	+	−
+	−	−	−	−	+	+
+	−	+	−	+	−	−
+	+	−	+	−	−	−
+	+	+	+	+	+	+

burned-out units. Then a fraction should be chosen that avoids these combinations. For example, if the combination $4 = +$, $5 = +$, and $6 = +$ should be avoided, then the fraction with $4 = 12$, $5 = 13$, and $6 = 23$ should not be chosen because it contains this combination. Any other fraction avoids this combination. If the number of undesirable combinations is small, it is easy to find a fraction that avoids these combinations by trial and error or by computer search. A more general theoretical treatment of this subject can be found in Li and Cheng (1993).

Because different sets of generators may give the same wordlength pattern, the defining generators given in the tables in Appendix 4A are not necessarily unique. To illustrate how these tables can be used to assign factors to columns, consider the eight-run designs. The seven degrees of freedom in an eight-run design can be represented by the seven columns in Table 4.8 denoted by $1, 2, 12, 3, 13, 23, 123$, where 12 is the column obtained by multiplying columns 1 and 2 entrywise; the other columns can be obtained similarly. For the 2_{IV}^{4-1} designs in Table 4A.1, the defining relation is $4 = 123$ (or $4 = -123$). Therefore, factor 4 is assigned to column 7 (or minus column 7), because column 7 is coded 123. Similarly, for the 2_{III}^{5-2} design with $4 = 12$ and $5 = 13$, factor 4 is assigned to column 4 and factor 5 to column 5; it turns out that factors 4 and 5 can be assigned to any two columns from 4 to 7 and still have the same wordlength pattern. For the 2_{III}^{6-3} design with $4 = 12$, $5 = 13$, and $6 = 23$, factors 4, 5, and 6 are assigned to columns 4, 5, 6 but any other assignment will also give the same wordlength pattern. For the saturated 2_{III}^{7-4} designs, the seven factors take up all seven columns.

4.6   BLOCKING IN FRACTIONAL FACTORIAL DESIGNS

The choice of optimal blocking schemes for 2^k designs was considered in Section 3.12. For 2^{k-p} fractional factorial designs, this problem is more complicated due to the presence of two defining contrast subgroups, one for defining the blocking scheme and the other for defining the fraction. The former is called the **block defining contrast subgroup**, as in Section 3.12. To distinguish the two defining contrast subgroups, the latter is referred to as the **treatment defining contrast subgroup** in the context of blocking 2^{k-p} designs. Without blocking the term "treatment" is omitted as we have done in this chapter. We now use an example to illustrate how these two groups act jointly in determining the confounding of effects. Consider the 16-run 2^{6-2} design in four blocks in Table 4B.2, with the treatment defining contrast subgroup

$$\mathbf{I} = \mathbf{1235} = \mathbf{1246} = \mathbf{3456} \tag{4.22}$$

and the block defining contrast subgroup

$$\mathbf{B_1} = \mathbf{134}, \mathbf{B_2} = \mathbf{234}, \text{ and } \mathbf{B_1 B_2} = \mathbf{12} .$$

By multiplying each defining word in (4.22) by $\mathbf{B_1} = \mathbf{134}$, we have

$$\mathbf{B_1} = \mathbf{134} = \mathbf{245} = \mathbf{236} = \mathbf{156},$$

where $\mathbf{245} = \mathbf{134} \times \mathbf{1235}$, $\mathbf{236} = \mathbf{134} \times \mathbf{1246}$, and $\mathbf{156} = \mathbf{134} \times \mathbf{3456}$. If one only looks at the block defining generators, it might be concluded that the first block effect $\mathbf{B_1}$ is only confounded with $\mathbf{134}$. Because of the treatment defining relation in (4.22), $\mathbf{134}$ is also aliased with the effects $\mathbf{245}$, $\mathbf{236}$, $\mathbf{156}$. Therefore, $\mathbf{B_1}$ is confounded with four effects:

$$\mathbf{B_1} = \mathbf{134} = \mathbf{245} = \mathbf{236} = \mathbf{156}.$$

Similarly, it can be shown that

$$\mathbf{B_2} = \mathbf{234} = \mathbf{145} = \mathbf{136} = \mathbf{256}$$

and

$$\mathbf{B_1 B_2} = \mathbf{12} = \mathbf{35} = \mathbf{46} = \mathbf{123456} .$$

Among the remaining 12 ($= 15 - 3$) degrees of freedom that do not involve the block effects, each of the six main effects is aliased with three-factor or higher interactions, e.g., $\mathbf{1} = \mathbf{235} = \mathbf{246} = \mathbf{13456}$. Thus, the six main effects are clear. Among the two-factor interactions, $\mathbf{12}, \mathbf{35}$, and $\mathbf{46}$ are confounded with the block effect $\mathbf{B_1 B_2}$ and are not eligible for estimation. The remaining 12 two-factor interactions are aliased in six pairs, i.e.,

$$\mathbf{13} = \mathbf{25} = \mathbf{2346} = \mathbf{1456},$$
$$\mathbf{14} = \mathbf{26} = \mathbf{2345} = \mathbf{1356},$$
$$\mathbf{15} = \mathbf{23} = \mathbf{2456} = \mathbf{1346},$$
$$\mathbf{16} = \mathbf{24} = \mathbf{2356} = \mathbf{1345},$$
$$\mathbf{34} = \mathbf{56} = \mathbf{1245} = \mathbf{1236},$$
$$\mathbf{36} = \mathbf{45} = \mathbf{1256} = \mathbf{1234}.$$

In summary, among the 15 degrees of freedom for the blocked design, 3 are allocated for block effects and 6 are for clear main effects. The remaining 6 degrees of freedom are six pairs of aliased two-factor interactions, where a two-factor interaction is estimable only if its alias is negligible.

In the context of blocking, a main effect or two-factor interaction is said to be **clear** if it is not aliased with any other main effects or two-factor interactions as well as not confounded with any block effects. The definition of **strongly clear** effect has the additional requirement that it is not aliased with any three-factor interactions. Because of the presence of two defining contrast subgroups, there is no natural extension of the minimum aberration criterion for blocked 2^{k-p} designs. We use the *total number of clear effects* to compare and rank order different blocked 2^{k-p} designs. Good designs according to this criterion and other considerations given in Sun, Wu, and Chen (1997) are listed in Tables 4B.1–4B.5 for 8–128 runs. A more complete collection of blocked designs can be found in Sun, Wu, and Chen (1997) and Sitter, Chen, and Feder (1997). In the "design generators" column of these tables, the defining generators for the treatment defining contrasts are given. In the "block generators" column, the defining generators for blocking are given. A list of clear effects is given in the last column.

The total number of clear effects should not be used as the sole criterion for selecting designs. For example, consider the two 2^{6-2} designs in four blocks in Table 4B.2. The first was shown previously to have six clear main effects. The second design is defined by

$$I = 125 = 1346 = 23456 \quad \text{and} \quad B_1 = 13, \ B_2 = 14, \ B_1 B_2 = 34.$$

It is left as an exercise to show that this design has three clear main effects **3, 4, 6** and six clear two-factor interactions **23, 24, 26, 35, 45, 56**. Although it has nine clear effects, three more than in the first design, three of its main effects are not clear. Thus, the choice between the designs depends on what set of main effects and two-factor interactions is believed to be more important. A similar conclusion was made in Section 4.5 for these two designs without blocking.

Finally we would like to show how the information given in the tables in Appendix 4B can be used for assigning factors to columns. Consider, for example, the 2^{4-1} design in Table 4B.1 with $I = 1234$ and $B_1 = 12$, $B_2 = 13$. Without the fourth factor defined as $4 = 123$, this design would be the same as the one displayed in Table 3.11. Thus, the only difference in the present case is to assign the column **123** in Table 3.11 to factor **4**. Other issues like graph representation, random choice of fraction, and avoidance of undesirable combinations can be dealt with as discussed in Section 4.5.

4.7 PRACTICAL SUMMARY

1. The main practical motivation for choosing a fractional factorial design is run size economy. Aliasing of effects is the price paid for economy.

2. A 2^{k-p} design is determined by its defining contrast subgroup, which consists of $2^p - 1$ defining words and the identity **I**. The length of its shortest word is the resolution of the design.

3. A main effect or two-factor interaction is said to be clear if none of its aliases are main effects or two-factor interactions and strongly clear if none of its aliases are main effects or two-factor or three-factor interactions.

4. Useful rules for resolution IV or V designs are:
   (i) In any resolution IV design, the main effects are clear.
   (ii) In any resolution V design, the main effects are strongly clear and the two-factor interactions are clear.
   (iii) Among the resolution IV designs with given k and p, those with the largest number of clear two-factor interactions are the best.

5. Projective property of resolution: The projection of a resolution R design onto any $R - 1$ factors is a full factorial in the $R - 1$ factors.

6. Unless there is specific information on the relative importance among the factorial effects, the minimum aberration criterion should be used for selecting good 2^{k-p} designs. Minimum aberration automatically implies maximum resolution. Minimum aberration 2^{k-p} designs are given in Appendix 4A. The minimum aberration criterion is sometimes supplemented by the criterion of maximizing the total number of clear main effects and two-factor interactions. Appendix 4A also includes designs that are better than the minimum aberration designs according to the latter criterion.

7. In a blocked 2^{k-p} design, there are the treatment defining contrast subgroup for defining the fraction and the block defining contrast subgroup for defining the blocking scheme. A main effect or two-factor interaction is said to be clear if it is also not aliased with any other main effects or two-factor interactions as well as not confounded with any block effects and strongly clear if it is also not aliased with any three-factor interactions. A blocked 2^{k-p} design with more clear effects is considered to be better. Optimal blocked designs according to this and related criteria are given in Appendix 4B.

8. The analysis of fractional factorial experiments is the same as in the full factorial experiments except that the observed significance of an effect should be attributed to the combination of the effect and all its aliased effects.

9. Optimal factor settings can be determined by the signs of significant effects and the corresponding interaction plots. A good illustration is given by the analysis of the leaf spring experiment in Section 4.3.

10. An alternative to the two-step procedure for nominal-the-best problems is to minimize the estimated mean-squared error

$$\widehat{MSE} = (\hat{y} - \text{target})^2 + \hat{\sigma}^2 = (\hat{y} - \text{target})^2 + \exp(\ln \hat{\sigma}^2).$$

By substituting the fitted models for $\hat{y}$ and $\ln \hat{\sigma}^2$ into the $\overline{MSE}$ formula, factor settings that minimize the $\overline{MSE}$ can be determined.

11. Two approaches to resolve ambiguities in aliased effects are:
    (i) Use *a priori* knowledge to dismiss some of the aliased effects. If a lower order effect is aliased with a higher order effect, the latter may be dismissed.
    (ii) Run a follow-up experiment.

12. Three methods for choosing a follow-up experiment are:
    (i) Choose the additional settings to make the aliased effects orthogonal. The settings for the remaining effects are chosen on an *ad hoc* basis. It is useful when a small number of effects needs to be de-aliased.
    (ii) Optimal design approach. Apply an optimal design criterion like the D or D_s criterion to a model consisting of the effects (including all its aliases) identified as significant in the original experiment. It works for any model and shape of experimental region. By using fast optimal design algorithms, it can be used to solve large design problems.
    (iii) Fold-over technique. Choose the second design by switching the signs of some or all columns in the first design. Its objectives are more narrow and the method is less flexible than the optimal design approach.

EXERCISES

1. Prove that, for a resolution R design, its projection onto any $R-1$ factors is a full factorial. (*Hint*: If the projection is not a full factorial, it must be a fractional factorial with some defining relation among the $R-1$ factors.)

2. Because the BCQ and DEQ in the leaf spring experiment are aliased, the two-step procedure in Section 4.3 has another version by substituting $x_B x_C x_Q$ in (4.9) by $x_D x_E x_Q$. Repeat the analysis in Section 4.3 using the following steps:
    (a) Produce the $D \times E \times Q$ interaction plot. Determine the level of E to reduce dispersion.
    (b) Perform step (i) of the two-step procedure.
    (c) Identify the adjustment factor.
    (d) Perform step (ii) of the two-step procedure.
    (e) Compare the results here with those in Section 4.3 in terms of the $\hat{\sigma}^2$ value, $\hat{y}$ value, and degree of extrapolation of the adjustment factor to move $\hat{y}$ on target.
    (f) If on physical grounds it is known that a heat time between 21 and 27 seconds has a significant effect on free height while the effect of

hold-down time on free height is based only on the model for $\hat{y}$, which version of the two-step procedure would be preferred?

3. (a) Choose factor settings over $-1 \leq x_B, x_C, x_D, x_E, x_Q \leq 1$ to minimize $\widehat{MSE} = (\hat{y} - 8)^2 + \hat{\sigma}^2$ based on models (4.8) and (4.9).

   (b) For the setting chosen in (a), compute its $\hat{y}$, $\hat{\sigma}^2$, and $\widehat{MSE}$ values. Compare them with those obtained in Section 4.3 and in Exercise 2. If you want to further reduce these values, do you need to relax the range of the factor levels? Which factor would you consider first?

   (c) Repeat the work in (a) by relaxing the range of one factor you chose in (b). Comment on the improvement and the need for extrapolation.

4. Repeat Exercise 3 by substituting the $x_B x_C x_Q$ term in (4.9) by $x_D x_E x_Q$. Apart from the difference in the choice of the adjustment factor (which is reflected in the choice of the factor to relax its range of levels), do you see much difference in the results here and those reported in Exercise 3?

5. Pretending that the pooled variance estimate $s^2 = 0.017$ is not available in the $\bar{y}$ analysis of the leaf spring experiment, use the IER or EER version of Lenth's method to test significance of the effects. Compare your results with those in Section 4.3.

6. A textile scientist has used a 2^4 design, with half the points run once and half the points run twice. Those points in the 2^{4-1} fraction with defining relation $\mathbf{I} = \mathbf{1234}$ were run twice and those points in the 2^{4-1} fraction with defining relation $\mathbf{I} = -\mathbf{1234}$ were run once. The order in which these tests were performed was determined by randomization. The scientist asks you to obtain estimates of all the main effects and interactions with their associated standard errors. How would you do this?

7. An experimenter obtained eight yields for the design given in Table 4.9.

Table 4.9   Yield Experiment

1	2	3	4	5	Yield
−	−	−	−	−	26
+	+	−	−	−	28
−	+	+	−	+	23
+	−	+	−	+	22
−	+	+	+	−	19
+	−	+	+	−	18
−	−	−	+	+	30
+	+	−	+	+	28

(a) Make two interaction plots for factors 3 and 4, one for 3 against 4, and the other for 4 against 3.

(b) Calculate the main effect of 3 and of 4 and the 3×4 interaction.

(c) If it is known that the standard deviation of each observation (i.e., each run) is 1.5, what is the standard error of the estimates in (b)?

(d) Based on the results in (a)–(c), which of the three effects in (b) are significant? (You can use the effect estimate ± 2 standard error as the 95% confidence interval to answer this question.) Comment on synergistic and antagonistic interactions.

8. For the follow-up experiment discussed in Section 4.4, among the 1024 candidate settings for two additional runs, choose the best settings according to the D and D_s criteria. If several settings attain the optimal value, list all of them.

9. For the leaf spring experiment, because the block effect is not of interest to the experimenter, it may be dropped in the D criterion. This will give rise to a new D_s criterion consisting of the effects B, DQ, BCQ, and DEQ in model (4.11). Among the 1024 candidate settings for two additional runs, choose the best settings according to the new D_s criterion. If several settings attain the optimal value, list all of them. Compare the settings with those found in Exercise 8.

10. For the leaf spring experiment, suppose that four runs are to be added in a follow-up experiment.

(a) Use existing optimal design software to find four additional settings to optimize the D and D_s criteria.

(b) Do any of the settings found in (a) satisfy the orthogonality constraint between the aliased effects BCQ and DEQ?

(c) Among the settings that satisfy the orthogonality constraint between BCQ and DEQ, how well do they perform in terms of the D criterion and, particularly, the D_s criterion?

11. (a) Show that d_2 defined in Section 4.4.3 has the generators $4 = -12$, $5 = -13$, $6 = -23$, $7 = 123$. (*Hint*: Rewrite the seven columns -1, -2, -3, -4, -5, -6, -7 as $1'$, $2'$, $3'$, $4'$, $5'$, $6'$, $7'$. Show that $4' = -1'2'$, etc.)

(b) Find the defining contrast subgroup of d_2 and compare it with that of d_1.

12. A 16-run experiment employs the minimum aberration 2_{III}^{9-5} design with $5 = 123$, $6 = 124$, $7 = 134$, $8 = 234$ and $9 = 1234$. Denote the design by d.

(a) Find the 31 defining words in its defining contrast subgroup. Which among them have length 3 and what do they have in common? Show that no effect is clear.

(b) The fold-over technique can be applied as follows. Suppose that all the main effects are to be de-aliased. Obtain the second design by switching all the signs in the nine columns for the nine factors in d. Find its defining generators and the 31 defining words in its defining contrast subgroup.

(c) By comparing the defining contrast subgroups for the two designs, obtain the defining contrast subgroup for the augmented 2^{9-4} design. Show that the augmented design has resolution IV. Identify all its clear effects.

(d) Suppose that the main effect **9** and all the two-factor interactions involving **9** are to be de-aliased. Obtain the second design by switching the signs in the column for factor 9 in d. Obtain the defining contrast subgroup for the augmented 2^{9-4} design. What is its resolution? Identify all its clear and strongly clear effects.

(e) In (b) and (d) above, the fold-over technique requires 16 additional runs while only nine effects are to be de-aliased. Describe how the optimal design approach can be used to de-alias the given set of effects with a smaller second design.

13. (a) Design an eight-run experiment to study the effect of the following five factors on yield: temperature (160 or 180°F), concentration (30 or 40%), catalyst (A or B), stirring rate (60 or 100 rpm), pH (low or high) with the prior knowledge that the combinations (180°F, 40%, 100 rpm) of (temperature, concentration, stirring rate) and (180°F, B, high) of (temperature, catalyst, pH) may lead to disastrous results and should be avoided. Write down the factor levels for each of the eight runs in a planning matrix.

(b) For the five factors in (a), find an eight-run design such that the catalyst-by-temperature interaction and the catalyst-by-concentration interaction are neither aliased with the main effects nor with each other.

(c) Prove that it is impossible to find a 2^{5-2} design such that the catalyst-by-temperature interaction and the pH-by-concentration interaction are neither aliased with the main effects nor with each other. (*Hint*: Use a graph representation of two-factor interactions that are not aliased with the main effects.)

14. A 2^{5-2} design is defined by **4 = 13** and **5 = 23**.
    (a) Find its defining words and resolution.
    (b) After the analysis, variable 5 turns out to be inert. It is assumed that all two-factor interactions involving variable 5 and all higher order interactions are negligible. In addition to estimating the four main

effects, there are still three degrees of freedom left. What two-factor interactions can be estimated with these three degrees of freedom?

15. (a) What is the resolution of each of the fractional factorial designs indicated below? Which design do you prefer? Justify your answers.

$$\text{(i)} \ 2^{6-2} \text{ with } 5 = 1234, \ 6 = 124,$$
$$\text{(ii)} \ 2^{6-2} \text{ with } 5 = 123, \ \ 6 = 124.$$

(b) For the design in (ii), if we further know that any two-factor interaction involving factor 6 (i.e., $16, 26, 36, 46, 56$) is negligible, which two-factor interactions are estimable under the usual assumptions that three-factor and higher interactions are negligible?

(c) Under the same assumptions as in (b), find a scheme to arrange the design in (ii) in two blocks each of size 8. Explain why your choice is the best.

16. Two choices of generators for a 2^{6-2} design are being considered:

$$A: \ \ 5 = 1234, \ \ 6 = 123,$$
$$B: \ \ 5 = 123, \ \ \ \ 6 = 234.$$

(a) Which design, A or B, would you recommend? Why?

(b) Show that it is impossible to have a 2_V^{6-2} design. (*Hint:* Count the degrees of freedom.)

17. Find the most economical design that can estimate the main effects of five factors (each with two levels) and all their two-factor interactions. What is the resolution of this design? If it is known that the level combination $(+, +, +, +, +)$ of the five factors can lead to disastrous results (e.g., explosion, burn-out), specify how the design should be chosen.

18. A 2^{k-p} design is said to be *saturated* if the number of factors equals the run size minus 1. Show that it can be formally written as a $2_{III}^{(2^r-1)-(2^r-r-1)}$ design. Denote by $1, 2, \ldots, r$ the r independent columns that generate all the 2^r runs. Show how the remaining columns in the $2_{III}^{(2^r-1)-(2^r-r-1)}$ design can be generated from $1, 2, \ldots, r$.

19. Suppose that a letter does not appear in any of the defining words of length 3 or 4 for a 2^{k-p} design. Prove that the main effect represented by this letter is strongly clear and all the two-factor interactions involving this letter are clear. (*Note:* Two such examples can be found in Sections 4.2 and 4.4.3.)

20. An experimenter considers running a 2^{7-3} design. He is contemplating two possibilities: design A with generators $I = 1235 = 1246 = 12347$ and design B with generators $I = 1235 = 1246 = 1347$. Which design is better? Why?

21. Consider the 2_{IV}^{16-10} design with generators $7 = 123, 8 = 124, 9 = 134, t_0 = 234, t_1 = 125, t_2 = 135, t_3 = 145, t_4 = 126, t_5 = 136, t_6 = 123456$.
   (a) Show that the following 18 two-factor interactions (2fi's) are clear:
   $1t_6, 2t_6, 3t_6, 4t_6, 5t_6, 6t_3, 6t_6, 7t_6, 8t_6, 9t_6, t_0t_6, t_1t_6, t_2t_6, t_3t_4, t_3t_5, t_3t_6, t_4t_6, t_5t_6$.
   (b) Represent the 18 2fi's in (a) in a graph.
   (c) Draw a graph to represent the 29 clear 2fi's for the second 2_{IV}^{16-10} design in Table 4A.5.
   (d) Show that the graph in (b) is isomorphic to a subgraph of the graph in (c). (*Note*: Two graphs are *isomorphic* if they are identical after relabeling the nodes of one graph.)
   (e) Based on the finding in (d), argue that the design considered in this exercise is inferior to the one in part (c) in terms of the ability to clearly estimate two-factor interactions.

22. Consider the 2_{IV}^{15-9} design with generators $7 = 123, 8 = 124, 9 = 134, t_0 = 234, t_1 = 125, t_2 = 135, t_3 = 126, t_4 = 146, t_5 = 123456$.
   (a) Show that it has the following 19 clear 2fi's: $1t_5, 2t_5, 3t_5, 4t_5, 5t_4, 5t_5, 6t_2, 6t_5, 7t_5, 8t_5, 9t_5, t_0t_5, t_1t_4, t_1t_5, t_2t_3, t_2t_4, t_2t_5, t_3t_5, t_4t_5$.
   (b) Represent the 19 2fi's in (a) in a graph.
   (c) Draw a graph to represent the 27 clear 2fi's for the second 2_{IV}^{15-9} design in Table 4A.5.
   (d) Show that the graph in (b) is not a subgraph of the graph in (c) even though the latter has eight more lines.
   (e) Argue that, except in rare situations, the design considered in this exercise is inferior to the one in part (c) in terms of the ability to clearly estimate two-factor interactions. Identify these rare situations.

23. An experimenter who wishes to use a 2^{8-2} design can only do 16 runs in a day and would like to include "day" as a blocking variable. What design would you recommend? Why? Give the treatment generators and the block generators for your design and the collection of clear effects.

24. It was shown in Section 4.2 that the 2_{III}^{6-2} design with $I = 125 = 1346 = 23456$ has nine clear effects: $3, 4, 6, 23, 24, 26, 35, 45, 56$. If this design is arranged in four blocks with $B_1 = 13$ and $B_2 = 14$, show that the same nine effects are still clear. (*Hint:* **13** and **14** do not appear in the five-letter word **23456**.)

25. An experimenter used the following design for studying five variables in eight runs in four blocks of size 2.

Run	1	2	3	B_1	B_2	4	5
1	−	−	−	+	+	−	−
2	+	−	−	−	−	−	+
3	−	+	−	−	+	+	+
4	+	+	−	+	−	+	−
5	−	−	+	+	−	+	+
6	+	−	+	−	+	+	−
7	−	+	+	−	−	−	−
8	+	+	+	+	+	−	+

(a) By reordering the runs, write down the design matrix in four blocks of size 2. This design matrix should contain five columns and eight rows. Indicate which two run numbers occur in each block.

(b) Explain why and under what conditions the main effect **3** is (or is not) confounded with a block effect.

(c) Explain why and under what conditions the main effect **4** is (or is not) confounded with a block effect.

26. (a) Prove that it is impossible to arrange the 2_V^{5-1} design with $I = 12345$ in eight blocks without confounding the main effects.

(b) Show that the 2_{IV}^{5-1} design with $I = 1235$ can be arranged in eight blocks with $B_1 = 14$, $B_2 = 24$, and $B_3 = 34$. It is somewhat counterintuitive that a resolution IV design can be arranged in the largest possible number (eight in this case) of blocks while a resolution V design cannot. Explain in intuitive terms why maximum resolution alone cannot guarantee maximal blocking. [*Note:* A theoretical characterization and explanation of this phenomenon can be found in Mukerjee and Wu (1999).]

27. In a resistance spot welding experiment, five factors were chosen to study their effects on the tensile strength, which is the maximum load a weld can sustain in a tensile test. The five factors are button diameter (A), welding time (B), holding time (C), electrode force (D), and machine type (E), each at two levels. The last factor is qualitative, while the others are quantitative. A 2_V^{5-1} design with $I = -ABCDE$ was used for

Table 4.10   Design Matrix and Tensile Strength Data, Spot Welding Experiment

Run	A	B	C	D	E	Tensile strength		
1	−	−	−	−	−	1330	1330	1165
2	+	+	−	−	−	1935	1935	1880
3	+	−	+	−	−	1770	1770	1770
4	−	+	+	−	−	1275	1275	1275
5	+	−	−	+	−	1880	1935	1880
6	−	+	−	+	−	1385	1440	1495
7	−	−	+	+	−	1220	1165	1440
8	+	+	+	+	−	2155	2100	2100
9	+	−	−	−	+	1715	1715	1660
10	−	+	−	−	+	1385	1550	1550
11	−	−	+	−	+	1000	1165	1495
12	+	+	+	−	+	1990	1990	1990
13	−	−	−	+	+	1275	1660	1550
14	+	+	−	+	+	1660	1605	1660
15	+	−	+	+	+	1880	1935	1935
16	−	+	+	+	+	1275	1220	1275

the experiment. Each run has three replicates. The data are given in Table 4.10.

(a) What effects are clear and strongly clear in this experiment?

(b) Analyze the location and dispersion effects separately, including the fitted models for each.

(c) For the dispersion effect model in (b), interpret the meaning of the significant effects. Use this model or an interaction effects plot to select optimal factor settings.

(d) Based on the location effect model in (b), select optimal factor settings in terms of maximizing the tensile strength.

(e) Is there any adjustment factor? Based on the results in (c) and (d), can you devise a two-step procedure for design optimization? (*Hint:* Because the response is a larger-the-better characteristic, you should consider the two-step procedure in Section 5.2.)

28. In a door panel stamping experiment, six factors were chosen to study their effects on the formability of a panel. One measure of formability is the thinning percentage of the stamped panel at a critical position. The six factors are concentration of lubricant (*A*), panel thickness (*B*), force on the outer portion of the panel (*C*), force on the inner portion of the panel (*D*), punch speed (*E*), and thickness of lubrication (*F*), each at two levels. The experiment was done in two days. "Day" was considered to be a blocking factor (*G*) in order to reduce day-to-day variation, with

Table 4.11   Design Matrix and Response Data, Stamping Experiment

Run	A	B	C	D	E	F	G	Thinning Percentage
1	−	+	+	−	−	−	−	8.17
2	−	+	+	+	+	−	−	6.85
3	−	+	−	+	−	−	−	9.16
4	−	+	−	−	+	−	−	8.50
5	−	−	−	+	−	+	−	5.97
6	−	−	+	+	+	+	−	4.87
7	−	−	+	−	−	+	−	3.00
8	−	−	−	−	+	+	−	4.43
9	+	−	−	−	+	−	−	6.30
10	+	−	+	−	−	−	−	3.00
11	+	−	−	+	−	−	−	5.53
12	+	−	+	+	+	−	−	3.00
13	+	+	+	−	−	+	−	7.51
14	+	+	+	+	+	+	−	7.18
15	+	+	−	−	+	+	−	8.50
16	+	+	−	+	−	+	−	7.18
17	+	+	−	−	−	−	+	9.49
18	+	+	−	+	+	−	+	8.17
19	+	+	+	−	+	−	+	6.85
20	+	+	+	+	−	−	+	7.18
21	+	−	−	−	−	+	+	5.53
22	+	−	+	−	+	+	+	4.10
23	+	−	+	+	−	+	+	4.87
24	+	−	−	+	+	+	+	5.53
25	−	−	+	+	−	−	+	2.67
26	−	−	−	−	−	−	+	6.30
27	−	−	−	+	+	−	+	5.20
28	−	−	+	−	+	−	+	2.23
29	−	+	−	+	+	+	+	8.83
30	−	+	−	−	−	+	+	8.83
31	−	+	+	+	−	+	+	7.84
32	−	+	+	−	+	+	+	7.51

"−" representing day 1 and "+" day 2. A 2_{IV}^{7-2} design with $\mathbf{I} = ABCDEF$ $= -ABFG = -CDEG$ was used for the experiment. The design matrix and the response y (thinning percentage) are given in Table 4.11.

(a) Assume that the block-by-treatment interactions are negligible. Show that all main effects and two-factor interactions among A, B, C, D, E, and F are clear.

(b) Because the logit transformation $\text{logit}(y) = \ln(y/(1-y))$ maps the interval $(0, 1)$ to the real line, it is often used for percentage data. Analyze $\text{logit}(y)$ values in terms of the factorial effects and the block

Table 4.12 Design Matrix and Closing Effort Data, Hatchback Experiment

Run	Factor			Car	Closing Effort
	L	P	F		
1	−	−	−	1	7.0
2	+	+	−	1	7.0
3	+	−	+	1	6.5
4	−	+	+	1	7.0
5	+	−	−	2	5.0
6	−	+	−	2	5.0
7	−	−	+	2	3.5
8	+	+	+	2	4.5
9	−	−	−	3	7.0
10	+	+	−	3	7.5
11	+	−	+	3	7.0
12	−	+	+	3	4.5
13	+	−	−	4	14.0
14	−	+	−	4	11.0
15	−	−	+	4	13.0
16	+	+	+	4	13.5

effect. Identify optimal factor settings in terms of reducing percentage thinning (i.e., improving formability).

(c) Repeat the analysis in (b) with the $\ln(y)$ values. Explain why the results of the analysis with a log transform are similar to those with a logit transform. (*Hint:* Look at the range of the percentage values in the experiment.)

29. Young (1996) reported an experiment using a replicated 2^3 design in four blocks. Because the effort to close the hatch in a new car design was unstable, the experiment studied the impact of nominal length (L), position (P), and firmness (F) on closing effort. The design and data appear in Table 4.12, where − and + represent the extremes of the variation observed for each of the three factors in the production system. Because there is known car-to-car variation, four runs were performed on the same car as shown in Table 4.12. Each car is treated as a block.

(a) Because implementing a full factorial design for each of the four blocks would require 32 ($= 8 \times 4$) runs, the chosen design is a half fraction of a full factorial design for each car. Find the generator that defines this half fraction.

(b) What model matrix should be used for effect estimation? Assuming no block-by-treatment interactions, how many degrees of freedom remain for error variance estimation? Obtain a variance estimate.

(c) Using the variance estimate s^2 in (b), apply the studentized maximum modulus test to test the significance of the seven factorial effects and three block effects.

(d) If the reliability of s^2 is in doubt, the IER version of Lenth's method can still be used to test the significance of the factorial effects. Compare the findings from the two approaches and comment.

(e) Discuss the consequences of ignoring blocking in the analysis.

APPENDIX 4A: TABLES OF 2^{k-p} FRACTIONAL FACTORIAL DESIGNS

Notes:

1. The first nine factors are denoted by $1, \ldots, 9$, the $(10 + i)$th factor is denoted by t_i, the $(20 + i)$th factor by u_i and the $(30 + i)$th factor by v_i, $i = 0, \ldots, 9$.

2. In a 2^{k-p} design, the first $k - p$ independent columns that generate the 2^{k-p} runs are denoted by $1, \ldots, k - p$. The remaining p columns are generated as interactions of the first $k - p$ columns. These p generators are listed under "Design Generators" in the tables given in Appendices 4A and 4B.

3. Two-factor interactions are abbreviated as 2fi's.

Table 4A.1   Eight-Run 2^{k-p} Fractional Factorial Designs ($k - p = 3$)

Number of Factors k	Fraction and Resolution	Design Generators	Clear Effects
4^a	2^{4-1}_{IV}	$4 = 123$	$1, 2, 3, 4$
5^b	2^{5-2}_{III}	$4 = 12, 5 = 13$	none
6	2^{6-3}_{III}	$4 = 12, 5 = 13, 6 = 23$	none
7	2^{7-4}_{III}	$4 = 12, 5 = 13, 6 = 23,$ $7 = 123$	none

^{a}The aliases are $1 = 234, 2 = 134, 3 = 124, 4 = 123, 12 = 34, 13 = 24, 14 = 23$.
^{b}The aliases are $1 = 24 = 35 = 12345, 2 = 14 = 345 = 1235, 3 = 15 = 245 = 1234, 4 = 12 = 235 = 1345, 5 = 13 = 234 = 1245, 23 = 45 = 125 = 134, 25 = 34 = 123 = 145$.

Table 4A.2 Sixteen-Run 2^{k-p} Fractional Factorial Designs ($k - p = 4$)
(k is the number of factors and F&R is the fraction and resolution)

k	F&R	Design Generators	Clear Effects
5	2_V^{5-1}	$5 = 1234$	all five main effects, all 10 2fi's
6	2_{IV}^{6-2}	$5 = 123, 6 = 124$	all six main effects
6^a	2_{III}^{6-2}	$5 = 12, 6 = 134$	$3, 4, 6, 23, 24,$ $26, 35, 45, 56$
7	2_{IV}^{7-3}	$5 = 123, 6 = 124, 7 = 134$	all seven main effects
8	2_{IV}^{8-4}	$5 = 123, 6 = 124, 7 = 134,$ $8 = 234$	all eight main effects
9	2_{III}^{9-5}	$5 = 123, 6 = 124, 7 = 134,$ $8 = 234, 9 = 1234$	none
10	2_{III}^{10-6}	$5 = 123, 6 = 124, 7 = 134,$ $8 = 234, 9 = 1234, t_0 = 34$	none
11	2_{III}^{11-7}	$5 = 123, 6 = 124, 7 = 134,$ $8 = 234, 9 = 1234, t_0 = 34,$ $t_1 = 24$	none
12	2_{III}^{12-8}	$5 = 123, 6 = 124, 7 = 134,$ $8 = 234, 9 = 1234, t_0 = 34,$ $t_1 = 24, t_2 = 14$	none
13	2_{III}^{13-9}	$5 = 123, 6 = 124, 7 = 134,$ $8 = 234, 9 = 1234, t_0 = 34,$ $t_1 = 24, t_2 = 14, t_3 = 23$	none
14	2_{III}^{14-10}	$5 = 123, 6 = 124, 7 = 134,$ $8 = 234, 9 = 1234, t_0 = 34,$ $t_1 = 24, t_2 = 14, t_3 = 23,$ $t_4 = 13$	none
15	2_{III}^{15-11}	$5 = 123, 6 = 124, 7 = 134,$ $8 = 234, 9 = 1234, t_0 = 34,$ $t_1 = 24, t_2 = 14, t_3 = 23,$ $t_4 = 13, t_5 = 12$	none

^{a} Its aliases are given in (4.3) of Section 4.2.

Table 4A.3 Thirty-Two Run 2^{k-p} Fractional Factorial Design ($k - p = 5$, $6 \leq k \leq 16$) (k is the number of factors and F&R is the fraction and resolution)

k	F&R	Design Generators	Clear Effects
6	2_{VI}^{6-1}	$6 = 12345$	all six main effects, all 15 2fi's
7	2_{IV}^{7-2}	$6 = 123$, $7 = 1245$	all seven main effects, $14, 15, 17, 24, 25, 27, 34,$ $35, 37, 45, 46, 47, 56, 57, 67$
8	2_{IV}^{8-3}	$6 = 123$, $7 = 124$, $8 = 1345$	all eight main effects, $15, 18, 25, 28, 35, 38, 45,$ $48, 56, 57, 58, 68, 78$
9	2_{IV}^{9-4}	$6 = 123$, $7 = 124$, $8 = 125$, $9 = 1345$	all nine main effects, $19, 29, 39, 49, 59, 69,$ $79, 89$
9	2_{IV}^{9-4}	$6 = 123$, $7 = 124$, $8 = 134$, $9 = 2345$	all nine main effects, $15, 19, 25, 29, 35, 39,$ $45, 49, 56, 57, 58, 59,$ $69, 79, 89$
10	2_{IV}^{10-5}	$6 = 123$, $7 = 124$, $8 = 125$, $9 = 1345$, $t_0 = 2345$	all 10 main effects
10	2_{III}^{10-5}	$6 = 12$, $7 = 134$, $8 = 135$, $9 = 145$, $t_0 = 345$	$3, 4, 5, 7, 8, 9, t_0, 23,$ $24, 25, 27, 28, 29, 2t_0,$ $36, 46, 56, 67, 68, 69,$ $6t_0$
11	2_{IV}^{11-6}	$6 = 123$, $7 = 124$, $8 = 134$, $9 = 125$, $t_0 = 135$, $t_1 = 145$	all 11 main effects
11	2_{III}^{11-6}	$6 = 12$, $7 = 13$, $8 = 234$, $9 = 235$, $t_0 = 245$, $t_1 = 1345$	$4, 5, 8, 9, t_0, t_1, 14,$ $15, 18, 19, 1t_0, 1t_1$
12	2_{IV}^{12-7}	$6 = 123$, $7 = 124$, $8 = 134$, $9 = 234$, $t_0 = 125$, $t_1 = 135$, $t_2 = 145$	all 12 main effects
12	2_{III}^{12-7}	$6 = 12$, $7 = 13$, $8 = 14$, $9 = 234$, $t_0 = 235$, $t_1 = 245$, $t_2 = 1345$	$5, 9, t_0, t_1, t_2,$ $15, 19, 1t_0, 1t_1, 1t_2$
13	2_{IV}^{13-8}	$6 = 123$, $7 = 124$, $8 = 134$, $9 = 234$, $t_0 = 125$, $t_1 = 135$, $t_2 = 235$, $t_3 = 145$	all 13 main effects
14	2_{IV}^{14-9}	$6 = 123$, $7 = 124$, $8 = 134$, $9 = 234$, $t_0 = 125$, $t_1 = 135$, $t_2 = 235$, $t_3 = 145$, $t_4 = 245$	all 14 main effects
15	2_{IV}^{15-10}	$6 = 123$, $7 = 124$, $8 = 134$, $9 = 234$, $t_0 = 125$, $t_1 = 135$, $t_2 = 235$, $t_3 = 145$, $t_4 = 245$, $t_5 = 345$	all 15 main effects
16	2_{IV}^{16-11}	$6 = 123$, $7 = 124$, $8 = 134$, $9 = 234$, $t_0 = 125$, $t_1 = 135$, $t_2 = 235$, $t_3 = 145$, $t_4 = 245$, $t_5 = 345$, $t_6 = 12345$	all 16 main effects

Table 4A.4 Thirty-Two Run 2^{k-p} Fractional Factorial Design ($k-p=5$, $17 \leq k \leq 31$) (k is the number of factors and F&R is the fraction and resolution)

k	F&R	Design Generators
17	2_{III}^{17-12}	$6=12$, $7=13$, $8=14$, $9=234$, $t_0=1234$, $t_1=15$, $t_2=235$, $t_3=1235$, $t_4=245$, $t_5=1245$, $t_6=345$, $t_7=1345$
18	2_{III}^{18-13}	$6=12$, $7=13$, $8=23$, $9=14$, $t_0=234$, $t_1=1234$, $t_2=15$, $t_3=235$, $t_4=1235$, $t_5=245$, $t_6=1245$, $t_7=345$, $t_8=1345$
19	2_{III}^{19-14}	$6=12$, $7=13$, $8=23$, $9=14$, $t_0=24$, $t_1=234$, $t_2=1234$, $t_3=15$, $t_4=235$, $t_5=1235$, $t_6=245$, $t_7=1245$, $t_8=345$, $t_9=1345$
20	2_{III}^{20-15}	$6=12$, $7=13$, $8=23$, $9=14$, $t_0=24$, $t_1=234$, $t_2=1234$, $t_3=15$, $t_4=25$, $t_5=235$, $t_6=1235$, $t_7=245$, $t_8=1245$, $t_9=345$, $u_0=1345$
21	2_{III}^{21-16}	$6=12$, $7=13$, $8=23$, $9=14$, $t_0=24$, $t_1=234$, $t_2=1234$, $t_3=15$, $t_4=25$, $t_5=235$, $t_6=1235$, $t_7=245$, $t_8=1245$, $t_9=345$, $u_0=1345$, $u_1=12345$
22	2_{III}^{22-17}	$6=12$, $7=13$, $8=23$, $9=14$, $t_0=24$, $t_1=134$, $t_2=234$, $t_3=1234$, $t_4=15$, $t_5=25$, $t_6=135$, $t_7=235$, $t_8=1235$, $t_9=145$, $u_0=245$, $u_1=1345$, $u_2=2345$
23	2_{III}^{23-18}	$6=12$, $7=13$, $8=23$, $9=14$, $t_0=24$, $t_1=134$, $t_2=234$, $t_3=1234$, $t_4=15$, $t_5=25$, $t_6=135$, $t_7=235$, $t_8=1235$, $t_9=145$, $u_0=245$, $u_1=1245$, $u_2=345$, $u_3=1345$
24	2_{III}^{24-19}	$6=12$, $7=13$, $8=23$, $9=14$, $t_0=24$, $t_1=134$, $t_2=234$, $t_3=1234$, $t_4=15$, $t_5=25$, $t_6=135$, $t_7=235$, $t_8=1235$, $t_9=145$, $u_0=245$, $u_1=1245$, $u_2=345$, $u_3=1345$, $u_4=2345$
25	2_{III}^{25-20}	$6=12$, $7=13$, $8=23$, $9=123$, $t_0=14$, $t_1=24$, $t_2=124$, $t_3=34$, $t_4=134$, $t_5=15$, $t_6=25$, $t_7=125$, $t_8=35$, $t_9=135$, $u_0=245$, $u_1=1245$, $u_2=345$, $u_3=1345$, $u_4=2345$, $u_5=12345$
26	2_{III}^{26-21}	$6=12$, $7=13$, $8=23$, $9=123$, $t_0=14$, $t_1=24$, $t_2=124$, $t_3=34$, $t_4=134$, $t_5=234$, $t_6=15$, $t_7=25$, $t_8=125$, $t_9=35$, $u_0=135$, $u_1=245$, $u_2=1245$, $u_3=345$, $u_4=1345$, $u_5=2345$, $u_6=12345$
27	2_{III}^{27-22}	$6=12$, $7=13$, $8=23$, $9=123$, $t_0=14$, $t_1=24$, $t_2=124$, $t_3=34$, $t_4=134$, $t_5=234$, $t_6=15$, $t_7=25$, $t_8=125$, $t_9=35$, $u_0=135$, $u_1=235$, $u_2=145$, $u_3=245$, $u_4=1245$, $u_5=345$, $u_6=1345$, $u_7=2345$
28	2_{III}^{28-23}	$6=12$, $7=13$, $8=23$, $9=123$, $t_0=14$, $t_1=24$, $t_2=124$, $t_3=34$, $t_4=134$, $t_5=234$, $t_6=15$, $t_7=25$, $t_8=125$, $t_9=35$, $u_0=135$, $u_1=235$, $u_2=145$, $u_3=245$, $u_4=1245$, $u_5=345$, $u_6=1345$, $u_7=2345$, $u_8=12345$
29	2_{III}^{29-24}	$6=12$, $7=13$, $8=23$, $9=123$, $t_0=14$, $t_1=24$, $t_2=124$, $t_3=34$, $t_4=134$, $t_5=234$, $t_6=1234$, $t_7=15$, $t_8=25$, $t_9=125$, $u_0=35$, $u_1=135$, $u_2=235$, $u_3=145$, $u_4=245$, $u_5=1245$, $u_6=345$, $u_7=1345$, $u_8=2345$, $u_9=12345$
30	2_{III}^{30-25}	$6=12$, $7=13$, $8=23$, $9=123$, $t_0=14$, $t_1=24$, $t_2=124$, $t_3=34$, $t_4=134$, $t_5=234$, $t_6=1234$, $t_7=15$, $t_8=25$, $t_9=125$, $u_0=35$, $u_1=135$, $u_2=235$, $u_3=1235$, $u_4=145$, $u_5=245$, $u_6=1245$, $u_7=345$, $u_8=1345$, $u_9=2345$, $v_0=12345$
31	2_{III}^{31-26}	$6=12$, $7=13$, $8=23$, $9=123$, $t_0=14$, $t_1=24$, $t_2=124$, $t_3=34$, $t_4=134$, $t_5=234$, $t_6=1234$, $t_7=15$, $t_8=25$, $t_9=125$, $u_0=35$, $u_1=135$, $u_2=235$, $u_3=1235$, $u_4=45$, $u_5=145$, $u_6=245$, $u_7=1245$, $u_8=345$, $u_9=1345$, $v_0=2345$, $v_1=12345$

Note: **No main effect or two-factor interaction is clear** for the designs in this table.

Table 4A.5   Sixty-Four Run 2^{k-p} Fractional Factorial Designs ($k-p=6$, $7 \le k \le 17$) (k is the number of factors and F&R is the fraction and resolution)

k	F&R	Design Generators	Clear Effects
7	2^{7-1}_{VII}	$7 = 123456$	all 21 2fi's
8	2^{8-2}_{V}	$7 = 1234, 8 = 1256$	all 28 2fi's
9	2^{9-3}_{IV}	$7 = 123, 8 = 1245, 9 = 1346$	all 2fi's except 12, 13, 17, 23, 27, 37
10	2^{10-4}_{IV}	$7 = 123, 8 = 1245, 9 = 1246,$ $t_0 = 1356$	all 2fi's except 12, 13, 17, 23, 27, 37, 56, 58, 59, 68, 69, 89
11	2^{11-5}_{IV}	$7 = 123, 8 = 124, 9 = 1345,$ $t_0 = 1346, t_1 = 1256$	the 10 2fi's involving t_1 and the 24 2fi's between any of the factors 1, 2, 3, 4, 7, 8 and any of the factors 5, 6, 9, t_0
12	2^{12-6}_{IV}	$7 = 123, 8 = 124, 9 = 1345,$ $t_0 = 1346, t_1 = 1256, t_2 = 23456$	the 36 2fi's between any of the factors 1, 2, 3, 4, 7, 8 and any of the factors 5, 6, 9, t_0, t_1, t_2
13	2^{13-7}_{IV}	$7 = 123, 8 = 124, 9 = 135,$ $t_0 = 145, t_1 = 236, t_2 = 2456,$ $t_3 = 3456$	the 20 2fi's between any of the factors 4, 5, 8, 9, t_0 and any of the factors 6, t_1, t_2, t_3
13	2^{13-7}_{IV}	$7 = 123, 8 = 124, 9 = 134,$ $t_0 = 2345, t_1 = 2346, t_2 = 156,$ $t_3 = 123456$	the 36 2fi's between any of the factors 2, 3, 4, 7, 8, 9 and any of the factors 5, 6, t_0, t_1, t_2, t_3
14	2^{14-8}_{IV}	$7 = 123, 8 = 124, 9 = 125,$ $t_0 = 2345, t_1 = 136, t_2 = 146,$ $t_3 = 156, t_4 = 3456$	$1t_0$, $1t_4$, $3t_0$, $3t_4$, $4t_0$, $4t_4$, $5t_0$, $5t_4$
14	2^{14-8}_{IV}	$7 = 123, 8 = 124, 9 = 134,$ $t_0 = 234, t_1 = 125, t_2 = 135,$ $t_3 = 145, t_4 = 2356$	the 25 2fi's that involve either factor 6 or t_4 or both
15	2^{15-9}_{IV}	$7 = 123, 8 = 124, 9 = 125,$ $t_0 = 2345, t_1 = 136, t_2 = 146,$ $t_3 = 156, t_4 = 3456, t_5 = 123456$	none
15	2^{15-9}_{IV}	$7 = 123, 8 = 124, 9 = 134,$ $t_0 = 234, t_1 = 125, t_2 = 135,$ $t_3 = 235, t_4 = 145, t_5 = 2456$	the 27 2fi's that involve either factor 6 or t_5 or both
16	2^{16-10}_{IV}	$7 = 123, 8 = 124, 9 = 134,$ $t_0 = 125, t_1 = 135, t_2 = 126,$ $t_3 = 136, t_4 = 1456, t_5 = 2456,$ $t_6 = 3456$	none
16	2^{16-10}_{IV}	$7 = 123, 8 = 124, 9 = 134,$ $t_0 = 234, t_1 = 125, t_2 = 135,$ $t_3 = 235, t_4 = 145, t_5 = 245,$ $t_6 = 3456$	the 29 2fi's that involve either factor 6 or t_6 or both
17	2^{17-11}_{IV}	$7 = 123, 8 = 124, 9 = 134,$ $t_0 = 234, t_1 = 125, t_2 = 135,$ $t_3 = 126, t_4 = 136, t_5 = 1456,$ $t_6 = 2456, t_7 = 3456$	none
17	2^{17-11}_{IV}	$7 = 123, 8 = 124, 9 = 134,$ $t_0 = 234, t_1 = 125, t_2 = 135,$ $t_3 = 235, t_4 = 145, t_5 = 245,$ $t_6 = 345, t_7 = 123456$	the 31 2fi's that involve either factor 6 or t_7 or both

Note: Because the designs in this table have at least resolution IV, **all their main effects are clear** and will not be repeated in the column "Clear Effects."

Table A4.6   Sixty-Four Run 2^{k-p} Fractional Factorial Designs ($k - p = 6$, $18 \leq k \leq 32$)
(k is the number of factors and F&R is the fraction and resolution)

k	F&R	Design Generators
18	2_{IV}^{18-12}	$7 = 123$, $8 = 124$, $9 = 134$, $t_0 = 234$, $t_1 = 125$, $t_2 = 135$, $t_3 = 235$, $t_4 = 126$, $t_5 = 136$, $t_6 = 1456$, $t_7 = 2456$, $t_8 = 3456$
19	2_{IV}^{19-13}	$7 = 123$, $8 = 124$, $9 = 134$, $t_0 = 234$, $t_1 = 125$, $t_2 = 135$, $t_3 = 235$, $t_4 = 126$, $t_5 = 136$, $t_6 = 236$, $t_7 = 1456$, $t_8 = 2456$, $t_9 = 3456$
20	2_{IV}^{20-14}	$7 = 123$, $8 = 124$, $9 = 134$, $t_0 = 234$, $t_1 = 125$, $t_2 = 135$, $t_3 = 235$, $t_4 = 126$, $t_5 = 136$, $t_6 = 236$, $t_7 = 1456$, $t_8 = 2456$, $t_9 = 3456$, $u_0 = 123456$
21	2_{IV}^{21-15}	$7 = 123$, $8 = 124$, $9 = 134$, $t_0 = 234$, $t_1 = 125$, $t_2 = 135$, $t_3 = 235$, $t_4 = 145$, $t_5 = 126$, $t_6 = 146$, $t_7 = 246$, $t_8 = 156$, $t_9 = 356$, $u_0 = 456$, $u_1 = 23456$
22	2_{IV}^{22-16}	$7 = 123$, $8 = 124$, $9 = 134$, $t_0 = 234$, $t_1 = 125$, $t_2 = 135$, $t_3 = 235$, $t_4 = 145$, $t_5 = 126$, $t_6 = 136$, $t_7 = 146$, $t_8 = 246$, $t_9 = 156$, $u_0 = 356$, $u_1 = 456$, $u_2 = 23456$
23	2_{IV}^{23-17}	$7 = 123$, $8 = 124$, $9 = 134$, $t_0 = 234$, $t_1 = 125$, $t_2 = 135$, $t_3 = 235$, $t_4 = 145$, $t_5 = 245$, $t_6 = 126$, $t_7 = 136$, $t_8 = 146$, $t_9 = 346$, $u_0 = 156$, $u_1 = 356$, $u_2 = 456$, $u_3 = 23456$
24	2_{IV}^{24-18}	$7 = 123$, $8 = 124$, $9 = 134$, $t_0 = 234$, $t_1 = 125$, $t_2 = 135$, $t_3 = 235$, $t_4 = 145$, $t_5 = 245$, $t_6 = 126$, $t_7 = 136$, $t_8 = 236$, $t_9 = 146$, $u_0 = 246$, $u_1 = 156$, $u_2 = 356$, $u_3 = 456$, $u_4 = 23456$
25	2_{IV}^{25-19}	$7 = 123$, $8 = 124$, $9 = 134$, $t_0 = 234$, $t_1 = 125$, $t_2 = 135$, $t_3 = 235$, $t_4 = 145$, $t_5 = 245$, $t_6 = 345$, $t_7 = 126$, $t_8 = 136$, $t_9 = 236$, $u_0 = 146$, $u_1 = 246$, $u_2 = 156$, $u_3 = 356$, $u_4 = 456$, $u_5 = 23456$
26	2_{IV}^{26-20}	$7 = 123$, $8 = 124$, $9 = 134$, $t_0 = 234$, $t_1 = 125$, $t_2 = 135$, $t_3 = 235$, $t_4 = 145$, $t_5 = 245$, $t_6 = 345$, $t_7 = 126$, $t_8 = 136$, $t_9 = 236$, $u_0 = 146$, $u_1 = 246$, $u_2 = 346$, $u_3 = 156$, $u_4 = 256$, $u_5 = 356$, $u_6 = 456$
27	2_{IV}^{27-21}	$7 = 123$, $8 = 124$, $9 = 134$, $t_0 = 234$, $t_1 = 125$, $t_2 = 135$, $t_3 = 235$, $t_4 = 145$, $t_5 = 245$, $t_6 = 345$, $t_7 = 12345$, $t_8 = 126$, $t_9 = 136$, $u_0 = 236$, $u_1 = 146$, $u_2 = 246$, $u_3 = 346$, $u_4 = 156$, $u_5 = 256$, $u_6 = 356$, $u_7 = 456$
28	2_{IV}^{28-22}	$7 = 123$, $8 = 124$, $9 = 134$, $t_0 = 234$, $t_1 = 125$, $t_2 = 135$, $t_3 = 235$, $t_4 = 145$, $t_5 = 245$, $t_6 = 345$, $t_7 = 12345$, $t_8 = 126$, $t_9 = 136$, $u_0 = 236$, $u_1 = 146$, $u_2 = 246$, $u_3 = 346$, $u_4 = 12346$, $u_5 = 156$, $u_6 = 256$, $u_7 = 356$, $u_8 = 456$
29	2_{IV}^{29-23}	$7 = 123$, $8 = 124$, $9 = 134$, $t_0 = 234$, $t_1 = 125$, $t_2 = 135$, $t_3 = 235$, $t_4 = 145$, $t_5 = 245$, $t_6 = 345$, $t_7 = 12345$, $t_8 = 126$, $t_9 = 136$, $u_0 = 236$, $u_1 = 146$, $u_2 = 246$, $u_3 = 346$, $u_4 = 12346$, $u_5 = 156$, $u_6 = 256$, $u_7 = 356$, $u_8 = 12356$, $u_9 = 456$
30	2_{IV}^{30-24}	$7 = 123$, $8 = 124$, $9 = 134$, $t_0 = 234$, $t_1 = 125$, $t_2 = 135$, $t_3 = 235$, $t_4 = 145$, $t_5 = 245$, $t_6 = 345$, $t_7 = 12345$, $t_8 = 126$, $t_9 = 136$, $u_0 = 236$, $u_1 = 146$, $u_2 = 246$, $u_3 = 346$, $u_4 = 12346$, $u_5 = 156$, $u_6 = 256$, $u_7 = 356$, $u_8 = 12356$, $u_9 = 456$, $v_0 = 12456$
31	2_{IV}^{31-25}	$7 = 123$, $8 = 124$, $9 = 134$, $t_0 = 234$, $t_1 = 125$, $t_2 = 135$, $t_3 = 235$, $t_4 = 145$, $t_5 = 245$, $t_6 = 345$, $t_7 = 12345$, $t_8 = 126$, $t_9 = 136$, $u_0 = 236$, $u_1 = 146$, $u_2 = 246$, $u_3 = 346$, $u_4 = 12346$, $u_5 = 156$, $u_6 = 256$, $u_7 = 356$, $u_8 = 12356$, $u_9 = 456$, $v_0 = 12456$, $v_1 = 13456$
32	2_{IV}^{32-26}	$7 = 123$, $8 = 124$, $9 = 134$, $t_0 = 234$, $t_1 = 125$, $t_2 = 135$, $t_3 = 235$, $t_4 = 145$, $t_5 = 245$, $t_6 = 345$, $t_7 = 12345$, $t_8 = 126$, $t_9 = 136$, $u_0 = 236$, $u_1 = 146$, $u_2 = 246$, $u_3 = 346$, $u_4 = 12346$, $u_5 = 156$, $u_6 = 256$, $u_7 = 356$, $u_8 = 12356$, $u_9 = 456$, $v_0 = 12456$, $v_1 = 13456$, $v_2 = 23456$

Note: The designs in this table have resolution IV; **all their main effects are clear but none of their two-factor interactions are clear.**

Table 4A.7   128-Run 2^{k-p} Minimum Aberration Fractional Factorial Design
($k-p=7$, $8 \le k \le 14$) (k is the number of factors and F&R is the fraction
and resolution)

k	F&R	Design Generators	Clear Effects
8	2_{VIII}^{8-1}	$8 = 1234567$	all 8 main effects, all 28 2fi's
9	2_{VI}^{9-2}	$8 = 13457$, $9 = 12467$	all 9 main effects, all 36 2fi's
10	2_{V}^{10-3}	$8 = 3456$, $9 = 13457$, $t_0 = 12467$	all 10 main effects, all 45 2fi's
11	2_{V}^{11-4}	$8 = 3456$, $9 = 13457$, $t_0 = 12467$, $t_1 = 2567$	all 11 main effects, all 55 2fi's
12	2_{IV}^{12-5}	$8 = 145$, $9 = 1236$, $t_0 = 2467$, $t_1 = 3567$, $t_2 = 123457$	all 12 main effects, all 2fi's except 14, 15, 18, 45, 48, 58
13	2_{IV}^{13-6}	$8 = 12345$, $9 = 1236$, $t_0 = 124567$, $t_1 = 134567$, $t_2 = 2347$, $t_3 = 567$	all 13 main effects, all 2fi's except 23, $2t_0$, $2t_1$, $3t_0$, $3t_1$, $t_0 t_1$, 56, 57, $5t_3$, 67, $6t_3$, $7t_3$
14	2_{IV}^{14-7}	$8 = 123$, $9 = 456$, $t_0 = 1245$, $t_1 = 1346$, $t_2 = 12467$, $t_3 = 13567$, $t_4 = 23457$	all 14 main effects, all 2fi's except 12, 13, 18, 23, 28, 38, 45, 46, 49, 56, 59, 69, $7t_2$, $7t_3$, $7t_4$, $t_2 t_3$, $t_2 t_4$, $t_3 t_4$

APPENDIX 4B: TABLES OF 2^{k-p} FRACTIONAL
FACTORIAL DESIGNS IN 2^q BLOCKS

Note. Two-factor interactions are abbreviated as 2fi's.

Table 4B.1   Eight-Run 2^{k-p} Fractional Factorial Designs in 2^q Blocks
($k-p=3$, $4 \le k \le 6$) (k = number of factors, 2^{k-p} = number of runs,
2^q = number of blocks)

k	p	q	Design Generators	Block Generators	Clear Effects
4	1	1	$4 = 123$	$B_1 = 12$	all four main effects
4	1	2	$4 = 123$	$B_1 = 12$, $B_2 = 13$	all four main effects
5	2	1	$4 = 12$, $5 = 13$	$B_1 = 23$	none
6	3	1	$4 = 12$, $5 = 13$, $6 = 23$	$B_1 = 123$	none

Table 4B.2   Sixteen-Run 2^{k-p} Fractional Factorial Designs in 2^q blocks ($k-p = 4$, $5 \leq k \leq 9$) (k = number of factors, 2^{k-p} = number of runs, 2^q = number of blocks)

k	p	q	Design Generators	Block Generators	Clear Effects
5	1	1	$5 = 1234$	$B_1 = 12$	all five main effects, all 2fi's except 12
5	1	2	$5 = 1234$	$B_1 = 12$, $B_2 = 13$	all five main effects, 14, 15, 24, 25, 34, 35, 45
5	1	3	$5 = 123$	$B_1 = 14$, $B_2 = 24$, $B_3 = 34$	all five main effects
6	2	1	$5 = 123, 6 = 124$	$B_1 = 134$	all six main effects
6	2	1	$5 = 12, 6 = 134$	$B_1 = 13$	3, 4, 6, 23, 24, 26, 35, 45, 56
6	2	2	$5 = 123, 6 = 124$	$B_1 = 134$, $B_2 = 234$	all six main effects
6	2	2	$5 = 12, 6 = 134$	$B_1 = 13$, $B_2 = 14$	3, 4, 6, 23, 24, 26, 35, 45, 56
6	2	3	$5 = 123, 6 = 124$	$B_1 = 13$, $B_2 = 23$, $B_3 = 14$	all six main effects
7	3	1	$5 = 123, 6 = 124$, $7 = 134$	$B_1 = 234$	all seven main effects
7	3	2	$5 = 123, 6 = 124$, $7 = 134$	$B_1 = 12$, $B_2 = 13$	all seven main effects
7	3	3	$5 = 123, 6 = 124$, $7 = 134$	$B_1 = 12$, $B_2 = 13$, $B_3 = 14$	all seven main effects
8	4	1	$5 = 123, 6 = 124$, $7 = 134, 8 = 234$	$B_1 = 12$	all eight main effects
8	4	2	$5 = 123, 6 = 124$, $7 = 134, 8 = 234$	$B_1 = 12$, $B_2 = 13$	all eight main effects
8	4	3	$5 = 123, 6 = 124$, $7 = 134, 8 = 234$	$B_1 = 12$, $B_2 = 13$, $B_3 = 14$	all eight main effects
9	5	1	$5 = 12, 6 = 13$, $7 = 14, 8 = 234$, $9 = 1234$	$B_1 = 23$	none
9	5	2	$5 = 12, 6 = 13$, $7 = 14, 8 = 234$, $9 = 1234$	$B_1 = 23$, $B_2 = 24$	none

Table 4B.3   Thirty-Two Run 2^{k-p} Fractional Factorial Designs in 2^q blocks ($k - p = 5$, $6 \leq k \leq 9$) (k = number of factors, 2^{k-p} = number of runs, 2^q = number of blocks)

k	p	q	Design Generators	Block Generators	Clear 2fi's
6	1	1	6 = 12345	$B_1 = 123$	all 15 2fi's
6	1	2	6 = 12345	$B_1 = 134$, $B_2 = 234$	all 2fi's except 12
6	1	3	6 = 12345	$B_1 = 135$, $B_2 = 235$, $B_3 = 145$	all 2fi's except 12, 34, 56
6	1	4	6 = 12345	$B_1 = 12$, $B_2 = 13$, $B_3 = 14$, $B_4 = 15$	none
7	2	1	6 = 123, 7 = 1245	$B_1 = 134$	all 2fi's except 12, 13, 16, 23, 26, 36
7	2	2	6 = 123, 7 = 1245	$B_1 = 134$, $B_2 = 234$	all 2fi's except 12, 13, 16, 23, 26, 36
7	2	3	6 = 123, 7 = 1245	$B_1 = 234$, $B_2 = 235$, $B_3 = 1345$	14, 15, 17, 24, 25, 27, 34, 35, 37, 46, 56, 67
7	2	4	6 = 123, 7 = 145	$B_1 = 12$, $B_2 = 13$, $B_3 = 14$, $B_4 = 15$	none
8	3	1	6 = 123, 7 = 124, 8 = 1345	$B_1 = 125$	15, 18, 25, 28, 35, 38, 45, 48, 56, 57, 58, 68, 78
8	3	2	6 = 123, 7 = 124, 8 = 1345	$B_1 = 13$, $B_2 = 14$	15, 18, 25, 28, 35, 38, 45, 48, 56, 57, 58, 68, 78
8	3	3	6 = 123, 7 = 124, 8 = 1345	$B_1 = 13$, $B_2 = 23$, $B_3 = 14$	15, 18, 25, 28, 35, 38, 45, 48, 56, 57, 58, 68, 78
8	3	4	6 = 123, 7 = 124, 8 = 135	$B_1 = 12$, $B_2 = 13$, $B_3 = 14$, $B_4 = 15$	none
9	4	1	6 = 123, 7 = 124, 8 = 134, 9 = 2345	$B_1 = 12$	15, 19, 25, 29, 35, 39, 45, 49, 56, 57, 58, 59, 69, 79, 89
9	4	2	6 = 123, 7 = 124, 8 = 134, 9 = 2345	$B_1 = 12$, $B_2 = 13$	15, 19, 25, 29, 35, 39, 45, 49, 56, 57, 58, 59, 69, 79, 89
9	4	3	6 = 123, 7 = 124, 8 = 134, 9 = 2345	$B_1 = 12$, $B_2 = 13$, $B_3 = 14$	15, 19, 25, 29, 35, 39, 45, 49, 56, 57, 58, 59, 69, 79, 89
9	4	4	6 = 123, 7 = 124, 8 = 135, 9 = 145	$B_1 = 12$, $B_2 = 13$, $B_3 = 14$, $B_4 = 15$	none

Note: **All the main effects are clear** for all of the designs in this table and will not be repeated in the column "Clear 2fi's."

Table 4B.4 **Sixty-Four Run 2^{k-p} Fractional Factorial Designs in 2^q Blocks** ($k - p = 6$, $7 \leq k \leq 9$) (k = number of factors, 2^{k-p} = number of runs, 2^q = number of blocks)

k	p	q	Design Generators	Block Generators	Clear 2fi's
7	1	1	7 = 123456	$B_1 = 123$	all 21 2fi's
7	1	2	7 = 123456	$B_1 = 123$, $B_2 = 145$	all 21 2fi's
7	1	3	7 = 123456	$B_1 = 123$, $B_2 = 145$, $B_3 = 246$	all 21 2fi's
7	1	4	7 = 12345	$B_1 = 12$, $B_2 = 34$, $B_3 = 135$, $B_4 = 16$	all 2fi's except 12, 16, 26, 34, 57
7	1	5	7 = 12345	$B_1 = 12$, $B_2 = 13$, $B_3 = 14$, $B_4 = 15$, $B_5 = 16$	none
8	2	1	7 = 1234, 8 = 1256	$B_1 = 135$	all 28 2fi's
8	2	2	7 = 1234, 8 = 1256	$B_1 = 135$, $B_2 = 246$	all 28 2fi's
8	2	3	7 = 1234, 8 = 1256	$B_1 = 146$, $B_2 = 246$, $B_3 = 13456$	all 2fi's except 12, 35
8	2	4	7 = 1234, 8 = 1256	$B_1 = 13$, $B_2 = 14$, $B_3 = 25$, $B_4 = 26$	12, 15, 16, 17, 18, 23, 24, 27, 28, 35, 36, 37, 38, 45, 46, 47, 48, 57, 58, 67, 68
8	2	5	7 = 123, 8 = 12456	$B_1 = 12$, $B_2 = 13$, $B_3 = 14$, $B_4 = 15$, $B_5 = 16$	none
9	3	1	7 = 123, 8 = 1245, 9 = 1346	$B_1 = 1256$	all 2fi's except 12, 13, 17, 23, 27, 37
9	3	2	7 = 123, 8 = 1245, 9 = 1346	$B_1 = 156$, $B_2 = 123456$	all 2fi's except 12, 13, 17, 23, 27, 37
9	3	3	7 = 123, 8 = 1245, 9 = 1346	$B_1 = 156$, $B_2 = 256$, $B_3 = 356$	all 2fi's except 12, 13, 17, 23, 27, 37
9	3	4	7 = 123, 8 = 1245, 9 = 1346	$B_1 = 12$, $B_2 = 13$, $B_3 = 14$, $B_4 = 56$	15, 16, 18, 19, 25, 26, 28, 29, 35, 36, 38, 39, 45, 46, 48, 49, 57, 58, 59, 67, 68, 69, 78, 79
9	3	5	7 = 123, 8 = 124, 9 = 13456	$B_1 = 12$, $B_2 = 13$, $B_3 = 14$, $B_4 = 15$, $B_5 = 16$	none

Note: **All the main effects are clear** for all of the designs in this table and will not be repeated in the column "Clear 2fi's."

Table 4B.5 128-Run 2^{k-p} Fractional Factorial Designs in 2^q Blocks
($k - p = 7$, $k = 8, 9$) (k = Lnumber of factors, 2^{k-p} = number of runs,
2^q = number of blocks)

k	p	q	Design Generators	Block Generators	Clear 2fi's
8	1	1	$8 = 1234567$	$B_1 = 1234$	all 28 2fi's
8	1	2	$8 = 1234567$	$B_1 = 1234, B_2 = 1256$	all 28 2fi's
8	1	3	$8 = 1234567$	$B_1 = 1234, B_2 = 1256,$ $B_3 = 1357$	all 28 2fi's
8	1	4	$8 = 123456$	$B_1 = 123, B_2 = 145,$ $B_3 = 246, B_4 = 17$	all 2fi's except 17
8	1	5	$8 = 123456$	$B_1 = 12, B_2 = 13,$ $B_3 = 45, B_4 = 46,$ $B_5 = 147$	all 2fi's except 12, 13, 23, 45, 46, 56, 78
8	1	6	$8 = 1234567$	$B_1 = 12, B_2 = 13,$ $B_3 = 14, B_4 = 15,$ $B_5 = 16, B_6 = 17$	none
9	2	1	$8 = 12345,$ $9 = 12367$	$B_1 = 1246$	all 36 2fi's
9	2	2	$8 = 12345,$ $9 = 12367$	$B_1 = 1246, B_2 = 1357$	all 36 2fi's
9	2	3	$8 = 12345,$ $9 = 12367$	$B_1 = 146, B_2 = 2346,$ $B_3 = 1357$	all 36 2fi's
9	2	4	$8 = 12345,$ $9 = 12367$	$B_1 = 12, B_2 = 134,$ $B_3 = 136, B_4 = 357$	all 2fi's except 12, 46
9	2	5	$8 = 12345,$ $9 = 12367$	$B_1 = 123, B_2 = 14,$ $B_3 = 25, B_4 = 16,$ $B_5 = 27$	all 2fi's except 14, 16, 25, 27, 38, 39, 46, 57, 89
9	2	6	$8 = 12345,$ $9 = 12367$	$B_1 = 12, B_2 = 13,$ $B_3 = 14, B_4 = 15,$ $B_5 = 16, B_6 = 17$	none

Note: **All the main effects are clear** for all of the designs in this table and will not be repeated in the column "Clear 2fi's."

REFERENCES

Atkinson, A.C. and Donev, A.N. (1992), *Optimum Experimental Designs*, New York: Oxford University Press.

Box, G.E.P. and Hunter, J.S. (1961), "The 2^{k-p} Fractional Factorial Designs," *Technometrics*, 3, 311–351 and 449–458.

Chen, J. (1992), "Some Results on 2^{n-k} Fractional Factorial Designs and Search for Minimum Aberration Designs," *Annals of Statistics*, 20, 2124–2141.

Chen, J. (1998), "Intelligent Search for 2^{13-6} and 2^{14-7} Minimum Aberration Designs, *Statistica Sinica*, 8, 1265–1270.

Chen, J. and Wu, C.F.J. (1991), "Some Results on s^{n-k} Fractional Factorial Designs with Minimum Aberration or Optimal Moments," *Annals of Statistics*, 19, 1028–1041.

Chen, J., Sun, D.X., and Wu, C.F.J. (1993), "A Catalogue of Two-Level and Three-Level Fractional Factorial Designs with Small Runs," *International Statistical Review*, 61, 131–145.

Daniel, C. (1962), "Sequences of Fractional Replicates in the 2^{p-q} Series," *Technometrics*, 58, 403–429.

Daniel, C. (1976), *Applications of Statistics to Industrial Experimentation*, New York: John Wiley & Sons.

ECHIP (1998), *ECHIP Version* 6.4, Hockessin, DE: ECHIP.

Fries, A. and Hunter, W.G. (1980), "Minimum Aberration 2^{k-p} Designs," *Technometrics*, 22, 601–608.

Kacker, R.N. and Tsui, K.L. (1990), "Interaction Graphs: Graphical Aids for Planning Experiments," *Journal of Quality Technology*, 22, 1–14.

Li, C.C. and Cheng, C.S. (1993), "Constructing Orthogonal Fractional Factorial Designs When Some Factor-Level Combinations Are Debarred," *Technometrics*, 35, 277–283.

Meyer, R.D., Steinberg, D.M., and Box, G. (1996), "Follow-up Designs to Resolve Confounding in Multifactor Experiments (with discussion), *Technometrics*, 38, 303–332.

Mukerjee, R. and Wu, C.F.J. (1999), "Blocking in Regular Fractional Factorials: a Projective Geometric Approach," *Annals of Statistics*, 27, 1256–1271.

Pignatiello, J.J., Jr. and Ramberg, J.S. (1985), "Discussion of 'Off-Line Quality Control, Parameter Design, and the Taguchi Method' by Kackar, R.N.," *Journal of Quality Technology*, 17, 198–206.

Pukelsheim, F. (1993), *Optimal Design of Experiments*, New York: John Wiley & Sons.

SAS Institute (1989), *SAS/QC Software: Reference, Version 6, First Edition*, Cary, NC: SAS Institute.

Sitter, R.R., Chen, J., and Feder, M. (1997), "Fractional Resolution and Minimum Aberration in Blocked 2^{n-p} Designs," *Technometrics*, 39, 382–390.

Sun, D.X., Wu, C.F.J., and Chen, Y. (1997), "Optimal Blocking Schemes for 2^{n} and 2^{n-k} Designs," *Technometrics*, 39, 298–307.

Tang, B. and Wu, C.F.J. (1996), "Characterization of the Minimum Aberration 2^{n-k} Designs in Terms of Their Complementary Designs," *Annals of Statistics*, 24, 2549–2559.

Wu, C.F.J. and Chen, Y. (1992), "A Graph-Aided Method for Planning Two-Level Experiments When Certain Interactions Are Important," *Technometrics*, 34, 162–175.

Young, J.C. (1996), "Blocking, Replication and Randomization—the Key to Effective Experiments: A Case Study," *Quality Engineering*, 9, 269–277.

CHAPTER 5

Full Factorial and Fractional Factorial Experiments at Three Levels

While two-level experiments require fewer runs and are more economical to conduct, there are practical considerations that make it desirable to study factors with more than two levels. In this chapter, experiments with three levels are considered. The mathematical structure for two-level designs and its ramification for data analysis have a natural extension to three-level designs. The main difference arises from the fact that each three-level factor has two degrees of freedom and there are two systems for parameterizing the interaction effects in three-level designs: the *orthogonal components system* and the *linear-quadratic system*. Standard analysis of variance is applicable to the former system while a new regression analysis strategy needs to be developed for the latter system. Strategies for handling multiple responses are also considered.

5.1 A SEAT-BELT EXPERIMENT

Consider an experiment to study the effect of four factors on the pull strength of truck seat belts following a crimping operation which joins an anchor and cable. The four factors are hydraulic pressure of the crimping machine (A), die flat middle setting (B), length of crimp (C), and anchor lot (D), each of which was studied at three levels as given in Table 5.1. The experimental plan is a three-level fractional factorial design in 27 runs and each run is replicated three times as shown in Table 5.2. The three levels of each factor are represented by 0, 1, and 2.

There are two quality characteristics: crimp tensile strength (lb), which has a minimum customer specification of 4000 lb, and flash (mm), which is excess metal from the crimping process that cannot exceed 14.00 mm. Strength

Table 5.1   **Factors and Levels, Seat-Belt Experiment**

	Level		
Factor	0	1	2
A. pressure (psi)	1100	1400	1700
B. die flat (mm)	10.0	10.2	10.4
C. crimp length (mm)	18	23	27
D. anchor lot (number)	P74	P75	P76

Table 5.2   **Design Matrix and Response Data, Seat-Belt Experiment**

Run	Factor				Strength			Flash		
	A	B	C	D						
1	0	0	0	0	5164	6615	5959	12.89	12.70	12.74
2	0	0	1	1	5356	6117	5224	12.83	12.73	13.07
3	0	0	2	2	3070	3773	4257	12.37	12.47	12.44
4	0	1	0	1	5547	6566	6320	13.29	12.86	12.70
5	0	1	1	2	4754	4401	5436	12.64	12.50	12.61
6	0	1	2	0	5524	4050	4526	12.76	12.72	12.94
7	0	2	0	2	5684	6251	6214	13.17	13.33	13.98
8	0	2	1	0	5735	6271	5843	13.02	13.11	12.67
9	0	2	2	1	5744	4797	5416	12.37	12.67	12.54
10	1	0	0	1	6843	6895	6957	13.28	13.65	13.58
11	1	0	1	2	6538	6328	4784	12.62	14.07	13.38
12	1	0	2	0	6152	5819	5963	13.19	12.94	13.15
13	1	1	0	2	6854	6804	6907	14.65	14.98	14.40
14	1	1	1	0	6799	6703	6792	13.00	13.35	12.87
15	1	1	2	1	6513	6503	6568	13.13	13.40	13.80
16	1	2	0	0	6473	6974	6712	13.55	14.10	14.41
17	1	2	1	1	6832	7034	5057	14.86	13.27	13.64
18	1	2	2	2	4968	5684	5761	13.00	13.58	13.45
19	2	0	0	2	7148	6920	6220	16.70	15.85	14.90
20	2	0	1	0	6905	7068	7156	14.70	13.97	13.66
21	2	0	2	1	6933	7194	6667	13.51	13.64	13.92
22	2	1	0	0	7227	7170	7015	15.54	16.16	16.14
23	2	1	1	1	7014	7040	7200	13.97	14.09	14.52
24	2	1	2	2	6215	6260	6488	14.35	13.56	13.00
25	2	2	0	1	7145	6868	6964	15.70	16.45	15.85
26	2	2	1	2	7161	7263	6937	15.21	13.77	14.34
27	2	2	2	0	7060	7050	6950	13.51	13.42	13.07

needs to be maximized to ensure that the seat belt does not come apart in an accident. Flash should be minimized because it causes cracks in a plastic boot covering this assembly which is put on during a later processing step. Originally, the recommended settings were pressure at 1700 lb and crimp length at 23 mm. The goal of the experiment was to identify factor settings which maximize the tensile strength and minimize the flash subject to the two constraints above. Overall, a mean strength of 7000 lb with standard deviation of 250 lb and mean flash of 13.50 mm with standard deviation of 0.2 mm are desired.

5.2 LARGER-THE-BETTER AND SMALLER-THE-BETTER PROBLEMS

The two response characteristics in the seat-belt experiment, strength and flash, differ substantially from the layer thickness in the epitaxial growth experiment (Chapter 3) and the free height in the leaf spring experiment (Chapter 4). Both layer thickness and free height had fixed target values and the purpose was to move the response y as closely as possible to a target value t. They were called nominal-the-best problems. In the seat-belt experiment, the strength should be as high as possible and the flash as low as possible. There is no fixed nominal value for either strength or flash. They are referred to as **larger-the-better** and **smaller-the-better** problems, respectively.

For nominal-the-best problems, step (ii) of its two-step procedure in (3.3) is easier to implement. One can often use physical knowledge or data analysis to identify an adjustment factor which has a big influence on $E(y)$. For larger-the-better problems, it is usually difficult to find a factor that can easily increase $E(y)$. If such a factor were available, it would have been used by the designer to solve the problem. There would have been no need to perform a statistically designed experiment. Because increasing the mean value is considered to be a more difficult task, it should be done in the first step. Therefore, the recommendation is to reverse the order of the two-step procedure in (3.3) for larger-the-better problems.

Two-Step Procedure for Larger-the-Better Problems

    (*i*) *Find factor settings that maximize $E(y)$.*

    (*ii*) *Find other factor settings that minimize $Var(y)$.*   (5.1)

Similarly, a two-step procedure for smaller-the-better problems can be stated as follows.

Two-Step Procedure for Smaller-the-Better Problems

    (*i*) *Find factor settings that minimize $E(y)$.*

    (*ii*) *Find other factor settings that minimize $Var(y)$.*   (5.2)

Since zero can be used as an ideal target value for smaller-the-better characteristics, one may consider minimizing $\sum_i (y_i - 0)^2 = \sum_i y_i^2$, the sum of squares of the y_i's, in a *single* step. That is, model $\sum_i y_i^2$ as a function of the factors and find factor settings that minimize $\sum_i y_i^2$.

For any of the two-step procedures to be effective, the factor employed in step (ii) should have a significant effect on the characteristic considered in step (ii) but *not* on the characteristic considered in step (i). This requirement may, however, not be met in some experiments. In these situations, an alternative is to analyze $\bar{y}$ and $\ln s^2$ separately and identify significant effects and optimum factor settings by jointly considering their effects on the characteristics to be optimized. The analysis will be less formal than the two-step procedures and subject-matter knowledge of the problem needs to be utilized.

5.3 3^k FULL FACTORIAL DESIGNS

In two-level experiments, only linear effects can be studied for quantitative factors, and comparisons between two factor levels can be made for qualitative factors. In many practical situations, three-level factors need to be considered. Three major reasons are as follows:

1. A quantitative factor like temperature may affect the response in a non-monotone fashion. Only two temperature settings preclude an analysis of the curvilinear relation between response and temperature. Three or more settings are required to detect a curvature effect.

2. A qualitative factor may have several levels, such as three machine types for manufacturing a part, three suppliers of the same component, or three hand positions in an ergonomic study. If the three machine types are used in current production, they should all be compared in the study. Unlike the quantitative situation, the response at one level of the qualitative factor cannot be used to infer about the response at another level (unless these two levels share an underlying latent variable). Therefore, one cannot compare three machine types based on a study of only two machine types. Moreover, it would be more time consuming and expensive to first study only two machine types in a two-level experiment and later add the third type in a follow-up experiment.

3. If x_0 is the nominal or current setting of a product or process, say the dimension of a part or injection pressure, it is common to study x_0 and two settings around x_0. This is particularly suitable if the goal is to improve the performance of the product/process by changing x_0, where either a lower or higher setting could realize this potential. It would be less time consuming to consider both higher and lower settings in a single experiment.

In order to understand three-level fractional factorial designs (of which the seat-belt experiment is a special case), we need to consider 3^k full factorial designs first. These designs study all the $3 \times \cdots \times 3 = 3^k$ combinations of k factors each at three levels. To illustrate the method to be given in this section, we use a simplified version of the seat-belt experiment in Table 5.2 by considering only the three factors A, B, and C. For these three factors, the 27-run experiment uses a full factorial 3^3 design with three replicates. Since a 3^3 design is a special case of a multi-way layout, the analysis of variance method introduced in Section 2.4 can be applied to this experiment. In this section, only the strength data are considered. Using analysis of variance, we can compute the sum of squares for main effects A, B, C, interactions $A \times B$, $A \times C$, $B \times C$, $A \times B \times C$, and the residual sum of squares. Details are given in Table 5.3. Each main effect has two degrees of freedom because each factor has three levels. Each two-factor interaction has $(3 - 1) \times (3 - 1) = 4$ degrees of freedom and the $A \times B \times C$ interaction has $(3 - 1) \times (3 - 1) \times (3 - 1) = 8$ degrees of freedom. Because there are three replicates, the residual degrees of freedom are 54 $[= 27 \times (3 - 1)]$.

Table 5.3 ANOVA Table, Simplified Seat-Belt Experiment

Source	Degrees of Freedom	Sum of Squares	Mean Squares	F	p Value
A	2	34,621,746	17,310,873	85.58	0.000
B	2	938,539	469,270	2.32	0.108
C	2	9,549,481	4,774,741	23.61	0.000
$A \times B$	4	3,298,246	824,561	4.08	0.006
AB	2	2,727,451	1,363,725	6.74	0.002
AB^2	2	570,795	285,397	1.41	0.253
$A \times C$	4	3,872,179	968,045	4.79	0.002
AC	2	2,985,591	1,492,796	7.38	0.001
AC^2	2	886,587	443,294	2.19	0.122
$B \times C$	4	448,348	112,087	0.55	0.697
BC	2	427,214	213,607	1.06	0.355
BC^2	2	21,134	10,567	0.05	0.949
$A \times B \times C$	8	5,206,919	650,865	3.22	0.005
ABC	2	4,492,927	2,246,464	11.11	0.000
ABC^2	2	263,016	131,508	0.65	0.526
AB^2C	2	205,537	102,768	0.51	0.605
AB^2C^2	2	245,439	122,720	0.61	0.549
residual	54	10,922,599	202,270		
total	80	68,858,056			

As in Section 2.4 we can use the F statistic to test the significance of each of the factors and of the four interactions. To test the main effect A for the strength data, we have

$$F_A = \frac{SS_A/2}{SS_r/54} = \frac{MS_A}{MS_r} = \frac{17{,}310{,}873}{202{,}270} = 85.58,$$

where SS_A denotes the sum of squares of factor A and SS_r the residual sum of squares. Similarly,

$$F_B = \frac{469{,}270}{202{,}270} = 2.32, \qquad F_C = \frac{4{,}774{,}741}{202{,}270} = 23.61,$$

$$F_{A \times B} = \frac{824{,}561}{202{,}270} = 4.08, \qquad F_{A \times C} = \frac{968{,}045}{202{,}270} = 4.79,$$

$$F_{B \times C} = \frac{112{,}087}{202{,}270} = 0.55, \qquad F_{A \times B \times C} = \frac{650{,}865}{202{,}270} = 3.22.$$

Based on the p values given in the last column of Table 5.3, A, C, $A \times B$, $A \times C$, and $A \times B \times C$ are significant.

The sum of squares for a two-factor interaction can be further decomposed into two components, each with two degrees of freedom. Consider, for example, factors A and B, whose levels are denoted by x_1 and x_2, respectively in Table 5.4. Its $A \times B$ interaction has two components denoted by AB and AB^2, where AB represents the contrasts among the response values whose x_1 and x_2 satisfy

$$x_1 + x_2 = 0, 1, 2 \ (mod \ 3),$$

and AB^2 represents the contrasts among the response values whose x_1 and x_2 satisfy

$$x_1 + 2x_2 = 0, 1, 2 \ (mod \ 3).$$

(In modulus 3 calculus, any multiple of 3 equals zero.) Because the contrasts compare values at three levels [i.e., $0, 1, 2 \ (mod \ 3)$], each of AB and AB^2 has two degrees of freedom. The nine level combinations of A and B can be

Table 5.4 Factor A and B Combinations (x_1 denotes the levels of factor A and x_2 denotes the levels of factor B)

x_1	x_2 0	1	2
0	$\alpha i(y_{00})$	$\beta k(y_{01})$	$\gamma j(y_{02})$
1	$\beta j(y_{10})$	$\gamma i(y_{11})$	$\alpha k(y_{12})$
2	$\gamma k(y_{20})$	$\alpha j(y_{21})$	$\beta i(y_{22})$

represented by the cells in the 3×3 square in Table 5.4. The letters $\alpha, \beta, \gamma, i, j, k$ represent groupings of the observations, e.g., α represents the group of observations y_{00}, y_{12}, and y_{21}. Therefore, the nine combinations in Table 5.4 are arranged in two mutually orthogonal Latin squares, one represented by α, β, γ and the other by i, j, k. Let

$$\bar{y}_\alpha = \tfrac{1}{3}(y_{00} + y_{12} + y_{21}),$$

$$\bar{y}_\beta = \tfrac{1}{3}(y_{01} + y_{10} + y_{22}),$$

$$\vdots$$

$$\bar{y}_k = \tfrac{1}{3}(y_{01} + y_{12} + y_{20}),$$

where each y_{ij} represents the average of n replicates in the (i, j) cell. The AB interaction component represents the contrasts among $\bar{y}_\alpha$, $\bar{y}_\beta$, and $\bar{y}_\gamma$, because $\bar{y}_\alpha$, $\bar{y}_\beta$, and $\bar{y}_\gamma$ represent the averages of the observations with $x_1 + x_2 = 0, 1, 2$ (*mod* 3), respectively. By treating the 3×3 table in Table 5.4 as a Latin square design in the treatments α, β, and γ, the sum of squares for the AB interaction component, denoted by SS_{AB}, can be interpreted as the treatment sum of squares for the Latin square design. By extending an ANOVA formula for latin squares (see Table 2.19 in Section 2.6),

$$SS_{AB} = 3n\left[(\bar{y}_\alpha - \bar{y}_.)^2 + (\bar{y}_\beta - \bar{y}_.)^2 + (\bar{y}_\gamma - \bar{y}_.)^2\right], \tag{5.3}$$

where $\bar{y}_. = (\bar{y}_\alpha + \bar{y}_\beta + \bar{y}_\gamma)/3$ and n is the number of replicates. For the simplified seat-belt experiment, $n = 9$ ($= 3 \times 3$) because factor C has three levels and there are three replicates for each level combination of factors A, B, and C. Also, $\bar{y}_\alpha = 6024.407$, $\bar{y}_\beta = 6177.815$, and $\bar{y}_\gamma = 6467.0$, so that $\bar{y}_. = 6223.074$ and

$$SS_{AB} = (3)(9)\left[(6024.407 - 6223.074)^2 + (6177.815 - 6223.074)^2\right.$$

$$\left. + (6467.0 - 6223.074)^2\right]$$

$$= 2{,}727{,}451. \tag{5.4}$$

Similarly, the AB^2 interaction component represents the contrasts among the three groups represented by the letters i, j, and k. The corresponding $\bar{y}_i$, $\bar{y}_j$, and $\bar{y}_k$ values represent the averages of observations with $x_1 + 2x_2 = 0, 1, 2$ (*mod* 3), respectively, and the formula for SS_{AB^2} can be defined in a similar manner.

The decomposition of the $A \times B \times C$ interaction can be similarly defined. It has the four components ABC, ABC^2, AB^2C, and AB^2C^2, which represent the contrasts among the three groups of (x_1, x_2, x_3) satisfying each of the four systems of equations,

$$x_1 + \ x_2 + \ x_3 = 0, 1, 2 \ (mod \ 3),$$

$$x_1 + \ x_2 + 2x_3 = 0, 1, 2 \ (mod \ 3),$$

$$x_1 + 2x_2 + \ x_3 = 0, 1, 2 \ (mod \ 3),$$

$$x_1 + 2x_2 + 2x_3 = 0, 1, 2 \ (mod \ 3).$$

(5.5)

Note that the coefficient of x_1 in the equation corresponds to the exponent of factor A. A similar correspondence holds for x_2 and B and for x_3 and C. For example, AB^2C corresponds to the equation $x_1 + 2x_2 + x_3 = 0, 1, 2 \ (mod \ 3)$. To avoid ambiguity, *the convention that the coefficient for the first nonzero factor is* 1 *will be used.* Therefore, ABC^2 is used instead of A^2B^2C, even though the two are equivalent. Their equivalence can be explained as follows. For A^2B^2C, the three groups satisfying $2x_1 + 2x_2 + x_3 = 2(x_1 + x_2 + 2x_3) = 0, 1, 2 \ (mod \ 3)$ can be equivalently defined by $x_1 + x_2 + 2x_3 = 0, 1, 2 \ (mod \ 3)$ by relabeling the groups.

Unlike for two-level experiments, we cannot use the $A \times B$ and AB notation interchangeably. *The $A \times B$ interaction represents four degrees of freedom while each of AB and AB^2 represents a component of $A \times B$ with two degrees of freedom.* Because AB and AB^2 are orthogonal components of $A \times B$ and ABC, ABC^2, AB^2C, and AB^2C^2 are orthogonal components of $A \times B \times C$, we refer to this system of parameterization as the **orthogonal components system**.

The sums of squares for the components AB, AB^2, AC, AC^2, BC, BC^2, $ABC, \dots, AB^2C^2$ are given in Table 5.3. With this more elaborate decomposition, an F test can be applied to each of the 10 sums of squares. Judging from the values of the F statistic and the corresponding p values given in Table 5.3, only AB, AC, and ABC appear to be significant. The question is whether the statistical significance of interaction components like AB can be given any intuitive interpretation. Recall that AB represents the contrasts among the $\bar{y}_\alpha$, $\bar{y}_\beta$, and $\bar{y}_\gamma$ values given above. From the distribution of the nine factor combinations in Table 5.4 among α, β, and γ, there is no obvious physical interpretation of the contrasts among the three groups represented by α, β, and γ. An alternative and more interpretable decomposition of the $A \times B$ interaction needs to be developed and will be considered in Section 5.6.

Another way to represent the 13 components of the factorial effects of the 3^3 design is given in Table 5.5. Let x_1, x_2, and x_3 represent the entries of the columns A, B, and C, respectively. Then the other interaction components can be obtained by using the equations that define the components. For example, each entry of the column AB^2C equals $x_1 + 2x_2 + x_3 \ (mod \ 3)$ as explained before. The main use of this matrix is in defining 27-run fractional factorial designs, which will be discussed in the next section. Note that the

Table 5.5 Main Effects and Interaction Components for the 3^3 Design

Run	A	B	C	AB	AB^2	AC	AC^2	BC	BC^2	ABC	ABC^2	AB^2C	AB^2C^2
1	0	0	0	0	0	0	0	0	0	0	0	0	0
2	0	0	1	0	0	1	2	1	2	1	2	1	2
3	0	0	2	0	0	2	1	2	1	2	1	2	1
4	0	1	0	1	2	0	0	1	1	1	1	2	2
5	0	1	1	1	2	1	2	2	0	2	0	0	1
6	0	1	2	1	2	2	1	0	2	0	2	1	0
7	0	2	0	2	1	0	0	2	2	2	2	1	1
8	0	2	1	2	1	1	2	0	1	0	1	2	0
9	0	2	2	2	1	2	1	1	0	1	0	0	2
10	1	0	0	1	1	1	1	0	0	1	1	1	1
11	1	0	1	1	1	2	0	1	2	2	0	2	0
12	1	0	2	1	1	0	2	2	1	0	2	0	2
13	1	1	0	2	0	1	1	1	1	2	2	0	0
14	1	1	1	2	0	2	0	2	0	0	1	1	2
15	1	1	2	2	0	0	2	0	2	1	0	2	1
16	1	2	0	0	2	1	1	2	2	0	0	2	2
17	1	2	1	0	2	2	0	0	1	1	2	0	1
18	1	2	2	0	2	0	2	1	0	2	1	1	0
19	2	0	0	2	2	2	2	0	0	2	2	2	2
20	2	0	1	2	2	0	1	1	2	0	1	0	1
21	2	0	2	2	2	1	0	2	1	1	0	1	0
22	2	1	0	0	1	2	2	1	1	0	0	1	1
23	2	1	1	0	1	0	1	2	0	1	2	2	0
24	2	1	2	0	1	1	0	0	2	2	1	0	2
25	2	2	0	1	0	2	2	2	2	1	1	0	0
26	2	2	1	1	0	0	1	0	1	2	0	1	2
27	2	2	2	1	0	1	0	1	0	0	2	2	1

three levels of a factor are denoted by 0, 1, and 2, because they appear naturally in the modulus 3 calculus that is employed in the construction of three-level designs.

For the 3^4 designs with 81 runs, the 80 degrees of freedom can be divided into 40 components, consisting of the four main effects A, B, C, D, the 12 components

$$AB, AB^2, AC, AC^2, AD, AD^2, BC, BC^2, BD, BD^2, CD, CD^2$$

for two-factor interactions, the 16 components

$$ABC, ABC^2, AB^2C, AB^2C^2, ABD, ABD^2, AB^2D, AB^2D^2,$$

$$ACD, ACD^2, AC^2D, AC^2D^2, BCD, BCD^2, BC^2D, BC^2D^2$$

for three-factor interactions, and the eight components

$$ABCD, ABCD^2, ABC^2D, ABC^2D^2,$$

$$AB^2CD, AB^2CD^2, AB^2C^2D \text{ and } AB^2C^2D^2$$

for the $A \times B \times C \times D$ interaction. Details of these components are left as an exercise.

5.4 3^{k-p} FRACTIONAL FACTORIAL DESIGNS

As for two-level fractional factorial designs, the main justification for using three-level fractional factorial designs is run size economy. Take, for example, the 3^4 designs in 81 runs. According to the discussion at the end of the previous section, out of the 80 degrees of freedom in the design, 48 (i.e., 24 components each having 2 degrees of freedom) are for estimating three-factor and four-factor interactions. According to the hierarchical ordering principle (whose extension to designs with more than two levels is obvious), these interactions are difficult to interpret and usually insignificant. Unless the experiment is not costly, it is not economical to use a 3^4 design. Usually, it is more efficient to use a one-third or one-ninth fraction of the 3^4 design.

Returning to the original seat-belt experiment, it employs a one-third fraction of the 3^4 design by choosing the column for factor D (lot number) to be equal to column A + column B + column C (*mod* 3); that is, every entry in column D is the sum of the first three entries modulus 3. For example, for run number 12, the levels for A, B, and C are 1, 0, and 2, respectively, so that $1 + 0 + 2 = 0$ (*mod* 3) or level 0 for D. Recall that *in modulus 3 calculus, any multiple of 3 equals zero*. As for two-level designs, this relationship can be represented by the notation

$$D = ABC.$$

If $x_1, \ldots, x_4$ are used to represent these four columns, then $x_4 = x_1 + x_2 + x_3$ (*mod* 3), or equivalently,

$$x_1 + x_2 + x_3 + 2x_4 = 0 \ (mod \ 3), \tag{5.6}$$

which can be represented by

$$\mathbf{I} = ABCD^2. \tag{5.7}$$

In (5.7), the squared power of D corresponds to the coefficient 2 of $2x_4$ in (5.6). The column of 0's, denoted by $\mathbf{I}$, is the identity in group theory since $0 + z = z$ for any z and plays the same role as the column of $+$'s in two-level designs. The relation in (5.7) is called the defining relation for the 3^{4-1} design used for the seat-belt experiment.

Its aliasing patterns can be deduced from the defining relation (5.6) or (5.7). For example, by adding $2x_1$ to both sides of (5.6), we have

$$2x_1 = 3x_1 + x_2 + x_3 + 2x_4 = x_2 + x_3 + 2x_4 \ (mod\ 3), \qquad (5.8)$$

which means that the three groups defined by $2x_1 = 0, 1, 2 \ (mod\ 3)$ [or equivalently by $x_1 = 0, 1, 2 \ (mod\ 3)$] are *identical* to the three groups defined by $x_2 + x_3 + 2x_4 = 0, 1, 2 \ (mod\ 3)$. Noting that the contrasts among the three groups in the former case define the main effect A and the contrasts among the three groups in the latter case define the interaction BCD^2, we say that A and BCD^2 are **aliased**. Similarly, adding x_1 to both sides of (5.6) leads to the aliasing of A and AB^2C^2D. Therefore, A has two aliases, BCD^2 and AB^2C^2D, or notationally $A = BCD^2 = AB^2C^2D$. By following the same derivation, it is easy to show that the following effects are aliased:

$$
\begin{aligned}
A &= BCD^2 &&= AB^2C^2D, \\
B &= ACD^2 &&= AB^2CD^2, \\
C &= ABD^2 &&= ABC^2D^2, \\
D &= ABC &&= ABCD, \\
AB &= CD^2 &&= ABC^2D, \\
AB^2 &= AC^2D &&= BC^2D, \\
AC &= BD^2 &&= AB^2CD, \qquad (5.9) \\
AC^2 &= AB^2D &&= BC^2D^2, \\
AD &= AB^2C^2 &&= BCD, \\
AD^2 &= BC &&= AB^2C^2D^2, \\
BC^2 &= AB^2D^2 &&= AC^2D^2, \\
BD &= AB^2C &&= ACD, \\
CD &= ABC^2 &&= ABD.
\end{aligned}
$$

These 13 sets of aliased effects represent the 13 orthogonal columns in the 3^{4-1} design. These 13 columns can be identified from the columns in Table 5.5 for the 3^3 design. The only extra work is to interpret each column as a triplet of aliased effects given in (5.9).

If three-factor interactions are assumed negligible, from the aliasing relations in (5.9), A, B, C, D, AB^2, AC^2, AD, BC^2, BD, and CD can be estimated. Using the terminology in Section 4.2, these main effects or components of two-factor interactions are called **clear** because they are not aliased with any other main effects or two-factor interaction components. A two-factor interaction, say $A \times B$, is called *clear* if both of its components,

AB and AB^2, are clear in the sense just defined. It is interesting that each of the six two-factor interactions has only one component that is clear; the other component is aliased with one component of another two-factor interaction. For example, for $A \times B$, AB^2 is clear but AB is aliased with CD^2. Similarly, a main effect or two-factor interaction component is said to be **strongly clear** if it is not aliased with any other main effects, two-factor or three-factor interaction components. A two-factor interaction is said to be *strongly clear* if both of its components are strongly clear.

Generally, a 3^{k-p} design is a fractional factorial design with k factors in 3^{k-p} runs. It is a (3^{-p})th fraction of the 3^k design. The fractional plan is defined by p independent generators. For example, the 3^{5-2} design with $D = AB$, $E = AB^2C$ has 27 runs and five factors. The one-ninth fraction is defined by $\mathbf{I} = ABD^2 = AB^2CE^2$, from which two additional relations can be obtained:

$$\mathbf{I} = (ABD^2)(AB^2CE^2) = A^2CD^2E^2 = AC^2DE$$

and

$$\mathbf{I} = (ABD^2)(AB^2CE^2)^2 = B^2C^2D^2E = BCDE^2.$$

Therefore the defining contrast subgroup for this design consists of the following defining relation:

$$\mathbf{I} = ABD^2 = AB^2CE^2 = AC^2DE = BCDE^2. \tag{5.10}$$

This subgroup has four words, whose lengths are 3, 4, 4, and 4. As in the two-level case (see Section 4.2), A_i is used to denote the number of words of length i in the subgroup and $W = (A_3, A_4, \ldots)$ to denote the wordlength pattern. Based on W, the definitions of **resolution** and **minimum aberration** are the same as given before in Section 4.2. For the 3^{5-2} design given above, $W = (1, 3, 0)$. Another 3^{5-2} design given by $D = AB$, $E = AB^2$ has the defining contrast subgroup

$$\mathbf{I} = ABD^2 = AB^2E^2 = ADE = BDE^2, \tag{5.11}$$

with the wordlength pattern $W = (4, 0, 0)$. According to the aberration criterion, the first design has less aberration than the second design. Moreover, it can be shown that the first design has minimum aberration.

How many factors can a 3^{k-p} design study? By writing $n = k - p$, this design has 3^n runs with the independent generators $x_1, x_2, \ldots, x_n$. By following the same derivation for the 3^3 design in Section 5.3, we can obtain altogether $(3^n - 1)/2$ orthogonal columns as different combinations of $\sum_{i=1}^{n} \alpha_i x_i$ with $\alpha_i = 0, 1, 2$, where at least one α_i should not be zero and the first nonzero α_i should be written as "1" to avoid duplication. For $n = 3$, the $(3^n - 1)/2 = 13$ columns were given in Table 5.5. Thus, the maximum number of factors (i.e., the maximum k value) in a 3^{k-p} design with $n = k - p$ is $(3^n - 1)/2$.

A general algebraic treatment of 3^{k-p} designs can be found in Kempthorne (1952).

In Tables 5A.1–5A.3 of Appendix 5A, selected 3^{k-p} designs are given for 9, 27, and 81 runs. For 9 and 27 runs, the collection of designs covers the whole range of factors. For 81 runs, it covers designs up to 10 factors. These designs are based on the computer search results in Chen, Sun, and Wu (1993) and the theoretical results in Suen, Chen, and Wu (1997). For each k and p, the minimum aberration designs are listed first. For $5 \leq k \leq 8$ and 27 runs, Table 5A.2 includes designs that do not have minimum aberration but possess other desirable properties to be discussed later. In each case only the design generators are given. The remaining defining relations can be obtained by multiplying the design generators as shown before. The last column gives the collection of main effects, two-factor interactions or two-factor interaction components that are clear. In practical applications, the 9-run and 27-run designs are used more often because they do not require many runs and can study up to 4 and 13 factors, respectively. The 81-run designs are used less often, except when it is not costly to conduct 81 runs. More economical designs for three-level factors include the 18-run and 36-run orthogonal arrays in Chapter 7 and the central composite designs in Chapter 9.

In Table 5A.2, there are two 3^{5-2} designs of resolution III. Their defining contrast subgroups are given in (5.10) and (5.11). From (5.10), it can be shown that for the first design (which has minimum aberration), only C, E, and the CE interaction component are clear. On the other hand, from (5.11), it can be shown that for the second design, C, $A \times C$, $B \times C$, $C \times D$, and $C \times E$ are clear. Thus, the second design is preferred if one of the five factors is far more important than the others and its interactions with the other factors are also likely to be important. Use of this design will ensure that these effects can be estimated clearly. On the other hand, the first design allows a richer combination of main effects and components of two-factor interactions to be estimated. See Sun and Wu (1994) for details. Sun and Wu also provides an explanation for why the non-minimum aberration designs for $k = 6, 7, 8$ in Table 5A.2 may be preferred.

The second 3^{5-2} design considered above does not include the important factor C in its defining relations. By removing C from the list of factors being considered for the 3^{5-2} design, its defining contrast subgroup given in (5.11) can be treated as the defining contrast subgroup for the reduced problem with four factors (i.e., A, B, D, E) and nine runs. It is then easy to show (by comparison with the minimum aberration 3^{4-2} design in Table 5A.1) that the defining contrast subgroup in (5.11) corresponds to the minimum aberration 3^{4-2} design. In general, if for reasons given above, one factor should not be included in the defining contrast subgroup for a 3^{k-p} design, then the best way to construct a design is to choose a minimum aberration $3^{(k-1)-p}$ design for the remaining $k-1$ factors and then use its defining contrast subgroup for the original k factors to construct a 3^{k-p} design. This method is exactly the same as the one described in Section 4.5 for 2^{k-p} designs. It can also be used for designs with factors having more levels and will not be repeated later.

The choice of a minimum aberration design and use of clear interactions is not as important for three-level experiments as for two-level experiments. For three-level experiments, the seriousness of effect aliasing is lessened by the use of a flexible data analysis strategy, to be presented in Section 5.6.

5.5   SIMPLE ANALYSIS METHODS: PLOTS AND ANALYSIS OF VARIANCE

A simple analysis starts by making a main effects plot and interaction plots to see what effects might be important. This step can be followed by a formal analysis like analysis of variance (as discussed in Section 5.3) and half-normal plots.

The strength data will be considered first. The location main effects and interaction plots are given in Figures 5.1 and 5.2. The main effects plot suggests that factor A is the most important followed by factors C and D. The interaction plots in Figure 5.2 suggest that there may be interactions because the lines are not parallel.

In Section 5.3, the 26 degrees of freedom in the experiment were grouped into 13 sets of effects. The corresponding ANOVA table in Table 5.6 gives the sum of squares for these 13 effects. Based on the p values in Table 5.6, clearly the factor A, C, and D main effects are significant. There are also

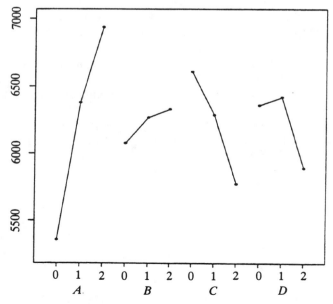

Figure 5.1. Main Effects Plot of Strength Location, Seat-Belt Experiment.

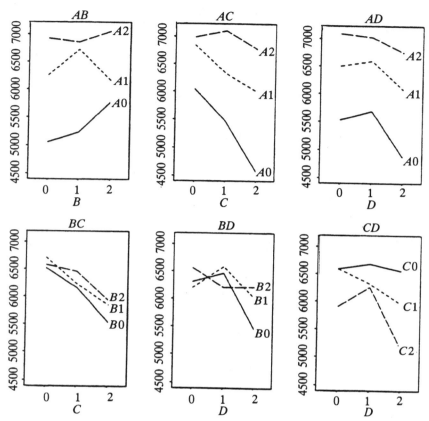

Figure 5.2. Interaction Plots of Strength Location, Seat-Belt Experiment.

two aliased sets of effects that are significant, $AB = CD^2$ and $AC = BD^2$. These findings are consistent with those based on the main effects plot and interaction plots. In particular, the significance of AB and CD^2 is supported by the $A \times B$ and $C \times D$ interaction plots, and the significance of AC and BD^2 is supported by the $A \times C$ and $B \times D$ interaction plots.

As explained in Chapter 4, aliased effects in two-level experiments can be de-aliased by performing a follow-up experiment. Alternately, *a priori* knowledge may suggest that one of the aliased effects can be dismissed. Fortunately for three-level designs, running a follow-up experiment for de-aliasing effects may be avoided by using an alternative system for parameterizing the interaction effects. A data analysis strategy based on this system will be presented in the next section and be used to reanalyze the data. One advantage of the new analysis method is its ability to simultaneously estimate some components of the $A \times B$ interaction and of the $C \times D$ interaction, even though the AB and CD^2 components are aliased.

Next, the analysis of dispersion (i.e., $\ln s^2$) for the strength data will be considered. The corresponding strength main effects plot and interaction

Table 5.6 ANOVA Table for Strength Location, Seat-Belt Experiment

Source	Degrees of Freedom	Sum of Squares	Mean Squares	F	p Value
A	2	34,621,746	17,310,873	85.58	0.000
B	2	938,539	469,270	2.32	0.108
$AB = CD^2$	2	2,727,451	1,363,725	6.74	0.002
AB^2	2	570,795	285,397	1.41	0.253
C	2	9,549,481	4,774,741	23.61	0.000
$AC = BD^2$	2	2,985,591	1,492,796	7.38	0.001
AC^2	2	886,587	443,294	2.19	0.122
$BC = AD^2$	2	427,214	213,607	1.06	0.355
BC^2	2	21,134	10,567	0.05	0.949
D	2	4,492,927	2,246,464	11.11	0.000
AD	2	263,016	131,508	0.65	0.526
BD	2	205,537	102,768	0.51	0.605
CD	2	245,439	122,720	0.61	0.549
residual	54	10,922,599	202,270		

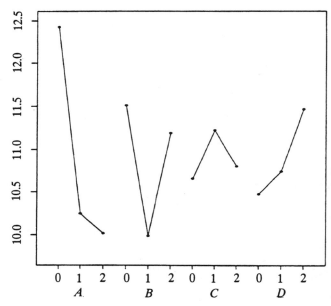

Figure 5.3. Main Effects Plot of Strength Dispersion, Seat-Belt Experiment.

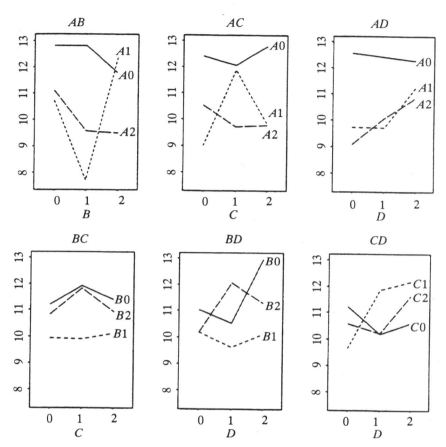

Figure 5.4. Interaction Plots of Strength Dispersion, Seat-Belt Experiment.

plots are displayed in Figures 5.3 and 5.4. Since there is no replication for the dispersion analysis, analysis of variance cannot be used to test effect significance. Instead, a half-normal plot can be drawn as follows. The 26 degrees of freedom in the experiment can be divided into 13 groups, each having two degrees of freedom. These 13 groups correspond to the 13 rows in Table 5.6 which consist of four main effects and nine interaction components. The two degrees of freedom in each group can be decomposed further into a linear effect and a quadratic effect. As explained in Section 1.8 these effects are defined by the contrast vectors $\frac{1}{\sqrt{2}}(-1, 0, 1)$ and $\frac{1}{\sqrt{6}}(1, -2, 1)$, respectively, where the values in the vectors are associated with the $\ln s^2$ values at the levels $(0, 1, 2)$ for the group. Because the linear and quadratic effects are standardized and orthogonal to each other, these 26 effect estimates can be plotted on the half-normal probability scale as in Figure 5.5. Informal

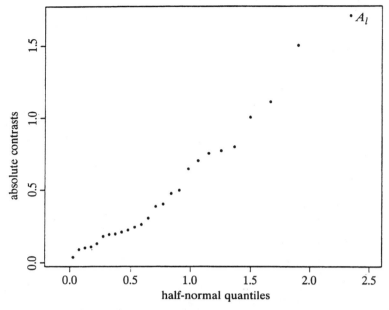

Figure 5.5. Half-Normal Plot of Strength Dispersion Effects, Seat-Belt Experiment.

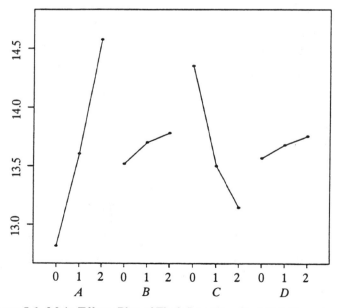

Figure 5.6. Main Effects Plot of Flash Location, Seat-Belt Experiment.

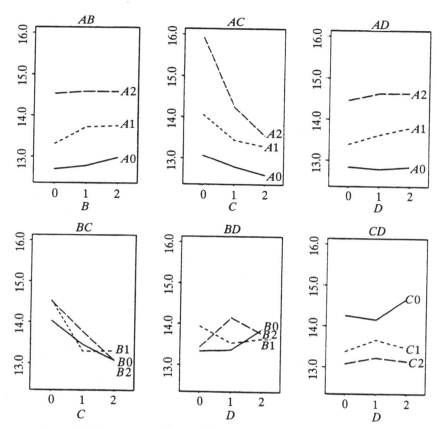

Figure 5.7. Interaction Plots of Flash Location, Seat-Belt Experiment.

analysis of the plot suggests that the factor A linear effect may be significant. This can be confirmed by using Lenth's method. The t_{PSE} value for the A linear effect is 3.99, which has a p value of 0.050 (EER). The details are left as an exercise.

A similar analysis can be performed to identify the flash location and dispersion effects. The main effects and interaction plots for flash location are displayed in Figures 5.6 and 5.7. The corresponding ANOVA table appears in Table 5.7. It identifies the A and C main effects and the BC^2 and AC^2 interaction components as significant. There is also one aliased set of effects that is significant, $AC = BD^2$. The significance is more likely attributed to AC than to BD^2, since both parent factors A and C of AC are significant while the parent factors B and D of BD are not. This argument is justified by the effect heredity principle (see Section 3.5).

Table 5.7   ANOVA Table for Flash Location, Seat-Belt Experiment

Source	Degrees of Freedom	Sum of Squares	Mean Squares	F	p Value
A	2	41.72712	20.86356	117.89	0.000
B	2	0.98174	0.49087	2.77	0.071
$AB = CD^2$	2	0.49137	0.24568	1.39	0.258
AB^2	2	0.05381	0.02690	0.15	0.859
C	2	20.86782	10.43391	58.96	0.000
$AC = BD^2$	2	4.22123	2.11062	11.93	0.000
AC^2	2	5.65452	2.82726	15.98	0.000
$BC = AD^2$	2	0.09154	0.04577	0.26	0.773
BC^2	2	1.71374	0.85687	4.84	0.012
D	2	0.47708	0.23854	1.35	0.268
AD	2	0.85537	0.42768	2.42	0.099
BD	2	0.13797	0.06898	0.39	0.679
CD	2	0.32670	0.16335	0.92	0.404
residual	54	9.55680	0.17698		

The main effects and interaction plots for flash dispersion are displayed in Figures 5.8 and 5.9. The half-normal plot of the 26 effects for flash dispersion can be drawn using the method described before for strength dispersion and is displayed in Figure 5.10. From the half-normal plot, only the A linear effect is identified as significant (which again can be confirmed by using Lenth's method).

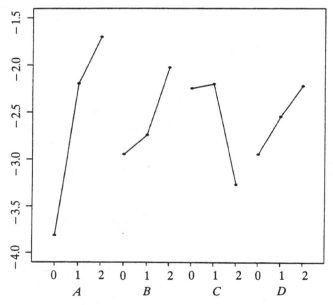

Figure 5.8. Main Effects Plot of Flash Dispersion, Seat-Belt Experiment.

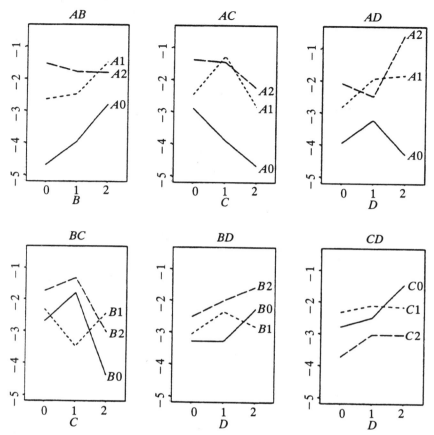

Figure 5.9. Interaction Plots of Flash Dispersion, Seat-Belt Experiment.

As in Section 4.2, we can determine the optimal factor settings that maximize the strength location by examining the main effects plot and interaction plots in Figures 5.1 and 5.2 that correspond to the significant effects identified in the ANOVA table in Table 5.6. The same method can be used to determine the optimal factor settings that minimize the strength dispersion, flash location, and flash dispersion, respectively. The details are left as an exercise. The most obvious findings are that level 2 of factor A should be chosen to maximize strength while level 0 of factor A should be chosen to minimize flash. There is an obvious conflict in meeting the two objectives. Trade-off strategies for handling multiple characteristics and conflicting objectives will be considered in Section 5.7. A joint analysis of strength and flash is deferred to that section.

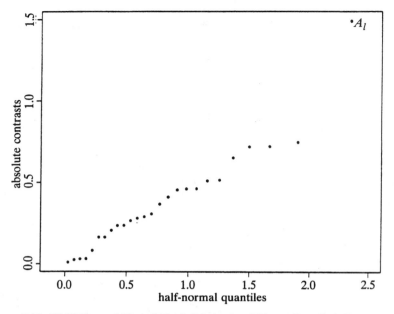

Figure 5.10. Half-Normal Plot of Flash Dispersion Effects, Seat-Belt Experiment.

5.6   AN ALTERNATIVE ANALYSIS METHOD

In the seat-belt experiment, the factors A, B, and C are quantitative. As in Section 1.8, the two degrees of freedom in a quantitative factor, say A, can be decomposed into the linear and quadratic components. Letting y_0, y_1, and y_2 represent the observations at level 0, 1, and 2, the *linear effect* is defined as

$$y_2 - y_0$$

and the *quadratic effect* as

$$(y_2 + y_0) - 2y_1,$$

which can be re-expressed as the difference between two consecutive linear effects $(y_2 - y_1) - (y_1 - y_0)$. Mathematically, the linear and quadratic effects are represented by two mutually orthogonal vectors:

$$A_l = \frac{1}{\sqrt{2}}(-1, 0, 1), \qquad A_q = \frac{1}{\sqrt{6}}(1, -2, 1). \qquad (5.12)$$

For the sake of brevity, they are also referred to as the l and q effects. The scaling constants $\sqrt{2}$ and $\sqrt{6}$ yield vectors with unit length. Using contrast vectors with unit length allows a meaningful statistical comparison of their corresponding estimates in a half-normal plot. The linear (or quadratic) effect is obtained by taking the inner product between A_l (or A_q) and the vector $\mathbf{y} = (y_0, y_1, y_2)$. For factor B, B_l and B_q are similarly defined. Then the four degrees of freedom in the $A \times B$ interaction can be decomposed into four mutually orthogonal terms: $(AB)_{ll}$, $(AB)_{lq}$, $(AB)_{ql}$, $(AB)_{qq}$, which are defined as follows: for $i, j = 0, 1, 2$,

$$(AB)_{ll}(i, j) = A_l(i)B_l(j), \qquad (AB)_{lq}(i, j) = A_l(i)B_q(j),$$
$$(AB)_{ql}(i, j) = A_q(i)B_l(j), \qquad (AB)_{qq}(i, j) = A_q(i)B_q(j). \tag{5.13}$$

They are called the **linear-by-linear, linear-by-quadratic, quadratic-by-linear,** and **quadratic-by-quadratic** interaction effects. They are also referred to as the $l \times l$, $l \times q$, $q \times l$, and $q \times q$ effects. It is easy to show that they are orthogonal to each other. Using the nine level combinations of factors A and B, $y_{00}, \ldots, y_{22}$ given in Table 5.4, the contrasts $(AB)_{ll}$, $(AB)_{lq}$, $(AB)_{ql}$, $(AB)_{qq}$ can be expressed as follows:

$$(AB)_{ll}: \tfrac{1}{2}\{(y_{22} - y_{20}) - (y_{02} - y_{00})\},$$

$$(AB)_{lq}: \frac{1}{2\sqrt{3}}\{(y_{22} + y_{20} - 2y_{21}) - (y_{02} + y_{00} - 2y_{01})\},$$

$$(AB)_{ql}: \frac{1}{2\sqrt{3}}\{(y_{22} + y_{02} - 2y_{12}) - (y_{20} + y_{00} - 2y_{10})\},$$

$$(AB)_{qq}: \tfrac{1}{6}\{(y_{22} + y_{20} - 2y_{21}) - 2(y_{12} + y_{10} - 2y_{11}) + (y_{02} + y_{00} - 2y_{01})\}.$$

An $(AB)_{ll}$ interaction effect measures the difference between the conditional linear B effects at levels 0 and 2 of factor A. A significant $(AB)_{ql}$ interaction effect means that there is curvature in the conditional linear B effect over the three levels of factor A. The other interaction effects $(AB)_{lq}$ and $(AB)_{qq}$ can be similarly interpreted. Extension of (5.13) to three-factor and higher order interactions is straightforward and is left as an exercise. This system of parameterizing main and interaction effects is referred to as the **linear-quadratic system.**

For designs of at least resolution V, all the main effects and two-factor interactions are clear. Then, further decomposition of these effects according to the linear-quadratic system allows all the effects (each with one degree of freedom) to be compared in a half-normal plot. Note that for effects to be compared in a half-normal plot, they should be uncorrelated and have the same variance.

For designs with resolution III or IV, a more elaborate analysis method is required to extract the maximum amount of information from the data. For illustration, consider the 3^{3-1} design with $C = AB$ whose design matrix is

Table 5.8 Design Matrix for the 3^{3-1} Design

Run	A	B	C
1	0	0	0
2	0	1	1
3	0	2	2
4	1	0	1
5	1	1	2
6	1	2	0
7	2	0	2
8	2	1	0
9	2	2	1

given in Table 5.8. Its main effects and two-factor interactions have the following aliasing relations:

$$A = BC^2, \quad B = AC^2, \quad C = AB, \quad AB^2 = BC = AC. \tag{5.14}$$

In addition to estimating the six degrees of freedom in the main effects A, B, and C, there are two degrees of freedom left for estimating the three aliased effects AB^2, BC, and AC, which, as discussed in Section 5.3, are difficult to interpret. Instead, consider using the remaining two degrees of freedom to estimate any pair of the $l \times l$, $l \times q$, $q \times l$, or $q \times q$ effects between A, B, and C. Suppose that the two interaction effects taken are $(AB)_{ll}$ and $(AB)_{lq}$. Then the eight degrees of freedom can be represented by the model matrix given in Table 5.9. Note that the scaling constants $\sqrt{2}$ and $\sqrt{6}$ in l and q are dropped in the table for simplicity. The columns in the model matrix can be easily obtained from the design matrix through the definitions of the various effects. By applying the formulas in (5.12) to column A in Table 5.8, we obtain columns A_l and A_q in Table 5.9. Columns B_l, B_q, C_l, and C_q are obtained similarly. Then columns $(AB)_{ll}$ and $(AB)_{lq}$ are obtained by multiplying A_l with B_l and A_l with B_q, respectively.

Table 5.9 A System of Contrasts for the 3^{3-1} Design

Run	A_l	A_q	B_l	B_q	C_l	C_q	$(AB)_{ll}$	$(AB)_{lq}$
1	-1	1	-1	1	-1	1	1	-1
2	-1	1	0	-2	0	-2	0	2
3	-1	1	1	1	1	1	-1	-1
4	0	-2	-1	1	0	-2	0	0
5	0	-2	0	-2	1	1	0	0
6	0	-2	1	1	-1	1	0	0
7	1	1	-1	1	1	1	-1	1
8	1	1	0	-2	-1	1	0	-2
9	1	1	1	1	0	-2	1	1

Because any component of $A \times B$ is orthogonal to A and to B, there are only four non-orthogonal pairs of columns whose correlations are:

$$Corr((AB)_{ll}, C_l) = -\sqrt{\tfrac{3}{8}}, \quad Corr((AB)_{ll}, C_q) = -\sqrt{\tfrac{1}{8}},$$
$$Corr((AB)_{lq}, C_l) = \sqrt{\tfrac{1}{8}}, \quad Corr((AB)_{lq}, C_q) = -\sqrt{\tfrac{3}{8}}. \tag{5.15}$$

Obviously, $(AB)_{ll}$ and $(AB)_{lq}$ can be estimated in addition to the three main effects. Because the last four columns are not mutually orthogonal, they cannot be estimated with full efficiency. The D_s criterion given in Section 4.4.2 can be used to measure the estimation efficiency. The details are left as an exercise. The estimability of $(AB)_{ll}$ and $(AB)_{lq}$ demonstrates an advantage of the linear-quadratic system over the orthogonal components system. For the same design, the AB interaction component cannot be estimated because it is aliased with the main effect C.

A geometric explanation of the estimability of some pairs of $(AB)_{ll}$, $(AB)_{lq}$, $(AB)_{ql}$, and $(AB)_{qq}$ is as follows. Each of the four one-dimensional vectors representing these four interaction effects has a 45° angle with the two-dimensional plane representing AB^2. Because AB^2 is not aliased with any main effect [see (5.14)], its plane is orthogonal to the six-dimensional subspace representing A, B, and C. It can be shown that any of the four pairs of interaction effects—$\{(AB)_{ll}, (AB)_{lq}\}$, $\{(AB)_{ll}, (AB)_{ql}\}$, $\{(AB)_{lq}, (AB)_{qq}\}$, and $\{(AB)_{ql}, (AB)_{qq}\}$—can be estimated in addition to A, B, and C.

For this design, we propose analyzing the data by using a stepwise regression or subset selection technique with 18 candidate variables: 6 for the l and q effects of A, B, C and 12 for the $l \times l$, $l \times q$, $q \times l$, and $q \times q$ effects of $A \times B$, $A \times C$, and $B \times C$. To avoid incompatible models, the *effect heredity principle* (see Section 3.5) can be invoked to rule out interactions whose parent factors are not significant.

If factors A, B, and C are all quantitative, a different but related strategy may be used. Instead of decomposing the effect into linear and quadratic components, regression analysis on the original scale of the factors can be performed. If the values of the factors are denoted by x_A, x_B, x_C, then x_A^2 can be used for the quadratic effect and $x_A x_B$ for the interaction effect. Although these effects are not orthogonal, regression analysis can be used because the orthogonality property is not required. This approach works particularly well for factors with unevenly spaced levels, such as factor C (crimp length) in the seat-belt experiment, because the linear and quadratic contrasts assume equally spaced levels. (Orthogonal contrasts for unequally spaced levels are developed in an exercise.)

For a qualitative factor like factor D (lot number) in the seat-belt experiment, the linear contrast $(-1, 0, 1)$ may make sense because it represents the comparison between levels 0 and 2. On the other hand, the "quadratic" contrast $(1, -2, 1)$, which compares level 1 with the average of levels 0 and 2, makes sense only if such a comparison is of practical interest. For example, if levels 0 and 2 represent two internal suppliers, then the "quadratic" contrast measures the difference between internal and external suppliers. When the quadratic contrast makes no sense, two out of the following three contrasts can be chosen to represent the two degrees of freedom for the main effect of a qualitative factor:

$$D_{01} = \begin{cases} -1 \\ 1 \\ 0 \end{cases} \text{ for level } \begin{matrix} 0 \\ 1 \\ 2 \end{matrix} \text{ of factor } D,$$

$$D_{02} = \begin{cases} -1 \\ 0 \\ 1 \end{cases} \text{ for level } \begin{matrix} 0 \\ 1 \\ 2 \end{matrix} \text{ of factor } D,$$

$$D_{12} = \begin{cases} 0 \\ -1 \\ 1 \end{cases} \text{ for level } \begin{matrix} 0 \\ 1 \\ 2 \end{matrix} \text{ of factor } D. \tag{5.16}$$

Mathematically, they are represented by the standardized vectors:

$$D_{01} = \frac{1}{\sqrt{2}}(-1,1,0), \quad D_{02} = \frac{1}{\sqrt{2}}(-1,0,1), \quad D_{12} = \frac{1}{\sqrt{2}}(0,-1,1). \tag{5.17}$$

These contrasts are not orthogonal to each other and have pairwise correlations of $\frac{1}{2}$ or $-\frac{1}{2}$. On the other hand, each of them is readily interpretable as a comparison between two of the three levels. The two contrasts should be chosen to be of interest to the investigator. For example, if level 0 is the main supplier and levels 1 and 2 are minor suppliers, then D_{01} and D_{02} should be used. The interaction between a quantitative factor and a qualitative factor, say $A \times D$, can be decomposed into four effects. As in (5.13), we define the four interaction effects as follows:

$$(AD)_{l,01}(i,j) = A_l(i)D_{01}(j), \qquad (AD)_{l,02}(i,j) = A_l(i)D_{02}(j),$$
$$(AD)_{q,01}(i,j) = A_q(i)D_{01}(j), \qquad (AD)_{q,02}(i,j) = A_q(i)D_{02}(j). \tag{5.18}$$

We include the parameterizations of (5.17) and (5.18) in the linear-quadratic system even though the terminology is no longer precise.

Since many of these contrasts are not mutually orthogonal, a general purpose analysis strategy cannot be based on the orthogonality assumption. Therefore, the following variable selection strategy is recommended.

Variable Selection Strategy

(i) For a quantitative factor, say A, use A_l and A_q for the A main effect.

(ii) For a qualitative factor, say D, use D_l and D_q if D_q is interpretable; otherwise, select two contrasts from D_{01}, D_{02}, and D_{12} for the D main effect.

(iii) For a pair of factors, say X and Y, use the products of the two contrasts of X and the two contrasts of Y [chosen in (i) or (ii)] as defined in (5.13) or (5.18) to represent the four degrees of freedom in the interaction $X \times Y$.

(iv) Using the contrasts defined in (i)–(iii) for all the factors and their two-factor interactions as candidate variables, perform a stepwise regression or subset selection procedure to identify a suitable model. To avoid incompatible models, use the effect heredity principle to rule out interactions whose parent factors are both not significant.

(v) If all the factors are quantitative, use the original scale, say x_A, to represent the linear effect of A, x_A^2 the quadratic effect, and $x_A^i x_B^j$ the interaction between x_A^i and x_B^j. This works particularly well if some factors have unevenly spaced levels.

Returning to the seat-belt experiment, although the original design has resolution IV, its capacity for estimating two-factor interactions is much better than what the definition of resolution IV would suggest. After estimating the four main effects, there are still 18 degrees of freedom available for estimating some components of the two-factor interactions. From (5.9), A, B, C, and D are estimable and only one of the two components in each of the six interactions $A \times B$, $A \times C$, $A \times D$, $B \times C$, $B \times D$, and $C \times D$ is estimable. Because of the difficulty of providing a physical interpretation of an interaction component, a simple and efficient modeling strategy that does not throw away the information in the interactions is to consider the contrasts (A_l, A_q), (B_l, B_q), (C_l, C_q), and (D_{01}, D_{02}, D_{12}) for the main effects and the 30 products between these four groups of contrasts for the interactions. Using these 39 contrasts as the candidate variables, a variable selection procedure was applied to the data.

Performing a stepwise regression (with F critical value of 4) on the strength data (response y_1), the following model with an R^2 of 0.811 was identified:

$$\hat{y}_1 = 6223.0741 + 1116.2859A_l - 190.2437A_q + 178.6885B_l$$
$$- 589.5437C_l + 294.2883(AB)_{ql} + 627.9444(AC)_{ll}$$
$$- 191.2855D_{01} - 468.4190D_{12} - 486.4444(CD)_{l,12}. \qquad (5.19)$$

Note that this model obeys effect heredity. The A, B, C, and D main effects and $A \times B$, $A \times C$, and $C \times D$ interactions are significant. In contrast, the simple analysis from the previous section identified the A, C, and D main effects and the AC ($= BD^2$) and AB ($= CD^2$) interaction components as significant.

Performing a stepwise regression on the flash data (response y_2), the following model with an R^2 of 0.857 was identified:

$$\hat{y}_2 = 13.6657 + 1.2408 A_l + 0.1857 B_l$$
$$- 0.8551 C_l + 0.2043 C_q - 0.9406 (AC)_{ll}$$
$$- 0.3775 (AC)_{ql} - 0.3765 (BC)_{qq} - 0.2978 (CD)_{l,12}. \quad (5.20)$$

Again, the identified model obeys effect heredity. The A, B, and C main effects and $A \times C$, $B \times C$, and $C \times D$ interactions are significant. In contrast, the simple analysis from the previous section identified the A and C main effects and the AC $(= BD^2)$, AC^2, and BC^2 interaction components as significant.

The difference in the results from applying the two approaches is due mainly to the fact that the simple analysis uses the orthogonal components system while the variable selection procedure uses the linear-quadratic system. There is an intricate relationship between the effects in the two systems, however. A secondary reason is that the variable selection procedure, as applied here, used a more generous criterion to include variables in the model. For example, the B main effect is included in both models but is not declared significant in the ANOVA analyses.

The 3^{k-p} designs can also be analyzed using a Bayesian variable selection procedure that will be presented in Chapter 8.

5.7 ANALYSIS STRATEGIES FOR MULTIPLE RESPONSES I: OUT-OF-SPEC PROBABILITIES

Models (5.19) and (5.20) were identified for strength and flash location in the previous section. Since the half-normal plots in Figures 5.5 and 5.10 revealed a single dispersion effect A_l for both strength and flash, the following models were fitted to $z_1 = \ln s_1^2$ for strength and $z_2 = \ln s_2^2$ for flash:

$$\hat{z}_1 = 10.8945 - 1.7027 A_l \quad (5.21)$$

and

$$\hat{z}_2 = -2.5704 + 1.4899 A_l. \quad (5.22)$$

After obtaining the location and dispersion models for each response characteristic, how should they be used in determining the optimal factor settings? If the same factor setting can be found that optimizes the location and minimizes the dispersion, then there is no need to choose a compromise setting. In most situations, however, a trade-off has to be made between optimizing the location and minimizing the dispersion. Some trade-off strategies are based on optimizing a single measure that incorporates both the location and dispersion effects. For nominal-the-best problems, the mean-squared error criterion is an obvious choice:

$$E(y - t)^2 = Var(y) + (E(y) - t)^2,$$

which can be estimated by

$$\widehat{MSE} = \exp(\hat{z}) + (\hat{y} - t)^2, \tag{5.23}$$

where $\hat{z}$ is based on the dispersion model for $z = \ln s^2$, $\hat{y}$ is based on the location model for $\bar{y}$, and t is the target value. For larger-the-better and smaller-the-better problems, the mean-squared error criterion does not have a natural extension because there is no target value. For such problems, a specification limit for the response characteristic is often available. Therefore, the probability that the response is outside the specification limit(s) can be used as a quality measure and will be referred to as the **out-of-spec probability**. For larger-the-better problems, the limit is a lower limit L; for smaller-the-better problems, the limit is an upper limit U; and for nominal-the-best problems, there are both lower and upper limits. Then, the out-of-spec probability is:

$$p = Prob(y \leq L),$$

$$p = Prob(y \geq U), \text{ or} \tag{5.24}$$

$$p = Prob(y \leq L \text{ or } y \geq U),$$

respectively. Minimizing the out-of-spec probability is consistent with the dual objectives of minimizing the dispersion and optimizing the location. For example, if the location effect of a larger-the-better characteristic is large and its dispersion effect is small, then the corresponding out-of-spec probability is small.

Multiple response characteristics occur commonly in real experiments. Besides the seat-belt experiment, the paint experiment in Section 6.6 has seven responses. Other examples are given in the exercises of this and later chapters. For multiple responses, using the out-of-spec probability has the additional advantage of providing a quality measure on the same scale for each response characteristic. This property will be better appreciated by noting that, for the seat-belt experiment, the strength is measured in thousands of pounds and the flash is measured in tens of millimeters. If the numerical values of strength and flash are used to measure quality, how should they be combined into a single quality measure and be weighted to reflect the relative importance of the two characteristics? In many situations, it is difficult to find such weights. As pointed out at the end of Section 5.5, maximizing the strength and minimizing the flash led to a conflicting choice of the level of factor A. Finding a compromise choice would require that the "largeness" of strength and the "smallness" of flash be measured on the same scale.

Next we will show how the out-of-spec probability can be estimated for a given factor setting based on the location and dispersion models. Take the settings $A = 1700$, $B = 10.0$, $C = 23$, and $D = $ P74, which correspond to the level combination $A_2 B_0 C_1 D_0$. For the strength data y_1, the predicted mean strength based on (5.19) is

$$\hat{y}_1 = 6223.0741 + 1116.2859(1/\sqrt{2}) - 190.2437(1/\sqrt{6})$$
$$+ 178.6885(-1/\sqrt{2}) - 589.5437(0/\sqrt{2}) + 294.2883(1/\sqrt{6})(-1/\sqrt{2})$$
$$+ 627.9444(1/\sqrt{2})(0/\sqrt{2}) - 191.2855(-1/\sqrt{2}) - 468.4190(0/\sqrt{2})$$
$$- 486.4444(0/\sqrt{2})(0/\sqrt{2}) = 6858.69 \tag{5.25}$$

and the predicted variance based on (5.21) is

$$\hat{\sigma}_1^2 = \exp(\hat{z}_1) = \exp\left[10.8945 - 1.7027(1/\sqrt{2})\right] = (127.14)^2. \tag{5.26}$$

Recall that the objective for strength is to achieve a mean of 7000 lb with a standard deviation of 250 lb. Therefore, we use 6500 lb, which is two standard deviations below the mean, as the lower specification limit in the following calculations. Based on (5.25), (5.26), and the normality assumption for y_1, the out-of-spec probability $p_1 = Prob(y_1 \leq 6500)$ for strength at $A_2 B_0 C_1 D_0$ can be estimated by

$$\hat{p}_1 = Prob(y_1 \leq 6500) = Prob\left(Z \leq \frac{6500 - 6858.69}{127.14}\right) = \Phi(-2.82) = 0.002,$$

where Z is the standard normal random variable and Φ is the normal cumulative distribution function. The objective for the flash response (y_2) is to achieve a mean of 13.5 mm with a standard deviation of 0.2 mm. We use 13.9 mm, which is two standard deviations above the mean, as the upper specification limit. Based on (5.20) and (5.22), similar calculations show that at $A_2 B_0 C_1 D_0$, the predicted mean flash is

$$\hat{y}_2 = 13.6657 + 1.2408(1/\sqrt{2}) + 0.1857(-1/\sqrt{2})$$
$$- 0.8551(0/\sqrt{2}) + 0.2043(-2/\sqrt{6}) - 0.9406(1/\sqrt{2})(0/\sqrt{2})$$
$$- 0.3775(1/\sqrt{6})(0/\sqrt{2}) - 0.3765(1/\sqrt{6})(-2/\sqrt{6})$$
$$- 0.2978(0/\sqrt{2})(0/\sqrt{2}) = 14.37, \tag{5.27}$$

the predicted variance is

$$\hat{\sigma}_2^2 = \exp(\hat{z}_2) = \exp\left[-2.5704 + 1.4899(1/\sqrt{2})\right] = (0.47)^2, \tag{5.28}$$

and the out-of-spec probability for flash can be estimated by

$$\hat{p}_2 = Prob(y_2 \geq 13.9) = Prob\left(Z \geq \frac{13.9 - 14.37}{0.47}\right) = 1 - \Phi(-1) = 0.84.$$

The same computation can be performed for each of the 81 ($= 3^4$) level combinations of factors A, B, C, and D. Table 5.10 gives the best 35 settings in terms of the $\hat{p}_1$. By comparing the $\hat{p}_1$ column and the mean strength

Table 5.10 Factor Settings Ranked by $\hat{p}_1$, Seat-Belt Experiment

| Setting | Factor | | | | $\hat{p}_1$ | $\hat{y}_1$ | $\hat{\sigma}_1$ |
	A	B	C	D			
1	2	2	2	1	0.00	7482.33	127.14
2	2	2	0	0	0.00	7384.20	127.14
3	2	2	1	1	0.00	7342.01	127.14
4	2	2	1	0	0.00	7281.31	127.14
5	2	1	2	1	0.00	7271.03	127.14
6	2	2	0	1	0.00	7201.69	127.14
7	2	2	2	0	0.00	7178.41	127.14
8	2	1	0	0	0.00	7172.90	127.14
9	2	2	0	2	0.00	7160.94	127.14
10	2	1	1	1	0.00	7130.70	127.14
11	2	1	1	0	0.00	7070.00	127.14
12	2	0	2	1	0.00	7059.72	127.14
13	2	1	0	1	0.00	6990.38	127.14
14	2	1	2	0	0.00	6967.10	127.14
15	2	0	0	0	0.00	6961.59	127.14
16	2	1	0	2	0.00	6949.64	127.14
17	2	0	1	1	0.00	6919.40	127.14
18	2	0	1	0	0.00	6858.69	127.14
19	2	2	1	2	0.01	6814.82	127.14
20	2	0	0	1	0.01	6779.07	127.14
21	1	0	0	0	0.02	6974.09	232.12
22	2	0	2	0	0.02	6755.80	127.14
23	2	0	0	2	0.03	6738.33	127.14
24	1	1	0	0	0.03	6930.54	232.12
25	1	2	0	0	0.05	6886.98	232.12
26	1	0	0	1	0.10	6791.57	232.12
27	1	0	0	2	0.14	6750.83	232.12
28	1	1	0	1	0.14	6748.02	232.12
29	1	1	0	2	0.19	6707.28	232.12
30	1	2	0	1	0.19	6704.46	232.12
31	2	1	1	2	0.21	6603.52	127.14
32	1	2	0	2	0.24	6663.72	232.12
33	1	0	1	1	0.31	6617.93	232.12
34	1	1	1	1	0.37	6574.37	232.12
35	1	0	1	0	0.40	6557.22	232.12

Table 5.11   Factor Settings Ranked by $\hat{p}_2$, Seat-Belt Experiment

Setting	A	B	C	D	$\hat{p}_2$	$\hat{y}_2$	$\hat{\sigma}_2$
1	0	0	2	0	0.00	12.43	0.16
2	0	0	2	2	0.00	12.29	0.16
3	0	1	1	0	0.00	12.37	0.16
4	0	1	1	1	0.00	12.37	0.16
5	0	1	1	2	0.00	12.37	0.16
6	0	2	2	2	0.00	12.55	0.16
7	0	0	2	1	0.00	12.58	0.16
8	0	1	2	2	0.00	12.60	0.16
9	0	0	1	0	0.00	12.62	0.16
10	0	0	1	1	0.00	12.62	0.16
11	0	0	1	2	0.00	12.62	0.16
12	0	2	2	0	0.00	12.70	0.16
13	0	1	2	0	0.00	12.75	0.16
14	0	0	0	1	0.00	12.77	0.16
15	0	2	2	1	0.00	12.85	0.16
16	0	2	1	0	0.00	12.88	0.16
17	0	2	1	1	0.00	12.88	0.16
18	0	2	1	2	0.00	12.88	0.16
19	0	1	2	1	0.00	12.90	0.16
20	0	0	0	0	0.00	12.92	0.16
21	0	2	0	1	0.00	13.03	0.16
22	0	0	0	2	0.00	13.07	0.16
23	0	1	0	1	0.00	13.09	0.16
24	0	2	0	0	0.00	13.18	0.16
25	0	1	0	0	0.00	13.24	0.16
26	0	2	0	2	0.00	13.33	0.16
27	1	0	2	2	0.00	13.02	0.28
28	0	1	0	2	0.00	13.39	0.16
29	1	0	2	0	0.00	13.17	0.28
30	1	1	1	0	0.01	13.25	0.28
31	1	1	1	1	0.01	13.25	0.28
32	1	1	1	2	0.01	13.25	0.28
33	1	2	2	2	0.01	13.28	0.28
34	1	0	2	1	0.02	13.32	0.28
35	1	1	2	2	0.02	13.34	0.28
36	2	0	2	2	0.04	13.10	0.47
37	1	2	2	0	0.04	13.43	0.28
38	1	1	2	0	0.07	13.49	0.28
39	1	0	1	0	0.07	13.49	0.28
40	1	0	1	1	0.07	13.49	0.28
41	1	0	1	2	0.07	13.49	0.28
42	2	0	2	0	0.08	13.25	0.47
43	1	2	2	1	0.12	13.58	0.28
44	2	2	2	2	0.13	13.36	0.47
45	2	0	2	1	0.14	13.40	0.47

Table 5.11 (*Continued*)

Setting	Factor				$\hat{p}_2$	$\hat{y}_2$	$\hat{\sigma}_2$
	A	B	C	D			
46	2	1	2	2	0.15	13.42	0.47
47	1	1	2	1	0.17	13.64	0.28
48	2	2	2	0	0.20	13.51	0.47
49	2	1	2	0	0.24	13.57	0.47
50	1	2	1	0	0.30	13.76	0.28
51	1	2	1	1	0.30	13.76	0.28
52	1	2	1	2	0.30	13.76	0.28
53	2	2	2	1	0.30	13.66	0.47
54	2	1	2	1	0.35	13.72	0.47
55	1	0	0	1	0.35	13.79	0.28

column ($\hat{y}_1$) in Table 5.10, it is clear that a low out-of-spec probability for strength is determined primarily by a large mean strength. Table 5.11 gives the best 55 settings in terms of the $\hat{p}_2$. By comparing the $\hat{p}_2$ column and the mean flash column ($\hat{y}_2$) in Table 5.11, it is seen that a low out-of-spec probability for flash is determined primarily by a small mean flash. In both cases, the variance has a negligible effect on the out-of-spec probability. Once $\hat{p}_1$ and $\hat{p}_2$ are obtained, we can use their weighted average

$$\hat{p} = \lambda \hat{p}_1 + (1 - \lambda) \hat{p}_2 \qquad (5.29)$$

as an overall quality index, where λ measures the relative importance of y_1 to y_2. The $\hat{p}$ value in (5.29) estimates the *overall out-of-spec probability*, $\lambda p_1 + (1 - \lambda) p_2$. Table 5.12 gives the best five settings in terms of the $\hat{p}$ value for $\lambda = \frac{2}{3}, \frac{1}{2}$, and $\frac{1}{3}$. The results in Tables 5.10–5.12 can be summarized as follows:

(i) By comparing Tables 5.10 and 5.11, there is generally a reverse relationship between small $\hat{p}_1$ and small $\hat{p}_2$. This is not surprising in view of the previous finding that higher strength favors the higher level of factor A while smaller flash favors the lower level of factor A and that A is the most significant effect for both response variables.

Table 5.12 **Factor Settings Ranked by $\hat{p}$ for Various λ, Seat-Belt Experiment**

Setting	$\lambda = \frac{2}{3}$					$\lambda = \frac{1}{2}$					$\lambda = \frac{1}{3}$				
	A	B	C	D	$\hat{p}$	A	B	C	D	$\hat{p}$	A	B	C	D	$\hat{p}$
1	2	0	2	0	0.04	2	0	2	0	0.05	2	0	2	0	0.06
2	2	0	2	1	0.05	2	0	2	1	0.07	2	0	2	1	0.09
3	2	2	2	0	0.07	2	2	2	0	0.10	1	1	1	1	0.13
4	2	1	2	0	0.08	2	1	2	0	0.12	2	2	2	0	0.14
5	2	2	2	1	0.10	2	2	2	1	0.15	1	0	1	1	0.15

(ii) Based on Table 5.12, the best five settings are not affected much by the choice of λ. For $\lambda = \frac{2}{3}, \frac{1}{2}$, the best five settings are the same and all have level 2 for factor A. Only for $\lambda = \frac{1}{3}$, which weights flash more heavily, do two new settings $A_1B_1C_1D_1$ and $A_1B_0C_1D_1$ emerge as the third and fifth best. Both have level 1 for factor A because smaller flash does not favor A_2.

(iii) Strength has a much bigger influence than flash on the overall out-of-spec probability. By comparing the three tables, there is a much higher correlation between the $\hat{p}_1$ rankings and the $\hat{p}$ rankings than between the $\hat{p}_2$ rankings and the $\hat{p}$ rankings. For example, the best five settings for $\lambda = \frac{2}{3}, \frac{1}{2}$ have the relatively high rankings 22, 12, 7, 14, and 1 in Table 5.10 but much lower rankings 42, 45, 48, 49, and 53 in Table 5.11. Only the two settings $A_1B_1C_1D_1$ and $A_1B_0C_1D_1$ (for $\lambda = \frac{1}{3}$) have the higher rankings of 31 and 40 in Table 5.10 (but the much lower rankings of 34 and 35 in Table 5.11).

For m response characteristics with $m > 2$, the extension is straightforward. For example, the **overall out-of-spec probability** is defined to be

$$p = \sum_1^m \lambda_i p_i, \tag{5.30}$$

where p_i is the out-of-spec probability for the ith response, λ_i measures the relative importance of the ith response, $0 \le \lambda_i \le 1$, and $\sum_1^m \lambda_i = 1$.

The trade-off strategy will not be effective if the specification limit(s) is (are) so loose that the out-of-spec probability is zero or nearly zero for most factor settings. In this situation, we recommend tightening the specification limit(s) to the extent that a meaningful comparison among factor settings can be made. Tightening the limit(s) has another useful application. Consider the two scenarios in Figure 5.11. Both have the same out-of-spec probability, but (b) has a much more skewed distribution than (a). Obviously (a) has a better overall quality than (b) but the difference cannot be detected by the use of the out-of-spec probability with the given specification limit. By moving the limit up, the out-of-spec probability will start to favor scenario (a). In fact, one can use the out-of-spec probabilities for several choices of specification limits to give a more realistic comparison of the two scenarios. Figure 5.12

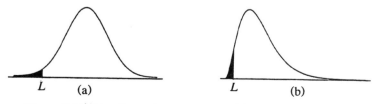

Figure 5.11. Two Scenarios with Lower (L) Specification Limit.

Figure 5.12. Two Scenarios with Lower (L) and Upper (U) Specification Limits (t is the target value).

depicts two scenarios for nominal-the-best problems. Both have the same out-of-spec probability, but (a) is more concentrated around the target value than (b). Obviously (a) is the preferred choice. By narrowing the lower and upper specification limits, the out-of-spec probability will start to favor scenario (a).

For nominal-the-best problems, a different but related trade-off strategy can be employed. Suppose that t_i is the target value and $t_i - \Delta_i$ and $t_i + \Delta_i$ are the lower and upper specification limits for y_i, $i = 1, 2$. Then

$$\lambda \frac{\widehat{MSE}_1}{\Delta_1^2} + (1 - \lambda) \frac{\widehat{MSE}_2}{\Delta_2^2} \qquad (5.31)$$

can be used as a single quality index, where $\widehat{MSE}$ is the estimated mean-squared error defined in (5.23) for y_i with target t_i, and λ measures the relative importance of y_1 to y_2. In (5.31), Δ_1 and Δ_2 serve as normalizing constants, so that $\widehat{MSE}_1/\Delta_1^2$ and $\widehat{MSE}_2/\Delta_2^2$ are comparable. How does the weighted mean-squared error approach in (5.31) compare with the overall out-of-spec probability approach in (5.29)? The latter better captures the out-of-spec probability while the former better captures the concentration of distribution around the target value. The choice depends on the objective of the experiment.

If the optimal factor settings can be determined solely by the factor main effects, an informal trade-off strategy can be developed as follows. For illustration, assume that the interaction effects are negligible for both strength and flash. Based on the main effects plot of strength location in Figure 5.1, A, C, and D are significant for strength and $A_2C_0D_0$ or $A_2C_0D_1$ should be chosen. (Note that the choice of the level of B is left open.) Based on the main effects plot of flash location in Figure 5.6, A and C are significant for flash and A_0C_2 should be chosen. The recommended settings for strength and flash are summarized in Table 5.13. If there is a conflict on the recommended setting, the investigator may use subject matter knowledge to make a choice. If an informal resolution cannot be made, a more formal strategy such as those considered above can be used. For factors that are not

Table 5.13   **Summary of Recommended Factor Level Settings, Seat-Belt Experiment**

Response	A	B	C	D
strength (y_1)	2		0	0 or 1
flash (y_2)	0		2	

Note: Numbers indicate the levels of the factor. A blank indicates no recommendation.

significant for any of the response characteristics, their levels can be determined by other considerations such as cost or ease of experimental setup. Returning to Table 5.13, the choice for B is left open, D should be chosen at 0 or 1, and there is a conflict on the choice of level for A and C. By putting all the recommended settings in the same table, this tabular approach makes it easier to find a compromise on conflicting recommendations. In spite of the appeal of simplicity, its use is severely limited by the assumption that the factor settings are determined solely by the main effects and the lack of a quantitative basis for making a compromise.

Another approach to handling multiple responses will be considered in Section 9.6.

5.8   BLOCKING IN 3^k AND 3^{k-p} DESIGNS

The simplest case is the arrangement of a 3^2 design in three blocks. Let the nine units of the 3^2 design be represented as in Table 5.4. Then the three blocks can be chosen to correspond to the units denoted by α, β, and γ. As shown in Section 5.3, the contrasts between these three blocks are represented by the interaction component AB. Therefore the block effect, which is denoted in this section by lowercase b_1, b_2, etc., is *confounded* with, the interaction component AB. Notationally this confounding relation is denoted by

$$b_1 = AB.$$

For this blocking scheme, the main effects A, B and the interaction component AB^2 are clear, but the interaction component AB is sacrificed for the block effect. Similar to the discussion in Section 3.12, the blocking schemes for 3^k full factorial designs can be chosen optimally according to the *minimum aberration* criterion.

Blocking for 3^{k-p} fractional factorial designs is more complicated due to the presence of the *treatment defining contrast subgroup* and the *block defining contrast subgroup*. Because of the similarity to the two-level case, the following discussion will not be as detailed as that in Section 4.6. Consider the arrangement of a 3^{4-1} design with $I = ABCD^2$ in nine blocks. Choose

$$b_1 = AB, \qquad b_2 = AC \qquad (5.32)$$

as the defining generators for the blocking scheme. Then the other two defining relations for blocking are

$$b_1 b_2 = (AB)(AC) = AB^2 C^2, \qquad b_1 b_2^2 = (AB)(A^2 C^2) = BC^2. \quad (5.33)$$

Since $b_1 b_2$ and $b_1 b_2^2$ represent the third and fourth components of the block effects, they are replaced by b_3 and b_4, respectively, to emphasize the fact that they represent main effects of the blocking variable. Because of the treatment defining relation, each of the four block effects b_1, b_2, b_3, and b_4 is confounded with three factorial effects. They can be obtained by multiplying

$$I = ABCD^2 = A^2 B^2 C^2 D \qquad (5.34)$$

by each of the relations in (5.32) and (5.33). For example, by multiplying (5.34) by $b_1 = AB$, we obtain

$$b_1 = AB = ABC^2 D = CD^2.$$

All the aliasing and confounding relations for this blocking scheme are summarized as follows:

$$
\begin{aligned}
\mathbf{I} \quad &= ABCD^2, \\
A \quad &= BCD^2 \quad = AB^2 C^2 D, \\
B \quad &= ACD^2 \quad = AB^2 CD^2, \\
C \quad &= ABD^2 \quad = ABC^2 D^2, \\
D \quad &= ABC \quad = ABCD, \\
b_1 = AB \quad &= CD^2 \quad = ABC^2 D, \\
b_2 = AC \quad &= BD^2 \quad = AB^2 CD, \\
BC \quad &= AD^2 \quad = AB^2 C^2 D^2, \\
AB^2 \quad &= AC^2 D \quad = BCD^2, \\
AC^2 \quad &= AB^2 D \quad = BC^2 D^2, \\
b_3 = AD \quad &= AB^2 C^2 \quad = BCD, \\
b_4 = BC^2 \quad &= AB^2 D^2 \quad = AC^2 D^2, \\
BD \quad &= AB^2 C \quad = ACD, \\
CD \quad &= ABC^2 \quad = ABD.
\end{aligned}
\qquad (5.35)
$$

From (5.35), we can see that the main effects A, B, C, and D and the interaction components AB^2, AC^2, BD, and CD are clear. In the context of blocking, clear and strongly clear effects are defined as in Section 5.4, with the additional requirement that they not be confounded with any block effects. The choice of "optimal" blocking schemes for 3^{k-p} designs is determined by maximizing the number of clear effects and following other optimality criteria which are too complicated to be discussed here. See Cheng and Wu (1999) for details.

In Appendix 5B, efficient blocking schemes are given for 9-, 27- and 81-run designs. In each case, the defining generators for the fractional factorial design and for the blocking scheme and the collection of clear main effects and two-factor interaction components are given. As in the two-level case, more than one scheme may be given if neither scheme dominates the other in terms of the combinations of clear effects. The choice should be made based on what effects in the particular experiment are likely to be important. The collection in these tables is condensed from those in Cheng and Wu (1999).

5.9   PRACTICAL SUMMARY

1. If the response characteristic is to be maximized, it is called a larger-the-better problem. If the response characteristic is to be minimized, it is called a smaller-the-better problem. When an adjustment factor is available, use procedures (5.1) and (5.2), respectively, for larger-the-better and smaller-the-better problems to perform a two-step optimization. If there is no adjustment factor, perform a single-step optimization by directly minimizing $\sum_i y_i^2$ for smaller-the-better problems.

2. Practical reasons for choosing a three-level factor:
   (i) to study the curvature effect of a quantitative factor,
   (ii) to compare several levels of a qualitative factor, and
   (iii) to compare the nominal or current setting and two settings above and below it.

3. In the orthogonal components system, the $A \times B$ interaction is decomposed into two orthogonal interaction components—AB and AB^2—and the $A \times B \times C$ interaction is decomposed into four orthogonal interaction components—ABC, ABC^2, AB^2C, and AB^2C^2. Each component has two degrees of freedom. The interaction components in this system are used in the defining relations for 3^{k-p} designs and in the analysis of variance.

4. A 3^{k-p} design is determined by its defining contrast subgroup, which consists of $(3^p - 1)/2$ defining words and the identity $\mathbf{I}$. The length of its shortest word is the resolution of the design.

5. A main effect or two-factor interaction component is said to be clear if none of its aliases are main effects or two-factor interaction components and strongly clear if none of its aliases are main effects or two-factor or three-factor interaction components. A two-factor interaction is clear (and respectively strongly clear) if both of its components are clear (and respectively strongly clear).

6. Unless there is specific information on the relative importance among the factorial effects, the minimum aberration criterion in Section 5.4 should be used for selecting good 3^{k-p} designs. Minimum aberration implies maximum resolution. Minimum aberration 3^{k-p} designs are given in Appendix 5A.

7. Use main effects plot and interaction plots and ANOVA tables for simple analysis.

8. In the linear-quadratic system, the linear and quadratic main effects are defined in (5.12) and the linear-by-linear, linear-by-quadratic, quadratic-by-linear, and quadratic-by-quadratic interaction effects are defined in (5.13). Their extensions to qualitative factors are given in (5.16)–(5.18). Their main advantage over the orthogonal components system is their ability to estimate some interaction effects even when the corresponding interaction components are aliased with the main effects. See the illustration and explanation in Section 5.6.

9. Use the variable selection strategy in Section 5.6 (which is based on the linear-quadratic system and the effect heredity principle) to identify the significant main effects and interaction effects in the linear-quadratic system.

10. There are two approaches to combine the location and dispersion effects into a single measure:
    (i) the mean-squared error in (5.23) and
    (ii) the out-of-spec probability in (5.24), which is estimated by substituting the location and dispersion models as illustrated in Section 5.7.

11. Trade-off strategies for multiple characteristics:
    (i) Use the overall out-of-spec probability in (5.29) and (5.30), which assumes that a lower and/or upper specification limit for each response characteristic is available.
    (ii) Use the overall mean-squared error in (5.31), which only works if all characteristics are nominal-the-best.
    (iii) Use informal resolution based on the summary table of recommended settings for all characteristics (see Table 5.13 for an illustration).

12. Optimal blocking schemes for 3^k and 3^{k-p} designs are given in Appendix 5B.

EXERCISES

1. Extend the matrix in Table 5.5 to the 3^4 design by finding an 81×40 matrix whose 40 columns represent the four main effects and 36 interaction components as described in Section 5.3.

2. Express the 26 effects in the linear-quadratic system for the 3^3 design in a 27×26 matrix and compare it with the matrix in Table 5.5.

3. (a) Use Lenth's method to test the significance of the 26 effects in the half-normal plot (see Figure 5.5) for strength dispersion. Confirm that the A linear effect is the only significant effect and find its p value.

   (b) Repeat (a) for flash dispersion. Its half-normal plot is given in Figure 5.10.

4. (a) Based on the significant effects identified in the ANOVA for strength location in Table 5.6, use the main effects plot for A, C, and D and the $A \times B$, $C \times D$, $A \times C$, and $B \times D$ interaction plots to determine optimal factor settings for maximizing strength location. Will the results be affected by further considering the minimization of dispersion? (Recall that only factor A is significant for strength dispersion.)

   (b) Based on the significant effects identified in the ANOVA for flash location in Table 5.7, use the main effects plot for A and C, and the $A \times C$, $B \times C$, and $B \times D$ interaction plots to determine optimal factor settings for minimizing flash location. Will the results be affected by further considering the minimization of dispersion? (Recall that only factor A is significant for flash dispersion.)

   (c) Are the optimal factor settings recommended in (a) and (b) consistent? If not, can you find compromise settings? Describe the difficulties you encountered in reaching a reasonable compromise. This experience will help you better appreciate the strategies suggested in Section 5.7.

5. Table 5.14 displays the design matrix and response data for a 2^3 confirmatory study of the seat-belt experiment. The factor levels are 1600 and 1800 psi and 22 and 24 mm using two anchor lots. Analyze the data. Are the results of your analysis consistent with the conclusions from the initial experiment?

6. For three unequally spaced levels x_1, x_2, and x_3 of a quantitative factor, create linear and quadratic orthogonal polynomials by starting with three columns with entries $(1, 1, 1)$, (x_1, x_2, x_3), (x_1^2, x_2^2, x_3^2) and using the Gram-Schmidt orthonormalization procedure.

Table 5.14 Design Matrix and Response Data, Confirmatory Seat-Belt Experiment

Run	Pressure	Length	Lot Number	Strength			Flash		
1	1600	22	P74	6887	6383	6888	12.57	13.06	13.50
2	1800	22	P74	7144	6064	6879	14.65	13.28	13.64
3	1600	24	P74	7215	6926	6873	13.70	13.73	14.08
4	1800	24	P74	6858	7101	6687	14.00	14.26	14.70
5	1600	22	P78	7073	7134	6293	13.69	14.82	13.44
6	1800	22	P78	6758	6395	6623	12.89	12.75	13.16
7	1600	24	P78	6829	6761	7008	13.75	14.12	14.45
8	1800	24	P78	6974	6833	6833	13.87	14.14	13.55

7. Take any two effects from $(AB)_{ll}$, $(AB)_{lq}$, $(AB)_{ql}$, and $(AB)_{qq}$ in the 3^{3-1} design with $C = AB$.

   (a) Compute the canonical correlation between the two-dimensional plane representing these two effects and the six-dimensional subspace representing A, B, and C.

   (b) Compute the D_s efficiency for estimating these two effects when the full model consists of the six main effects A_l, A_q, B_l, B_q, C_l, C_q and the two chosen interaction effects. (*Hint*: To use the formula in (4.13) for the D_s efficiency, normalize the eight column vectors corresponding to the eight effects to have unit length. A D_s value of 1 indicates full efficiency and a value close to 1 indicates high efficiency.)

   (c) In view of the results in (a) and (b), comment on the estimability and estimation efficiency of the two given interaction effects.

   (d) Based on the results in (c) for the six pairs chosen from $(AB)_{ll}$, $(AB)_{lq}$, $(AB)_{ql}$, and $(AB)_{qq}$, which pairs have higher efficiencies? Is there a geometric interpretation for the higher efficiency?

8. For the 3_{IV}^{7-3} design with $E = ABC$, $F = ABD$, and $G = AB^2C^2D$:

   (a) Find its defining contrast subgroup and all the aliasing relations.

   (b) Identify the clear main effects and clear two-factor interaction components.

9. In an experiment to improve the adhesion of plastic pin covers, an ultrasonic welder bonded the cover to the module. The response was pin pull strength which was to be maximized. Six factors were studied in a 3^{6-3} design with two replicates. The factors were pressure of type 1 (A), pressure of type 2 (B), speed (C), stroke stop (D), weld (E), and hold time (F). Because the engineers thought that $A \times C$ and $A \times E$ might be important, the AC, AC^2, AE, and AE^2 components were assigned to columns orthogonal to the six main effects. The design matrix and response data appear in Table 5.15.

Table 5.15   Design Matrix and Response Data, Ultrasonic Bonding Experiment

Run	A	C	E	B	D	F	Strength	
1	0	0	0	0	0	0	18.25	17.25
2	0	0	1	1	1	1	4.75	7.5
3	0	0	2	2	2	2	11.75	11.25
4	0	1	0	1	1	1	13.0	8.75
5	0	1	1	2	2	2	12.5	11.0
6	0	1	2	0	0	0	9.25	13.0
7	0	2	0	2	2	2	21.0	15.0
8	0	2	1	0	0	0	3.5	5.25
9	0	2	2	1	1	1	4.0	8.5
10	1	0	0	0	1	2	6.75	15.75
11	1	0	1	1	2	0	5.0	13.75
12	1	0	2	2	0	1	17.25	13.5
13	1	1	0	1	2	0	13.5	21.25
14	1	1	1	2	0	1	9.0	10.25
15	1	1	2	0	1	2	15.0	9.75
16	1	2	0	2	0	1	10.5	8.25
17	1	2	1	0	1	2	11.0	11.5
18	1	2	2	1	2	0	19.75	14.25
19	2	0	0	0	2	1	17.0	20.0
20	2	0	1	1	0	2	17.75	17.5
21	2	0	2	2	1	0	13.0	12.0
22	2	1	0	1	0	2	8.75	12.25
23	2	1	1	2	1	0	12.25	9.0
24	2	1	2	0	2	1	13.0	11.25
25	2	2	0	2	1	0	10.0	10.0
26	2	2	1	0	2	1	14.5	17.75
27	2	2	2	1	0	2	8.0	11.0

(a) Identify the defining relation for the 3^{6-3} design and find its defining contrast subgroup. (*Hint:* Express each of the factors B, D, and F in terms of A, C, and E.)

(b) Based on the results in (a), find all the aliases of the AC, AC^2, AE, and AE^2 interaction components. Show that none of them are clear. This implies that the original design does not meet the requirement that $A \times C$ and $A \times E$ be estimated clearly of other two-factor interactions.

(c) Use the simple analysis methods and the variable selection procedure to analyze the data. (Because the factor levels are not given, you may assume that they are all evenly spaced quantitative factors.)

(d) Noting that strength is a larger-the-better characteristic and that there are only two replicates per run, what quality index would you use for optimization?

(e) Make recommendations on factor settings that maximize strength.

Table 5.16   Design Matrix and Response Data, Control Value Experiment

Run	A	B	C	Push-Off Force				
1	0	0	0	121.7	118.1	119.7	97.8	98.2
2	0	1	1	131.4	119.5	136.4	139.9	128.0
3	0	2	2	71.0	73.0	41.4	79.5	62.2
4	1	0	1	129.8	130.6	119.4	116.7	131.0
5	1	1	2	67.2	47.3	25.6	37.2	57.3
6	1	2	0	142.0	98.7	97.6	128.7	101.6
7	2	0	2	56.2	67.8	115.5	64.2	58.9
8	2	1	0	169.0	174.9	115.1	108.5	137.8
9	2	2	1	177.2	121.8	112.9	124.8	134.3

10. In an experiment to improve the leakage of a spark control valve, the end cap on the valve must be air tight and withstand a push-off force of 40 lb. However, it was thought that a higher force, such as 110 lb, was needed to prevent leaks. A 3^{3-1} design with $C = AB$ was used to study three factors: weld time (A) at 0.3, 0.5, and 0.7 seconds; pressure (B) at 15, 20, and 25 psi; and moisture (C) at 0.8, 1.2, and 1.8 percent. The design matrix and push-off force data appear in Table 5.16. The five observations per run represent the results from testing five control valves.

   (a) Use the simple analysis methods to analyze the $\bar{y}$ (i.e., location) and $\ln s^2$ (i.e., dispersion) data.

   (b) How many degrees of freedom are available for estimating interaction effects? Use the variable selection procedure to reanalyze the $\bar{y}$ and $\ln s^2$ data. Are any interaction effects in the linear-quadratic system identified as significant? Compare the results with those in (a) based on the interaction plots.

   (c) Make recommendations on factor settings that best achieve the objectives of maximizing location and minimizing dispersion.

   (d) Alternately, the two criteria in (c) can be combined into a single measure in terms of the out-of-spec probability by using 110 lb as the lower specification limit. Find optimal factor settings that minimize the out-of-spec probability. Compare the results with those in (c).

11. By working out all the aliasing relations involving the main effects and two-factor interaction components, find all the clear effects for the 3^{6-2}_{IV} design in Table 5A.3.

12. Repeat the previous problem for the 3^{7-3}_{IV} design in Table 5A.3.

13. Show that no 3^{6-3} design can allow any two-factor interaction component to be estimated clearly.

14. Suppose that A_i denotes the event that the ith response characteristic y_i is out of specification, $i = 1, 2$. Then the probability that either y_1 or y_2 is out of specification is $Prob(A_1 \cup A_2) = Prob(A_1) + Prob(A_2) - Prob(A_1 \cap A_2) = p_1 + p_2 - Prob(A_1 \cap A_2)$. Furthermore, if y_1 and y_2 are assumed to be independent, then $Prob(A_1 \cup A_2) = p_1 + p_2 - p_1 p_2$. Discuss the difficulty of this approach and its similarities with and differences from the overall out-of-spec probability approach in (5.29).

15. A 3^{6-3} design was used in an experiment to study a core drill process which produced rotor cores from glass-ceramic slabs. The core drill process was a new machining process which needed to be optimized. The goal was to minimize damage and stress to the rotor cores. Six factors (each at three levels) were considered: drill grit size (A), spindle rotation rate (B), feed rate (C), spindle amplitude (D), wax type (E), and type of

Table 5.17   Design Matrix and Response Data, Core Drill Experiment

Run	A	B	C	D	E	F	ID Finish	OD Finish	ID Force	OD Force	Chipping
1	0	0	0	0	0	0	47.6	68.0	10.8	17.3	7.5
2	0	0	1	1	1	1	26.1	27.1	5.4	17.1	11.3
3	0	0	2	2	2	2	27.3	47.3	4.9	12.0	7.5
4	0	1	0	1	1	2	35.8	38.2	8.2	13.9	13.2
5	0	1	1	2	2	0	39.5	96.5	4.0	13.1	5.4
6	0	1	2	0	0	1	20.4	33.1	4.6	6.6	7.2
7	0	2	0	2	2	1	34.7	71.9	6.4	19.9	12.2
8	0	2	1	0	0	2	61.7	37.5	3.6	6.0	9.6
9	0	2	2	1	1	0	26.1	25.6	2.8	9.1	13.9
10	1	0	0	1	2	2	98.7	79.7	25.0	31.0	8.3
11	1	0	1	2	0	0	54.8	48.3	17.4	19.1	4.9
12	1	0	2	0	1	1	25.8	32.5	15.6	20.5	17.7
13	1	1	0	2	0	1	57.0	60.2	12.2	14.1	13.7
14	1	1	1	0	1	2	45.8	36.7	17.1	22.1	14.4
15	1	1	2	1	2	0	34.4	38.7	10.2	13.3	10.8
16	1	2	0	0	1	0	57.8	39.1	21.1	26.3	21.0
17	1	2	1	1	2	1	79.2	42.8	14.0	16.9	16.5
18	1	2	2	2	0	2	72.1	42.1	5.8	10.5	8.8
19	2	0	0	2	1	1	20.6	33.5	5.0	16.6	3.3
20	2	0	1	0	2	2	15.5	16.8	4.6	17.5	7.0
21	2	0	2	1	0	0	17.2	29.0	2.7	6.1	2.2
22	2	1	0	0	2	0	50.4	49.8	10.3	19.1	7.6
23	2	1	1	1	0	1	20.8	25.2	4.7	10.4	4.2
24	2	1	2	2	1	2	23.0	16.4	2.5	6.7	5.2
25	2	2	0	1	0	2	24.7	23.2	10.8	14.1	1.3
26	2	2	1	2	1	0	24.0	17.8	2.7	6.7	15.0
27	2	2	2	0	2	1	16.8	23.2	3.8	8.2	7.9

slab (F). The responses are inner diameter (ID) and outer diameter (OD) surface finish (microinches) and ID and OD z-axis force (pounds) and degree of chipping. To achieve the experimental goal, each of the five responses should ideally be minimized. The design matrix and response data appear in Table 5.17. The response data are averages over two rotor cores.

(a) Identify the defining relation for the 3^{6-3} design and find its defining contrast subgroup.

(b) Use the simple analysis methods and the variable selection procedure to analyze each of the five response characteristics.

(c) Find optimal factor settings for each response characteristic.

(d) What trade-off strategy would you use to make compromises on the recommendations made in (c)? Because there is a single replicate per run, the out-of-spec probability approach cannot be used. Is the informal method mentioned at the end of Section 5.7 the most practical approach you can take? Give justification for whatever strategy you adopt.

16. In the milling operation of an aluminum cast tape drive deck, the goal was to minimize microfinish and polishing time. The four factors (each at three levels) are coolant mix (A), type of milling operation (B), feed rate (C), and coolant application (D). The experiment used a 3^{4-2} design. The design matrix and response data appear in Table 5.18.

(a) Because of the small number (i.e., two) of replicates, would you still recommend the ln s^2 analysis? If not, what measure would you use that can suitably combine the location and dispersion effects? Analyze this measure for each of the two response characteristics.

(b) Based on the analysis results in (a) for microfinish and polishing time, find factor settings that best achieve the objectives of minimizing microfinish and polishing time. You may use both the formal and informal trade-off strategies in Section 5.7.

Table 5.18 Design Matrix and Response Data, Casting Experiment

Run	Factor A	B	C	D	Microfinish		Polish Time	
1	0	0	0	0	34.0	31.8	49	54
2	0	1	1	1	102.8	98.4	96	67
3	0	2	2	2	57.6	57.2	50	44
4	1	0	1	2	31.8	24.8	30	48
5	1	1	2	0	162.2	150.8	74	111
6	1	2	0	1	32.4	27.5	37	51
7	2	0	2	1	31.9	37.7	88	96
8	2	1	0	2	53.1	57.2	101	50
9	2	2	1	0	53.8	36.9	45	54

APPENDIX 5A: TABLES OF 3^{k-p} FRACTIONAL FACTORIAL DESIGNS

Table 5A.1   Nine-Run 3^{k-p} Fractional Factorial Designs ($k-p=2$, $k=3$ and 4)
(k is the number of factors and F&R is the fraction and resolution)

k	F&R	Design Generators	Clear Effects
3	3_{III}^{3-1}	$C = AB$	none
4	3_{III}^{4-2}	$C = AB$, $D = AB^2$	none

Table 5A.2   Twenty-Seven Run 3^{k-p} Fractional Factorial Designs ($k-p=3$, $4 \le k \le 13$)
(k is the number of factors and F&R is the fraction and resolution)

k	F&R	Design Generators	Clear Effects
4	3_{IV}^{4-1}	$D = ABC$	$A, B, C, D, AB^2, AC^2,$ AD, BC^2, BD, CD
5	3_{III}^{5-2}	$D = AB$, $E = AB^2C$	C, E, CE
5	3_{III}^{5-2}	$D = AB$, $E = AB^2$	$C, A \times C, B \times C,$ $C \times D, C \times E$
6	3_{III}^{6-3}	$D = AB$, $E = AB^2C$, $F = AB^2C^2$	none
6	3_{III}^{6-3}	$D = AB$, $E = AC$, $F = BC$	none
7	3_{III}^{7-4}	$D = AB$, $E = AC^2$, $F = BC^2$, $G = AB^2C^2$	none
7	3_{III}^{7-4}	$D = AB^2$, $E = ABC$, $F = AC^2$, $G = BC^2$	none
8	3_{III}^{8-5}	$D = AB$, $E = ABC$, $F = AB^2C$, $G = AC^2$, $H = BC^2$	none
8	3_{III}^{8-5}	$D = AB^2$, $E = ABC$, $F = AB^2C$, $G = AC^2$, $H = BC^2$	none
9	3_{III}^{9-6}	$D = AB$, $E = ABC$, $F = AB^2C$, $G = AC^2$, $H = BC^2$, $J = AB^2C^2$	none
10	3_{III}^{10-7}	$D = AB$, $E = AC$, $F = BC$, $G = ABC$, $H = AC^2$, $J = BC^2$, $K = ABC^2$	none
11	3_{III}^{11-8}	$D = AB$, $E = AC$, $F = BC$, $G = ABC$, $H = AC^2$, $J = BC^2$, $K = ABC^2$, $L = AB^2$	none
12	3_{III}^{12-9}	$D = AB$, $E = AC$, $F = BC$, $G = ABC$, $H = AC^2$, $J = BC^2$, $K = ABC^2$, $L = AB^2$, $M = AB^2C$	none
13	3_{III}^{13-10}	$D = AB$, $E = AC$, $F = BC$, $G = ABC$, $H = AC^2$, $J = BC^2$, $K = ABC^2$, $L = AB^2$, $M = AB^2C$, $N = AB^2C^2$	none

Table 5A.3 Eighty-One Run 3^{k-p} Fractional Factorial Designs ($k - p = 4$, $5 \leq k \leq 10$) (k is the number of factors and F&R is the fraction and resolution)

k	F&R	Design Generators	Clear Effects
5	3_V^{5-1}	$E = ABCD$	all five main effects, all 10 2fi's
6	3_{IV}^{6-2}	$E = ABC, F = AB^2D$	all six main effects, $C \times D, C \times F, D \times E, E \times F$, AC^2, AD^2, AE, AF, BC^2, BD, BE, BF^2, CE, DF
7	3_{IV}^{7-3}	$E = ABC, F = ABD,$ $G = AB^2C^2D$	all seven main effects, $AB^2, AC^2, AG^2, BD^2, BG$, CD, CG^2, DG, EG^2, FG, AF, BE, CE, DF, EF
8	3_{IV}^{8-4}	$E = ABC, F = ABD,$ $G = AB^2C^2D, H = AB^2CD^2$	all eight main effects, AG^2, AH^2, BG, BH, CD, CE, DF, EF
9	3_{IV}^{9-5}	$E = ABC, F = ABD, G = AB^2C^2D,$ $H = AB^2CD^2, J = AC^2D^2$	all nine main effects
10	3_{IV}^{10-6}	$E = ABC, F = ABD, G = AB^2C^2D,$ $H = AB^2CD^2, J = AC^2D^2, K = BC^2D^2$	all 10 main effects

APPENDIX 5B: TABLES OF 3^{k-p} FRACTIONAL FACTORIAL DESIGNS IN 3^q BLOCKS

Table 5B.1 Twenty-Seven Run 3^{k-p} Designs in 3^q Blocks ($k - p = 3$, $3 \leq k \leq 5$) (k = number of factors, 3^{k-p} = number of runs, 3^q = number of blocks, and two-factor interactions denoted by 2fi's)

k	p	q	Design Generators	Block Generators	Clear Effects
3	0	1		$b_1 = ABC$	all three main effects, all 6 2fi's
3	0	2		$b_1 = ABC,$ $b_2 = AB^2$	all three main effects, AB, AC, BC
4	1	1	$D = ABC$	$b_1 = AB$	all four main effects, $AB^2, AC^2, AD, BC^2, BD, CD$
4	1	1	$D = ABC$	$b_1 = AB^2$	all four main effects, AC^2, AD, BC^2, BD, CD
4	1	2	$D = ABC$	$b_1 = AB,$ $b_2 = AC$	all four main effects, AB^2, AC^2, BD, CD
5	2	1	$D = AB,$ $E = AB^2C$	$b_1 = AC$	C, E, CE
5	2	1	$D = AB,$ $E = AB^2$	$b_1 = AC$	$C, AC^2, BC, BC^2,$ $C \times D, C \times E$
5	2	2	$D = AB,$ $E = AB^2C$	$b_1 = AC,$ $b_2 = BC$	C, E, CE
5	2	2	$D = AB,$ $E = AC$	$b_1 = AB^2,$ $b_2 = ABC$	BC, BE^2, CD^2, DE

Table 5B.2 Twenty-Seven Run 3^{k-p} Designs in 3^q Blocks
($k - p = 3$, $6 \le k \le 12$) (k = number of factors, 3^{k-p} = number of runs, 3^q = number of blocks)

k	p	q	Design Generators	Block Generators
6	3	1	$D = AB$, $E = AB^2C$, $F = AB^2C^2$	$b_1 = AC$
6	3	2	$D = AB$, $E = AB^2C$, $F = AB^2C^2$	$b_1 = AC$, $b_2 = BC$
7	4	1	$D = AB$, $E = AC^2$, $F = BC^2$, $G = AB^2C^2$	$b_1 = ABC$
7	4	2	$D = AB$, $E = AC^2$, $F = BC^2$, $G = AB^2C^2$	$b_1 = AB^2$, $b_2 = BC$
8	5	1	$D = AB$, $E = ABC$, $F = AB^2C$, $G = AB^2$, $H = BC^2$	$b_1 = AB^2C^2$
8	5	2	$D = AB$, $E = ABC$, $F = AB^2C$, $G = AB^2$, $H = BC^2$	$b_1 = AB^2$, $b_2 = BC$
9	6	1	$D = AB$, $E = ABC$, $F = AB^2C$, $G = AB^2$, $H = BC^2$, $J = AB^2C^2$	$b_1 = AB^2$
9	6	2	$D = AB$, $E = ABC$, $F = AB^2C$, $G = AB^2$, $H = BC^2$, $J = AB^2C^2$	$b_1 = AB^2$, $b_2 = BC$
10	7	1	$D = AB$, $E = AC$, $F = BC$, $G = ABC$, $H = AC^2$, $J = BC^2$, $K = ABC^2$	$b_1 = AB^2$
11	8	1	$D = AB$, $E = AB^2$, $F = AC$, $G = BC$, $H = ABC$, $J = AB^2C$, $K = AC^2$, $L = BC^2$	$b_1 = ABC^2$
12	9	1	$D = AB$, $E = AB^2$, $F = AC$, $G = BC$, $H = ABC$, $J = AB^2C$, $K = AC^2$, $L = BC^2$, $M = ABC^2$	$b_1 = AB^2C^2$

Notes: 1. **No main effect or two-factor interaction is clear** for the designs in this table.
2. For $q = 2$ and $(k, p) = (10, 7)$, $(11, 8)$, and $(12, 9)$, no blocking scheme exists without confounding the main effects and block effects.

Table 5B.3   Eighty-One Run 3^{k-p} Designs in 3^q Blocks
($k - p = 4$, $4 \le k \le 10$) (k = number of factors, 3^{k-p} = number of runs,
3^q = number of blocks, and two-factor interactions denoted by 2fi's)

k	p	q	Design Generators	Block Generators	Clear Effects
4	0	1		$b_1 = ABCD$	all four main effects, all 12 2fi's
4	0	2		$b_1 = ABC$, $b_2 = AB^2D$	all four main effects, all 12 2fi's
4	0	3		$b_1 = AB$, $b_2 = AC$, $b_3 = AD$	all four main effects, AB^2, AC^2, AD^2, BC, BD, CD
5	1	1	$E = ABCD$	$b_1 = AB^2C$	all five main effects, all 20 2fi's
5	1	2	$E = ABCD$	$b_1 = AB$, $b_2 = AC^2D$	all five main effects, all 2fi's except AB
5	1	3	$E = ABCD$	$b_1 = AB$, $b_2 = AC$, $b_3 = AD$	all five main effects, AB^2 AC^2, AD^2, AE, BC, BD, BE^2, CD, CE^2, DE^2
6	2	1	$E = ABC$, $F = AB^2D$	$b_1 = BCD$	all six main effects, AC^2, AD^2, AE, AF, BC^2, BD, BE, BF^2, $C \times D$, CE, $C \times F$, $D \times E$, DF, $E \times F$
6	2	1	$E = AB$ $F = AB^2CD$	$b_1 = AC^2D$	C, D, F, all 2fi's except $A \times B$, $A \times E$, $B \times E$
6	2	2	$E = ABC$, $F = AB^2D$	$b_1 = ACD$, $b_2 = AB$	all six main effects, AC^2, AD^2, AE, AF, BC^2, BD, BE, BF^2, $C \times D$, CE, $C \times F$, $D \times E$, DF, $E \times F$
6	2	2	$E = AB$, $F = AB^2CD$	$b_1 = AB^2$, $b_2 = AC^2D$	C, D, F, all 2fi's except $A \times B$, $A \times E$, $B \times E$
6	2	3	$E = ABC$, $F = ABD$	$b_1 = AB^2$, $b_2 = AC$, $b_3 = AD$,	all six main effects, AC^2, AD^2, AE, AF, BC^2, BD^2, BE, BF, CD, CF^2, DE^2, EF
7	3	1	$E = ABC$, $F = ABD$, $G = ACD$	$b_1 = BCD$	all seven main effects, AB^2, AC^2, AD^2, AE, AF, AG, BE, BF, BG^2, CE, CF^2, CG, DE^2, DF, DG, EF, EG, FG

Table 5B.3 (*Continued*)

k	p	q	Design Generators	Block Generators	Clear Effects
7	3	2	$E = ABC, F = ABD,$ $G = ACD$	$b_1 = BC,$ $b_2 = BD$	all seven main effects, AB^2, $AC^2, AD^2, AE, AF, AG,$ $BE, BF, BG^2, CE, CF^2,$ $CG, DE^2, DF, DG, DJ,$ EF, EG, FG
7	3	3	$E = ABC,$ $F = ABD$ $G = ACD$	$b_1 = AB,$ $b_2 = AC,$ $b_3 = AD$	all seven main effects, AB^2, $AC^2, AD^2, BE, BF, CE,$ CG, DF, DG, EF, EG, FG
8	4	1	$E = ABC, F = ABD,$ $G = ACD, H = BCD$	$b_1 = AB^2C^2D$	all eight main effects, AE, $AF, AG, AH^2, BE, BF,$ $BG^2, BH, CE, CF^2, CG,$ CH, DE^2, DF, DG, DH
8	4	2	$E = ABC, F = ABD,$ $G = ACD, H = BCD$	$b_1 = AB^2,$ $b_2 = AB^2C^2D$	all eight main effects, AE, $AF, AG, AH^2, BE, BF,$ $BG^2, BH, CE, CF^2, CG,$ CH, DE^2, DF, DG, DH
8	4	3	$E = ABC, F = ABD,$ $G = ACD, H = BCD$	$b_1 = AB,$ $b_2 = AC,$ $b_3 = BD$	all eight main effects, AF, $AG, AH^2, BE, BG^2, BH,$ CE, CF^2, CH, DE^2, DF DG
9	5	1	$E = ABC, F = ABD,$ $G = AB^2C^2D,$ $H = AB^2CD^2,$ $J = AC^2D^2$	$b_1 = BC^2D^2$	all nine main effects
9	5	1	$E = AB, F = AB^2C,$ $G = AB^2D, H = ACD,$ $J = BCD$	$b_1 = AC^2D$	$C, D, F, G, H, J,$ $CF, CG^2, CJ, DF^2, DG, DJ,$ FH, GH, HJ
9	5	2	$E = ABC, F = ABD,$ $G = AB^2C^2D,$ $H = AB^2CD^2,$ $J = AC^2D^2$	$b_1 = ACD,$ $b_2 = AB$	all nine main effects
9	5	2	$E = AB, F = AB^2C,$ $G = AB^2D, H = ACD,$ $J = BCD$	$b_1 = AC,$ $b_2 = AD$	$C, D, F, G, H, J,$ $CF, CG^2, CJ, DF^2, DG, DJ,$ FH, GH, HJ
9	5	3	$E = ABC, F = ABD,$ $G = AB^2C^2D,$ $H = AB^2CD^2,$ $J = AC^2D^2$	$b_1 = AB,$ $b_2 = BC$ $b_3 = BD$	all nine main effects
9	5	3	$E = AB, F = AB^2C,$ $G = AB^2D, H = ACD,$ $J = BCD$	$b_1 = AB^2,$ $b_2 = AC,$ $b_3 = AD$	$C, D, F, G, H,$ $J, CF, CJ, DG,$ FH, GH, HJ

Table 5B.3   (*Continued*)

k	p	q	Design Generators	Block Generators	Clear Effects
10	6	1	$E = ABC$, $F = ABD$, $G = AB^2C^2D$, $H = AB^2CD^2$, $J = AC^2D^2$, $K = BC^2D^2$	$b_1 = AB$	all 10 main effects
10	6	2	$E = ABC$, $F = ABD$, $G = AB^2C^2D$, $H = AB^2CD^2$, $J = AC^2D^2$, $K = BC^2D^2$	$b_1 = AB$, $b_2 = AC$	all 10 main effects
10	6	3	$E = AB$, $F = AC$, $G = AB^2D$, $H = BCD$ $J = AC^2D$, $K = ABCD^2$	$b_1 = AB^2$, $b_2 = ABC$, $b_3 = AD$	D, G, H, J, K (the remaining five main effects are still estimable)

REFERENCES

Chen, J., Sun, D.X., and Wu, C.F.J. (1993), "A Catalogue of Two-Level and Three-Level Fractional Factorial Designs with Small Runs," *International Statistical Review*, 61, 131–145.

Cheng, S. W. and Wu, C.F.J. (1999), "Choice of Optimal Blocking Schemes in Two-Level and Three-Level Designs," Technical Report No. 347, Department of Statistics, University of Michigan.

Kempthorne, O. (1952), *Design and Analysis of Experiments*, New York: John Wiley & Sons.

Suen, C., Chen, H., and Wu, C.F.J. (1997), "Some Identities on q^{n-m} Designs with Application to Minimum Aberration Designs," *Annals of Statistics*, 25, 1176–1188.

Sun, D.X. and Wu, C.F.J. (1994), "Interaction Graphs for Three-Level Fractional Factorial Designs," *Journal of Quality Technology*, 26, 297–307.

CHAPTER 6

Other Design and Analysis Techniques for Experiments at More Than Two Levels

Various methods for designing and analyzing experiments with factors at more than two levels are presented in this chapter. The method of replacement is used to construct 2^m4^n (i.e., mixed two-level and four-level) designs from 2^{k-p} designs and optimal 2^m4^n designs are chosen according to the minimum aberration criterion. An analysis strategy for 2^m4^n designs is presented. A class of 36-run designs is constructed as a product of 2^{k-p} and 3^{k-p} designs. Then 25-run fractional factorial designs at five levels and 49-run fractional factorial designs at seven levels are considered. Finally, the method of sliding levels is considered, which can be used to handle factors that are related, i.e., the levels of one factor depend on the levels of another factor.

6.1 A ROUTER BIT EXPERIMENT BASED ON A MIXED TWO-LEVEL AND FOUR-LEVEL DESIGN

Consider an experiment to improve router-bit life reported in Phadke (1986). In a routing process, 8×4 inch printed wiring boards are cut from an 18×24 inch panel which is $\frac{1}{16}$ of an inch thick. When a router bit becomes dull, it produces boards with rough edges, which requires an extra cleaning process. Also, frequently changing router bits is expensive in terms of downtime. Failure is determined by evidence of an excessive amount of dust, where router-bit life is measured in inches cut in the x-y plane. A 32-run design was used to study the nine factors given in Table 6.1. The two levels of stacking height (G) correspond to a stack of three and four panels, respectively. The routing machine has four spindles (E) and each spindle works on its own stack of 18×24 inch panels. The spindles are synchronized so that they have

256

Table 6.1   Factors and Levels, Router Bit Experiment (current settings indicated by ∗)

Factor	Level			
A. suction (in. of Hg)	1	2*		
B. x-y feed (in./min)	60*	80		
C. in-feed (in./min)	10*	50		
D. bit type	1	2	3	4*
E. spindle position	1	2	3	4
F. suction foot	SR	BB*		
G. stacking height (in.)	3/16	1/4*		
H. Slot depth (mils)	60*	100		
J. speed (rpm)	30,000	40,000*		

Table 6.2   Design Matrix and Lifetime Data, Router Bit Experiment (1 lifetime unit = 100 inches)

Run	A	B	C	D	E	F	G	H	J	Lifetime
1	−	−	−	0	0	−	−	−	−	3.5
2	−	−	−	1	1	+	+	−	−	0.5
3	−	−	−	2	3	−	+	+	−	0.5
4	−	−	−	3	2	+	−	+	−	17.0
5	−	+	+	2	0	+	+	−	−	0.5
6	−	+	+	3	1	−	−	−	−	2.5
7	−	+	+	0	3	+	−	+	−	0.5
8	−	+	+	1	2	−	+	+	−	0.5
9	+	−	+	3	0	−	+	+	−	17.0
10	+	−	+	2	1	+	−	+	−	2.5
11	+	−	+	1	3	−	−	−	−	0.5
12	+	−	+	0	2	+	+	−	−	3.5
13	+	+	−	1	0	+	−	+	−	0.5
14	+	+	−	0	1	−	+	+	−	2.5
15	+	+	−	3	3	+	+	−	−	0.5
16	+	+	−	2	2	−	−	−	−	3.5
17	−	−	−	0	0	−	−	−	+	17.0
18	−	−	−	1	1	+	+	−	+	0.5
19	−	−	−	2	3	−	+	+	+	0.5
20	−	−	−	3	2	+	−	+	+	17.0
21	−	+	+	2	0	+	+	−	+	0.5
22	−	+	+	3	1	−	−	−	+	17.0
23	−	+	+	0	3	+	−	+	+	14.5
24	−	+	+	1	2	−	+	+	+	0.5
25	+	−	+	3	0	−	+	+	+	17.0
26	+	−	+	2	1	+	−	+	+	3.5
27	+	−	+	1	3	−	−	−	+	17.0
28	+	−	+	0	2	+	+	−	+	3.5
29	+	+	−	1	0	+	−	+	+	0.5
30	+	+	−	0	1	−	+	+	+	3.5
31	+	+	−	3	3	+	+	−	+	0.5
32	+	+	−	2	2	−	−	−	+	17.0

the same horizontal $(x\text{-}y)$ feed (B), vertical feed (in-feed) (C), and rotational speed (J). The experimenters are not interested in differences between the spindles since all the spindles will be used in production. Settings for the other factors need to be found which work well for all the spindles. The design matrix and lifetime data are presented in Table 6.2. Notice that the factors A, B, C, and J all have the same levels for the four spindles. In Table 6.2, the low and high levels of the two-level factors are represented by $-$ and $+$ and the four levels of D and E are represented by 0, 1, 2, and 3.

The experiment was stopped after 1700 inches. Eight of the 32 router bits (at runs $4, 9, 17, 20, 22, 25, 27, 32$) had not failed when the experiment was stopped; such data are called right-censored data. Actually, the router bits were inspected every 100 inches, so that the data are interval censored. For illustrative purposes, the midpoints of the intervals for the interval-censored data and the right-censoring times for the right-censored data will be used. Strategies for handling censored data will be discussed in Chapter 12. The data in Table 6.2 are given in units of 100 inches, e.g., 3.5 for 350 inches. The experimenters were interested in the relative importance of the nine main effects and the following four two-factor interactions: BJ, CJ, GJ, and BG. The experimental objective was to select factor levels which would improve router-bit life.

Before proceeding to the analysis of this experiment we need to show how fractional factorial designs with mixed two-level and four-level factors are constructed, which will be done in the next two sections.

6.2 METHOD OF REPLACEMENT AND CONSTRUCTION OF $2^m 4^n$ DESIGNS

The simplest way to construct designs with mixed two-level and four-level factors is to start with a 2^{k-p} design and replace three of its columns by a four-level column. To illustrate this method, consider the 8×7 matrix on the right side of Table 6.3. It is a 2^{7-4} design with eight runs and seven factors

Table 6.3 $OA(8, 2^4 4^1)$ **Constructed from** $OA(8, 2^7)$

Run	A	3	13	23	123		Run	1	2	12	3	13	23	123
1	0	−	+	+	−		1	−	−	+	−	+	+	−
2	0	+	−	−	+		2	−	−	+	+	−	−	+
3	1	−	+	−	+		3	−	+	−	−	+	−	+
4	1	+	−	+	−	←	4	−	+	−	+	−	+	−
5	2	−	−	+	+		5	+	−	−	−	−	+	+
6	2	+	+	−	−		6	+	−	−	+	+	−	−
7	3	−	−	−	−		7	+	+	+	−	−	−	−
8	3	+	+	+	+		8	+	+	+	+	+	+	+

Table 6.4 Generating a Four-Level Factor from Three Two-Level Factors

α	β	$\alpha\beta$		New Factor
$-$	$-$	$+$		0
$-$	$+$	$-$		1
$+$	$-$	$-$	$\longrightarrow$	2
$+$	$+$	$+$		3

represented by the columns **1, 2, 12, 3, 13, 23, 123**. For consistency of notation its levels are represented by $-$ and $+$. Noting that column **12** represents the interaction between column **1** and column **2**, we can replace the three columns in this matrix by a four-level column according to the rule given in Table 6.4. The generated column is denoted by A in the left matrix of Table 6.3. This matrix has one four-level column A and four two-level columns **3, 13, 23,** and **123**. A pair of columns in a matrix are called **orthogonal** if all possible combinations of levels in the two columns appear equally often. Column A is orthogonal to any of the two-level columns in the left matrix of Table 6.3 because each of the combinations $(0, -)$, $(1, -)$, $(2, -)$, $(3, -)$, $(0, +)$, $(1, +)$, $(2, +)$, $(3, +)$ appears once among the eight runs. This combinatorial definition of orthogonality has a statistical justification in terms of the orthogonality in effect estimation. When restricted to a pair of orthogonal columns, the design becomes a full factorial design for the two factors represented by the columns. Because in a full factorial design the main effects of one factor are orthogonal to the main effects of the other factor (i.e., in the ANOVA table, the sums of squares for the two factors are statistically uncorrelated), the effect estimates for these two columns are uncorrelated. In geometric terms, the subspaces represented by the two main effects are orthogonal.

The reader may wonder why a mixed notation is used in Tables 6.2 and 6.3 with $-$ and $+$ for two-level factors and $0, 1, 2, 3$ for four-level factors. An alternative would be to use 0 and 1 for two-level factors. Then the $\alpha\beta$ column in Table 6.4 would be $(0, 1, 1, 0)$, which is the sum (mod 2) of $\alpha = (0, 0, 1, 1)$ and $\beta = (0, 1, 0, 1)$. In data analysis the $\alpha\beta$ column can be used to estimate the interaction between the factors represented by α and β. Treating 0 and 1 in the $\alpha\beta$ column as low and high levels in the computation of the interaction effect would result in an effect estimate with the opposite sign from the usual definition of the interaction effect as given in Chapter 3. To avoid confusion in the sign interpretation of the interaction effect estimate or the level interpretation of 0 and 1, we choose not to use the $(0, 1)$ notation for two-level factors in 2^m4^n designs.

Both matrices in Table 6.3 are special cases of a very useful class of designs called orthogonal arrays (Rao 1947, 1973).

Orthogonal Array

An orthogonal array $OA(N, s_1^{m_1}, \ldots, s_\gamma^{m_\gamma}, t)$ of strength t is an $N \times m$ matrix, $m = m_1 + \cdots + m_\gamma$, in which m_i columns have $s_i(\geq 2)$ symbols or levels such that, for any t columns, all possible combinations of symbols appear equally often in the matrix.

For $\gamma = 1$, all factors have the same number of levels. These arrays are called **symmetrical orthogonal arrays**. For $\gamma > 1$, they are called **asymmetrical (or mixed-level) orthogonal arrays**. When there is no confusion, we may refer to an orthogonal array without specifying its strength, i.e., using $OA(N, s_1^{m_1}, \ldots, s_\gamma^{m_\gamma})$ for simplicity of notation. Usually when we refer to an array of strength two, we do not specify its strength and use the simpler notation.

The 2^{k-p} designs and the 3^{k-p} designs considered in Chapters 4 and 5 are special cases of orthogonal arrays. It is easy to see that the resolution of a 2^{k-p} or 3^{k-p} design is the same as the strength of the design plus 1 when viewed as an orthogonal array. For example, a resolution III design is an orthogonal array of strength two. Using the orthogonal array notation, the 2^{7-4} design on the right of Table 6.3 can be denoted by $OA(8, 2^7)$. The left design of Table 6.3 is an $OA(8, 2^4 4^1)$. Both have strength two.

A simple and general method for constructing designs with two-level and four-level factors consists of the following steps:

Method of Replacement

(i) Choose a two-level orthogonal array $OA(N, 2^m)$ which has three columns of the form $(\alpha, \beta, \alpha\beta)$, where $\alpha\beta$ is the interaction column between columns α and β, as in Table 6.4.

(ii) Replace $(\alpha, \beta, \alpha\beta)$ by a four-level column according to the rule in Table 6.4.

(iii) Repeat step (ii) for other sets of columns $(\alpha_i, \beta_i, \alpha_i \beta_i)$ in the $OA(N, 2^m)$.

It can be shown that the newly constructed four-level column is orthogonal to the remaining two-level columns in the $OA(N, 2^m)$. The proof is left as an exercise. If the number of mutually exclusive sets of columns $(\alpha_i, \beta_i, \alpha_i \beta_i)$ in step (iii) is n, n four-level columns are generated. Because each of the remaining $m - 3n$ columns still has two levels, the constructed design is an orthogonal array with $m - 3n$ two-level columns and n four-level columns and is denoted by $OA(N, 2^{m-3n} 4^n)$.

This construction method is referred to as the *method of replacement*. The best way to implement the method is to use a saturated two-level fractional factorial design in step (i). Recall from Section 4.2 that this design has 2^k

runs and can accommodate $2^k - 1$ factors. Denote it by $OA(N, 2^{N-1})$ with $N = 2^k$. To illustrate the construction method, take $k = 4$. The corresponding 16-run saturated design has 15 columns represented by **1, 2, 12, 3, 13, 23, 123, 4, 14, 24, 124, 34, 134, 234, 1234**. These 15 columns can be grouped into five mutually exclusive sets of the form $(\alpha, \beta, \alpha\beta)$:

$$(1, 2, 12), (3, 4, 34), (13, 24, 1234), (23, 124, 134), (123, 14, 234).$$

By successively replacing each of the five sets of columns by a four-level column, we can obtain the following orthogonal arrays:

$$OA(16, 2^{12}4^1), OA(16, 2^9 4^2), OA(16, 2^6 4^3), OA(16, 2^3 4^4), OA(16, 4^5).$$

The last array has five factors all at four levels. If there is no confusion about the run size, we may drop the reference to OA and simply refer to them as 2^m4^n designs with $n = 1, 2, 3, 4, 5$ and $m + 3n = 15$.

In general, for even k, all the $2^k - 1$ columns in the $OA(N, 2^{N-1})$ with $N = 2^k$ can be grouped into $(2^k - 1)/3$ sets of the form $(\alpha, \beta, \alpha\beta)$. By applying the method of replacement to some or all of these sets of columns, we can obtain up to $(2^k - 1)/3$ four-level columns with the remaining columns still at two levels, which results in the construction of the $OA(2^k, 2^m4^n)$ with $n = 1, \ldots, (2^k - 1)/3$ and $m + 3n = 2^k - 1$. For odd k, only $2^k - 5$ columns can be grouped into $(2^k - 5)/3$ sets of the form $(\alpha, \beta, \alpha\beta)$, which leads to the construction of $OA(2^k, 2^m4^n)$ with $n = 1, \ldots, (2^k - 5)/3$ and $m + 3n = 2^k - 1$. A proof of this result and an algorithm for finding mutually exclusive sets of the form $(\alpha, \beta, \alpha\beta)$ can be found in Wu (1989). For odd k, Mukerjee and Wu (1995) proved that an $OA(2^k, 2^1 4^{(2^k - 2)/3})$ does not exist, thus confirming that $(2^k - 5)/3$, the number of four-level columns in the constructed designs, is the maximum. For $k = 5$, $(2^k - 5)/3 = 9$, which says that the largest number of four-level factors

Table 6.5 Mutually Exclusive Sets of Columns of the Form $(\alpha, \beta, \alpha\beta)$ in the 2^k Designs

k	Sets of Columns
3	(1, 2, 12)
4	(1, 2, 12), (3, 4, 34), (13, 24, 1234), (23, 124, 134), (123, 14, 234)
5	(1, 2, 12), (4, 5, 45), (234, 1235, 145), (1234, 135, 245), (134, 15, 345), (14, 25, 1245), (124, 235, 1345), (24, 35, 2345), (34, 125, 12345)
6	(1, 2, 12), (3, 4, 34), (13, 24, 1234), (23, 124, 134), (123, 14, 234), (5, 6, 56), (15, 26, 1256), (125, 16, 256), (25, 126, 156), (35, 46, 3456), (345, 36, 456), (45, 346, 356), (135, 246, 123456), (12345, 136, 2456), (245, 12346, 1356), (235, 1246, 13456), (1345, 236, 12456), (1245, 1346, 2356), (1235, 146, 23456), (2345, 1236, 1456), (145, 2346, 12356)

in a 32-run orthogonal array is 9. By counting the available degrees of freedom, one might have thought that it could have accommodated 10 four-level factors.

Table 6.5 gives the grouping of columns in an $OA(N, 2^{N-1})$ with $N = 2^k$ into the maximum number of mutually exclusive sets of the form $(\alpha, \beta, \alpha\beta)$ for $k = 3, 4, 5, 6$. By applying the method of replacement to the columns in this table, one can easily construct various $2^m 4^n$ designs for 8, 16, 32, 64 runs. Because the groupings given in Table 6.5 are not unique and, for a particular grouping scheme, there are different ways to choose a subset of columns in the table, a natural question to ask is how to select good $2^m 4^n$ designs and what criterion should be used. This will be dealt with in the next section.

6.3 MINIMUM ABERRATION $2^m 4^n$ DESIGNS
with $n = 1, 2$

The minimum-aberration criterion for selecting good 2^{k-p} designs has a natural extension to $2^m 4^n$ designs. To illustrate this extension, consider the problem of selecting an $OA(16, 2^4 4^1)$. As in Table 6.3, denote the four-level column by $A = (1, 2, 12)$, which is obtained from the three two-level columns 1, 2, and 12 in the original $OA(16, 2^{15})$ by the method of replacement. How should the four two-level columns in the $OA(16, 2^4 4^1)$ be chosen from the remaining 12 two-level columns in the $OA(16, 2^{15})$? First, consider the design

$$d_1 : A, 3, 4, 23, 134,$$

which consists of the four-level column A and four two-level columns denoted by 3, 4, 23, and 134. [Recall that the 15 columns in the $OA(16, 2^{15})$ can be represented by all possible combinations of the numbers 1, 2, 3, 4.] Let B, C, D, E represent the four two-level factors, i.e., $B = 3$, $C = 4$, $D = 23$, $E = 134$. It is easy to show that these four factors are independent and there are no aliasing relationships among them. By comparing column A and the three two-level columns 1, 2, 12 in Table 6.4, it can be seen that factor A has three degrees of freedom represented by the contrasts between the two levels in each of the three columns 1, 2, and 12. Denote these three degrees of freedom by $a_1 = 1$, $a_2 = 2$, and $a_3 = 12$ (with $a_3 = a_1 a_2$). Then it is easy to show that this one-quarter fractional factorial design d_1 has the following defining relation:

$$\mathbf{I} = a_1 BCE = a_2 BD = a_3 CDE, \tag{6.1}$$

which is the defining contrast subgroup of design d_1. For comparison, consider the alternative design

$$d_2 : A, 3, 4, 34, 124.$$

By performing a similar calculation, its defining contrast subgroup can be shown to be

$$I = BCD = a_3CE = a_3BDE. \tag{6.2}$$

How can d_1 and d_2 be compared in terms of their respective defining contrast subgroups? Before answering this question, we need to classify the defining relations for 2^m4^1 designs into two types. The first type involves only the two-level factors and is called *type 0*. The second type involves one of the a_i's in the four-level factor A and some two-level factors and is called *type 1*. Note that any two a_i's that appear in a defining relation can be replaced by the third a_i because of the relation $I = a_1a_2a_3$. In typical situations, a defining relation of type 1 is less serious than a defining relation of type 0 of the same length. It can be justified as follows. Because factor A has three degrees of freedom (which are represented by its components a_1, a_2, a_3), it does not often happen that all the a_i's are significant. In particular, if A is a quantitative factor, the argument in the next section [after Equation 6.5)] shows that one of the a_i's can be treated as a cubic effect which is less likely to be important. Therefore, *a priori* knowledge may allow the experimenter to choose the least significant a_i component to be included in a defining relation of type 1.

For a design d, let $A_{i0}(d)$ and $A_{i1}(d)$ be the number of type 0 and type 1 words of length i in the defining contrast subgroup of d. The vector

$$W(d) = \{A_i(d)\}_{i \geq 3}, \tag{6.3}$$

where $A_i(d) = (A_{i0}(d), A_{i1}(d))$ is the *wordlength pattern* of d. A design d with positive $A_{2j}(d)$ is not useful because two of its main effects are aliased. This is why $A_2(d)$ is not included in $W(d)$. The **resolution** of d is defined to be the smallest i such that $A_{ij}(d)$ is positive for at least one j. In view of the previous discussion on the two types of defining relations, it is more important to have a smaller A_{i0} than a smaller A_{i1} for the same i. This consideration leads to the following criterion.

Minimum Aberration 2^m4^1 Design

Let d_1 and d_2 be two 2^m4^1 designs of the same run size and r be the smallest i such that $A_i(d_1) \neq A_i(d_2)$. If $A_{r0}(d_1) < A_{r0}(d_2)$ or $A_{r0}(d_1) = A_{r0}(d_2)$ but $A_{r1}(d_1) < A_{r1}(d_2)$, then d_1 is said to have less aberration than d_2. A design d has minimum aberration if no other design has less aberration than d.

It is easily verified that for design d_1 in (6.1), $A_{30} = 0$, $A_{31} = 1$, and $A_{41} = 2$; for d_2 in (6.2), $A_{30} = A_{31} = A_{41} = 1$ and d_1 has minimum aberration.

This approach (Wu and Zhang, 1993) can be extended to 2^m4^2 designs. Denote the two four-level factors in these designs by $A = (\mathbf{1}, \mathbf{2}, \mathbf{12})$ and

$B = (3, 4, 34)$. Then there are three types of defining relations. *Type 0* is defined as before. *Type 1* involves one component of factor A or B and some two-level factors, while *type 2* involves one component of factor A, one component of factor B, and some two-level factors. For a design d, let $A_{ij}(d)$ be the number of type j words of length i in the defining contrast subgroup of d and $W(d) = \{A_i(d)\}_{i \geq 3}$ and $A_i(d) = (A_{i0}(d), A_{i1}(d), A_{i2}(d))$. As previously argued, for the same length, a defining relation of type 0 is most serious while a defining relation of type 2 is least serious. This leads to the following criterion.

Minimum Aberration $2^m 4^2$ Design

Let d_1 and d_2 be two $2^m 4^2$ designs of the same run size and r be the smallest i such that $A_i(d_1) \neq A_i(d_2)$. Suppose that one of the following three conditions holds: (i) $A_{i0}(d_1) < A_{i0}(d_2)$; (ii) $A_{i0}(d_1) = A_{i0}(d_2)$, $A_{i1}(d_1) < A_{i1}(d_2)$; (iii) $A_{i0}(d_1) = A_{i0}(d_2)$, $A_{i1}(d_1) = A_{i1}(d_2)$, $A_{i2}(d_1) < A_{i2}(d_2)$. Then d_1 is said to have less aberration than d_2. A design d has minimum aberration if no other design has less aberration than d.

Consider the following two 16-run $2^3 4^2$ designs:

$$d_3: A, B, 14, 23, 234, \qquad d_4: A, B, 14, 23, 1234,$$

where the three two-level columns are represented by the last three words. They have resolution III and the following wordlength patterns:

$$W(d_3) = ((0, 1, 3), (0, 1, 2), (0, 0, 0)),$$
$$W(d_4) = ((1, 0, 3), (0, 0, 3), (0, 0, 0)).$$

It can be shown that d_3 has minimum aberration.

Minimum aberration $2^m 4^1$ and $2^m 4^2$ designs with 16, 32, and 64 runs are given in Tables 6A.1–6A.3 of Appendix 6A and 6B.1–6B.3 of Appendix 6B, respectively, for selected values of m. For each m, the design generators (which define the four-level and two-level factors), its resolution, and the collection of clear main effects and two-factor interactions are given. These designs are based on the numerical and theoretical results given in Wu and Zhang (1993). (The definitions of clear main effects and two-factor interactions are deferred to the next section.) The information in these tables can be readily used in planning an experiment. Take, for example, the 32-run design with $m = 5$ in Table 6A.2. It has five independent columns denoted by **1, 2, 3, 4,** and **5**. The four-level column A is based on **(1, 2, 12)**. The five two-level columns are denoted by **3, 4, 5, 6** (= **245**) and **7** (= **1345**), with factor **6** assigned to the interaction column **245** and factor **7** assigned to the interaction column **1345**. The defining contrast subgroup of the design can be obtained from its generators. From the subgroup it can be determined that the design has resolution IV, all its main effects are clear, and the two-factor interactions 34, 35, 47, and 57 are clear. To avoid confusion between "twelve" and the 12 interaction, we use t_0, t_1, t_2, etc., to denote the Arabic numbers 10, 11, 12, etc., as was done in Chapter 4.

6.4 AN ANALYSIS STRATEGY FOR 2^m4^n EXPERIMENTS

Before discussing an analysis strategy, we must address the important question of how to decompose the three degrees of freedom in a four-level factor. For a quantitative factor A, orthogonal polynomials can be used again to obtain the linear, quadratic, and cubic contrasts:

$$A_l = (-3, -1, 1, 3),$$
$$A_q = (1, -1, -1, 1), \tag{6.4}$$
$$A_c = (-1, 3, -3, 1),$$

where $-3, -1, 1, 3$ in A_l are the weights given to observations at levels 0, 1, 2, and 3, with the weights in A_q, A_c being similarly interpreted. The three contrasts are mutually orthogonal vectors in (6.4). For qualitative factors, however, it is often difficult to interpret them, especially A_l and A_c, which have unequal weights.

A better system of decomposition is the following:

$$A_1 = (-1, -1, 1, 1),$$
$$A_2 = (1, -1, -1, 1), \tag{6.5}$$
$$A_3 = (-1, 1, -1, 1),$$

where only weights of $+1$ and -1 are attached to observations at each of the four levels. The three contrasts are mutually orthogonal. For qualitative factors, each of them can be interpreted as a comparison between two groups of levels. For example, A_1 corresponds to the difference $(y_2 + y_3) - (y_0 + y_1)$ between levels $2, 3$ and levels $0, 1$. Interestingly enough, they also have a natural interpretation for quantitative factors with evenly spaced levels. The contrast A_1 obviously represents a linear effect and indeed has a high correlation of 0.89 with A_l; A_2 represents a quadratic effect because $(y_3 + y_0) - (y_2 + y_1) = (y_3 - y_2) - (y_1 - y_0)$ is the difference between two linear effects at the high levels (2 and 3) and the low levels (0 and 1); strictly speaking, A_3 does not measure the cubic effect but it is highly correlated with A_c in (6.4) with correlation 0.89. This suggests that the effect $(y_3 + y_1) - (y_2 + y_0)$ can be interpreted as being nearly cubic. Note that the A_i vectors are defined without the normalizing constant $\frac{1}{2}$, as would normally be done. The main reason is to keep them consistent with the weights $(-1, 1)$ used in computing the main effects of two-level factors. This will avoid a possible confusion in data analysis that must deal with both four-level and two-level factors.

Another advantage of the (A_1, A_2, A_3) system stems from the fact that the three vectors are of the form $(\alpha, \beta, \alpha\beta)$. By comparing (6.5) with Table 6.4, it is seen that A_1 corresponds to α, A_2 corresponds to $\alpha\beta$, and A_3 corresponds to β. This relationship makes it easier to relate and trace each A_i to a factorial effect in the original two-level design from which the $2^m 4^n$ design is generated.

Since the (A_1, A_2, A_3) system enjoys definite advantages over the (A_l, A_q, A_c) system, its use is recommended for analyzing both types of factors. The interaction between two four-level factors A and B can be easily obtained by multiplying (A_1, A_2, A_3) and (B_1, B_2, B_3), which gives the following nine contrasts: for $i, j = 1, 2, 3$, and $x, y = 0, 1, 2, 3$,

$$(AB)_{ij}(x, y) = A_i(x)B_j(y). \tag{6.6}$$

As in Section 5.6, we call $(AB)_{11}$ the linear-by-linear interaction effect, $(AB)_{12}$ the linear-by-quadratic interaction effect, $(AB)_{23}$ the quadratic-by-cubic interaction effect, and so forth. Higher order interactions as well as interactions between a four-level factor and a two-level factor are similarly defined.

Because the decompositions for main effects and interactions in (6.5) and (6.6) produce a system of orthogonal contrasts, each with one degree of freedom, the analysis can proceed as before as long as the aliasing relations between effects are available. These aliasing relations can be obtained from those of the original two-level design. In obtaining these relations, one must recognize that the three contrasts in (6.5) for the main effect of a four-level factor and the interactions between two four-level factors or between a two-level factor and a four-level factor can be traced to the factorial effects in the original two-level design. For example, if A_1, A_2, and A_3 in (6.5) represent the three degrees of freedom of factor A in Table 6.3, they can be traced to columns **1**, **12**, and **2**, respectively, in the $OA(8, 2^7)$ of Table 6.3.

The definition of clear and strongly clear effects for two-level experiments has a straightforward extension here. A main effect contrast of a two-level or four-level factor or a two-factor interaction contrast between factors of either type is said to be **clear** if it is not aliased with any other main effect contrasts or two-factor interaction contrasts. [Note that a contrast represents one degree of freedom such as those defined in (6.5) and (6.6).] It is said to be **strongly clear** if it satisfies the further requirement that it is not aliased with any three-factor interaction contrasts. One can argue as in Section 6.3 that some interaction components between two four-level factors are less likely to be significant than those between a two-level factor and a four-level factor, which are in turn less likely to be significant than those between two two-level factors. Its major implication is in the further classification of non-clear effects: those whose only aliases are interactions between two four-level factors are least serious and may be regarded almost as important as clear effects. Similarly, non-clear effects that are not aliased with any interactions between two two-level factors are less serious than those aliased with interactions between two two-level factors. Although no formal definition is given here to reflect this subtle difference, its distinction can be useful in data analysis, as can be seen in the analysis in the next section.

The decomposition of main effects and interactions in (6.5) and (6.6) produces a system of orthogonal contrasts, each of which can be traced to a factorial effect in the original two-level design. Therefore, the analysis of $2^m 4^n$ experiments can proceed as in the two-level case except for the interpretation of effect aliasing. This suggests the following analysis strategy.

An Analysis Strategy for $2^m 4^n$ Experiments   (6.7)

Step 1. Suppose an experiment has 2^k runs. Its $2^k - 1$ degrees of freedom (i.e., $2^k - 1$ orthogonal contrasts) can be represented as products of the k generating columns $\mathbf{1}, \mathbf{2}, \ldots, \mathbf{k}$, where $\mathbf{1}$ is the column of half $-$'s and half $+$'s, $\mathbf{2}$ is the column of quarter $-$'s, quarter $+$'s, quarter $-$'s, and quarter $+$'s, and so on. Express the main effect contrasts in (6.5) of any four-level factor and the main effect of any two-level factor as products of $\mathbf{1}, \mathbf{2}, \ldots, \mathbf{k}$. By multiplying the main effect contrasts, interaction effect contrasts such those defined in (6.6) can also be represented as products of $\mathbf{1}, \mathbf{2}, \ldots, \mathbf{k}$. Two contrasts are aliased if they correspond to the same algebraic expression. All the aliasing relations can be obtained from examining these algebraic expressions.

Step 2. Because each contrast corresponds to a factorial effect in a 2^k design, the standard analysis methods discussed in Chapters 3 and 4 can be used. A half-normal plot or a formal test can be used to identify significant effects. In interpreting effect significance, the A_i contrast defined in (6.5) should be interpreted as a linear effect for $i = 1$, a quadratic effect for $i = 2$, and a cubic effect for $i = 3$ for a quantitative factor and as contrasts of means for a qualitative factor. (Here A stands for a generic factor and can be substituted with other factor names.)

6.5   ANALYSIS OF THE ROUTER BIT EXPERIMENT

The strategy in (6.7) is illustrated with the analysis of the router bit experiment.

Step 1. The effects in the 32-run design for the experiment can be associated with the 31 effects in a 2^5 full factorial design. Let **1–5** denote the generating columns for the 32-run design, where **1** is the column of 16 $-$'s and 16 $+$'s, **2** is the column of the pattern 8 $-$'s and 8 $+$'s repeated twice, and so on. Then, the nine factor main effects can be represented as

$$A = 2, \qquad B = 3, \qquad C = -23,$$
$$D = (D_1, D_2, D_3) = (234, -345, -25),$$
$$E = (E_1, E_2, E_3) = (4, -5, -45), \tag{6.8}$$
$$F = -35, \qquad G = -2345, \qquad H = -24, \qquad J = 1.$$

The representation is illustrated with the identities $E_1 = 4$ and $E_2 = -5$ (which imply $E_3 = E_1 E_2 = -45$). From (6.5), E_1 assigns $-$ to its 0 and 1 levels and $+$ to its 2 and 3 levels. By applying this rule to the E column in Table 6.2, a vector consisting of a repeated pattern of $(- - + +)$ is obtained, which equals **4**. Similarly, by assigning $-$ to the 1 and 2 levels of the E column in Table 6.2 and $+$ to its 0 and 3 levels, E_2 is seen to be a vector consisting of a repeated pattern of $(+ -)$, which equals -5.

From (6.8), it can be deduced that

$$D_1 = -BH, \qquad D_2 = AG, \qquad D_3 = -CF$$

and

$$E_1 = -AH, \qquad E_2 = BF, \qquad E_3 = -CG. \tag{6.9}$$

In deriving (6.9), interactions involving the four-level factors D and E were not included, because, as argued before, these interactions are usually less likely to be significant than those involving only two-level factors. Otherwise, there would be more aliases. For example, $D_2 = AG = BE_3 = E_1 F$. Moreover, AB, AC, BC, FG, FH, and GH are completely aliased with main effects, since $C = -AB$ and $H = -FG$.

Because $J = 1$ and none of the other factors involve column **1**, it is easy to see that J is strongly clear and any two-factor interaction involving J is clear. Three two-factor interactions have aliases that are interactions involving D or E:

$$AF = BD_3 = -CE_2 = -D_2 H = D_1 E_3 = E_1 G,$$
$$CH = AD_1 = BE_1 = D_2 E_2 = E_3 F = D_3 G, \tag{6.10}$$
$$BG = AE_3 = -CD_2 = D_3 E_1 = D_1 F = -E_2 H.$$

By multiplying each term in (6.10) by J, it is easily seen that AFJ, CHJ, and BGJ are only aliased with three-factor interactions involving D and E.

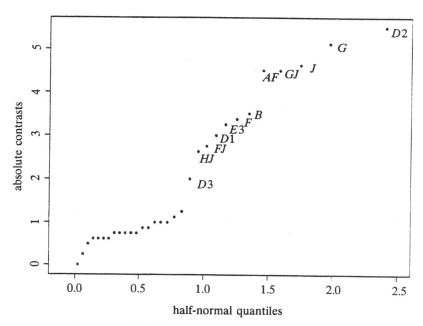

Figure 6.1. Half-Normal Plot, Router Bit Experiment.

Therefore, we have identified all 31 effects: 13 main effects $A, B, C, F, G, H, J, D_1, D_2, D_3, E_1, E_2, E_3$; 12 clear two-factor interactions $AJ, BJ, CJ, FJ, GJ, HJ, D_1J, D_2J, D_3J, E_1J, E_2J, E_3J$; three aliased two-factor interactions AF, BG, CH; and three aliased three-factor interactions AFJ, BGJ, CHJ.

Step 2. The half-normal plot of the 31 effects is displayed in Figure 6.1. From the half-normal plot, D_2, G, J, GJ, and AF are identified as important. Recall that GJ is clear but AF is aliased with five two-factor interactions [see (6.10)]. Based on the effect heredity principle in Section 3.5, only D_2H and E_1G may be important because none of the other four aliased interactions have parent factor effects among the significant factor effects, D_2, G, or J. Between D_2H and E_1G, more importance may be attached to D_2H. As discussed before, the four spindles are synchronized. It is hard to imagine that the effect of G will vary substantially among spindles, thus ruling out the E_1G effect. If it is desirable to de-alias some of the aliased interactions, a follow-up experiment should be conducted by following the strategies given in Section 4.4.

Based on the previous discussion, the lifetime data y is fitted by a regression model consisting of the effects D_2, G, J, GJ, and D_2H. The

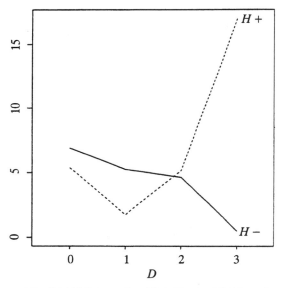

Figure 6.2. $D \times H$ Interaction Plot, Router Bit Experiment.

model for predicted lifetime is

$$\hat{y} = 5.8125 + 2.75D_2 - 2.56G + 2.31J - 2.25GJ + 2.25D_2H. \quad (6.11)$$

To determine the setting of the four-level factor D, an examination of the $D \times H$ interaction plot (Figure 6.2) is recommended. Based on the plot, level 3 for D and level $+$ for H should be chosen. For G and J, using (6.11), level $-$ for G and level $+$ for J are recommended. Among the recommendations, J at $+$ is the current setting (which is marked with an asterisk in Table 6.1); D at 3, G at $-$, and H at $+$ are new. Based on model (6.11) the predicted lifetime at the current setting ($D = 3$, $G = +$, $H = -$, $J = +$) is

$$\hat{y} = 5.8125 + 2.75(+1) - 2.56(+1) + 2.31(+1)$$

$$-2.25(+1)(+1) + 2.25(+1)(-1) = 3.8125 \quad (6.12)$$

and at the new setting ($D = 3$, $G = -$, $H = +$, $J = +$) is

$$\hat{y} = 5.8125 + 2.75(+1) - 2.56(-1) + 2.31(+1) - 2.25(-1)(+1)$$

$$+2.25(+1)(+1) = 17.9325. \quad (6.13)$$

[Note that for the computations in (6.12) and (6.13), D_2 in (6.11) equals $(+1, -1, -1, +1)$ at the four levels $0, 1, 2, 3$ of factor D.] The improvement is about 470%. There is a downside, however, to the recommended change from $+$ to $-$ for G. Recall from the description in Section 6.1 that $+$ corresponds to four printed boards while $-$ corresponds to three printed boards. Therefore, the recommended change would result in a 25% loss in productivity. Based on the coefficient of G in (6.11), the change would increase the predicted lifetime by 9.62 $[= (2)(2.56) + (2)(2.25)]$, which is about 54% of the predicted lifetime from (6.13) at the new setting. Whether the recommendation should be adopted will depend on the relative importance among productivity, router-bit life, and process downtime. If the 25% loss cannot be tolerated, then only the recommended changes for factors D and H should be adopted. At the revised settings ($D = 3$, $G = +$, $H = +$, $J = +$), the predicted lifetime based on (6.11) is 8.3125, which is still a substantial improvement of 218%. All the predicted savings based on a fitted model should be confirmed (or disproved) by performing a confirmation experiment at the new settings.

The design for this experiment was chosen to ensure that the four designated interactions BG, BJ, CJ, and GJ can be estimated clearly of the main effects and two-factor interactions not involving D or E. Consequently it is a secondary concern whether the design has minimum aberration. Although the 32-run minimum aberration $2^7 4^2$ design in Table 6B.2 of Appendix 6B is generally a better design, it does not meet the specific requirements of the experiment.

6.6 A PAINT EXPERIMENT BASED ON A MIXED TWO-LEVEL AND THREE-LEVEL DESIGN

An experiment was performed to improve a painting process of charcoal grill parts (Hale-Bennett and Lin, 1997). The paint was too thick and within- and between-part variation was unacceptable. The parts are painted in a booth in which 12 paint guns are positioned, guns 1–6 on the right side and guns 7–12 on the left side. Guns 7, 8, 9, and 11 oscillate, while the rest of the guns are fixed. Two parts are painted at a time and are hung on the top and bottom hooks of a rack, respectively. Optimal paint thickness is 1.5 mils for the outside and side of the grill and 1.0 mil for the inside of the grill. A mixed-level 36-run design was used to study six factors: gun 1 position, gun 8 position, gun 9 position at two levels denoted by $-$ and $+$, and powder pressure, gun 7 position, and gun 11 position at three levels denoted by $-$, 0, and $+$. These factors are denoted by A–F. Powder pressure, gun 7 position, and gun 11 position were thought to be the more important variables, so they were studied at more levels. The actual levels were not reported because of confidentiality; the three levels are quantitative and assumed to be evenly spaced. The design matrix and paint thickness data appear in Tables 6.6 and 6.7. The runs were performed in random order. The responses for top and bottom are from grills hung on the top and bottom hooks of the paint rack. Responses for inside, outside, and side are from the three critical locations on a grill. For paint thickness, the mean and range based on 16 grills are

Table 6.6 Design Matrix, Paint Experiment

Run	A Gun 1	B Gun 8	C Gun 9	D Pressure	E Gun 7	F Gun 11
1	−	−	+	−	−	0
2	+	−	−	−	−	0
3	−	+	−	−	−	0
4	+	+	+	−	−	0
5	−	−	+	0	−	+
6	+	−	−	0	−	+
7	−	+	−	0	−	+
8	+	+	+	0	−	+
9	−	−	+	+	−	−
10	+	−	−	+	−	−
11	−	+	−	+	−	−
12	+	+	+	+	−	−
13	−	−	+	−	0	+
14	+	−	−	−	0	+
15	−	+	−	−	0	+
16	+	+	+	−	0	+
17	−	−	+	0	0	−
18	+	−	−	0	0	−
19	−	+	−	0	0	−
20	+	+	+	0	0	−
21	−	−	+	+	0	0
22	+	−	−	+	0	0
23	−	+	−	+	0	0
24	+	+	+	+	0	0
25	−	−	+	−	+	−
26	+	−	−	−	+	−
27	−	+	−	−	+	−
28	+	+	+	−	+	−
29	−	−	+	0	+	0
30	+	−	−	0	+	0
31	−	+	−	0	+	0
32	+	+	+	0	+	0
33	−	−	+	+	+	+
34	+	−	−	+	+	+
35	−	+	−	+	+	+
36	+	+	+	+	+	+

Table 6.7 Thickness Data, Paint Experiment

| | Inside | | | | Outside | | | | Side | | | | |
| | Top | | Bottom | | Top | | Bottom | | Top | | Bottom | | |
Run	Mean	Range	Mean	Range	Mean	Range	Mean	Range	Mean	Range	Mean	Range	Weight
1	0.775	0.3	0.725	0.2	0.850	0.4	1.075	0.4	1.15	0.1	0.95	0.1	0.118
2	0.550	0.4	0.675	0.6	0.875	0.4	0.850	0.1	1.05	0.1	0.95	0.1	0.100
3	0.550	0.6	0.575	0.3	0.825	0.7	0.950	0.5	0.95	0.1	1.40	0.2	0.108
4	0.600	0.4	0.675	0.2	0.900	0.4	1.025	0.4	0.90	0.0	0.95	0.1	0.102
5	0.900	0.2	1.025	0.3	1.425	0.4	1.375	0.7	1.65	0.1	1.90	0.2	0.166
6	0.875	0.1	0.950	0.7	1.950	1.8	1.375	0.9	1.95	0.1	1.85	0.1	0.145
7	1.000	0.3	0.875	0.5	1.250	0.2	1.550	0.4	1.75	0.1	1.65	0.1	0.182
8	1.000	0.7	0.950	0.3	1.300	0.2	1.575	0.7	1.80	0.2	1.80	0.1	0.166
9	1.400	0.2	1.250	0.2	1.525	0.4	1.850	0.4	1.95	0.1	2.60	0.6	0.242
10	1.225	0.5	1.100	0.4	1.500	0.6	1.775	0.8	1.60	0.4	1.45	0.1	0.205
11	1.225	0.3	1.225	0.2	1.525	0.2	1.625	0.4	1.80	0.0	2.45	0.1	0.209
12	1.475	0.4	1.150	0.4	1.750	0.3	2.200	0.6	2.65	0.1	2.35	0.1	0.220
13	0.600	0.3	0.575	0.1	0.975	0.4	1.075	0.4	1.10	0.4	1.10	0.2	0.126
14	0.600	0.3	0.700	0.3	1.125	0.7	1.025	0.3	1.30	0.2	1.35	0.1	0.112
15	0.625	0.7	0.600	0.4	0.975	0.4	0.850	0.7	1.20	0.1	1.25	0.1	0.124
16	0.500	0.2	0.600	0.5	1.025	0.5	0.900	0.8	1.30	0.0	1.20	0.4	0.122
17	0.925	0.3	0.925	0.2	1.600	0.6	1.500	0.7	2.10	0.0	1.75	0.1	0.166
18	1.025	0.1	1.000	0.5	1.525	0.2	1.425	1.0	1.50	0.1	1.30	0.0	0.176
19	0.875	0.2	1.025	0.1	1.200	0.2	1.475	0.4	1.45	0.1	1.55	0.1	0.186
20	0.850	0.3	0.850	0.5	1.375	0.7	2.075	0.7	2.00	0.2	2.05	0.1	0.150
21	1.200	0.4	1.175	0.2	1.875	0.7	2.175	0.6	2.50	0.4	2.85	0.5	0.236
22	1.250	0.3	0.200	0.2	1.525	0.1	2.175	1.0	2.05	0.3	2.20	0.0	0.230
23	1.150	0.4	0.925	0.3	1.675	0.3	1.750	1.2	1.90	0.0	2.20	0.0	0.224
24	1.150	0.2	1.300	0.2	1.625	0.3	2.050	0.4	1.60	0.0	1.95	0.3	0.220
25	0.500	0.4	0.575	0.3	1.075	0.3	0.900	0.3	1.25	0.3	1.10	0.2	0.110
26	0.550	0.4	0.525	0.3	0.875	0.4	0.875	0.2	0.85	0.1	1.10	0.0	0.104
27	0.575	0.2	0.650	0.3	1.000	0.2	0.950	0.3	0.95	0.3	1.10	0.0	0.105
28	0.600	0.4	0.625	0.4	0.925	0.4	1.150	0.6	1.15	0.1	1.10	0.2	0.120
29	0.900	0.7	1.050	0.6	1.400	0.8	1.775	0.2	2.10	0.2	1.95	0.3	0.184
30	1.025	0.6	1.000	0.3	1.225	0.4	1.450	0.3	1.25	0.1	1.40	0.0	0.184
31	0.850	0.4	0.750	0.8	1.375	1.1	2.325	1.1	2.30	0.8	2.35	0.5	0.180
32	0.975	0.4	0.875	0.2	1.400	0.7	1.750	0.4	2.05	0.3	2.15	0.1	0.176
33	1.100	0.2	1.150	0.2	1.850	1.6	1.475	1.6	2.00	0.4	2.35	0.1	0.220
34	1.200	0.4	1.150	0.7	1.600	0.7	1.700	0.6	1.90	0.2	2.15	0.1	0.224
35	1.150	0.5	1.150	0.3	1.725	1.7	1.675	0.7	2.80	0.0	2.20	0.2	0.210
36	1.300	0.5	1.350	0.3	1.550	0.8	2.075	0.7	2.00	0.2	2.35	0.3	0.226

reported. The weight response is the amount of paint used to paint the grills. The goal of the experiment was to find process conditions that can achieve the nominal paint thickness on average with smaller variation. An important benefit from the variation reduction would be the savings in the the amount of paint used.

6.7 DESIGN AND ANALYSIS OF 36-RUN EXPERIMENTS AT TWO AND THREE LEVELS

The design in Table 6.6 is the first example in the book of a mixed two-level and three-level design. It can be constructed as follows. By examining columns A, B, and C of the design matrix, it is easily seen that these two-level factors form a 2^{3-1} design with $C = AB$. The 2^{3-1} design is used for the first four rows and repeated for the next eight groups of four rows. By coding the $-$, 0, and $+$ levels of the three-level factors in Table 6.6 as 1, 2, and 0, it can be easily shown that column F = column D + column E (*mod* 3) and thus the three-level factors form a 3^{3-1} design with $F = DE$. (Note that this coding does not affect the analysis to be presented later. It merely serves to show a simple relation between D, E, and F.) Each of the nine combinations of the 3^{3-1} design appears in four consecutive entries in Table 6.6. The 36-run design in Table 6.6 consists of the 4×9 combinations of the 2^{3-1} design and the 3^{3-1} design and is called a $2^{3-1} \times 3^{3-1}$ design. Because it is an orthogonal array of strength two, it can also be referred to as an $OA(36, 2^3 3^3)$.

The defining relation of the $2^{3-1} \times 3^{3-1}$ design can be obtained from those of its two component designs: $\mathbf{I} = ABC$ and $\mathbf{I} = DEF^2$. By multiplying these two defining words, two more defining words are obtained:

$$ABC(DEF^2) = ABCDEF^2,$$

$$ABC(DEF^2)^2 = ABCD^2E^2F. \tag{6.14}$$

Thus the defining contrast subgroup of the $2^{3-1} \times 3^{3-1}$ design is

$$\mathbf{I} = ABC = DEF^2 = ABCDEF^2 = ABCD^2E^2F. \tag{6.15}$$

Because the $2^{3-1} \times 3^{3-1}$ design is a *product* of the 2^{3-1} design and the 3^{3-1} design, the aliasing relations of its effects have a simple structure. From $\mathbf{I} = ABC$, we have the aliasing relations among two-level factor effects:

$$A = BC, \qquad B = AC, \qquad C = AB.$$

From $\mathbf{I} = DEF^2$, we have the aliasing relations among three-level factor effects:

$$D = EF^2, \qquad E = DF^2, \qquad F = DE, \qquad DE^2 = DF = EF.$$

Noting that a two-level factor appears with a three-level factor only in one of the two six-letter words in (6.15), any two-factor interaction between a two-level factor and a three-level factor is clear. By ignoring higher order effects among aliased effects, the 35 degrees of freedom in the $2^{3-1} \times 3^{3-1}$ design can be used to estimate the following effects: six main effects (A, B, C, D, E, F), DE^2 (and its aliases), nine clear interactions $(AD, AE, AF, BD, BE, BF, CD, CE, CF)$, ADE^2, BDE^2, and CDE^2 (and their respective aliases). In counting the degrees of freedom, it should be noted that an interaction between a two-level factor and a three-level factor has two degrees of freedom.

Because these effects are orthogonal to each other, they can be analyzed by ANOVA. For simplicity's sake, we only consider the analysis of the top

Table 6.8   ANOVA Table for Top Inside Paint Thickness, Paint Experiment

Source	Degrees of Freedom	Sum of Squares	Mean Squares	F	p Value
A	1	0.005625	0.005625	2.3143	0.179
B	1	0.000625	0.000625	0.2571	0.630
C	1	0.005625	0.005625	2.3143	0.179
D	2	2.539201	1.269601	522.3500	0.000
E	2	0.038993	0.019497	8.0214	0.020
F	2	0.006076	0.003038	1.2500	0.352
$G = DE^2$	2	0.019306	0.009653	3.9714	0.080
$A \times D$	2	0.017812	0.008906	3.6643	0.091
$A \times E$	2	0.023229	0.011615	4.7786	0.057
$A \times F$	2	0.000729	0.000365	0.1500	0.864
$B \times D$	2	0.001979	0.000990	0.4071	0.683
$B \times E$	2	0.020104	0.010052	4.1357	0.074
$B \times F$	2	0.021979	0.010990	4.5214	0.063
$C \times D$	2	0.011563	0.005781	2.3786	0.174
$C \times E$	2	0.045729	0.022865	9.4071	0.014
$C \times F$	2	0.005104	0.002552	1.0500	0.406
Residual	6	0.014583	0.002431		

inside mean response, which is given in the second column of Table 6.7. An ANOVA can be performed by pooling the sum of squares for ADE^2, BDE^2, and CDE^2 with six degrees of freedom as the residual sum of squares. The ANOVA table displayed in Table 6.8 indicates that D, E, and $C \times E$ are significant. Analysis of the other responses and their implications for achieving the experimental goals are left as an exercise.

Because D, E, and $C \times E$ each have two degrees of freedom, it is desirable to know which of their components contribute to their significance. To perform a more refined analysis than an ANOVA, we need to associate each of the 35 degrees of freedom with an effect. First note that the 3^{3-1} design can accommodate an additional factor, say G, which has the following values: $(0, -, +, +, 0, -, -, +, 0)$. In this case, a simple analysis could be done by calculating 35 orthogonal effects and then using a half-normal plot to identify the significant effects. Linear and quadratic contrasts can be used for the three-level factors, say D_l and D_q. Any interaction between a two-level factor and a three-level factor can be decomposed into two components. For example, the $A \times D$ interaction can be decomposed into AD_l and AD_q as defined in (6.16), where l and q stand for "linear" and "quadratic." Therefore, nine contrasts for main effects and 18 contrasts for two-factor interactions have been identified. The eight remaining contrasts are the main effects for G and the interactions between all the two-level factors and G. In the paint experiment, $G = DE^2$ is a two-factor interaction component so that $A \times G$, $B \times G$, and $C \times G$ are aliased with three-factor interaction components. Recall that the contrasts need to be scaled to have unit length. The

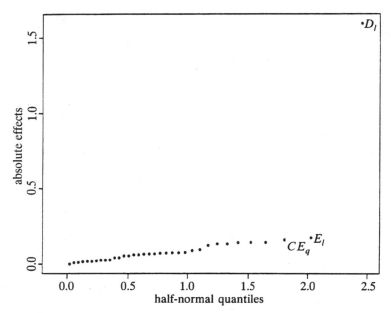

Figure 6.3. Half-Normal Plot of Top Inside Mean Paint Thickness, Paint Experiment.

scaling factors for this 36-run design are $\sqrt{36}$ for the two-level factor main effects, $\sqrt{24}$ for both the linear component of the three-level factor main effects and the linear component of the two-level factor by three-level factor interaction, and $\sqrt{72}$ for both the quadratic component of the three-level factor main effects and the quadratic component of the two-level factor by three-level factor interaction.

The half-normal plot of paint thickness effects is given in Figure 6.3. Clearly, the pressure (D) linear effect is significant. After dropping D_l, the half-normal plot of the remaining effects, as displayed in Figure 6.4, shows that E_l and CE_q are significant; this is consistent with the ANOVA table results in Table 6.8. This is the first use in the book of a two-stage version of the half-normal plot technique. It works generally as follows:

    (i) Use the half-normal plot or Lenth's method to identify the top cluster of effects.

    (ii) Apply the same procedure on the remaining effects and identify the second cluster of effects.

    (iii) Possibly repeat the same procedure on the remaining effects.

If a formal procedure like Lenth's method is used to identify significant effects, the critical value decreases as the number of effects under test

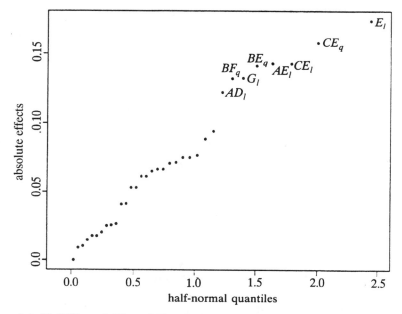

Figure 6.4. Half-Normal Plot of Top Inside Mean Paint Thickness (D_l removed), Paint Experiment.

decreases. This sequential test procedure is referred to as a **step-down multiple comparison** procedure.

The linear-quadratic system and the variable selection strategy for 3^{k-p} designs (see Section 5.6) can be adapted for mixed two-level and three-level designs. For example, an interaction between a two-level factor and a three-level factor, say $A \times D$, can be decomposed into two effect components AD_l and AD_q: for $i = -, +$ and $j = -, 0, +$,

$$AD_l(i, j) = A(i)D_l(j), \qquad AD_q(i, j) = A(i)D_q(j), \qquad (6.16)$$

where

$$A(-, +) = (-1, 1), \quad D_l(-, 0, +) = (-1, 0, 1), \quad D_q(-, 0, +) = (1, -2, 1).$$

If D is a qualitative factor, $A \times D$ can be decomposed into AD_{01} and AD_{02} which are defined in a similar manner as (6.16) with D_{01} and D_{02} given in (5.17). For the paint experiment, the variable selection strategy in Section 5.6 can be applied to the following candidate variables:

(i) A, B, C, the linear and quadratic components of D, E, F, AD, AE, $AF, BD, BE, BF, CD, CE, CF$,

(ii) the $l \times l$, $l \times q$, $q \times l$ and $q \times q$ effects between D, E, and F.

The variables in (i) are orthogonal to each other while those in (ii) are not. The non-orthogonality can be seen by computing their pairwise correlations as was done in Section 5.6. Note that AB, BC, and AC are not included in the list of candidate variables because they are each fully aliased with a main effect. On the other hand the interaction effects between three-level factors are included in (ii) because they are not fully aliased with any main effect. A stepwise regression, performed on the 39 candidate variables, identified D_l, E_l, $(DF)_{qq}$, and CE_q as significant. Fitting these effects gives a model for predicted top inside paint thickness:

$$\hat{y} = 0.9181 + 0.3250\, D_l - 0.0613 E_l + 0.0172(DF)_{qq} + 0.0188\, CE_q. \quad (6.17)$$

Thus, in addition to D_l, E_l, and CE_q, stepwise regression identified $(DF)_{qq}$. Note that D_l by itself has an R^2 of 0.91. By adding the three other terms in (6.17), R^2 only climbs to 0.94.

Besides the $2^{3-1} \times 3^{3-1}$ design used in the paint experiment, there are five other 36-run designs that can be constructed from 2^{k-p} and 3^{k-p} designs: the full factorial $2^2 \times 3^2$ design and the fractional factorial $2^2 \times 3^{3-1}$, $2^2 \times 3^{4-2}$, $2^{3-1} \times 3^2$, and $2^{3-1} \times 3^{4-2}$ designs. For each fractional factorial design its defining relation and effect aliasing can be inferred by following the derivations in (6.14) and (6.15). Only the $2^{3-1} \times 3^{4-2}$ design is considered here. Denote the 2^{3-1} design by $C = AB$ and the 3^{4-2} design by $F = DE$ and $G = DE^2$. Then it can be shown that all the two-factor interactions between the two-level factors A, B, C and the three-level factors D, E, F, G are clear, which account for 24 degrees of freedom. The remaining 11 degrees of freedom are for the seven main effects, none of which are clear. The proof is left as an exercise. Because this estimability property is a result of the product structure of the design, it also holds for the other 36-run designs.

How useful are these 36-run designs for practical experimentation? Except possibly in parameter design experiments, higher priority is usually given to the estimation of main effects than to two-factor interactions. The reversed priority given by these 36-run designs may not be justifiable unless they are the only designs available. In Chapter 7, a 36-run orthogonal array $OA(36, 2^{11}3^{12})$ will be presented, which can accommodate more two-level and three-level factors and does not give undue priority to the estimation of a particular class of interactions. On the other hand, the aliasing relationships among its effects are much more complex. Further comparison of the two types of 36-run designs will be given in Chapter 7.

6.8 r^{k-p} FRACTIONAL FACTORIAL DESIGNS FOR ANY PRIME NUMBER r

The 2^{k-p} and 3^{k-p} series of designs considered in Chapters 4 and 5 can be extended to cover fractional factorial designs at r levels, for any prime number r. For $r \geq 5$, these designs are more useful for qualitative factors with a large number of levels or for computer experiments which typically study more complex relations. As will be shown later, for $k - p = 2$, they are equivalent to hyper-Graeco-Latin squares of order r and are useful for studying several treatment factors in the presence of two blocking factors arranged in an $r \times r$ square. In the next two subsections, we consider the two smallest and most useful designs in this extended class: 25-run fractional factorial designs at five levels and 49-run fractional factorial designs at seven levels. In Section 6.8.3, a general construction method is given.

6.8.1 25-Run Fractional Factorial Designs at Five Levels

It is easy to verify that the 25×6 matrix in Table 6C of Appendix 6C is an orthogonal array. The 25 runs are generated by the 5×5 combinations in the first two columns of the matrix. The remaining four columns are generated from the first two columns by using the following method. Let x_1 and x_2 denote the entries of column 1 and column 2, respectively. Then the third through sixth columns are obtained by

$$x_3 = x_1 + x_2 \quad (mod \ 5),$$

$$x_4 = x_1 + 2x_2 \ (mod \ 5),$$

$$x_5 = x_1 + 3x_2 \ (mod \ 5),$$

$$x_6 = x_1 + 4x_2 \ (mod \ 5).$$

(6.18)

This construction is analogous to the one for 3^{k-p} designs. The main difference is the use of the modulus 5 calculus instead of the modulus 3 calculus.

For the full factorial 5^2 design with factors A and B, the last four columns of the matrix in Table 6C can be used as the four components of the $A \times B$ interaction. As an obvious extension of the AB and AB^2 interaction components for three-level designs (see Section 5.3), we denote these four components by AB, AB^2, AB^3, and AB^4. The power of B indicates the coefficient of x_2 in (6.18). Each component represents the contrasts among the five groups of (x_1, x_2) satisfying the equation in (6.18) that defines the component and thus has four degrees of freedom. The sum of squares for $A \times B$ can be decomposed into four terms that represent the sums of squares for AB, AB^2, AB^3, and AB^4, respectively.

The matrix in Table 6C can be used to accommodate up to four additional factors at five levels by assigning factors C, D, E, F to columns 3, 4, 5, and 6, respectively. The corresponding defining relations are

$$C = AB, \qquad D = AB^2, \qquad E = AB^3, \qquad F = AB^4.$$

Thus, the first k columns in Table 6C give a $5^{k-(k-2)}$ design. For $k = 3, \ldots, 6$, we obtain 5^{3-1}, 5^{4-2}, 5^{5-3}, and 5^{6-4} designs. The defining contrast subgroup for each of the designs can be obtained from the equations in (6.18). For the 5^{3-1} design, the defining equation $x_3 = x_1 + x_2 (mod\ 5)$ is equivalent to $x_1 + x_2 + 4x_3 = 0(mod\ 5)$. In terms of A, B, and C, this equivalence can be restated as

$$C = AB \text{ is equivalent to } \mathbf{I} = ABC^4,$$

where $\mathbf{I}$ denotes the column of zeros and plays the role of the identity in group theory. From the defining relation $\mathbf{I} = ABC^4$, the following aliasing relations among main effects and two-factor interactions can be inferred:

$$A = BC^4, \quad B = AC^4, \quad C = AB, \quad AB^2 = BC = AC^3,$$
$$AB^3 = BC^3 = AC, \quad AB^4 = BC^2 = AC^2. \tag{6.19}$$

The derivations are left as an exercise.

For the 5^{4-2} design, the two defining relations are $\mathbf{I} = ABC^4 = AB^2D^4$. By multiplying ABC^4 and AB^2D^4 and using the convention that the first letter must have unitary power (which was used in Chapter 5 for 3^{k-p} designs), we have

$$\mathbf{I} = ABC^4 \times AB^2D^4 = A^2B^3C^4D^4 = AB^4C^2D^2,$$

whose last identity follows from the equivalence between $2x_1 + 3x_2 + 4x_3 + 4x_4 = 0(mod\ 5)$ and $3(2x_1 + 3x_2 + 4x_3 + 4x_4) = x_1 + 4x_2 + 2x_3 + 2x_4 = 0(mod\ 5)$. Similarly, we have

$$\mathbf{I} = ABC^4 (AB^2D^4)^2 = A^3C^4D^3 = AC^3D,$$

$$\mathbf{I} = ABC^4 (AB^2D^4)^3 = A^4B^2C^4D^2 = AB^3CD^3,$$

$$\mathbf{I} = ABC^4 (AB^2D^4)^4 = B^4C^4D = BCD^4.$$

Therefore, the 5^{4-2} design has the defining contrast subgroup:

$$\mathbf{I} = ABC^4 = AB^2D^4 = AB^4C^2D^2 = AC^3D = AB^3CD^3 = BCD^4.$$

The defining contrast subgroups for the 5^{5-3} and 5^{6-4} designs can be obtained in a similar manner and their derivation is left as an exercise.

The 5^{3-1} design is equivalent to a 5×5 Latin square if the first two columns of Table 6C represent the rows and columns of the square and its third column represents the Latin letters. By taking its fourth column for the Greek letters, the 5^{4-2} design is equivalent to a 5×5 Graeco-Latin square. By taking the remaining two columns of Table 6C for the lowercase letters and the Arabic numerals, the 5^{6-4} design is a hyper-Graeco-Latin square of order 5, which is displayed in Table 6.9.

The data analysis strategies for 5^{k-p} designs are similar to those given in Chapter 5 for three-level designs. Consider the 5^{3-1} design first. According

Table 6.9 5^{6-4} Design Represented as a Hyper-Graeco-Latin Square

	0	1	2	3	4
0	$A\alpha a1$	$B\beta b2$	$C\gamma c3$	$D\delta d4$	$E\epsilon e5$
1	$B\gamma d5$	$C\delta e1$	$D\epsilon a2$	$E\alpha b3$	$A\beta c4$
2	$C\epsilon b4$	$D\alpha c5$	$E\beta d1$	$A\gamma e2$	$B\delta a3$
3	$D\beta e3$	$E\gamma a4$	$A\delta b5$	$B\epsilon c1$	$C\alpha d2$
4	$E\delta c2$	$A\epsilon d3$	$B\alpha e4$	$C\beta a5$	$D\gamma b1$

to (6.19), its 24 degrees of freedom are divided into six groups, each with four degrees of freedom. Each of the last three groups consists of three aliased two-factor interaction components. To overcome the difficulties associated with aliasing and the physical interpretation of interaction components like AB^2, BC^2, or AC^3, an effective approach is to extend the linear-quadratic system of contrasts given in Section 5.6 to five-level designs. For a quantitative factor, say A, its linear, quadratic, cubic, and quartic effects can be represented by the following contrast vectors:

$$A_1 = \frac{1}{\sqrt{10}}(-2, -1, 0, 1, 2),$$

$$A_2 = \frac{1}{\sqrt{14}}(2, -1, -2, -1, 2),$$

$$A_3 = \frac{1}{\sqrt{10}}(-1, 2, 0, -2, 1),$$

$$A_4 = \frac{1}{\sqrt{70}}(1, -4, 6, -4, 1).$$

Similarly, the 16 degrees of freedom in the $A \times B$ interaction can be represented by the linear-by-linear, linear-by-quadratic, ..., quartic-by-quartic contrast vectors, which are defined to be

$$(AB)_{mn}(i, j) = A_m(i) B_n(j),$$

where $m, n = 1, 2, 3, 4$ and $i, j = 0, 1, 2, 3, 4$. The detailed development is very similar to the one given in Section 5.6 and will not be given here. For a qualitative factor, say D, its four degrees of freedom can be represented by four contrasts such as $D_{01}, D_{02}, D_{03}, D_{04}$, where

$$D_{0n}(i) = \begin{cases} -1, & \text{for } i = 0, \\ 1, & \text{for } i = n, \\ 0, & \text{otherwise.} \end{cases}$$

Again, these are extensions of the D_{01} and D_{02} contrasts given in Section 5.6. Once the contrasts for main effects and interactions are defined, a data analysis strategy following the variable selection procedure given in Section 5.6 should be used.

6.8.2 49-Run Fractional Factorial Designs at Seven Levels

The 49 runs in the 49×8 matrix of Table 6D of Appendix 6D are generated by the 7×7 combinations of its first two columns. Let x_1 and x_2 denote the entries of column **1** and column **2**, respectively. Then the remaining six columns are generated by

$$x_{j+2} = x_1 + jx_2 (mod\ 7), \quad j = 1, \ldots, 6.$$

By assigning these columns to additional factors, $7^{k-(k-2)}$ fractional factorial designs with $k = 3, \ldots, 8$ are obtained. The defining contrast subgroups and aliasing relations of effects for these designs can be derived by using modulus 7 calculus in exactly the same manner as for the five-level designs discussed in Section 6.8.1.

For a quantitative factor at seven levels, its main effect can be decomposed into six orthogonal contrasts corresponding to orthogonal polynomials of order 1–6. The orthogonal polynomials are listed in Appendix G. In modeling real data, it rarely makes sense to use a fifth- or a sixth-degree polynomial because of the concern with overfitting and the difficulty in interpreting their significance. It is usually prudent to use lower order polynomials, say less than fifth degree, in defining a system of contrasts for main effects and interactions. Once the system is given, the data analysis strategy can again follow the general steps given in Section 5.6.

6.8.3 General Construction

Generally an r^{k-p} design, r being a prime number, can be easily constructed. Denote $k - p$ by n. Let $x_1, \ldots, x_n$ be n independent and orthogonal columns, each taking the values $0, \ldots, r-1$. These n columns generate the r^n design. From them we can generate $(r^n - 1)/(r - 1)$ columns defined as follows:

$$\sum_{i=1}^{n} c_i x_i (mod\ r) \tag{6.20}$$

where c_i is any integer between 0 and $r - 1$ and the first nonzero c_i must be 1. Because the contrasts defined by $\sum_1^n c_i x_i$ are the same as those defined by any of its multiples $\lambda(\sum_{i=1}^n c_i x_i)$, the restriction that the first nonzero c_i be 1 ensures uniqueness. Because there are $r^n - 1$ nonzero combinations of the c_i's, the uniqueness restriction leads to $(r^n - 1)/(r - 1)$ distinct combinations in (6.20). By assigning factors to the columns defined in (6.20), we can obtain $r^{k-(k-n)}$ designs with $k = n, \ldots, (r^n - 1)/(r - 1)$.

After 7, the next prime number is 11. Except possibly in computer experiments or large field trials, 121 run designs at 11 levels are seldom used. If r is a prime power, the construction method given above can be extended by employing Galois field theory. A general treatment can be found in Kempthorne (1952). An alternative to the use of Galois field theory is the method of replacement discussed in Section 6.2. Instead of using the Galois field $GF(2^2)$ for constructing 4^{k-p} designs, it was shown there that such designs can be obtained by applying the method of replacement to 2^{k-p} designs. Because of the transparent relationship between the two classes of designs, many nice properties of the 2^{k-p} designs can be exploited in deriving the aliasing relations of the 4^{k-p} designs and in choosing minimum aberration 4^{k-p} designs. More generally, if r is a prime power, i.e., $r = s^m$ with s being a prime, r^{k-p} designs can be constructed from s^{k-p} designs by an extension of the method of replacement. Details can be found in Wu, Zhang, and Wang (1992).

6.9 RELATED FACTORS: METHOD OF SLIDING LEVELS AND NESTED EFFECTS ANALYSIS

Two factors are said to be *related* when the desirable range of one factor depends on the level of the other factor. For example, in a chemical reaction, the desirable reaction time range shifts downward as temperature increases because the combination of high temperature and long reaction time can be disastrous. To avoid these extreme combinations, one can slide the levels of factor A according to the levels of factor B. Figure 6.5 depicts a nine-run experiment using "sliding" levels. Other examples of experiments with **sliding levels** are aperture and exposure time in an integrated circuit (IC) fabrication

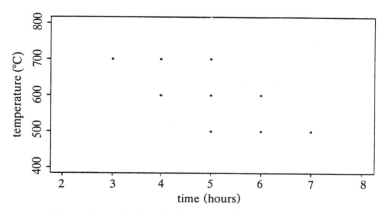

Figure 6.5. A Design for Factors with Sliding Levels.

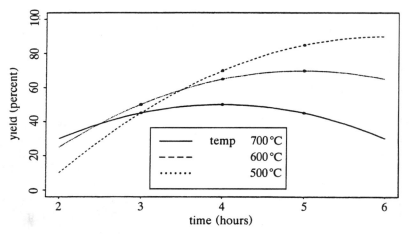

Figure 6.6. Interaction from a Design without Sliding Levels.

experiment, where exposure time is decreased as aperture is increased; pulse rate and weld time in a welding process experiment, where weld time is decreased as pulse rate is increased; and gas and air quantity in a light bulb sealing process experiment, where the air quantity is increased as the gas quantity is decreased.

One rationale for using sliding levels is to avoid a bad region in the experimental space. For the chemical reaction example, the range is chosen so that the best reaction time is thought to be contained within it, thereby avoiding temperature–reaction time combinations which are known to be bad, such as a low temperature–short reaction time or high temperature–long reaction time.

A second rationale for using sliding levels is that it can reduce or even eliminate interactions. Figure 6.6 displays the interaction between temperature and reaction time which would be evident from a full factorial design (temperature at 500, 600, and 700°C and reaction time at 3, 4, and 5 hours); there is an interaction because the curves are not parallel. Figure 6.7 displays how the interaction is removed by using a design with sliding levels, where the range of the reaction time is 5–7 hours for temperature at 500°C, 4–6 hours for temperature at 600°C, and 3–5 hours for temperature at 700°C. The interaction shown in Figure 6.7(a) is removed by coding the (low, middle, high) levels of both factors as $(-1, 0, 1)$ as in Figure 6.7(b).

Consider the choice of levels for two factors, A and B, in which there are three sliding levels for factor B at each level of factor A. This amounts to deciding on the location (i.e., the middle level) and scale (i.e., how far apart the high and low levels are) of the levels. Thus, the use of sliding levels is a centering and scaling transformation, where the amount of centering and scaling is dependent on the other factor. It can be seen that an interaction in

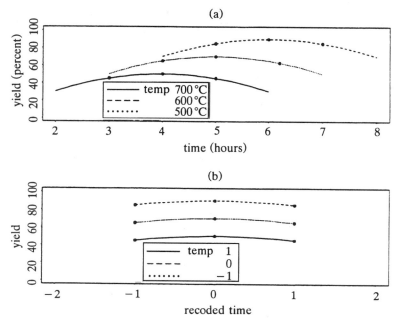

Figure 6.7. Design I for Factors with Sliding levels: (a) Response and Design in Original Factors; (b) Response and Design in Centered and Scaled Factors.

the original factors can be eliminated only if the relation between the mean response, $E(y)$, and factors A and B satisfy

$$E(y) = f_1[(x_A - c_A)/s_A] + f_2\{[x_B - c_B(x_A)]/s_B(x_A)\}, \qquad (6.21)$$

where the c's and s's denote the centering and scaling constants (with those for factor B depending on factor A) and f_1 and f_2 are two arbitrary functions. Figure 6.7, as previously discussed, illustrates that a proper choice of sliding levels can eliminate an interaction provided the mean response satisfies the form in (6.21). That is, if the nine-run design (Design I) in Figure 6.7(a) is used and the three levels of each factor are coded as $(-1, 0, +1)$, then the relation between yield and the coded factors is as displayed in Figure 6.7(b).

For the same example, improperly locating the sliding levels will not eliminate the interaction, however. In Figure 6.8(a), a different nine-run design (Design II) is used. After recoding the three levels of time as $(-1, 0, +1)$, it is seen from Figure 6.8(b) that the interaction is not eliminated. By using a similar argument, it can be shown that the interaction will not be eliminated for an improper choice of scale, e.g., where the reaction time scale depends on the temperature level.

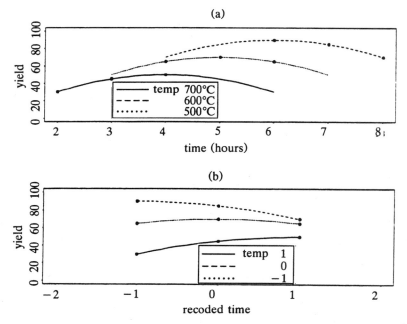

Figure 6.8. Design II for Factors with Sliding Levels: (a) Response and Design in Original Factors; (b) Response and Design in Centered and Scaled Factors.

Nested Effects Modeling

The previous discussion on the difficulty of completely eliminating interactions suggests that interactions should be considered in modeling and analysis. Sliding level designs can be viewed as nested designs because the slid factor is nested within its related factor. This suggests the use of *nested effects modeling*.

In terms of the sliding level designs, the effect of the slid factor should be studied separately at each level of its related factor. For factor B's levels depending on A's, two cases need to be considered: A is either qualitative or quantitative. When factor A is qualitative, we propose analyzing the effect of recoded factor B at each level of factor A, e.g., the effect of the two-level factor B at the ith level of A, denoted by $B|A_i$. If B is quantitative with more than two levels, then the linear, quadratic, etc., effects of B at the ith level of A, denoted by $B_l|A_i$, $B_q|A_i$, etc., are analyzed. In this way, the potential interaction between the related factors is accounted for. Besides the nested B effects such as $B_l|A_i$ and $B_q|A_i$, the factor A effect needs to be fitted. For example, if A is a qualitative factor and has three levels, we can consider the two contrasts $A_{0,1}$ and $A_{0,2}$, where $A_{i,j}$ denotes the contrast between levels i and j of factor A. See Table 6.10 for the corresponding covariates, where $A_{0,2}$ is the same as A_l. Note that the meaning of $A_{i,j}$ is different from the usual one in factorial design, because the levels of

Table 6.10 Covariates for Factor B Nested within Factor A and Unrelated Factor C in a 3^{3-1} Design with $C = AB$

Factor A B C	$A_{0,1}$	$A_{0,2}$	A_l	A_q	$B_l|A_0$	$B_q|A_0$	$B_l|A_1$	$B_q|A_1$	$B_l|A_2$	$B_q|A_2$	C_l	C_q
0 0 0	-1	-1	-1	1	-1	1	0	0	0	0	-1	1
0 1 1	-1	-1	-1	1	0	-2	0	0	0	0	0	-2
0 2 2	-1	-1	-1	1	1	1	0	0	0	0	1	1
1 0 1	1	0	0	-2	0	0	-1	1	0	0	0	-2
1 1 2	1	0	0	-2	0	0	0	-2	0	0	1	1
1 2 0	1	0	0	-2	0	0	1	1	0	0	-1	1
2 0 2	0	1	1	1	0	0	0	0	-1	1	1	1
2 1 0	0	1	1	1	0	0	0	0	0	-2	-1	1
2 2 1	0	1	1	1	0	0	0	0	1	1	0	-2

B at level i of A are different from those at level j of A. That is, the factor A contrasts represent differences between groups of B levels. When A is quantitative, we can replace $A_{0,1}$ and $A_{0,2}$ by the "linear" and "quadratic" effects of A, A_l and A_q, as given in Table 6.10. Again, A_l and A_q should be interpreted differently from the usual ones; they are the linear and quadratic contrasts of the three groups.

Note that the (coded) effects of the related factors (e.g., factors A and B in Table 6.10) need not be orthogonal to those of other unrelated factors (e.g., factor C in Table 6.10) in the design. For example, A_l and A_q in Table 6.10 are orthogonal to C_l and C_q, but $B_l|A_i$ and $B_q|A_i$ are not. This suggests using multiple regression techniques to fit the model and assess the significance of these effects.

The nested effects modeling strategy works for general forms of centering and scaling constants $c_B(x_A)$ and $s_B(x_A)$ in (6.21). Any difference in $c_B(x_A)$ and $s_B(x_A)$ as a function of the x_A levels can be modeled by an appropriate choice of the $B|A$ terms. The nested effects modeling strategy will be illustrated with the following light bulb experiment.

6.9.1 Analysis of Light Bulb Experiment

An experiment on a light bulb sealing process was performed to improve a cosmetic problem that was frequently occurring [Taguchi (1987), Section 17.6].

A 16-run 2^{10-6} design with 10 replicates was used to study 10 factors (A–J) of which four pairs of factors, (C, D), (E, F), (G, H), and (I, J), were related. Factor A is stem diameter whose levels are smaller and larger. Factor B is sealing machine speed whose levels are 4.8 and 5.4 rpm. Table 6.11 shows the design matrix that was used; the factors A–J were assigned to columns **6, 7, 10, 11, 13, 14, 1, 2, 4**, and **8**, respectively; the two levels for each

Table 6.11 Design Matrix and Covariates, Light Bulb Experiment

Run	G + CD	H	GH + EF	I	A	B	J	C	D	IJ	E	F	Covariate $H\mid G_-$	$H\mid G_+$	$J\mid I_-$	$J\mid I_+$
1	−	−	+	−	+	+	−	+	+	+	+	−	−	0	−	0
2	−	−	+	−	+	+	+	−	+	−	+	+	−	0	+	0
3	−	−	+	+	−	−	−	+	−	−	+	+	−	0	0	−
4	−	−	+	+	−	−	+	−	+	+	−	−	−	0	0	+
5	−	+	−	−	+	−	−	−	+	+	−	+	+	0	−	0
6	−	+	−	−	+	−	+	+	−	−	+	−	+	0	+	0
7	−	+	−	+	−	+	−	−	+	−	+	−	+	0	0	−
8	−	+	−	+	−	+	+	+	−	+	−	+	+	0	0	+
9	+	−	−	−	−	+	−	+	+	+	+	−	0	−	−	0
10	+	−	−	−	−	+	+	−	−	−	−	+	0	−	+	0
11	+	−	−	+	+	−	−	+	+	−	−	+	0	−	0	−
12	+	−	−	+	+	−	+	−	−	+	+	−	0	−	0	+
13	+	+	+	−	−	−	−	−	−	+	+	+	0	+	−	0
14	+	+	+	−	−	−	+	+	+	−	−	−	0	+	+	0
15	+	+	+	+	+	+	−	−	−	−	−	−	0	+	0	−
16	+	+	+	+	+	+	+	+	+	+	+	+	0	+	0	+

Table 6.12 Gas and Air Levels, Light Bulb Experiment

		Air	
	Gas	1	2
1	S_g	S_a	$S_a - 1$
2	$S_g + 1$	$S_a + 1$	S_a

(S_g and S_a denote standard gas and air settings)

factor are denoted by $(-, +)$. The pairs of factors consisted of the gas quantity and air quantity used at the four sealing stages in the process. Sliding levels were used with air quantity being increased as gas quantity is increased from S_g to $S_g + 1$ (see Table 6.12). In addition to the main effects for the 10 factors, the interactions GH and IJ, should also be considered since the third and the fourth stages are the most critical in the sealing process. Table 6.11 shows that GH and IJ which are associated with columns 3 and 12, can be estimated by this experiment. For this design, CD and EF are aliased with G and GH, respectively, as indicated by the heading in the first and third columns of Table 6.11, and therefore are assumed to be negligible. The experimental response was light bulb appearance, which was classified into one of three ordered categories: defective, slight cosmetic problem (but passed inspection), and perfect. We scored the categories as 1, 2, and 3 (see Table 6.13 for the appearance data) and analyzed the scores,

Table 6.13 Appearance Data, Light Bulb Experiment

Run	Appearance									
1	1	1	1	1	1	2	2	2	2	3
2	1	1	1	1	1	1	2	2	2	2
3	1	1	1	1	2	2	2	2	3	3
4	1	1	1	2	2	2	2	2	3	3
5	1	1	2	2	2	3	3	3	3	3
6	1	1	1	1	1	1	2	2	2	3
7	1	1	1	3	3	3	3	3	3	3
8	1	1	2	2	3	3	3	3	3	3
9	1	1	1	1	2	2	2	2	3	3
10	1	1	1	1	1	2	2	2	2	2
11	1	1	1	1	2	2	2	2	3	3
12	1	1	1	2	2	2	3	3	3	3
13	1	1	2	2	2	2	3	3	3	3
14	1	1	1	1	1	1	1	1	2	2
15	1	1	2	2	2	3	3	3	3	3
16	1	1	2	3	3	3	3	3	3	3

where higher scores are desired. The 10 observations per run are treated as replicates. (A formal analysis methodology for ordered categorical data will be considered in Chapter 13.) The model consists of an intercept, the main effects for factors $A-G$ and I and the nested effects, $H|G_-$, $H|G_+$, $J|I_-$, and $J|I_+$, where the subscripts $(-, +)$ are the $(-, +)$ factor levels. Using Table 6.11, the corresponding covariates for these effects can be obtained by replacing $(-, +)$ with $(-1, +1)$. These contrasts are orthogonal so that the estimated effects are uncorrelated. The least squares estimates, standard errors, t statistics, and p values for these effects presented in Table 6.14 indicate significant $H|G_-$, $J|I_-$, and I effects (at the 0.05 level). It is interesting to see that H is significant only at the $-$ level of G and J is significant only at the $-$ level of I. The predicted score for appearance has the form

$$\hat{y} = 1.92 + 0.23I + 0.24H|G_- - 0.29J|I_-. \qquad (6.22)$$

Using the signs and magnitudes of the estimated effects in (6.22), $G_-H_+I_+$ is recommended. Because the estimated effect for $H|G_-$ is positive, G_-H_+ is chosen. Similarly, I_+ is chosen because the estimated effect for I is positive. The setting of J is not determined because J appears in $J|I_-$ and I_- is not the chosen setting.

Finally, the relationship between two related factors can be classified either as *symmetrical* or *asymmetrical*. The window forming experiment reported in Phadke et al. (1983) provides good examples for both types. In the experiment, exposure time and aperture have a symmetric relationship

Table 6.14   Least Squares Estimates, Standard Errors, t Statistics, and p Values, Light Bulb Experiment

Effect	Least Squares Estimate	Standard Error	t	p Value
intercept	1.92	0.06	32.0	0.000
A	0.07	0.06	1.1	0.250
B	0.03	0.06	-0.5	0.600
C	-0.09	0.06	-1.6	0.120
D	-0.01	0.06	-0.1	0.920
E	0.04	0.06	0.7	0.470
F	0.07	0.06	1.1	0.250
G	0.01	0.06	0.1	0.920
$H \mid G_-$	0.24	0.08	2.8	0.006
$H \mid G_+$	0.13	0.08	1.5	0.140
I	0.23	0.06	3.9	0.000
$J \mid I_-$	-0.29	0.08	-3.4	0.001
$J \mid I_+$	0.08	0.08	0.9	0.380

because their product is related to total light energy. Consequently, one can slide the levels of aperture as the exposure time varies, which was done in the experiment; alternatively, one can slide the levels of exposure time as the aperture varies. Although the two approaches result in different factor level combinations, they would both avoid the undesirable combinations of (high, high) and (low, low) aperture and exposure time. On the other hand, photoresist viscosity and spin speed have an asymmetric relationship. As viscosity increases, the spin speed should be increased so that the photoresist can be spread uniformly on a wafer. It would be unnatural to first specify spin speed and then slide the viscosity levels. Asymmetrical relationships also occur if one factor is qualitative and the other quantitative. Since the qualitative factor cannot be slid because it does not have enough levels, the quantitative factor is often nested within the qualitative factor, assuming it is physically feasible. The relationship between factors, whether it is symmetric or asymmetric, has an impact on experimental design. For symmetric relationships, the choice of the factor to be nested is not crucial, whereas for asymmetric relationships, there is only one way to slide the factor levels. Further discussion and examples of the method of sliding levels can be found in Hamada and Wu (1995).

6.10   PRACTICAL SUMMARY

1. Mixed two-level and four-level designs, abbreviated as $2^m 4^n$ designs, can be constructed by the method of replacement from 2^{k-p} designs. The nature of construction makes it easy to discern the aliasing relations for the former designs from those for the latter designs.

2. Optimal 2^m4^1 and 2^m4^2 designs can be selected according to the minimum aberration criteria given in Section 6.3. These designs for selected values of m and run size can be found in Appendices 6A and 6B.

3. A main effect contrast of a two-level or four-level factor or a two-factor interaction contrast between factors of either type is clear if it is not aliased with any other main effect contrasts or two-factor interaction contrasts and strongly clear if it satisfies the further requirement that it is not aliased with any three-factor interaction contrasts.

4. Use the (A_1, A_2, A_3) system in (6.5) and (6.6) to parameterize the main effects and two-factor interactions for four-level factors. Because the contrasts in the system are uncorrelated and each of them has one degree of freedom, they can be analyzed in the same way as in 2^{k-p} experiments. An analysis strategy is given in (6.7) and illustrated in Section 6.5.

5. To study several two-level and three-level factors, use the following 36-run designs: $2^{3-1} \times 3^{4-2}$, $2^{3-1} \times 3^{3-1}$, $2^{3-1} \times 3^2$, $2^2 \times 3^{4-2}$, $2^2 \times 3^{3-1}$, $2^2 \times 3^2$. Among them, designs with more factors are more economical. Experiments based on these designs can be analyzed by using ANOVA and half-normal plots. To further analyze the three-level factors, use the linear-quadratic system and the variable selection strategy. An example is given in Section 6.7.

6. Construction of fractional factorial designs at r levels, where r is prime, is given in Section 6.8. These designs can be interpreted as hyper-Graeco-Latin squares of order r. Smaller designs, such as the 25-run designs at five levels and the 49-run designs at seven levels, are the more useful and can be found in Appendices 6C and 6D.

7. Two factors are related if the desirable range of one factor depends on the level of the other factor. To avoid bad regions in the experimental space, levels of related factors can be chosen by the method of sliding levels, i.e., to slide the levels of one factor according to the levels of the other factor. Then use nested effects modeling for data analysis. It also eliminates a spurious interaction between two related factors.

EXERCISES

1. Show that the newly constructed four-level column using the procedure in Section 6.2 is orthogonal to the remaining two-level columns of the $OA(N, 2^m)$.

2. The minimum aberration criterion given in Section 6.3 is called type 0 because it first compares the values of A_{r0} in its definition. For 2^n4^1 designs, if the roles of A_{r0} and A_{r1} are reversed in the definition, the

resulting criterion is called minimum aberration of type 1. A third alternative is to apply the usual minimum aberration criterion to the vector of sums $(A_{30} + A_{31}, A_{40} + A_{41}, \ldots)$.

(a) Describe situations in which either of these two alternative definitions may make more sense.

(b) For $2^7 4^1$ designs with 16 runs, show that the design with A, $3, 4, 14, 23, 34, 123, 134$ has minimum aberration of type 1 while the corresponding design in Table 6A.1 has minimum aberration of type 0 but not of type 1. Which design is better in terms of the sums $(A_{30} + A_{31}, A_{40} + A_{41}, \ldots)$? [Further discussion on these alternative criteria can be found in Wu and Zhang (1993).]

3. For $2^m 4^n$ designs with general n, extend the definition of type j defining relations to $0 \leq j \leq n$. Based on the extended definitions, define the minimum aberration criterion for $2^m 4^n$ designs.

4. Design a follow-up experiment to de-alias the three aliased interaction effects, AF, $D_2 H$, and $E_1 G$ in the router-bit experiment. Use a regression model consisting of these three aliases and the effects identified in the analysis in Section 6.4 and the D_s criterion in Section 4.4 to choose four additional runs for a follow-up experiment. Can you extend the foldover technique to $2^m 4^n$ experiments to solve the same problem? Which approach do you prefer?

5. Confirm the aliasing relations in (6.9) and (6.10).

6. Compare the design chosen for the router-bit experiment and the minimum aberration $2^7 4^2$ design given in Table 6B.2 in terms of clear main effects and two-factor interactions and aliases that involve D or E. If it is necessary that BG, BJ, CJ, and GJ be estimated clearly of main effects and other two-factor interactions between two-level factors, which design is preferred?

7. Determine the p values of the 31 effects in Figure 6.1 using Lenth's method. Identify the important effects and compare the results with those found in Section 6.5 based on visual judgment.

8. In the analysis of the router-bit experiment, the (A_1, A_2, A_3) system was used for factor D. Because D is a qualitative factor, an alternative is to use the pairwise contrasts among the four levels of D: $D_{0,1}, D_{0,2}, D_{0,3}$. Reanalyze the data by using this system for D and compare the results with those in Section 6.5.

9. Derive the defining contrast subgroup for the 32-run design with $m = 5$ in Table 6A.2. Identify each of the 31 degrees of freedom for the design,

and use this to determine the clear main effects and two-factor interactions.

10. Derive the aliasing relations in (6.19) for the 5^{3-1} design with $C = AB$.

11. By multiplying both sides of the aliasing relation $A = BC^4$ in (6.19) by factor C, we can obtain the relation $AC = BC^5 = B$, which contradicts the aliasing relation $AC^4 = B$ in (6.19). Point out the logical flaw in the derivation. [*Hint*: This common error can be avoided by using algebraic equations in $x_1, x_2, x_3 (mod\ 5)$ to verify derivations.]

12. Find the defining contrast subgroup for the 5^{5-3} design with $C = AB$, $D = AB^2$, and $E = AB^3$.

13. From the $OA(49, 7^8)$ in Table 6D, construct a 7×7 Graeco-Latin square and a 7×7 hyper-Graeco-Latin square. [*Hint:* A similar construction for 5×5 squares was given in Section 6.8.1.]

14. (a) Analyze the mean and range of the remaining six response characteristics in the paint experiment.

    (b) Because the paint thickness has a target value of 1.5 mils for the outside and side and 1.0 mil for the inside, the six remaining responses can be classified as nominal-the-best problems. By converting the range values to standard deviation values based on the normality assumption, a mean-squared error can be obtained for each of the six responses. By treating the mean-squared errors as the responses, analyze the data.

    (c) Find optimal factor settings by applying the trade-off strategies in Section 5.7 to the analysis results in (a) and (b).

15. Use Lenth's method to obtain the p values for the three largest effects identified in the half-normal plot for the paint experiment in Figures 6.3 and 6.4: D_l, E_l, and CE_q.

16. Consider the $2^{3-1} \times 3^{4-2}$ design with $C = AB$, $F = DE$, and $G = DE^2$.

    (a) Find its defining contrast subgroup.

    (b) From (a), identify all its clear effects (which account for 24 degrees of freedom).

    (c) For each of the remaining 11 degrees of freedom, find all the aliased main effects and two-factor interactions.

17. The belief that interactions are eliminated by sliding levels has two ramifications.

    (a) In the analysis of such factors, only main effects need to be considered. In analyzing only main effects, it is possible to declare no effect for the slid factor B when an AB interaction exists. Demonstrate this with the information displayed in Figure 6.9.

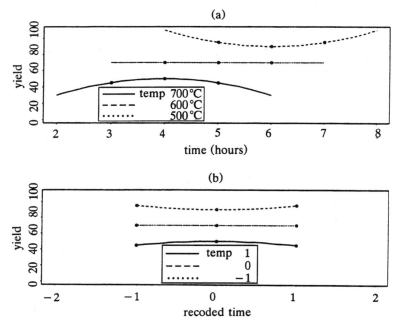

Figure 6.9. Missed Time Effect from Main-Effects-Only Analysis: (a) Response and Design in Original Factors; (b) Response and Design in Centered and Scaled Factors.

    (b) Believing that interactions do not need to be estimated, the design matrix columns associated with them might be used to study other experimental factors. Discuss the consequences of doing this.

18. (a) For the design of the light bulb experiment (Table 6.11), find the defining generators and all the aliasing relations among the main effects and two-factor interactions. Was the design well chosen?

    (b) For the same experiment, there is a 16-run design in which all main effects and the four designated interactions can be estimated. Find the design and compare it with the one in (a).

19. To improve the stability of reel motor in tape cartridge drives, a 32-run $2^3 4^2$ experiment was used to study five factors: file reel (A) at 1, 2, 3, and 4; machine reel (B) at 1, 2, 3, and 4; armature inertia (C) at normal and increased 30%; motor power card (D) at two settings; and temperature (E) at 30 and 40°C. The engineers anticipated important $A \times C$ and $B \times C$ interactions. The responses were stabilization times in milliseconds (ms) for the file and machine reel motor. The experimental goal was to increase stability, i.e., minimize stabilization time. The design matrix and stabilization times appear in Table 6.15.

Table 6.15 Design Matrix and Response Data, Reel Motor Experiment

Run	C	A	E	D	B	File Reel		Machine Reel	
1	0	0	0	0	0	38.34	38.70	26.19	26.01
2	0	0	0	1	1	39.87	40.23	21.24	21.33
3	0	0	1	0	2	38.97	39.51	22.05	22.05
4	0	0	1	1	3	38.25	38.70	21.87	21.96
5	0	1	0	0	1	34.29	34.65	21.15	20.97
6	0	1	0	1	0	40.14	40.41	20.52	20.43
7	0	1	1	0	3	39.51	40.32	20.52	20.70
8	0	1	1	1	2	32.67	32.58	20.70	20.52
9	0	2	0	0	2	38.07	37.80	22.95	22.95
10	0	2	0	1	3	55.08	55.80	22.50	22.50
11	0	2	1	0	0	43.29	44.01	22.77	22.68
12	0	2	1	1	1	46.98	46.80	21.06	20.97
13	0	3	0	0	3	34.65	34.11	22.23	22.32
14	0	3	0	1	2	27.90	29.70	23.67	23.31
15	0	3	1	0	1	37.35	37.53	21.42	21.24
16	0	3	1	1	0	30.33	30.51	22.32	22.50
17	1	0	0	0	0	41.04	40.95	21.96	22.23
18	1	0	0	1	1	40.05	41.40	21.24	21.60
19	1	0	1	0	2	41.31	42.21	24.39	24.21
20	1	0	1	1	3	37.17	38.70	21.42	24.03
21	1	1	0	0	1	38.07	37.80	21.15	21.06
22	1	1	0	1	0	38.34	39.87	21.69	21.60
23	1	1	1	0	3	50.40	49.23	21.24	21.06
24	1	1	1	1	2	39.33	40.50	20.52	20.52
25	1	2	0	0	2	43.38	43.20	24.48	24.30
26	1	2	0	1	3	60.30	58.50	21.78	21.60
27	1	2	1	0	0	44.10	42.30	21.24	21.15
28	1	2	1	1	1	55.98	58.68	22.68	22.41
29	1	3	0	0	3	35.91	36.90	22.68	22.41
30	1	3	0	1	2	33.03	32.40	22.32	22.32
31	1	3	1	0	1	35.19	36.72	21.42	21.33
32	1	3	1	1	0	36.00	36.09	22.41	22.32

(a) Identify the defining relation for the $2^3 4^2$ design. Find its resolution and clear effects. Can the six degrees of freedom in the $A \times C$ and $B \times C$ interactions be estimated clear of the main effects and other two-factor interactions?

(b) Find a 32-run minimum aberration $2^3 4^2$ design and compare it with the design in (a) in terms of the number of clear effects and the estimability of the $A \times C$ and $B \times C$ interactions.

(c) For each of the responses, file stabilization time and machine stabilization time, analyze the data, including a fitted model and optimal factor settings.

(d) Use an appropriate trade-off strategy to find factor settings that best achieve the dual objectives of minimizing the file and machine stabilization times.

(e) Based on the residuals from the two models fitted in (c), find the correlation between the two responses. Will a significant correlation affect the results obtained so far? How would you reanalyze the data by taking correlation into account?

20. In a plastic-forming process, a plastic ring is hot-formed to retain a metal insert at the top of a brake servo valve body. The experimental goal was to maximize the failure load of the hot-formed ring. A 16-run $2^4 4^1$ experiment was used to study five factors: temperature (A) at 180, 200, 220, and 240°C; hold time (B) at 8 and 15 seconds; batch number (C) at 1 and 2; maximum force (D) at 6 and 7 kN; force apply rate (E) at 4 and 1 kN/s. The design matrix and load data appear in Table 6.16.

(a) Identify the defining relation for the $2^4 4^1$ design. Find its resolution and clear effects.

(b) Show that it is not a minimum aberration design. Find a minimum aberration $2^4 4^1$ design. Compare the two designs in terms of resolution, clear effects, and possibly other properties. Which one would you recommend?

Table 6.16   Design Matrix and Response Data, Plastic-Forming Experiment

Run	Factor					Load	
	A	B	C	D	E		
1	0	0	0	0	0	1.9	1.8
2	0	0	1	1	1	2.4	2.3
3	1	0	0	0	0	2.2	2.3
4	1	0	1	1	1	2.6	2.3
5	2	0	0	0	1	2.5	2.4
6	2	0	1	1	0	3.2	2.8
7	3	0	0	0	1	2.9	3.0
8	3	0	1	1	0	3.6	3.5
9	0	1	0	1	0	2.7	2.8
10	0	1	1	0	1	2.7	2.6
11	1	1	0	1	0	3.0	2.8
12	1	1	1	0	1	3.1	2.8
13	2	1	0	1	1	3.0	3.4
14	2	1	1	0	0	3.2	3.3
15	3	1	0	1	1	3.8	4.4
16	3	1	1	0	0	4.2	4.2

(c) Analyze the data, including a fitted model and optimal factor settings.

21. An experiment was performed to improve a process producing weatherstrips which form a weather seal between the glass window and car door. To make weatherstrips, rubber, extruded onto a metal strip, forms lips which are glued together and then cured in an oven. Because of excessive scrap due to adhesion failure, the goal was to maximize adhesion and reduce adhesion variability. A 3^{4-1} design was used to study four factors: oven temperature (A) in degrees Fahrenheit; line speed (B) in feet per minute (fpm); adhesive type (C) (of types 1–3); and adhesive thickness (D) at 3, 5, 7 mils. Because slower speed and higher temperature may lead to overheating (and similarly, faster speed and lower temperature to underheating), the oven temperature should be increased as the line speed is made faster. The sliding levels are shown in Table 6.17. The

Table 6.17 Line Speed and Oven Temperature Levels, Weatherstrip Experiment

Line Speed	Oven Temperature		
	1	2	3
30	390	400	410
45	420	435	450
60	430	450	470

sliding levels were used to avoid unnecessarily making scrap. Samples from both dies (inboard and outboard) that produce the weatherstrips were taken. The design matrix and adhesion data appear in Table 6.18.

(a) Analyze $\bar{y}$ and $\ln s^2$ for the inboard and outboard data. Find factor settings to meet the experimental goal.

(b) Is there any correlation between the inboard and outboard responses? How would it affect the analysis and results in (a)?

22. (a) Suppose that the 16-run $2^4 4^1$ design given in Table 6A.1 is to be arranged in two blocks. Find an optimal blocking scheme and justify your choice in terms of the optimality measure you use.

(b) In general, describe an optimal blocking scheme to arrange any $2^m 4^1$ design in two blocks.

Table 6.18   Design Matrix and Response Data, Weatherstrip Experiment

	Factor				Adhesion					
Run	A	B	C	D	Inboard			Outboard		
1	0	0	0	0	52	57	43	67	77	64
2	0	0	1	1	32	24	25	31	64	31
3	0	0	2	2	118	118	118	94	118	118
4	0	1	0	1	25	24	29	38	23	26
5	0	1	1	2	58	44	49	81	88	72
6	0	1	2	0	118	118	118	107	104	76
7	0	2	0	2	28	29	46	41	33	31
8	0	2	1	0	13	14	13	14	15	30
9	0	2	2	1	32	27	26	36	32	30
10	1	0	0	0	49	43	42	41	46	44
11	1	0	1	1	27	25	27	46	31	27
12	1	0	2	2	118	92	118	118	90	118
13	1	1	0	1	25	41	24	24	23	25
14	1	1	1	2	6	11	31	31	66	43
15	1	1	2	0	92	118	118	118	118	118
16	1	2	0	2	23	29	31	28	28	31
17	1	2	1	0	20	21	23	41	45	41
18	1	2	2	1	118	118	118	118	118	118
19	2	0	0	0	24	23	22	33	30	35
20	2	0	1	1	14	17	14	13	13	13
21	2	0	2	2	75	89	77	31	32	32
22	2	1	0	1	22	22	22	22	23	22
23	2	1	1	2	16	17	14	32	29	48
24	2	1	2	0	118	118	90	118	118	118
25	2	2	0	2	30	29	26	21	21	18
26	2	2	1	0	9	11	11	8	10	10
27	2	2	2	1	88	72	72	41	45	44

23. Find an optimal blocking scheme to arrange the 32-run $2^3 4^2$ design in Table 6B.2 in four blocks. Justify your choice in terms of the optimality measure you use.

APPENDIX 6A: TABLES OF $2^m 4^1$ MINIMUM ABERRATION DESIGNS

In this appendix, the only four-level factor is denoted by $A = (1, 2, 12)$, which is obtained by the method of replacement from the three columns 1, 2 and 12; m is the number of two-level factors.

Table 6A.1   Minimum Aberration $2^m 4^1$ Designs with 16 Runs, $3 \le m \le 12$

m	Resolution	Design Generators	Clear Effects
3	IV	$A, 3, 4, 5 = 134$	$A, 3, 4, 5$
4	III	$A, 3, 4, 5 = 23, 6 = 134$	$4, 6$
5	III	$A, 3, 4, 5 = 23, 6 = 24,$ $7 = 134$	7
6	III	$A, 3, 4, 5 = 23, 6 = 24,$ $7 = 134, 8 = 1234$	*none*
7	III	$A, 3, 4, 5 = 13, 6 = 14,$ $7 = 23, 8 = 24, 9 = 124$	*none*
8	III	$A, 3, 4, 5 = 13, 6 = 14,$ $7 = 23, 8 = 24, 9 = 123,$ $t_0 = 124$	*none*
9	III	$A, 3, 4, 5 = 13, 6 = 23,$ $7 = 34, 8 = 123, 9 = 134,$ $t_0 = 234, t_1 = 1234$	*none*
10	III	$A, 3, 4, 5 = 13, 6 = 14,$ $7 = 23, 8 = 34, 9 = 123,$ $t_0 = 134, t_1 = 234, t_2 = 1234$	*none*
11	III	$A, 3, 4, 5 = 13, 6 = 14,$ $7 = 23, 8 = 24, 9 = 34,$ $t_0 = 123, t_1 = 134, t_2 = 234,$ $t_3 = 1234$	*none*
12	II	A, remaining 12 columns	none

Table 6A.2   Minimum Aberration $2^m 4^1$ Designs with 32 Runs, $4 \le m \le 9$

m	Resolution	Design Generators	Clear Effects
4	V	$A, 3, 4, 5, 6 = 1345$	all five main effects, all 10 2fi's
5	IV	$A, 3, 4, 5, 6 = 245, 7 = 1345$	all six main effects, $34, 35, 47, 57$
6	IV	$A, 3, 4, 5, 6 = 235, 7 = 245,$ $8 = 1345$	all seven main effects, 58
7	IV	$A, 3, 4, 5, 6 = 234, 7 = 235,$ $8 = 245, 9 = 1345$	all eight main effects
8	III	$A, 3, 4, 5, 6 = 13, 7 = 145,$ $8 = 234, 9 = 235, t_0 = 12345$	$4, 5, 7, 8, 9, t_0$
9	III	$A, 3, 4, 5, 6 = 13, 7 = 14,$ $8 = 234, 9 = 235, t_0 = 245,$ $t_1 = 1345$	$5, 8, 9, t_0, t_1$

Table 6A.3   Minimum Aberration 2^m4^1 Designs with 64 Runs, $5 \le m \le 9$

m	Resolution	Design Generators	Clear Effects
5	VI	$A, 3, 4, 5, 6, 7 = 123456$	all six main effect, all 15 2fi's
6	V	$A, 3, 4, 5, 6, 7 = 1345,$ $8 = 2456$	all seven main effects, all 21 2fi's
7	IV	$A, 3, 4, 5, 6, 7 = 1345,$ $8 = 2346, 9 = 12356$	all eight main effects, all 2fi's except 37, 38, 39, 78, 79, 89
8	IV	$A, 3, 4, 5, 6, 7 = 356,$ $8 = 1345, 9 = 2456,$ $t_0 = 12346$	all nine main effects, all 2fi's except 35, 36, 37, $56, 57, 67, 48, 49, 4t_0,$ $89, 8t_0, 9t_0$
9	IV	$A, 3, 4, 5, 6, 7 = 356,$ $8 = 1235, 9 = 1345,$ $t_0 = 2456, t_1 = 12346$	all ten main effects, $34, 39, 3t_0, 3t_1, 45,$ $46, 47, 59, 5t_0, 5t_1, 69,$ $6t_0, 6t_1, 79, 7t_0, 7t_1$

APPENDIX 6B: TABLES OF 2^m4^2 MINIMUM ABERRATION DESIGNS

In this appendix the two four-level factors are denoted by $A = (1, 2, 12)$ and $B = (3, 4, 34)$, which are obtained by the method of replacement from the columns 1, 2 and 12 and from the columns 3, 4 and 34; m is the number of two-level factors.

Table 6B.1 2^m4^2 Designs with 16 Runs, $1 \le m \le 8$

m	Resolution	Design Generators
1	III	$A, B, 5 = 14$
2	III	$A, B, 5 = 14, 6 = 23$
3	III	$A, B, 5 = 14, 6 = 23, 7 = 234$
4	III	$A, B, 5 = 14, 6 = 23, 7 = 124, 8 = 234$
5	III	$A, B, 5 = 14, 6 = 23, 7 = 24, 8 = 124,$ $9 = 234$
6	III	$A, B, 5 = 13, 6 = 14, 7 = 23, 8 = 24,$ $9 = 134, t_0 = 234$
7	III	$A, B, 5 = 13, 6 = 14, 7 = 23, 8 = 24,$ $9 = 124, t_0 = 134, t_1 = 234$
8	III	$A, B, 5 = 13, 6 = 14, 7 = 23, 8 = 24,$ $9 = 124, t_0 = 134, t_1 = 234, t_2 = 1234$

Note: **No main effect or two-factor interaction is clear** for the designs in this table.

Table 6B.2 2^m4^2 **Designs with 32 Runs,** $2 \leq m \leq 10$

m	Resolution	Design Generators	Clear Effects
2	IV	$A, B, 5, 6 = 235$	all four main effects
3	IV	$A, B, 5, 6 = 235, 7 = 1245$	all five main effect
4	IV	$A, B, 5, 6 = 235, 7 = 1245,$ $8 = 1345$	all six main effects
5	III	$A, B, 5, 6 = 14, 7 = 235,$ $8 = 1245, 9 = 1345$	$5, 7, 8, 9$
6	III	$A, B, 5, 6 = 14, 7 = 234,$ $8 = 235, 9 = 1245, t_0 = 1345$	$5, 8, 9, t_0$
7	III	$A, B, 5, 6 = 13, 7 = 14,$ $8 = 234, 9 = 235, t_0 = 1245,$ $t_1 = 1345$	$5, 9, t_0, t_1$
8	III	$A, B, 5, 6 = 13, 7 = 14,$ $8 = 234, 9 = 235, t_0 = 1234,$ $t_1 = 1245, t_2 = 1345$	$5, 9, t_1, t_2$
9	III	$A, B, 5, 6 = 13, 7 = 14,$ $8 = 15, 9 = 234, t_0 = 235,$ $t_1 = 1234, t_2 = 1245, t_3 = 1345$	t_0, t_2, t_3
10	III	$A, B, 5, 6 = 13, 7 = 14,$ $8 = 15, 9 = 234, t_0 = 235,$ $t_1 = 345, t_2 = 1234, t_3 = 1245,$ $t_4 = 1345$	t_0, t_3, t_4

Table 6B.3 2^m4^2 **Designs with 64 Runs,** $3 \leq m \leq 7$

m	Resolution	Design Generators	Clear Effects
3	V	$A, B, 5, 6, 7 = 123456$	all five main effects, all 10 2fi's
4	IV	$A, B, 5, 6, 7 = 1356,$ $8 = 2456$	all six main effects, $A5, A6, B5, B6, 56$ $57, 58, 67, 68$
5	IV	$A, B, 5, 6, 7 = 1356,$ $8 = 2456, 9 = 2346$	all seven main effects, $A5, B5, 56, 57, 67, 68, 79$
6	IV	$A, B, 5, 6, 7 = 1356,$ $8 = 2456, 9 = 2346, t_0 = 1235$	all eight main effects, $56, 57, 68, 79$
7	IV	$A, B, 5, 6, 7 = 1356,$ $8 = 2456, 9 = 2346, t_0 = 1235,$ $t_1 = 1246$	all nine main effects, $56, 57, 68, 79, 7t_0, 7t_1$

APPENDIX 6C: $OA(25, 5^6)$

Table 6C

Run	1	2	3	4	5	6
1	0	0	0	0	0	0
2	0	1	1	2	3	4
3	0	2	2	4	1	3
4	0	3	3	1	4	2
5	0	4	4	3	2	1
6	1	0	1	1	1	1
7	1	1	2	3	4	0
8	1	2	3	0	2	4
9	1	3	4	2	0	3
10	1	4	0	4	3	2
11	2	0	2	2	2	2
12	2	1	3	4	0	1
13	2	2	4	1	3	0
14	2	3	0	3	1	4
15	2	4	1	0	4	3
16	3	0	3	3	3	3
17	3	1	4	0	1	2
18	3	2	0	2	4	1
19	3	3	1	4	2	0
20	3	4	2	1	0	4
21	4	0	4	4	4	4
22	4	1	0	1	2	3
23	4	2	1	3	0	2
24	4	3	2	0	3	1
25	4	4	3	2	1	0

APPENDIX 6D: $OA(49, 7^8)$

Table 6D

Run	1	2	3	4	5	6	7	8
1	0	0	0	0	0	0	0	0
2	0	1	1	2	3	4	5	6
3	0	2	2	4	6	1	3	5
4	0	3	3	6	2	5	1	4
5	0	4	4	1	5	2	6	3
6	0	5	5	3	1	6	4	2
7	0	6	6	5	4	3	2	1
8	1	0	1	1	1	1	1	1
9	1	1	2	3	4	5	6	0
10	1	2	3	5	0	2	4	6

Table 6D (*Continued*)

Run	1	2	3	4	5	6	7	8
11	1	3	4	0	3	6	2	5
12	1	4	5	2	6	3	0	4
13	1	5	6	4	2	0	5	3
14	1	6	0	6	5	4	3	2
15	2	0	2	2	2	2	2	2
16	2	1	3	4	5	6	0	1
17	2	2	4	6	1	3	5	0
18	2	3	5	1	4	0	3	6
19	2	4	6	3	0	4	1	5
20	2	5	0	5	3	1	6	4
21	2	6	1	0	6	5	4	3
22	3	0	3	3	3	3	3	3
23	3	1	4	5	6	0	1	2
24	3	2	5	0	2	4	6	1
25	3	3	6	2	5	1	4	0
26	3	4	0	4	1	5	2	6
27	3	5	1	6	4	2	0	5
28	3	6	2	1	0	6	5	4
29	4	0	4	4	4	4	4	4
30	4	1	5	6	0	1	2	3
31	4	2	6	1	3	5	0	2
32	4	3	0	3	6	2	5	1
33	4	4	1	5	2	6	3	0
34	4	5	2	0	5	3	1	6
35	4	6	3	2	1	0	6	5
36	5	0	5	5	5	5	5	5
37	5	1	6	0	1	2	3	4
38	5	2	0	2	4	6	1	3
39	5	3	1	4	0	3	6	2
40	5	4	2	6	3	0	4	1
41	5	5	3	1	6	4	2	0
42	5	6	4	3	2	1	0	6
43	6	0	6	6	6	6	6	6
44	6	1	0	1	2	3	4	5
45	6	2	1	3	5	0	2	4
46	6	3	2	5	1	4	0	3
47	6	4	3	0	4	1	5	2
48	6	5	4	2	0	5	3	1
49	6	6	5	4	3	2	1	0

REFERENCES

Hale-Bennett, C. and Lin, D.K.J. (1997), "From SPC to DOE: a Case Study at Meco, Inc.," *Quality Engineering*, 9, 489–502.

Hamada, M. and Wu, C.F.J. (1995), "The Treatment of Related Experimental Factors by Sliding Levels," *Journal of Quality Technology* 27, 45–55.

Kempthorne, O. (1952), *Design and Analysis of Experiments*, New York: John Wiley Sons.

Mukerjee, R. and Wu, C.F.J. (1995), "On the Existence of Saturated and Nearly Saturated Asymmetrical Orthogonal Arrays," *Annals of Statistics*, 23, 2102–2115.

Phadke, M.S. (1986), "Design Optimization Case Studies," *AT&T Technical Journal*, 65, 51–68.

Phadke, M.S., Kackar, R.N., Speeney, D.V., and Grieco, M.J. (1983), "Off-Line Quality Control in Integrated Circuit Fabrication Using Experimental Design," *The Bell System Technical Journal*, 62, 1273–1309.

Rao, C.R. (1947), "Factorial Experiments Derivable from Combinatorial Arrangements of Arrays." *Journal of the Royal Statistical Society*, 9, 128–139.

Rao, C.R. (1973), "Some Combinatorial Problems of Arrays and Applications to Design of Experiments," in *A Survey of Combinatorial Theory* (Ed. J.N. Srivastava et al.), North-Holland: Amsterdam, pp. 349–359.

Taguchi, G. (1987), *System of Experimental Design*, Unipub/Kraus International Publications: White Plains, NY.

Wu, C.F.J. (1989), "Construction of $2^m 4^n$ Designs Via a Grouping Scheme," *Annals of Statistics*, 17, 1880–1885.

Wu, C.F.J. and Zhang, R. (1993), "Minimum Aberration Designs with Two-Level and Four-Level Factors," *Biometrika*, 80, 203–209.

Wu, C.F.J., Zhang, R., and Wang, R. (1992), "Construction of Asymmetrical Orthogonal Arrays of the Type $OA(s^k, s^m(s^{r_1})^{n_1} \cdots (s^{r_t})^{n_t})$," *Statistica Sinica*, 2, 203–219.

Nonregular Designs: Construction and Properties

The experimental designs employed in the previous chapters are constructed through defining relations among factors. For these designs, any two factorial effects either can be estimated independently of each other or are fully aliased. These designs are called **regular designs**. Designs that do not possess this property are called **nonregular designs**, which include many mixed-level orthogonal arrays. For reasons of run size economy or flexibility, nonregular designs may be used. A collection of useful nonregular designs and their statistical properties are given in this chapter. Several simple and effective methods for constructing nonregular designs are presented. Analysis strategies for nonregular designs will be the focus of the next chapter.

7.1 TWO EXPERIMENTS: WELD-REPAIRED CASTINGS AND BLOOD GLUCOSE TESTING

Hunter, Hodi, and Eager (1982) used a 12-run design to study the effects of seven factors on the fatigue life of weld-repaired castings. The factors and levels are listed in Table 7.1. The seven factors were assigned using the first seven columns of the 12-run design shown in Table 7.2. The response is the logged lifetime of the casting. The goal of the experiment was to identify the factors that affect the casting lifetime. These factors could then be adjusted to increase the lifetime.

Consider an experiment to study the effect of one two-level factor and seven three-level factors on blood glucose readings made by a clinical laboratory testing device. The experiment used an 18-run mixed-level orthogonal array. The factors and their levels are given in Table 7.3. Here we consider only one aspect of the study which was to identify factors that affect the mean reading. Note that factor F combines two variables, sensitivity and absorption, so that the levels of F are pairings of sensitivity and absorption

Table 7.1 Factors and Levels, Cast Fatigue Experiment

Factor	Level −	Level +
A. initial structure	as received	β treat
B. bead size	small	large
C. pressure treat	none	HIP
D. heat treat	anneal	solution treat/age
E. cooling rate	slow	rapid
F. polish	chemical	mechanical
G. final treat	none	peen

Table 7.2 Design Matrix and Lifetime Data, Cast Fatigue Experiment

Run	A	B	C	D	E	F	G	8	9	10	11	Logged Lifetime
1	+	+	−	+	+	+	−	−	−	+	−	6.058
2	+	−	+	+	+	−	−	−	+	−	+	4.733
3	−	+	+	+	−	−	−	+	−	+	+	4.625
4	+	+	+	−	−	−	+	−	+	+	−	5.899
5	+	+	−	−	−	+	−	+	+	−	+	7.000
6	+	−	−	−	+	−	+	+	−	+	+	5.752
7	−	−	+	−	+	+	−	+	+	+	+	5.682
8	−	−	+	−	+	+	−	+	+	+	−	6.607
9	−	+	−	+	+	−	+	+	+	−	−	5.818
10	+	−	+	+	−	+	+	+	−	−	−	5.917
11	−	+	+	−	+	+	+	−	−	−	+	5.863
12	−	−	−	−	−	−	−	−	−	−	−	4.809

Table 7.3 Factors and Levels, Blood Glucose Experiment

Factor	Level 0	Level 1	Level 2
A. wash	no	yes	
B. microvial volume (ml)	2.0	2.5	3.0
C. caras H_2O level (ml)	20	28	35
D. centrifuge RPM	2100	2300	2500
E. centrifuge time (min)	1.75	3	4.5
F. (sensitivity, absorption)	(0.10, 2.5)	(0.25, 2)	(0.50, 1.5)
G. temperature (°C)	25	30	37
H. dilution ratio	1:51	1:101	1:151

Table 7.4 Design Matrix and Response Data, Blood Glucose Experiment

Run	\multicolumn{8}{c}{Factor}							Mean Reading
	A	G	B	C	D	E	F	H	
1	0	0	0	0	0	0	0	0	97.94
2	0	0	1	1	1	1	1	1	83.40
3	0	0	2	2	2	2	2	2	95.88
4	0	1	0	0	1	1	2	2	88.86
5	0	1	1	1	2	2	0	0	106.58
6	0	1	2	2	0	0	1	1	89.57
7	0	2	0	1	0	2	1	2	91.98
8	0	2	1	2	1	0	2	0	98.41
9	0	2	2	0	2	1	0	1	87.56
10	1	0	0	2	2	1	1	0	88.11
11	1	0	1	0	0	2	2	1	83.81
12	1	0	2	1	1	0	0	2	98.27
13	1	1	0	1	2	0	2	1	115.52
14	1	1	1	2	0	1	0	2	94.89
15	1	1	2	0	1	2	1	0	94.70
16	1	2	0	2	1	2	0	1	121.62
17	1	2	1	0	2	0	1	2	93.86
18	1	2	2	1	0	1	2	0	96.10

levels. Because the 18-run design cannot accommodate eight three-level factors, it was necessary to combine the two variables into a single factor. The design matrix and response data (Hamada and Wu, 1992) are given in Table 7.4.

The analysis of these two experiments will be considered in the next chapter.

7.2 SOME ADVANTAGES OF NONREGULAR DESIGNS OVER THE 2^{k-p} AND 3^{k-p} SERIES OF DESIGNS

Most designs considered in the previous chapters have a run size which is either a power of 2 or of 3 because they are fractions of 2^k, 3^k, or $2^m 4^n$ full factorial designs. The two experiments in Section 7.1 have 12 and 18 runs, respectively, neither of which is a power of 2 or of 3. Why were they chosen for the experiments? The first and most obvious reason is **run size economy**. To appreciate this point, note that in the series of 2^k or 2^{k-p} designs, the run size must be 4, 8, 16, 32, ... There are large gaps in the run sizes. For example, if 8–11 factors are to be studied, then a minimum of 16 runs is required. On the other hand, one can use the design matrix in Table 7.2 to study 8–11 factors simply by assigning the first k columns to the k factors,

$8 \leq k \leq 11$. Thus, if the runs are expensive to conduct, using the 12-run design in Table 7.2 would save four runs that would have been required in a 2^{k-p} design. In some experimental situations, each run can be costly or time consuming to conduct, e.g., making a prototype or performing an intensive simulation or numerical computation.

Because the cast fatigue experiment has seven factors, it would be possible to use a 2_{III}^{7-4} design. If there were sufficient resources to conduct 12 runs (but not 16 runs), then a 12-run experiment would give more information than an eight-run experiment. Another reason for preferring the 12-run experiments is that the 2_{III}^{7-4} design is a saturated design which leaves no degrees of freedom for estimating interactions or error.

By applying the previous argument to the series of 3^k or 3^{k-p} designs, there are even larger gaps in their run sizes, 9, 27, 81, ... Recall that the 9-, 27-, and 81-run designs can accommodate a maximum of 4, 13, and 40 three-level factors, respectively. If an experiment has 5–7 three-level factors, then it would require a 27-run design from the series. On the other hand, one can use the 18-run design matrix in Table 7.4 for the experiment by assigning 5, 6, or 7 columns from the columns $B-H$ to the factors. This design requires 9 fewer runs as compared to the 27-run design in the 3^{k-p} series. The design matrix in Table 7.4 has the additional advantage that it can also accommodate 1 two-level factor. This demonstrates the second advantage of nonregular designs, namely, they are more **flexible** in accommodating various combinations of factors with different numbers of levels.

The 3^{k-p} designs in Chapter 5 can be treated as regular designs or nonregular designs, depending on the parameterization system. If the orthogonal components system is adopted, a 3^{k-p} design is a regular design because any pair of main effects or interaction components in the system are either orthogonal or fully aliased. If the linear-quadratic system is used, as shown in Section 5.6, some effects in the system have absolute correlation between 0 and 1 and so the corresponding 3^{k-p} design should be treated as a nonregular design. Experiments based on these designs can be analyzed using the variable selection strategy in Section 5.6 or the Bayesian variable selection strategy in Section 8.5.

A general discussion of the construction and properties of nonregular designs, which include the 12- and 18-run designs, will be given in this chapter.

7.3 A LEMMA ON ORTHOGONAL ARRAYS

Because nonregular designs are special cases of orthogonal arrays defined in Section 6.2, they share the desirable property that any two factorial effects represented by the columns of an orthogonal array can be estimated and interpreted independently of each other. The following lemma provides a simple and effective guide on the run size of an orthogonal array, whose definition was given in Section 6.2.

Lemma 7.1. For an orthogonal array $OA(N, s_1^{m_1} \cdots s_\gamma^{m_\gamma}, t)$, its run size N must be divisible by the least common multiple of $\prod_{i=1}^{\gamma} s_i^{k_i}$ for all possible combinations of k_i with $k_i \leq m_i$ and $\sum_{i=1}^{\gamma} k_i = t$, $i = 1, \ldots, \gamma$.

Proof. According to the definition of orthogonality, in any t columns of the array with k_i columns having s_i symbols, $k_i \leq m_i$, $\sum_{i=1}^{\gamma} k_i = t$, each of the $\prod_{i=1}^{\gamma} s_i^{k_i}$ combinations of symbols appears equally often. This implies that N is divisible by the least common multiple of $\prod_{i=1}^{\gamma} s_i^{k_i}$ over these choices of t columns. Q.E.D.

The lemma will be used repeatedly in the discussion of various orthogonal arrays. Because orthogonal arrays of strength two can accommodate more columns than arrays of higher strength and the latter arrays can often be obtained by taking a subset of columns in the former arrays, to save space, *we will only consider orthogonal arrays of strength two in this chapter and omit the reference to strength.*

For a comprehensive treatment on the theory and collection of orthogonal arrays, see Hedayat, Sloane, and Stufken (1999) and Dey and Mukerjee (1999).

7.4 PLACKETT-BURMAN DESIGNS AND HALL'S DESIGNS

Two-level orthogonal arrays are denoted by $OA(N, 2^m)$. From Lemma 7.1, N must be a multiple of 4. These arrays can be constructed from Hadamard matrices. A **Hadamard matrix** of order N, denoted by H_N, is an $N \times N$ orthogonal matrix with entries 1 or -1. We can assume without loss of generality that its first column consists of 1's. Then the remaining $N-1$ columns are orthogonal to the first column and must have half 1's and half -1's. By removing the first column from a Hadamard matrix, the remaining $N \times (N-1)$ matrix is an $OA(N, 2^{N-1})$. This shows that the construction of $OA(N, 2^{N-1})$ is equivalent to the construction of H_N. Using the terminology introduced in Chapter 3, the $OA(N, 2^{N-1})$ can be used as a design matrix while the H_N can be used as a model matrix whose first column corresponds to the intercept term in a regression model. Throughout the book we assume that the first column of H_N consists of 1's. For notational simplicity, we often use $+$ and $-$ in the tables to represent 1 and -1, respectively.

A large collection of $OA(N, 2^{N-1})$ was given in Plackett and Burman (1946) for $N = 4k$ but not a power of 2. In the statistical literature, they are often referred to as the **Plackett-Burman designs**. The smallest among them are the 12-, 20-, and 24-run designs given in Appendix 7A. These designs can be generated from the first row by *cyclic* arrangement. The second row is generated from the first row by moving the entries of the first row one position to the right and placing the last entry in the first position. The third row is generated from the second row by using the same method, and the

Table 7.5 Generating Row Vectors for Plackett-Bruman Designs of Run Size N

N	Vector
12	$+ + - + + + - - - + -$
20	$+ + - - + + + + - + - + - - - - + + -$
24	$+ + + + + - + - + + - - + + - - + - + - - - -$
36	$- + - + + + - - - + + + + + - + + + - - + - - - - + - + - + + - - + -$
44	$+ + - - + - + - - + + + - + + + + + - - - + - + + + - - - - - + - - -$
	$+ + - + - + + -$

process continues until $N - 1$ rows are generated. A row of -1's is then added as the last row, completing the design. The method of cyclic generation can also be applied columnwise. For example, the 11×11 submatrix of the design matrix in Table 7.2 is generated by cyclic arrangement of its first column. It is known that $OA(N, 2^{N-1})$ for any $N = p^m + 1$, where p is a prime number, can be constructed. For $m = 1$, the design can be constructed by a cyclic arrangement of its first row. In Table 7.5, the generating vectors of the Plackett-Burman designs for $N = 12, 20, 24, 36,$ and 44 are given. The full designs for $N = 36$ and $N = 44$ are not given in Appendix 7A because they can be easily constructed from the generating vectors given in Table 7.5. For $N = 28$, the cyclic arrangement cannot be applied and the design is given in Appendix 7A. A survey on the construction, properties, and applications of Hadamard matrices can be found in Hedayat and Wallis (1978).

An integer N is a *Hadamard number* if an H_N exists. For any Hadamard number N, H_{2N} can be constructed as follows,

$$H_{2N} = \begin{pmatrix} H_N & H_N \\ H_N & -H_N \end{pmatrix}. \qquad (7.1)$$

It is straightforward to verify that the matrix in (7.1) is indeed a Hadamard matrix. By removing its first column, we then obtain an $OA(2N, 2^{2N-1})$ for any Hadamard number N.

As will be illustrated in Section 8.1 for the 12-run design, many Plackett-Burman designs have complex aliasing relations. Traditionally, they are used for estimating the main effects only. Here we will use a modification of the projective rationale first introduced in Section 4.2 to explain why it is possible to estimate some interactions if the number of significant main effects is small. Take the 12-run design in Appendix 7A as an example. Suppose the analysis of main effects suggests that there are only three significant main effects. Can the interactions among the three factors be estimated? This can be answered by projecting the design onto the three factors. It is easy to verify that the projection onto any set of three columns of the design consists of the sum of a 2^3 design and a 2^{3-1} design with $\mathbf{I} = -ABC$. This suggests

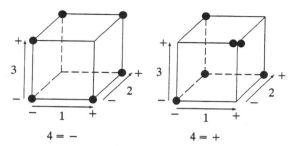

Figure 7.1. Projection of $OA(12, 2^{11})$ in Appendix 7A.1 Onto Its First Four Columns.

that the projected design will allow all the main effects and interactions among the three factors to be estimated.

Suppose four main effects are identified as significant. It can be verified that any projection onto four columns of the design is equivalent to the one given in Figure 7.1, where the design is projected onto the first four columns. (Two designs or arrays are said to be **equivalent** or **isomorphic** if one design can be obtained from the other by row permutations, column permutations, or relabeling of levels.) From Figure 7.1, it is clear that the projected design needs five additional runs to make it a 2^4 design. It only requires one additional run at $(-, +, -, -)$ to make it a 2_{IV}^{4-1} design with $\mathbf{I} = -ABCD$. Because there are 10 degrees of freedom for the four main effects and their six two-factor interactions, complex aliasing suggests that these 10 effects may all be estimated without additional runs. It turns out by simple computation that indeed they can all be estimated if higher-order interactions are negligible. These 10 effects are not orthogonal, however. To measure the estimation efficiency of a set of non-orthogonal effects, we can use the following D_s efficiency criterion. Let $\mathbf{x}_1, \mathbf{x}_2, \ldots, \mathbf{x}_m$ denote the standardized contrast vectors of m non-orthogonal effects, i.e., $\|\mathbf{x}_i\| = 1$. Then the D_s *efficiency* for the estimate associated with $\mathbf{x}_i$ is

$$\mathbf{x}_i^T \mathbf{x}_i - \mathbf{x}_i^T \mathbf{X}_{(i)} \left(\mathbf{X}_{(i)}^T \mathbf{X}_{(i)} \right)^{-1} \mathbf{X}_{(i)}^T \mathbf{x}_i = 1 - \mathbf{x}_i^T \mathbf{X}_{(i)} \left(\mathbf{X}_{(i)}^T \mathbf{X}_{(i)} \right)^{-1} \mathbf{X}_{(i)}^T \mathbf{x}_i, \quad (7.2)$$

where $\mathbf{X}_{(i)} = [\mathbf{x}_1, \ldots, \mathbf{x}_{i-1}, \mathbf{x}_{i+1}, \ldots, \mathbf{x}_m]$. Note that (7.2) is a special case of the D_s criterion given in (4.13). If x_i were orthogonal to the remaining $m - 1$ vectors, the second term in (7.2) would be zero and the D_s value in (7.2) would be 1. Therefore, 1 is the upper bound for the D_s efficiency. The closer the D_s efficiency is to 1, the more accurately the effect for x_i can be estimated. For each of the 10 effects considered above, the D_s efficiency is 0.62.

In contrast with the geometric projection property of regular 2^{k-p} designs, this estimability of interactions without relying on geometric projection is referred to as having a **hidden projection** property (Wang and Wu, 1995). This hidden projection property for four factors holds generally as shown by the following theorem.

Theorem 7.1. Let X be any $OA(N, 2^k)$ with $k \geq 4$ and N not a multiple of 8. Then the projection of X onto any four factors has the property that all the main effects and two-factor interactions among the four factors are estimable when the higher order interactions are negligible.

The result and its proof are due to Cheng (1995). Theorem 7.1 is also applicable to the Plackett-Burman designs with 20, 28, 36, and 44 runs.

Returning to the 12-run design, it can be shown by straightforward calculations that its projection onto five factors allows all the main effects and five or six two-factor interactions among the five factors to be estimated. Unless the number of interactions to be estimated is large, this property ensures their estimability without adding runs. Details on this result and the following results for the 20-run design can be found in Lin and Draper (1992), Wang and Wu (1995), and the exercises.

The hidden projection approach also gives encouraging results for the 20-run design. Its projection onto five factors allows all the main effects and q two-factor interactions to be estimated, where $q = 7, 9, 10$ depending on the type of the projection.) there are altogether 10 two-factor interactions among the five factors. For projection onto six factors, all the main effects and q two-factor interactions are estimable, where $q = 10, 11, 12, 13$, depending on the type of the projection. For larger designs, their hidden projection properties can be similarly studied.

The four tables in Appendix 7B are the 16-run nonregular two-level orthogonal arrays constructed by Hall (1961). It is known that the saturated 2_{III}^{15-11} design (see Section 4.2) is a regular $OA(16, 2^{15})$ and is called a type I design by Hall. The four designs in Appendix 7B are given in the ascending order of their "nonregularity." The type II design is the closest to being regular in that it has the smallest number of pairs of main effects and two-factor interactions that are partially aliased. (Any two factorial effects are said to be *partially aliased* if they are neither fully aliased nor orthogonal.) On the other hand, the type V design is the most nonregular because it has the largest number of partially aliased pairs. Details on the notion of closeness to "regularity" and the aliasing relations of these designs can be found in Sun and Wu (1993). Although these four designs have rarely been used in practical experimentation, they may be considered as alternatives to the regular 16-run 2^{k-p} designs, especially if the complex aliasing of the former designs is considered to be an advantage. The advantages of complex aliasing for data analysis will be discussed and demonstrated in Chapter 8.

7.5　A COLLECTION OF USEFUL MIXED-LEVEL ORTHOGONAL ARRAYS

Appendix 7C gives a collection of nonregular mixed-level orthogonal arrays that have 12–54 runs and can accommodate factors at two to six levels. The source of each array is given at the bottom of the table. Because of their run size economy and flexibility in accommodating different combinations of factor levels, they are extremely useful. In the following, we will describe some major properties of the arrays in the collection and discuss how their use may arise in practice. Some of these arrays can be constructed by using the methods to be described in the next two sections.

1. $OA(12, 3^1 2^4)$. Note that a 12-run regular design is a product of the 3^1 design and the 2_{III}^{3-1} design, which is denoted by $3 \times 2^{3-1}$ or $OA(12, 3^1 2^3)$ in the orthogonal array notation. From Lemma 7.1, the run size N in an $OA(N, 3^1 2^3)$ must be at least 12 because 12 is the least common multiple of 4 ($= 2 \times 2$) and 6 ($= 2 \times 3$). Therefore $OA(12, 3^1 2^4)$ is a nonregular design, which is useful if there are four factors at two levels and the three levels of the other factor cannot be reduced to two (otherwise an eight-run 2^{k-p} design should be used). It, however, leaves five ($= 11 - 6$) degrees of freedom unused. It is impossible to add a fifth two-level factor without affecting orthogonality. (The proof is left as an exercise.) If the requirement of orthogonality can be relaxed slightly, then more two-level columns can be added. In Table 7C.1, the two added columns are denoted by 6′ and 7′. The only non-orthogonal pairs of columns are (4, 6′) and (5, 7′). Because the number of non-orthogonal pairs of columns is small, the expanded array is called a **nearly orthogonal array** and is denoted by $OA'(12, 3^1 2^6)$, where ′ is used to indicate the fact that it is nearly orthogonal. For several definitions of nearly orthogonal arrays, see Wang and Wu (1992).

2. $OA(18, 2^1 3^7)$ and $OA(18, 6^1 3^6)$. Note that an 18-run regular design is a $2^1 \times 3^{4-2}$ [i.e., an $OA(18, 2^1 3^4)$] or a 6×3 design [i.e., an $OA(18, 6^1 3^1)$]. From Lemma 7.1, the run size N in an $OA(N, 2^1 3^4)$ or $OA(N, 6^1 3^1)$ must be at least 18. Therefore $OA(18, 2^1 3^7)$ and $OA(18, 6^1 3^6)$ are nonregular designs, which are useful when the number of three-level factors is 5–7 for the former and 2–6 for the latter. For the $OA(18, 2^1 3^7)$ in Table 7C.2, the interaction between column 1 and column 2, which equals *column 1 + column 2(mod 3)*, is orthogonal to columns 3–8. Because of this orthogonality property, we can replace columns 1 and 2 by a six-level column, which is denoted by column 1′ in the table. Column 1′ and columns 3–8 make up the $OA(18, 6^1 3^6)$, which is a saturated design because $17 = (6 - 1) + 6(3 - 1)$. As discussed in Section 7.2, these arrays are extremely economical for studying five to seven three-level factors. The hidden projective rationale can be used to explain why some interactions in these arrays can also be studied. Consider, for simplicity, the $OA(18, 3^7)$, which is the last seven columns of the $OA(18, 2^1 3^7)$. Its

projection onto any three factors allows the three main effects and the $l \times l$, $l \times q$, and $q \times l$ components of the three interactions to be estimated. Altogether these effects have 15 degrees of freedom. (The details are left as an exercise.) In contrast, if the nine-run 3^{3-1} design with $\mathbf{I} = ABC$ is used, only two interaction components can be estimated (see Section 5.6). For projection onto four factors, it can be shown that the four main effects and the $l \times l$ components of the six interactions can be estimated. Altogether these effects have 14 degrees of freedom. (The details are left as an exercise.) In contrast, the saturated nine-run 3^{4-2} design does not allow any interaction to be estimated. An alternative is to use the 27-run 3^{4-1} design with $\mathbf{I} = ABCD$, which would require nine more runs. Details on the hidden projection properties of the 18-run array can be found in Wang and Wu (1995).

3. $OA(20, 5^1 2^8)$. Note that a 20-run regular design is a $5 \times 2^{3-1}$ design [i.e., an $OA(20, 5^1 2^3)$]. Again Lemma 7.1 can be used to show that 20 is the minimal run size for an $OA(N, 5^1 2^3)$. Therefore $OA(20, 5^1 2^8)$ is a nonregular design, which is useful if one factor must have five levels and the number of two-level factors is at least 4. It, however, leaves seven $(= 19 - 12)$ degrees of freedom unused. Two more columns ($10'$ and $11'$ in Table 7C.3) can be added so that the expanded array is a nearly orthogonal array $OA'(20, 5^1 2^{10})$.

4. $OA(24, 3^1 2^{16})$ and $OA(24, 6^1 2^{14})$. In the discussion on the 12-run array, it was pointed out that no 12-run orthogonal array can accommodate one three-level factor and more than four two-level factors. Using this fact and Lemma 7.1, it is easy to see that 24 is the smallest run size when one three-level (or six-level) factor and five or more two-level factors are to be studied. Since a 24-run regular design is a $3 \times 2^{7-4}$ design [i.e., $OA(24, 3^1 2^7)$] or a $6 \times 2^{3-1}$ design (i.e., $OA(24, 6^1 2^3)$], $OA(24, 3^1 2^{16})$ and $OA(24, 6^1 2^{14})$ are nonregular designs. These designs are useful if the number of two-level factors is at least 8 for the former design and at least 4 for the latter design. For each array two more two-level columns can be added so that the expanded array is a nearly orthogonal array. See Tables 7C.4 and 7C.5 for details.

5. $OA(36, 2^{11} 3^{12})$. It was pointed out in Section 6.7 that a 36-run regular design can accommodate at most three two-level factors and four three-level factors, i.e., it is a $2^{3-1} \times 3^{4-2}$ design. To study more factors, $OA(36, 2^{11} 3^{12})$ will be the best choice. Two versions are given in Tables 7C.6 and 7C.7. They are arranged to minimize the number of level changes for two-level and three-level factors, respectively. Recall that when the level of a factor is hard to change, the number of its level changes should be minimized.

6. $OA(36, 3^7 6^3)$ and $OA(36, 2^8 6^3)$. These arrays in Tables 7C.8 and 7C.9 are useful for studying two to three six-level factors as well as several two-level or three-level factors.

7. $OA(48, 2^{14} 4^{12})$. This array in Table 7C.10 is useful only if the 32-run $2^m 4^n$ designs given in Chapter 6 cannot handle the numbers of two-level and

four-level factors, e.g., $n = 10, 11, 12$, or $n = 9$ and $m \geq 5$, or $n = 8$ and $m \geq 8$. This follows from the fact that the 32-run designs with the largest number of four-level factors are $OA(32, 2^4 4^9)$ and $OA(32, 2^7 4^8)$.

8. $OA(50, 2^1 5^{11})$. Since the 25-run 5^{k-p} designs given in Section 6.8 can accommodate up to 6 five-level factors, this array is useful when there are 7–11 five-level factors. In addition it can study 1 two-level factor. By slightly relaxing the orthogonality requirement, two additional two-level columns (columns 13′ and 14′ in Table 7C.11) can be added so that the expanded array is a nearly orthogonal array $OA'(50, 2^3 5^{11})$.

9. $OA(54, 2^1 3^{25})$ and $OA(54, 6^1 3^{24})$. Note that a 54-run regular design is a $2^1 \times 3^{13-10}$ design [i.e., an $OA(54, 2^1 3^{13})$]. Therefore, $OA(54, 2^1 3^{25})$ is a non-regular design that can accommodate many more three-level factors than the corresponding 54-run regular design. The $OA(54, 6^1 3^{24})$ is a saturated array. Both are useful for studying more than 13 three-level factors and 1 additional two-level or six-level factor.

There are many methods for constructing mixed-level orthogonal arrays with different combinations of factor levels and collections of these arrays are available in the literature. See, for example, Dey (1985), Wang and Wu (1991), and Hedayat, Sloane, and Stufken (1999). If the run size required for an orthogonal array (as dictated by the formula in Lemma 7.1) is too large, a nearly orthogonal array with a much smaller run size may be considered. Many nearly orthogonal arrays can be found in Wang and Wu (1992) and Nguyen (1996).

7.6 CONSTRUCTION OF MIXED-LEVEL ORTHOGONAL ARRAYS BASED ON DIFFERENCE MATRICES

The construction method given in this section employs two mathematical concepts: difference matrices and Kronecker sums. Let G be an additive group of p elements denoted by $\{0, 1, \ldots, p - 1\}$, where the sum of any two elements x and y is $x + y \pmod{p}$ for prime p, and is defined as in a Galois field for prime power p. A $\lambda p \times r$ matrix with elements from G denoted by $D_{\lambda p, r; p}$ is called a **difference matrix** if, among the differences modulus p of the corresponding elements of any of its two columns, each element of G occurs exactly λ times. For $p = 2$, $D_{n, n; 2}$ is a Hadamard matrix of order n. For two matrices $A = [a_{ij}]$ of order $n \times r$ and B of order $m \times s$, both with entries from G, define their *Kronecker sum* to be

$$A * B = [B^{a_{ij}}]_{1 \leq i \leq n, 1 \leq j \leq r}, \tag{7.3}$$

where

$$B^k = (B + k\mathbf{J}) \bmod p \tag{7.4}$$

is obtained from adding $k \, (mod \, p)$ to the elements of B and $\mathbf{J}$ is the $m \times s$ matrix of 1's. Let $A = [A_1, \ldots, A_n]$ and $B = [B_1, \ldots, B_n]$ be two partitioned matrices such that for each i, both A_i and B_i have entries from an additive group G_i with p_i elements. The *generalized Kronecker sum* of A and B is defined as

$$A \otimes B = [A_1 * B_1, \ldots, A_n * B_n].$$

If A is an orthogonal array with the partition

$$A = [OA(N, p_1^{s_1}), \ldots, OA(N, p_n^{s_n})]$$

and

$$B = [D_{M, k_1; \, p_1}, \ldots, D_{M, k_n; \, p_n}],$$

where $D_{M, k_i; \, p_i}$ is a difference matrix and N and M are both multiples of the p_i's, then their generalized Kronecker sum

$$A \otimes B = [OA(N, p_1^{s_1}) * D_{M, k_1; \, p_1}, \ldots, OA(N, p_n^{s_n}) * D_{M, k_n; p_n}] \quad (7.5)$$

is an orthogonal array $OA(NM, p_1^{k_1 s_1} \cdots p_n^{k_n s_n})$. A proof can be found in Wang and Wu (1991).

General Method for Constructing Asymmetrical Orthogonal Arrays (7.6)

The method consists of the following two or three steps:

   (i) Construct the orthogonal array $A \otimes B$ as in (7.5).
   (ii) Let $\mathbf{L} = 0_N * OA(M, q_1^{r_1} \cdots q_m^{r_m})$ be a matrix consisting of N copies of the orthogonal array $OA(M, q_1^{r_1} \cdots q_m^{r_m})$ as its rows, where 0_N is the $N \times 1$ vector of zeros. By adding the columns of $\mathbf{L}$ to $A \otimes B$, the resulting matrix $[A \otimes B, \mathbf{L}]$ is an orthogonal array $OA(NM, p_1^{k_1 s_1} \cdots p_n^{k_n s_n} q_1^{r_1} \cdots q_m^{r_m})$.
   (iii) (Optional) Use the method of replacement to replace two-level columns in the array in step (ii) by four-level columns and, more generally, p-level columns by p^r-level columns, $r > 1$.

The orthogonality between $A \otimes B$ and $\mathbf{L}$ can be shown as follows. By rearranging the rows of A, each column of $OA(N, p_i^{s_i}) * D_{M, k_i; \, p_i}$ in $A \otimes B$ can be written as $(w^0, \ldots, w^{p_i - 1})^T$, and each column of $\mathbf{L}$ can be written as $(v, v, \ldots, v)^T$, where w and v are $(NM/p_i) \times 1$ vectors, v consists of each of the q_i symbols the same number of times, and $w^j = w + j \pmod{p_i}$. The orthogonality of these two vectors is now obvious.

The *adding columns* technique in step (ii) enhances the variety of arrays and the flexibility in the choice of factor levels because $OA(M, q_1^{r_1} \ldots q_m^{r_m})$ is usually available for a variety of q_i and r_i. This flexibility can also be achieved by the method of replacement.

For the convenience of array construction, a collection of difference matrices is given in Appendix 7D. The method in (7.6) is illustrated with the construction of several orthogonal arrays in the following examples.

Example 7.1. 18-Run Arrays

Step (i). Using $D_{6,6;3}$ in Appendix 7D, $OA(3, 3^1) = (0, 1, 2)^T$, (7.3), and (7.4), it can be shown that

$$OA(3, 3^1) * D_{6,6;3} = \begin{bmatrix} D_{6,6;3} \\ D_{6,6;3} + 1(mod\ 3) \\ D_{6,6;3} + 2(mod\ 3) \end{bmatrix},$$

which is an $OA(18, 3^6)$ and is equivalent to the last six columns of the matrix in Table 7C.2.

Step (ii). Let L_6 denote a six-run orthogonal array. Then

$$\left[OA(3, 3^1) * D_{6,6;3}, 0_3 * L_6 \right]$$

is an orthogonal array, which is equivalent to $OA(18, 2^1 3^7)$ and $OA(18, 6^1 3^6)$ in Table 7C.2 if L_6 is chosen to be $OA(6, 2^1 3^1)$ and $OA(6, 6^1)$, respectively.

Example 7.2. 36-Run Arrays.

Step (i). Using $D_{12,12;3}$ in Appendix 7D and $OA(3, 3^1)$, $OA(3, 3^1) * D_{12,12;3}$ is an $OA(36, 3^{12})$.

Step (ii). Let L_{12} denote a 12-run orthogonal array. Then

$$\left[OA(3, 3^1) * D_{12,12;3}, 0_3 * L_{12} \right]$$

is an orthogonal array, which is equivalent to $OA(36, 2^{11} 3^{12})$ (see Table 7C.6 or 7C.7), $OA(36, 3^{12}(12)^1)$, $OA(36, 3^{12} 2^2 6^1)$, $OA(36, 3^{13} 4^1)$, and $OA(36, 2^4 3^{13})$ if L_{12} is chosen to be $OA(12, 2^{11})$, $OA(12, (12)^1)$, $OA(12, 2^2 6^1)$, $OA(12, 3^1 4^1)$, and $OA(12, 2^4 3^1)$, respectively.

These examples also demonstrate the versatility of the method in constructing a variety of mixed-level orthogonal arrays.

The method in (7.6) can also be used to construct arrays in Tables 7C.10, 7C.11, and 7C.12. Details for constructing these and many other arrays can be found in Wang and Wu (1991). It can also be extended (Wang and Wu, 1992) to construct several classes of nearly orthogonal arrays (which were briefly mentioned in Section 7.5).

7.7 CONSTRUCTION OF MIXED-LEVEL ORTHOGONAL ARRAYS THROUGH THE METHOD OF REPLACEMENT

In Section 6.2, the method of replacement was used to replace three two-level columns by one four-level column. By *reversing* the assignment rule in Table 6.4, one four-level column can be replaced by three two-level columns. If the original four-level column, denoted by A, is orthogonal to the remaining columns in an array, each of the three two-level columns, denoted by $(\alpha, \beta, \alpha\beta)$ in Table 6.4, is also orthogonal to the remaining columns. This can be easily proved as follows. Denote a remaining column by a. Because A is orthogonal to a, each level combination between A and a appears the same number of times in the array, say v. It is obvious then that each level combination between α and a appears $2v$ times in the array, which proves the orthogonality between α and a. A similar argument shows that β and $\alpha\beta$ are orthogonal to a.

The method of replacement can also be applied to any nonregular two-level orthogonal array that contains three columns of the form $(\alpha, \beta, \alpha\beta)$. Take, for example, the $OA(2N, 2^{2N-1})$ that is obtained from H_{2N} in (7.1) by deleting its first column. Let $\mathbf{1}$ denote the column of 1's in H_N in (7.1) and $\mathbf{c}$ be any other column in H_N. Then H_{2N} contains the following three columns:

$$\begin{pmatrix} \mathbf{c} \\ \mathbf{c} \end{pmatrix}, \quad \begin{pmatrix} \mathbf{c} \\ -\mathbf{c} \end{pmatrix}, \quad \begin{pmatrix} \mathbf{1} \\ -\mathbf{1} \end{pmatrix},$$

which are of the form $(\alpha, \beta, \alpha\beta)$. By replacing these three columns by a four-level column, we can generate an $OA(2N, 4^1 2^{2N-4})$. The smallest such array is $OA(24, 4^1 2^{20})$. The method can be extended to more than four levels. For example, seven two-level columns of the form $(\alpha, \beta, \alpha\beta, \gamma, \alpha\gamma, \beta\gamma, \alpha\beta\gamma)$ in an array can be replaced by an eight-level column and vice versa *without affecting orthogonality*. Because the seven columns are generated by three independent columns, they have eight distinct level combinations, which correspond to the eight levels. By applying this method to the $OA(16, 2^{15})$, we can obtain an $OA(16, 8^1 2^8)$. Mixed eight-, four-, and two-level designs can also be generated by using the method repeatedly. For example, in the $OA(32, 2^{31})$, we can replace the seven columns $1, 2, 12, 3, 13, 23, 123$ by an eight-level column, the three columns $4, 5, 45$ by a four-level column, and the three columns $14, 25, 1245$ by another four-level column, resulting in an $OA(32, 8^1 4^2 2^{18})$.

**Table 7.6   Generating a Nine-Level Factor from
Four Three-Level Factors and Vice Versa**

A	B	AB	AB^2	New Factor
0	0	0	0	0
0	1	1	2	1
0	2	2	1	2
1	0	1	1	3
1	1	2	0	4
1	2	0	2	5
2	0	2	2	6
2	1	0	1	7
2	2	1	0	8

(with $\longleftrightarrow$ between AB^2 and New Factor columns, opposite rows 4–5)

Similarly, a nine-level column and four three-level columns of the form (A, B, AB, AB^2) can be replaced by each other without affecting orthogonality. The replacement rule is given in Table 7.6.

More generally, a t-level column, where $t = s^u$ and s is a prime power, can be replaced by K s-level columns, where $K = (s^u - 1)/(s - 1)$ and the K columns are generated by u independent columns, and vice versa. The replacement does not affect the orthogonality property of the original array. It also preserves the degrees of freedom of the two arrays because both the t-level column and the K s-level columns have $s^u - 1$ degrees of freedom.

Even when t is not a prime power, the method can still be utilized. Suppose $t = \prod_{i=1}^{j} s_i^{u_i}$, where s_i are distinct prime numbers. Then the t levels can be replaced by K_i s_i-level columns, where $K_i = (s_i^{u_i} - 1)/(s_i - 1)$, $i = 1, \ldots, j$. More generally, the t levels can be replaced by *any* orthogonal array consisting of t rows and an appropriate number of columns. By following the argument given at the beginning of the section, it is easy to show that the replacement does not affect orthogonality. The simplest case is $t = 6$. A six-level column is replaced by a two-level column and a three-level column as shown in Table 7.7. The method can be used to generate $OA(36, 2^1 3^8 6^2)$ and $OA(36, 2^2 3^9 6^1)$ from $OA(36, 3^7 6^3)$ (given in Table 7C.8) and $OA(36, 2^9 3^1 6^2)$ and $OA(36, 2^{10} 3^2 6^1)$ from $OA(36, 2^8 6^3)$ (given in Table 7C.9).

**Table 7.7   Replacing a Six-Level Column by One Two-Level
Column and One Three-Level Column**

0	0	0
1	0	1
2	0	2
3	1	0
4	1	1
5	1	2

(with $\longrightarrow$ in the middle between the first column and the last two columns)

Unlike the previous cases, this replacement will result in a loss of degrees of freedom because the six-level column has five degrees of freedom while the two replacing columns have only three degrees of freedom.

The reverse of the replacement rule in Table 7.7 does not always preserve orthogonality. Counterexamples abound. For example, the first two columns of $OA(12, 3^1 2^4)$ in Table 7C.1 cannot be replaced by a six-level column without affecting orthogonality. If, however, the interaction between the two-level column and the three-level column is orthogonal to the remaining columns, then orthogonality is unaffected by replacing the two columns by the six-level column in Table 7.7. As an application, consider the $OA(18, 2^1 3^7)$ in Table 7C.2. The interaction between its first two columns is orthogonal to the remaining three-level columns. By replacing these two columns by a six-level column, we obtain the $OA(18, 6^1 3^6)$ in Table 7C.2. For further information on the various replacement rules, see Addelman (1962), Dey (1985), and Wu, Zhang, and Wang (1992).

7.8 ORTHOGONAL MAIN-EFFECT PLANS THROUGH COLLAPSING FACTORS

The method of collapsing factors (Addelman, 1962) is very useful for constructing plans that are not orthogonal arrays but are smaller in run size than what would be required by the use of orthogonal arrays. For example, it is known from Lemma 7.1 that no $OA(9, 2^1 3^3)$ exists and the smallest run size for the given combination of factor levels is 18. By assigning the three levels of column A in the $OA(9, 3^4)$ in Table 7.8 to two levels according to the rule $0 \rightarrow 0, 1 \rightarrow 1, 2 \rightarrow 1$, a two-level column is obtained and is denoted by A' in the right table of Table 7.8. The right table has one two-level column and three three-level columns. It is not an orthogonal array because its first

Table 7.8 Construction of $OME(9, 2^1 3^3)$ from $OA(9, 3^4)$ through Collapsing Factor

Run	A	B	C	D		Run	A'	B	C	D
1	0	0	0	0		1	0	0	0	0
2	0	1	2	1		2	0	1	2	1
3	0	2	1	2		3	0	2	1	2
4	1	0	1	1	$\longrightarrow$	4	1	0	1	1
5	1	1	0	2		5	1	1	0	2
6	1	2	2	0		6	1	2	2	0
7	2	0	2	2		7	1	0	2	2
8	2	1	1	0		8	1	1	1	0
9	2	2	0	1		9	1	2	0	1

Table 7.9 Factors A' and B in $OME(9, 2^1 3^3)$ Appear in Proportional Frequencies, where n_{ij} Appear as Cell Counts with $n_{i.}$ and $n_{.j}$ as Margin Totals

A'	B 0	1	2	Row Sum
0	1	1	1	3
1	2	2	2	6
Column Sum	3	3	3	

column has three 0's and six 1's. Its three three-level columns are still orthogonal. Although A' is not orthogonal to the other columns, its levels appear with the levels of any other column in proportional frequencies (see Table 7.9). These frequencies are not constant as would be required by an orthogonal array but satisfy the following definition of proportional frequencies. For two factors denoted by A and B, let r and s be the number of levels of A and B, respectively. Let n_{ij} be the number of observations at level i of A and level j of B in the two-factor design. We say that n_{ij} are in *proportional frequencies* if they satisfy

$$n_{ij} = \frac{n_{i.} n_{.j}}{n_{..}}, \qquad (7.7)$$

where $n_{i.} = \sum_{j=1}^{s} n_{ij}$, $n_{.j} = \sum_{i=1}^{r} n_{ij}$, and $n_{..} = \sum_{i=1}^{r} \sum_{j=1}^{s} n_{ij}$.

The main advantage of having proportional frequencies is that they ensure that the main effect estimates of factor A and of factor B are uncorrelated. Under the assumption of a linear model with main effects for A and B but no interactions (see Section 2.3 for the two-way layout), it is known that the main effect estimates of factor A and of factor B are uncorrelated if and only if n_{ij} are in proportional frequencies. A proof can be found in John (1971).

In general, we can replace a factor with t levels in an orthogonal array by a new factor with s levels, $s < t$, by using a many-to-one correspondence between the t levels and the s levels. Two examples of many-to-one correspondences are given in Tables 7.8 and 7.10. The technique is called the **method of collapsing factors**. Because the cell frequencies between any two

Table 7.10 Collapsing a Four-Level Factor to a Three-Level Factor

0		0		0
1		1		1
2	$\longrightarrow$	1	or	2
3		2		2

factors of an orthogonal array are equal, it is easy to see that the collapsed factor and any remaining factor in the orthogonal array have proportional frequencies. From the result in the preceding paragraph, the main effect estimates of the collapsed factor and of any other factor in the original array are uncorrelated. If t is a multiple of s, the method of collapsing factors reduces to the method of replacement as described in Section 7.7, and the newly constructed plan is still an orthogonal array. If t is not a multiple of s, the newly constructed plan is not an orthogonal array but all its main effect estimates are still orthogonal (i.e., uncorrelated). Such plans are therefore called **orthogonal main-effect plans** and are abbreviated as *OME*s. For example, the right table of Table 7.8 is an $OME(9, 2^1 3^3)$ but not an $OA(9, 2^1 3^3)$. It is obvious that orthogonal main-effect plans are more general than orthogonal arrays. To avoid confusion, we shall use the term orthogonal main-effect plans in the book only to refer to plans that are *not* orthogonal arrays.

The method of collapsing factors can be repeatedly applied to different columns of an orthogonal array to generate a variety of orthogonal main-effect plans. For example, by collapsing another three-level column in the $OA(9, 3^4)$ to a two-level column, we can generate an $OME(9, 2^2 3^2)$. Recall from Lemma 7.1 that an $OA(N, 2^2 3^2)$ would require $N = 36$. This again demonstrates the run size saving from using an orthogonal main-effect plan. By repeating the same procedure, an $OME(9, 3^1 2^3)$ can be generated. This plan is, however, not as good as the $OME(8, 3^1 2^4)$ to be given later in this section. Collapsing should not be applied to the last remaining three-level column because the resulting $OME(9, 2^4)$ is inferior to the $OA(8, 2^4)$.

When should an orthogonal main-effect plan be used in practice? Its most obvious advantage is *run size economy*. There is presumably some loss of estimation efficiency from using an orthogonal main-effect plan. From the definitions, the only difference between an orthogonal main-effect plan and an orthogonal array is that the former has some factors and pairs of factors which do not have equal frequencies. Because all main effect estimates in an orthogonal main-effect plan are orthogonal, from the estimation point of view, the only difference between them is in the efficiency of main effect estimates, not the covariance between them. It turns out that the efficiency is generally very high unless the number of levels of the collapsed factor is much smaller than that of the original factor. Consider the simplest case of collapsing a three-level factor to a two-level factor. Assume the total run size is $3m$. For the collapsed factor, the run size allocation between the two levels is $2m$ and m. The variance of its main effect estimate is

$$Var(\bar{y}_1 - \bar{y}_0) = \sigma^2 \left(\frac{1}{m} + \frac{1}{2m} \right) = \frac{3}{2m} \sigma^2.$$

Had it been possible to find a balanced allocation of $(3/2)m$ runs to each level of the two-level factor, the variance of its main effect estimate $(4/3m)\sigma^2$

would be the smallest. By taking the ratio of the two variances, the relative efficiency of the main effect estimate of the collapsed factor is $\frac{8}{9}$, which is quite high. Because it may not be possible to find an orthogonal main-effect plan with balanced assignment on all its two-level factors, $\frac{8}{9}$ is a lower bound on its relative efficiency and the actual efficiency may be higher.

Given the high efficiency of main effect estimates in an orthogonal main-effect plan, its only real disadvantage is the *imbalance* of run size allocation between levels. It is often unacceptable to practitioners to assign one-third of the total runs to one level and two-thirds to the other level unless there is an *a priori* belief or reason that one level should be studied much more intensively. This may happen if there is very little known about this level (e.g., a new treatment or machine) or a higher precision is required because of its wider use in practice. The level receiving two-thirds of the runs should be the one that needs more intensive study. In Table 7.8, it is level 1. If level 0 turns to be more important, the collapsing should assign $0 \to 0$, $1 \to 0, 2 \to 1$.

Another popular version of the method is to collapse a four-level factor to a three-level factor. By following the assignment rule in Table 7.10, an $OME(8, 3^1 2^4)$ can be generated from the $OA(8, 4^1 2^4)$ which was given in Table 6.4. The advantage of using the $OME(8, 3^1 2^4)$ is again run size economy. From Lemma 7.1, an $OA(N, 3^1 2^4)$ would require $N = 12$. With a slight loss of estimation efficiency, use of the $OME(8, 3^1 2^4)$ would save four runs. Similarly, from the $OA(16, 4^2 2^9)$ given in Section 6.2, one can generate an $OME(16, 4^1 3^1 2^9)$ and an $OME(16, 3^2 2^9)$.

What are the relative efficiencies of the main effect estimates of the collapsed three-level factor? Consider the first version of the assignment rule in Table 7.10. The middle level receives one-half of the runs while each of the other two levels receives one-quarter of the runs. Assuming the factor is quantitative and the total number of runs is $4m$, its linear effect estimate has variance

$$Var(\bar{y}_2 - \bar{y}_0) = \sigma^2 \left(\frac{1}{m} + \frac{1}{m} \right) = \frac{2}{m}\sigma^2$$

while its quadratic effect estimate has variance

$$Var\left(\frac{1}{2}\bar{y}_0 + \frac{1}{2}\bar{y}_2 - \bar{y}_1 \right) = \sigma^2 \left(\frac{1}{4m} + \frac{1}{4m} + \frac{1}{2m} \right) = \frac{\sigma^2}{m}.$$

Had it been possible to assign $\frac{4}{3}m$ runs to each of the three levels, the linear effect estimate would have the variance $(3/2m)\sigma^2$ and the quadratic effect estimate would have the variance $(9/8m)\sigma^2$. Therefore the relative efficiencies are $\frac{3}{4}$ for the linear effect and $\frac{9}{8}$ for the quadratic effect. Note that the quadratic effect has efficiency greater than 1 because the middle level receives one-half of the runs. This high efficiency is achieved at the expense of the linear effect whose efficiency is much lower. This version of the assignment rule should be used if it is more important to estimate the quadratic effect precisely. For the second version of the assignment rule in Table 7.10, level 2 receives one-half of the runs. A similar argument shows that its relative efficiencies are 1 for the linear effect and $\frac{9}{11}$ for the quadratic effect. (These derivations are left as an exercise.) The second version should be used if it is more important to estimate the linear effect precisely.

Many more orthogonal main-effect plans can be generated from orthogonal arrays by the method of collapsing factors. A collection of the most useful plans for $N \leq 36$ is given in Table 7E.1. In general, orthogonal main-effect plans constructed through collapsing fewer factors are preferred over those through collapsing more factors because the collapsing of each factor would result in the loss of degrees of freedom and of estimation efficiency. A good example is the comparison between the $OME(8, 3^1 2^4)$ and the $OME(9, 3^1 2^3)$ discussed before. The former is preferred because it has one fewer run and can accommodate one more two-level factor. It is obtained from collapsing one factor while the latter is obtained from collapsing three factors. This rule should be considered when choosing an OME plan from Table 7E.1.

The $OME(16, 3^7)$ in Table 7E.2 is interesting because it cannot be obtained from collapsing factors and is one degree of freedom short of being saturated (i.e., $2 \times 7 = 14 = 15 - 1$). Its ingenious construction is due to Starks (1964).

More information on the method of collapsing factors can be found in Addelman (1962). A more extensive collection of orthogonal main-effect plans is available in Table 3.4 of Dey (1985).

7.9   PRACTICAL SUMMARY

1. Regular designs are those constructed through defining relations among factors. They include the r^{n-k} designs with r being a prime power and the mixed-level designs in Chapter 6. In a regular design, any pair of factorial effects are either orthogonal or fully aliased. Nonregular designs are those that do not possess this property. They include designs considered in this chapter.

2. The main reasons for using nonregular designs are run size economy and flexibility in the choice of factor level combinations. A third benefit for using these designs will be discussed in point 6.

3. The Plackett-Burman designs are obtained from Hadamard matrices by removing its first column of $+$'s. For run size N, they can be used to study up to $N - 1$ two-level factors. They are given in Appendix 7A for $N = 12, 20, 24, 28, 36, 40, 44, 48$.

4. Hall's designs of Types II to V in Appendix 7B are 16-run nonregular designs. The commonly used 16-run designs are the regular 2^{n-k} designs with $n - k = 4$, which are Hall's designs of Type I.

5. A collection of useful mixed-level orthogonal arrays are given in Appendix 7C and their properties discussed in Section 7.5.

6. Many of the designs in points 3–5 enjoy a hidden projection property. Even though they do not have the geometric projection property as required by a resolution V design, their projections onto a small number (e.g., 3–5) of factors allow the main effects as well as some (or all) of the two-factor interactions to be estimated. Their run size economy does not hinder their ability in estimating two-factor interactions.

   (i) Theorem 7.1 ensures this hidden projection property for projection onto four factors.

   (ii) The hidden projection property is discussed in detail in Section 7.4 for the 12-run and 20-run Plackett-Burman designs and in Section 7.5 for $OA(18, 2^1 3^7)$.

7. Nearly orthogonal arrays can be used to accommodate more factors beyond what orthogonal arrays can do. In these arrays the number of non-orthogonal pairs of columns is small. They are indicated by the notation OA'. Several such arrays are given in Appendix 7C.

8. A general method for constructing mixed-level orthogonal arrays is given in Section 7.6, which is based on the difference matrix and generalized Kronecker sum. It leads to the construction of several arrays in Appendix 7C.

9. Additional mixed-level orthogonal arrays can be constructed by applying the method of replacement in Section 7.7 to existing orthogonal arrays.

10. Orthogonal main-effect plans, abbreviated as $OMEs$, are arrays whose main-effect estimates are orthogonal (i.e., uncorrelated). Their main difference from orthogonal arrays is that some factors and pairs of factors in an OME do not have equal frequencies. They should only be used as an economic alternative to orthogonal arrays, i.e., when an OME is available for a given run size and factor level combination while the corresponding orthogonal array is not.

11. The method of collapsing factors is the most effective way for constructing orthogonal main-effect plans. For collapsing a three-level factor to a two-level factor and a four-level factor to a three-level factor, the efficiency of its main effect estimates is high. Its only disadvantage is the imbalance of run size allocation between levels, which may not be acceptable on practical grounds. A collection of useful $OMEs$ is given in Appendix 7E.

EXERCISES

1. (a) Show that the projections of the 12-run Plackett-Burman design onto any four columns are equivalent.
   (b) Consider the main effects for the four columns in (a) and their six interactions. Show that the D_s efficiency for each of the 10 effects is 0.62.

2. (a) Show that there are two non-equivalent 12×5 submatrices of the 12-run Plackett-Burman design.
   (b) Show that one of them, called design 5.1, has two repeated runs (i.e., two identical row vectors). How many degrees of freedom does it have for estimating effects? Show that it can estimate five main effects as well as five two-factor interactions (out of 10).
   (c) Show that the second type, called design 5.2, has two mirror image runs (i.e., one row vector having opposite signs from the other). How many degrees of freedom does it have for estimating effects? Show that it can estimate five main effects as well as six two-factor interactions.
   (d) For planning a five-factor experiment, which of designs 5.1 and 5.2 would you prefer?

Table 7.11 A 10-Factor 12-Run Experiment with Six Added Runs

Run	\multicolumn{10}{c} Factor										y
	1	2	3	4	5	6	7	8	9	10	
1	+	+	−	+	+	+	−	−	−	+	70.19
2	−	+	+	−	+	+	+	−	−	−	57.12
3	+	−	+	+	−	+	+	+	−	−	63.17
4	−	+	−	+	+	−	+	+	+	−	99.43
5	−	−	+	−	+	+	−	+	+	+	90.72
6	−	−	−	+	−	+	+	−	+	+	110.37
7	+	−	−	−	+	−	+	+	−	+	120.36
8	+	+	−	−	−	+	−	+	+	−	40.15
9	+	+	+	−	−	−	+	−	+	+	81.38
10	−	+	+	+	−	−	−	+	−	+	88.89
11	+	−	+	+	+	−	−	−	+	−	9.63
12	−	−	−	−	−	−	−	−	−	−	36.25
13	−		−				−	+		+	104.18
14	+		−				+	−		−	50.28
15	−	+					−	−		+	71.74
16	−	+					−	+		−	43.66
17	+	+					−	+		+	67.76
18	−	+					+	+		+	129.10

3. Consider the constructed data (Lin and Draper, 1992) in Table 7.11, which has two parts: the first 12 runs for 10 factors and the added 6 runs for 5 out of the 10 factors.

   (a) Analyze the first 12 runs by using a main-effect model. Show that factors **1, 3, 7, 8,** and **10** have significant main effects.

   (b) In order to further study the interactions among the five factors identified in (a), additional runs need to be conducted. Explain how runs 13–18 in the table were chosen. (*Hint:* Part of the 18 runs for the five factors forms a resolution V design.)

   (c) Use the full data to analyze the main effects and two-factor interactions of the five identified factors. Show that one two-factor interaction is significant.

   (d) As shown in the previous exercise, based on the first 12 runs, the 5 identified factors and 5 or 6 of their two-factor interactions are estimable. Use forward regression to analyze the first 12 runs with the 5 identified factors and their 10 two-factor interactions as candidate variables. Show that no two-factor interaction is found to be significant (because the data are too noisy).

   (e) If the purpose of adding runs is to identify the significant interaction found in (c), only one of the six added runs in the table is needed. Show that, by applying forward regression to the data consisting of runs 1–12 and 14, the five main effects and the interaction [identified in (c)] are significant. Based on this result, comment on the advisability of adding six runs in the original constructed example.

4. It is known (Wang and Wu, 1995) that there are nine non-equivalent 20×5 submatrices of the 20-run Plackett-Burman design.

   (a) Use a random search to find one 20×5 submatrix that has no repeated runs. Show that, if this submatrix is used to study 5 factors, all the 5 main effects and 10 two-factor interactions can be estimated. Compute the D_s efficiency for each effect estimate.

   (b) To follow the geometric projection approach, one would need to find additional runs to supplement the 20×5 submatrix in (a) so that the augmented design has resolution V, thus allowing all the 5 main effects and 10 two-factor interactions to be estimated. How many runs are required? What is the D_s efficiency for each effect estimate from the augmented design?

   (c) Based on the run size and efficiency comparison in (a) and (b), comment on the pros and cons of the hidden projection approach and the geometric projection approach.

   (d) Use a random search to find two other 20×5 submatrices that allow 9 and respectively 7 two-factor interactions (in addition to the 5 main effects) to be estimated.

(e) Among the three submatrices in (a) and (d) which one would you prefer for studying 5 factors?

5. Mukerjee and Wu (1995) used an elaborate argument to prove that it is impossible to further increase the number of three-level columns in the $OA(36, 2^{11}3^{12})$ in the sense that the maximum m in $OA(36, 2^{35-2m}3^m)$ is 12. Can you find a simpler and more direct proof?

6. Prove the following statements (Wang and Wu, 1992):
   (a) The maximum for m in an $OA(12, 3^1 2^m)$ is 4.
   (b) The maximum for m in an $OA(20, 5^1 2^m)$ is 8.
   (c) The maximum for m in an $OA(24, 3^1 2^m)$ is 16.

7. (a) Find all the defining words of length 3 for Hall's design of type II. (*Hint:* There are 10 such words.)
   (b) Find all the pairs of main effects and two-factor interactions in Hall's design of type II that are partially aliased.

8. (a) Are all the projections of the $OA(18, 3^7)$ onto any four columns equivalent?
   (b) Take the first four columns for projection. Consider the estimation of the linear and quadratic effects of the four factors and their six $l \times l$ interaction effects. Show that these 14 effect estimates are linearly independent. Compute the D_s efficiency for each of the 14 effect estimates.

9. Use the method in (7.6) with the difference matrix $D_{10, 10; 5}$ to construct $OA(50, 2^1 5^{11})$ (given in Table 7C.11) and $OA(50, 10^1 5^{10})$. Identify the orthogonal array in step (ii) of (7.6) for each of the two constructed arrays.

10. (a) Use the method in (7.6) with the difference matrix $D_{6, 6; 3}$ to construct $OA(54, 2^1 3^{25})$ and $OA(54, 6^1 3^{24})$ (both given in Table 7C.12). Identify the orthogonal array in step (ii) of (7.6) for each of the two constructed arrays.
   (b) For each of the constructed arrays in (a), identify four three-level columns of the form (A, B, AB, AB^2) and replace them by a nine-level column. Show that this leads to the construction of $OA(54, 2^1 9^1 3^{21})$ and $OA(54, 6^1 9^1 3^{20})$.

11. Use the method of replacement in Section 7.7 to construct $OA(32, 8^1 4^m 2^{24-3m})$ for $3 \leq m \leq 8$. (*Hint:* After replacing the 7 columns

1, 2, 12, 3, 13, 23, 123 by an eight-level column, group the remaining 24 columns in eight mutually exclusive sets of the form $(\alpha, \beta, \alpha\beta)$.)

12. For the second collapsing rule in Table 7.10, prove that the relative efficiencies are 1 for the linear effect and $\frac{9}{11}$ for the quadratic effect.

13. Consider the $OA(18, 6^1 3^6)$ given in Table 7C.2. By collapsing its six-level factor to a four-level factor, an $OME(18, 4^1 3^6)$ can be generated.
    (a) Find the smallest N in an $OA(N, 4^1 3^6)$. Comment on the run size saving from using an $OME(18, 4^1 3^6)$.
    (b) Give the best assignment rule for collapsing if, among the linear, quadratic, and cubic contrasts [defined in (6.5)] of the four-level factor, the linear contrast should be estimated more precisely.
    (c) Repeat question (b) if the quadratic contrast should be estimated more precisely.
    (d) For each of the rules in (b) and (c) give the relative efficiencies for the linear, quadratic, and cubic contrasts of the four-level factor.

14. (a) An investigator is contemplating an experiment with 1 three-level factor, 1 four-level factor, and 10 two-level factors. If an orthogonal array is to be chosen, what is the minimum run size required? Find an orthogonal array with the minimum run size. (*Hint:* Use Lemma 7.1 and count degrees of freedom.)
    (b) If budgetary constraints only allow at most 18 runs to be conducted, it would not be possible to use an orthogonal array. As a compromise, the investigator is willing to reduce the number of two-level factors from 10 to 9. Find instead an orthogonal main-effect plan from Table 7E.1.

15. Seven two-level factors and six three-level factors are to be studied in an experiment.

    (a) What is the minimum run size required if an orthogonal array is to be used? Find such an array and arrange it so that the number of level changes in one three-level factor is minimized.
    (b) In order to reduce the run size, an orthogonal main-effect plan is considered as an alternative. What is the minimum run size required if an orthogonal main-effect plan is to be used? Find two such plans from Table 7E.1. Are the run size savings substantial?

16. Prove that, if the interaction between a two-level column and a three-level column is orthogonal to the remaining columns of an orthogonal array,

then orthogonality is unaffected by replacing the two columns by a six-level column using the reverse of the rule in Table 7.7.

17. (a) Verify that the four sets of columns $(2, 3, 14)$, $(2, 4, 15)$, $(2, 5, 16)$, and $(2, 6, 17)$ in Table 7C.4 are of the form $(\alpha, \beta, \alpha\beta)$.

    (b) Obtain an $OA(24, 3^1 4^1 2^{13})$ by replacing columns 2, 3, and 14 in Table 7C.4 by a four-level column according to the rule in Table 6.4.

    (c) By further replacing columns 2, 4, and 15 by a four-level column using the same replacement rule, an array with $(24, 3^1 4^2 2^{11})$ can be obtained. Is this array orthogonal? If not, explain why the method of replacement cannot be used here to ensure orthogonality. How good is the array as a nearly orthogonal array in terms of the number of non-orthogonal pairs and correlations?

    (d) Construct arrays with $(24, 3^1 4^3 2^9)$ and $(24, 3^1 4^4 2^7)$ by further replacing columns $(2, 5, 16)$ and $(2, 6, 17)$ by two four-level columns. Comment on the non-orthogonality of the arrays. Discuss when such arrays may be used in terms of run size economy, flexibility in factor level combinations, and non-orthogonality.

[*Note:* This method of construction is called the *idle column technique* by Taguchi (1987, Chapter 10). Its main difference from the method of replacement is that it allows sets of columns of the form $(\alpha, \beta_i, \alpha\beta_i)$ with the *same* α in every set to be replaced by four-level columns. The common column α is referred to as the "idle column."]

APPENDIX 7A: PLACKETT-BURMAN DESIGNS $OA(N, 2^{N-1})$ WITH $12 \leq N \leq 48$ AND $N = 4k$ BUT NOT A POWER OF 2

1. N = 12

Run	1	2	3	4	5	6	7	8	9	10	11
1	+	+	−	+	+	+	−	−	−	+	−
2	−	+	+	−	+	+	+	−	−	−	+
3	+	−	+	+	−	+	+	+	−	−	−
4	−	+	−	+	+	−	+	+	+	−	−
5	−	−	+	−	+	+	−	+	+	+	−
6	−	−	−	+	−	+	+	−	+	+	+
7	+	−	−	−	+	−	+	+	−	+	+
8	+	+	−	−	−	+	−	+	+	−	+
9	+	+	+	−	−	−	+	−	+	+	−
10	−	+	+	+	−	−	−	+	−	+	+
11	+	−	+	+	+	−	−	−	+	−	+
12	−	−	−	−	−	−	−	−	−	−	−

2. $N = 20$

Run	1	2	3	4	5	6	7	8	9	10	11	12	13	14	15	16	17	18	19
1	+	+	−	−	+	+	+	+	−	+	−	+	−	−	−	−	+	+	−
2	−	+	+	−	−	+	+	+	+	−	+	−	+	−	−	−	−	+	+
3	+	−	+	+	−	−	+	+	+	+	−	+	−	+	−	−	−	−	+
4	+	+	−	+	+	−	−	+	+	+	+	−	+	−	+	−	−	−	−
5	−	+	+	−	+	+	−	−	+	+	+	+	−	+	−	+	−	−	−
6	−	−	+	+	−	+	+	−	−	+	+	+	+	−	+	−	+	−	−
7	−	−	−	+	+	−	+	+	−	−	+	+	+	+	−	+	−	+	−
8	−	−	−	−	+	+	−	+	+	−	−	+	+	+	+	−	+	−	+
9	+	−	−	−	−	+	+	−	+	+	−	−	+	+	+	+	−	+	−
10	−	+	−	−	−	−	+	+	−	+	+	−	−	+	+	+	+	−	+
11	+	−	+	−	−	−	−	+	+	−	+	+	−	−	+	+	+	+	−
12	−	+	−	+	−	−	−	−	+	+	−	+	+	−	−	+	+	+	+
13	+	−	+	−	+	−	−	−	−	+	+	−	+	+	−	−	+	+	+
14	+	+	−	+	−	+	−	−	−	−	+	+	−	+	+	−	−	+	+
15	+	+	+	−	+	−	+	−	−	−	−	+	+	−	+	+	−	−	+
16	+	+	+	+	−	+	−	+	−	−	−	−	+	+	−	+	+	−	−
17	−	+	+	+	+	−	+	−	+	−	−	−	−	+	+	−	+	+	−
18	−	−	+	+	+	+	−	+	−	+	−	−	−	−	+	+	−	+	+
19	+	−	−	+	+	+	+	−	+	−	+	−	−	−	−	+	+	−	+
20	−	−	−	−	−	−	−	−	−	−	−	−	−	−	−	−	−	−	−

3. $N = 24$.

Run	1	2	3	4	5	6	7	8	9	10	11	12	13	14	15	16	17	18	19	20	21	22	23
1	+	+	+	+	+	−	+	−	+	+	−	−	+	+	−	−	+	−	+	−	−	−	−
2	−	+	+	+	+	+	−	+	−	+	+	−	−	+	+	−	−	+	−	+	−	−	−
3	−	−	+	+	+	+	+	−	+	−	+	+	−	−	+	+	−	−	+	−	+	−	−
4	−	−	−	+	+	+	+	+	−	+	−	+	+	−	−	+	+	−	−	+	−	+	−
5	−	−	−	−	+	+	+	+	+	−	+	−	+	+	−	−	+	+	−	−	+	−	+
6	+	−	−	−	−	+	+	+	+	+	−	+	−	+	+	−	−	+	+	−	−	+	−
7	−	+	−	−	−	−	+	+	+	+	+	−	+	−	+	+	−	−	+	+	−	−	+
8	+	−	+	−	−	−	−	+	+	+	+	+	−	+	−	+	+	−	−	+	+	−	−
9	−	+	−	+	−	−	−	−	+	+	+	+	+	−	+	−	+	+	−	−	+	+	−
10	−	−	+	−	+	−	−	−	−	+	+	+	+	+	−	+	−	+	+	−	−	+	+
11	+	−	−	+	−	+	−	−	−	−	+	+	+	+	+	−	+	−	+	+	−	−	+
12	+	+	−	−	+	−	+	−	−	−	−	+	+	+	+	+	−	+	−	+	+	−	−
13	−	+	+	−	−	+	−	+	−	−	−	−	+	+	+	+	+	−	+	−	+	+	−
14	−	−	+	+	−	−	+	−	+	−	−	−	−	+	+	+	+	+	−	+	−	+	+
15	+	−	−	+	+	−	−	+	−	+	−	−	−	−	+	+	+	+	+	−	+	−	+
16	+	+	−	−	+	+	−	−	+	−	+	−	−	−	−	+	+	+	+	+	−	+	−
17	−	+	+	−	−	+	+	−	−	+	−	+	−	−	−	−	+	+	+	+	+	−	+
18	+	−	+	+	−	−	+	+	−	−	+	−	+	−	−	−	−	+	+	+	+	+	−
19	−	+	−	+	+	−	−	+	+	−	−	+	−	+	−	−	−	−	+	+	+	+	+
20	+	−	+	−	+	+	−	−	+	+	−	−	+	−	+	−	−	−	−	+	+	+	+
21	+	+	−	+	−	+	+	−	−	+	+	−	−	+	−	+	−	−	−	−	+	+	+
22	+	+	+	−	+	−	+	+	−	−	+	+	−	−	+	−	+	−	−	−	−	+	+
23	+	+	+	+	−	+	−	+	+	−	−	+	+	−	−	+	−	+	−	−	−	−	+
24	−	−	−	−	−	−	−	−	−	−	−	−	−	−	−	−	−	−	−	−	−	−	−

Note: An $OA(24, 2^{23})$ can also be constructed from H_{12} using (7.1).

4. $N = 28$. It has the form

$$\begin{bmatrix} X & Y & Z \\ Z & X & Y \\ Y & Z & X \\ - & - & - \end{bmatrix}$$

where the last row consists of all $-$'s and the generating matrices X, Y, and Z are defined as follows:

$$X = \begin{bmatrix}
+ & - & + & + & + & + & - & - & - \\
+ & + & - & + & + & + & - & - & - \\
- & + & + & + & + & + & - & - & - \\
- & - & - & + & - & + & + & + & + \\
- & - & - & + & + & - & + & + & + \\
- & - & - & - & + & + & + & + & + \\
+ & + & + & - & - & - & + & - & + \\
+ & + & + & - & - & - & + & + & - \\
+ & + & + & - & - & - & - & + & +
\end{bmatrix},$$

$$Y = \begin{bmatrix}
- & + & - & - & - & + & - & - & + \\
- & - & + & + & - & - & + & - & - \\
+ & - & - & - & + & - & - & + & - \\
- & - & + & - & + & - & - & - & + \\
+ & - & - & - & - & + & + & - & - \\
- & + & - & + & - & - & - & + & - \\
- & - & + & - & - & + & - & + & - \\
+ & - & - & + & - & - & - & - & + \\
- & + & - & - & + & - & + & - & -
\end{bmatrix},$$

$$Z = \begin{bmatrix}
+ & + & - & + & - & + & + & - & + \\
- & + & + & + & + & - & + & + & - \\
+ & - & + & - & + & + & - & + & + \\
+ & - & + & + & + & - & + & - & + \\
+ & + & - & - & + & + & + & + & - \\
- & + & + & + & - & + & - & + & + \\
+ & - & + & + & - & + & + & + & - \\
+ & + & - & + & + & - & - & + & + \\
- & + & + & - & + & + & + & - & +
\end{bmatrix}.$$

5. $N = 36$. It can be obtained from a cyclic arrangement of the row vector given in Table 7.5 and adding a row of $-$'s as the last row.

6. $N = 40$. It can be constructed from H_{20} using (7.1).

7. $N = 44$. It can be obtained from a cyclic arrangement of the row vector given in Table 7.5 and adding a row of $-$'s as the last row.

8. $N = 48$. It can be constructed from H_{24} using (7.1).

APPENDIX 7B: HALL'S 16-RUN ORTHOGONAL ARRAYS OF TYPES II TO V

1. Type II

Run	1	2	3	4	5	6	7	8	9	10	11	12	13	14	15
1	−	−	−	−	−	−	−	−	−	−	−	−	−	−	−
2	−	−	−	−	−	−	−	+	+	+	+	+	+	+	+
3	−	−	−	+	+	+	+	−	−	−	−	+	+	+	+
4	−	−	−	+	+	+	+	+	+	+	+	−	−	−	−
5	−	+	+	−	−	+	+	−	−	+	+	−	−	+	+
6	−	+	+	−	−	+	+	+	+	−	−	+	+	−	−
7	−	+	+	+	+	−	−	−	−	+	+	+	+	−	−
8	−	+	+	+	+	−	−	+	+	−	−	−	−	+	+
9	+	−	+	−	+	−	+	−	+	−	+	−	+	−	+
10	+	−	+	−	+	−	+	+	−	+	−	+	−	+	−
11	+	−	+	+	−	+	−	−	+	−	+	+	−	+	−
12	+	−	+	+	−	+	−	+	−	+	−	+	−	+	+
13	+	+	−	−	+	+	−	−	+	+	−	+	−	−	+
14	+	+	−	−	+	+	−	+	−	−	+	−	+	+	−
15	+	+	−	+	−	−	+	−	+	+	−	−	+	+	−
16	+	+	−	+	−	−	+	+	−	−	+	+	−	−	+

2. Type III

Run	1	2	3	4	5	6	7	8	9	10	11	12	13	14	15
1	−	−	−	−	−	−	−	−	−	−	−	−	−	−	−
2	−	−	−	−	−	−	−	+	+	+	+	+	+	+	+
3	−	−	−	+	+	+	+	−	−	−	−	+	+	+	+
4	−	−	−	+	+	+	+	+	+	+	+	−	−	−	−
5	−	+	+	−	−	+	+	−	−	+	+	−	−	+	+
6	−	+	+	−	−	+	+	+	+	−	−	+	+	−	−
7	−	+	+	+	+	−	−	−	−	+	+	+	+	−	−
8	−	+	+	+	+	−	−	+	+	−	−	−	−	+	+
9	+	−	+	−	+	−	+	−	+	−	+	−	+	−	+
10	+	−	+	−	+	−	+	+	−	+	−	+	−	+	−
11	+	−	+	+	−	+	−	−	+	+	−	−	+	+	−
12	+	−	+	+	−	+	−	+	−	−	+	+	−	−	+
13	+	+	−	−	+	+	−	−	+	+	−	+	−	−	+
14	+	+	−	−	+	+	−	+	−	−	+	−	+	+	−
15	+	+	−	+	−	−	+	−	+	−	+	+	−	+	−
16	+	+	−	+	−	−	+	+	−	+	−	−	+	−	+

3. Type IV

Run	1	2	3	4	5	6	7	8	9	10	11	12	13	14	15
1	−	−	−	−	−	−	−	−	−	−	−	−	−	−	−
2	−	−	−	−	−	−	−	+	+	+	+	+	+	+	+
3	−	−	−	+	+	+	+	−	−	−	−	+	+	+	+
4	−	−	−	+	+	+	+	+	+	+	+	−	−	−	−
5	−	+	+	−	−	+	+	−	−	+	+	−	−	+	+
6	−	+	+	−	−	+	+	+	+	−	−	+	+	−	−
7	−	+	+	+	+	−	−	−	−	+	+	+	+	−	−
8	−	+	+	+	+	−	−	+	+	−	−	−	−	+	+
9	+	−	+	−	+	−	+	−	+	−	+	−	+	−	+
10	+	−	+	−	+	+	−	−	+	+	−	+	−	+	−
11	+	−	+	+	−	−	+	+	−	+	−	−	+	+	−
12	+	−	+	+	−	+	−	+	−	−	+	+	−	−	+
13	+	+	−	−	+	−	+	+	−	+	−	+	−	−	+
14	+	+	−	−	+	+	−	+	−	−	+	−	+	+	−
15	+	+	−	+	−	−	+	−	+	−	+	+	−	+	−
16	+	+	−	+	−	+	−	−	+	+	−	−	+	−	+

4. Type V

Run	1	2	3	4	5	6	7	8	9	10	11	12	13	14	15
1	−	−	−	−	−	−	−	−	−	−	−	−	−	−	−
2	−	−	−	−	−	−	−	+	+	+	+	+	+	+	+
3	−	−	−	+	+	+	+	−	−	−	−	+	+	+	+
4	−	−	−	+	+	+	+	+	+	+	+	−	−	−	−
5	−	+	+	−	−	+	+	−	−	+	+	−	−	+	+
6	−	+	+	−	−	+	+	+	+	−	−	+	+	−	−
7	−	+	+	+	+	−	−	−	+	−	+	−	+	−	+
8	−	+	+	+	+	−	−	+	−	+	−	+	−	+	−
9	+	−	+	−	+	−	+	−	−	+	+	+	+	−	−
10	+	−	+	−	+	−	+	+	+	−	−	−	−	+	+
11	+	−	+	+	−	+	−	−	+	+	−	+	−	−	+
12	+	−	+	+	−	+	−	+	−	−	+	−	+	+	−
13	+	+	−	−	+	+	−	−	+	−	+	+	−	+	−
14	+	+	−	−	+	+	−	+	−	+	−	−	+	−	+
15	+	+	−	+	−	−	+	−	+	+	−	−	+	+	−
16	+	+	−	+	−	−	+	+	−	−	+	+	−	−	+

APPENDIX 7C: SOME USEFUL MIXED-LEVEL ORTHOGONAL ARRAYS

Table 7C.1 $OA(12, 3^1 2^4)$ (based on columns 1–5)

Run	1	2	3	4	5	6'	7'
1	0	0	0	0	0	0	0
2	0	0	1	0	1	0	1
3	0	1	0	1	1	1	1
4	0	1	1	1	0	1	0
5	1	0	0	1	1	1	0
6	1	0	1	1	0	1	1
7	1	1	0	0	1	0	0
8	1	1	1	0	0	0	1
9	2	0	0	1	0	0	1
10	2	0	1	0	1	1	0
11	2	1	0	0	0	1	1
12	2	1	1	1	1	0	0

Note: The seven columns 1–5, 6', and 7' form a nearly orthogonal array $OA'(12, 3^1 2^6)$. The only non-orthogonal pairs of columns are $(4, 6')$ and $(5, 7')$.
Source: Wang and Wu (1992).

Table 7C.2 $OA(18, 2^1 3^7)$ (based on columns 1–8) and $OA(18, 6^1 3^6)$ (based on columns 1' and 3–8)

Run	1'	1	2	3	4	5	6	7	8
1	0	0	0	0	0	0	0	0	0
2	0	0	0	1	1	1	1	1	1
3	0	0	0	2	2	2	2	2	2
4	1	0	1	0	0	1	1	2	2
5	1	0	1	1	1	2	2	0	0
6	1	0	1	2	2	0	0	1	1
7	2	0	2	0	1	0	2	1	2
8	2	0	2	1	2	1	0	2	0
9	2	0	2	2	0	2	1	0	1
10	3	1	0	0	2	2	1	1	0
11	3	1	0	1	0	0	2	2	1
12	3	1	0	2	1	1	0	0	2
13	4	1	1	0	1	2	0	2	1
14	4	1	1	1	2	0	1	0	2
15	4	1	1	2	0	1	2	1	0
16	5	1	2	0	2	1	2	0	1
17	5	1	2	1	0	2	0	1	2
18	5	1	2	2	1	0	1	2	0

Note: The interaction between columns 1 and 2, i.e., the column vector $(0\,0\,0\,1\,1\,1\,2\,2\,2\,1\,1\,1\,2\,2\,2\,0\,0\,0)^T$, is orthogonal to columns 3–8.
Source: Masuyama (1957), Taguchi (1987).

Table 7C.3 $OA(20, 5^1 2^8)$ **(based on columns 1–9)**

Run	1	2	3	4	5	6	7	8	9	10'	11'
1	0	0	0	0	0	0	0	0	0	0	0
2	0	0	1	1	1	1	0	1	1	1	0
3	0	1	0	1	0	0	1	1	1	0	1
4	0	1	1	0	1	1	1	0	0	1	1
5	1	0	0	0	1	1	0	1	1	0	0
6	1	0	1	0	0	0	1	1	0	1	1
7	1	1	0	1	0	1	0	0	0	1	1
8	1	1	1	1	1	0	1	0	1	0	0
9	2	0	0	1	0	1	1	1	1	1	1
10	2	0	1	1	0	0	0	0	0	0	0
11	2	1	0	0	1	0	1	0	1	1	0
12	2	1	1	0	1	1	0	1	0	0	1
13	3	0	0	0	1	0	1	1	0	0	1
14	3	0	1	1	1	0	0	0	1	1	1
15	3	1	0	0	0	1	0	0	1	1	0
16	3	1	1	1	0	1	1	1	0	0	0
17	4	0	0	1	1	1	1	0	0	1	0
18	4	0	1	0	0	1	1	0	1	0	1
19	4	1	0	1	1	0	0	1	0	0	1
20	4	1	1	0	0	0	0	1	1	1	0

Note: Columns 1–9, 10', and 11' form a nearly orthogonal array $OA'(20, 5^1 2^{10})$. The only non-orthogonal pairs of columns are (6, 10'), (8, 10'), (9, 10'), (7, 11'), (8, 11'), (9, 11').

Source: Wang and Wu (1992).

Table 7C.4 $OA(24, 3^1 2^{16})$ **(based on columns 1–17)**

Run	1	2	3	4	5	6	7	8	9	10	11	12	13	14	15	16	17	18'	19'
1	0	0	0	0	0	0	0	0	0	0	0	0	0	0	0	0	0	0	0
2	0	0	0	1	0	1	1	0	1	1	0	0	1	0	1	0	1	0	1
3	0	0	1	0	1	1	0	1	0	1	0	0	1	1	0	1	1	1	1
4	0	0	1	1	1	0	0	0	1	0	1	0	1	1	1	1	0	1	0
5	1	0	0	0	1	1	0	0	1	1	1	1	0	0	0	1	1	1	0
6	1	0	0	1	1	0	1	1	0	1	1	0	0	0	1	1	0	1	1
7	1	0	1	0	0	1	1	1	1	0	1	0	0	1	0	0	1	0	0
8	1	0	1	1	0	0	0	1	1	1	0	1	0	1	1	0	0	0	1
9	2	0	0	0	1	0	1	1	1	0	0	1	1	0	0	1	0	0	1
10	2	0	0	1	0	1	0	1	0	0	1	1	1	0	1	0	1	1	0
11	2	0	1	0	0	0	1	0	0	1	1	1	1	1	0	0	0	1	1
12	2	0	1	1	1	1	1	0	0	0	0	1	0	1	1	1	1	0	0
13	0	1	1	1	1	1	1	1	1	1	1	1	1	0	0	0	0	0	0
14	0	1	1	0	1	0	0	1	0	0	1	1	0	0	1	0	1	0	1
15	0	1	0	1	0	0	1	0	1	0	1	1	0	1	0	1	1	1	1
16	0	1	0	0	0	1	1	1	0	1	0	1	0	1	1	1	0	1	0
17	1	1	1	1	0	0	1	1	0	0	0	0	1	0	0	1	1	1	0

Table 7C.4 (*Continued*)

Run	1	2	3	4	5	6	7	8	9	10	11	12	13	14	15	16	17	18'	19'
18	1	1	1	0	0	1	0	0	1	0	0	1	1	0	1	1	0	1	1
19	1	1	0	1	1	0	0	0	0	1	0	1	1	1	0	0	1	0	0
20	1	1	0	0	1	1	1	0	0	0	1	0	1	1	1	0	0	0	1
21	2	1	1	1	0	1	0	0	0	1	1	0	0	0	0	1	0	0	1
22	2	1	1	0	1	0	1	0	1	1	0	0	0	0	1	0	1	1	0
23	2	1	0	1	1	1	0	1	1	0	0	0	0	1	0	0	0	1	1
24	2	1	0	0	0	0	0	1	1	1	1	0	1	1	1	1	1	0	0

Notes: (i) Column $(11 + i)$ is the interaction between column 2 and column i, where $i = 3, 4, 5, 6$. (ii) $OA(24, 3^1 4^1 2^{13})$ can be obtained by replacing columns 2, 3, and 14 by a four-level column with $(0, 0) \to 0$, $(0, 1) \to 1$, $(1, 0) \to 2$, and $(1, 1) \to 3$. (iii) Columns 1–17, 18', and 19' form a nearly orthogonal array $OA'(24, 3^1 2^{18})$. The only non-orthogonal pairs of columns are $(16, 18')$ and $(17, 19')$.

Source: Wang and Wu (1992).

Table 7C.5 $OA(24, 6^1 2^{14})$ **(based on columns 1–15)**

Run	1	2	3	4	5	6	7	8	9	10	11	12	13	14	15	16'	17'
1	0	0	0	0	0	0	0	0	0	0	0	0	0	0	0	0	0
2	1	0	0	1	0	1	1	0	1	1	0	0	1	0	1	0	1
3	1	0	1	0	1	1	0	1	0	1	0	0	1	1	0	1	0
4	0	0	1	1	1	0	0	0	1	0	1	0	1	1	1	1	1
5	2	0	0	0	1	1	0	0	1	1	1	1	0	0	0	1	1
6	3	0	0	1	1	0	1	1	0	1	1	0	0	0	1	0	0
7	3	0	1	0	0	1	1	1	1	0	1	0	0	1	0	1	1
8	2	0	1	1	0	0	0	1	1	1	0	1	0	1	1	0	0
9	4	0	0	0	1	0	1	1	1	0	0	1	1	0	0	0	1
10	5	0	0	1	0	1	0	1	0	0	1	1	1	0	1	1	0
11	5	0	1	0	0	0	1	0	0	1	1	1	1	1	0	0	1
12	4	0	1	1	1	1	1	0	0	0	1	0	1	1	1	1	0
13	0	1	1	1	1	1	1	1	1	1	1	1	1	0	0	0	0
14	1	1	1	0	1	0	0	1	0	0	1	1	0	0	1	0	1
15	1	1	0	1	0	0	1	0	1	0	1	1	0	1	0	1	0
16	0	1	0	0	0	1	1	1	0	1	0	1	0	1	1	1	1
17	2	1	1	1	0	0	1	1	0	0	0	0	1	0	0	1	1
18	3	1	1	0	0	1	0	0	1	0	0	1	1	0	1	0	0
19	3	1	0	1	1	0	0	0	1	0	1	0	1	1	0	1	1
20	2	1	0	0	1	1	1	0	0	0	1	0	1	1	1	0	0
21	4	1	1	1	0	1	0	0	0	1	1	0	0	0	0	0	1
22	5	1	1	0	1	0	1	0	1	1	0	0	0	0	1	1	0
23	5	1	0	1	1	1	0	1	1	0	0	0	0	1	0	0	1
24	4	1	0	0	0	0	0	1	1	1	1	0	1	1	1	1	0

Notes: (i) Column $(11 + i)$ is the interaction between column 2 and column i, where $i = 3, 4$. (ii) Columns 1–15, 16', and 17' form a nearly orthogonal array $OA'(24, 6^1 2^{16})$. Its only non-orthogonal pairs of columns are $(14, 16')$, $(15, 17')$.

Source: Agrawal and Dey (1982), Nguyen (1996).

Table 7C.6 $OA(36, 2^{11}3^{12})$

Run	1	2	3	4	5	6	7	8	9	10	11	12	13	14	15	16	17	18	19	20	21	22	23
1	0	0	0	0	0	0	0	0	0	0	0	0	0	0	0	0	0	0	0	0	0	0	0
2	0	0	0	0	0	0	0	0	0	0	0	1	1	1	1	1	1	1	1	1	1	1	1
3	0	0	0	0	0	0	0	0	0	0	0	2	2	2	2	2	2	2	2	2	2	2	2
4	0	0	0	0	0	1	1	1	1	1	1	0	0	0	0	1	1	1	1	2	2	2	2
5	0	0	0	0	0	1	1	1	1	1	1	1	1	1	1	2	2	2	2	0	0	0	0
6	0	0	0	0	0	1	1	1	1	1	1	2	2	2	2	0	0	0	0	1	1	1	1
7	0	0	1	1	1	0	0	0	1	1	1	0	0	1	2	0	1	2	2	0	1	1	2
8	0	0	1	1	1	0	0	0	1	1	1	1	1	2	0	1	2	0	0	1	2	2	0
9	0	0	1	1	1	0	0	0	1	1	1	2	2	0	1	2	0	1	1	2	0	0	1
10	0	1	0	1	1	0	1	1	0	0	1	0	0	2	1	0	2	1	2	1	0	2	1
11	0	1	0	1	1	0	1	1	0	0	1	1	1	0	2	1	0	2	0	2	1	0	2
12	0	1	0	1	1	0	1	1	0	0	1	2	2	1	0	2	1	0	1	0	2	1	0
13	0	1	1	0	1	1	0	1	0	1	0	0	1	2	0	2	1	0	2	2	1	0	1
14	0	1	1	0	1	1	0	1	0	1	0	1	2	0	1	0	2	1	0	0	2	1	2
15	0	1	1	0	1	1	0	1	0	1	0	2	0	1	2	1	0	2	1	1	0	2	0
16	0	1	1	1	0	1	1	0	1	0	0	0	1	2	1	0	0	2	1	2	2	1	0
17	0	1	1	1	0	1	1	0	1	0	0	1	2	0	2	1	1	0	2	0	0	2	1
18	0	1	1	1	0	1	1	0	1	0	0	2	0	1	0	2	2	1	0	1	1	0	2
19	1	0	1	1	0	0	1	1	0	1	0	0	1	0	2	2	2	0	1	1	0	1	2
20	1	0	1	1	0	0	1	1	0	1	0	1	2	1	0	0	0	1	2	2	1	2	0
21	1	0	1	1	0	0	1	1	0	1	0	2	0	2	1	1	1	2	0	0	2	0	1
22	1	0	1	0	1	1	1	0	0	0	1	0	1	1	2	2	0	1	0	0	2	2	1
23	1	0	1	0	1	1	1	0	0	0	1	1	2	2	0	0	1	2	1	1	0	0	2
24	1	0	1	0	1	1	1	0	0	0	1	2	0	0	1	1	2	0	2	2	1	1	0
25	1	0	0	1	1	1	0	1	1	0	0	0	2	1	0	1	2	2	0	2	0	1	1
26	1	0	0	1	1	1	0	1	1	0	0	1	0	2	1	2	0	0	1	0	1	2	2
27	1	0	0	1	1	1	0	1	1	0	0	2	1	0	2	0	1	1	2	1	2	0	0
28	1	1	1	0	0	0	0	1	1	0	1	0	2	1	1	1	0	0	2	1	2	0	2
29	1	1	1	0	0	0	0	1	1	0	1	1	0	2	2	2	1	1	0	2	0	1	0
30	1	1	1	0	0	0	0	1	1	0	1	2	1	0	0	0	2	2	1	0	1	2	1
31	1	1	0	1	0	1	0	0	0	1	1	0	2	2	2	1	2	1	1	0	1	0	0
32	1	1	0	1	0	1	0	0	0	1	1	1	0	0	0	2	0	2	2	1	2	1	1
33	1	1	0	1	0	1	0	0	0	1	1	2	1	1	1	0	1	0	0	2	0	2	2
34	1	1	0	0	1	0	1	0	1	1	0	0	2	0	1	2	1	2	0	1	1	2	0
35	1	1	0	0	1	0	1	0	1	1	0	1	0	1	2	0	2	0	1	2	2	0	1
36	1	1	0	0	1	0	1	0	1	1	0	2	1	2	0	1	0	1	2	0	0	1	2

Note: The two-level factors (e.g., columns 1 and 2) have smaller numbers of level changes so that this design is more useful when hard-to-change factors have two levels.

Source: Seiden (1954), Taguchi (1987).

Table 7C.7 $OA(36, 2^{11}3^{12})$

Run	1	2	3	4	5	6	7	8	9	10	11	12	13	14	15	16	17	18	19	20	21	22	23
1	0	0	0	1	1	0	0	1	0	2	2	0	1	0	1	0	0	0	1	1	1	0	1
2	0	0	0	0	2	0	2	0	2	0	0	1	1	1	0	1	0	0	0	1	1	1	0
3	0	0	1	0	0	2	1	2	0	0	1	0	0	1	1	0	1	0	0	0	1	1	1
4	0	0	2	2	0	1	0	0	1	1	0	0	1	0	1	1	0	1	0	0	0	1	1
5	0	1	2	2	0	0	1	1	2	0	2	2	1	1	0	1	1	0	1	0	0	0	1
6	0	1	2	1	2	1	2	2	2	2	1	0	1	1	1	0	1	1	0	1	0	0	0
7	0	1	0	0	2	2	0	2	1	1	2	2	0	1	1	1	0	1	1	0	1	0	0
8	0	1	1	2	1	2	2	0	0	2	0	2	0	0	1	1	1	0	1	1	0	1	0
9	0	2	1	2	1	0	0	2	2	1	1	1	0	0	0	1	1	1	0	1	1	0	1
10	0	2	1	0	0	1	2	1	1	2	2	1	1	0	0	0	1	1	1	0	1	1	0
11	0	2	2	1	2	2	1	1	0	1	0	1	0	1	0	0	0	1	1	1	0	1	1
12	0	2	0	1	1	1	1	0	1	0	1	2	0	0	0	0	0	0	0	0	0	0	0
13	1	1	1	2	2	1	1	2	1	0	0	1	1	0	1	0	0	0	1	1	1	0	1
14	1	1	1	1	0	1	0	1	0	1	1	2	1	1	0	1	0	0	0	1	1	1	0
15	1	1	2	1	1	0	2	0	1	1	2	1	0	1	1	0	1	0	0	0	1	1	1
16	1	1	0	0	1	2	1	1	2	2	1	1	1	0	1	1	0	1	0	0	0	1	1
17	1	2	0	0	1	1	2	2	0	1	0	0	1	1	0	1	1	0	1	0	0	0	1
18	1	2	0	2	0	2	0	0	0	0	2	1	1	1	1	0	1	1	0	1	0	0	0
19	1	2	1	1	0	0	1	0	2	2	0	0	0	1	1	1	0	1	1	0	1	0	0
20	1	2	2	0	2	0	0	1	1	0	1	0	0	0	1	1	1	0	1	1	0	1	0
21	1	0	2	0	2	1	1	0	0	2	2	2	0	0	0	1	1	1	0	1	1	0	1
22	1	0	2	1	1	2	0	2	2	0	0	2	1	0	0	0	1	1	1	0	1	1	0
23	1	0	0	2	0	0	2	2	1	2	1	2	0	1	0	0	0	1	1	1	0	1	1
24	1	0	1	2	2	2	2	1	2	1	2	0	0	0	0	0	0	0	0	0	0	0	0
25	2	2	2	0	0	2	2	0	2	1	1	2	1	0	1	0	0	0	1	1	1	0	1
26	2	2	2	2	1	2	1	2	1	2	2	0	1	1	0	1	0	0	0	1	1	1	0
27	2	2	0	2	2	1	0	1	2	2	0	2	0	1	1	0	1	0	0	0	1	1	1
28	2	2	1	1	2	0	2	2	0	0	2	2	1	0	1	1	0	1	0	0	0	1	1
29	2	0	1	1	2	2	0	0	1	2	1	1	1	1	0	1	1	0	1	0	0	0	1
30	2	0	1	0	1	0	1	1	1	1	0	2	1	1	1	0	1	1	0	1	0	0	0
31	2	0	2	2	1	1	2	1	0	0	1	1	0	1	1	1	0	1	1	0	1	0	0
32	2	0	0	1	0	1	1	2	2	1	2	1	0	0	1	1	1	0	1	1	0	1	0
33	2	1	0	1	0	2	2	1	1	0	0	0	0	0	1	1	1	0	1	1	0	1	1
34	2	1	0	2	2	0	1	0	0	1	1	0	1	0	0	0	1	1	1	0	1	1	0
35	2	1	1	0	1	1	0	0	2	0	2	0	1	0	0	0	1	1	1	0	1	1
36	2	1	2	0	0	0	0	2	0	2	0	1	0	0	0	0	0	0	0	0	0	0	0

Note: The three-level factors (e.g., columns 1 and 2) have smaller numbers of level changes so that this design is more useful when hard-to-change factors have three levels.

Source: Seiden (1954), Taguchi (1987).

Table 7C.8 $OA(36, 3^7 6^3)$

Run	1	2	3	4	5	6	7	8	9	10
1	0	0	0	0	0	0	0	0	0	0
2	0	1	1	2	1	0	1	2	2	2
3	0	2	2	2	2	1	2	1	1	0
4	0	3	3	1	1	1	2	0	0	1
5	0	4	4	1	0	2	1	2	1	2
6	0	5	5	0	2	2	0	1	2	1
7	1	0	1	0	0	1	1	1	1	1
8	1	1	2	0	1	2	2	0	0	2
9	1	2	0	1	2	1	0	2	2	2
10	1	3	4	2	0	0	2	1	2	1
11	1	4	5	1	2	0	1	0	0	0
12	1	5	3	2	1	2	0	2	1	0
13	2	0	2	1	1	0	0	2	1	1
14	2	1	0	2	2	2	1	1	0	1
15	2	2	1	1	0	2	2	0	2	0
16	2	3	5	2	0	1	0	2	0	2
17	2	4	3	0	2	0	2	1	1	2
18	2	5	4	0	1	1	1	0	2	0
19	3	0	3	2	2	1	1	0	2	2
20	3	1	5	0	0	1	2	2	1	0
21	3	2	4	0	1	0	0	1	0	2
22	3	3	0	1	1	2	1	1	1	0
23	3	4	2	2	0	2	0	0	2	1
24	3	5	1	1	2	0	2	2	0	1
25	4	0	4	2	2	2	2	2	0	0
26	4	1	3	1	0	0	0	1	2	0
27	4	2	5	2	1	0	1	0	1	1
28	4	3	1	0	2	2	0	0	1	2
29	4	4	0	0	1	1	2	2	2	1
30	4	5	2	1	0	1	1	1	0	2
31	5	0	5	1	1	2	2	1	2	2
32	5	1	4	1	2	1	0	0	1	1
33	5	2	3	0	0	2	1	2	0	1
34	5	3	2	0	2	0	1	2	2	0
35	5	4	1	2	1	1	0	1	0	0
36	5	5	0	2	0	0	2	0	1	2

Source: Finney (1982).

Table 7C.9 $OA(36, 2^8 6^3)$

Run	1	2	3	4	5	6	7	8	9	10	11
1	0	0	0	0	0	0	0	0	0	0	0
2	0	1	1	0	1	1	0	1	1	1	0
3	0	2	2	1	1	0	1	0	0	1	1
4	0	3	3	1	0	0	1	1	1	1	0
5	0	4	4	0	1	1	1	0	1	0	1

Table 7C.9 (*Continued*)

Run	1	2	3	4	5	6	7	8	9	10	11
6	0	5	5	1	0	1	0	1	0	0	1
7	1	0	1	0	0	0	0	0	1	1	1
8	1	1	2	1	0	1	1	0	1	0	0
9	1	2	5	1	1	0	0	1	1	0	1
10	1	3	0	0	1	1	1	1	0	0	1
11	1	4	3	1	1	1	0	0	0	1	0
12	1	5	4	0	0	0	1	1	0	1	0
13	2	0	2	0	0	1	1	1	0	1	1
14	2	1	5	0	1	0	1	1	0	0	0
15	2	2	1	1	0	1	1	0	1	0	0
16	2	3	4	1	1	1	0	0	0	1	0
17	2	4	0	1	0	0	0	1	1	1	1
18	2	5	3	0	1	0	0	0	1	0	1
19	3	0	3	1	1	0	1	1	1	0	0
20	3	1	0	1	1	0	1	0	0	1	1
21	3	2	4	0	0	0	0	0	0	0	0
22	3	3	1	1	0	1	0	1	0	0	1
23	3	4	5	0	0	1	1	0	1	1	1
24	3	5	2	0	1	1	0	1	1	1	0
25	4	0	4	1	1	1	1	1	1	0	1
26	4	1	3	0	0	1	0	0	0	0	1
27	4	2	0	0	1	1	0	1	1	1	0
28	4	3	5	0	0	0	1	0	1	1	0
29	4	4	2	1	0	0	0	1	0	0	0
30	4	5	1	1	1	0	1	0	0	1	1
31	5	0	5	1	1	1	0	0	0	1	0
32	5	1	4	1	0	0	0	1	1	1	1
33	5	2	3	0	0	1	1	1	0	1	1
34	5	3	2	0	1	0	0	0	1	0	1
35	5	4	1	0	1	0	1	1	0	0	0
36	5	5	0	1	0	1	1	0	1	0	0

Source: Finney (1982).

Table 7C.10 $OA(48, 2^{11}4^{12})$

Run	1	2	3	4	5	6	7	8	9	10	11	12	13	14	15	16	17	18	19	20	21	22	23
1	0	0	0	0	0	0	0	0	0	0	0	0	0	0	0	0	0	0	0	0	0	0	0
2	0	0	0	3	3	3	1	1	1	2	2	2	0	0	0	0	0	1	1	1	1	1	1
3	0	0	0	1	1	1	2	2	2	3	3	3	0	0	1	1	1	1	1	1	0	0	0
4	0	1	3	2	3	1	3	2	0	1	0	2	1	0	1	0	1	0	1	0	1	0	1
5	0	1	3	1	2	3	0	3	2	2	1	0	1	0	1	1	0	0	0	1	1	1	0
6	0	1	3	3	1	2	2	0	3	0	2	1	1	0	0	1	1	1	0	0	0	1	1
7	0	3	2	1	0	2	1	3	0	1	2	3	0	1	1	0	1	0	0	1	0	1	1
8	0	3	2	2	1	0	0	1	3	3	1	2	0	1	1	1	0	1	0	0	1	0	1

Table 7C.10　(*Continued*)

Run	1	2	3	4	5	6	7	8	9	10	11	12	13	14	15	16	17	18	19	20	21	22	23
9	0	3	2	0	2	1	3	0	1	2	3	1	0	1	0	1	1	0	1	0	1	1	0
10	0	2	1	3	2	0	1	2	3	1	3	0	1	1	0	1	0	0	1	1	0	0	1
11	0	2	1	0	3	2	3	1	2	0	1	3	1	1	0	0	1	1	0	1	1	0	0
12	0	2	1	2	0	3	2	3	1	3	0	1	1	1	1	0	0	1	1	0	0	1	0
13	1	1	1	1	1	1	1	1	1	1	1	1	0	0	0	0	0	0	0	0	0	0	0
14	1	1	1	2	2	2	0	0	0	3	3	3	0	0	0	0	0	1	1	1	1	1	1
15	1	1	1	0	0	0	3	3	3	2	2	2	0	0	1	1	1	1	1	1	0	0	0
16	1	0	2	3	2	0	2	3	1	0	1	3	1	0	1	0	1	0	1	0	1	0	1
17	1	0	2	0	3	2	1	2	3	3	0	1	1	0	1	1	0	0	0	1	1	1	0
18	1	0	2	2	0	3	3	1	2	1	3	0	1	0	0	1	1	1	0	0	0	1	1
19	1	2	3	0	1	3	0	2	1	0	3	2	0	1	1	0	1	0	0	1	0	1	1
20	1	2	3	3	0	1	1	0	2	2	0	3	0	1	1	1	0	1	0	0	1	0	1
21	1	2	3	1	3	0	2	1	0	3	2	0	0	1	0	1	1	0	1	0	1	1	0
22	1	3	0	2	3	1	0	3	2	0	2	1	1	1	0	1	0	0	1	1	0	0	1
23	1	3	0	1	2	3	2	0	3	1	0	2	1	1	0	0	1	1	0	1	1	0	0
24	1	3	0	3	1	2	3	2	0	2	1	0	1	1	1	0	0	1	1	0	0	1	0
25	2	2	2	2	2	2	2	2	2	2	2	2	0	0	0	0	0	0	0	0	0	0	0
26	2	2	2	1	1	1	3	3	3	0	0	0	0	0	0	0	0	1	1	1	1	1	1
27	2	2	2	3	3	3	0	0	0	1	1	1	0	0	1	1	1	1	1	1	0	0	0
28	2	3	1	0	1	3	1	0	2	3	2	0	1	0	1	0	1	0	1	0	1	0	1
29	2	3	1	3	0	1	2	1	0	0	3	2	1	0	1	1	0	0	0	1	1	1	0
30	2	3	1	1	3	0	0	2	1	2	0	3	1	0	0	1	1	1	0	0	0	1	1
31	2	1	0	3	2	0	3	1	2	3	0	1	0	1	1	0	1	0	0	1	0	1	1
32	2	1	0	0	3	2	2	3	1	1	3	0	0	1	1	1	0	1	0	0	1	0	1
33	2	1	0	2	0	3	1	2	3	0	1	3	0	1	0	1	1	0	1	0	1	1	0
34	2	0	3	1	0	2	3	0	1	3	1	2	1	1	0	1	0	0	1	1	0	0	1
35	2	0	3	2	1	0	1	3	0	2	3	1	1	1	0	0	1	1	0	1	1	0	0
36	2	0	3	0	2	1	0	1	3	1	2	3	1	1	1	0	0	1	1	0	0	1	0
37	3	3	3	3	3	3	3	3	3	3	3	3	0	0	0	0	0	0	0	0	0	0	0
38	3	3	3	0	0	0	2	2	2	1	1	1	0	0	0	0	0	1	1	1	1	1	1
39	3	3	3	2	2	2	1	1	1	0	0	0	0	0	1	1	1	1	1	1	0	0	0
40	3	2	0	1	0	2	0	1	3	2	3	1	1	0	1	0	1	0	1	0	1	0	1
41	3	2	0	2	1	0	3	0	1	1	2	3	1	0	1	1	0	0	0	1	1	1	0
42	3	2	0	0	2	1	1	3	0	3	1	2	1	0	0	1	1	1	0	0	0	1	1
43	3	0	1	2	3	1	2	0	3	2	1	0	0	1	1	0	1	0	0	1	0	1	1
44	3	0	1	1	2	3	3	2	0	0	2	1	0	1	1	1	0	1	0	0	1	0	1
45	3	0	1	3	1	2	0	3	2	1	0	2	0	1	0	1	1	0	1	0	1	1	0
46	3	1	2	0	1	3	2	1	0	2	0	3	1	1	0	1	0	0	1	1	0	0	1
47	3	1	2	3	0	1	0	2	1	3	2	0	1	1	0	0	1	1	0	1	1	0	0
48	3	1	2	1	3	0	1	0	2	0	3	2	1	1	0	0	1	1	0	0	1	0	

Source: Suen (1989).

Table 7C.11 $OA(50, 2^1 5^{11})$ **(based on columns 1 to 12)**

Run	1	2	3	4	5	6	7	8	9	10	11	12	13′	14′
1	0	0	0	0	0	0	0	0	0	0	0	0	0	1
2	0	1	0	1	2	3	4	1	2	3	4	0	1	0
3	0	2	0	2	4	1	3	0	2	4	1	3	0	0
4	0	3	0	3	1	4	2	2	0	3	1	4	1	1
5	0	4	0	4	3	2	1	2	1	0	4	3	1	0
6	1	0	0	1	0	2	2	3	4	4	3	1	1	0
7	1	1	0	2	2	0	1	4	3	1	3	4	0	1
8	1	2	0	3	4	3	0	4	1	2	2	1	1	1
9	1	3	0	4	1	1	4	3	3	2	0	2	0	0
10	1	4	0	0	3	4	3	1	4	1	2	2	0	1
11	0	0	1	1	1	1	1	1	1	1	1	1	0	1
12	0	1	1	2	3	4	0	2	3	4	0	1	1	0
13	0	2	1	3	0	2	4	1	3	0	2	4	0	0
14	0	3	1	4	2	0	3	3	1	4	2	0	1	1
15	0	4	1	0	4	3	2	3	2	1	0	4	1	0
16	1	0	1	2	1	3	3	4	0	0	4	2	1	0
17	1	1	1	3	3	1	2	0	4	2	4	0	0	1
18	1	2	1	4	0	4	1	0	2	3	3	2	1	1
19	1	3	1	0	2	2	0	4	4	3	1	3	0	0
20	1	4	1	1	4	0	4	2	0	2	3	3	0	1
21	0	0	2	2	2	2	2	2	2	2	2	2	0	1
22	0	1	2	3	4	0	1	3	4	0	1	2	1	0
23	0	2	2	4	1	3	0	2	4	1	3	0	0	0
24	0	3	2	0	3	1	4	4	2	0	3	1	1	1
25	0	4	2	1	0	4	3	4	3	2	1	0	1	0
26	1	0	2	3	2	4	4	0	1	1	0	3	1	0
27	1	1	2	4	4	2	3	1	0	3	0	1	0	1
28	1	2	2	0	1	0	2	1	3	4	4	3	1	1
29	1	3	2	1	3	3	1	0	0	4	2	4	0	0
30	1	4	2	2	0	1	0	3	1	3	4	4	0	1
31	0	0	3	3	3	3	3	3	3	3	3	3	0	1
32	0	1	3	4	0	1	2	4	0	1	2	3	1	0
33	0	2	3	0	2	4	1	3	0	2	4	1	0	0
34	0	3	3	1	4	2	0	0	3	1	4	2	1	1
35	0	4	3	2	1	0	4	0	4	3	2	1	1	0
36	1	0	3	4	3	0	0	1	2	2	1	4	1	0
37	1	1	3	0	0	3	4	2	1	4	1	2	0	1
38	1	2	3	1	2	1	3	2	4	0	0	4	1	1
39	1	3	3	2	4	4	2	1	1	0	3	0	0	0
40	1	4	3	3	1	2	1	4	2	4	0	0	0	1
41	0	0	4	4	4	4	4	4	4	4	4	4	0	1
42	0	1	4	0	1	2	3	0	1	2	3	4	1	0
43	0	2	4	1	3	0	2	4	1	3	0	2	0	0
44	0	3	4	2	0	3	1	1	4	2	0	3	1	1
45	0	4	4	3	2	1	0	1	0	4	3	2	1	0

Table 7C.11 (*Continued*)

Run	1	2	3	4	5	6	7	8	9	10	11	12	13'	14'
46	1	0	4	0	4	1	1	2	3	3	2	0	1	0
47	1	1	4	1	1	4	0	3	2	0	2	3	0	1
48	1	2	4	2	3	2	4	3	0	1	1	0	1	1
49	1	3	4	3	0	0	3	2	2	1	4	1	0	0
50	1	4	4	4	2	3	2	0	3	0	1	1	0	1

Note: Columns 1–12, 13', and 14' form a nearly orthogonal array $OA'(50, 2^3 5^{11})$. The only non-orthogonal pairs of columns are (1, 13'), (1, 14'), (13', 14').

Source: Masuyama (1957), Nguyen (1996).

Table 7C.12 $OA(54, 2^1 3^{25})$ (**based on columns 1–26**) **and** $OA(54, 6^1 3^{24})$
(**based on columns 1', 3–26**)

Run	1'	1	2	3	4	5	6	7	8	9	10	11	12	13	14	15	16	17	18	19	20	21	22	23	24	25	26
1	0	0	0	0	0	0	0	0	0	0	0	0	0	0	0	0	0	0	0	0	0	0	0	0	0	0	0
2	1	0	1	0	0	0	0	1	2	0	1	2	1	2	0	1	2	1	2	0	1	2	1	2	0	1	2
3	2	0	2	0	0	0	0	2	1	1	0	2	2	1	1	0	2	2	1	1	0	2	2	1	1	0	2
4	3	1	0	0	0	0	0	2	1	2	1	0	2	1	2	1	0	2	1	2	1	0	2	1	2	1	
5	4	1	1	0	0	0	0	2	0	2	1	1	2	0	2	1	1	2	0	2	1	1	2	0	2	1	1
6	5	1	2	0	0	0	0	1	1	2	2	0	1	1	2	2	0	1	1	2	2	0	1	1	2	2	0
7	0	0	0	0	1	1	2	0	0	0	0	0	1	1	1	1	2	2	2	2	1	1	1	1	1		
8	1	0	1	0	1	1	2	1	2	0	1	2	2	0	1	2	0	0	1	2	0	1	2	0	1	2	0
9	2	0	2	0	1	1	2	2	1	1	0	2	0	2	2	1	0	1	0	0	2	1	0	2	2	1	0
10	3	1	0	0	1	1	2	0	2	1	2	1	1	0	2	0	2	2	1	0	1	0	1	0	2	0	2
11	4	1	1	0	1	1	2	2	0	2	1	1	0	1	0	2	2	1	2	1	0	0	0	1	0	2	2
12	5	1	2	0	1	1	2	2	0	1	1	2	2	0	0	1	0	0	1	2	2	2	0	0	1		
13	0	0	0	0	2	2	1	0	0	0	0	0	2	2	2	2	1	1	1	1	2	2	2	2			
14	1	0	1	0	2	2	1	1	2	0	1	2	0	1	2	0	1	2	0	1	2	0	0	1	2	0	1
15	2	0	2	0	2	2	1	2	1	1	0	2	1	0	0	2	1	0	2	2	1	0	1	0	0	2	1
16	3	1	0	0	2	2	1	0	2	1	2	1	2	1	0	1	0	1	0	2	0	2	2	1	0	1	0
17	4	1	1	0	2	2	1	2	0	2	1	1	1	2	1	0	0	0	1	0	2	2	1	2	1	0	0
18	5	1	2	0	2	2	1	1	1	2	2	0	0	0	1	1	2	2	2	0	0	1	0	0	1	1	2
19	0	0	0	1	0	1	1	1	1	1	1	1	1	1	1	1	1	1	1	1	1	0	0	0	0	0	
20	1	0	1	1	0	1	1	2	0	1	2	0	2	0	1	2	0	2	0	1	2	0	1	2	0	1	2
21	2	0	2	1	0	1	1	0	2	2	1	0	0	2	2	1	0	0	2	2	1	0	2	1	1	0	2
22	3	1	0	1	0	1	1	1	0	2	0	2	1	0	2	0	2	1	0	2	0	2	0	2	1	2	1
23	4	1	1	1	0	1	1	0	1	0	2	2	0	1	0	2	2	0	1	0	2	2	2	0	2	1	1
24	5	1	2	1	0	1	1	2	2	0	0	1	2	2	0	0	1	2	2	0	0	1	1	1	2	2	0
25	0	0	0	1	1	2	0	1	1	1	1	1	2	2	2	2	2	2	0	0	0	0	1	1	1	1	
26	1	0	1	1	1	2	0	2	0	1	2	0	0	1	2	0	1	1	2	0	1	2	0	1	2	0	
27	2	0	2	1	1	2	0	0	2	2	1	0	1	0	0	2	1	2	1	1	0	2	0	2	2	1	0
28	3	1	0	1	1	2	0	1	0	2	0	2	2	1	0	1	0	0	2	1	2	1	1	0	2	0	2
29	4	1	1	1	1	2	0	0	1	0	2	2	1	2	1	0	0	2	0	2	1	1	0	1	0	2	2
30	5	1	2	1	1	2	0	2	2	0	0	1	0	0	1	1	2	1	1	2	2	0	2	2	0	0	1
31	0	0	0	1	2	0	2	1	1	1	1	1	0	0	0	0	2	2	2	2	2	2	2	2	2	2	
32	1	0	1	1	2	0	2	2	0	1	2	0	1	2	0	1	2	0	1	2	0	1	0	1	2	0	1
33	2	0	2	1	2	0	2	0	2	2	1	0	2	1	1	0	2	1	0	0	2	1	1	0	0	2	1
34	3	1	0	1	2	0	2	1	0	2	0	2	0	2	1	2	1	2	1	0	1	0	2	1	0	1	0
35	4	1	1	1	2	0	2	0	1	0	2	2	2	0	2	1	1	1	2	1	0	0	1	2	1	0	0
36	5	1	2	1	2	0	2	2	2	0	0	1	1	1	2	2	0	0	0	1	1	2	0	0	1	1	2
37	0	0	0	2	0	2	2	2	2	2	2	2	2	2	2	2	2	2	2	2	2	2	0	0	0	0	0

Table 7C.12 (*Continued*)

Run	1'	1	2	3	4	5	6	7	8	9	10	11	12	13	14	15	16	17	18	19	20	21	22	23	24	25	26
38	1	0	1	2	0	2	2	0	1	2	0	1	0	1	2	0	1	0	1	2	0	1	1	2	0	1	2
39	2	0	2	2	0	2	2	1	0	0	2	1	1	0	0	2	1	1	0	0	2	1	2	1	1	0	2
40	3	1	0	2	0	2	2	2	1	0	1	0	2	1	0	1	0	2	1	0	1	0	0	2	1	2	1
41	4	1	1	2	0	2	2	1	2	1	0	0	1	2	1	0	0	1	2	1	0	0	2	0	2	1	1
42	5	1	2	2	0	2	2	0	0	1	1	2	0	0	1	1	2	0	0	1	1	2	1	1	2	2	0
43	0	0	0	2	1	0	1	2	2	2	2	2	0	0	0	0	1	1	1	1	1	1	1	1	1	1	1
44	1	0	1	2	1	0	1	0	1	2	0	1	1	2	0	1	2	2	0	1	2	0	2	0	1	2	0
45	2	0	2	2	1	0	1	1	0	0	2	1	2	1	1	0	2	0	2	2	1	0	0	2	2	1	0
46	3	1	0	2	1	0	1	2	1	0	1	0	0	2	1	2	1	1	0	2	0	2	1	0	2	0	2
47	4	1	1	2	1	0	1	1	2	1	0	0	2	0	2	1	1	0	1	0	2	2	0	1	0	2	2
48	5	1	2	2	1	0	1	0	0	1	1	2	1	1	2	2	0	2	2	0	0	1	2	2	0	0	1
49	0	0	0	2	2	1	0	2	2	2	2	2	1	1	1	1	1	0	0	0	0	0	2	2	2	2	2
50	1	0	1	2	2	1	0	0	1	2	0	1	2	0	1	2	0	1	2	0	1	2	0	1	2	0	1
51	2	0	2	2	2	1	0	1	0	0	2	1	0	2	2	1	0	2	1	1	0	2	1	0	0	2	1
52	3	1	0	2	2	1	0	2	1	0	1	0	1	0	2	0	2	0	2	1	2	1	2	1	0	1	0
53	4	1	1	2	2	1	0	1	2	1	0	0	0	1	0	2	2	2	0	2	1	1	1	2	1	0	0
54	5	1	2	2	2	1	0	0	0	1	1	2	2	2	0	0	1	1	1	2	2	0	0	0	1	1	2

Notes: (i) Columns 5 and 6 are the interaction columns of columns 3 and 4 (i.e., if A = column 3 and B = column 4, then AB = column 5 and AB^2 = column 6).
(ii) By replacing columns 3–6 by a nine-level column (using the rule in Table 7.6), $OA(54, 2^1 9^1 3^{21})$ and $OA(54, 6^1 9^1 3^{20})$ are obtained.

Source: Dey (1985).

APPENDIX 7D: SOME USEFUL DIFFERENCE MATRICES

Table 7D.1 $D_{6,6;3}$

0	0	0	0	0	0
0	1	2	0	1	2
0	2	1	1	0	2
0	0	2	1	2	1
0	2	0	2	1	1
0	1	1	2	2	0

Table 7D.2 $D_{12,12;3}$

0	0	0	1	1	0	0	1	0	2	2	0
0	0	0	0	2	0	2	0	2	0	0	1
0	0	1	0	0	2	1	2	0	0	1	0
0	0	2	2	0	1	0	0	1	1	0	0
0	1	2	2	0	0	1	1	2	0	2	2
0	1	2	1	2	1	2	2	2	2	1	0
0	1	0	0	2	2	0	2	1	1	2	2
0	1	1	2	1	2	2	0	0	2	0	2
0	2	1	2	1	0	0	2	2	1	1	1
0	2	1	0	0	1	2	1	1	2	2	1
0	2	2	1	2	2	1	1	0	1	0	1
0	2	0	1	1	1	1	0	1	0	1	2

Table 7D.3 $D_{12,12;4}$

00	00	00	00	00	00	00	00	00	00	00	00
00	00	00	01	01	01	11	11	11	10	10	10
00	00	00	11	11	11	10	10	10	01	01	01
00	11	01	10	01	11	01	10	00	11	00	10
00	11	01	11	10	01	00	01	10	10	11	00
00	11	01	01	11	10	10	00	01	00	10	11
00	01	10	11	00	10	01	00	11	01	11	10
00	01	10	10	11	00	11	01	00	10	01	11
00	01	10	00	10	11	00	11	01	11	10	01
00	10	11	01	10	00	01	11	10	01	00	11
00	10	11	00	01	10	10	01	11	11	01	00
00	10	11	10	00	01	11	10	01	00	11	01

Note: Its four elements are denoted by 00, 01, 10, and 11 with their addition following the modulus 2 addition on each component.

Table 7D.4 $D_{5,5;5}$

0	0	0	0	0
0	1	2	3	4
0	2	3	4	1
0	3	4	1	2
0	4	1	2	3

Table 7D.5 $D_{10,10;5}$

0	4	1	1	4	0	1	4	4	1
0	0	3	4	3	4	1	0	1	4
0	1	0	2	2	1	4	4	1	0
0	2	2	0	1	1	0	1	4	4
0	3	4	3	0	4	4	1	0	1
0	3	2	2	3	0	3	2	2	3
0	4	4	0	2	2	3	0	3	2
0	0	1	3	1	3	2	2	3	0
0	1	3	1	0	3	0	3	2	2
0	2	0	4	4	2	2	3	0	3

Table 7D.6 $D_{12,6;6}$

0	0	0	0	0	0
0	1	3	2	4	0
0	2	0	1	5	2
0	3	1	5	4	2
0	4	3	5	2	1
0	5	5	3	1	1
0	0	2	3	2	3
0	1	2	4	0	5
0	2	5	2	3	4
0	3	4	1	1	4
0	4	1	0	3	5
0	5	4	4	5	3

APPENDIX 7E: SOME USEFUL ORTHOGONAL MAIN-EFFECT PLANS

Table 7E.1 Orthogonal Main-Effect Plans Obtained by Collapsing Factors

OME Plan	Number of Runs	Original OA	Source
$3^1 2^4$	8	$4^1 2^4$	§6.2
$3^2 2^2, 3^3 2^1$	9	3^4	§5.3
$4^{n_1} 3^{n_2} 2^{15-3(n_1+n_2)}$ $(n_1+n_2 = 1,\ldots,5)$	16	4^5	§6.2
$m^1 2^8$ $(m=5,6,7)$	16	$8^1 2^8$	§7.7
$3^5 2^2, 3^6 2^1$	16	$OME(16, 3^7)^a$	table 7E.2
$5^1 3^6, 4^1 3^6$	18	$6^1 3^6$	table 7C.2
$2^{1+n_1} 3^{n_2}$ $(n_1+n_2 = 7)$	18	$2^1 3^7$	table 7C.2
$3^1 2^8$	20	$5^1 2^8$	table 7C.3
$3^1 2^{20}$	24	$4^1 2^{20}$	§7.7
$4^1 3^1 2^{12}$	24	$(12)^1 2^{12}$	b
$5^1 2^{14}$	24	$6^1 2^{14}$	table 7C.5
$5^{n_1} 4^{n_2} 3^{n_3} 2^{n_4}$ $(\Sigma n_i = 6)$	25	5^6	table 6C.1
$3^{n_1} 2^{n_2}$ $(n_1+n_2 = 13, n_1 \geq 3)$	27	3^{13}	table 5A.2
$m^1 3^{n_1} 2^{n_2}$ $(n_1+n_2 = 9, m = 5,\ldots,9)$	27	$9^1 3^9$	c
$m^1 2^{16}$ $(5 \leq m \leq 15)$	32	$(16)^1 2^{16}$	d
$4^{n_1} 3^{n_2} 2^{31-3(n_1+n_2)}$ $(n_1+n_2 = 1,\ldots,9)$	32	$4^9 2^4$	§6.2
$m^1 4^{n_1} 3^{n_2} 2^{3n_3}$ $(\Sigma n_i = 8, m = 5,6,7,8)$	32	$8^1 4^8$	e
$3^{n_1} 2^{11+n_2}$ $(n_1+n_2 = 12, n_1 \geq 1)$	36	$2^{11} 3^{12}$	table 7C.6
$6^{n_1} 5^{n_2} 4^{n_3} 3^{n_4} 2^{n_5}$ $(\Sigma n_i = 10, n_4+n_5 \geq 7)$	36	$3^7 6^3$	table 7C.8
$6^{n_1} 5^{n_2} 4^{n_3} 3^{n_4} 2^{8+n_5}$ $(\Sigma n_i = 3)$	36	$2^8 6^3$	table 7C.9

Note: The orthogonal main-effect (*OME*) plan is obtained by collapsing factors in the orthogonal array (*OA*), whose source is indicated by a section number, e.g., §6.2, or a table number, e.g., Table 7C.2.

[a] This is the only case in which the original plan is an *OME*, not an *OA*.

[b] Obtained by (7.6) using $D_{12, 12; 2}$ and $OA(12, (12)^1)$.

[c] Obtained from $OA(27, 3^{13})$ by the replacement rule in Table 7.6.

[d] Obtained from $OA(32, 2^{31})$ by replacing 15 two-level columns of the form (1, 2, 3, 4, 12, 13, 14, 23, 24, 34, 123, 124, 134, 234, 1234) by a 16-level column.

[e] Obtained from $OA(32, 2^7 4^8)$ in Section 6.2 by replacing its seven two-level columns by an eight-level column.

Table 7E.2   OME(16, 3^7)

Run	1	2	3	4	5	6	7
1	−1	0	0	1	0	1	1
2	1	−1	0	0	1	0	1
3	1	1	−1	0	0	1	0
4	0	1	1	−1	0	0	1
5	1	0	1	1	−1	0	0
6	0	1	0	1	1	−1	0
7	0	0	1	0	1	1	−1
8	0	0	0	0	0	0	0
9	1	0	0	−1	0	−1	−1
10	−1	1	0	0	−1	0	−1
11	−1	−1	1	0	0	−1	0
12	0	−1	−1	1	0	0	−1
13	−1	0	−1	−1	1	0	0
14	0	−1	0	−1	−1	1	0
15	0	0	−1	0	−1	−1	1
16	0	0	0	0	0	0	0

Source: Starks (1964).

REFERENCES

Addelman, S. (1962), "Orthogonal Main-Effect Plans for Asymmetrical Factorial Experiments," *Technometrics*, 4, 21–46.

Agrawal, V. and Dey, A. (1982), "A Note on Orthogonal Main-Effect Plans for Asymmetrical Factorials," *Sankhya, Series B*, 44, 278–282.

Cheng, C.S. (1995), "Some Projection Properties of Orthogonal Arrays," *Annals of Statistics*, 23, 1223–1233.

Dey, A. (1985), *Orthogonal Fractional Factorial Designs*, New York: Halsted Press.

Dey, A. and Mukerjee, R. (1999) *Fractional Factorial Plans*, New York: Wiley.

Finney, D.J. (1982), "Some Enumerations for the 6×6 Latin Squares," *Utilitas Mathematica*, 21A, 137–153.

Hall, M.J., Jr. (1961), "Hadamard Matrices of Order 16," *Research Summary*, 1, 21–26, Pasadena, CA: Jet Propulsion Laboratory.

Hamada, M. and Wu, C.F.J. (1992), "Analysis of Designed Experiments with Complex Aliasing," *Journal of Quality Technology*, 24, 130–137.

Hedayat, A. and Wallis, W.D. (1978), "Hadamard Matrices and Their Applications," *Annals of Statistics*, 6, 1184–1238.

Hedayat, A., Sloane, N., and Stufken, J. (1999), *Orthogonal Arrays: Theory and Applications*, New York: Springer.

Hunter, G.B., Hodi, F.S., and Eager, T.W. (1982), "High-Cycle Fatigue of Weld Repaired Cast Ti-6Al-4V," *Metallurgical Transactions*, 13A, 1589–1594.

John, P.W.M. (1971), *Statistical Design and Analysis of Experiments*, New York: Macmillan.

Lin, D.K.J. and Draper, N.R. (1992), "Projection Properties of Plackett and Burman Designs," *Technometrics*, 34, 423–428.

Masuyama, M. (1957), "On Difference Sets for Constructing Orthogonal Arrays of Index Two and of Strength Two," *Report on Statistical Application and Research, JUSE*, 5, 27–34.

Mukerjee, R. and Wu, C.F.J. (1995), "On the Existence of Saturated and Nearly Saturated Asymmetrical Orthogonal Arrays," *Annals of Statistics*, 23, 2102–2115.

Nguyen, N.K. (1996), "A Note on the Construction of Near-Orthogonal Arrays with Mixed Levels and Economic Run Size," *Technometrics*, 38, 279–283.

Plackett, R.L. and Burman, J.P. (1946), "The Design of Optimum Multifactorial Experiments," *Biometrika*, 33, 305–325.

Seiden, E. (1954), "On the Problem of Construction of Orthogonal Arrays," *Annals of Mathematical Statistics*, 25, 151–156.

Starks, T.H. (1964), "A Note on Small Orthogonal Main-Effect Plans for Factorial Experiments," *Technometrics*, 6, 220–222.

Suen, C. (1989), "A Class of Orthogonal Main-Effect Plans," *Journal of Statistical Planning and Inference*, 21, 391–394.

Sun, D.X. and Wu, C.F.J. (1993), "Statistical Properties of Hadamard Matrices of Order 16," in *Quality Through Engineering Design* (Ed. W. Kuo), New York: Elsevier, pp. 169–179.

Taguchi, G. (1987), *System of Experimental Design*, White Plains: Unipub/Kraus International Publications.

Wang, J.C. and Wu, C.F.J. (1991), "An Approach to the Construction of Asymmetrical Orthogonal Arrays," *Journal of the American Statistical Association*, 86, 450–456.

Wang, J.C. and Wu, C.F.J. (1992), "Nearly Orthogonal Arrays with Mixed Levels and Small Runs," *Technometrics*, 34, 409–422.

Wang, J.C. and Wu, C.F.J. (1995), "A Hidden Projection Property of Plackett-Burman and Related Designs," *Statistica Sinica*, 5, 235–250.

Wu, C.F.J., Zhang, R., and Wang, R. (1992), "Construction of Asymmetrical Orthogonal Arrays of the Type $OA(s^k, s^m(s^{r_1})^{n_1} \cdots (s^{r_i})^{n_i})$," *Statistica Sinica*, 2, 203–219.

CHAPTER 8

Experiments with Complex Aliasing

In many nonregular designs, the aliasing of effects has a complex pattern. In this chapter, we will consider frequentist and Bayesian analysis of experiments based on designs with complex aliasing. Both approaches will allow the main effects as well as some interactions to be studied. Their application to two real experiments identifies important interactions that were missed in previous analyses which focused on main effects only. The Bayesian approach employs a Gibbs sampler to perform an efficient stochastic search of the model space. Both approaches can also be applied to the analysis of supersaturated experiments. A class of supersaturated designs is also given.

8.1 PARTIAL ALIASING OF EFFECTS AND THE ALIAS MATRIX

The cast fatigue experiment and the blood glucose experiment as reported in Section 7.1 are based on designs whose main effects and two-factor interactions have complex aliasing relations. For example, the first experiment is based on the 12-run Plackett-Burman design (which was discussed in Section 7.4). It has only seven factors, with the last four columns (labeled 8–11) unused. Since the design is an orthogonal array, we can denote it notationally by $OA(12, 2^7)$, where 12 denotes the run size, 7 the number of factors, and 2 the number of levels. Because the 11 columns in Table 7.2 are mutually orthogonal, the main effect estimates associated with the columns are uncorrelated and are unbiased estimates of the main effects if there are no interactions. In the presence of two-factor interactions, are the main effect estimates still unbiased? A two-factor interaction between factors i and j can be represented by a column vector obtained by taking the entrywise product of columns i and j. For example, in the cast fatigue experiment, the $B \times C$ interaction is represented by the vector $(-, -, +, +, -, +, +, -, -, -, +, +)^T$. Let

$$\hat{A} = \bar{y}(A = +) - \bar{y}(A = -) \qquad (8.1)$$

denote the main effect of A, where $\bar{y}(A = +)$ denotes the average of observations at the $+$ level of A. (The definitions for $\hat{B}$ to $\hat{G}$ are similar.) In the absence of significant interaction effects, $\hat{A}$ is an unbiased estimate of A. Assuming there are no three and higher-order interactions, it can be shown that

$$E(\hat{A}) = A - \tfrac{1}{3}BC - \tfrac{1}{3}BD - \tfrac{1}{3}BE + \tfrac{1}{3}BF - \tfrac{1}{3}BG$$
$$+ \tfrac{1}{3}CD - \tfrac{1}{3}CE - \tfrac{1}{3}CF + \tfrac{1}{3}CG + \tfrac{1}{3}DE$$
$$+ \tfrac{1}{3}DF - \tfrac{1}{3}DG - \tfrac{1}{3}EF - \tfrac{1}{3}EG - \tfrac{1}{3}FG,$$

$$E(\hat{B}) = B - \tfrac{1}{3}AC - \tfrac{1}{3}AD - \tfrac{1}{3}AE + \tfrac{1}{3}AF - \tfrac{1}{3}AG$$
$$- \tfrac{1}{3}CD - \tfrac{1}{3}CE - \tfrac{1}{3}CF + \tfrac{1}{3}CG + \tfrac{1}{3}DE$$
$$- \tfrac{1}{3}DF - \tfrac{1}{3}DG + \tfrac{1}{3}EF + \tfrac{1}{3}EG - \tfrac{1}{3}FG,$$

$$E(\hat{C}) = C - \tfrac{1}{3}AB + \tfrac{1}{3}AD - \tfrac{1}{3}AE - \tfrac{1}{3}AF + \tfrac{1}{3}AG$$
$$- \tfrac{1}{3}BD - \tfrac{1}{3}BE - \tfrac{1}{3}BF + \tfrac{1}{3}BG - \tfrac{1}{3}DE$$
$$- \tfrac{1}{3}DF - \tfrac{1}{3}DG + \tfrac{1}{3}EF - \tfrac{1}{3}EG + \tfrac{1}{3}FG,$$

$$E(\hat{D}) = D - \tfrac{1}{3}AB + \tfrac{1}{3}AC + \tfrac{1}{3}AE + \tfrac{1}{3}AF - \tfrac{1}{3}AG$$
$$- \tfrac{1}{3}BC + \tfrac{1}{3}BE - \tfrac{1}{3}BF - \tfrac{1}{3}BG - \tfrac{1}{3}CE \qquad (8.2)$$
$$- \tfrac{1}{3}CF - \tfrac{1}{3}CG - \tfrac{1}{3}EF - \tfrac{1}{3}EG + \tfrac{1}{3}FG,$$

$$E(\hat{E}) = E - \tfrac{1}{3}AB - \tfrac{1}{3}AC + \tfrac{1}{3}AD - \tfrac{1}{3}AF - \tfrac{1}{3}AG$$
$$- \tfrac{1}{3}BC + \tfrac{1}{3}BD + \tfrac{1}{3}BF + \tfrac{1}{3}BG - \tfrac{1}{3}CD$$
$$+ \tfrac{1}{3}CF - \tfrac{1}{3}CG - \tfrac{1}{3}DF - \tfrac{1}{3}DG - \tfrac{1}{3}FG,$$

$$E(\hat{F}) = F + \tfrac{1}{3}AB - \tfrac{1}{3}AC + \tfrac{1}{3}AD - \tfrac{1}{3}AE - \tfrac{1}{3}AG$$
$$- \tfrac{1}{3}BC - \tfrac{1}{3}BD + \tfrac{1}{3}BE - \tfrac{1}{3}BG - \tfrac{1}{3}CD$$
$$+ \tfrac{1}{3}CE + \tfrac{1}{3}CG - \tfrac{1}{3}DE + \tfrac{1}{3}DG - \tfrac{1}{3}EG,$$

$$E(\hat{G}) = G - \tfrac{1}{3}AB + \tfrac{1}{3}AC - \tfrac{1}{3}AD - \tfrac{1}{3}AE - \tfrac{1}{3}AF$$
$$+ \tfrac{1}{3}BC - \tfrac{1}{3}BD + \tfrac{1}{3}BE - \tfrac{1}{3}BF - \tfrac{1}{3}CD$$
$$- \tfrac{1}{3}CE + \tfrac{1}{3}CF - \tfrac{1}{3}DE + \tfrac{1}{3}DF - \tfrac{1}{3}EF.$$

A general approach for deriving these formulas will be given later in this section. From (8.2) all the two-factor interactions not involving A are "partial" aliases of A with aliasing coefficient $\tfrac{1}{3}$ or $-\tfrac{1}{3}$. Two effects are said to be **partially aliased** if the absolute value of their aliasing coefficient is strictly between 0 and 1. If the experiment had used all 11 columns, then there would be $\binom{10}{2} = 45$ two-factor interactions that are partially aliased with A. "Complexity" refers to the enormous number of two-factor interactions that are partially aliased with the main effect. For the 12-run design, i.e., the 12×11 matrix in Table 7.2, each main effect has the same property as factor A in that it is partially aliased with the 45 two-factor interactions not involving the main effect and the aliasing coefficients are $\frac{1}{3}$ or $-\frac{1}{3}$. By comparison, the factorial effects in 2^{k-p} or 3^{k-p} designs, as discussed in Chapters 4 and 5, are either orthogonal (i.e., uncorrelated) or fully aliased (i.e., with aliasing coefficient 1 or -1) with each other, and the number of (full) aliases is much smaller. Before we discuss how complex aliasing can affect the analysis of such experiments, we give a general method for computing aliasing coefficients.

Let $\mathbf{y}$ be an $N \times 1$ vector of observations, $\mathbf{X}_1$ and $\mathbf{X}_2$ be $N \times k_1$ and $N \times k_2$ matrices, and $\boldsymbol{\beta}_1$ and $\boldsymbol{\beta}_2$ be $k_1 \times 1$ and $k_2 \times 1$ vectors of parameters. Assume the true model is

$$\mathbf{y} = \mathbf{X}_1\boldsymbol{\beta}_1 + \mathbf{X}_2\boldsymbol{\beta}_2 + \boldsymbol{\epsilon}, \tag{8.3}$$

with $E(\boldsymbol{\epsilon}) = \mathbf{0}$ and $Var(\boldsymbol{\epsilon}) = \sigma^2\mathbf{I}$, where $\mathbf{I}$ is the identity matrix, and the working model is

$$\mathbf{y} = \mathbf{X}_1\boldsymbol{\beta}_1 + \boldsymbol{\epsilon}, \tag{8.4}$$

where $\mathbf{X}_1$ is nonsingular. The primary parameters of interest are $\boldsymbol{\beta}_1$, which are estimated by the ordinary least squares estimator

$$\hat{\boldsymbol{\beta}}_1 = \left(\mathbf{X}_1^T\mathbf{X}_1\right)^{-1}\mathbf{X}_1^T\mathbf{y} \tag{8.5}$$

under the working model (8.4). Then, under the true model (8.3),

$$\begin{aligned} E\left(\hat{\boldsymbol{\beta}}_1\right) &= \left(\mathbf{X}_1^T\mathbf{X}_1\right)^{-1}\mathbf{X}_1^T E(\mathbf{y}) \\ &= \left(\mathbf{X}_1^T\mathbf{X}_1\right)^{-1}\mathbf{X}_1^T(\mathbf{X}_1\boldsymbol{\beta}_1 + \mathbf{X}_2\boldsymbol{\beta}_2) \\ &= \boldsymbol{\beta}_1 + \left(\mathbf{X}_1^T\mathbf{X}_1\right)^{-1}\mathbf{X}_1^T\mathbf{X}_2\boldsymbol{\beta}_2. \end{aligned} \tag{8.6}$$

Thus, $(\mathbf{X}_1^T\mathbf{X}_1)^{-1}\mathbf{X}_1^T\mathbf{X}_2\boldsymbol{\beta}_2$ is the bias of $\hat{\boldsymbol{\beta}}_1$ under the more general model (8.3). We call

$$L = \left(\mathbf{X}_1^T\mathbf{X}_1\right)^{-1}\mathbf{X}_1^T\mathbf{X}_2 \tag{8.7}$$

the **alias matrix** because it provides the aliasing coefficients for the estimate $\hat{\boldsymbol{\beta}}_1$. In the example above, $\mathbf{X}_1$ consists of the 7 columns in Table 7.2, representing the main effects, $\mathbf{X}_2$ the $\binom{7}{2} = 21$ columns representing the two-factor interactions, $\boldsymbol{\beta}_1$ the main effects, and $\boldsymbol{\beta}_2$ the two-factor interactions. If many of the entries of the alias matrix L are nonzero, we call the aliasing **complex**. Since an $A \times B$ interaction is orthogonal to A and to B (assuming A and B are orthogonal), the coefficients in the matrix L between $A \times B$ and A and $A \times B$ and B must be zero. The 12-run Plackett-Burman design has the most extreme aliasing pattern in that all the coefficients between a factor X and a two-factor interaction not involving X are nonzero.

8.2 TRADITIONAL ANALYSIS STRATEGY: SCREENING DESIGN AND MAIN EFFECT ANALYSIS

For experiments based on designs with complex aliasing, it is difficult to disentangle the large number of aliased effects and to interpret their significance. For this reason, such designs were traditionally used for screening factors only, that is, to estimate factor main effects but not their interactions. These designs are therefore referred to as **screening designs**. The validity of the main effect estimates for screening designs depends on the assumption that the interaction effects are negligible. In many practical situations this assumption is often questionable, which suggests the need for strategies that allow interactions to be entertained. Such strategies will be described in the next two sections.

Consider the cast fatigue experiment. If a half-normal plot, as displayed in Figure 8.1, is used to detect effect significance, only factor F appears significant. Its significance can also be confirmed by using Lenth's method. (The details are left as an exercise.) Alternatively we can use regression analysis to identify important factors. The model with F alone has an

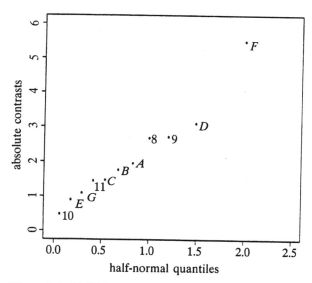

Figure 8.1. Half-Normal Plot, Cast Fatigue Experiment.

$R^2 = 0.45$ and the model with F and D has a higher R^2 value of 0.59. Adding other factors does not substantially improve the model fit or its R^2 value. Thus, the main effect analysis does not succeed in explaining the variation in the data very well.

8.3   SIMPLIFICATION OF COMPLEX ALIASING VIA EFFECT SPARSITY

The complex aliasing pattern in (8.2) can be greatly simplified under the assumption of *effect sparsity* that the number of relatively important two-factor interactions is small. By employing the effect sparsity principle and the hierarchical ordering principle (see Section 3.5), one can assume that only few main effects and even fewer two-factor interactions are relatively important. With this assumption, the complex aliasing pattern of a design can be greatly simplified. This simplification and the partial aliasing of effects make it possible to estimate some interactions. The following hypothetical example illustrates why a few interactions can be estimated in addition to the significant main effects identified in the initial screening of main effects. Suppose that in the cast fatigue experiment only A, F, G, AB, and FG are significant. If the seven main effect estimates [see (8.1)] were obtained assuming no additional interactions, their expectations in (8.2) are simplified as follows,

$$E(\hat{A}) = A - \tfrac{1}{3}FG,$$

$$E(\hat{B}) = -\tfrac{1}{3}FG,$$

$$E(\hat{C}) = -\tfrac{1}{3}AB + \tfrac{1}{3}FG,$$

$$E(\hat{D}) = -\tfrac{1}{3}AB + \tfrac{1}{3}FG, \qquad (8.8)$$

$$E(\hat{E}) = -\tfrac{1}{3}AB - \tfrac{1}{3}FG,$$

$$E(\hat{F}) = F + \tfrac{1}{3}AB,$$

$$E(\hat{G}) = G - \tfrac{1}{3}AB.$$

Effect sparsity reduces the 16 terms in the original equation (8.2) to one or two terms on the right side of equation (8.8). Moreover, it can be shown that A, F, G, AB, and FG are estimable since the corresponding model matrix has full rank. Thus, when the number of significant main effects and interactions is small, the selected main effects and interactions can be estimated, as this example illustrates.

In contrast with designs having complex aliasing, the 2^{k-p} designs have straightforward aliasing patterns; i.e., any two effects are either orthogonal or fully aliased. While the number of aliased effects with any main effect remains small, there is also a disadvantage. Suppose that in the cast fatigue experiment, the FG interaction and A main effect turn out to be important from an experiment using a 12-run Plackett-Burman design. If an 8-run 2_{III}^{7-4} design had been used instead with $A = FG$ as one of its defining relations, then it would have been impossible to disentangle FG from its alias A. On the other hand, both A and FG are estimable from the Plackett-Burman design. Moreover, this problem may not disappear by doubling the run size from 8 to 16. The heat exchanger experiment (given in the exercises) provides such an example. If a 16-run 2_{III}^{10-6} design had been used to study the 10 factors in the experiment with $EG = H$ as one of its defining relations, then it would have been impossible to disentangle EG from H. In contrast, the 12-run Plackett-Burman design that was actually used allows E, G, H, EG, and EH to be estimated simultaneously. Thus, this new perspective of Plackett-Burman designs and other designs with complex aliasing allows their known deficiency to be turned into an advantage.

Further Analysis of Cast Fatigue Experiment

Recall that the main effect analysis in Section 8.2 identified F as significant. By entertaining all the interactions with F, we found a significant polish by final treat (FG) interaction. Adding FG to F doubles the R^2 to 0.89! Moreover, adding D to the model (F, FG) only increases the R^2 to 0.92. Thus, the heat treat factor D appears not to be significant. The improved analysis may also lead to a better understanding of the process and some benefits as discussed below. Based on F and FG, the model for predicted fatigue lifetimes is

$$\hat{y} = 5.7 + 0.458F - 0.459FG, \tag{8.9}$$

where F and G take 1 or -1 corresponding to the "$+$" and "$-$" levels, respectively. By changing G from peening ("$+$") to none ("$-$") with F set at mechanical ("$+$"), the predicted life is

$$5.7 + 0.458 - 0.459(-1) = 5.7 + 0.92 = 6.62,$$

which represents a 16% increase ($= 0.92/5.70$). Similarly, by changing F from chemical ("$-$") to mechanical ("$+$") with G set at no peening ("$-$"), the predicted life is increased by 38%($= 1.84/4.78$). In practice this prediction needs to be verified by a confirmation experiment. Nevertheless, the potential for a dramatic improvement would not have been possible without discovering the FG interaction. If eliminating the peening step does not affect other important characteristics of the castings, it has the additional benefit of simplifying the original process.

Next, we present a general analysis strategy for designs with complex aliasing. Later, we will return to this example to discuss additional reasons for supporting the model (8.9).

8.4 AN ANALYSIS STRATEGY FOR DESIGNS WITH COMPLEX ALIASING

In this section, an analysis strategy, based on Hamada and Wu (1992), is presented for experimental designs with complex aliasing. It is based on two supporting principles, effect sparsity and effect heredity (see Section 3.5). Recall that effect heredity will exclude models that contain an interaction but not any of its parent main effects. In general the strategy works well when effect sparsity and effect heredity hold and the correlations between partially aliased effects (i.e., aliasing coefficients) are small to moderate.

The analysis of experiments with complex aliasing can be viewed as a variable selection problem, but the methods in Section 1.5 need to be modified. For example, backward selection cannot be used here because the full model is not estimable since there are more effects than runs. Also, all-subsets regression may be computationally infeasible because of the large number of effects and may likely identify models which violate effect heredity. The following strategy attempts to circumvent these problems. It has two versions, which are called Method I and Method II. We begin with Method I.

Method I

Step 1. For each factor X, entertain X and all its two-factor interactions XY with other factors. Use a stepwise regression procedure (see Section 1.5) to identify significant effects from the candidate variables and denote the selected model by M_X. Repeat this for each of the factors and then choose the best model. Go to step 2.

Step 2. Use a stepwise regression procedure to identify significant effects among the effects identified in the previous step as well as all the main effects. Go to step 3.

Step 3. Using effect heredity, entertain (i) the effects identified in step 2 and (ii) the two-factor interactions that have at least one component factor appearing among the main effects in (i). Also, consider (iii) interactions suggested by the experimenter. Use a stepwise regression procedure to identify significant effects among the effects in (i)–(iii). Go to step 2.

Step 4. Iterate between steps 2 and 3 until the selected model stops changing.

Since the model search in the procedure is guided by effect heredity, the problem of obtaining uninterpretable models is lessened. This problem cannot be completely avoided, however, because effect heredity is not enforced throughout the procedure. Its occurrence may suggest some limitation of information in the data, which will be discussed in Section 8.4.1.

Effect sparsity suggests that only a few iterations will be required. If all two-factor interactions are entertained indiscriminately in step 3, it is possible to get a good fitting model consisting only of interaction terms and no main effects; hence, nonsensical models may be obtained without assuming effect heredity. Stepwise regression is suggested in steps 2 and 3 for identifying effects. Step 2 is motivated by the possibility of missing main effects in step 1 because of the existence of interactions and complex aliasing.

Some explanation for handling factors with more than two levels is needed. For quantitative factors with three levels, the linear and quadratic effects can be estimated by using the linear and quadratic contrasts A_l and A_q defined in (5.12). For qualitative factors, we can use the same contrasts in (5.12) if A_l can be interpreted as a comparison between the first and third levels and A_q's interpretation as a comparison between the average of the first and third levels and the second level makes sense. Alternatively, for qualitative factors, we can use two of the three contrasts D_{01}, D_{02}, D_{12} defined in (5.16). For interactions between two three-level factors, the four components $l \times l$, $l \times q$, $q \times l$, and $q \times q$ can be entertained.

For four-level factors as discussed in Section 6.4, the contrasts A_1, A_2, A_3 given in (6.5) can be interpreted as linear, quadratic, and cubic effects for the quantitative factors and as comparisons between averages of two groups of levels for qualitative factors. They should be used for main effects and their products, as defined in (6.6) for interactions. Initially, only some of the interaction terms thought to be interpretable or more likely to occur should be included in the model. This will free up degrees of freedom for entertaining other effects.

If a more extensive search of models is desired, the final model obtained by Method I can be viewed as a good starting model. Other models can be entertained by adding or deleting effects one at a time. For example, if (A, B, BC) is chosen by stepwise regression, (A, B, AB) with a slightly smaller R^2 may be preferred if it provides a better interpretation.

The iterative search in Method I can be easily implemented computationally but does not provide a very extensive search for models. Method II provides a more complete search for models subject to the constraints imposed by the principles of effect sparsity and effect heredity.

Method II

Suppose effect sparsity suggests that no meaningful model can have more than h effects. Search over all models that have no more than h effects (plus an intercept term) and satisfy the effect heredity requirement. Choose the best model (or models) according to a sensible model selection criterion (e.g., the C_p criterion given in Section 1.5).

Method II provides a more extensive search than Method I if h is not smaller than the number of effects in the model selected by Method I. Method II is also conceptually very simple to understand.

If an objective estimate of σ^2 is not available so that the C_p criterion is not applicable, the following modified version of Method II may be considered.

Method IIA

Let h_0 be the number of effects in the final model chosen by Method I. Entertain all models that have h_0 effects (plus an intercept term) and satisfy the effect heredity requirement. Choose the one with the smallest residual sum of squares (RSS).

Usually the final model chosen by Method I should be close to the best chosen by Method II, which justifies the choice of h_0 in Method IIA. Once h_0 is fixed, the second term in the C_p criterion becomes a constant and its minimization is equivalent to the minimization of the RSS criterion.

The constraint imposed by effect heredity substantially reduces the model space searched in Method II. Consider, for example, the 12-run Plackett-Burman design with 11 factors and $h = 5$ in Method II. The total number of main effects and two-factor interactions is $11 + \binom{11}{2} = 66$. The total number of models with five terms (and the intercept) is $\binom{66}{5} = 8,936,928$. Among these models, there are 125,202 models satisfying the effect heredity requirement. This is about 1.4% of the total number of models. The enumeration is briefly shown as follows. Let M_i denote the class of models with $5 - i$ main effects and i two-factor interactions, where the interactions must have at least one parent factor among the $5 - i$ main effects. The size of M_i can be enumerated as follows,

$$\#M_0: \quad \binom{11}{5} = 462,$$

$$\#M_1: \quad \binom{11}{4} \times \left(\binom{11}{2} - \binom{7}{2} \right) = 330 \times 34 = 11220,$$

$$\#M_2: \quad \binom{11}{3} \times \binom{27}{2} = 165 \times 351 = 57915,$$

$$\#M_3: \quad \binom{11}{2} \times \binom{19}{3} = 55 \times 969 = 53295,$$

$$\#M_4: \quad \binom{11}{1} \times \binom{10}{4} = 11 \times 210 = 2310.$$

These five numbers sum to 125,202. The savings is even more dramatic for larger problems. Take, for example, the 20-run Plackett-Burman design with

19 factors (see Section 7.4) and $h = 4$ in Method II. The total number of main effects and two-factor interactions is $19 + \binom{19}{2} = 190$. The total number of models with four terms (and the intercept) is $\binom{190}{4} = 52,602,165$. By following the previous method of enumeration, it can be shown that, among these models, only 170,544 (about 0.32%) satisfy effect heredity.

Although effect heredity helps reduce the computational work of Method II, it can still be formidable as the examples in the previous paragraph show. As the number of factors and h increase, the search will become computationally prohibitive. Instead of conducting a search over all possible models (subject to the imposed constraints), an alternative is to use an efficient stochastic search such as the Bayesian variable selection procedure based on Gibbs sampling, which will be considered in the next section. When there is a need to distinguish the analysis strategies in this section from the Bayesian strategy, the former will be referred to as *frequentist* strategies.

We now return to the analysis of the cast fatigue data. Application of the analysis strategy leads to insights beyond that found in a previous analysis.

Cast Fatigue Experiment Revisited

Using Method I, the model (F, FG) is found in step 1 and does not change in further steps. It can be verified that Method IIA with $h_0 = 2$ also identifies the same model. (The details are left as an exercise.). The original analysis by Hunter et al. (1982), which considered main effects only, identified the F and D main effects. They noted a discrepancy between their findings and previous work, namely, the sign of the heat treat (factor D) effect was reversed. They further suggested that the cause was an interaction between heat treat and cooling rate (DE) and claimed that the design did not generate enough information to determine this. Because the DE interaction is orthogonal to both D and E main effects, these three effect estimates are uncorrelated, and therefore the DE interaction does not affect the sign of the factor D main effect. The design's aliasing patterns with some additional calculations can be used to explain this apparent reversal. From Table 8.1, which gives estimates of the 11 main effects and their partial aliases with FG, we observe that $\hat{D}$ $(= -0.516)$ actually estimates $D + \frac{1}{3}FG$. Because the estimated effects in Table 8.1 are twice the least squares estimates (LSEs), $\widehat{FG} = -0.918$ is twice the LSE from equation (8.9), giving $\frac{1}{3}\widehat{FG} = -0.306$. Consequently, the D estimate can be adjusted as follows:

$$\hat{D} - \tfrac{1}{3}\widehat{FG} = -0.516 - (-0.306) = -0.210.$$

While the corrected estimate remains negative, a 95% confidence interval for D, $(-0.526, 0.106) = (-0.210 \pm 0.316)$, shows that a positive D main effect is possible.

Table 8.1 Estimates and Alias Patterns, Cast Fatigue Experiment

Effect	Estimated Effect	Alias Pattern
A	0.326	$A - \frac{1}{3}FG$
B	0.294	$B - \frac{1}{3}FG$
C	-0.246	$C + \frac{1}{3}FG$
D	-0.516	$D + \frac{1}{3}FG$
E	0.150	$E - \frac{1}{3}FG$
F	0.915	F
G	0.183	G
8	0.446	$-\frac{1}{3}FG$
9	0.453	$-\frac{1}{3}FG$
10	0.081	$-\frac{1}{3}FG$
11	-0.242	$+\frac{1}{3}FG$

Note that the magnitude of the estimated effects of A, B, and C (not significant from Figure 8.1) can be explained from Table 8.1 by an FG interaction; that is,

$$\left(-\tfrac{1}{3}\widehat{FG}, -\tfrac{1}{3}\widehat{FG}, \tfrac{1}{3}\widehat{FG}\right) = (0.306, 0.306, -0.306)$$

are close to their estimates, 0.326, 0.294, and -0.246, respectively. Also, the effects $A-E$ and 8–11 have the same sign as their FG aliases, which lend further support to the existence of a significant FG interaction. Interestingly, Hunter et al. (1982) suggested an FG interaction to explain a discrepancy between the estimated effect for factor G and other results obtained from optical micrographs. The statistical analysis presented here provides quantitative support of this engineering conjecture.

In the following, the analysis strategy is illustrated with the blood glucose experiment given in Section 7.1. Again, it succeeds in identifying a better model than what was found using the main effect analysis.

Blood Glucose Experiment

First, consider the main effects analysis. Based on the half-normal plot given in Figure 8.2, effects E_q and F_q are identified. The R^2 value for the model with these two effects is only 0.36!

Now, consider the use of Method I. For the three-level factors, we decompose their main effects into linear and quadratic effects. In step 1, the model consisting of B_l, $(BH)_{lq}$, $(BH)_{qq}$ with an R^2 of 0.85 was identified. With one more term than the main effect model (consisting of E_q and F_q), this model has a substantially higher R^2 value. In subsequent steps, the same

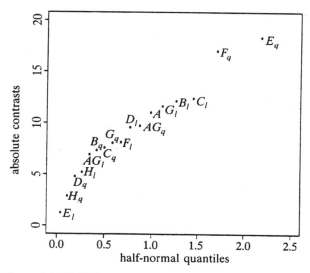

Figure 8.2. Half-Normal Plot, Blood Glucose Experiment.

model was identified whose predicted mean reading is

$$\hat{y} = 96.36 - 2.87B_l - 6.67(BH)_{lq} - 5.45(BH)_{qq}. \qquad (8.10)$$

(In (8.10), the linear and quadratic terms use the unscaled $(-1, 0, 1)$ and $(1, -2, 1)$ vectors so that $(BH)_{lq}$ is formed by multiplying B_l and H_q and $(BH)_{qq}$ is formed by multiplying B_q and H_q.) Since factors B and H are used in calibrating the machine, identifying the $B \times H$ interaction components is important for setting up the machine to give unbiased readings.

These data provide an example of the need for entertaining all the interactions XY with X in step 1 of Method I. If, in step 1, only the main effects were entertained, then E_q and F_q would have been identified. Among the interactions involving E or F, only $(EF)_{ll}$ would have been identified as an additional term. The resulting model consisting of E_q, F_q, and $(EF)_{ll}$ has a smaller R^2 value of 0.68.

8.4.1 Some Limitations

Methods I and II work well when there are only a few significant interactions that are partially aliased with the main effects. Otherwise some incompatible models may be identified, as the next example shows. In the example, both the interactions BC and CD are partially aliased with the main effect A.

Table 8.2   Design Matrix and Response Data, Plackett-Burman Design Example Experiment

Run	A	B	C	D	E	F	G	H	J	K	L	Response
1	+	+	−	+	+	+	−	−	−	+	−	−1.5418
2	+	−	+	+	+	−	−	−	+	−	+	1.3085
3	−	+	+	+	−	−	−	+	−	+	+	1.7652
4	+	+	+	−	−	−	+	−	+	+	−	9.5982
5	+	+	−	−	−	+	−	+	+	−	+	−5.5487
6	+	−	−	−	+	−	+	+	−	+	+	−2.5779
7	−	−	−	+	−	+	+	−	+	+	+	−1.9475
8	−	−	+	−	+	+	−	+	+	+	−	2.1151
9	−	+	−	+	+	−	+	+	+	−	−	−4.8022
10	+	−	+	+	−	+	+	+	−	−	−	2.0413
11	−	+	+	−	+	+	+	−	−	−	+	5.9876
12	−	−	−	−	−	−	−	−	−	−	−	−5.6237

Example

Suppose that a 12-run Plackett-Burman design is used to study a process where the true model is $y = 2A + 4C + 2BC - 2CD + \epsilon$. The model matrix and response data are displayed in Table 8.2; normally distributed error ϵ with mean 0 and standard deviation 0.5 was used to simulate the response data. Method I identifies only C and CD in step 1, whose corresponding R^2 is 0.81. In step 2, by entertaining C, CD, and the other main effects, A might be identified, leading to the model (C, CD, A) with an R^2 of 0.867; however, model (C, CD, BC), which would have been identified next in step 1 if one more effect had been added, has an R^2 of 0.877. There are other models which explain the data as well; the three main effect models (C, G), (C, H), and (C, K) have R^2 of 0.79, 0.78, and 0.77, respectively, compared with 0.81 for (C, CD). Also, the three main effect models (C, G, H), (C, G, K), and (C, H, K) are nearly the same with R^2 of 0.863, 0.844, and 0.844, respectively, compared with model (C, CD, A), whose R^2 is 0.867. Thus, the analysis procedure can find several incompatible models that are equally plausible.

If the analysis finds several incompatible models, it is a strong indication that the information provided by the data and design are limited and no analysis method can distinguish between them. One way to resolve the ambiguity is to conduct additional runs with the goal of discriminating between these models. The strategies in Section 4.4 for generating additional runs can be applied here to discriminate between incompatible models. Other approaches to the design of follow-up experiments can be found in Meyer, Steinberg, and Box (1996) and the ensuing discussions.

8.5 A BAYESIAN VARIABLE SELECTION STRATEGY FOR DESIGNS WITH COMPLEX ALIASING

The two methods considered in the previous section work well if the models under search are restricted to have a small number of terms and the number of candidate terms to choose from is moderate. As the number of candidate terms to choose from and/or the number of terms allowed in the model increases, both methods encounter some problems. The search over the model space is too incomplete for Method I to be effective while the required computations for Method II become prohibitively expensive even with the help of effect heredity to reduce the number of models. An efficient model search procedure is considered in this section that can handle larger and more complex problems than Methods I and II. It is done in a Bayesian framework, which has several advantages. First, the stochastic search is easily implemented by Gibbs sampling (as presented later in this section); the search moves from one model to another in the model space and visits the most likely models the most often. Its computational efficiency is derived from the well-established properties of Gibbs sampling. Second, the Bayesian approach can easily incorporate the effect sparsity and effect heredity principles through the choice of priors. One feature that distinguishes designed experiments from regular regression problems is the emphasis on hierarchical relationships between factorial effects. For example, the significance of a two-factor interaction may be related to the significance of its two parent main effects. This relationship can be easily modeled by a prior. The Bayesian variable selection strategy to be presented here is based on Chipman, Hamada, and Wu (1997).

We begin with a brief description of the Bayesian framework. Let y denote the data, $f(y|\theta)$, their likelihood given the vector of parameters θ, and $\pi(\theta)$ the prior distribution for θ. Then the posterior distribution for θ is

$$\pi(\theta|y) = f(y|\theta)\pi(\theta) / \int f(y|\theta)\pi(\theta)\,d\theta. \tag{8.11}$$

Using the posterior distribution, inference about θ can then be made.

Consider variable selection in the Bayesian framework (George and McCulloch, 1993) for the general linear model:

$$y = X\beta + \epsilon, \tag{8.12}$$

where X is the $N \times (p+1)$ model matrix, β is the $(p+1)$ vector containing the intercept and factorial effects, and $\epsilon \sim MN(0, \sigma^2 I_{N \times N})$. A $p+1$ vector δ of 0's and 1's is introduced to indicate the significance of the effects. The ith entry $\delta_i = 0$ indicates that β_i is small and therefore not significant. On the other hand, $\delta_i = 1$ indicates that β_i is large and therefore significant. Thus, for variable selection, the vector of $2p + 3$ $[= 2(p + 1) + 1]$ parameters is $\theta = (\beta, \delta, \sigma^2)$. Because δ specifies a model by the i's with $\delta_i = 1$, the posterior for δ is of particular interest. In this Bayesian framework, the models with the highest posterior probability are identified as important.

To complete the Bayesian formulation, priors for θ need to be specified. For β, the following *normal mixture* prior is used:

$$\pi(\beta_i | \delta_i) = \begin{cases} N(0, \sigma^2\tau_i^2) & \text{if } \delta_i = 0, \\ N(0, \sigma^2(c_i\tau_i)^2) & \text{if } \delta_i = 1. \end{cases} \tag{8.13}$$

When $\delta_i = 0$, the constant τ_i needs to be specified so that β_i is tightly centered around 0 and therefore does not have a large effect. The constant c_i needs to be chosen with $c_i \gg 1$ to indicate the possibility of a large β_i when $\delta_i = 1$. Thus, the constants τ_i and c_i should be chosen to represent respectively a "small" effect and how many times larger a "large" effect should be. An *inverted gamma* (*IG*) distribution is used for the prior for σ^2. That is,

$$\sigma^2 \sim IG(\nu/2, \nu\lambda/2),$$

whose density is

$$\pi(\sigma^2) \propto (\sigma^2)^{-(\nu/2+1)} \exp\{-\nu\lambda/(2\sigma^2)\}.$$

It can be shown that $\nu\lambda/\sigma^2 \sim \chi_\nu^2$, a chi-squared distribution with ν degrees of freedom.

8.5.1 Bayesian Model Priors

Because δ specifies a model, the prior for δ specifies a prior for that model. First consider using an independence prior,

$$\pi(\delta) = \prod_{i=1}^{p+1} p_i^{\delta_i}(1 - p_i)^{1-\delta_i}, \tag{8.14}$$

where $p_i = Prob(\delta_i = 1)$. The independence prior implies that the significance of a covariate does not depend on whether another covariate is significant or not. In view of the effect heredity principle, the independence prior is not appropriate for modeling interactions. Hierarchical priors (Chipman, 1996) for δ which incorporate the effect heredity principle should be used and are described next.

For three two-level factors A, B, and C, consider three main effects A, B, C and three two-factor interactions AB, AC, and BC. Then $\delta = (\delta_A, \delta_B, \delta_C, \delta_{AB}, \delta_{AC}, \delta_{BC})$. Based on the effect heredity principle, the significance of AB depends on whether the main effects A and B are included in the model or not. If neither main effect is significant, then the interaction seems less plausible and more difficult to explain. This belief can be incorporated into the prior for $\delta = (\delta_A, \ldots, \delta_{BC})$ as follows:

$$Prob(\delta) = Prob(\delta_A)Prob(\delta_B)Prob(\delta_C)$$
$$\times Prob(\delta_{AB}|\delta_A, \delta_B)Prob(\delta_{AC}|\delta_A, \delta_C)Prob(\delta_{BC}|\delta_B, \delta_C). \quad (8.15)$$

In (8.15), two principles are used to obtain the simplified form. First, the *conditional independence principle* assumes that conditional on first-order terms, the second-order terms ($\delta_{AB}, \delta_{AC}, \delta_{BC}$) are independent. Independence is also assumed among the main effects. The *inheritance principle* assumes that the significance of a term depends only on those terms from which it was formed, implying $Prob(\delta_{AB}|\delta_A, \delta_B, \delta_C) = Prob(\delta_{AB}|\delta_A, \delta_B)$.

The exact nature of this dependence on "parent" terms is defined by the components of the joint probability in (8.15). For example, the probability that "*AB* is active" takes on four different values:

$$Prob(\delta_{AB} = 1|\delta_A, \delta_B) = \begin{cases} p_{00} & \text{if } (\delta_A, \delta_B) = (0,0), \\ p_{01} & \text{if } (\delta_A, \delta_B) = (0,1), \\ p_{10} & \text{if } (\delta_A, \delta_B) = (1,0), \\ p_{11} & \text{if } (\delta_A, \delta_B) = (1,1). \end{cases} \quad (8.16)$$

Here, we choose p_{00} small (e.g., 0.01), p_{01} and p_{10} larger (e.g., 0.10), and p_{11} largest (e.g., 0.25). This represents the belief that a two-factor interaction without parents is quite unlikely, one with a single parent is more likely, and one with both parents is most likely. The term **relaxed weak heredity** is used to refer to this prior, and setting $p_{00} = 0$ yields **strict weak heredity**. The case $p_{00} = p_{01} = p_{10} = 0$ is referred to as **strong heredity** in which an interaction is only likely to be significant if both parents are significant. The strong heredity principle (Chipman, 1996) is also called the marginality principle by McCullagh and Nelder (1989). Probabilities of less than 0.5 for both main effects and interactions represent the belief that relatively few terms are active, i.e., effect sparsity holds.

This prior may be generalized to polynomials and interactions involving polynomials. Consider the case, with the fourth-order term A^2B^2, the third-order terms AB^2, A^2B, the second-order terms A^2, AB, B^2, and the first-order terms A and B. We consider the parents of a term to be those terms of the next smallest order which can form the original term when multiplied by a main effect. Here, A^2B^2 has parents A^2B (since multiplication by B produces A^2B^2) and AB^2. Some terms (such as A^2) will have only one parent (e.g., A). We assume that the significance of a term depends only on its parents, an assumption called the *immediate inheritance principle*. The conditional independence principle (which now says that terms of a given order are independent given all lower order terms) is applied as before. The analysis of the blood glucose experiment (to be presented later) will use these polynomial priors.

The covariates corresponding to the polynomials need some explanation. For example, A and A^2, the linear and quadratic polynomials, are defined in terms of x_A, the levels of factor A in their original scale. In particular, A is calculated by centering x_A by its mean; A^2 is calculated by first squaring A (i.e., squaring centered x_A) and then centering. Similarly, AB is the centered product of A (centered x_A) and B (centered x_B). These covariates are used rather than those of the linear-quadratic system because they do not require the levels of the quantitative factors to be evenly spaced. Another reason for using these polynomials is that it is easier to define their priors and to use the inheritance principle. In contrast, the quadratic effect in the linear-quadratic system involves both the x and x^2 terms, which would lead to a messier expression for the probability of the quadratic effect conditional on the linear effect.

For qualitative factors, say three-level factor D, the covariates D_{01}, D_{02}, and D_{12} as defined in (5.17) can be used in variable selection. Other covariates discussed in Section 2.3 for main effects and interaction effects involving qualitative factors can also be used.

8.5.2 Gibbs Sampling

The evaluation of the posterior for the $2p + 3$ dimensional vector $\boldsymbol{\theta}$ in (8.11) for variable selection can be conveniently implemented by using **Gibbs sampling** (Geman and Geman, 1984). Gibbs sampling is a simple *Markov Chain Monte Carlo* (MCMC) technique for drawing samples from a posterior distribution. In general, an MCMC constructs a Markov chain $\boldsymbol{\theta}^1, \boldsymbol{\theta}^2, \ldots$ whose limiting distribution is the posterior (Smith and Roberts, 1993). In Gibbs sampling, *full conditional distributions* are employed, namely $f(\theta_i | \boldsymbol{\theta}_{(-i)}, \mathbf{y})$, where

$$\boldsymbol{\theta}_{(-i)} = \left(\theta_1, \ldots, \theta_{i-1}, \theta_{i+1}, \ldots, \theta_{2p+3} \right).$$

Starting with an arbitrarily chosen value $\boldsymbol{\theta}^0$, $\boldsymbol{\theta}^1$ is obtained by the following sequence of random draws:

$$\theta_1^1 \sim f\left(\theta_1 \big| \boldsymbol{\theta}_{(-1)}^0, \mathbf{y} \right),$$

$$\theta_2^1 \sim f\left(\theta_2 \big| \theta_1^1, \theta_3^0, \ldots, \theta_{2p+3}^0, \mathbf{y} \right),$$

$$\theta_3^1 \sim f\left(\theta_3 \big| \theta_1^1, \theta_2^1, \theta_4^0, \ldots, \theta_{2p+3}^0, \mathbf{y} \right), \qquad (8.17)$$

$$\vdots$$

$$\theta_{2p+3}^1 \sim f\left(\theta_{2p+3} \big| \theta_1^1, \theta_2^1, \ldots, \theta_{2p+2}^1, \mathbf{y} \right).$$

In practice, the Markov chain is run "long enough" until it converges. Cowles and Carlin (1996) provided an overview of convergence and mixing diagnostics for MCMC.

Although the description of Gibbs sampling indicated the use of full conditionals of each of the $2p + 3$ parameters, parameters can be grouped if the full conditionals have a simple form. This occurs for the variable selection problem as formulated above, namely, the joint distribution for $f(\theta, y)$ can be expressed as

$$
\begin{aligned}
f(\theta, y) &= f(y | \beta, \sigma^2, \delta) \pi(\beta, \sigma^2, \delta) \\
&= f(y | \beta, \sigma^2, \delta) \pi(\beta | \sigma^2, \delta) \pi(\sigma^2 | \delta) \pi(\delta) \\
&= f(y | \beta, \sigma^2) \pi(\beta | \sigma^2, \delta) \pi(\sigma^2) \pi(\delta).
\end{aligned}
\tag{8.18}
$$

The third line of (8.18) follows from its second line by assuming that the distribution of y depends only on β and σ^2 and that the prior for σ^2 does not depend on δ. From (8.18), the full conditionals can be derived. For example,

$$
f(\beta | \delta, \sigma^2, y) \propto f(y | \beta, \sigma^2) \pi(\beta | \sigma^2, \delta),
\tag{8.19}
$$

which simplifies to a multivariate normal density. The other full conditionals can be determined similarly so that the Gibbs sampling algorithm can be shown to consist of a multivariate normal draw for $(\beta | \sigma^2, \delta)$, an inverse gamma draw for $(\sigma^2 | \beta, \delta)$, and $p + 1$ Bernoulli draws for $(\delta_i | \beta, \sigma^2, \{\delta_j\}_{j \neq i})$. Their derivations are given in Appendix 8A.

For linear models, β and σ^2 in the Bayesian formulation can be eliminated by an appropriate integration. This leads to a more efficient Gibbs sampler that makes draws only for the δ. See Chipman (1998) for details. Nevertheless, there are several reasons for presenting the version of the Gibbs sampler given above. First, it serves to illustrate the Bayesian computation in Gibbs sampling. Second, the simplified Gibbs sampler as described in Chipman (1998) does not work for nonlinear or nonnormal problems. The version given in this subsection can be extended in a straightforward manner for censored data in Chapter 12 and ordinal data and generalized linear models in Chapter 13.

8.5.3 Choice of Prior Tuning Constants

Several prior tuning constants need to be specified. The normal mixture prior on β has constants τ and c, and the inverse gamma prior for σ^2 has constants ν and λ. Since this methodology is used as a tool rather than strictly for Bayesian philosophical reasons, we view these as tuning constants as well as representations of prior information.

As in Box and Meyer (1986), we use $c_i = 10$, which indicates that an important effect is an order of magnitude larger than an unimportant one. For the choice of τ, we take, as in George and McCulloch (1993),

$$\tau_j = \Delta y / (3 \Delta X_j), \tag{8.20}$$

where Δy represents a "small" change in y and ΔX_j represents a "large" change in X_j. This implies that if $\delta_j = 0$, even a large change in X_j is unlikely to produce anything more than a small change in y. Here ΔX_j is defined as

$$\Delta X_j = \max(X_j) - \min(X_j). \tag{8.21}$$

A value of Δy must still be chosen. When expert opinions are not available,

$$\Delta y = \sqrt{Var(y)} / 5 \tag{8.22}$$

is found to work well in practice, where $Var(y)$ is the sample variance of the response (without any regression). This choice corresponds to the belief that after a model is fitted to the data, σ will be roughly 20% of the unadjusted standard deviation.

Choice of τ can be quite influential for the posterior, since it defines the magnitude of an important coefficient. Several values near the first guess in (8.20) can be used to assess the sensitivity of posteriors to these constants. Sensitivity to τ may be viewed as an advantage of the procedure because it allows the user to choose different models by fine-tuning τ. Unless there is a strong belief or past knowledge to suggest the choice of τ, the appropriateness of a chosen τ may be judged by the models (or posterior model probabilities) it generates. Too few or too many active terms may be considered inappropriate. Often, the experimenter has much better prior knowledge about model size than about the numeric value of a prior tuning constant like τ.

An improper prior (i.e., $\nu = 0$) for σ is not appropriate, since it would allow σ to be close to 0. When the number of covariates exceeds the number of observations, this prior will result in a posterior with very small σ values and many terms in the model. An informative prior on σ can correct this deficiency and is recommended. The assumption that $\sigma \approx \sqrt{Var(y)} / 5$ suggests that a prior on σ with a mean equal to $\sqrt{Var(y)} / 5$ be used. Among these priors, the desirable spread may be attained by selecting a prior with an upper quantile (say 99%) that is near $\sqrt{Var(y)}$. This approach often yields a value of ν near 2, which corresponds to a relatively uninformative prior. The value of λ changes from experiment to experiment because it depends on the scale of the response measurement; consequently, there is no rule of thumb for choosing λ. One approach that seems to work well in practice is to fine-tune λ so that the selected models do not have too many or too few terms.

8.5.4 Blood Glucose Experiment Revisited

We now turn to the analysis of the blood glucose experiment. Because its three-level factors are quantitative, polynomials for main effects and interactions should be used. There are 15 degrees of freedom for the main effects and an additional 98 two-factors interactions to be considered. Because none of them are fully aliased, all 113 candidate variables can be entertained. Linear and quadratic terms will be used throughout, with interactions having four components: linear-by-linear, linear-by-quadratic, quadratic-by-linear, and quadratic-by-quadratic. Choice of $(\nu, \lambda) = (2, 1.289)$ is based on the recommendation given in Section 8.5.3. Values of $c = 10$ and $\tau = 2\tau^*$ with τ^* given in (8.20) are used. The hierarchical priors used are

$$Pr(\delta_A = 1) = 0.25,$$

$$Pr(\delta_{A^2} = 1 \mid \delta_A) = \begin{cases} 0.01 & \text{if } \delta_A = 0, \\ 0.25 & \text{if } \delta_A = 1, \end{cases}$$

$$Pr(\delta_{AB^2} = 1 \mid \delta_{AB}, \delta_{B^2}) = \begin{cases} 0.01 & \text{if } \delta_{AB} = \delta_{B^2} = 0, \\ 0.10 & \text{if } \delta_{AB} \neq \delta_{B^2}, \\ 0.25 & \text{if } \delta_{AB} = \delta_{B^2} = 1. \end{cases}$$

$\hspace{10cm}$ (8.23)

The priors for δ_{AB}, δ_{A^2B}, and $\delta_{A^2B^2}$ are defined similarly as in (8.23) since each effect has two parents.

Because there are only 18 observations and 113 terms from which to choose, it is likely that there will be many models that fit the data well, and probably quite a few parsimonious ones. The hierarchical priors are useful here because they focus attention on good models that also make sense.

The complexity of the problem is apparent in the simulation, which takes much longer to mix sufficiently. The model probabilities obtained from chains of length 10,000 using multiple starts varied too much. A chain of length 50,000 was found sufficient and every fifth draw was kept to save storage space. Each of the stored 10,000 draws from the posterior distribution consists of 113 δ's, 113 corresponding β's, and σ. The model probabilities given in Tables 8.3 and 8.4 for the relaxed weak heredity and strict weak heredity priors, respectively, are the relative frequencies of the models appearing in the stored 10,0000 draws, i.e., the number of draws with δ's corresponding to a model divided by 10,000.

When relaxed weak heredity priors are used, the most probable model contains two terms, BH^2 and B^2H^2. This model clearly violates even weak heredity, so there must be a good reason for its large mass. The 10 most probable models are given in Table 8.3. The prevalence of models that contain both BH^2 and B^2H^2 may suggest a nonlinear relation involving these two variables (B = volume of blood, H = dilution ratio).

**Table 8.3   Posterior Model Probabilities (Relaxed Weak Heredity Prior),
Blood Glucose Experiment**

Model	Prob	R^2
BH^2, B^2H^2	0.183	0.7696
B, BH^2, B^2H^2	0.080	0.8548
B, BH, BH^2, B^2H^2	0.015	0.8601
F, BH^2, B^2H^2	0.014	0.7943
GE, BH^2, B^2H^2	0.013	0.8771
AH^2, BH^2, B^2H^2	0.009	0.8528
G^2D, BH^2, B^2H^2	0.009	0.8517
A, BH^2, B^2H^2	0.008	0.7938
B, B^2, BH^2, B^2H^2	0.008	0.8864
H, BH^2, B^2H^2	0.008	0.7855

**Table 8.4   Posterior Model Probabilities (Strict Weak Heredity Prior),
Blood Glucose Experiment**

Model	Prob	R^2
B, BH, BH^2, B^2H^2	0.146	0.8601
$B, BH, B^2H, BH^2, B^2H^2$	0.034	0.8828
H, H^2, BH^2, B^2H^2	0.033	0.7903
H, BH, BH^2, B^2H^2	0.031	0.7908
F, F^2, DF, D^2F, EF	0.024	0.8835
$H, H^2, AH^2, BH^2, B^2H^2$	0.017	0.8735
B, B^2, BH, BH^2, B^2H^2	0.013	0.8917
B, H, BH, BH^2, B^2H^2	0.013	0.8760
B, H, H^2, BH^2, B^2H^2	0.008	0.8756
E, E^2, CE, EF	0.008	0.6979

Rerunning the algorithm with strict weak heredity [with 0 replacing 0.01 in (8.23)] gives the results in Table 8.4. The best model in Table 8.4 is a superset of the best model in Table 8.3, with the appropriate terms for weak heredity added (namely B and BH). The fit of this model, indicated by an R^2 of 0.86, is quite good. Other models involving EF also appear possible but less likely. In Section 8.4, the model consisting of B_l, $(BH)_{lq}$, and $(BH)_{qq}$ was identified by Method I and is the second most likely under relaxed weak heredity (denoted by BH^2, B^2H^2 in Table 8.3). Under strict weak heredity (Table 8.4), the best model requires the addition of $(BH)_{ll}$ and B_l (denoted by BH and B). The additional information gained from comparing posteriors originating from weak and strict forms of the prior tells us that it is the higher order interactions between B and H that really drive the model and may indicate that caution be exercised in identifying a single "best" model.

The two model probability posteriors (under relaxed and strict weak heredity) are quite different. Although the change in priors for δ from a probability of 0.01 (relaxed) to 0 (strict) seems small, this represents a substantial change. The strict prior states that numerous models are impossible, while the relaxed prior merely downweights them. Some of these downweighted models fit the data well, and so there is a large shift in the posterior when the priors are changed.

If hierarchical priors are replaced with an independence prior, the posterior becomes much less peaked. For example, if the interactions and higher order terms have $Prob(\delta_i = 1) = 0.1$ independently of other model terms, the most probable model is visited only three times in 10,000, or a posterior probability of 0.0003. Moreover, there are numerous models that explain the data well, of which many are nonsensical. This explains why it is not sufficient to set the independent probabilities small.

8.5.5   Other Applications

The Bayesian variable selection strategy has applications beyond the analysis of orthogonal experiments with complex aliasing. Because it does not require that the factor main effects be orthogonal, it is also applicable to non-orthogonal designs which include the class of nearly orthogonal arrays given in Chapter 7. Non-orthogonality can also occur in experiments whose factors are *hard to set*. Because of the difficulties in setting the levels of the factors (e.g., temperature in an old furnace or the exact proportion of ingredients specified), the experiment may not be carried out as planned and the orthogonality of the original design may be destroyed. Even when the factors are not hard to set, miscommunication or careless implementation can lead to choosing wrong factor settings. A real experiment in wood-pulp processing is given in the exercises, which has four hard-to-set factors.

Because the Bayesian strategy does not require that the dimension of β be smaller than the run size of the experiment, it can also be applied to the analysis of experiments based on supersaturated designs, which will be considered in the next section.

8.6   SUPERSATURATED DESIGNS: DESIGN CONSTRUCTION AND ANALYSIS

A **supersaturated design** is a design in which the degrees of freedom for all its main effects and the intercept term exceed the total number of distinct factor level combinations of the design. Consider, for example, the 14×23 design matrix in Table 8.5. The 24 degrees of freedom for the 23 main effects (of two-level factors) and the intercept term exceed the run size 14. Therefore, the data from the experiment cannot be used to estimate all the factor effects *simultaneously*. The main attraction for using supersaturated designs

Table 8.5 Supersaturated Design Matrix and Adhesion Data, Epoxy Experiment

Run	1	2	3	4	5	6	7	8	9	10	11	12	13	14	15	16	17	18	19	20	21	22	23	Adhesion
1	+	+	+	−	−	−	+	+	+	+	+	−	+	−	−	+	−	−	+	−	−	−	+	133
2	+	−	−	−	−	−	+	+	+	−	−	−	+	+	+	−	+	−	−	+	+	−	−	62
3	+	+	−	+	+	−	−	−	−	−	+	−	+	+	+	+	+	−	−	−	−	+	+	45
4	+	+	−	+	−	+	−	−	−	+	+	−	+	−	+	−	+	+	+	−	−	−	−	52
5	−	−	+	+	+	+	−	+	+	−	−	−	+	−	+	+	−	−	+	−	+	+	+	56
6	−	−	+	+	+	+	+	−	+	+	+	−	−	+	+	+	+	+	+	+	+	−	−	47
7	−	−	−	−	+	−	−	+	−	+	−	+	+	+	−	+	+	+	+	+	−	−	+	88
8	−	+	+	−	−	+	−	+	−	+	−	−	−	−	−	−	−	+	−	+	+	+	−	193
9	−	−	−	−	−	+	+	−	−	−	+	+	+	−	−	+	+	+	−	−	−	+	+	32
10	+	+	+	+	−	+	+	+	−	−	−	+	−	+	+	+	−	+	−	+	−	−	+	53
11	−	+	−	+	+	−	−	+	+	−	+	−	−	+	−	−	+	+	−	−	−	+	+	276
12	+	−	−	−	+	+	+	−	+	+	+	+	+	−	−	−	−	−	+	−	+	+	+	145
13	+	+	+	+	+	−	+	−	+	−	−	+	−	−	−	−	−	+	−	+	+	−	+	130
14	−	−	+	−	−	−	−	−	−	−	+	+	+	−	+	−	−	−	−	+	−	+	−	127

is their run size economy, i.e., they can be used to study more factors than what the run size can afford. Obviously they should be used primarily for screening factor main effects, and the limitations as discussed in Section 8.4.1 also apply here. The key question is how such data should be analyzed. Using the effect sparsity principle, one can argue that the number of relatively important factor effects is small (even though their identity is not known until the analysis is done) and thus can all be estimated. The supersaturated nature of the design is not a hindrance in data analysis because both the frequentist and Bayesian analysis strategies build up models by variable selection. Neither approach starts with a model consisting of all the factor effects. The supersaturated nature of the design also implies that not all of its columns are orthogonal to each other.

Supersaturated designs can be constructed by the following simple method (Lin, 1993). By comparing the 14×23 matrix in Table 8.5 and the 28×24 matrix in Table 8.6, the former is obtained from the latter by taking those runs in the larger matrix whose levels for the column labeled "Extra" are +. The 14 runs in Table 8.5 correspond to runs 1, 3, 4, 6, 8–10, 13, 17, 22–25, and 28 in Table 8.6. More generally, the method starts with an $N \times (N-1)$ Plackett-Burman design, where $N = 4k$ is the run size. Then, choose a column of the matrix and use it to split the design into two halves, each corresponding to the + and − levels of the column. Each half is an $(N/2) \times (N-2)$ supersaturated design, where $N/2$ is the run size and $N-2$ is the maximum number of factors the design can study. Because the original Plackett-Burman design has the orthogonality property that for any pair of columns each of the $(+, +)$, $(+, -)$, $(-, +)$, and $(-, -)$ combinations appear equally often (see Section 7.4), each column of the constructed

Table 8.6 Original Epoxy Experiment Based on 28-Run Plackett-Burman Design

Run	1	2	3	4	5	6	7	8	9	10	11	12	13	14	15	16	17	18	19	20	21	22	23	Extra	Adhesion
1	+	+	+	−	−	−	+	+	+	+	+	−	+	−	−	+	−	−	+	−	−	−	−	+	133
2	−	+	−	−	−	−	−	+	+	+	+	−	+	−	−	+	+	+	+	+	−	+	+	−	49
3	+	−	−	−	−	−	−	+	+	+	−	−	−	+	+	+	−	+	−	−	+	+	−	+	62
4	+	+	−	+	+	−	−	−	−	−	+	−	+	+	+	+	+	−	−	−	−	+	+	+	45
5	+	+	−	−	+	+	+	−	−	−	−	−	−	−	−	−	+	+	−	+	+	+	−	−	88
6	+	+	−	+	−	+	−	−	−	+	+	−	+	−	+	−	+	+	+	−	−	−	−	+	52
7	−	−	+	+	+	+	+	+	−	+	−	+	+	−	−	−	+	−	−	−	−	−	−	−	300
8	−	−	+	+	+	+	−	+	+	−	−	−	+	−	+	+	−	−	+	−	+	+	+	+	56
9	−	−	+	+	+	+	−	+	+	+	−	−	+	+	+	+	+	+	+	+	−	−	−	+	47
10	−	−	−	−	+	−	−	+	−	+	−	+	+	+	−	+	+	+	+	+	−	−	+	+	88
11	+	−	+	−	−	+	−	−	+	−	−	−	+	+	+	−	+	+	+	−	−	−	+	−	116
12	−	+	+	+	−	−	−	+	−	−	+	+	−	+	+	+	−	+	+	−	−	+	+	−	83
13	−	+	+	−	−	+	−	+	−	+	−	−	−	−	−	−	−	+	−	+	+	+	+	+	193
14	−	−	−	+	−	−	−	−	+	+	−	−	−	−	+	−	−	−	−	+	−	−	+	−	230
15	+	−	+	−	+	−	+	−	−	+	−	−	−	+	+	+	−	+	+	+	−	−	+	−	51
16	−	+	+	−	+	−	−	−	+	−	+	+	+	−	+	+	−	+	−	+	−	−	−	−	82
17	−	−	−	−	−	+	+	−	−	−	+	+	−	−	+	+	+	−	−	−	−	+	+	+	32
18	+	−	+	+	−	−	−	−	+	−	+	+	+	+	−	+	−	+	+	+	+	+	+	−	58
19	+	−	−	+	+	−	+	+	−	−	+	−	−	−	−	+	−	+	−	−	+	−	−	−	201
20	+	+	+	−	+	+	−	−	+	+	+	+	−	+	+	−	+	−	−	−	+	−	+	−	56
21	−	+	−	+	−	+	+	−	+	−	−	+	+	+	−	−	−	+	+	−	+	−	+	−	97
22	+	+	+	+	−	+	+	+	−	−	−	+	−	+	+	+	−	+	−	+	−	−	+	+	53
23	−	+	−	+	+	−	−	+	+	−	+	−	−	+	−	−	+	+	−	−	−	+	+	+	276
24	+	−	−	−	+	+	+	−	+	+	+	+	+	−	−	−	−	+	−	+	+	+	+	+	145
25	+	+	+	+	+	−	+	−	+	−	−	+	−	−	−	−	+	−	+	+	−	+	−	+	130
26	−	+	−	−	+	+	+	+	−	−	+	−	+	+	+	−	−	−	+	+	−	+	−	−	55
27	+	−	−	+	−	+	−	+	+	+	+	+	−	+	−	+	−	−	+	+	−	+	−	−	160
28	−	−	+	−	−	−	−	−	−	−	+	+	+	−	+	−	−	−	−	+	−	+	−	+	127

supersaturated design has the *balance* property in that − and + appear equally often. It is a simple and general method for constructing supersaturated designs and works as long as a Plackett-Burman design exists. Construction and properties of Plackett-Burman designs can be found in Chapter 7. Other methods for constructing supersaturated designs can be found in Wu (1993), Nguyen (1996), and Tang and Wu (1997).

Returning to the analysis of the experiment in Table 8.5, it was based on a larger experiment (Williams, 1968) given in Table 8.6 whose goal was to develop an epoxide adhesive system for bonding a polyester cord. The 23 factors span the adhesive system (factor 4 is curing agent concentration), processing conditions (factors 12 and 15 are applied stretch in ovens and cord travel in first and second ovens) and adhesion testing (factor 19 is cure temperature). The response is strip adhesion, which needs to be maximized.

Table 8.7 Posterior Model Probabilities, Epoxy Experiment
Using Supersaturated Design

Model	Prob	R^2
4 12 15 19	0.0266	0.955
4 10 12 15 19	0.0259	0.973
4 10 11 12 15 19	0.0158	0.987
4 12 15 19 20	0.0120	0.969
4 11 12 15 19	0.0082	0.966
4 7 10 11 12 15 19	0.0076	0.998
1 4 12 15 19	0.0074	0.970
4 12 13 15 19	0.0071	0.964
1 4 10 12 15 19	0.0066	0.982
1 4 12 15 16 19	0.0064	0.978

As previously remarked, the analysis should be restricted to main effects only. Because supersaturated designs are used primarily as screening designs, this is a reasonable approach. Entertaining the main effects only, a stepwise regression (Section 1.5) leads to the model with factors 4, 12, 15, and 19 and an R^2 value of 0.95.

Now consider an analysis of the epoxy experiment using Bayesian variable selection in Section 8.5. The procedures recommended in Section 8.5.3 were used to choose the regression coefficient priors and the hierarchical priors were used in the variable selection. The values used were $\lambda = 67$, $\nu = 2$, $\tau = 2.31$, $c = 10$, which appear to produce reasonable results with no modifications. The joint posteriors, which are based on every fifth draw from 50,000 cycles of the chain, are given in Table 8.7. The results suggest that factors 4, 12, 15, 19 (and perhaps 10) are active, as was found using stepwise regression. Although the Bayesian methodology did not find substantially different results, it provides a more thorough search and has flexible priors whose advantages can be exploited in larger experiments with more candidate models. Moreover, by using the posterior model probabilities, it provides a quantitative basis for comparing and selecting models.

After identifying the model with factor effects 4, 12, 15, and 19, one may wonder how the remaining nine $(=14-5)$ degrees of freedom can be utilized. In addition to giving an error estimate through the residual sum of squares, it may provide information on the magnitude of the interactions among these four factors. Because the interactions in the supersaturated designs constructed by Lin's method are in general not fully aliased with the main effects, they can be further studied. [The same does not hold for designs constructed by Wu's (1993) method.] By using the effect heredity principle, interactions which have at least one parent term among 4, 12, 15, and 19 can be further entertained by using the frequentist or Bayesian strategies considered in the previous two sections. The details are left as an exercise.

8.7 PRACTICAL SUMMARY

1. Two factorial effects are said to be partially aliased if they are neither fully aliased nor orthogonal. In nonregular designs some effects are partially aliased and the aliasing pattern is complex. Because of the concern with complex aliasing, experiments based on designs with complex aliasing were used primarily for screening purpose, i.e., to screen out unimportant main effects.

2. In many real experiments the number of important interactions is very small. Because many interactions that are not significant can be set to zero, the complex aliasing pattern is greatly simplified, thus making it possible to study some interactions in such experiments. Two strategies for entertaining main effects and interactions are summarized below.

3. A frequentist strategy for analyzing experiments with complex aliasing is given in Section 8.4. It has three versions, I, II, and IIA. Method I is computationally simpler but its model search is far from complete. Method II conducts a more complete model search but is computationally more intensive. Method IIA is a compromise between I and II and can be used as a follow-up to Method I. All three methods are easy to program. If the number of significant interactions is very small, Method I is very effective.

4. A Bayesian strategy for analyzing experiments with complex aliasing is given in Section 8.5. It is based on a prior chosen to reflect effect sparsity and effect heredity. It employs Gibbs sampling to conduct an efficient stochastic model search. Its programming requires more effort but the model search is more efficient than the frequentist approach.

5. The Bayesian strategy can be applied to other problems, including designs with non-orthogonal factors, hard-to-set factors, and supersaturated designs.

6. A supersaturated design is a design in which the degrees of freedom for all its main effects and the intercept term exceed the total number of its distinct factor level combinations. A class of supersaturated designs is given in Section 8.6. The main reason for using supersaturated designs is run size economy. They are used primarily for screening purposes.

7. The frequentist and Bayesian strategies work well when the number of significant interactions that are partially aliased with the main effects is small to moderate. Otherwise the model search can lead to numerous incompatible models. When this happens, it is an indication that the information in the experiment is limited, and a follow-up experiment needs to be conducted to provide more discriminating data.

EXERCISES

1. Work out the complex aliasing relations for the $OA(18, 2^1 3^7)$ in Table 7.4 by using the linear-quadratic system and the orthogonal components system.

2. Let A and B be two factors in a two-level orthogonal array. Prove that both A and B are orthogonal to their interaction $A \times B$.

3. (a) Suppose seven factors A, B, C, D, E, F, G are assigned to the first seven columns of Hall's design of Type V in Appendix 7B. Use the alias matrix in (8.7) to obtain all the main effect and two-factor interaction aliases of each main effect estimate. Comment on the mixed nature of the aliases, i.e., some are full aliases and others are partial aliases.

   (b) Repeat (a) for factors assigned to the first seven columns of Hall's design of Type II.

   (c) By comparing the numbers of partial aliases in (a) and (b), confirm that Type V design is more nonregular than Type II design (which was suggested at the end of Section 7.4).

4. Reanalyze the cast fatigue experiment by following the Bayesian variable selection approach. Use both relaxed weak heredity prior and strictly weak heredity prior. Compare the chosen models with that by the frequentist approach.

5. Consider an experiment to improve a heat exchanger's reliability which studied 10 factors ($A–H, J, K$). Factors B, D, G, and H are concerned with the product design, factors C and E with material selection, and factors A, F, J, and K with the manufacturing process. A unit fails when it develops a tube wall crack. The lifetime is measured in $\times 100$ cycles. Each run was checked after cycles 42, 56.5, 71, 82, 93.5, 105, and 116 and stopped after 128 ($\times$ 100) cycles. The design matrix (based on the 12-run Plackett-Burman design) and interval-censored data (Hamada and Wu, 1991) appear in Table 8.8. Analyze the data using Method I or II and compare the results with those based on a Bayesian analysis. Use the interval midpoints for the interval-censored data and 128 for the right-censored datum as the response values.

6. Phadke et al. (1983) reported on an experiment to optimize contact window formation in an integrated circuit fabrication process, one of the first experiments performed in the U.S. which G. Taguchi was involved with. An $OA(18, 2^1 3^7)$ was used to study nine factors ($A–H, J$) listed in Table 8.9. There were two pairs of related factors, (photoresist viscosity,

Table 8.8   Design Matrix and Lifetime Data, Heat Exchanger Experiment

Run	F	B	A	C	D	E	G	H	J	K	Lifetime
						Factor					
1	−	−	−	−	−	−	−	−	−	−	(93.5, 105)
2	−	−	−	−	−	+	+	+	+	+	(42, 56.5)
3	−	−	+	+	+	−	−	+	+	+	(128, ∞)
4	−	+	−	+	+	+	+	−	−	+	(56.5, 71)
5	−	+	+	−	+	−	+	−	+	−	(56.5, 71)
6	−	+	+	+	−	+	−	+	−	−	(0, 42)
7	+	−	+	+	−	+	+	−	+	−	(56.5, 71)
8	+	−	+	−	+	+	−	−	−	+	(42, 56.5)
9	+	−	−	+	+	−	+	+	−	−	(82, 93.5)
10	+	+	+	−	−	−	+	+	−	+	(82, 93.5)
11	+	+	−	+	−	−	−	−	+	+	(82, 93.5)
12	+	+	−	−	+	+	−	+	+	−	(42, 56.5)

spin speed) and (aperture, exposure time). Their respective sliding levels are given in Tables 8.10 and 8.11. The design matrix is given in Table 8.12. An added complication is that two two-level factors B and D were combined into a single three-level factor BD, where the three levels of this combined factor correspond to $B_0 D_0$, $B_1 D_0$, and $B_0 D_1$ and are denoted by 0, 1, 2 in the column "BD" of Table 8.12. In Table 8.12, for $BD = 0, 2$, viscosity = 204, and spin speed = 2000, 3000, 4000, respectively, for $C = 0, 1, 2$; for $BD = 1$, viscosity = 206, and spin speed = 3000, 4000, 5000, respectively, for $C = 0, 1, 2$. Two responses were measured, pre-etch line width and post-etch line width, whose data appear in Tables 8.13 and 8.14. (Another response, window size, will be given in an exercise in Chapter 12.) There are either 5 or 10 replicates for each run in Tables 8.13 and 8.14. Five replicates represent line widths measured on five locations (center, east, west, north, and south) of one wafer. The next five values in the same row represent corresponding values on the

Table 8.9   Factors and Levels, Window Forming Experiment

Factor	Level		
A. mask dimension (μm)	2	2.5	
B. viscosity	204	206	
C. spin speed (rpm)	low	normal	high
D. bake temperature (°C)	90	105	
E. bake time (min)	20	30	40
F. aperture	1	2	3
G. exposure time (%)	20 over	normal	20 under
H. developing time (s)	30	45	60
J. plasma etch time (min)	14.5	13.2	15.8

Table 8.10 Spin Speed and Viscosity Levels, Window Forming Experiment

| Viscosity | Spin Speed | | |
	Low	Normal	High
204	2000	3000	4000
206	3000	4000	5000

Table 8.11 Exposure and Aperture Levels, Window Forming Experiment

| Aperture | Exposure | | |
	20% Over	Normal	20% Under
1	96	120	144
2	72	90	108
3	40	50	60

Table 8.12 Design Matrix, Window Forming Experiment

| Run | Factor | | | | | | | |
	A	BD	C	E	F	G	H	J
1	0	0	0	0	0	0	0	0
2	0	0	1	1	1	1	1	1
3	0	0	2	2	2	2	2	2
4	0	1	0	0	1	1	2	2
5	0	1	1	1	2	2	0	0
6	0	1	2	2	0	0	1	1
7	0	2	0	1	0	2	1	2
8	0	2	1	2	1	0	2	0
9	0	2	2	0	2	1	0	1
10	1	0	0	2	2	1	1	0
11	1	0	1	0	0	2	2	1
12	1	0	2	1	1	0	0	2
13	1	1	0	1	2	0	2	1
14	1	1	1	2	0	1	0	2
15	1	1	2	0	1	2	1	0
16	1	2	0	2	1	2	0	1
17	1	2	1	0	2	0	1	2
18	1	2	2	1	0	1	2	0

second wafer. For runs 5, 15, and 18, one wafer was broken, which explains why only five values are recorded. The analysis of this experiment is complicated by the presence of two pairs of related factors, the complex aliasing of the 18-run array and the creation of a three-level factor BD from two two-level factors. Use the tools you have learned in this chapter and Chapter 6 to conduct a comprehensive analysis of the data. Identify factor settings to minimize the dispersion of line width.

Table 8.13   Pre-Etch Line-Width Data, Window Forming Experiment

Run	Line Width									
1	2.43	2.52	2.63	2.52	2.50	2.36	2.50	2.62	2.43	2.49
2	2.76	2.66	2.74	2.60	2.53	2.66	2.73	2.95	2.57	2.64
3	2.82	2.71	2.78	2.55	2.36	2.76	2.67	2.90	2.62	2.43
4	2.02	2.06	2.21	1.98	2.13	1.85	1.66	2.07	1.81	1.83
5	1.87	1.78	2.07	1.80	1.83					
6	2.51	2.56	2.55	2.45	2.53	2.68	2.60	2.85	2.55	2.56
7	1.99	1.99	2.11	1.99	2.00	1.96	2.20	2.04	2.01	2.03
8	3.15	3.44	3.67	3.09	3.06	3.27	3.29	3.49	3.02	3.19
9	3.00	2.91	3.07	2.66	2.74	2.73	2.79	3.00	2.69	2.70
10	2.69	2.50	2.51	2.46	2.40	2.75	2.73	2.75	2.78	3.03
11	3.20	3.19	3.32	3.20	3.15	3.07	3.14	3.14	3.13	3.12
12	3.21	3.32	3.33	3.23	3.10	3.48	3.44	3.49	3.25	3.38
13	2.60	2.56	2.62	2.55	2.56	2.53	2.49	2.79	2.50	2.56
14	2.18	2.20	2.45	2.22	2.32	2.33	2.20	2.41	2.37	2.38
15	2.45	2.50	2.51	2.43	2.43					
16	2.67	2.53	2.72	2.70	2.60	2.76	2.67	2.73	2.69	2.60
17	3.31	3.30	3.44	3.12	3.14	3.12	2.97	3.18	3.03	2.95
18	3.46	3.49	3.50	3.45	3.57					

Table 8.14   Post-Etch Line-Width Data, Window Forming Experiment

Run	Line Width									
1	2.95	2.74	2.85	2.76	2.70	3.03	2.95	2.75	2.82	2.85
2	3.05	3.18	3.20	3.16	3.06	3.25	3.15	3.09	3.11	3.16
3	3.69	3.57	3.78	3.55	3.40	3.92	3.62	3.71	3.71	3.53
4	2.68	2.62	2.90	2.45	2.70	2.29	2.31	2.77	2.46	2.49
5	1.75	1.15	2.07	2.12	1.53					
6	3.42	2.98	3.22	3.13	3.17	3.34	3.21	3.23	3.25	3.28
7	2.62	2.49	2.53	2.41	2.51	2.76	2.94	2.68	2.62	2.51
8	4.13	4.38	4.41	4.03	4.03	4.00	4.02	4.18	3.92	3.91
9	3.94	3.82	3.84	3.57	3.71	3.44	3.30	3.41	3.28	3.20
10	3.17	2.85	2.84	3.06	2.94	3.70	3.34	3.45	3.41	3.29
11	4.01	3.91	3.92	3.80	3.90	3.67	3.31	2.86	3.41	3.23
12	4.04	3.80	4.08	3.81	3.94	4.51	4.37	4.45	4.24	4.48
13	3.40	3.12	3.11	3.25	3.06	3.22	3.03	2.89	2.92	2.98
14	3.18	3.03	3.40	3.17	3.32	3.18	2.83	3.17	3.07	3.02
15	2.86	2.46	2.30	2.60	2.55					
16	2.85	2.14	1.22	2.80	3.03	3.40	2.97	2.96	2.87	2.88
17	4.06	3.87	3.90	3.94	3.87	4.02	3.49	3.51	3.69	3.47
18	4.49	4.28	4.34	4.39	4.25					

7. Confirm the significance of the factor F effect in the main effect analysis of the cast fatigue experiment by using Lenth's method and finding its p value.

8. Show that Method I and Method IIA identify the same model for the cast fatigue experiment.

9. Apply Method II to the cast fatigue experiment and the blood glucose experiment with $h = 3, 4, 5$. Compare the models identified by Methods I and II.

10. By following the enumeration method in Section 8.4, show that, for the 20-run Plackett-Burman design and $h = 4$, the total number of models satisfying effect heredity is 170,544.

11. Show that for the 12-run Plackett-Burman design and $h = 4$, the total number of models satisfying effect heredity is 15,510, which is 2.15% of the total number of models, $\binom{66}{4} = 720,720$.

12. Reanalyze the epoxy experiment in Table 8.5 by including two-factor interactions as candidate variables. Use Method I, Method IIA, and the Bayesian variable selection strategy. Compare the results from using the three methods. Is any significant interaction uncovered?

13. The other 14 runs (with "Extra"$= -$) in Table 8.6 can be treated as a 14×23 supersaturated design. Analyze these data by using the methods considered in Section 8.5 and in the previous exercise. Is the conclusion different from that based on Table 8.5.

14. Analyze the experiment based on the 28×24 Plackett-Burman design in Table 8.6 and compare the results with those based on its two half designs as previously analyzed.

15. Silicon nitride (Si_3N_4) ceramic is a strong, wear-resistant, oxidation-resistant, and thermal-shock-resistant material widely used in high-temperature structural applications such as cutting tools, bearings, and heat engine components. An experiment (Yuan, 1998) was performed to find settings that maximize the strength of Si_3N_4. One two-level factor and six three-level factors were chosen for study: heating schedule (A), where "2 steps" includes a presintering stage over "1 step," the total amount of four sintering additives CaO, Y_2O_3, MgO, and Al_2O_3 (B), the amount of CaO (C), the ratio of the mol% of Y_2O_3 to the ratio of the mol% of MgO (D), the ratio of the mol% of Y_2O_3 to the mol% of Al_2O_3 (E), sintering temperature (F), and sintering time (G). The levels of these factors are given in Table 8.15. An $OA(18, 2^1 3^6)$ was used for this

Table 8.15 Levels of Factors, Ceramics Experiment

Level	A	B	C	D	E	F	G
0	2 steps	14 mol%	0.0 mol%	1:1	2:1	1800°C	1 h
1	1 step	16 mol%	1.0 mol%	1:2	1:1	1850°C	2 h
2		18 mol%	2.0 mol%	1:6	1:4	1900°C	3 h

Table 8.16 Design Matrix and Strength Data, Ceramics Experiment

Run	A	B	C	D	E	F	G	Strength				
1	0	1	1	0	2	0	2	996.8	783.6	796.9		
2	0	1	0	1	1	2	0	843.8	816.2	714.3	824.4	
3	0	1	2	2	0	1	1	647.1	667.9	534.3	617.7	
4	0	2	1	0	1	2	1	616.3	552.3	552.6	596.0	
5	0	2	0	1	0	1	2	517.8	526.1	498.1	499.5	
6	0	2	2	2	2	0	0	1002.0	1097.0	882.9	940.1	
7	0	0	1	1	2	1	0	806.5	933.5	964.9	1046.0	
8	0	0	0	2	1	0	1	801.5	803.2	846.2	756.4	
9	0	0	2	0	0	2	2	739.2	863.3	797.0	929.6	
10	1	1	1	2	0	2	0	615.0	627.5	583.9	597.1	563.9
11	1	1	0	0	2	1	1	795.9	854.0	937.0	999.2	724.8
12	1	1	2	1	1	0	2	850.9	921.8	990.6	943.5	840.9
13	1	2	1	1	0	0	1	513.0	665.9	718.9	646.4	
14	1	2	0	2	2	2	2	831.3	981.4	912.5	950.7	987.3
15	1	2	2	0	1	1	0	806.1	908.1	627.6	855.0	
16	1	0	1	2	1	1	2	727.3	643.9	584.0	643.4	602.1
17	1	0	0	0	0	0	0	836.8	716.3	862.9	796.2	
18	1	0	2	1	2	2	1	1001.0	937.6	955.3	995.8	1009.0

experiment. Each run has three to five replicates. The design matrix and measured strength data are given in Table 8.16. Analyze the data by using Method I and the Bayesian variable selection strategy. Based on the model identified in your analysis, recommend optimal factor settings for strength maximization.

16. Consider an experiment (Chipman, Hamada, and Wu, 1997) involving 11 factors in a wood pulp production process. The process consisted of chemical and mechanical treatments; factors $A-G$ involved the chemical treatment, and factors $H-K$ involved the mechanical treatment. The planned experiment was a 20-run Plackett-Burman design with a center point replicated twice (i.e., the total run size was 22). Data from only 19 runs were available since difficulties were encountered in performing

Table 8.17   Design Matrix and Response Data, Wood Pulp Experiment

Run	A	B	C	D	E	F	G	H	I	J	K	Burst Index
						Factor						
1	1	−1	1	1	−0.33	−1	1	1	0.74	1	−0.89	1.61
2	−1	−1	1	1	1.63	1	−1	1	−1.02	1	−0.76	1.97
3	−1	1	−1	−1	−1.04	−1	1	1	−0.55	1	1.85	1.48
4	1	1	−1	1	1.82	−1	−1	1	0.35	1	1.03	0.55
5	1	1	−1	−1	0.31	1	1	1	−0.67	1	−1.08	0.55
6	−1	−1	−1	−1	1.00	1	−1	1	0.75	−1	−0.87	1.59
7	1	1	−1	1	−0.57	1	−1	−1	−1.19	−1	2.26	1.64
8	−1	−1	−1	−1	−0.32	−1	−1	−1	−1.16	−1	−0.79	1.50
9	1	1	1	−1	1.69	−1	1	−1	−1.20	−1	−0.87	1.97
10	−1	−1	−1	1	1.32	−1	1	1	−1.17	−1	2.17	1.67
11	1	−1	1	−1	1.57	−1	−1	−1	−1.41	1	1.12	1.52
12	−1	1	1	1	1.61	−1	1	−1	−0.77	−1	−0.40	4.37
13	−1	1	1	−1	−1.06	1	1	1	0.45	−1	2.32	2.38
14	1	1	1	1	−0.76	1	−1	1	−0.62	−1	−0.83	2.04
15	−1	−1	1	1	−0.33	1	1	−1	−1.69	1	−1.38	2.24
16	1	−1	−1	1	1.36	1	1	−1	3.35	−1	0.66	1.76
17	−1	1	−1	1	−0.23	−1	−1	−1	1.45	1	−0.65	1.73
18	0	0	0	0	−0.10	0	0	0	−0.09	0	0.39	1.74
19	0	0	0	0	0.05	0	0	0	0.58	0	0.16	1.76

three of the runs from the Plackett-Burman design portion. Several of the factors were hard to set, notably, factors E, I, and K (wood-to-liquid ratio, slurry concentrations at two stages). The planned levels were ±1 in runs 1–17 and 0 in runs 18–19, but the actual values of factors E, I, and K were different. The actual factor levels and burst index (one of the observed quality characteristics) are given in Table 8.17.

(a) Analyze the experiment by using Method I. Rank the selected models according to the C_p criterion.

(b) Analyze the experiment by using the Bayesian variable selection strategy with weak heredity prior. Rank the selected models according to the posterior model probability.

(c) Compare the results in (a) and (b).

APPENDIX 8A: FURTHER DETAILS FOR THE FULL CONDITIONAL DISTRIBUTIONS

In this appendix, we provide more details for the full conditional distributions used in Bayesian variable selection. Recall from (8.18) that $f(\boldsymbol{\theta}, \mathbf{y})$ can be expressed as

$$f(\boldsymbol{\theta}, \mathbf{y}) = f(\mathbf{y}|\boldsymbol{\beta}, \sigma^2)\pi(\boldsymbol{\beta}|\sigma^2, \boldsymbol{\delta})\pi(\sigma^2)\pi(\boldsymbol{\delta}). \qquad (8.24)$$

Now consider the forms of the likelihood and priors used in (8.24). From $(\mathbf{y}|\boldsymbol{\beta}, \sigma^2) \sim MN(\mathbf{X}\boldsymbol{\beta}, \sigma^2 \mathbf{I})$, where MN stands for multivariate normal, the likelihood $f(\mathbf{y}|\boldsymbol{\beta}, \sigma^2)$ has the form

$$f(\mathbf{y}|\boldsymbol{\beta}, \sigma^2) \propto \sigma^{-N} \exp\{-(\mathbf{y} - \mathbf{X}\boldsymbol{\beta})^T(\mathbf{y} - \mathbf{X}\boldsymbol{\beta})/(2\sigma^2)\}, \qquad (8.25)$$

where N is the number of observations. The priors $\pi(\cdot)$ used in Section 8.5.1 are summarized next. The normal mixture prior for $(\boldsymbol{\beta}|\sigma^2, \boldsymbol{\delta})$ in (8.13) can be expressed as

$$\pi(\boldsymbol{\beta}|\sigma^2, \boldsymbol{\delta}) \propto \sigma^{-(p+1)} \exp\{-\boldsymbol{\beta}^T \mathbf{D}_\delta^{-1} \boldsymbol{\beta}/(2\sigma^2)\}, \qquad (8.26)$$

where $\mathbf{D}_\delta$ is the diagonal matrix

$$diag\left\{\left(c_1^{\delta_1}\tau_1\right)^2, \ldots, \left(c_{p+1}^{\delta_{p+1}}\tau_{p+1}\right)^2\right\}.$$

That is, $(\boldsymbol{\beta}|\sigma^2, \boldsymbol{\delta}) \sim MN(\mathbf{0}, \sigma^2 \mathbf{D}_\delta)$. The inverse gamma prior for σ^2 has the form

$$\pi(\sigma^2) \propto (\sigma^2)^{-(\nu/2+1)} \exp\{-\nu\lambda/(2\sigma^2)\}, \qquad (8.27)$$

i.e., $\sigma^2 \sim IG(\nu/2, \nu\lambda/2)$. The prior for $\boldsymbol{\delta}$, $\pi(\boldsymbol{\delta})$, is a multinomial on 2^{p+1} values.

The full conditional distributions for the Gibbs algorithm are easily determined from the joint distribution in (8.24). The form of the full conditional distribution for $\boldsymbol{\beta}$ can be obtained by dropping the terms in (8.24) not involving $\boldsymbol{\beta}$:

$$f(\boldsymbol{\beta}|\sigma^2, \boldsymbol{\delta}, \mathbf{y}) \propto f(\mathbf{y}|\boldsymbol{\beta}, \sigma^2)\pi(\boldsymbol{\beta}|\sigma^2, \boldsymbol{\delta}). \qquad (8.28)$$

Since both the likelihood and $\boldsymbol{\beta}$ prior are multivariate normal, it can be shown by combining (8.25) and (8.26) that (8.28) has the following form:

$$f(\boldsymbol{\beta}|\sigma^2, \boldsymbol{\delta}, \mathbf{y}) \propto \exp\{-(1/2\sigma^2)[\boldsymbol{\beta}^T(\mathbf{D}_\delta^{-1} + \mathbf{X}^T\mathbf{X})\boldsymbol{\beta} - \boldsymbol{\beta}^T\mathbf{X}^T\mathbf{y} - \mathbf{y}^T\mathbf{X}\boldsymbol{\beta}]\}. \qquad (8.29)$$

By simplifying (8.29), it can be seen that

$$(\boldsymbol{\beta}|\sigma^2, \boldsymbol{\delta}, \mathbf{y}) \sim MN(\sigma^{-2}\mathbf{A}_\delta \mathbf{X}^T\mathbf{y}, \mathbf{A}_\delta), \qquad (8.30)$$

where

$$\mathbf{A}_\delta = \sigma^2(\mathbf{X}^T\mathbf{X} + \mathbf{D}_\delta^{-1})^{-1}.$$

By dropping the $\pi(\boldsymbol{\delta})$ term in (8.24), the form of the full conditional distribution for σ^2 is

$$f(\sigma^2|\boldsymbol{\beta},\boldsymbol{\delta},\mathbf{y}) \propto f(\mathbf{y}|\boldsymbol{\beta},\sigma^2)\pi(\boldsymbol{\beta}|\sigma^2,\boldsymbol{\delta})\pi(\sigma^2). \tag{8.31}$$

By simplifying (8.31), it can be shown that the full conditional distribution for σ^2 is

$$f(\sigma^2|\boldsymbol{\beta},\boldsymbol{\delta},\mathbf{y}) \propto (\sigma^2)^{-((N+p+1+\nu)/2+1/2)}$$

$$\times \exp\left\{-\left[\nu\lambda + (\mathbf{y}-\mathbf{X}\boldsymbol{\beta})^T(\mathbf{y}-\mathbf{X}\boldsymbol{\beta}) + \boldsymbol{\beta}^T\mathbf{D}_{\boldsymbol{\delta}}^{-1}\boldsymbol{\beta}\right]/(2\sigma^2)\right\}.$$

That is,

$$(\sigma^2|\beta,\delta,\mathbf{y})$$

$$\sim IG\left(\tfrac{1}{2}(N+p+1+\nu), \tfrac{1}{2}\left[\nu\lambda + (\mathbf{y}-\mathbf{X}\boldsymbol{\beta})^T(\mathbf{y}-\mathbf{X}\boldsymbol{\beta}) + \boldsymbol{\beta}^T\mathbf{D}_{\boldsymbol{\delta}}^{-1}\boldsymbol{\beta}\right]\right).$$

$$\tag{8.32}$$

By dropping $f(\mathbf{y}|\boldsymbol{\beta},\sigma^2)$ and $\pi(\sigma^2)$ in (8.24), the full conditional distribution for $\boldsymbol{\delta}$ is

$$f(\boldsymbol{\delta}|\boldsymbol{\beta},\sigma^2,\mathbf{y}) \propto \pi(\boldsymbol{\beta}|\sigma^2,\boldsymbol{\delta})\pi(\boldsymbol{\delta}). \tag{8.33}$$

In Gibbs sampling, each element δ_i is sampled conditional on the remaining elements $\boldsymbol{\delta}_{(-i)}$, $\boldsymbol{\beta}$, σ^2, and $\mathbf{y}$. The draw for δ_i is Bernoulli so that the probability of $\delta_i = 1$ conditionally on all other variables needs to be determined. Using (8.33), the full conditional distribution for $\delta_i = 1$ has the form

$$f\left(\delta_i|\boldsymbol{\delta}_{(-i)},\boldsymbol{\beta},\sigma^2,\mathbf{y}\right) \propto \pi\left(\boldsymbol{\beta}|\delta_i,\boldsymbol{\delta}_{(-i)},\sigma^2\right)\pi\left(\delta_i,\boldsymbol{\delta}_{(-i)}\right).$$

Thus

$$Pr\left(\delta_i = 1|\boldsymbol{\delta}_{(-i)},\boldsymbol{\beta},\sigma^2,\mathbf{y}\right)$$

$$= \frac{\pi\left(\boldsymbol{\beta}|\delta_i=1,\boldsymbol{\delta}_{(-i)},\sigma^2\right)\pi\left(\delta_i=1,\boldsymbol{\delta}_{(-i)}\right)}{\pi\left(\boldsymbol{\beta}|\delta_i=1,\boldsymbol{\delta}_{(-i)},\sigma^2\right)\pi\left(\delta_i=1,\boldsymbol{\delta}_{(-i)}\right) + \pi\left(\boldsymbol{\beta}|\delta_i=0,\boldsymbol{\delta}_{(-i)},\sigma^2\right)\pi\left(\delta_i=0,\boldsymbol{\delta}_{(-i)}\right)}$$

$$= \frac{\pi\left(\delta_i=1,\boldsymbol{\delta}_{(-i)}\right)}{\pi\left(\delta_i=1,\boldsymbol{\delta}_{(-i)}\right) + \pi\left(\delta_i=0,\boldsymbol{\delta}_{(-i)}\right)\left(\pi\left(\boldsymbol{\beta}|\delta_i=0,\boldsymbol{\delta}_{(-i)},\sigma^2\right)/\pi\left(\boldsymbol{\beta}|\delta_i=1,\boldsymbol{\delta}_{(-i)},\sigma^2\right)\right)}.$$

$$\tag{8.34}$$

Therefore, a draw for δ_i requires the evaluation of the joint prior for $\boldsymbol{\delta}$ at $(\delta_i = 1, \boldsymbol{\delta}_{(-i)})$ and $(\delta_i = 0, \boldsymbol{\delta}_{(-i)})$, and the calculation of the ratio $\pi(\boldsymbol{\beta} \mid \delta_i = 0, \boldsymbol{\delta}_{(-i)}, \sigma^2) / \pi(\boldsymbol{\beta} \mid \delta_i = 1, \boldsymbol{\delta}_{(-i)}, \sigma^2)$. Given the model prior $\pi(\boldsymbol{\delta})$ (which can be chosen as in Section 8.5.1), the computation of the joint prior for $\boldsymbol{\delta}$ is straightforward. The ratio $\pi(\boldsymbol{\beta} \mid \delta_i = 0, \boldsymbol{\delta}_{(-i)}, \sigma^2) / \pi(\boldsymbol{\beta} \mid \delta_i = 1, \boldsymbol{\delta}_{(-i)}, \sigma^2)$ can be determined from (8.26).

REFERENCES

Box, G. and Meyer, R.D. (1986), "Dispersion Effects from Fractional Designs," *Technometrics*, 28, 19–27.

Chipman, H. (1996), "Bayesian Variable Selection with Related Predictors," *Canadian Journal of Statistics*, 24, 17–36.

Chipman, H. (1998), "Fast Model Search for Designed Experiments with Complex Aliasing" in *Quality Improvement through Statistical Methods* (Eds. B. Abraham, and N.U. Nair), Boston: Birkhauser, pp. 207–220.

Chipman, H., Hamada, M., and Wu, C.F.J. (1997), "A Bayesian Variable-Selection Approach for Analyzing Designed Experiments with Complex Aliasing," *Technometrics*, 39, 372–381.

Cowles, M.K. and Carlin, B.P. (1996), "Markov Chain Monte Carlo Convergence Diagnostics: A Comparative Review," *Journal of the American Statistical Association*, 91, 883–904.

Geman, S. and Geman, D. (1984), "Stochastic Relaxation, Gibbs Distributions, and the Bayesian Restoration of Images," *IEEE Transactions on Pattern Analysis and Machine Intelligence*, 6, 721–741.

George, E.I. and McCulloch, R.E. (1993), "Variable Selection via Gibbs Sampling," *Journal of the American Statistical Association*, 88, 881–889.

Hamada, M. and Wu, C.F.J. (1991), "Analysis of Censored Data from Highly Fractionated Experiments," *Technometrics*, 33, 25–38.

Hamada, M. and Wu, C.F.J. (1992), "Analysis of Designed Experiments with Complex Aliasing," *Journal of Quality Technology*, 24, 130–137.

Hunter, G.B., Hodi, F.S., and Eager, T.W. (1982), "High-Cycle Fatigue of Weld Repaired Cast Ti-6Al-4V," *Metallurgical Transactions*, 13A, 1589–1594.

Lin, D.K.J. (1993), "A New Class of Supersaturated Designs," *Technometrics*, 35, 28–31.

McCullagh, P. and Nelder, J.A. (1989), *Generalized Linear Models*, 2nd ed., London: Chapman & Hall.

Meyer, R.D., Steinberg, D.M., and Box, G. (1996), "Follow-up Designs to Resolve Confounding in Multifactor Experiments (with discussions)," *Technometrics*, 38, 303–332.

Nguyen, N.K. (1996), "An Algorithmic Approach to Constructing Supersaturated Designs," *Technometrics*, 38, 69–73.

Phadke, M.S., Kackar, R.N., Speeney, D.V., and Grieco, M.J. (1983), "Off-Line Quality Control in Integrated Circuit Fabrication Using Experimental Design," *The Bell System Technical Journal*, 62, 1273–1309.

Smith, A.F.M. and Roberts, G.O. (1993), "Bayesian Computation via the Gibbs Sampler and Related Markov Chain Monte Carlo Methods," *Journal of the Royal Statistical Society, Series B*, 55, 3–23.

Tang, B. and Wu, C.F.J. (1997), "A Method for Constructing Supersaturated Designs and Its Es^2 Optimality," *Canadian Journal of Statistics*, 25, 191–201.

Williams, K.R. (1968), "Designed Experiments," *Rubber Age*, 100, 65–71.

Wu, C.F.J. (1993), "Construction of Supersaturated Designs through Partially Aliased Interactions," *Biometrika*, 80, 661–669.

Yuan, C.G. (1998), "The Optimization of Mechanical Properties of Si_3N_4 Ceramics Using Experimental Designs," unpublished Ph.D. dissertation, Department of Materials Science and Engineering, University of Michigan.

CHAPTER 9

Response Surface Methodology

Study of a process or system often focuses on the relationship between the response and the input factors. Its purpose can be to optimize the response or to understand the underlying mechanism. If the input factors are quantitative and there are only a few of them, response surface methodology is an effective tool for studying this relationship. A sequential experimentation strategy is considered, which facilitates an efficient search of the input factor space by using a first-order experiment followed by a second-order experiment. Analysis of a second-order experiment can be done by approximating the response surface relationship with a fitted second-order regression model. Second-order designs that allow the second-order regression models to be efficiently estimated are considered. These include central composite designs, Box-Behnken designs, and uniform shell designs. Desirability functions are used to develop strategies to handle multiple response characteristics.

9.1 A RANITIDINE SEPARATION EXPERIMENT

Morris et al. (1997) reported on an experiment that studied important factors in the separation of ranitidine and related products by capillary electrophoresis. Ranitidine hydrochloride is the active ingredient in Zantac, a popular treatment for ulcers. Electrophoresis is the process used to separate ranitidine and four related compounds. From screening experiments, the investigators identified three factors as important: pH of the buffer solution, the voltage used in electrophoresis, and the concentration of α-CD, a component of the buffer solution. The factor levels are given in Table 9.1. The (coded) design matrix and response data for the ranitidine experiment are given in Table 9.2. The response chromatographic exponential function (CEF) is a quality measure in terms of separation achieved and time of final separation; the goal is to *minimize* CEF.

The experiment used a central composite design which consists of three parts. The first eight points in Table 9.2 form a 2^3 design. Because they are

Table 9.1　Factors and Levels, Ranitidine Experiment

Factor	Levels
A pH	2, 3.42, 5.5, 7.58, 9
B. voltage (kV)	9.9, 14, 20, 26, 30.1
C. α-CD (mM)	0, 2, 5, 8, 10

Table 9.2　Design Matrix and Response Data, Ranitidine Experiment

Run	Factor A	B	C	CEF	ln CEF
1	−1	−1	−1	17.293	2.850
2	1	−1	−1	45.488	3.817
3	−1	1	−1	10.311	2.333
4	1	1	−1	11757.084	9.372
5	−1	−1	1	16.942	2.830
6	1	−1	1	25.400	3.235
7	−1	1	1	31697.199	10.364
8	1	1	1	12039.201	9.396
9	0	0	−1.67	7.474	2.011
10	0	0	1.67	6.312	1.842
11	0	−1.68	0	11.145	2.411
12	0	1.68	0	6.664	1.897
13	−1.68	0	0	16548.749	9.714
14	1.68	0	0	26351.811	10.179
15	0	0	0	9.854	2.288
16	0	0	0	9.606	2.262
17	0	0	0	8.863	2.182
18	0	0	0	8.783	2.173
19	0	0	0	8.013	2.081
20	0	0	0	8.059	2.087

on the corners of the 2^3 cube, they are called *cube points* or *corner points*. The next six points form three pairs of points along the three coordinate axes and are therefore called *axial points* or *star points*. The last six runs are at the center of the design region and are called *center points*. These 15 distinct points of the design are represented in the three-dimensional space as in Figure 9.1. As will be shown in Section 9.7, this design is a *second-order design* in the sense that it allows all the linear and quadratic components of the main effects and the linear-by-linear interactions to be estimated. This experiment was part of a systematic study that involved other experiments using response surface methodology. Before returning to the analysis of this experiment, we need to develop some details for response surface methodology in the next few sections.

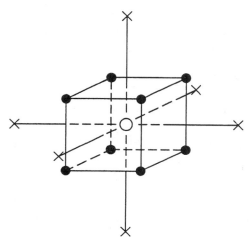

Figure 9.1. A Central Composite Design in Three Dimensions [cube point (dot), star point (cross), center point (circle)].

9.2 SEQUENTIAL NATURE OF RESPONSE SURFACE METHODOLOGY

Suppose that a scientific or engineering investigation is concerned with a process or system that involves a response y that depends on the *input factors* (also called *input variables* or *process variables*) $X_1, X_2, \ldots, X_k$. Their relationship can be modeled by

$$y = f(X_1, X_2, \ldots, X_k) + \epsilon, \tag{9.1}$$

where the form of the true response function f is unknown and ϵ is an error term that represents the sources of variability not captured by f. We will assume that the ϵ over different runs are independent and have mean zero and variance σ^2. The X_i's are expressed in the original scale such as minutes, degrees Celcius (°C), or milligrams (mg). As in regression analysis (see Section 1.4), it is convenient and computationally efficient to convert X to **coded variables** $x_1, x_2, \ldots, x_k$, which are dimensionless and have mean 0 and the same standard deviation. For example, each factor in the ranitidine experiment has five symmetrically spaced levels in the original scale, which can be expressed as $x_i - \alpha c_i$, $x_i - c_i$, x_i, $x_i + c_i$, $x_i + \alpha c_i$, with $\alpha > 1$. By taking the linear transformation $x = (X - x_i)/c_i$, the coded variable x has the five levels $-\alpha, -1, 0, 1, \alpha$. Another transformation is to center and then standardize the original variable as in regression analysis, e.g.,

$$x_{ij} = \frac{X_{ij} - \overline{X}_{\cdot j}}{s_j}, \tag{9.2}$$

where X_{ij} is the value of the jth variable for the ith run and $\overline{X}_{.j}$ and s_j are the mean and standard deviation of the X_{ij}'s over i.

From now on, we shall assume that the input factors are in coded form and express (9.1) as

$$y = f(x_1, x_2, \dots, x_k) + \epsilon. \tag{9.3}$$

Because the relationship between the response y and the x_i's can be graphically depicted as a surface lying over the region for the x_i's, the study of this relationship is henceforth called a **response surface study**.

Often the purpose of the investigation is to maximize or minimize the response or to achieve a desired value of the response. Because f is unknown and y is observed with random error, we need to run experiments to obtain data about f. Success of the investigation depends critically on how well f can be approximated. **Response surface methodology** (Box and Wilson, 1951) is a strategy to achieve this goal that involves experimentation, modeling, data analysis, and optimization.

If there are many factors whose importance cannot be ruled out at the beginning of the response surface study, a **screening experiment** should be conducted to eliminate the unimportant ones. Such an experiment is based on a highly fractionated design like the 2^{k-p}, 3^{k-p}, Plackett-Burman designs, and nonregular orthogonal arrays discussed in the previous chapters. A screening experiment would not be necessary, however, if the number of factors is already small at the outset of the study.

Once a small number of important factors is identified, subsequent experiments can be conducted more efficiently and require fewer runs. The remaining investigation is divided into two phases. In the *first phase*, the main goal is to determine if the current conditions or levels of the input factors are close to the optimum (i.e., maximum or minimum) of the response surface or far from it. When the experimental region is far from the optimum region of the surface, a first-order approximation of the surface should be adequate and the following **first-order model** should be used:

$$y = \beta_0 + \sum_{i=1}^{k} \beta_i x_i + \epsilon, \tag{9.4}$$

where β_i represents the slope or linear effect of the coded variable x_i. A design or experiment that allows the coefficients in (9.4) to be estimated is called a **first-order design** or **first-order experiment**, respectively. Examples include resolution III 2^{k-p} designs and Plackett-Burman designs. Runs at the center point of the experimental region are added to a first-order experiment so that a curvature check of the underlying surface can be made. A search must be conducted over the region for the x_i's to determine if a first-order experiment should be continued or, in the presence of curvature, be replaced by a more elaborate second-order experiment. Two search methods will be considered in Section 9.3: **steepest ascent search** and **rectangular grid search**.

When the experimental region is near or within the optimum region, the *second phase* of the response surface study begins. Its main goal is to obtain an accurate approximation to the response surface in a small region around the optimum and to identify optimum process conditions. Near the optimum of the response surface, the curvature effects are the dominating terms and the response surface should be approximated by the **second-order model,**

$$y = \beta_0 + \sum_{i=1}^{k} \beta_i x_i + \sum_{i<j}^{k} \beta_{ij} x_i x_j + \sum_{i=1}^{k} \beta_{ii} x_i^2 + \epsilon, \tag{9.5}$$

where β_i represents the linear effect of x_i, β_{ij} represents the linear-by-linear interaction between x_i and x_j, and β_{ii} represents the quadratic effect of x_i. A design or experiment that allows the coefficients in (9.5) to be estimated is called a **second-order design** or **second-order experiment,** respectively. Analysis and optimization of the fitted second-order model will be considered in Section 9.4. Several classes of second-order designs will be considered in Sections 9.7 and 9.8. These two phases of the response surface study are illustrated in Figure 9.2.

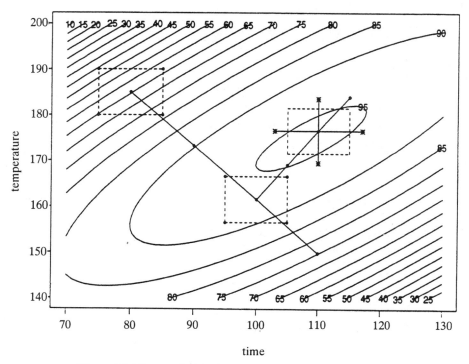

Figure 9.2. Sequential Exploration of the Response Surface.

Response surface methodology is useful for solving different types of problems. Its most common usage is response *optimization* such as the minimization of the CEF value in the ranitidine experiment and the maximization of yield in a chemical reaction. It is also useful for *mapping the response surface* over a region of interest. The following casting experiment provides a good illustration. To reduce the variation in the core hardness of a 2.8-liter crankshaft, a first-order experiment was conducted and a particular setting of the levels of copper, silicon, and manganese was identified as having a small hardness variation. Further experimentation in a small region around this setting was desired because for other casting mixes the required hardness would be different and an exploration of the response surface relationship would enable the process engineers to predict in advance the changes in the levels of copper, silicon, and manganese required for a desired (and different) hardness. In the same experiment, it was discovered that the copper additions were not well controlled during production, thereby causing a low pearlite level in some of the castings. Sensitivity of hardness to copper level became the focus. An accurate response surface approximation could provide valuable information on the sensitivity. The sensitivity of the response to the uncontrollable variation in the input factors is the subject of robust parameter design, which will be taken up in the next chapter.

9.3   FROM FIRST-ORDER EXPERIMENTS TO SECOND-ORDER EXPERIMENTS: STEEPEST ASCENT SEARCH AND RECTANGULAR GRID SEARCH

In a response surface study, the move from a first-order experiment to a second-order experiment often involves an iterative search of the design region and sequential experimentation. In this section, two methods of iterative search will be considered.

As was remarked in the previous section, a first-order experiment becomes ineffective when its experimental region is close to the optimum of the response surface. Near the optimum, the curvature effects of the response surface dominate the linear effects and a second-order experiment should be conducted instead. How does the experimenter know when to switch from a first-order experiment to a second-order experiment? A simple method is to check for an overall curvature effect by adding center runs to a first-order experiment.

9.3.1   Curvature Check

Suppose that the first-order experiment is based on a two-level orthogonal design with run size n_f and n_c center point runs are added to the first-order experiment, where "f" stands for the factorial design and "c" stands for center point. Denote the sample average over the factorial runs by $\bar{y}_f$ and the sample average at the center points by $\bar{y}_c$. Code the low and high levels of the factorial design by $-$ and $+$ and the level of the center point by 0. Under the second-order model in (9.5), it is easy to show that

$$E(\bar{y}_c) = \beta_0 \tag{9.6}$$

and

$$E(\bar{y}_f) = \beta_0 + \sum_{i=1}^{k} \beta_{ii}. \tag{9.7}$$

In deriving (9.6), note that $x_i = 0$ at the center point. In deriving (9.7), note that the orthogonality of the two-level design ensures that the coefficients β_i and β_{ij}, $i < j$, in model (9.5) sum to zero over the factorial runs. From (9.6) and (9.7), we have

$$E(\bar{y}_f - \bar{y}_c) = \sum_{i=1}^{k} \beta_{ii},$$

that is, we can use the difference $\bar{y}_f - \bar{y}_c$ to test if the **overall curvature**

$$\beta_{11} + \cdots + \beta_{kk}$$

is zero. Because the variance of $\bar{y}_f - \bar{y}_c$ is $\sigma^2(1/n_f + 1/n_c)$, we can use the following t test to test the significance of the overall curvature at level α:

$$\frac{|\bar{y}_f - \bar{y}_c|}{s(1/n_f + 1/n_c)^{1/2}} > t_{n_c - 1, \alpha/2}, \tag{9.8}$$

where s^2 is the sample variance based on the n_c center runs. If the curvature check is not significant, the search may continue with the use of another first-order experiment and steepest ascent. Otherwise, it should be switched to a second-order experiment.

In using the overall curvature to measure the curvature effects, we assume that the β_{ii} have the same sign. If they have different signs, it is possible that some of the large β_{ii} cancel each other, thus yielding a very small $\sum_{i=1}^{k} \beta_{ii}$. Then the test statistic in (9.8) will not be able to detect the presence of these curvature effects. This rarely occurs in practice unless the first-order experiment is conducted near a saddle point of the response surface. If it is important to detect the individual curvature effects β_{ii}, a second-order experiment should be conducted.

It is clear from the previous discussion that adding center points to a first-order experiment serves two purposes:

(i) It allows the check of the overall curvature effect.

(ii) It provides an unbiased estimate of the process error variance.

9.3.2 Steepest Ascent Search

The linear effects β_i in the first-order model (9.4) can be estimated by least squares estimates in regression analysis or by main effects in two-level designs. The fitted response is given by

$$\hat{y} = \hat{\beta}_0 + \sum_{i=1}^{k} \hat{\beta}_i x_i. \tag{9.9}$$

Taking the partial derivative of $\hat{y}$ with respect to x_i,

$$\frac{\partial \hat{y}}{\partial x_i} = \hat{\beta}_i, \qquad i = 1, \dots, k, \tag{9.10}$$

gives the direction of **steepest ascent**,

$$\lambda\left(\hat{\beta}_1, \dots, \hat{\beta}_k\right), \qquad \lambda > 0. \tag{9.11}$$

If a linear approximation to the unknown response surface is adequate, the response value will increase along the steepest ascent direction at least locally near the point of approximation. If the response is to be minimized, then the *steepest descent* direction $-\lambda(\hat{\beta}_1, \dots, \hat{\beta}_k)$ should be considered instead.

The search for a higher response continues by drawing a line from the center point of the design in the steepest ascent direction. Several runs should be taken along the steepest ascent path. The location on the path where a maximum response is observed will be taken as the center point for the next experiment. The design is again a first-order design plus n_c center runs. If the curvature check is significant, the design should be augmented by star points to make it a second-order design. The procedure is illustrated in the following with a yield maximization experiment.

To illustrate the steepest ascent method, consider an experiment to maximize the yield of a chemical reaction whose factors are time and temperature. The first-order design is a 2^2 design with two center points. The $-$ and $+$ levels are $75, 85$ minutes for time and $180, 190°C$ for temperature. The center points, denoted by $(0, 0)$, correspond to 80 minutes and $185°C$. The design matrix and yield data appear as runs 1–6 in Table 9.3. The first-order design appears in Figure 9.3 as the dotted box and center point against the contours of the unknown response surface.

To analyze the first-order experiment, the following model is fitted:

$$y = \beta_0 + \beta_1 x_1 + \beta_2 x_2 + \beta_{12} x_1 x_2 + (\beta_{11} + \beta_{22}) x_1^2, \tag{9.12}$$

where x_1 and x_2 are the time and temperature covariates from runs 1–6 of Table 9.3. Note that $x_1^2 = x_2^2$ so that the coefficient for x_1^2 is $\beta_{11} + \beta_{22}$, which is the measure of overall curvature. Fitting the regression model (9.12) yields

Table 9.3   Design Matrix and Yield Data for First-Order Design

Run	Time	Temperature	Yield
1	−1	−1	65.60
2	−1	1	45.59
3	1	−1	78.72
4	1	1	62.96
5	0	0	64.78
6	0	0	64.33
7	2	2(−1.173)	89.73
8	4	4(−1.173)	93.04
9	6	6(−1.173)	75.06

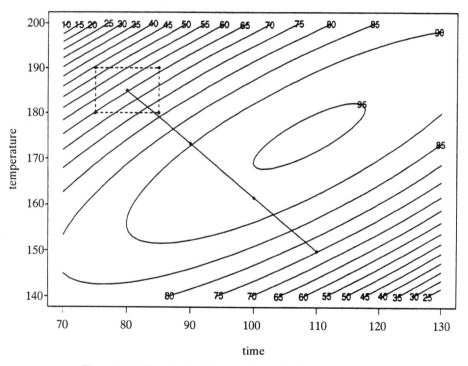

Figure 9.3. First-Order Experiment with Steepest Ascent.

the results in Table 9.4. Note that the t statistic in Table 9.4 has one degree of freedom because model (9.12) with five terms is fitted to six runs.

From Table 9.4, there is no indication of interaction and curvature, which suggests that a steepest ascent search should be conducted. From (9.11), the steepest ascent direction is proportional to (7.622, −8.942), or equivalently, (1, −1.173), that is, for one unit increase in time, there is a 1.173 unit decrease in temperature. Increasing time in steps of 2 units or 10 minutes is

Table 9.4   Least Squares Estimates, Standard Errors, t Statistics, and p Values, Initial First-Order Experiment

Effect	Estimate	Standard Error	t	p Value
intercept	64.5550	0.2250	286.91	0.00
β_1	7.6225	0.1591	47.91	0.01
β_2	-8.9425	0.1591	-56.21	0.01
β_{12}	1.0625	0.1591	6.68	0.09
$\beta_{11} + \beta_{22}$	-1.3375	0.2756	-4.85	0.13

Table 9.5   Design Matrix and Yield Data for Second-Order Design

Run	Time	Temperature	Yield
1	-1	-1	91.21
2	-1	1	94.17
3	1	-1	87.46
4	1	1	94.38
5	0	0	93.04
6	0	0	93.06
7	-1.41	0	93.56
8	1.41	0	91.17
9	0	-1.41	88.74
10	0	1.41	95.08

used because a step of 1 unit would give a point near the south-east corner of the first-order design. The results for three steps appear as runs 7–9 in Table 9.3. Times $(2, 4, 6)$ correspond to 90, 100, and 110 minutes and temperatures $(-1.173)(2, 4, 6)$ correspond to 173.27, 161.54, and 149.81°C. Because 100 minutes and 161.54°C correspond to the maximum yield along the path, they would be a good center point for the next experiment. (Note that the $-$ and $+$ values correspond to ± 5°C and ± 5 minutes for temperature and time, respectively.) Thus, a first-order experiment can be run as indicated by runs 1–6 in Table 9.5, where run 5 is from the second step of the steepest ascent search and run 6 is a replicate at the center point. If much time had elapsed between running the steepest ascent search and the second first-order experiment, a potential block effect in terms of the time-to-time difference may suggest running several new center points and discarding the steepest ascent run. This first-order design appears in Figure 9.4 as the dotted box and center point against the contours of the unknown response surface.

Fitting the regression model (9.12) to the second first-order experiment leads to the results in Table 9.6. The results indicate that there are significant interaction and curvature effects. This suggests augmenting the first-order design so that runs at the axial points of a central composite design are

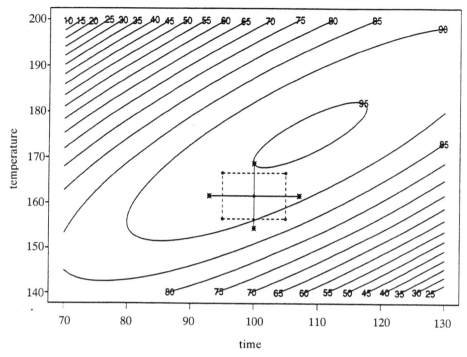

Figure 9.4. Second-Order Experiment.

Table 9.6   Least Squares Estimates, Standard Errors, t Statistics, and p Values for Follow-up First-Order Experiment

Effect	Estimate	Standard Error	t	p Value
intercept	93.0500	0.0100	9305.00	0.000
β_1	−0.8850	0.0071	−125.17	0.005
β_2	2.4700	0.0071	349.31	0.002
β_{12}	0.9900	0.0071	140.01	0.005
$\beta_{11} + \beta_{22}$	−1.2450	0.0122	−101.65	0.006

performed. See runs 7–10 in Table 9.5 and the end points of the cross in Figure 9.4. The axial points correspond to 92.95 and 107.05 minutes for time and 154.49 and 168.59°C for temperatures, i.e., $\pm \sqrt{2}$ in coded units. (The choice of $\pm \sqrt{2}$ can be justified by the rotatability criterion in Section 9.7.) Fitting a second-order model (9.5) (with $k = 2$) yields the results in Table 9.7 and the estimated response surface

$$\hat{y} = 93.05 - 0.87x_1 + 2.36x_2 + 0.99x_1x_2 - 0.43x_1^2 - 0.65x_2^2. \quad (9.13)$$

**Table 9.7   Least Squares Estimates, Standard Errors, t Statistics,
and p Values for Second-Order Experiment**

Effect	Estimate	Standard Error	t	p Value
intercept	93.0500	0.2028	458.90	0.000
β_1	-0.8650	0.1014	-8.53	0.001
β_2	2.3558	0.1014	23.24	0.000
β_{12}	0.9900	0.1434	6.90	0.002
β_{11}	-0.4256	0.1341	-3.17	0.034
β_{22}	-0.6531	0.1341	-4.87	0.008

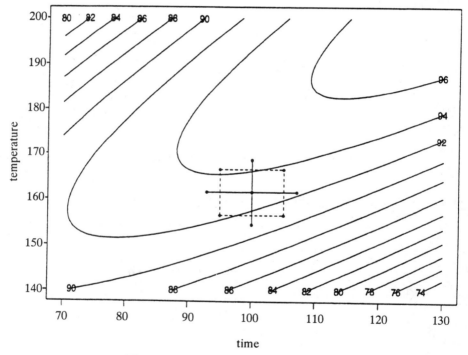

Figure 9.5. Estimated Response Surface.

The contours of the estimated response surface are displayed in Figure 9.5, which suggests that moving in a north-east direction would increase the yield, namely by increasing the time and temperature.

Further analysis of a second-order response surface such as (9.13) will be taken up in Section 9.4.

9.3.3   Rectangular Grid Search

Steepest ascent search is not the only method for conducting an iterative search. A related but different method is to start with a first-order experiment with widely spaced factor levels, use the sign and magnitude of the main effect for each factor to narrow the search to a smaller range, and then perform a first- or second-order experiment over the smaller region of factor levels. The search may continue over a smaller subregion or move to a neighboring region using the same strategy. This search strategy is illustrated in Figure 9.6. The initial design has three levels coded -1, 0, and $+1$ with the center point indicated by 0. Because the response at point A is lower, the factor range is narrowed to the right half with coded levels 0, 0.5, and 1, where 0.5 indicates the center point of the second design. Important information can be ascertained from comparing the responses over the two designs. The main effect for the first design (indicated by the slope of the line AB) is larger than the main effect for the second design (indicated by the slope of the line CB). The curvature [based on (9.8)] for the first design is more significant than the curvature for the second design. Both suggest that the second design is closer to the peak of the unknown response curve. Because

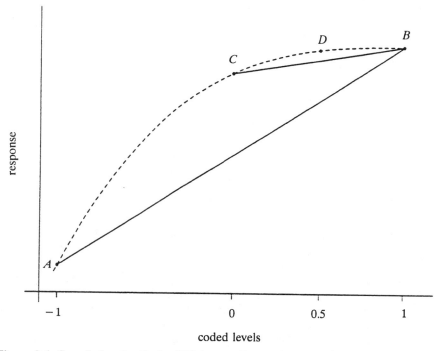

Figure 9.6. Search for the Peak of Unknown Response Curve (dashed curve): initial design over $[-1, 1]$, second design over $[0, 1]$.

Table 9.8   Factors and Levels, Ranitidine Screening Experiments

Factor	First Experiment			Second Experiment		
	−	0	+	−	0	+
A. methanol (%)	4	12	20	4	8	12
B. α-CD (mM)	1	5.5	10	1	3.25	5.5
C. pH	2	5.5	9	2	3.75	5.5
D. voltage (kV)	10	20	30	10	15	20
E. temperature (°C)	24	42	60	24	33	42
F. NaH$_2$PO$_4$ (mM)	20	35	50	20	27.5	35

Table 9.9   Design Matrix, Ranitidine Screening Experiment

Run	Factor					
	A	*B*	*C*	*D*	*E*	*F*
1	+	+	+	+	+	+
2	−	+	−	+	+	−
3	+	−	−	−	+	−
4	−	−	−	+	−	+
5	−	+	+	−	−	−
6	+	−	+	+	−	−
7	+	+	−	−	−	+
8	−	−	+	−	+	+
9	+	+	−	+	−	−
10	+	−	−	+	+	+
11	−	+	+	+	−	+
12	−	+	−	−	+	+
13	+	+	+	−	+	−
14	+	−	+	−	−	+
15	−	−	−	−	−	−
16	−	−	+	+	+	−
17	0	0	0	0	0	0
18	0	0	0	0	0	0
19	0	0	0	0	0	0
20	0	0	0	0	0	0

this strategy searches over different rectangular subregions of the factor region, it will be called the **rectangular grid search** method.

This strategy was employed in two first-order (i.e., screening) experiments that preceded the ranitidine experiment considered in Section 9.1. The first screening experiment was based on a 16-run 2_{IV}^{6-2} design (with $I = 1235 = 2346 = 1456$) plus four runs at the center point. The factor levels and design matrix are given in Tables 9.8 and 9.9, respectively. Morris et al. (1997) reported a very high overall curvature, $\bar{y}_f - \bar{y}_c = 16,070 - 11 = 16,059$, for this experiment. This large positive value suggests that one or more factors would attain the minimum within the range. Because the response CEF is to be

Table 9.10 Selected Analysis Results, Ranitidine Screening Experiments

Factor	First Experiment		Second Experiment	
	t	p Value	t	p Value
pH	0.650	0.56	−3.277	0.02
voltage	−5.456	0.01	−0.398	0.71
α-CD	2.647	0.08	−0.622	0.56

minimized, the factor range is narrowed to the left or right half, depending on whether the leftmost or rightmost level has the lower CEF. Take, for example, the factor methanol in Table 9.8. Its leftmost level 4 has a lower CEF. Therefore, in the second screening experiment the center level 12 becomes the new rightmost level, the leftmost level 4 remains the same, and the new center level is 8. The factor levels for the second screening experiment are given in the right half of Table 9.8.

As reported by Morris et al. (1997), methanol and buffer concentration (NaH_2PO_4) are not significant at the 10% level in both screening experiments. The factor temperature is significant at the 10% level for the second experiment but was dropped for further study because the existing instrument was unable to adjust the temperature effectively over a sufficient length of the capillary tube. For the remaining three factors, pH, voltage, and α-CD, the t statistics and p values are given in Table 9.10 for both experiments. Because the t statistics for the second experiment have a negative sign, the CEF is a decreasing function over the ranges of the three factors. This suggests that the factor range should be extended to the right by taking the rightmost level of the second screening design (in right half of Table 9.8) to be the center point of a new central composite design, which explains how the design for the experiment in Section 9.1 was chosen. For example, for pH, the center point (5.5) in Table 9.1 is the rightmost level in Table 9.8, the left star point (2) in Table 9.1 is the leftmost level in Table 9.8, and 3.42, 7.58, and 9 are three additional levels in the central composite design. For α-CD, the center point (5) in Table 9.1 is slightly lower than 5.5, which is the rightmost level in Table 9.8.

9.4 ANALYSIS OF SECOND-ORDER RESPONSE SURFACES

Analysis of a second-order experiment begins with the fitting of the second-order model in (9.5). By using the least squares method to estimate the parameters in the model, the fitted second-order model is

$$\hat{y} = \hat{\beta}_0 + \sum_{i=1}^{k} \hat{\beta}_i x_i + \sum_{i<j}^{k} \hat{\beta}_{ij} x_i x_j + \sum_{i=1}^{k} \hat{\beta}_{ii} x_i^2, \qquad (9.14)$$

which can be expressed in matrix form as

$$\hat{y} = \hat{\beta}_0 + \mathbf{x}^T \mathbf{b} + \mathbf{x}^T \mathbf{B} \mathbf{x}, \tag{9.15}$$

where $\mathbf{x}^T = (x_1, \ldots, x_k)$, $\mathbf{b}^T = (\hat{\beta}_1, \ldots, \hat{\beta}_k)$, and $\mathbf{B}$ is the $k \times k$ symmetric matrix

$$\mathbf{B} = \begin{bmatrix} \hat{\beta}_{11} & \frac{1}{2}\hat{\beta}_{12} & \cdots & \frac{1}{2}\hat{\beta}_{1k} \\ \frac{1}{2}\hat{\beta}_{12} & \hat{\beta}_{22} & \cdots & \frac{1}{2}\hat{\beta}_{2k} \\ \vdots & \vdots & \cdots & \vdots \\ \frac{1}{2}\hat{\beta}_{1k} & \frac{1}{2}\hat{\beta}_{2k} & \cdots & \hat{\beta}_{kk} \end{bmatrix}. \tag{9.16}$$

Differentiating $\hat{y}$ in (9.15) with respect to $\mathbf{x}$ and setting it to $\mathbf{0}$,

$$\frac{\partial \hat{y}}{\partial \mathbf{x}} = \mathbf{b} + 2\mathbf{B}\mathbf{x} = \mathbf{0}, \tag{9.17}$$

leads to the solution

$$\mathbf{x}_s = -\tfrac{1}{2}\mathbf{B}^{-1}\mathbf{b}, \tag{9.18}$$

which is called the *stationary point* of the quadratic surface (or system) represented by (9.15).

The nature of the quadratic surface around the stationary point is better understood in a new coordinate system. Let $\mathbf{P}$ be the $k \times k$ matrix whose columns are the standardized eigenvectors of $\mathbf{B}$. Then

$$\mathbf{P}^T \mathbf{B} \mathbf{P} = \mathbf{\Lambda},$$

where $\mathbf{\Lambda} = diag(\lambda_1, \ldots, \lambda_k)$ is a diagonal matrix and λ_i are the eigenvalues of $\mathbf{B}$ associated with the ith column of $\mathbf{P}$. The orthogonality of the eigenvectors ensures that $\mathbf{P}^T \mathbf{P} = \mathbf{P}\mathbf{P}^T = \mathbf{I}$. By *translating* the model (9.15) to a new center, namely the stationary point, and *rotating* to new axes associated with the eigenvectors, i.e.,

$$\mathbf{z} = \mathbf{x} - \mathbf{x}_s, \mathbf{v} = \mathbf{P}^T \mathbf{z},$$

we have

$$\begin{aligned} \hat{y} &= \hat{\beta}_0 + (\mathbf{z} + \mathbf{x}_s)^T \mathbf{b} + (\mathbf{z} + \mathbf{x}_s)^T \mathbf{B}(\mathbf{z} + \mathbf{x}_s) \\ &= \left(\hat{\beta}_0 + \mathbf{x}_s^T \mathbf{b} + \mathbf{x}_s^T \mathbf{B}\mathbf{x}_s \right) + \mathbf{z}^T \mathbf{B}\mathbf{z} + \mathbf{z}^T \mathbf{b} + 2\mathbf{z}^T \mathbf{B}\mathbf{x}_s \\ &= \hat{y}_s + \mathbf{z}^T \mathbf{B}\mathbf{z}, \end{aligned} \tag{9.19}$$

where $\hat{y}_s$ is the fitted response at $\mathbf{x}_s$ and the last equality follows from $\mathbf{z}^T(\mathbf{b} + 2\mathbf{B}\mathbf{x}_s) = \mathbf{0}$ via (9.18). By rotating $\mathbf{z}$ to $\mathbf{v} = (v_1, \ldots, v_k)^T$ and noting that $\mathbf{z} = \mathbf{P}\mathbf{P}^T\mathbf{z} = \mathbf{P}\mathbf{v}$,

$$\hat{y} = \hat{y}_s + \mathbf{v}^T\mathbf{P}^T\mathbf{B}\mathbf{P}\mathbf{v}$$

$$= \hat{y}_s + \mathbf{v}^T\mathbf{\Lambda}\mathbf{v}$$

$$= \hat{y}_s + \sum_{i=1}^{k} \lambda_i v_i^2. \tag{9.20}$$

This method of translation and rotation to simplify a quadratic model into the form (9.20) is called *canonical analysis*. The behavior of $\hat{y}$ around the stationary point $\hat{y}_s$ can be nicely described in terms of the eigenvalues λ_i's of $\mathbf{B}$ and the *canonical variables* v_i's. From the signs of the λ_i's, we can classify the second-order response surfaces into two broad types:

(i) When the λ_i's are of the same sign, the contours around the stationary point are elliptical. It is called an *elliptic system*. When the signs are

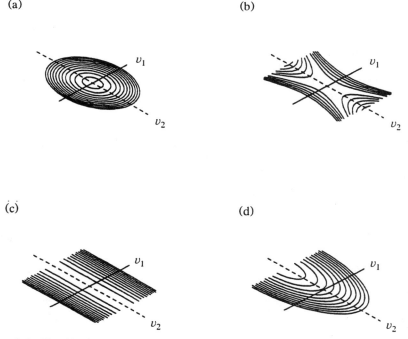

Figure 9.7. Classification of Second-Order Responses Surfaces: (a) Elliptic, (b) Hyperbolic, (c) Stationary Ridge, (d) Rising Ridge.

negative, the stationary point is a point of *maximum* response. When the signs are positive, the stationary point is a point of *minimum* response.

(ii) When the λ_i's are of mixed signs, the contours around the stationary point are hyperbolic. It is called a *hyperbolic system*. The stationary point is a *saddle* point.

The two systems are illustrated in Figures 9.7(a) and (b).

9.4.1 Ridge Systems

As one of the eigenvalues in (9.20) becomes very small relative to the others, a different type of system emerges. For illustration we consider a two-dimensional system with $\lambda_1 < 0$, $\lambda_2 < 0$. As λ_2 approaches 0, the elliptic contours become more and more elongated along the v_2 axis. A small change in the response value will require a big move along the v_2 axis (see Figure 9.8). So far we have assumed that all the eigenvalues are nonzero to ensure that the parameters in the second-order model are estimable. In the limiting case of $\lambda_2 = 0$, the contours around the stationary point have a different form. As can be seen in Figure 9.7(c), the contours along the v_2 axis are parallel lines. If the stationary point is inside the experimental region, it is called a **stationary ridge system** because the response is a constant value along the ridge [the dashed line in Figure 9.7(c)] which passes through the stationary point. While the maximum response along the v_1 axis is attained only at the stationary point, it is attained along the whole line in the v_2 direction. It is called a *line maximum*, in contrast with a point maximum, as in the elliptic case. A stationary ridge system can also be viewed as a limiting form of a hyperbolic system. This can be seen by following the same argument with $\lambda_1 < 0$, $\lambda_2 > 0$, and λ_2 approaching 0. That is, when the contours along the v_2 axis in Figure 9.7(b) are being pushed away from the stationary point and the contours along the v_1 axis become flatter, Figure 9.7(b) becomes more and more like Figure 9.7(c).

A stationary ridge system has the benefit that there is considerable *flexibility* in locating optimum operating conditions along a whole line (or a plane if there are two zero eigenvalues) around the stationary point. This flexibility may lead to other operating conditions that are better in terms of other characteristics such as ease of operation, less cost, or shorter reaction time. In practice, a system having exactly the stationary ridge form is rarely encountered. Random error and mild lack of fit can change the shape of the contours and make them look more like elongated elliptic or hyperbolic contours. Instead of being able to identify the optimum operating conditions along an entire line, only an interval around the stationary point will be identified.

If the stationary point in the elongated elliptic system in Figure 9.8 is far from the experimental region, the contours in the experimental region, say, in

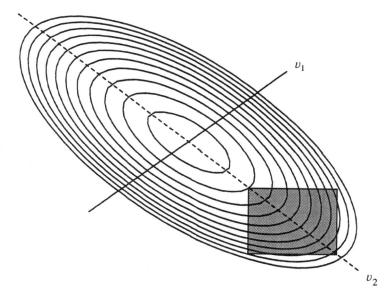

Figure 9.8. An Elliptic System Elongated Along the v_2 Axis. (Shaded experimental region is in the southeast corner.)

the southeast corner of Figure 9.8, can be described as a **rising ridge system**. Within this region (shaded in Figure 9.8), the contours can be described as in Figure 9.7(d). It is called a rising ridge because the response rises slowly along the v_2 axis toward the stationary point, which lies outside the region. If the response at the stationary point is a minimum, it is called a **falling ridge**. The contours in Figure 9.7(d) can also arise if the experimental region is in the northwest corner of an elongated hyperbolic system [which can be obtained by elongating the contours in Figure 9.7(b) along the v_2 axis]. Therefore a rising ridge system can be viewed as a limiting form of an elliptic system [see Figure 9.7(a)] or a hyperbolic system [see Figure 9.7(b)] as λ_1 remains fixed, λ_2 becomes very small, and at the same time the stationary point is moved toward infinity. Detection of a rising ridge suggests that the current experimental region is far from the optimum region, and *movement outside the current region* for additional experimentation is warranted.

A more general discussion on ridge systems can be found in Box and Draper (1987).

9.5 ANALYSIS OF THE RANITIDINE EXPERIMENT

Now we consider the analysis of data from the ranitidine experiment reported in Table 9.2. First, a blockage occurred in the separation column for run 7; consequently the observation is treated as faulty and is dropped in the

Table 9.11 Least Squares Estimates, Standard Errors, t Statistics, and p Values, Ranitidine Experiment (Run 7 dropped)

Effect	Estimate	Standard Error	t	p Value
intercept	2.1850	0.5785	3.78	0.00
β_1	1.1169	0.4242	2.63	0.03
β_2	0.7926	0.4242	1.87	0.09
β_3	0.0101	0.4262	0.02	0.98
β_{11}	2.7061	0.3788	7.14	0.00
β_{22}	−0.0460	0.3786	−0.12	0.91
β_{33}	−0.1295	0.3850	−0.34	0.74
β_{12}	1.4667	0.5890	2.49	0.03
β_{13}	−0.1918	0.5890	−0.33	0.75
β_{23}	0.2028	0.5890	0.34	0.74

subsequent analysis. Apparently, there were problems with the separation process around run 7 so that the region around run 7 would not be of further interest to the experimenters. Because the CEF values of the remaining runs cover several orders of magnitude (i.e., the ratio of the largest to the smallest is around 5×10^3), they are transformed to the natural log scale so that the ln CEF is analyzed as the response. The second-order model (9.5) was fitted yielding the results in Table 9.11. These results indicate that only factors pH and voltage are important. The estimated response surface using only the significant effects in Table 9.11 is displayed in Figure 9.9.

Because it was difficult to locate the minimum from the plot, a follow-up second-order experiment in pH and voltage was performed by Morris et al. (1997). Since pH was the most significant variable, the range of pH was narrowed. Thus, the factors were pH (A) and voltage (B), whose five levels were (4.19, 4.50, 5.25, 6.00, 6.31) and (20.0, 26.0, 28.5), respectively. The scaled values are $(-1.41, -1, 0, 1, 1.41)$. The design matrix and ln CEF response data are given in Table 9.12, where the run order was randomized. Fitting the second-order model (9.5) yields the results given Table 9.13. The estimated response surface (based on those effects in Table 9.13 with p values ≤ 0.20) is displayed in Figure 9.10.

Next we consider the canonical analysis for the final ranitidine second-order experiment. Based on Table 9.13,

$$\hat{y} = 2.0244 + 1.9308x_1 - 1.3288x_2 + 1.4838x_1^2 + 0.7743x_2^2 - 0.6680x_1x_2$$

$$(9.21)$$

and **b** and **B** of (9.15) are

$$\mathbf{b} = (1.9308, -1.3288)^T$$

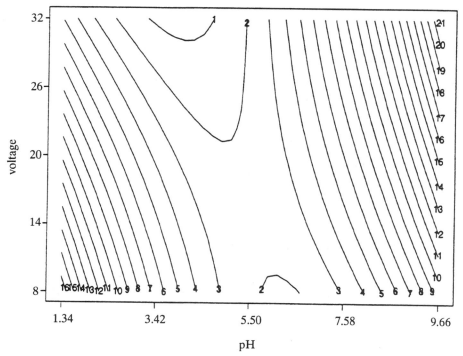

Figure 9.9. Estimated Response Surface, Ranitidine Experiment (run 7 dropped).

**Table 9.12   Design Matrix and Response Data, Final
Second-Order Ranitidine Experiment**

Run	Factor A	B	ln CEF
1	0	-1.41	6.943
2	1	-1	6.248
3	-1.41	0	2.100
4	0	0	2.034
5	0	0	2.009
6	0	0	2.022
7	1	1	3.252
8	1.41	0	9.445
9	0	1.41	1.781
10	0	0	1.925
11	-1	-1	2.390
12	-1	1	2.066
13	0	0	2.113

Table 9.13 Least Squares Estimates, Standard Errors, t Statistics, and p Values, Final Second-Order Ranitidine Experiment

Effect	Estimate	Standard Error	t	p Value
intercept	2.0244	0.5524	3.66	0.0080
β_1	1.9308	0.4374	4.41	0.0031
β_2	-1.3288	0.4374	-3.04	0.0189
β_{12}	-0.6680	0.6176	-1.08	0.3153
β_{11}	1.4838	0.4703	3.15	0.0160
β_{22}	0.7743	0.4703	1.65	0.1437

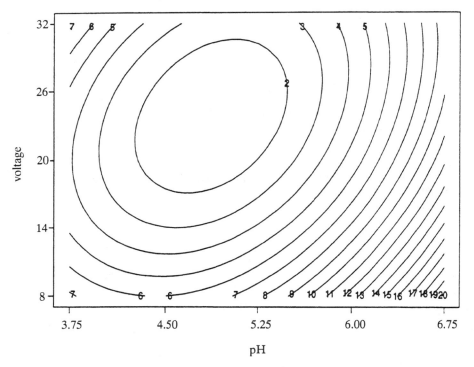

Figure 9.10. Estimated Response Surface, Final Second-Order Ranitidine Experiment.

and

$$\mathbf{B} = \begin{bmatrix} 1.4838 & -0.3340 \\ -0.3340 & 0.7743 \end{bmatrix}.$$

Using (9.18) yields the stationary point $\mathbf{x}_s = (-0.5067, 0.6395)^T$. Then performing a decomposition of $\mathbf{B}$ yields eigenvalues

$$\Lambda = diag(1.6163, 0.6418)$$

and eigenvectors

$$\mathbf{P} = \begin{bmatrix} -0.9295 & -0.3687 \\ 0.3687 & -0.9295 \end{bmatrix}.$$

Since both λ_1 and λ_2 are positive, $y_s = 1.1104$ is the minimum value which is achieved at $\mathbf{x}_s$ (pH of 4.87 and voltage of 23.84). These results are consistent with the estimated response surface in Figure 9.10.

9.6 ANALYSIS STRATEGIES FOR MULTIPLE RESPONSES II: CONTOUR PLOTS AND THE USE OF DESIRABILITY FUNCTIONS

In Section 5.7 the out-of-spec probabilities were used as a trade-off measure among multiple responses. Its use requires a variance model and its estimate for the computation of the out-of-spec probabilities. This requirement may not be met in response surface studies, where the main interest lies in the mean predicted response $\hat{y}$ and the experiment is rarely replicated at every design point. In this section alternative methods that do not require a variance estimate are considered.

If there are few responses and only two or three important input variables, the problem can be solved by overlaying and examining the **contour plots** for the responses over a few pairs of input variables. Otherwise a more formal method is required. If, among the responses, one is of primary importance, it can be formulated as a **constrained optimization** problem by optimizing this response subject to constraints imposed on the other responses. For example, in an experiment to separate components of an amphetamine discussed in the exercises, there were four responses: $R_s A$ (resolution of the amphetamine components), $R_s tot$ (total resolution), time (analysis time), and power (power generated during separation). The experimental goal was to maximize $R_s tot$ with the constraints $R_s A > 1.5$, time < 20 minutes, and power < 1.0 W. Typically one can use standard optimization packages, such as linear or quadratic programming, to solve the constrained optimization problem. This approach is not appropriate for investigations whose goal is to find an optimal balance among several response characteristics. For example, in the development of a tire tread compound, the following response characteristics are all important: (i) an abrasion index to be maximized, (ii) elongation at break to be within specified limits, and (iii) hardness to be within specified limits.

An alternative approach to provide a trade-off among several responses is to transform each predicted response $\hat{y}$ to a desirability value d, where $0 \le d \le 1$. The value of the **desirability function** d increases as the "desirability" of the corresponding response increases. In the following, a class of desirability functions (Derringer and Suich, 1980) is described for three types of problems.

First we consider the *nominal-the-best* problem. Denote the target value for a response by t. It is reasonable to assume that the product is most desirable at $y = t$. Choose a lower value L and an upper value U such that the product is considered unacceptable if $y < L$ or $y > U$. If lower and upper specification limits are available from product or process design, they should be chosen to be L and U. (Recall that we need to have specification limits to define the out-of-spec probabilities in Section 5.7.) Otherwise substantive knowledge may allow the investigator to choose such L and U values. The desirability function is then defined as

$$d = \begin{cases} \left| \dfrac{\hat{y} - L}{t - L} \right|^{\alpha_1}, & L \le \hat{y} \le t, \\[3mm] \left| \dfrac{\hat{y} - U}{t - U} \right|^{\alpha_2}, & t \le \hat{y} \le U, \end{cases} \qquad (9.22)$$

with $d = 0$ for $\hat{y} < L$ or $\hat{y} > U$. The choice of α_1 and α_2 in (9.22) is more subjective than the choice of L and U. A small α value would suggest that the response does not have to be very close to the target t. On the other hand, a large α value would imply the importance of being close to t. If the penalties for being above or below the target are very different, this difference can be reflected by choosing different values for α_1 and α_2 in (9.22). For example, in bottling soft drinks under-fill has a more severe consequence than over-fill because the potential damage of under-fill on product reputation greatly outweighs the minor cost in the additional amount of sugar and water.

Next we consider the *smaller-the-better* problem with a being the smallest possible value for the response y. (Recall that in Section 5.2 a was chosen to be 0.) Treat a as the target value and choose U to be a value above which the product is considered to be unacceptable. Then we can choose the right half of the d function in (9.22) as the desirability function, that is, let

$$d = \left| \frac{\hat{y} - U}{a - U} \right|^{\alpha}, \qquad a \le \hat{y} \le U, \qquad (9.23)$$

and $d = 0$ for $\hat{y} > U$.

For the *larger-the-better* problem, there is no fixed ideal target as infinity is theoretically the ideal target. Suppose that the investigator can choose a value L below which the product is considered to be unacceptable and a value U above which the product is considered to be nearly perfect. Then the desirability function can be defined as

$$d = \left| \frac{\hat{y} - L}{U - L} \right|^{\alpha}, \qquad L \le \hat{y} \le U, \qquad (9.24)$$

with $d = 0$ for $\hat{y} < L$ and $d = 1$ for $\hat{y} > U$.

In some practical situations, the L and U values in the Derringer-Suich formulation cannot be properly chosen. (Several such examples can be found in the experiments given in the exercises.) A different desirability function could then be defined over the whole real line or half line. Use of the exponential function appears to be ideal for this purpose. For the nominal-the-best problem with the target value t, we suggest the following double-exponential function as its desirability function,

$$d = \begin{cases} \exp\{-c_1|\hat{y} - t|^{\alpha_1}\}, & -\infty < \hat{y} \leq t, \\ \exp\{-c_2|\hat{y} - t|^{\alpha_2}\}, & t \leq \hat{y} < \infty. \end{cases} \tag{9.25}$$

Similarly, for the smaller-the-better problem with a being its smallest possible value, its desirability function can be defined as

$$d = \exp\{-c|\hat{y} - a|^{\alpha}\}, \qquad a \leq \hat{y} < \infty. \tag{9.26}$$

In (9.25) and (9.26) no L or U value needs to be specified. For the larger-the-better problem, we need to assume that an L value can be chosen as before. Then its desirability function can be defined as

$$d = 1 - \exp\{-c\hat{y}^{\alpha}\}/\exp\{-cL^{\alpha}\}, \qquad L \leq \hat{y} < \infty, \tag{9.27}$$

with $d = 0$ for $\hat{y} < L$.

In (9.25)–(9.27), the constants c and α can be used to fine-tune the scale and shape of the desirability function. A smaller α value makes the desirability function drop more slowly from its peak. A smaller c increases the spread of the desirability function by lowering the whole exponential curve between 0 and 1.

Suppose that a fitted response model and a desirability function d_i are chosen for the ith response, $i = 1, \ldots, m$. Then an **overall desirability function** d can be defined as the geometric mean of the d_i's, $i = 1, \ldots, m$,

$$d = \{d_1 d_2 \cdots d_m\}^{1/m}. \tag{9.28}$$

It can be extended to

$$d = d_1^{w_1} d_2^{w_2} \cdots d_m^{w_m} \tag{9.29}$$

to reflect the possible difference in the importance of the different responses, where the weights w_i satisfy $0 < w_i < 1$ and $w_1 + \cdots + w_m = 1$. The d function in (9.29) is equivalent to taking the weighted average of the log d_i's. Another possibility is to take the arithmetic mean or weighted average of the d_i's. The d function in (9.28) and (9.29) has the desirable property that if any $d_i = 0$ (i.e., if one of the responses is unacceptable), then $d = 0$ (i.e., the overall product or process is unacceptable). It is for these reasons that the geometric mean rather than the arithmetic mean is recommended.

Any setting for the input factors that maximizes the d value in (9.28) or (9.29) is chosen to be one which achieves an optimal balance over the m responses. After finding such a setting, it should be verified whether all constraints on the responses are satisfied. Some constraints may not be explicitly spelled out at the beginning of the study. Sometimes only when a surprising choice of process conditions is identified as "optimal" would the investigator realize that certain constraints on the responses should have been imposed. Running a confirmation experiment at the identified conditions will provide assurance that all responses are in a desirable region.

9.7   CENTRAL COMPOSITE DESIGNS

The design employed in the ranitidine experiment in Section 9.1 is a special case of **central composite designs**, which are the most commonly used second-order designs in response surface studies. Suppose that the k input factors in coded form are denoted by $x = (x_1, \ldots, x_k)$. A central composite design (Box and Wilson, 1951) consists of the following three parts:

(i) n_f *cube points* (or *corner points*) with $x_i = -1, 1$ for $i = 1, \ldots, k$. They form the *factorial portion* of the design.

(ii) n_c *center points* with $x_i = 0$ for $i = 1, \ldots, k$.

(iii) $2k$ *star points* (or *axial points*) of the form $(0, \ldots, x_i, \ldots, 0)$ with $x_i = \alpha, -\alpha$ for $i = 1, \ldots, k$.

For the ranitidine experiment, the cube points are the 2^3 design, $n_c = 6$ and $\alpha = 1.67$ or 1.68.

The central composite design can be used in a *single* experiment or in a *sequential* experiment. As shown in Section 9.3, the cube points and some of the center points form a first-order design. If the experimental data suggest that there is an overall curvature, the design can be expanded to a second-order design by adding the star points and some additional center points. On the other hand, if the experimental region is close to the optimum region, a sequential experiment will be unnecessary and the central composite design should be used in a single experiment. The option of using a central composite design in a sequential fashion is a definite advantage.

In choosing a central composite design, there are three issues:

(i) choosing the factorial portion of the design,

(ii) determining the α value for the star points, and

(iii) determining the number of center points.

These issues will be addressed in the remaining part of this section.

Choice of the Factorial Portion

How small can a 2^{k-p} design used for the factorial portion be so that the resulting central composite design is a second-order design? First we should note that the total number of parameters in the second-order model in (9.5) is $(k+1)(k+2)/2$. Therefore the total number of *distinct* design points in a central composite design, $N = n_f + 2k + 1$, must be at least $(k+1)(k+2)/2$. Before stating the following theorem on the estimability of parameters in (9.5), recall that in a 2^{k-p} design, there are $2^{k-p} - 1$ sets of aliased effects.

Theorem 9.1. In any central composite design whose factorial portion is a 2^{k-p} design that does not use any main effect as a defining relation, the following parameters in (9.5) are estimable: β_0, β_i, β_{ii}, $i = 1, \ldots, k$, and one β_{ij} selected from each set of aliased effects for $i < j$. It is not possible to estimate more than one β_{ij} from each set of aliased effects.

Its proof can be obtained by straightforward but tedious calculations (Hartley, 1959). The theorem requires that the defining relation of the 2^{k-p} design in the factorial portion does not contain any words of length one. It is interesting to note that even defining words of length two are allowed and that words of length four are worse than words of length three. This somewhat counter-intuitive result can be explained by the geometry of central composite designs. Because the linear main effects are not sacrificed in the defining relation, the linear and quadratic effects can be estimated by exploiting the information in the star points. In this regard note that a central composite design has three or five levels for each factor. On the other hand, the star points do not contain information on the interactions β_{ij} and thus cannot be used to de-alias any aliased β_{ij}'s. Because defining words of length four lead to the aliasing of some β_{ij}'s, they should be avoided.

Any resolution III design whose defining relation does not contain words of length four is said to have **resolution III***. Theorem 9.1 implies that *any central composite design whose factorial portion has resolution III* is a second-order design.*

Theorem 9.1 can be used to find central composite designs with economical run sizes. A collection of such designs is given in Table 9A.1. Their properties are briefly discussed as follows.

1. For $k = 2$, using the resolution II design with $\mathbf{I} = AB$ for the factorial portion leads to a second-order design according to Theorem 9.1. If more runs can be afforded, the 2^2 design can be chosen for the factorial portion.

2. For $k = 3$, either the 2_{III}^{3-1} or the 2^3 design can be chosen for the factorial portion.

3. For $k = 4$, either the 2_{III}^{4-1} design with $\mathbf{I} = ABD$ or the 2^4 design can be chosen for the factorial portion. According to Theorem 9.1, the 2_{IV}^{4-1} design is not a good choice.

4. For $k = 5$, the 2_V^{5-1} design is a good choice for the factorial portion. Because it has resolution V, there is really no need to use the 2^5 design.

5. For $k = 6$, the 2^{6-2} design given in the table has resolution III* because it has no word of length four. The minimum aberration 2^{6-2} design has resolution IV and is therefore not a good choice for the factorial portion. Use of the 2_{VI}^{6-1} design would require 16 additional runs and is hardly justified.

6. For $k = 7$, the 2^{7-2} design given in the table has resolution III*. The minimum aberration 2^{7-2} design has resolution IV and is therefore not a good choice for the factorial portion.

Smaller central composite designs can be found by using the Plackett-Burman designs for the factorial portion. Five such designs are given in Table 9A.1. Their properties are briefly discussed as follows.

1. For $k = 4$, an intermediate run size between the 8-run and 16-run designs is available by using the 12-run Plackett-Burman design. It can be shown by computer enumeration that all the 12×4 submatrices of the 12-run Plackett-Burman design (see Appendix 7A) are isomorphic. (Two matrices are *isomorphic* if one can be obtained from the other by permutations of rows, columns, and levels.) Therefore we can choose the first four columns for the factorial portion. It turns out that, among its 12 rows, two are identical (or called repeated runs). One of them can be removed and only an 11×4 matrix is needed for the factorial portion.

2. For $k = 5$, the same technique leads to the construction of a very economical design. Among all the 12×5 submatrices of the 12-run Plackett-Burman design, there are two non-isomorphic types. One type has two identical rows. By removing one of the rows, we can use an 11×5 matrix for the factorial portion, resulting in a 22-run central composite design, which has only one more run over the minimal required size of 21. The second type has no identical rows and therefore has one more run than the previous design.

3. For $k = 7$, the smallest resolution III* design has 32 runs. Instead of using a 2_{III}^{7-2} design, we can use a 24×7 submatrix of the 24-run Plackett-Burman design for the factorial portion. Among the different types of the 24×7 submatrices, one type has two pairs of identical rows. By removing one row from each pair, we can use a 22×7 submatrix for the factorial portion. The resulting central composite design has 37 runs, which is one more than the minimal required size of 36. Another type has only one pair of identical rows. By removing one row from the submatrix, we can use a 23×7 submatrix for the factorial portion.

Further discussion of small composite designs based on the Plackett-Burman designs can be found in Draper and Lin (1990).

If the α value in a central composite design is not chosen to be $\sqrt{k}$, it is known that the second-order design property of the central composite design is not affected by the removal of the center point from the design (Box and Hunter, 1957). For the estimability of the parameters in the second-order model in (9.5), one can use the cube and star points of the central composite design. Such a design is referred to as a *composite design* and its run size is $n_f + 2k$. By comparing this value with the minimal run size $(k + 1)(k + 2)/2$ in Table 9A.1, it is seen that the smallest designs in the table for $k = 2, 3, 5, 6, 7$ have the minimal run size and are saturated. For $\alpha = \sqrt{k}$, the center point cannot be removed, however. In general we recommend that the center point be retained because it provides information on the error variance and helps stabilize the variance of the predicted response $\hat{y}$.

Most of the designs in Table 9A.1 are saturated or nearly saturated. There are other classes of saturated or nearly saturated second-order designs. [See Section 6 of Draper and Lin (1990) for a review.] Because small composite designs are simple to construct and reasonably balanced, they are generally a good choice. In specific situations, however, other designs may have some advantages and should be adopted. Two such classes of designs will be considered in the next section.

Choice of α

In general, α should be chosen between 1 and $\sqrt{k}$ and rarely outside this range. First we consider the two extreme choices.

For $\alpha = 1$, the star points are placed at the center of the faces of the cube. Take the case of $k = 3$. The six star points are at the center of the six faces of the 2^3 cube. The design is therefore called the *face center cube*. This choice has two advantages. First they are the only central composite designs that require three levels. If one or more of the factors are qualitative and can only take three levels, central composite designs with five levels would not work. For example, the factors in the amphetamine experiment in the exercises have three levels and α is chosen to be 1. Second, they are effective designs if the design region is a cube. If the choice of factor range is independent of each other, the region is naturally cuboidal. Many of the factorial experiments considered in the previous chapters have cuboidal regions. For a cuboidal region, the efficiency of the parameter estimates is increased by pushing the star points toward the extreme, namely, the faces of the cube.

The choice of $\alpha = \sqrt{k}$ makes the star points and cube points lie on the same sphere. The design is often referred to as a *spherical design*. It is more suitable for a spherical design region, which is not as commonly encountered as a cuboidal region. By placing the star points at the extreme of the region, the efficiency of the parameter estimates is increased. However, for large k, this choice should be taken with caution because the star points are too far from the center point and no information is available for the response surface in the intermediate range of the factors, especially along the axes.

In general the choice of α depends on the geometric nature of and the practical constraints on the design region. If it is desired to collect information closer to the faces of the cube, a smaller α value should be chosen. If estimation efficiency is a concern, the star points should be pushed toward the extremes of the region. When converted into the original scale of the factor, a particular α value may not be feasible and a neighboring value may have to be chosen.

Barring the previous concerns, a general criterion for choosing α is needed. The notion of rotatability (Box and Hunter, 1957) provides one such criterion. Denote the predicted response at $\mathbf{x} = (x_1, \ldots, x_k)$ by $\hat{y}(\mathbf{x})$. A design is called **rotatable** if $Var(\hat{y}(\mathbf{x}))$ depends only on $\|\mathbf{x}\| = (x_1^2 + \cdots + x_k^2)^{1/2}$, that is, if the accuracy of prediction of the response is the same on a sphere around the center of the design. For a central composite design whose factorial portion is a 2^{k-p} design of resolution V, it can be shown that rotatability is guaranteed by choosing

$$\alpha = \sqrt[4]{n_f}. \tag{9.30}$$

For the ranitidine experiment, $n_f = 8$ and $\sqrt[4]{8} = 1.682$ but $\alpha = 1.68$ ($= 3.5/2.08$) was chosen for factor A and $\alpha = 1.67$ ($= 5/3$) was chosen for factors B and C in the experiment in order to have integer values for the voltage and α-CD levels and the stated levels of pH to the second decimal place (see Table 9.1).

Unless some practical considerations dictate the choice of α, (9.30) serves as a useful guide even when the factorial portion does not have resolution V. The rotatability criterion should not, however, be strictly followed. Its definition depends on how the x_i variables are coded. For different coding schemes, the contours of $Var(\hat{y}(\mathbf{x}))$ in the original scale are different. For example, if linear transformations with different scales are used in coding different variables, rotatability implies that the variance contours are elliptic in the original scale and the shape of the ellipse depends on the scales chosen for the variables. This lack of invariance with respect to the coding scheme makes it less attractive as a theoretical criterion.

Because rotatability often requires that the factorial portion has resolution V, use of a smaller composite design whose factorial portion has a lower resolution has not been popular. However, some of the central composite designs whose factorial portions have resolution V or higher have excessive degrees of freedom. The smallest case is $k = 6$. The smallest design that satisfies the rotatability requirement is a 2_{VI}^{6-1} design for the factorial portion. The total run size of the central composite design would be 45, which has 17 additional degrees of freedom over the minimal run size. We do not need so many degrees of freedom for error variance estimation or model checking. In fact, even the 29-run central composite design given in Table 9A.1 can provide sufficient information on the error variance. It is unlikely that, in any model fitting, all the 28 terms in the second-order model (for $k = 6$) are significant. After fitting the model and dropping the terms that are not significant, there should be enough degrees of freedom left for error estimation. The 29-run design is not rotatable but the additional $16 (= 45 - 29)$ runs required for rotatability can hardly be justified. Because of the lack of invariance of rotatability, it should not be imposed as the primary criterion. **Near rotatability** will be a more reasonable and practical criterion. The argument made here can also be used to justify the other small composite designs in Table 9A.1 with $k \geq 5$.

Number of Runs at the Center Point

For $\alpha = \sqrt{k}$ (which is suitable for spherical regions), it was pointed out before that at least one center point is required for the estimability of the parameters in the second-order model. Otherwise the variance of $\hat{y}(\mathbf{x})$ becomes infinite. In order to stabilize the prediction variance, the rule of thumb is to take *three to five runs at the center point* when α is close to $\sqrt{k}$. When α is close to 1 (which is suitable for cuboidal regions), *one or two runs at the center point* should be sufficient for the stability of the prediction variance. Between these two extremes, two to four runs should be considered. If the primary purpose of taking replicate runs at the center point is to estimate the error variance, then more than four or five runs may be required. For example, the ranitidine experiment had six runs at the center point.

Further discussion on the choice of α and the number of runs at the center point can be found in Box and Draper (1987).

9.8 BOX-BEHNKEN DESIGNS AND UNIFORM SHELL DESIGNS

Box and Behnken (1960) developed a family of three-level second-order designs by combining two-level factorial designs with balanced or partially balanced incomplete block designs in a particular manner. The method is illustrated with the construction of its smallest design for three factors (i.e., $k = 3$). A balanced incomplete block design with three treatments and three blocks is given by

Block	Treatment 1	Treatment 2	Treatment 3
1	×	×	
2	×		×
3		×	×

Take the three treatments as the three input factors x_1, x_2, x_3 in a response surface study. Replace the two crosses in each block by the two columns of the two-level 2^2 design and insert a column of zeros wherever a cross does not appear. Repeating this procedure for the next two blocks and adding a few runs at the center point lead to the construction of the Box-Behnken design for $k = 3$:

x_1	x_2	x_3
-1	-1	0
-1	1	0
1	-1	0
1	1	0
-1	0	-1
-1	0	1
1	0	-1
1	0	1
0	-1	-1
0	-1	1
0	1	-1
0	1	1
0	0	0

Its total run size is $12 + n_c$, where n_c denotes the number of center runs. Similar construction leads to the designs given in Table 9B.1 for $k = 4, 5$. The $(\pm 1, \pm 1)$ in the matrices denotes the four runs of the 2^2 design as previously explained.

One advantage of the Box-Behnken designs is that they only require *three* *levels* for each factor. In contrast, the central composite designs require five levels except when α is chosen to be 1. All the design points (except at the center) have length two, i.e., they all lie on the same sphere. These designs are particularly suited for spherical regions. Because of the spherical property, there should be at least three to five runs at the center point. The design for $k = 4$ is known to be rotatable and the other designs are nearly rotatable.

The designs for $k = 4, 5$ have the property that they can be arranged in orthogonal blocks. For $k = 4$, the design can be arranged in three blocks, which are separated by the dashed lines in Table 9B.1. Each block consists of nine distinct points, including a center point. For this design, the second-order model (9.5) can be expanded to include the block effects without affecting the parameter estimates in (9.5) (i.e., the parameter effects and block effests are orthogonal). Any design with this property is said to be **orthogonally blocked**. Similarly, the design for $k = 5$ in Table 9B.1 can be arranged in two orthogonal blocks, each of which must contain at least one center point. Orthogonal blocking is a desirable property and more important than the rotatability requirement if the experiment should be arranged in blocks and the block effects are likely to be large. Conditions that guarantee orthogonal blocking can be found in Box and Draper (1987).

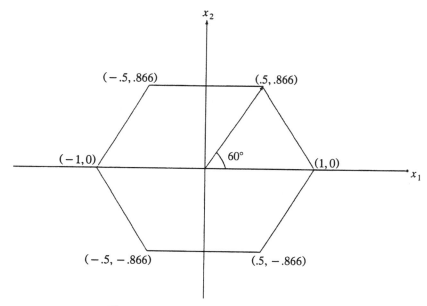

Figure 9.11. Uniform Shell Design, $k = 2$.

By comparing the run sizes of the designs in Tables 9A.1 and 9B.1, the Box-Behnken design for $k = 3$ is quite economical; the design for $k = 4$ has 27 design points, which are 12 more than the minimal required size of 15; the design for $k = 5$ is even larger, with 21 more than the minimal required size of 21. This comparison will get worse for the Box-Behnken designs with $k > 5$, which explains why they are not included in Table 9B.1. As argued near the end of Section 9.7, we do not recommend second-order designs which are much larger than the minimal required size.

Uniform shell designs are the other family of second-order designs to be considered in this section. In the case of $k = 2$, it can be described by a regular hexagon plus the center point in Figure 9.11. The six points are uniformly spaced on the circle. Because of its geometric appeal, this design is recommended over the central composite design with $\alpha = \sqrt{2}$ and its factorial portion defined by $\mathbf{I} = AB$.

For general k, Doehlert (1970) and Doehlert and Klee (1972) constructed a family of designs with $k^2 + k + 1$ points which consist of $k^2 + k$ points uniformly spaced on the k-dimensional sphere and one center point. For $k > 5$, its run size is much larger than the minimal run size $(k + 1)(k + 2)/2$. For reasons stated near the end of Section 9.7, we do not recommend using these designs for $k > 5$. In Table 9C.1 uniform shell designs with $2 \le k \le 5$ are given. The design in Table 9C.1 for $k = 3$ happens to be identical to the

Box-Behnken design for $k = 3$. Its 12 non-center points are the midpoints of the edges of the 2^3 cube and form a cuboctahedron.

In using the designs in Table 9C.1, the following points should be considered:

(i) To save space, only $k(k + 1)/2$ points are given in the table. The complete design is obtained by taking the negatives of these points and adding the center point.

(ii) The design points are arranged to *minimize* the number of the levels. The number of levels for each factor is given at the bottom of the design.

(iii) These designs can be rotated without affecting the uniformity property. Rotation will likely increase the number of levels for some factors.

These designs are suited for spherical regions. Its main attractive feature is the uniformity property. For each design point, define its covering region as the collection of points on the sphere that are closer to this design point than to the other design points. Uniformity then ensures that the covering region of each design point has the *same* spherical area. As for any spherical designs (which include the central composite designs with $\alpha = \sqrt{k}$ and the Box-Behnken designs), a disadvantage is the *lack of coverage* of the region between the origin and the spherical shell. A remedy is to choose two concentric spherical shells of distinctly different radii and place one set of $k^2 + k$ points on each shell. For better spacing of the points, the points on one shell should be rotated to have a different orientation from the other, assuming that the resulting increase of factor levels will not cause any practical problem. The drawback is the doubling of the run size.

9.9 PRACTICAL SUMMARY

1. If there are too many input factors, a response surface study should be preceded by a screening experiment to eliminate the unimportant factors.

2. Response surface methodology can be implemented in two phases:
   (i) When the experimental region is far from the optimum region, conduct a first-order experiment based on a 2^{k-p} design or a Plackett-Burman design. A first-order model in (9.4) should be fitted. It may be repeated over different regions based on the search methods to be discussed later.
   (ii) When the experimental region is near or within the optimum region, conduct a second-order experiment based on a second-order design. A second-order model in (9.5) should be fitted.

3. To determine if a first-order experiment has moved to the curved part of the response surface, use the overall curvature test in (9.8) based on the first-order experiment plus several runs at the center point.

4. A search should be conducted over the input factor space to determine if a first-order experiment should be continued or be replaced by a second-order experiment. Two search methods are discussed in the next two points.

5. Once a first-order model is fitted [see (9.9)], a search for higher y values can be conducted in the steepest ascent direction (9.11). Several runs are taken along the steepest ascent path. This may be repeated until the curvature test suggests that there is a significant curvature effect and the first-order experiment should be augmented by a second-order experiment.

6. Another search method is the rectangular grid search. It uses the sign and magnitude of the main effect for each factor to narrow the search to a smaller range or to move the search to a neighboring region. An illustration is given in Section 9.3.3.

7. Canonical analysis in (9.20) is used to classify second-order response surfaces:

    (i) When all the eigenvalues are of the same sign, it is an elliptic system. For a negative sign, the stationary point is a point of maximum response; for a positive sign, the stationary point is a point of minimum response.

    (ii) When the eigenvalues are of mixed signs, it is a hyperbolic system and the stationary point is a saddle point.

    (iii) When one of the eigenvalues is zero or near zero, it is a ridge system. When the stationary point is inside the experimental region, it is a stationary ridge system. When the stationary point is far from the experimental region, it is a rising (or falling) ridge system.

8. Desirability functions are defined in (9.22)–(9.27) for nominal-the-best, smaller-the-better, and larger-the-better problems. By combining the desirability values for different response characteristics, an overall desirability is defined in (9.28) and (9.29). Factor settings that maximize the overall desirability are identified as those with an optimal balance over the different responses. Unlike the out-of-spec probability approach considered in Section 5.7, the desirability function approach does not require replicated runs or fitting of a variance model.

9. A second-order design allows all the regression coefficients in the second-order model to be estimated. Efficient second-order designs are discussed in the next two points and given in Appendices 9A–9C.

10. Central composite designs are the most commonly used second-order designs. They consist of three parts: cube points, center points, and star points.

    (i) The cube points are based on a first-order design (e.g., a 2^{k-p} or Plackett-Burman design). Run size economy can be achieved by choosing a small first-order design as is done in Table 9A.1.

    (ii) The distance α of the star points from the origin can be determined by the rotatability criterion in (9.30) or by the geometric nature of or practical constraints on the design region. If the run size of a rotatable design is too large, smaller designs with near rotatability and other properties (see Section 9.7) should be considered.

    (iii) Number of center runs: one to two for α close to 1, three to five for α close to $\sqrt{k}$, and two to four between these two extremes, where k is the number of input factors. For error variance estimation, more than four or five runs are required.

11. Box-Behnken designs and uniform shell designs are second-order designs suitable for spherical regions. They should be considered only when they enjoy some definite advantages over the central composite designs. The Box-Behnken designs require only three levels for each factor and can be arranged in orthogonal blocks for $k = 4, 5$. Orthogonal blocking is a desirable property if the experiment must be conducted in blocks and the block effects are potentially significant. Uniform shell designs place the design points more uniformly in the design region.

EXERCISES

1. An experimenter has been given the opportunity to make eight experimental runs to investigate the importance of two factors that affect yield. The two designs under consideration are Design A, a twice replicated 2^2, and Design B, a 2^2 with four additional center points.

    (a) Discuss the advantages and disadvantages of each design. Be sure to include a discussion of error variance estimation, degrees of freedom, linearity test, curvature check, and other relevant points.

    (b) Suppose that only four runs can be made per day; reevaluate the merits of the two designs.

2. Each of the two original screening experiments on the separation of ranitidine was performed in two blocks. Find an optimal blocking scheme to arrange the 2^{6-2} design given in Section 9.3 in two blocks.

3. The log transformation of the CEF values in the ranitidine experiment in Section 9.1 is a member of the power transformation family (see Section 2.5). Is there another member of the power transformation family that is more appropriate to use than the log transformation?

4. The original ranitidine experiment was conducted in two blocks over two days. The first block (i.e., the first day) consists of the eight cube points and four center points (runs 1–8, 15–18 in Table 9.2); the second block (i.e., the second day) consists of the six star points and two center points (runs 9–14, 19–20 in Table 9.2).

   (a) Does the design achieve orthogonal blocking? If not, how seriously are the parameter estimates affected by the presence of the block effect?

   (b) Reanalyze the data by incorporating the block effect. Does the general conclusion change?

5. In the original ranitidine experiment, the rectangular grid search led to the choice of the factor range in Table 9.1, which is the same as the factor range used in the first screening design. (See the left half of Table 9.8.) This may be disappointing as it did not lead the investigation to a new experimental region. Provide some explanation on why this happened in the experiment. (*Hint*: Consider effect significance in the second screening experiment and aliased interactions in the screening designs.)

6. Verify Theorem 9.1 for $k = 4$ and when the factorial portion is defined by $I = ABD$ and $I = ABCD$, respectively. For the latter design, identify all the estimable β_{ij}'s.

7. Prove Theorem 9.1 in the special case of $k = 2$ and $I = AB$ for the factorial portion by working out the $X^T X$ matrix, where X is in the $E(y) = X\beta$ representation of the model (9.5). Explain why the linear and quadratic effects are estimable based on the entries in the $X^T X$ matrix.

8. Prove that no $2_{III^*}^{5-2}$ design exists.

9. Prove that no $2_{III^*}^{7-3}$ design exists.

10. Is the central composite design (with $k = 2$, $\alpha = \sqrt{2}$, and $I = A$ for the factorial portion) a second-order design? Among the seven-run designs over a circular region, would you recommend this design?

11. Prove that the model matrix for the second-order model of a composite design without the center point is singular if $\alpha = \sqrt{k}$.

12. For the central composite design with $k = 2$ and $I = AB$ for the factorial portion, prove that it is impossible to choose an α value to make it rotatable.

Table 9.14   Factors and Levels, Amphetamine Experiment

Factor	Level −1	Level 0	Level 1
A. buffer concentration (mM)	50	100	150
B. buffer pH	2.5	3.0	3.5
C. chiral selector concentration (mM)	10	25	40
D. capillary temperature (°C)	20	30	40
E. applied voltage (kV)	15	20	25

13. For the central composite design with $k = 2$, $\alpha = \sqrt{2}$, and a 2^2 design for the factorial portion, there are two options for stabilizing the variance of $\hat{y}$ over a circular region: (i) choose three replicated runs at the center point, or (ii) choose one center run and two other runs near but not at the center point. Which option do you prefer? Where should the two runs be placed near the center for the second option?

14. An experiment was performed to optimize the chiral separation of amphetamine by capillary electrophoresis (Varesio et al., 1997). The process separates two compounds (enantiomers). The goal was to find conditions which allowed the separation with sufficient resolution. A central composite design with two center points was used to study five factors $A–E$ listed in Table 9.14. The design matrix and response data are displayed in Table 9.15. The four responses are $R_s A$ (the resolution of the amphetamine enantiomers), $R_s tot$ (total resolution), time (analysis time), and power (power generated during electrophoresis). In terms of the responses, the goal was to maximize $R_s tot$ with constraints $R_s A > 1.5$, time < 20 minutes, and power < 1.0 W.

    (a) Describe how the cube points and the star points in the design are chosen. Given that each factor has only three levels, can different star points be chosen?

    (b) Analyze the data and identify the factor settings that optimize the goal. Explain how you would handle the constrained maximization problem.

15. Hackler et al. (1956) reported an experiment on whiteware. The factors were the ratios clay/flux (A), clay/flint (B), and nepheline syenite/feldspar (C). The design matrix and response data are given in Table 9.16. The properties of fired whiteware depend on firing treatment and relative quantities of the raw materials used. The responses are percent of absorption of water, an average of two replicates. Lower

Table 9.15   Design Matrix and Response Data, Amphetamine Experiment

Run	A	B	C	D	E	R_sA	R_stot	Time	Power
			Factor						
1	−1	−1	−1	−1	1	1.29	11.11	26.15	0.55
2	1	−1	−1	−1	−1	1.08	10.09	46.42	0.45
3	−1	1	−1	−1	−1	0.72	6.72	30.52	0.13
4	1	1	−1	−1	1	1.29	12.25	23.98	1.18
5	−1	−1	1	−1	−1	1.77	9.47	52.30	0.19
6	1	−1	1	−1	1	2.24	12.99	39.31	1.25
7	−1	1	1	−1	1	1.84	9.82	27.63	0.39
8	1	1	1	−1	−1	1.96	10.28	65.19	0.35
9	−1	−1	−1	1	−1	0.72	6.08	43.11	0.21
10	1	−1	−1	1	1	0.76	9.08	17.72	1.85
11	−1	1	−1	1	1	0.66	6.53	13.24	0.58
12	1	1	−1	1	−1	0.75	6.57	27.39	0.48
13	−1	−1	1	1	1	1.77	8.95	28.18	0.63
14	1	−1	1	1	−1	1.49	9.50	51.44	0.54
15	−1	1	1	1	−1	1.21	5.56	30.30	0.15
16	1	1	1	1	1	1.85	9.93	24.14	1.38
17	0	0	0	0	0	1.55	9.40	31.94	0.58
18	0	0	0	0	0	1.73	10.06	31.56	0.57
19	−1	0	0	0	0	1.36	8.19	27.12	0.31
20	1	0	0	0	0	1.74	11.24	36.86	0.82
21	0	−1	0	0	0	1.85	11.49	35.98	0.70
22	0	1	0	0	0	1.40	8.67	28.59	0.52
23	0	0	−1	0	0	0.95	8.56	25.30	0.60
24	0	0	1	0	0	1.78	9.95	36.94	0.55
25	0	0	0	−1	0	1.77	11.85	43.25	0.48
26	0	0	0	1	0	1.44	8.56	29.42	0.64
27	0	0	0	0	−1	1.36	8.46	44.28	0.31
28	0	0	0	0	1	1.70	11.32	26.60	0.90

absorption is desirable. Cone 02, cone 1, and cone 3 are three firing temperatures in increasing order. The cones begin to melt when a certain temperature has been reached.

(a) Describe the design chosen for this experiment.

(b) Analyze the data and draw contour plots. Identify a region in which the observed absorption can be further reduced.

16. An experiment was reported in Mielnik (1993) on drilling holes, a widely used manufacturing process. The two factors studied were feed in inches/revolution and drill diameter in inches. The responses were thrust in lbf (pound-force) and torque in ft-lb. Thrust is the axial force needed to push the drill into the workpiece and torque is that required to

Table 9.16   Design Matrix and Response Data, Whiteware Experiment

| Run | Factor | | | Absorption | | |
	A	B	C	Cone 02	Cone 1	Cone 3
1	0.4	3	3	14.84	8.81	6.36
2	0.4	3	5	13.54	7.26	3.92
3	0.4	5	3	13.64	7.46	3.76
4	0.4	5	5	12.86	6.84	3.80
5	0.6	3	3	11.71	6.42	3.83
6	0.6	3	5	13.31	7.78	5.50
7	0.6	5	3	11.33	5.97	3.58
8	0.6	5	5	11.36	6.04	3.62
9	0.5	4	4	13.04	7.56	4.19
10	0.3	4	4	15.77	8.82	4.55
11	0.7	4	4	10.70	5.34	3.42
12	0.5	2	4	13.12	7.96	4.88
13	0.5	6	4	12.78	7.11	4.11
14	0.5	4	2	12.81	7.06	4.18
15	0.5	4	6	12.38	6.82	3.59

Table 9.17   Design Matrix and Response Data, Drill Experiment

Run	Feed	Diameter	Thrust	Torque
1	0.006	0.250	230	1.0
2	0.006	0.406	375	2.1
3	0.013	0.406	570	3.8
4	0.013	0.250	375	2.1
5	0.009	0.225	280	1.0
6	0.005	0.318	255	1.1
7	0.009	0.450	580	3.8
8	0.017	0.318	565	3.4
9	0.009	0.318	400	2.2
10	0.009	0.318	400	2.1
11	0.009	0.318	380	2.1
12	0.009	0.318	380	1.9

rotate the drill. Having both responses as small as possible is desirable. Note that drill diameters and feed rates can only be changed in discrete steps so that the chosen design is as close to a rotatable central composite design as possible. The planning matrix with actual levels and the responses appear in Table 9.17.

(a) Describe the design chosen for the experiment.

(b) Analyze the thrust and torque data separately.

(c) Use contour plots and desirability functions to identify factor settings that achieve the experimental goal. Justify your choice of desirability functions.

Table 9.18 Design Matrix and Response Data, Ammonia Experiment

Run	A	B	C	D	DMP
			Factor		
1	−1	−1	−1	−1	58.2
2	1	−1	−1	−1	23.4
3	−1	1	−1	−1	21.9
4	1	1	−1	−1	21.8
5	−1	−1	1	−1	14.3
6	1	−1	1	−1	6.3
7	−1	1	1	−1	4.5
8	1	1	1	−1	21.8
9	−1	−1	−1	1	46.7
10	1	−1	−1	1	53.2
11	−1	1	−1	1	23.7
12	1	1	−1	1	40.3
13	−1	−1	1	1	7.5
14	1	−1	1	1	13.3
15	−1	1	1	1	49.3
16	1	1	1	1	20.1
17	0	0	0	0	32.8
18	−1.4	0	0	0	31.1
19	1.4	0	0	0	28.1
20	0	−1.4	0	0	17.5
21	0	1.4	0	0	49.7
22	0	0	−1.4	0	49.9
23	0	0	1.4	0	34.2
24	0	0	0	−1.4	31.1
25	0	0	0	1.4	41.1

17. In an experiment on the ammoniation of 1, 2-propanediol, the chemical reaction yields primarily 1, 2-propanediamine and 2, 5-dimethylpiperazine (Chang et al., 1960). The factors were ammonia concentration (A), maximum temperature (B), water concentration (C), and initial hydrogen pressure (D). The design matrix and response data appear in Table 9.18. The planning matrix can be recovered by multiplying by a scaling factor and adding a centering factor: A (51, 102), B (20, 250), C (200, 300), and D (350, 850), where the numbers in parentheses represent scaling and centering constants, respectively. For example, for factor A, multiply by 51 and then add 102, so that coded level $+1$ corresponds to $(+1)(51) + 102 = 153$. The response is percent yield of 2,5-dimethylpiperazine (DMP), where a high DMP yield is desirable. Runs 1–17 are for the initial first-order design and runs 18–25 are the star points to complete the second-order design.

(a) Describe the design chosen for the experiment.

(b) Analyze the data, describe the fitted response surface, and identify factor settings that maximize the response.

Table 9.19 Design Matrix and Response Data, TAB Laser Experiment

Run	Force	Power	Time	Strength Mean	Std. Dev.
1	40	5.5	100	30.5	4.9
2	40	3.5	100	32.0	3.8
3	20	5.5	100	31.0	3.8
4	20	3.5	100	31.1	3.5
5	40	5	150	31.6	3.0
6	40	5	150	31.5	3.6
7	20	5	150	33.4	5.0
8	20	5	50	32.9	5.0
9	30	5.5	150	31.3	2.9
10	30	5.5	50	30.4	4.1
11	30	3.5	150	29.8	6.4
12	30	3.5	50	22.0	9.1
13	30	5	100	32.4	2.8
14	30	5	100	28.7	3.7
15	30	5	100	29.6	3.1
16	30	5	150	29.7	3.9

18. Tape-automated bonding (TAB) is used to package integrated circuits. An experiment on bonding strength (Wesling, 1994) studied the following factors involving a laser bonder: bond force (grams, stylus force), laser power (watts), and laser time (mseconds). The response was strength of the bonds (average and standard deviation strength based on 24 leads). The goal was to maximize strength. The design matrix and response data appear in Table 9.19.

(a) Describe the design chosen for the experiment.

(b) Analyze the data and identify factor settings that maximize the bonding strength. Your analysis should include mean and log variance modeling, possible use of a desirability function or out-of-spec probability for larger-the-better responses, and the incorporation of standard deviation in the maximization of strength.

19. An experiment was performed to study the effect of factors on the ability of cement to contain waste material (Heimann et al., 1992). Factors were matrix type (%Al_2O_3) at 3.75, 7.11, 36.73; concentration of waste element (mol/l) at 0.1, 0.4, 1.0; curing temperature of cement paste (°C) at 25, 50, 75; and curing time of cement paste (days) at 7, 36, 60. Matrix type refers to different types of cement. The responses are normalized mass loss (NML) of chromium (Cr) and aluminum (Al). Experimenters were interested in waste containment, i.e., to minimize NML. The design matrix and response data appear in Table 9.20.

Table 9.20 Design Matrix and Response Data, Cement Experiment

Run	Matrix Type	Concentration	Curing Temperature	Curing Time	NML Cr	Al
1	−1	−1	0	0	0.1	0.0007
2	1	−1	0	0	0.042	0.044
3	−1	1	0	0	0.16	0.0058
4	1	1	0	0	0.071	0.0058
5	0	0	−1	−1	0.23	0.0041
6	0	0	1	−1	0.2	0.0036
7	0	0	−1	1	0.15	0.3
8	0	0	1	1	0.19	0.0043
9	0	0	0	0	0.21	0.0036
10	−1	0	0	−1	0.24	0.0044
11	1	0	0	−1	0.19	0.0041
12	−1	0	0	1	0.24	0.004
13	1	0	0	1	0.2	0.0038
14	0	−1	−1	0	0.066	0.044
15	0	1	−1	0	0.14	0.0061
16	0	−1	1	0	0.046	0.029
17	0	1	1	0	0.16	0.0062
18	0	0	0	0	0.22	0.0039
19	0	−1	0	−1	0.35	0.0039
20	0	1	0	−1	0.033	0.01
21	0	−1	0	1	0.24	0.0044
22	0	1	0	1	0.191	0.0036
23	−1	0	−1	0	0.055	0.043
24	1	0	−1	0	0.16	0.0075
25	−1	0	1	0	0.063	0.037
26	1	0	1	0	0.14	0.0061
27	0	0	0	0	0.21	0.0036

(a) Describe the design chosen for the experiment. Would a central composite design be a better or worse choice?

(b) Analyze the Cr and Al losses separately.

(c) Use contour plots and desirability functions to identify factor settings that achieve the goal.

20. Find optimal factor settings for the seat-belt experiment in Chapter 5 by using desirability functions, pretending that the variance models are not available. Compare the results with those reported in Tables 5.10–5.12. Discuss the sensitivity of the optimal factor settings to the choice of the desirability function and the pros and cons of the desirability function approach and the out-of-spec probability approach as applied to this experiment.

Table 9.21 Design Matrix, Bulking Process Experiment

Position	Temperature	Tension	Slug Weight
0.000	384	15.71	14.70
0.026	399	9.71	8.25
0.087	408	12.00	23.03
0.237	360	12.00	19.98
0.251	414	15.71	9.49
0.259	375	6.00	13.31
0.324	384	8.29	28.37
0.391	366	15.71	6.67
0.415	381	9.71	0.00
0.458	375	18.00	21.64
0.480	390	12.00	15.00
0.502	405	6.00	8.36
0.545	399	14.29	30.00
0.569	414	8.29	23.33
0.635	396	15.71	1.63
0.701	405	18.00	16.69
0.709	366	8.29	20.50
0.723	420	12.00	10.02
0.873	371	12.00	6.97
0.934	381	14.29	21.75
0.960	396	8.29	15.30

21. Doehlert (1970) described an experiment that studied different types of synthetic yarn in a bulking process to determine the impact of four factors: position of the yarn sample in the feed package, temperature of the bulking tube, tension in the yarn fed into the bulking tube, and weight of the slug riding in the tube on the yarn. The planning matrix is given in Table 9.21. Show that it employed a uniform shell design with $k = 4$ by translating and rescaling the factor levels.

22. Repeat Exercise 15(d) in Chapter 5 by applying the desirability function approach to the five responses in Table 5.17. (Recall that the out-of-spec probability approach could not be used because there are no replicated runs.) Discuss the sensitivity of the optimal factor settings to the choice of the desirability function. Compare the results with those obtained in Exercise 15(d) in Chapter 5.

23. Repeat Exercise16(a) and (b) in Chapter 5 by applying the desirability function approach to the two responses in Table 5.18. (Even though the out-of-spec probability approach can be used, it would not be very efficient because there are only two replicates per run.) Discuss the sensitivity of the optimal factor settings to the choice of the desirability function. Compare the results with those obtained in Exercise 16(b) in Chapter 5.

APPENDIX 9A: TABLE OF CENTRAL COMPOSITE DESIGNS

Table 9A.1 Central Composite Designs for $2 \le k \le 6$ ($k =$ number of factors, $N = n_f + 2k + 1 =$ number of distinct design points, where $n_f =$ number of cube points, $2k =$ number of star points, $(k + 1)(k + 2)/2 =$ smallest number of points required for the second-order model to be estimable)

k	$(k+1)(k+2)/2$	N	n_f	Factorial Portion (cube points)
2	6	7	2	$2^{2-1}(\mathbf{I} = AB)$
2	6	9	4	2^2
3	10	11	4	2_{III}^{3-1} $(\mathbf{I} = ABC)$
3	10	15	8	2^3
4	15	17	8	2_{III}^{4-1} $(\mathbf{I} = ABD)$
4	15	20	11	11×4 submatrix of 12-run PB design[a]
4	15	25	16	2^4
5	21	22	11	11×5 submatrix of 12-run PB design[b]
5	21	23	12	12×5 submatrix of 12-run PB design[c]
5	21	27	16	2_V^{5-1} $(\mathbf{I} = ABCDE)$
6	28	29	16	$2_{III^*}^{6-2}$ $(\mathbf{I} = ABE = CDF = ABCDEF)$
7	36	37	22	22×7 submatrix given in Table 9A.2
7	36	38	23	23×7 submatrix given in Table 9A.3
7	36	47	32	$2_{III^*}^{7-2}(\mathbf{I} = ABCDF = DEG)$

Note: For $\alpha \ne \sqrt{k}$, the center point can be removed from the central composite designs without affecting its second-order design property. The number of design points will be $N - 1 = n_f + 2k$.

[a]Obtained by taking the first four columns and deleting row 3 of the 12-run Plackett-Burman design in Appendix 7A.

[b]Obtained by taking columns 1, 2, 3, 4, and 10 and deleting row 3 of the 12-run Plackett-Burman design in Appendix 7A.

[c]Obtained by taking the first five columns of the 12-run Plackett-Burman design in Appendix 7A.

Table 9A.2 22×7 Matrix

1	2	3	4	5	6	7
1	1	1	-1	1	-1	1
1	1	1	1	-1	1	1
1	1	-1	-1	1	1	-1
1	1	1	1	1	-1	-1
1	-1	-1	1	-1	-1	1
-1	1	1	-1	-1	1	1
1	-1	1	-1	1	1	-1
-1	1	-1	1	1	-1	-1

Table 9A.2 (*Continued*)

1	2	3	4	5	6	7
1	1	-1	1	-1	-1	1
1	-1	1	-1	-1	1	-1
-1	-1	1	-1	1	-1	1
-1	1	-1	1	-1	1	-1
1	1	-1	-1	1	-1	-1
1	-1	1	1	-1	-1	-1
-1	-1	-1	-1	-1	-1	-1
-1	1	1	-1	-1	-1	1
1	-1	-1	-1	-1	1	1
1	-1	-1	1	1	1	1
-1	-1	-1	1	1	1	1
-1	-1	1	1	1	1	-1
-1	-1	1	1	1	-1	1
-1	1	1	1	-1	1	-1

Table 9A.3 **23 × 7 Matrix**

1	2	3	4	5	6	7
1	1	1	-1	1	1	1
1	1	-1	1	-1	1	-1
1	1	1	-1	1	-1	-1
1	1	-1	1	1	-1	1
1	-1	1	1	-1	1	1
-1	1	1	-1	-1	1	-1
1	-1	-1	-1	1	-1	-1
-1	1	-1	1	1	-1	1
1	1	1	1	-1	1	-1
1	-1	1	-1	-1	-1	1
-1	-1	-1	-1	1	1	-1
-1	1	-1	1	-1	-1	-1
1	-1	-1	1	-1	-1	-1
-1	-1	1	-1	-1	-1	1
-1	1	-1	-1	-1	1	1
1	-1	-1	-1	-1	1	1
-1	1	-1	-1	1	1	1
1	-1	-1	1	1	1	1
-1	-1	1	1	1	1	-1
-1	-1	1	1	1	-1	1
-1	-1	1	1	1	1	-1
-1	1	1	1	-1	-1	1
-1	-1	-1	-1	-1	-1	-1

APPENDIX 9B: TABLE OF BOX-BEHNKEN DESIGNS

Table 9B.1 Box-Behnken Designs for $3 \leq k \leq 5$ (k = number of factors, N = run size, n_c = number of center runs)

k	N	Design Points
3	$12 + n_c$ $n_c \geq 1$	$\begin{bmatrix} \pm 1 & \pm 1 & 0 \\ \pm 1 & 0 & \pm 1 \\ 0 & \pm 1 & \pm 1 \\ 0 & 0 & 0 \end{bmatrix}$
4	$24 + n_c,$ $n_c \geq 3$	$\left[\begin{array}{cccc} \pm 1 & \pm 1 & 0 & 0 \\ 0 & 0 & \pm 1 & \pm 1 \\ 0 & 0 & 0 & 0 \\ \hdashline \pm 1 & 0 & 0 & \pm 1 \\ 0 & \pm 1 & \pm 1 & 0 \\ 0 & 0 & 0 & 0 \\ \hdashline \pm 1 & 0 & \pm 1 & 0 \\ 0 & \pm 1 & 0 & \pm 1 \\ 0 & 0 & 0 & 0 \end{array}\right]$
5	$40 + n_c,$ $n_c \geq 2$	$\left[\begin{array}{ccccc} \pm 1 & \pm 1 & 0 & 0 & 0 \\ 0 & 0 & \pm 1 & \pm 1 & 0 \\ 0 & \pm 1 & 0 & 0 & \pm 1 \\ \pm 1 & 0 & \pm 1 & 0 & 0 \\ 0 & 0 & 0 & \pm 1 & \pm 1 \\ 0 & 0 & 0 & 0 & 0 \\ \hdashline 0 & \pm 1 & \pm 1 & 0 & 0 \\ \pm 1 & 0 & 0 & \pm 1 & 0 \\ 0 & 0 & \pm 1 & 0 & \pm 1 \\ \pm 1 & 0 & 0 & 0 & \pm 1 \\ 0 & \pm 1 & 0 & \pm 1 & 0 \\ 0 & 0 & 0 & 0 & 0 \end{array}\right]$

Note: Each $(\pm 1, \pm 1)$ in the design matrix denotes the four points in the 2^2 designs with $(-1, -1)$, $(-1, 1)$, $(1, -1)$, and $(1, 1)$. Dashed lines indicate how designs can be blocked.

APPENDIX 9C: TABLE OF UNIFORM SHELL DESIGNS

Table 9C.1   Uniform Shell Designs for $2 \leq k \leq 5$ (k = number of factors, N = number of distinct design points)

k	N	Design Points				
		0.000	0.000			
		1.000	0.000			
		−1.000	0.000			
2	7	0.500	0.866			
		−0.500	0.866			
		0.500	−0.866			
		−0.500	−0.866			
		5	3			
		1.000	0.000	1.000		
		0.000	1.000	1.000		
		1.000	1.000	0.000		
3	13	−1.000	1.000	0.000		
		0.000	1.000	−1.000		
		1.000	0.000	−1.000		
		add origin and negatives				
		3	3	3		
		−1.118	−0.500	−0.500	−0.500	
		−1.118	0.500	−0.500	0.500	
		−1.118	−0.500	0.500	0.500	
		−1.118	0.500	0.500	−0.500	
		0.000	1.000	0.000	1.000	
4	21	0.000	0.000	1.000	1.000	
		0.000	1.000	1.000	0.000	
		0.000	−1.000	1.000	0.000	
		0.000	0.000	1.000	−1.000	
		0.000	1.000	0.000	−1.000	
		add origin and negatives				
		3	5	5	5	
		0.000	1.000	0.000	1.000	0.000
		0.000	0.000	1.000	1.000	0.000
		0.000	1.000	1.000	0.000	0.000
		0.866	0.500	0.500	0.500	−0.707
		0.866	0.500	0.500	0.500	0.707
		0.000	−1.000	1.000	0.000	0.000
		0.000	0.000	1.000	−1.000	0.000
5	31	0.866	−0.500	0.500	−0.500	−0.707
		0.866	−0.500	0.500	−0.500	0.707
		0.000	1.000	0.000	−1.000	0.000
		0.866	0.500	−0.500	−0.500	−0.707
		0.866	0.500	−0.500	−0.500	0.707
		0.866	−0.500	−0.500	0.500	−0.707
		0.866	−0.500	−0.500	0.500	0.707
		0.000	0.000	0.000	0.000	1.414
		add origin and negatives				
		3	5	5	5	5

Notes: (1) For $3 \leq k \leq 5$, only $k(k + 1)/2$ design points are given in the table; the remaining points are obtained by taking their negatives and adding the origin. (2) The integers at the bottom of each design denote the number of factor levels.

REFERENCES

Box, G.E.P. and Behnken, D.W. (1960), "Some New Three Level Designs for the Study of Quantitative Variables," *Technometrics*, 2, 455–475.

Box, G.E.P. and Draper, N.R. (1987), *Empirical Model-Building and Response Surfaces*, New York: John Wiley & Sons,.

Box, G.E.P. and Hunter, J.S. (1957), "Multifactor Experimental Designs for Exploring Response Surfaces," *Annals of Mathematical Statistics*, 28, 195–241.

Box, G.E.P. and Wilson, K.B. (1951), "On the Experimental Attainment of Optimum Conditions," *Journal of the Royal Statistical Society, Series B*, 13, 1–45.

Chang, C.D., Kononenko, O.K., and Franklin, R.E., Jr. (1960), "Maximum Data through a Statistical Design," *Industrial and Engineering Chemistry*, 52, 939–942.

Derringer, G. and Suich, R. (1980), "Simultaneous Optimization of Several Response Variables," *Journal of Quality Technology*, 12, 214–219.

Doehlert, D.H. (1970), "Uniform Shell Designs," *Journal of the Royal Statistical Society, Series C*, 19, 231–239.

Doehlert, D.H. and Klee, V.L. (1972), "Experimental Designs through Level Reduction of the *d*-dimensional Cuboctahedron," *Discrete Mathematics*, 2, 309–334.

Draper, N.R. and Lin, D.K.J. (1990), "Small Response-Surface Designs," *Technometrics*, 32, 187–194.

Hackler, W.C., Kriegel, W.W., and Hader, R.J. (1956), "Effect of Raw-Material Ratios on Absorption of Whiteware Compositions," *Journal of the American Ceramic Society*, 39, 20–25.

Hartley, H.O. (1959), "Smallest Composite Designs for Quadratic Response Surfaces," *Biometrics*, 15, 611–624.

Heimann, R.B, Conrad, D., Florence, L.Z., Neuwirth, M., Ivey, D.G., Mikula, R.J., and Lam, W.W. (1992), "Leaching of Simulated Heavy Metal Waste Stabilized/Solidified in Different Cement Matrices," *Journal of Hazardous Materials*, 31, 39–57.

Mielnik, E.M. (1993), "Design of a Metal-Cutting Drilling Experiment: A Discrete Two-Variable Problem," *Quality Engineering*, 6, 71–98.

Morris, V.M., Hargreaves, C., Overall, K., Marriott, P.J., and Hughes, J.G. (1997), "Optimization of the Capillary Electrophoresis Separation of Ranitidine and Related Compounds," *Journal of Chromatography A*, 766, 245–254.

Varesio, E., Gauvrit, J.-Y., Longeray, R., Lanteri, P., and Veuthey, J.-L. (1997), "Central Composite Design in the Chiral Analysis of Amphetamines by Capillary Electrophoresis," *Electrophoresis*, 18, 931–937.

Wesling, P. (1994), "TAB Inner-Lead Bond Process Characterization for Single-Point Laser Bonding," *IEEE Transactions on Components, Packaging and Manufacturing Technology—Part A*, 17, 142–148.

CHAPTER 10

Introduction to Robust Parameter Design

Robust parameter design (parameter design for short) is a statistical/engineering methodology that aims at reducing the performance variation of a system (i.e., a product or process) by choosing the setting of its control factors to make it less sensitive to noise variation. The input variables are divided into two broad categories: *control factors* and *noise factors*. Control factors are variables whose values remain fixed once they are chosen. They include the design parameters in product or process design. Noise factors are variables which are hard to control during the normal process or use conditions. In conducting a parameter design experiment, the settings of the noise factors are varied systematically to represent their variation in normal conditions. Fractional factorial designs or orthogonal arrays are often employed in conducting the experiments. Two experimentation strategies (which use cross arrays and single arrays) and two modeling strategies (location and dispersion modeling and response modeling) are considered. Taguchi's signal-to-noise ratio and its limitations for parameter design optimization are also discussed.

10.1 A ROBUST PARAMETER DESIGN PERSPECTIVE OF THE LAYER GROWTH AND LEAF SPRING EXPERIMENTS

The original epitaxial layer growth experiment in Chapter 3 and the leaf spring experiment in Chapter 4 were actually robust parameter design experiments. Their description in the parameter design context will be given next.

10.1.1 Layer Growth Experiment Revisited

The experimental factors for the original layer growth experiment appear in Table 10.1. There are eight control factors ($A-H$) and two noise factors

436

Table 10.1   Factors and Levels, Layer Growth Experiment

	Level	
Control Factor	−	+
A. susceptor-rotation method	continuous	oscillating
B. code of wafers	668G4	678D4
C. deposition temperature(°C)	1210	1220
D. deposition time	short	long
E. arsenic flow rate(%)	55	59
F. hydrochloric acid etch temperature(°C)	1180	1215
G. hydrochloric acid flow rate(%)	10	14
H. nozzle position	2	6

	Level	
Noise Factor	−	+
L. location	bottom	top
M. facet	1 2	3 4

Table 10.2   Cross Array and Thickness Data, Layer Growth Experiment

								Noise Factor							
Control Factor								L-Bottom				L-Top			
A	B	C	D	E	F	G	H	M-1	M-2	M-3	M-4	M-1	M-2	M-3	M-4
−	−	−	+	−	−	−	−	14.2908	14.1924	14.2714	14.1876	15.3182	15.4279	15.2657	15.4056
−	−	−	+	+	+	+	+	14.8030	14.7193	14.6960	14.7635	14.9306	14.8954	14.9210	15.1349
−	−	+	−	−	−	+	+	13.8793	13.9213	13.8532	14.0849	14.0121	13.9386	14.2118	14.0789
−	−	+	−	+	+	−	−	13.4054	13.4788	13.5878	13.5167	14.2444	14.2573	14.3951	14.3724
−	+	−	−	−	+	−	+	14.1736	14.0306	14.1398	14.0796	14.1492	14.1654	14.1487	14.2765
−	+	−	−	+	−	+	−	13.2539	13.3338	13.1920	13.4430	14.2204	14.3028	14.2689	14.4104
−	+	+	+	−	+	+	−	14.0623	14.0888	14.1766	14.0528	15.2969	15.5209	15.4200	15.2077
−	+	+	+	+	−	−	+	14.3068	14.4055	14.6780	14.5811	15.0100	15.0618	15.5724	15.4668
+	−	−	−	+	+	−	−	13.7259	13.2934	12.6502	13.2666	14.9039	14.7952	14.1886	14.6254
+	−	−	+	−	−	+	+	13.8953	14.5597	14.4492	13.7064	13.7546	14.3229	14.2224	13.8209
+	−	+	+	−	+	−	+	14.2201	14.3974	15.2757	15.0363	14.1936	14.4295	15.5537	15.2200
+	−	+	+	+	−	+	−	13.5228	13.5828	14.2822	13.8449	14.5640	14.4670	15.2293	15.1099
+	+	−	+	−	−	+	+	14.5335	14.2492	14.6701	15.2799	14.7437	14.1827	14.9695	15.5484
+	+	−	+	+	+	−	−	14.5676	14.0310	13.7099	14.6375	15.8717	15.2239	14.9700	16.0001
+	+	+	−	−	−	−	−	12.9012	12.7071	13.1484	13.8940	14.2537	13.8368	14.1332	15.1681
+	+	+	−	+	+	+	+	13.9532	14.0830	14.1119	13.5963	13.8136	14.0745	14.4313	13.6862

(L and M). On a susceptor there are four facets, so that factor M has four levels. On each facet, there are top and bottom positions (factor L). In the original experiment, Kackar and Shoemaker (1986) reported results for six facets; for convenience of illustration, we will consider only four facets here. Because it is desirable to have uniform thickness over facets and locations, factors L and M are treated as noise factors.

The experimental plan, as displayed in Table 10.2, used a 2_{IV}^{8-4} design for the eight control factors with the defining generators $D = -ABC$, $F = ABE$,

$G = ACE$, and $H = BCE$. A 2×4 design was used for the two noise factors. Thus, there are eight observations per control factor setting. This experimental layout is called a cross array and will be discussed further in Section 10.5.

Recall that the nominal value for thickness is 14.5 μm with specification limits of 14.5 ± 0.5 μm. Thus, the goal of the experiment was to minimize the epitaxial layer nonuniformity over facets 1–4 and top/bottom positions while maintaining an average thickness of 14.5 μm.

10.1.2   Leaf Spring Experiment Revisited

The factors for the leaf spring experiment are listed in Table 10.3. The quench oil temperature Q is not controllable in normal production. With extra effort in the experiment, Q can be chosen in two ranges of values, 130–150 and 150–170, and is therefore treated as a noise factor. There are four control factors (B-E) and one noise factor (Q). The design and free height data are presented in Table 10.4. A 2_{IV}^{4-1} design with $\mathbf{I} = BCDE$ was chosen for the control factors. For each control factor setting, two noise factor settings are chosen. Each control and noise factor setting is then replicated three times.

Table 10.3   Factors and Levels, Leaf Spring Experiment

	Level	
Control Factor	−	+
B. high heat temperature (°F)	1840	1880
C. heating time (seconds)	23	25
D. transfer time (seconds)	10	12
E. hold down time (seconds)	2	3
	Level	
Noise Factor	−	+
Q. quench oil temperature (°F)	130–150	150–170

Table 10.4   Cross Array and Height Data, Leaf Spring Experiment

| Control Factor | | | | Noise Factor | | | | | |
B	C	D	E	Q^-			Q^+		
−	+	+	−	7.78	7.78	7.81	7.50	7.25	7.12
+	+	+	+	8.15	8.18	7.88	7.88	7.88	7.44
−	−	+	+	7.50	7.56	7.50	7.50	7.56	7.50
+	−	+	−	7.59	7.56	7.75	7.63	7.75	7.56
−	+	−	+	7.94	8.00	7.88	7.32	7.44	7.44
+	+	−	−	7.69	8.09	8.06	7.56	7.69	7.62
−	−	−	−	7.56	7.62	7.44	7.18	7.18	7.25
+	−	−	+	7.56	7.81	7.69	7.81	7.50	7.59

Recall that the height of an unloaded spring (free height) is the response whose target value is 8 inches. The goal of the experiment was to minimize the variation about the target.

10.2 STRATEGIES FOR REDUCING VARIATION

Before introducing robust parameter design as a strategy for reducing variation, we shall review several alternative strategies in this section. Further discussion can be found in MacKay and Steiner (1997).

1. *Sampling Inspection.* Before a product is delivered to the customer or to the next process step, sampling inspection can be applied to ensure that with high probability the delivered product meets some prescribed standard. This is often viewed as a last resort and not a proactive quality improvement tool. It is often time consuming and costly. There is, however, no alternative if the quality of the delivered product can only be assured by inspecting out defective items.

2. *Control Charting and Process Monitoring.* A control chart such as the Shewhart Chart can be used to monitor the process. If the process is out of control (as identified by a plotted point outside the control limits), it is an indication that the process variation cannot be explained by *common causes*, and actions may be required to identify *special causes* and to root out the underlying problems. It is a more proactive approach as it does not wait until the end of the process to inspect out the defects. If the process is stable but does not meet the specification requirements or competitive benchmarking, experiments may be used to change the process conditions so that further improvement can be achieved.

For more information on sampling inspection and control charts, see Montgomery (1991).

3. *Blocking.* As explained in Chapter 1, blocking is effective when the blocking variable is chosen so that the block-to-block variation is large (i.e., heterogeneous) and the within-block variation is small (i.e., homogeneous). The block effect is usually of no direct interest to the investigators. Blocking is used to reduce variation that could affect the comparison of the treatments. Its success depends on how much the block-to-block variation accounts for the total variation.

4. *Covariate Adjustment.* There may be covariates whose values are not controlled at the planning stage of the experiment. If the response depends on the covariate, covariate adjustment can be done at the analysis stage to account for the variation due to the covariate which might otherwise confound the comparison of the treatments. Analysis of covariance is used to perform the adjustment. Details on the method and examples of covariates can be found in Section 2.9. The effectiveness of covariate adjustment for variation reduction depends on whether a significant covariate can be identified.

Blocking and covariate adjustment are passive measures that can be effective in reducing the variability in the comparison of treatments. Usually they are not used to reduce the intrinsic process variation or to remove the root causes.

5. *Reducing the Variation in the Noise Factors.* The response variation is often caused by the variation in the input variables of a product or process. Measures can then be taken to reduce the variation at the source. An old machine may be replaced by a new one which can produce parts with much smaller variation. A more expensive and better quality supplier may be found. More operators may be assigned to the station to ensure a more frequent inspection of items. A control device may be installed in the room to reduce the fluctuation of the room temperature. If the source of variation is the deviation from component target values, the product designer may decide to tighten the component tolerance so that the product performance has a smaller variation. As one can see from these examples, most of the counter-measures are quite costly. An alternative is to use robust parameter design to reduce variation by changing the settings of the control factors to make the product or process less sensitive to the noise variation. In order to under-stand this strategy, we need to explain the nature of noise factors in the next section.

10.3   NOISE (HARD-TO-CONTROL) FACTORS

Noise factors are factors whose values are hard to control during normal process or use conditions. The terms "noise" and "hard to control" will be used interchangeably in this book. There are many types of noise factors, some of which are not mutually exclusive. The major ones are discussed here.

1. *Variation in Process Parameters.* If a process parameter cannot be con-trolled precisely during process runs, it should be treated as a noise factor. The quench oil temperature in the leaf spring experiment is a good example. Furnace temperature in the heat treatment illustrates the dual nature of a process parameter. If it can be controlled precisely, it is treated as a control factor in the experiment. Otherwise, it should be treated as a noise factor.

2. *Variation in Product Parameters.* In product design, the components of a product are assigned target values and tolerances. For example, a mechanical part has a nominal value of 3 mm and ± 0.1 mm as its tolerance, or a circuit's output has 10 volts as its target and ± 0.2 volts as its tolerance. Because of the variation in manufacturing, the deviation of the component values from the target should be treated as noise variation. In studying the product performance, the variation transmitted by the variation within the tolerances of the components should be taken into account. If the target values can be varied, they should be treated as control factors. In circuit design, they are also called nominal values, and the determination of nominal values is

referred to as nominal design. This type of noise factor is called *internal noise* by Taguchi (1986) because the noise factor and the control factor represent two different aspects (i.e., tolerance and nominal value) of the same variable.

3. *Environmental Variation.* Humidity, temperature, and dust levels in a room are common examples of environmental factors. Unless one is willing to install an expensive device to control the room temperature (or humidity), a more economical approach is to live with the variation and employ parameter design methodology to make the product/process less sensitive to the variation.

4. *Load Factors.* These refer to the external load to which the product or process is subjected. For example, in designing a washer, the laundry load must be considered. In a different market (e.g., different countries or regions), it may be customary to put a heavier load than what is specified in the manual. If this is the case, then it is advisable to design for the heavier load. Other examples include the passenger load or luggage load of a vehicle, amount of food in a refrigerator, and the number of times a refrigerator door is opened.

5. *Upstream Variation.* In a process, the variation of a part coming from the previous process step (or from a supplier) can have a large effect on the current process step. If the upstream variation cannot be controlled, it should be treated as a noise factor. For example, stamped door panels may have a significant amount of variation during stamping, which can make it hard to assemble the door panels to the body frame. The engineers should try to reduce the variation in stamping as much as possible. The residual variation then should be taken into account by the subsequent assembly operation.

6. *Downstream or User Conditions.* In designing the current process step, consideration may be given to how the parts will be processed in later steps. For example, in the design of the ink printing step in circuit fabrication, a particular chemical may give less variation in the printed lines but may be more susceptible to baking which will be applied in a later step. The conditions under which a product will be used should be considered in the design of the product. If 20% of the vehicles will be sold to regions with bumpy road conditions, stronger shock absorbers may be installed.

7. *Unit-to-Unit and Spatial Variation.* Because uniformity of performance over many units is a major goal in quality improvement, the effect of unit-to-unit variation should be reduced. These include person-to-person variation, variation among the animals in the same litter, and batch-to-batch, lot-to-lot, and part-to-part variation. In agricultural experiments, the plot-to-plot or subplot-to-subplot variation can be attributed to the variation in soil nutrients, water conditions, etc. If the unit-to-unit variation is related to the spatial location of the unit, it is also referred to as spatial variation. Examples include the locations and facets in the layer growth experiment, four points on a door panel, and positions of parts in a furnace. For the last case, the temperature profile within the furnace may affect the result of the heat

treatment but is very hard or expensive to measure (unless sensors are installed in the furnace). Instead of measuring the temperature, a more economical approach is to treat the location as a *surrogate variable* for the temperature and use parameter design to achieve uniformity over the locations.

8. *Variation over Time.* This refers to a shorter time scale like morning to afternoon, days within the same week, or several weeks. Obviously uniformity over time is required in quality improvement. Time is often used as a surrogate variable for the many variables that are changing over time which may be unknown to the investigator or hard to measure individually. These latent variables may be the change of air quality from morning to afternoon, climatic conditions over days, and raw material changes over weeks. It can be time consuming and expensive to keep track of all these changes. If uniformity over time can be achieved, then there is no need to take such expensive measures.

9. *Degradation.* This refers to the slow deterioration of conditions over a long time period. It may be part wear, decay of bioactivity of an ingredient, or drift of an instrument. More active measures such as frequent calibration of instruments or replacement of worn parts may be taken. An alternative is to make the system insensitive to degradation of its components or subsystems.

In "traditional" design of experiments, replicates are taken over units or over time and uniformity over the replicates is an objective of experimentation. In this sense the traditional approach only uses time or location, which are surrogate variables, for variation reduction. Because robust parameter design seeks to identify variables that *directly* affect the performance variation, it is a *proactive* approach that should be viewed as an *extension and improvement over traditional practice.*

10.4   VARIATION REDUCTION THROUGH ROBUST PARAMETER DESIGN

As discussed in Section 10.2, the response variation can be reduced by tightening the noise variation, but the required measure can be costly. **Robust parameter design**, which was pioneered by Genichi Taguchi (see Taguchi, 1986, 1987), is an alternative strategy which seeks to change the control factor settings to reduce the response variation by exploiting the *interaction* between control and noise factors. Because control factors are usually much easier to change, it is a more economical and expedient approach than directly reducing the noise variation.

To facilitate a more formal discussion, denote the control factors by $\mathbf{x}$, the noise factors by $\mathbf{z}$, and the relationship between them and the response y by

$$y = f(\mathbf{x}, \mathbf{z}). \qquad (10.1)$$

Obviously the variation in $\mathbf{z}$ will be transmitted to y, which suggests tightening the variation in $\mathbf{z}$ to reduce the variation in y. Alternately, if $\mathbf{x}$ and $\mathbf{z}$ interact in their joint effect on y, then the variation in y can also be reduced by changing the settings of $\mathbf{x}$ since different settings of $\mathbf{x}$ may have different effects on the relationship between y and $\mathbf{z}$. This explains why robust parameter design works in reducing the variation in y. A special case of (10.1) will serve to explain this point better. Consider the following model:

$$y = \mu + \alpha x_1 + \beta z + \gamma x_2 z + \epsilon$$
$$= \mu + \alpha x_1 + (\beta + \gamma x_2)z + \epsilon \tag{10.2}$$

with two control factors x_1 and x_2 and one noise factor z. The model has a significant x_1 main effect and a significant interaction between x_2 and z, and ϵ represents the remaining variation not captured by z. One can then choose the value of x_2 to make the coefficient $\beta + \gamma x_2$ of z in (10.2) as small as possible, which will dampen the effect of z on y. In data analysis this can be achieved by making the **control-by-noise interaction plots** and choosing the control factor setting with the *flattest* response between y and z. For the layer growth experiment, such plots are given later in Figures 10.5 and 10.6. For example, the $H \times L$ plot in Figure 10.5 suggests that the $+$ level of H gives a flatter response and should be chosen for robustness. Further discussion of the control-by-noise interaction plots will be given in Section 10.5.2.

Robustness can also be achieved by exploiting the *nonlinearity* in the relationship (10.1) if the control factors are the nominal values of the components of a product and the noise factors are the tolerances around the nominal values. This type of noise factor was discussed in the previous section. Let $\mathbf{x}$ denote the component values and the relationship be represented by $y = f(\mathbf{x})$. Assume that each component value x_i has a distribution with mean x_{i0} and variance σ_i^2. Then the variation of $\mathbf{x}$ is *transmitted* to that of y through f. Therefore, σ^2, the variance of y, is approximately proportional to the squared slope of f at x_{i0},

$$\sigma^2 \approx \sum_i \left(\frac{\partial f}{\partial x_i} \Big|_{x_{i0}} \right)^2 \sigma_i^2. \tag{10.3}$$

If f is nonlinear, then a nominal value with a flatter slope is preferred. Figure 10.1 depicts one such scenario. Two nominal values a and b of transistor gain, each with the same tolerance, are being compared. The slope of $y = f(x)$ (which is the output voltage) is flatter at b than at a. Instead of tightening the tolerance around a, one can move the nominal value from a to b to reduce the output voltage variation around $f(b)$. For this problem it is much easier and cheaper to change the nominal value of the transistor gain than to tighten its tolerance around a. To distinguish between these two strategies, the former is called **parameter design** and the latter **tolerance design** by Taguchi.

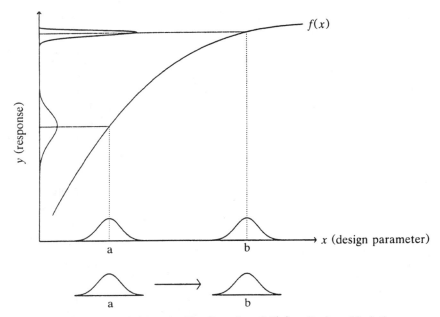

Figure 10.1. Exploiting the Nonlinearity of $f(x)$ to Reduce Variation.

Exploitation of nonlinearity can be viewed as a special case of exploiting control-by-noise interactions discussed at the beginning of this section. Take the nominal values in Figure 10.1 as a control factor and the tolerance around the nominal values as a noise factor. The slope of f at a (and respectively, b) can be used to measure the effect of the noise variation on y at the control factor setting a (and respectively, b). A significant difference between $f'(a)$ and $f'(b)$ implies nonlinearity of f but can also be interpreted as a significant control-by-noise interaction with the control and noise factors defined above.

10.5   EXPERIMENTATION AND MODELING STRATEGIES I: CROSS ARRAY

Once the noise factors are identified, they should be *systematically varied* in a parameter design experiment to reflect the noise variation that occurs in normal process or use conditions. Because noise factors are hard to control in normal conditions, it may require extra effort to vary them in a parameter design experiment. For example, the quench oil temperature in the leaf spring experiment is not controllable in normal production conditions. In order to study the robustness to the temperature variation, it is necessary to use two or more levels of the temperature for the experiment. Controlling the temperature in the two ranges 130–150°F and 150–170°F will require more effort and time. Another example is the robustness of a casting process to the size of the incoming parts, which is treated as a noise factor. A robust process should work well for parts of different sizes so that it is not necessary to adjust the process settings for different sizes. For the purpose of running a parameter design experiment, however, it is required that the incoming parts be sorted out according to their sizes so that two or more levels of the noise factor can be generated. In this situation, it is time consuming to sort and store the parts and may be disruptive to production.

Referring to the layer growth experiment, the experimental layout consists of a 2_{IV}^{8-4} design for the control factors and a 2×4 design for the noise factors. The former is called a control array and the latter is called a noise array. Generally, a **control array** refers to a design matrix for the control factors and a **noise array** refers to a design matrix for the noise factors. Each level combination in the control array is crossed with all the level combinations in the noise array. A **cross array** then consists of all the level combinations between those in the control array and those in the noise array. In mathematical terms, a cross array is a product between a control array and a noise array. If N_1 and N_2 denote the run sizes of the control and noise arrays, respectively, then $N = N_1 N_2$ is the run size for the cross array. For the layer growth experiment, $N_1 = 16$, $N_2 = 8$, and $N = 16 \times 8 = 128$. In Taguchi's terminology, a cross array is called an *inner-outer array*.

Usually orthogonal arrays are chosen for the control array and for the noise array. Because the level combinations in an orthogonal array represent points which are reasonably uniform over the region of the noise factors, running an experiment based on a cross array can be interpreted as doing a systematic Monte Carlo over the noise variation.

If some of the noise factors have more than three levels, the run size of the orthogonal array for the noise factors may be too large. An alternative is to employ a smaller plan with uniformly spread points for the noise factors. These plans include Latin hypercube sampling (Koehler and Owen, 1996) and "uniform" designs based on number-theoretic methods (Fang and Wang, 1994). Since the noise array is chosen to represent the noise variation, uniformity may be considered to be a more important requirement than orthogonality.

How should data from a cross array experiment be modeled and analyzed? Two approaches will be considered in the following subsections.

10.5.1 Location and Dispersion Modeling

The **location and dispersion modeling** approach builds models for measures of location and dispersion separately in terms of the control factor main effects and interactions. At each control factor setting, $\bar{y}_i$ and $\ln s_i^2$ (the sample mean and log sample variance over the noise replicates) are used as the measures of location and dispersion and are computed as

$$\bar{y}_i = \frac{1}{n_i} \sum_{j=1}^{n_i} y_{ij}, \qquad s_i^2 = \frac{1}{n_i - 1} \sum_{j=1}^{n_i} (y_{ij} - \bar{y}_i)^2, \qquad (10.4)$$

where n_i is the number of noise replicates for the ith control factor setting.

From the identified location and dispersion models, the **location factors** are defined to be those factors appearing in the location model; similarly, the **dispersion factors** are those factors appearing in the dispersion model. As in Section 3.2, any location factor that is not a dispersion factor is called an **adjustment factor**. For nominal-the-best problems, recommended settings for the control factors can be obtained by using the following procedure:

Two-Step Procedure for Nominal-the-Best Problem

$$(i) \ \textit{Select the levels of the dispersion factors to minimize dispersion.}$$
$$(ii) \ \textit{Select the level of the adjustment factor to bring the location} \qquad (10.5)$$
$$\textit{on target}$$

If the factor employed in step (ii) was also a dispersion factor, changing its setting could affect both the location and dispersion. One possible consequence is that the location may be adjusted to target but the dispersion has increased, which requires the readjustment of the dispersion factors and an iteration between the two steps. If one adjustment factor is not sufficient to bring the location on target, it may require two or more adjustment factors in step (ii) to do the job.

For the larger-the-better or smaller-the-better problems, as remarked in Section 5.2, it is usually advisable to interchange the two steps in the previous procedure.

Two-Step Procedure for Larger-the-Better and Smaller-the-Better Problems

$$(i) \ \textit{Select the levels of the location factors to maximize}$$
$$\textit{(or minimize) the location.}$$
$$(ii) \ \textit{Select the levels of the dispersion factors that are not} \qquad (10.6)$$
$$\textit{location factors to minimize dispersion}$$

The location-dispersion modeling approach is easy to understand and to implement. It is a natural approach for experiments based on cross arrays. In the cross array format, $\bar{y}_i$ and s_i^2 are computed over the *same* noise array. Any difference between the $\bar{y}_i$ (or s_i^2) values can then be attributed to the difference in the control factor settings, which justifies the modeling approach.

Layer Growth Experiment

Table 10.5 presents the $\bar{y}_i$ and $\ln s_i^2$ for the 16 control factor combinations of the layer growth experiment. To identify the important location and dispersion effects (from the eight main effects A–H and seven two-factor interactions involving factor A), half-normal plots of the location and dispersion effects, using the $\bar{y}_i$ and $z_i = \ln s_i^2$ as responses, were generated (see Figure

Table 10.5 Means, Log Variances, and Signal-to-Noise Ratios, Layer Growth Experiment

Control Factor											
A	B	C	D	E	F	G	H	$\bar{y}_i$	$\ln s_i^2$	$\ln \bar{y}_i^2$	$\hat{\eta}_i$
−	−	−	+	−	−	−	−	14.79	−1.018	5.389	6.41
−	−	−	+	+	+	+	+	14.86	−3.879	5.397	9.28
−	−	+	−	−	−	+	+	14.00	−4.205	5.278	9.48
−	−	+	−	+	+	−	−	13.91	−1.623	5.265	6.89
−	+	−	−	−	+	−	+	14.15	−5.306	5.299	10.60
−	+	−	−	+	−	+	−	13.80	−1.236	5.250	6.49
−	+	+	+	−	+	+	−	14.73	−0.760	5.380	6.14
−	+	+	+	+	−	−	+	14.89	−1.503	5.401	6.90
+	−	−	−	−	+	+	−	13.93	−0.383	5.268	5.65
+	−	−	−	+	−	−	+	14.09	−2.180	5.291	7.47
+	−	+	+	−	+	−	+	14.79	−1.238	5.388	6.63
+	−	+	+	+	−	+	−	14.33	−0.868	5.324	6.19
+	+	−	+	−	−	+	+	14.77	−1.483	5.386	6.87
+	+	−	+	+	+	−	−	14.88	−0.418	5.400	5.82
+	+	+	−	−	−	−	−	13.76	−0.418	5.243	5.66
+	+	+	−	+	+	+	+	13.97	−2.636	5.274	7.91

10.2). From these plots, D is an important location factor and A and H are important dispersion factors. The corresponding location and dispersion models are

$$\hat{y} = 14.352 + 0.402\, x_D$$

and

$$\hat{z} = -1.822 + 0.619 x_A - 0.982 x_H. \tag{10.7}$$

Using the two-step procedure in (10.5), the recommended levels to reduce variation are A at the "−" level and H at the "+" level. Then D can be used to bring the mean thickness on target.

By solving

$$\hat{y} = 14.352 + 0.402 x_D = 14.5,$$

$x_D = 0.368$, that is, the deposition time D should be equal to

$$\text{short time} + \frac{0.368-(-1)}{1-(-1)}(\text{long time} - \text{short time})$$

$$= 0.316 \text{ short time} + 0.684 \text{ long time},$$

which is roughly longer than the short time by two-thirds of the time span between the short and long times.

location

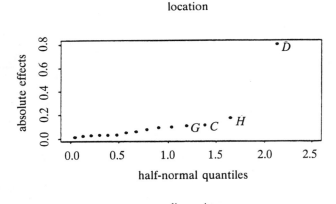

dispersion

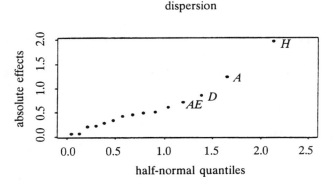

Figure 10.2. Half-Normal Plots of Location and Dispersion Effects, Layer Growth Experiment.

Leaf Spring Experiment

First, $\bar{y}_i$ and $z_i = \ln s_i^2$ for the eight control factor settings were computed as shown in Table 10.6. The calculations were performed using the six observations, which consist of three replicates each at the two levels of the noise factor Q.

Half-normal plots of the location and dispersion effects as shown in Figure 10.3 can be used to identify the important location and dispersion effects. From these plots, B, C, and E are important location factors and C is an important dispersion factor.

The corresponding location and dispersion models are

$$\hat{y} = 7.6360 + 0.1106x_B + 0.0881x_C + 0.0519x_E$$

Table 10.6   Means and Log Variances, Leaf Spring Experiment

Control Factor					
B	C	D	E	$\bar{y}_i$	$\ln s_i^2$
−	+	+	−	7.540	− 2.4075
+	+	+	+	7.902	− 2.6488
−	−	+	+	7.520	− 6.9486
+	−	+	−	7.640	− 4.8384
−	+	−	+	7.670	− 2.3987
+	+	−	−	7.785	− 2.9392
−	−	−	−	7.372	− 3.2697
+	−	−	+	7.660	− 4.0582

location

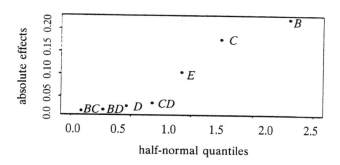

dispersion

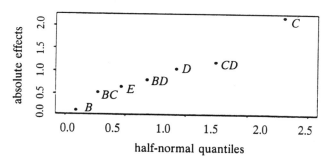

Figure 10.3. Half-Normal Plots of Location and Dispersion Effects, Leaf Spring Experiment.

and

$$\hat{z} = -3.6886 + 1.0901 x_C. \tag{10.8}$$

Using the two-step procedure in (10.5), set factor C at its $-$ level to reduce dispersion and then use factors B and E to bring the location on target. With $x_C = -1$, the equation for $\hat{y}$ becomes

$$\hat{y} = 7.5479 + 0.1106 x_B + 0.0519 x_E, \tag{10.9}$$

from which it is clearly seen that the target 8.0 cannot be attained unless x_B and x_E can be chosen beyond $+1$. Several options are available and considered in the following.

One option is to assume that B and E cannot be increased beyond $+1$. Then, for B and E at their $+$ levels and C at its $-$ level, which is denoted by $B_+ C_- E_+$, $\hat{y} = 7.71$ and $\hat{\sigma}^2 = \exp \hat{z} = 0.0084$, which give a mean-squared error (MSE) of

$$(7.71 - 8.00)^2 + 0.0084 = 0.0925.$$

Recognizing that choosing C at its $-$ level has the adverse effect of moving the mean $\bar{y}$ away from the target 8, the second option is to choose C at its $+$ level. At $B_+ C_+ E_+$, $\hat{y} = 7.89$ is closer to 8, but $\hat{\sigma}^2 = \exp \hat{z} = 0.0744$ is also higher. These settings yield an MSE of

$$(7.89 - 8.00)^2 + 0.0744 = 0.0865,$$

which is smaller than 0.0925 obtained by the two-step procedure because the decrease in squared bias dominates the increase in variance.

The third option is to assume that B and E can be increased beyond $+1$ to bring the mean on target. From (10.9), the coded values of B and E can be chosen to be 2.78 so that $\hat{y} = 8.00$. The corresponding uncoded values are high heat temperature at 1916°F and hold down time at 3.89 seconds. At a heating time of 23 seconds (i.e., C at its $-$ level), $\hat{\sigma}^2 = \exp \hat{z} = 0.0084$ gives an MSE of

$$(8.00 - 8.00)^2 + 0.0084 = 0.0084,$$

which is much smaller than under the two other scenarios. This tremendous reduction is achieved under the strong and often untenable assumption that the models hold beyond the experimental region. Because of the uncertainty from extrapolating outside the experimental region where there are no data to test the validity of the fitted model, a follow-up experiment needs to be done to verify the extrapolated predictions for larger levels of factors B and E.

Among the three scenarios discussed above, the first one, which follows the two-step procedure but cannot bring the location on target, is the worst. This illustrates the possibility that the *two-step procedure may fail* to find optimal control settings if *its second step cannot bring the location on target*. When the deviation of the location from target is significant, it may overwhelm the gain made in the first step by reducing dispersion. For this reason, B and E should not be treated as effective adjustment factors even though they appear in the $\hat{y}$ model but not in the $\ln s^2$ model of (10.8).

10.5.2 Response Modeling

A disadvantage of the location-dispersion modeling approach is that modeling location and dispersion in terms of the control factors may mask some important relationships between control and noise factors. For example, factors A and H are identified as important dispersion factors in (10.7), but the model contains no information about which noise factor interacts with A or H to make them important dispersion factors. Moreover, the dispersion measure is more likely to have a nonlinear relationship with the control factors even when the original response y has a linear relationship with the control and noise factors. This can make it more difficult to model the dispersion measure. A simple example that illustrates this problem is given as an exercise.

In the **response modeling** approach (Welch et al., 1990; Shoemaker, Tsui, and Wu, 1991), instead of computing the variance over the noise replicates, the response y is modeled as a function of both the control and noise factors. Call the fitted model $\hat{y}$ the *response model*. The analysis then consists of two parts:

(i) Make **control-by-noise interaction plots** for the significant interaction effects in the response model. From these plots, control factor settings, at which y has a flatter relationship with the noise factor, are chosen as *robust settings*.

(ii) Based on the fitted model $\hat{y}$, compute $Var(\hat{y})$ with respect to the variation among the noise factors in $\hat{y}$. Call $Var(\hat{y})$ the **transmitted variance model**. Because $Var(\hat{y})$ is a function of the control factors, we can use it to identify control factor settings with small transmitted variance.

The control factor settings determined by (i) and (ii) are not necessarily the same. In many situations, they lead to consistent or identical recommendations, but their equivalence requires some assumptions on the terms in the response model [see Hou and Wu (1997) for precise conditions for two-level designs].

The layer growth experiment will be used to illustrate the response modeling approach. A response model analysis of the leaf spring experiment is left as an exercise.

Layer Growth Experiment

To identify the important effects, a half-normal plot of all the estimable effects is generated, as displayed in Figure 10.4. Because the noise factor M has four levels, by using the system of contrasts in (6.5), we have the following three orthogonal contrasts for the M main effect:

$$M_l = (M_1 + M_2) - (M_3 + M_4),$$

$$M_q = (M_1 + M_4) - (M_2 + M_3),$$

$$M_c = (M_1 + M_3) - (M_2 + M_4),$$

where M_i stands for the effect of facet i. Each of M_l, M_q, and M_c can be viewed as a two-level factor with $-$ on two facets and $+$ on the other two facets. For example, the x_{M_l} variable in (10.10) is set at $+1$ for M_1 and M_2 and at -1 for M_3 and M_4; similarly, x_{M_q} is $+1$ for M_1 and M_4 and -1 for

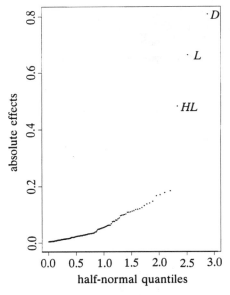

Figure 10.4. Half-Normal Plot of Response Model Effects, Layer Growth Experiment.

M_2 and M_3. From the plot, the effects D, L, and HL are identified as the most important. Note that the next largest effects appear as a cluster of four in Figure 10.4, which are H, M_l, CM_l, and AHM_q. Including these four effects gives the model

$$\hat{y} = 14.352 + 0.402x_D + 0.087x_H + 0.330x_L - 0.090x_{M_l}$$

$$- 0.239x_Hx_L - 0.083x_Cx_{M_l} - 0.082x_Ax_Hx_{M_q}. \qquad (10.10)$$

Model (10.10) contains three control-by-noise interaction effects. As discussed at the beginning of the subsection (and also in Section 10.4), control-by-noise interaction plots can be used to choose control factor settings to reduce the response variation due to noise. The interaction plots are given in Figures 10.5 and 10.6 for these three effects From Figure 10.5, recommended levels are the $+$ level for H and the $-$ level for C. Figure 10.6 suggests setting A at the $-$ level since the two middle response curves at A_- are much flatter than the other two curves corresponding to A_+. The deposition time (D) is an adjustment factor because it appears as a significant effect in (10.10) but does not interact with the noise factors M and L. This is also confirmed by the plot in Figure 10.7.

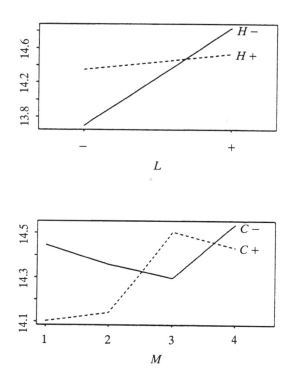

Figure 10.5. $H \times L$ and $C \times M$ Interaction Plots, Layer Growth Experiment.

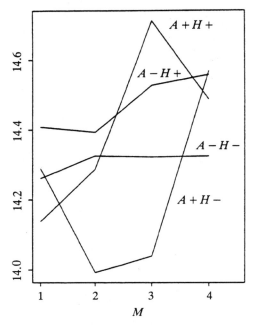

Figure 10.6. $A \times H \times M$ Interaction Plot, Layer Growth Experiment.

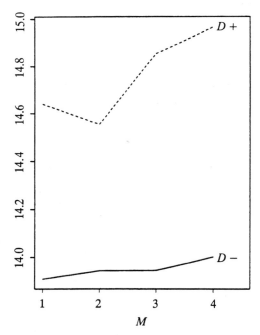

Figure 10.7. $D \times M$ Interaction Plot, Layer Growth Experiment.

The recommended settings H_+ and A_- are consistent with those based on the dispersion model in (10.7). The response model in (10.10) provides the additional information that H interacts with the noise factor L to reduce the variation of y due to L and that A interacts with the noise factor M to reduce the variation of y due to M. This information sheds new insight on how the deposition process works. The nozzle position (factor H) moves upward or downward so it makes physical sense to expect the nozzle position to have an impact on the thickness uniformity over the top and bottom positions (factor L) at each facet. Similarly, it makes physical sense to expect the rotation method (factor A) to have an impact on the thickness uniformity over the facets (factor M) but not over the top/bottom positions. For the layer growth experiment, examination of the control-by-noise interaction plots leads to consistent recommendations for the control factor settings.

In general, this may not be the case. For example, if a control factor C interacts with two noise factors N_1 and N_2, the $C \times N_1$ interaction plot may suggest the " $-$ " level of factor C while the $C \times N_2$ interaction plot may suggest the " $+$ " level of factor C. Using the transmitted variance model $Var(\hat{y})$ would avoid making conflicting recommendations on the control factor settings because the variance computation combines the various effects into a single formula. It has the additional advantage of providing an objective and formal rule for choosing factor settings, especially if visual judgment of the interaction plots is considered to be subjective.

Next we will use the response model in (10.10) to illustrate the computation of the transmitted variance model. By treating the noise factors L, M_l, and M_q as *random variables*, the variance of $\hat{y}$ with respect to them can be computed. For simplicity, we assume that each control or noise factor in (10.10) takes two values, -1 and $+1$, and L, M_l, and M_q are independent random variables. This leads to

$$x_L^2 = x_{M_l}^2 = x_{M_q}^2 = x_A^2 = x_C^2 = x_H^2 = 1,$$

$$E(x_L) = E(x_{M_l}) = E(x_{M_q}) = 0, \tag{10.11}$$

$$Cov(x_L, x_{M_l}) = Cov(x_L, x_{M_q}) = Cov(x_{M_l}, x_{M_q}) = 0.$$

From (10.10) and (10.11), it is easy to verify that

$$Var(\hat{y}) = (0.330 - 0.239x_H)^2 \, Var(x_L) + (-0.090 - 0.083x_C)^2 \, Var(x_{M_l})$$

$$+ (0.082x_A x_H)^2 \, Var(x_{M_q})$$

$$= \text{constant} + (0.330 - 0.239x_H)^2 + (-0.090 - 0.083x_C)^2$$

$$= \text{constant} - 2(0.330)(0.239)x_H + 2(0.090)(0.083)x_C$$

$$= \text{constant} - 0.158x_H + 0.015x_C. \tag{10.12}$$

Using (10.12), the recommended settings are the $+$ level for H and the $-$ level for C. This confirms the recommendation using the interaction plots in Figure 10.5. The factor A effect is significant in the log variance model in (10.7) and also according to the $A \times H \times M_q$ interaction plot. However, factor A does not appear in (10.12) since $x_A^2 = 1$ for $x_A = \pm 1$. This discrepancy can be explained by the way $Var(\hat{y})$ is computed. The $x_A x_H x_{M_q}$ term in (10.10) has M_q as its noise factor but the noise main effect x_{M_q} does not appear in (10.10). It requires the presence of both $x_A x_H x_{M_q}$ and x_{M_q} to compute their covariance so that x_A can be included in the transmitted variance model $Var(\hat{y})$. The omission of A in $Var(\hat{y})$ demonstrates a possible disadvantage of the transmitted variance approach. Another explanation of the observed discrepancy is given in Exercise 10.19.

10.6 EXPERIMENTATION AND MODELING STRATEGIES II: SINGLE ARRAY AND RESPONSE MODELING

Recall that the run size of a cross array experiment is $N = N_1 N_2$, where N_1 and N_2 are the run sizes of the control and noise arrays, respectively. If N is large and the cost of running the experiment is proportional to N, the cross array format may be too costly. An alternative is to use a **single array** for both the control and noise factors. The required run size can be much smaller. To illustrate this point, consider a modified version of the leaf spring experiment. It has three control factors,

- **(A)** heating time,
- **(B)** transfer time, and
- **(C)** hold down time,

and two noise factors,

- **(a)** high heat temperature and
- **(b)** quench oil temperature.

The last two are treated as noise factors if the high heat temperature and quench oil temperature are not controllable under normal process conditions. In order to estimate all the two-factor interactions, the cross array format must employ a 2^3 design for the control array and a 2^2 design for the noise array. (Note that a 2^{3-1} design for the three control factors would not allow any two-factor interactions among them to be estimated.) The resulting cross array given in Table 10.7 employs a 2^5 design for the five factors and has $8 \times 4 = 32$ runs. If the single array format is adopted, a 16-run 2_V^{5-1} design with $\mathbf{I} = ABCab$ can be used. Because it has resolution V, all the two-factor interactions can still be estimated. Therefore the single-array

Table 10.7 32-Run Cross Array

Runs	A	B	a b C	+ +	+ −	− +	− −
1–4	+	+	+	●	○	○	●
5–8	+	+	−	○	●	●	○
9–12	+	−	+	○	●	●	○
13–16	+	−	−	●	○	○	●
17–20	−	+	+	○	●	●	○
21–24	−	+	−	●	○	○	●
25–28	−	−	+	●	○	○	●
29–32	−	−	−	○	●	●	○

format can cut the total runs by half without sacrificing its ability to estimate any of the two-factor interactions. Because the cost of this experiment is roughly proportional to the total run size, using the 16-run single array would cut the cost approximately in half. The 16 runs in the single array are represented by those with filled-in dots in Table 10.7. (The remaining 16 runs indicated by empty circles in the table form another single array.)

Returning to the original leaf spring experiment, the cross array in Table 10.4 used a 2_{IV}^{5-1} design with $\mathbf{I} = BCDE$ that does not involve the noise factor Q. Using a single array based on a 2_V^{5-1} design with $\mathbf{I} = BCDEQ$ would have been better because a resolution V design would allow all the main effects and two-factor interactions to be clearly estimated. In contrast, none of the control-by-control interactions in the cross array are clear.

The router bit experiment in Chapter 6 provides another, albeit more complex, example of the need to use a single array for economic reasons. Factor E, spindle, should be treated as a noise factor because comparison between the four spindles is of no interest and the improvement of router bit life should be achieved over the four spindles. Using the classification of noise factors in Section 10.3, it is a type of unit-to-unit variation. The original experimental layout given in Table 6.2 employs a single array. The layout can be rearranged as in Table 10.8 by giving first the combinations of the eight control factors followed by the noise factor setting that goes with each control factor setting. It can be seen from the table that there are still 32 distinct combinations of the eight control factors and for each combination there is only one noise factor setting (i.e., one spindle). Clearly it is not a cross array. To synchronize the movement of the four spindles, the setting of each of the factors B (X–Y feed), C (in-feed), and J (rotational speed) must be the same for each of the four spindles. In the original experiment the setting of factor A (suction) was also chosen to be the same for the four spindles. Out of the 32 runs, there are only eight distinct combinations of factors A, B, C, and J. For each of the eight combinations, four different

Table 10.8   Cross Array Format, Router Bit Experiment

Run	A	B	C	J	D	F	G	H	0	1	2	3
	Control Factor								Noise Factor E			

Run	A	B	C	J	D	F	G	H	0	1	2	3
1	0	0	0	0	0	0	0	0	x			
2	0	0	0	0	1	1	1	0		x		
3	0	0	0	0	2	0	1	1				x
4	0	0	0	0	3	1	0	1			x	
5	0	1	1	0	2	1	1	0	x			
6	0	1	1	0	3	0	0	0		x		
7	0	1	1	0	0	1	0	1				x
8	0	1	1	0	1	0	1	1			x	
9	1	0	1	0	3	0	1	1	x			
10	1	0	1	0	2	1	0	1		x		
11	1	0	1	0	1	0	0	0				x
12	1	0	1	0	0	1	1	0			x	
13	1	1	0	0	1	1	0	1	x			
14	1	1	0	0	0	0	1	1		x		
15	1	1	0	0	3	1	1	0				x
16	1	1	0	0	2	0	0	0			x	
17	0	0	0	1	0	0	0	0	x			
18	0	0	0	1	1	1	1	0		x		
19	0	0	0	1	2	0	1	1				x
20	0	0	0	1	3	1	0	1			x	
21	0	1	1	1	2	1	1	0	x			
22	0	1	1	1	3	0	0	0		x		
23	0	1	1	1	0	1	0	1				x
24	0	1	1	1	1	0	1	1			x	
25	1	0	1	1	3	0	1	1	x			
26	1	0	1	1	2	1	0	1		x		
27	1	0	1	1	1	0	0	0				x
28	1	0	1	1	0	1	1	0			x	
29	1	1	0	1	1	1	0	1	x			
30	1	1	0	1	0	0	1	1		x		
31	1	1	0	1	3	1	1	0				x
32	1	1	0	1	2	0	0	0			x	

combinations of factors D, F, G, and H are assigned to the four spindles. Because the routing process must involve all the four spindles, the total cost of the experiment is proportional to 8 instead of 32 runs.

If a cross array format were adopted for the experiment, what design would be chosen? From Tables 6A.1–6A.3, the smallest orthogonal array for the eight control factors is $OA(16, 4^1 2^7)$, which has resolution III. By crossing this orthogonal array with the four spindles represented as an $OA(4, 4^1)$, a 64-run cross array is obtained. As discussed in the previous paragraph, the total cost of the experiment would be proportional to 16 instead of 64 runs.

Therefore using a cross array would require at least twice the run size of the single array. The cost in terms of material or time is also doubled. This explains why the single array format was adopted in the original experiment. Optimality criteria and selection of "optimal" single arrays will be addressed in Sections 10.8 and 10.9.

How should data from a single array experiment be analyzed? If it cannot be rearranged as a cross array (e.g. Table 10.8), obviously the location-dispersion modeling approach cannot be utilized because the approach requires that the same noise replicates be used for each control factor combination. The response modeling approach is recommended for single array experiments. For the router bit experiment the data analysis presented in Chapter 6 follows essentially the response modeling approach. By recognizing E as a noise factor, further analysis can be done, such as making control-by-noise interaction plots. These plots do not shed any new insight, however. Details of the analysis are left as an exercise.

10.7 CROSS ARRAYS: ESTIMATION CAPACITY AND OPTIMAL SELECTION

In order to compare the pros and cons of cross arrays and single arrays, we need first to understand what effects in a cross array are estimable in terms of the estimable effects in the control array and the noise array that make up the cross array. The following example illustrates a general theorem to be presented later. To avoid confusion, we shall use capital letters for control factors and small letters for noise factors in this section.

A main effect or two-factor interaction is said to be **eligible** if it is not aliased with any other main effects. If a design has at least resolution III, all its main effects are eligible. Suppose that a two-factor interaction is not eligible, i.e., it is aliased with at least one main effect. According to the hierarchical ordering principle, the latter is more likely to be important than the former, which makes the two-factor interaction ineligible for estimation. Eligibility is a weaker property than clear estimation because it only ensures that an effect is estimable if its two-factor interaction aliases (if they exist) are negligible. Therefore, in assessing a design, eligibility should be used as a secondary criterion after clear and strongly clear estimation.

Example 10.1. Suppose there are three control factors A, B, and C and three noise factors a, b, and c. The control array is given by the 2^{3-1}_{III} design with $I = ABC$ and the noise array is given by the 2^{3-1}_{III} design with $I = abc$. Then the cross array is a 16-run 2^{6-2}_{III} design with the defining relation

$$I = ABC = abc = ABCabc. \tag{10.13}$$

In the control array the three main effects A, B, and C are eligible and in the noise array the three noise main effects a, b, and c are eligible. It is easy to see from the defining relation in (10.13) that, among its 15 degrees of freedom, the six main effects A, B, C, a, b, and c are eligible and all the nine control-by-noise interactions Aa, Ab, Ac, Ba, Bb, Bc, Ca, Cb, and Cc are clear. None of the control-by-noise interactions are strongly clear because they are aliased with some three-factor interactions. For example, $Aa = BCa = Abc$ follows from (10.13).

Because a cross array can be viewed as a tensor product of the control array and the noise array, the collection of its estimable effects also has a similar product structure, namely, all the "products" between the estimable effects in the control array and the estimable effects in the noise array are estimable in the cross array. The result in Example 10.1 provides a simple illustration of the general result given in Theorem 10.1. In the theorem, a set of effects are called *estimable* if they can all be estimated in a given array under the assumption that other effects are negligible. If α_i is an effect in the control array and β_j is an effect in the noise array, $\alpha_i \beta_j$, being their symbolic product, is the *generalized interaction* between the two effects. For example, the four-factor interaction $ABab$ is the generalized interaction between AB and ab.

Theorem 10.1 Suppose a 2^{k-p} design d_C is chosen for the control array, a 2^{m-q} design d_N is chosen for the noise array, and a cross array, denoted by $d_C \otimes d_N$, is constructed from d_C and d_N.

(i) If $\alpha_1, \ldots, \alpha_A$ are the estimable factorial effects (among the control factors) in d_C and $\beta_1, \ldots, \beta_B$ are the estimable factorial effects (among the noise factors) in d_N, then $\alpha_i, \beta_j, \alpha_i \beta_j$ for $i = 1, \ldots, A$, $j = 1, \ldots, B$ are estimable in $d_C \otimes d_N$.

(ii) All the km control-by-noise interactions (i.e., two-factor interactions between a control factor main effect and a noise factor main effect) are clear in $d_C \otimes d_N$.

Proof. Let $u_1, \ldots, u_{2^p-1}$ denote the defining words for d_C and $v_1, \ldots, v_{2^q-1}$ denote the defining words for d_N. Then $d_C \otimes d_N$, which is a $2^{(k+m)-(p+q)}$ design, has the $2^{p+q} - 1$ defining words:

$$u_i, v_j, u_i v_j, \qquad i = 1, \ldots, 2^p - 1, \qquad j = 1, \ldots, 2^q - 1. \qquad (10.14)$$

To prove (i), we need to show that no pair of the effects among α_i, β_j, $\alpha_i \beta_j$ are aliased, i.e., none of their pairwise products appear among the defining words in (10.14). First we consider α_i and α_j. Because α_i and α_j are estimable in d_C, $\alpha_i \alpha_j$ cannot be equal to any of the u_i's. Because the v_j's are for the noise array d_N, $\alpha_i \alpha_j$ cannot be equal to any of the v_j's or $u_i v_j$'s. Similarly we can prove that β_i and β_j are not aliased. Consider now α_i and $\alpha_j \beta_k$. Their product $\alpha_i \alpha_j \beta_k$ cannot appear among the defining words in (10.14) because $\alpha_i \alpha_j$ is not equal to any of the u_i's. A similar argument shows that none of the following three pairs of effects are aliased: (α_i, β_j), $(\beta_i, \alpha_j \beta_k)$, and $(\alpha_i \beta_j, \alpha_{i'} \beta_{j'})$.

To prove (ii), we note that any of the km control-by-noise interactions can only appear in $u_i v_j$ among the words in (10.14). Because d_C and d_N must have at least resolution III, u_i and v_j have at least length 3. This and the fact that u_i only involves control factors and v_j only involves noise factors imply that $u_i v_j$ has at least length 6. Therefore any control-by-noise interaction can be aliased with four-factor or higher order interactions. It can also be aliased with some three-factor interactions that involve at least one control factor and one noise factor. (One such example was discussed in Example 10.1.) This proves that a control-by-noise interaction is clear in $d_C \otimes d_N$. Q.E.D.

Theorem 10.1 can be easily extended to r^{k-p} designs with r being a prime power. The detailed development follows the same proof but is more tedious. Theorem 10.1(i) is still valid when d_C and d_N are chosen to be very general arrays; that is, they do not even have to be orthogonal arrays. The result holds as long as some estimability assumption is satisfied. For a proof and the required model assumptions, see the Appendix of Shoemaker, Tsui, and Wu (1991). The conclusion in Theorem 10.1(ii) is particularly interesting. As demonstrated in Section 10.5, the control-by-noise interactions play a very important role in determining robust factor settings. An advantage of the cross array is that it allows *all* the control-by-noise interactions to be estimated under the assumption that three-factor and higher order interactions are negligible.

According to Theorem 10.1 and its generalization, all the control-by-noise interactions are clear in a cross array $d_C \otimes d_N$ no matter how the control array d_C and the noise array d_N are chosen (as long as they have at least resolution III). The choice of d_C depends on what factorial effects among the control factors should be estimable. In general, a *minimum aberration design* (or an efficient design according to the aberration criterion) *should be chosen for the control array*. Because the estimability of factorial effects among the noise factors is not a major concern, choice of the noise array may be dictated by the need to estimate some of the $\alpha_i \beta_j$'s in Theorem 10.1(i). If such a need cannot be clearly stated, a *minimum aberration design for the noise array* will be a safe choice. If three or more levels are required, other designs discussed in Chapters 5–9 may be used for the control array and the noise array. They include three-level, mixed-level, nonregular, and second-order designs.

10.8 CHOOSING BETWEEN CROSS ARRAYS AND SINGLE ARRAYS

How do the single arrays stack up against the cross arrays in terms of their estimation capacities? Here we use the term *estimation capacity* to refer to the overall performance of an array in estimating different effects such as the number of effects that are eligible, clear, or strongly clear. We will use the following comparison of the cross array in Example 10.1 with two competing single arrays to illustrate some general points to be given later.

Example 10.2. An alternative to the cross array in Example 10.1 is to choose a single array given by the minimum aberration 2_{IV}^{6-2} design with $I = ABCa = ABbc = abcC$. Because it has resolution IV, all its main effects are clear. The two-factor interactions are eligible but not clear and are divided into seven groups of aliased effects as follows:

$$AB = Ca = bc, \qquad AC = Ba, \qquad BC = Aa,$$
$$Ab = Bc, \qquad Ac = Bb, \qquad Cb = ac, \qquad Cc = ab. \qquad (10.15)$$

The two remaining degrees of freedom are for two groups of aliased three-factor interactions. Even though this design has a higher resolution than the cross array in Example 10.1, it is clearly an inferior choice for parameter design experiments because it does not allow any control-by-noise interactions to be clearly estimated. Its advantage is the ability to estimate the main effects clearly. This, however, does not outweigh its deficiency because the control-by-noise interactions provide the opportunities for robustness while the control main effects are used primarily to adjust the location on target in the two-step procedure described in Section 10.5.1.

Example 10.2 demonstrates the inadequacy of the resolution and minimum aberration criteria for parameter design experiments. These criteria are based on the hierarchical ordering principle, which treats effects of the same order equally. This ordering of effects does not work for parameter design experiments. For example, as argued previously, control-by-noise interactions are more important than control-by-control interactions. The following principle provides a more appropriate ordering of the lower order effects in parameter design experiments.

Effect Ordering Principle in Parameter Design

The importance of effects should be arranged in the following descending order of importance:

    (i)   control-by-noise interactions, control main effects, and noise main effects;

   (ii)   control-by-control interactions and control-by-control-   (10.16) by-noise interactions;

  (iii)   noise-by-noise interactions.

The control-by-noise interactions and control main effects are the most important because they are employed in the two-step procedure for parameter design optimization. The noise main effects are also included in (i). Even though the noise main effects cannot be adjusted in parameter design optimization, their magnitude can make it difficult to interpret the significance of other effects that are aliased with them. It was argued previously that the control-by-noise interactions are slightly more important than the control main effects. In general, we would prefer to put them in the same category. The effects in (ii) are one or two orders higher than those in (i) and are obtained by multiplying the first two effects in (i) by a control factor. Therefore one can use the effect ordering principle to argue that (ii) is less important than (i). The noise-by-noise interactions in (iii) are the least important because they are less likely to be important than the noise main effects, and the purpose here is not to estimate the magnitude of the noise effects but to exploit large control-by-noise interactions to dampen the effect of noise on the response. Note that the control-by-control-by-noise interactions in (ii) are three-factor interactions, which were assumed negligible or difficult to interpret in the previous chapters. In the context of parameter design, a control-by-control-by-noise interaction can be viewed as a *second-order effect* because its significance would suggest that a pair of control factors should be jointly adjusted to dampen the effect of the noise factor (in the three-factor interaction) on the response. One good example is the $A \times H \times M_q$ interaction in the response model for the layer growth experiment.

This principle should, however, not be followed indiscriminately, especially when a different ordering is more appropriate for the application under consideration.

Based on the effect ordering principle, we can find a better single array than the minimum aberration design in Example 10.2.

Example 10.3. Choose a single array with the defining relations $\mathbf{I} = ABCa = abc = ABCbc$. It is easy to see that the nine effects A, B, C, Ab, Ac, Bb, Bc, Cb, and Cc are clear; a, b, and c are eligible; and $Aa = BC$, $Ba = AC$, and $Ca = AB$ are eligible in three pairs of aliased effects. Even though it has resolution III, its estimation capacity is definitely superior to that of the minimum aberration design in Example 10.2. It is worth noting that the only three-letter word in its defining relations is the noise-by-noise-by-noise interaction abc, which is not important according to the effect ordering principle. This suggests that, in the construction of "optimal" single arrays, more noise factors should be included in the defining words of shorter length.

A comparison of the estimation capacities of the three designs in Examples 10.1–10.3 is given in Table 10.9. Design 10.2 is obviously the worst because it has only six clear effects. Both designs 10.1 and 10.3 have nine clear effects. The choice depends on what effects are considered to be important. Design 10.3 has a slight advantage because it allows three control main effects and six control-by-noise interactions to be clearly estimated. In many practical situations it is unlikely that all the nine control-by-noise interactions are important. If the six clear interactions in design 10.3 can be

Table 10.9 Estimation Capacity Comparison of the Three Designs in Examples 10.1–10.3

Design	Eligible Effects	Clear Effects
10.1	A, B, C, a, b, c	$Aa, Ab, Ac, Ba, Bb, Bc, Ca, Cb, Cc$
10.2	all 15 two-factor interactions	A, B, C, a, b, c
10.3	$a, b, c, Aa, Ba, Ca, AB, AC, BC$	$A, B, C, Ab, Bb, Cb, Ac, Bc, Cc$

chosen to represent the important ones, then the three control main effects can be used to adjust the location of the response. This *flexibility* in the selection of clear effects is an advantage of design 10.3. Design 10.1 does not have this flexibility because none of its main effects are clear. As explained before, eligible effects can be used as a secondary criterion after clear effects. One cannot, however, use the total number of eligible effects to assess a design because some of the eligible effects may be aliased with each other. For example, design 10.1 has six eligible effects while design 10.3 has nine. Because the nine eligible effects in design 10.3 include three pairs of aliased effects (which correspond to six degrees of freedom), there is little difference between the two designs in terms of eligible effects. A better criterion is to use the degrees of freedom for the eligible effects.

The run size savings from using single arrays is even more dramatic for experiments with more than two levels, as the following example shows.

Example 10.4. Suppose that there are three control factors A, B, and C, each at three levels, and two noise factors a and b, each at three levels. The control array is given by the 3_{III}^{3-1} design with $\mathbf{I} = ABC$ and the noise array is given by the 3^2 design. Then the cross array is an 81-run 3_{III}^{5-1} design with $\mathbf{I} = ABC$. Following an argument similar to the one employed in Section 5.6, it can be shown that in the control array the linear and quadratic components A_l, A_q, B_l, B_q, C_l, and C_q of the three main effects and the linear-by-linear interactions $(AB)_{ll}$ and $(AC)_{ll}$ are estimable. It is obvious that in the noise array all the main effects and two-factor interactions are estimable. According to an extension of Theorem 10.1, the 81-run cross array allows a total of 80 effects to be estimated, including the linear and quadratic components of the five main effects and the linear-by-linear, linear-by-quadratic, quadratic-by-linear, and quadratic-by-quadratic components of the six control-by-noise interactions $A \times a$, $A \times b$, $B \times a$, $B \times b$, $C \times a$, and $C \times b$. Most of the remaining 46 degrees of freedom in the cross array are allocated for effects of no interest or significance. A more economical approach is to choose a smaller single array for the five factors. For example, if only the main effects and the linear-by-linear components of the 10 two-factor interactions are of interest, a 27-run central composite design can be chosen. Smaller central composite designs with 22 or 23 runs are also available. These three designs are given in Table 9A.1.

The cost of a parameter design experiment cannot be solely determined by its total run size. A more realistic cost structure should have two components c_1, the cost per control replicate, and c_2, the cost per noise replicate. For example, in the layer growth experiment, c_1 represents the cost of changing the control factor settings and c_2 represents the cost of a wafer (since each noise replicate is a wafer). For a single array or a cross array, let N_1 be the number of its control factor combinations and N_2 be the number of noise replicates for each control factor combination. Then the total cost is given by

$$C = c_1 N_1 + c_2 N_1 N_2. \qquad (10.17)$$

Two cases are considered:
(i) If $c_1 \gg c_2 N_2$, then

$$C = c_1 N_1 \left(1 + \frac{c_2 N_2}{c_1} \right) \approx c_1 N_1, \qquad (10.18)$$

which suggests that N_1 should be minimized. A cross array is usually preferred because N_1, the size of its control array, is small. The layer growth experiment provides an excellent example for this case. Obviously c_1 is much larger than c_2 because changing the process conditions (i.e., the control factor settings) is much more time consuming and definitely more costly than the cost of wafers. Using a cross array in the original experiment was dictated by the nature of the process but can also be justified by cost considerations.
(ii) If $c_1 \ll c_2 N_2$, then

$$C = c_2 N_1 N_2 \left(1 + \frac{c_1}{c_2 N_2} \right) \approx c_2 N_1 N_2, \qquad (10.19)$$

which suggests that the total run size $N_1 N_2$ should be minimized. This usually favors the choice of single arrays because of their run size economy. The condition $c_1 \ll c_2 N_2$ is satisfied if c_1 is close to c_2 and N_2 is large. A good example is provided by the gear experiment in the exercises (see Table 10.10). Its N_2 equals 8 and $c_1 \approx c_2$ is a reasonable assumption because changing either the control settings or the noise settings would require a new set-up of the experiment.

Between these two extreme cases, the choice between single or cross arrays is not as clear cut. Other considerations should then be taken into account. A good example is the leaf spring experiment. Because changing the quench oil temperature requires a rerun of the process, its cost c_2 is close to c_1. This and the fact that $N_2 = 2$ suggests that neither term in the cost C of (10.17) dominates the other.

If a cross array is too costly to run as in situation (ii) above, an alternative to using a single array is to create a **compound noise factor** with two levels that represent two extreme conditions of the noise factors (Taguchi, 1987). For example, consider a case where there are four noise factors, each at two levels. Instead of using an eight-run array for the noise factors, one can choose two settings with opposite signs [e.g., $(-, -, +, -)$ and $(+, +, -, +)$] from the $16 \ (= 2^4)$ possible settings and call them the two levels of a compound noise factor. Its obvious advantage is the savings in run size. If N_1 represents the size of the control array, a cross array would require $8N_1$ runs while the use of a compound noise factor would require $2N_1$ runs. Since run size savings can also be achieved by using a single array, why should a compound noise factor be considered? The main reason is that the latter retains the product structure of the cross array, i.e., its experimental format is still a product between the control array and the compound noise factor. It, however, requires the assumption that the two levels of the compound noise factor represent the extreme noise conditions for *all* control factor combinations. This may not hold in a practical application or may be difficult to verify, thus making it hard for the investigator to correctly identify the two levels of the compound noise factor. If the investigator knew how the noise factors and the control factors jointly influence the response variation, it would have been unnecessary to run a parameter design experiment. Lack of such knowledge suggests that a good compound noise factor (if it exists) may be difficult to identify. The injection molding experiment in Chapter 11 uses a compound noise factor to represent the variation among four noise factors (see Table 11.2). Further discussion and comparison of the various options will be given in Section 11.6 in the context of that experiment.

10.9   OPTIMAL SELECTION OF SINGLE ARRAYS

From the discussion in Section 10.8, it is clear that the minimum aberration criterion cannot be used for choosing good single arrays; several alternate criteria can be recommended based on the effect ordering principle, however. One such criterion was used to screen out poor arrays. From the remaining arrays, those with a larger number of clear control-by-noise interactions, clear control main effects, clear noise main effects, and clear control-by-control interactions are judged to be good arrays. A collection of such arrays is given in Appendix 10A. For the specific selection criterion and a more complete collection of arrays, see Wu and Zhu (2000). Because any cross array can be expressed as a single array, cross arrays are treated as special cases of single arrays. In Appendix 10A, both are considered.

Tables 10A.1–10A.5 contain useful single arrays for 8, 16, 32, and 64 runs. To distinguish between control and noise factors, control factors are denoted by capital letters and noise factors by lowercase letters. The m generating columns for a 2^m-run design are denoted by the Arabic numbers $1, 2, \ldots, m$. Then all the $2^m - 1$ columns can be represented by the products of $1, 2, \ldots, m$.

To save space, only the generators of the arrays are given. The other defining words of the arrays can be obtained by multiplying the generators. If applicable, the last column of each table gives the collection of clear main effects and interactions of different types. When the total number of factors is close to the run size, it makes no sense to run a parameter design experiment since few or no control-by-noise interactions will be clear or eligible for estimation. For completeness, some of these nearly saturated arrays (e.g., $k = 13$ for the 16-run arrays in Table 10A.3) are included in the tables.

For each combination of k, n, c in the tables, where $k =$ total number of factors, $c =$ number of control factors, and $n =$ number of noise factors, one or more arrays are included. When two or three arrays are judged to be comparable, they are all included. Consider, for example, the two arrays in Table 10A.4 with $k = 8$, $c = 5$, and $n = 3$. The first array has five clear control main effects, nine clear control-by-noise interactions, and four clear control-by-control interactions while the second (which is a cross array) has 15 clear control-by-noise interactions. There is no clear-cut choice between the two arrays. Their relative merits are similar to the comparisons made in Section 10.8 between designs 10.1 and 10.3. That is, design 10.3 is slightly preferred because it has more flexibility in the choice of clear effects. In practice the choice should depend on the importance attached to different types of effects.

Some of the single arrays in the tables are uniformly better than the corresponding cross arrays in terms of the number of clear main effects and two-factor interactions. Obviously the resolution V designs in Table 10A.2 (for $k = 5$) and Table 10A.5 (for $k = 8$) and the resolution VI or VII designs in Table 10A.4 (for $k = 6$) and Table 10A.5 (for $k = 7$) are better than their corresponding cross arrays. Consider the case of $k = 5$, $c = 3$, and $n = 2$ in Table 10A.2. The corresponding cross array uses a 2_{III}^{3-1} design (with $\mathbf{I} = ABC$) for the control array and a 2^2 design for the noise array. It can be easily verified that this array has no clear control main effects or control-by-control interactions. In contrast, the resolution V single array in Table 10A.2 allows all the main effects and two-factor interactions to be clearly estimated. The domination over cross arrays is also observed for single arrays with resolution IV or III. Consider the single array in Table 10A.4 with $k = 7$, $c = 6$, and $n = 1$. It has nine clear control-by-control interactions while the corresponding cross array does not have any clear control-by-control interactions. This domination property holds for the arrays in Table 10A.4 with $(k, c, n) =$ (7, 5, 2), (7, 4, 3), (8, 7, 1), (8, 6, 2) (first array), (9, 8, 1), (9, 7, 2) (first array), and (9, 6, 3) (second array). As k becomes larger, at least for $n = 1$ (i.e., one noise factor), cross arrays start to dominate, as indicated by the arrays marked by ∗ in Table 10A.4 with $k = 11, 12, 13$ and $n = 1$. For 64-run arrays, it is clear from Table 10A.5 that single arrays are uniformly better than cross arrays.

A related approach is to use the concept of mixed resolution and the *D* optimal criterion to find optimal single arrays. See Borkowski and Lucas (1997).

10.10   SIGNAL-TO-NOISE RATIO AND ITS LIMITATIONS FOR PARAMETER DESIGN OPTIMIZATION

A variation of the location-dispersion modeling approach is the signal-to-noise ratio approach proposed by Taguchi (1986). Let μ and σ^2 be the mean and variance of the response at a given control factor setting and define

$$\eta = \ln \frac{\mu^2}{\sigma^2}. \tag{10.20}$$

Suppose that a cross array is employed for the experiment. The sample analog of η,

$$\hat{\eta} = \ln \frac{\bar{y}^2}{s^2}, \tag{10.21}$$

is called the **signal-to-noise ratio (SN ratio)** for the given control factor setting, where $\bar{y}$ and s^2 are the sample mean and variance as defined in (10.4) for the setting. A regression model can be used to model $\hat{\eta}$ as a function of the control factors. The fitted model is referred to as the *signal-to-noise ratio model* (*SN ratio model*). The following is a variation of the two-step procedure in (10.5):

Taguchi's Two-Step Procedure for Nominal-the-Best Problem

> (*i*)   *Select the levels of the significant control factors in the SN ratio model to maximize $\hat{\eta}$.*
> (*ii*)   *Select the level of a location factor that does not appear in (i) to bring the location on target.*   (10.22)

This procedure differs from (10.5) in the use of the SN ratio in step (i). The two procedures are not in complete opposition because a large $\hat{\eta}$ in (10.21) typically implies a small s^2 (which was advocated by the procedure in (10.5)) and/or a large $\bar{y}$. On the other hand, a large $\bar{y}$ is not required in step (i) of (10.5), which raises some question about the appropriateness of the SN ratio approach.

Although the SN ratio $\hat{\eta}$ is a natural measure in some disciplines (e.g., electrical circuit theory), it has little or no meaning for many other problems. It typically requires that the system be a linear input-output system so that the variance and squared mean are increased or decreased in the same proportion. Because of this stringent condition, it lacks a strong physical justification in most practical situations.

It can be justified, however, if the following relationship holds:

$$Var(y) \propto [E(y)]^2, \qquad (10.23)$$

which can be represented by the model

$$y = \mu(\mathbf{x}_1, x_2)\epsilon(\mathbf{x}_1), \qquad (10.24)$$

where $E(\epsilon) = 1$ and $Var(\epsilon) = P(\mathbf{x}_1)$. This model implies that y has mean $\mu(\mathbf{x}_1, x_2)$ and variance $\mu^2(\mathbf{x}_1, x_2)P(\mathbf{x}_1)$ so that

$$\eta = \ln \frac{[E(y)]^2}{Var(y)} = \ln \frac{1}{P(\mathbf{x}_1)}. \qquad (10.25)$$

Maximizing the signal-to-noise ratio η is therefore equivalent to minimizing the squared coefficient of variation $P(\mathbf{x}_1)$.

Under model (10.24), Taguchi's two-step procedure can be restated as follows:

> (*i*) Select $\mathbf{x}_1^*$ to minimize $P(\mathbf{x}_1)$.
> (*ii*) Select x_2^* so that $\mu(\mathbf{x}_1^*, x_2^*)$ equals the target value. (10.26)

By comparing (10.5) and (10.26), the factors in $\mathbf{x}_1^*$ play the role of the dispersion factors and the factor x_2 plays the role of the adjustment factor in (10.5). The relationship in (10.23) and (10.24) suggests that, when $\mu \rightarrow \lambda\mu$, $\sigma \rightarrow \lambda\sigma$, so that $\eta \rightarrow \eta$ remains the same. In this context, the SN ratio η is called a *performance measure independent of adjustment* because it is unaffected by the location adjustment in step (ii) of (10.26). This property often holds if a *scale* factor is used for adjustment. Deposition time in the layer growth experiment and mold size in injection molding are good examples of scale factors. For a theory and discussion of performance measure independent of adjustment, see León, Shoemaker, and Kacker (1987) and León and Wu (1992).

The model in (10.24) is rather specialized and cannot adequately describe many practical situations. This raises questions about the performance of the SN ratio approach for general models. Consider the model

$$y = \mu + \alpha x_2 + \beta z + \gamma x_1 z, \qquad (10.27)$$

which is a simplification of model (10.2) by dropping the error term ϵ. The variation of y in (10.27) is transmitted by the noise variation z; x_1 is a dispersion factor and x_2 is an adjustment factor. A numerical and simulation study in Bérubé and Wu (1998) shows that, under (10.27), the SN ratio approach correctly identifies x_2 as an adjustment factor *only* when

$$\left|\frac{\gamma}{\beta}\right| \gg \left|\frac{\alpha}{\mu}\right| \quad \text{and} \quad \left|\frac{\alpha}{\mu}\right| \approx 0. \qquad (10.28)$$

In the majority of situations where (10.28) is violated, the SN ratio approach fails to identify x_2 as an adjustment factor and Taguchi's two-step procedure does not work. Extensive studies in Bérubé and Wu show that the SN ratio approach does not perform well for other classes of models, including

$$y = \mu(\mathbf{x}_1, x_2) + \left(\mu(\mathbf{x}_1, x_2) \right)^\lambda \epsilon(\mathbf{x}_1),$$

where $E(\epsilon) = 0$ and $Var(\epsilon) = P(\mathbf{x}_1)$. Its validity deteriorates as the true model deviates from the assumption in (10.24).

The third problem with the SN ratio approach arises from the complexity in modeling the $\hat{\eta}$ statistic. As remarked at the beginning of Section 10.5.2, the sample variance s^2 tends to have a more complex relationship with the control factors than the original response y. Because $\hat{\eta}$ involves both s^2 and $\bar{y}$, it has an even more complex relationship. If there is one group of factors (i.e., the dispersion factors) that affect s^2 and another factor (i.e., the adjustment factor) that affects $\bar{y}$, this relationship may not be revealed in the analysis of $\hat{\eta}$ because the latter combines the location and dispersion measures into a single quantity. Except in situations like model (10.24), it is generally more difficult to find a parsimonious model for $\hat{\eta}$ than for y, $\bar{y}$, or $\ln s^2$. If the variation in $\ln s^2$ dominates the variation in $\bar{y}$, analysis of $\hat{\eta}$ and of $\ln s^2$ will lead to essentially the same findings. There will be little difference between the two-step procedures in (10.5) and in (10.22). Otherwise the recommendations from using the two procedures can be quite different.

The previous discussion suggests that the SN ratio approach has a limited justification, which is often hard to verify by data or be supported on physical grounds. It is thus not recommended for parameter design optimization. In case it is used in data analysis, it should be supplemented by the location-dispersion modeling and the response modeling approaches. The results from the three approaches should then be compared. If there is a serious disagreement between them, it may raise doubt about the validity of the SN ratio analysis.

Another approach to the understanding of the SN ratio analysis is through data transformation. Because η is approximately equivalent to the negative of the log variance of $\ln y$, the SN ratio analysis can be viewed as a special case of the location-dispersion analysis within the family of the power transformations of y [see (2.34)]. Details on the modeling, analysis, and optimization strategies can be found in Nair and Pregibon (1986) and Box (1988).

10.10.1 SN Ratio Analysis of Layer Growth Experiment

Here we shall use the layer growth experiment to illustrate the SN ratio approach. For each of the 16 control factor settings in the experiment, the SN ratio statistic $\hat{\eta}$ in (10.21) is computed and given in the last column of

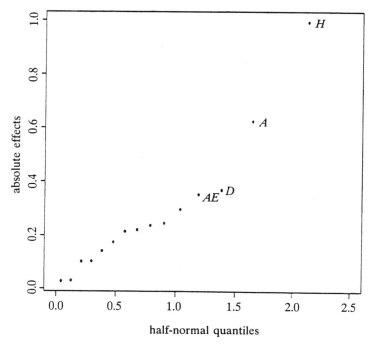

Figure 10.8. Half-Normal Plots of Effects Based on SN Ratio, Layer Growth Experiment.

Table 10.5. Because $\hat{\eta}$ in step (i) of the two-step procedure in (10.22) should be maximized, we need to identify significant effects for $\hat{\eta}$. The required model fitting procedure is the same as in Section 10.5.1 except that the response is $\hat{\eta}$ instead of $\ln s^2$. A half-normal plot of the effects for $\hat{\eta}$ is given in Figure 10.8. Noting the close similarity between Figure 10.8 and the bottom of Figure 10.2 (which is for $\ln s^2$), the SN ratio analysis leads to the same conclusion as for the dispersion analysis based on $\ln s^2$. The similarity can be explained as follows. By examining the columns for $\ln s_i^2$ and $\ln \bar{y}_i^2$ in Table 10.5, it is seen that the variation among the former is much larger than the variation among the latter. This and the relationship

$$\hat{\eta}_i = \ln \bar{y}_i^2 - \ln s_i^2$$

suggest that, for the layer growth experiment, the variation in the SN ratio $\hat{\eta}$ is dominated by the $\ln s^2$ term.

Even though the SN ratio approach and the location-dispersion modeling approach lead to the same results for the experiment, the latter approach is preferred for its simplicity and other advantages as explained in this section.

10.11 FURTHER TOPICS

This chapter covers some basic techniques in robust parameter design. As research on robust parameter design will continue to grow, we will not attempt to give a comprehensive review of the progress made so far. For a summary of opinions and research prior to 1992, see the panel discussion edited by Nair (1992). Only a few selected topics are briefly reviewed here.

1. *Exploitation of Noise Variation.* If there is a positive correlation between control runs with the same noise setting, Bérubé and Nair (1998) showed that a substantial efficiency gain in the estimation of the dispersion effects can be achieved by exploiting this correlation. Steinberg and Bursztyn (1998) showed that the test for dispersion effect is more powerful if the noise factor levels are systematically varied and included in the model rather than treated as random variation. Other work on the methods of estimating dispersion effects include Box and Meyer (1986), Nair and Pregibon (1988), Rosenbaum (1994), and Bergman and Hynén (1997). If the noise factor levels are hard to control even in a parameter design experiment (e.g., ambient temperature may not be controllable), the experiment may still be conducted in a variety of noise conditions (e.g., warmer and cooler room temperatures). By recording the actual values of the noise factors and treating them as covariates, a more elaborate modeling strategy that incorporates the noise effects can be used (Freeny and Nair, 1992).

2. *Modeling of Location and Dispersion.* Motivated by response surface methodology, Vining and Myers (1990) proposed a dual response modeling approach, which is similar to the location-dispersion modeling discussed in Section 10.5.1. Engel (1992) proposed a joint modeling strategy for location and dispersion. The regression parameters in the location model can be efficiently estimated by weighted least squares with weights estimated from the estimated dispersion effects. Related extensions were considered in Engel and Huele (1996) and Wolfinger and Tobias (1998).

3. *Split-Unit Structure.* The layer growth experiment provides a good example of a split-unit error structure. Each control run corresponds to a separate process condition. For each process condition, the noise conditions are represented by the wafers on different facets and locations. Therefore each process condition (i.e., control run) can be treated as a whole unit, and the wafers mounted on the rotating device can be treated as sub-units within the whole unit. Discussion of split-unit experiments in the context of parameter design can be found in Box and Jones (1992) and Section 11.6. Extensions and adaptation of the minimum aberration criterion for split-unit experiments are considered in Bingham and Sitter (1999a, 1999b).

4. *Parameter Design for General Loss Functions.* The two-step procedure for nominal-the-best problems in (10.5) was based on the quadratic loss

function. Suppose that the true quality loss is highly skewed. For example, a smaller windshield has to be scrapped while a larger windshield can be reworked; the cost of scrap is much higher than the cost of rework. In this situation, is it still appropriate to bring the location on target as in step (ii) of (10.5)? Intuitively, the location should be moved to the side of the target with the smaller loss. A systematic development of a two-step procedure for general loss functions can be found in Moorhead and Wu (1998).

5. *Parameter Design and Tolerance Design.* By exploiting the control-by-noise interactions, parameter design can reduce the response variation by changing the control factor settings. If further reduction in the response variation is desired, a more expensive measure to reduce the noise variation directly may have to be taken. This step is called *tolerance design* by Taguchi (1986) and was briefly discussed in point 5 of Section 10.2. The procedure of *performing a parameter design followed by a tolerance design* is called a *two-stage design*. An economical alternative (Li and Wu, 1999) is to perform parameter design and tolerance design in a *single* step and is referred to as an *integrated design*. See also Chan and Xiao (1995).

10.12   PRACTICAL SUMMARY

1. Prior to using robust parameter design, strategies for reducing variation can be tried which include (i) sampling inspection, (ii) control chart and process monitoring, (iii) blocking, (iv) covariate adjustment, and (v) reducing noise variation.

2. There are two broad types of factors in a system (i.e., a product or process): control factors and noise factors. Control factors are variables whose values remain fixed once they are chosen. Noise factors are variables which are hard to control during normal process or use conditions. It is important to recognize the nature of noise factors for a given application. Nine types of noise factors are discussed in Section 10.3.

3. The strategy of robust parameter design is to choose control factor settings that make a system less sensitive to noise variation. Even though noise factors are hard to change and control during normal process or use conditions, in a parameter design experiment, they should be systematically varied to represent typical noise conditions.

4. There are two experimental formats for parameter design experiments: cross arrays and single arrays. Comparisons between these two formats are discussed in Section 10.8. Of particular value is the discussion in (10.17)–(10.19) in terms of the costs per control replicate and per noise replicate.

5. For a cross array experiment, there are two analysis strategies:
   (i) location-dispersion modeling in Section 10.5.1 and the two-step procedure in (10.5) or (10.6);
   (ii) response modeling in Section 10.5.2 which exploits the control-by-noise interaction plots, the response model, and the transmitted variance model.

6. A third option for analyzing a cross array experiment is to use the signal-to-noise (SN) ratio in (10.21) and the two-step procedure in (10.22). Its validity is based on the model assumption in (10.23) or (10.24). Because of its limited justification, use of the SN ratio is not recommended. Its usage should be supplemented by the location-dispersion modeling and the response modeling. If there is a serious disagreement among the results from following these approaches, the validity of the SN ratio analysis may be questionable.

7. For a single array experiment that cannot be expressed as a cross array, the location-dispersion modeling approach is not applicable. The response modeling approach should be used.

8. For two-level factors, Theorem 10.1(i) provides a fundamental result on the estimable effects in a cross array by relating them to the estimable effects in the control array and in the noise array. Theorem 10.1(ii) shows that all control-by-noise interactions in a cross array are clear. Based on the theorem, an optimal design, such as a minimum aberration design, should be chosen for the control array. Choice of an optimal noise array is less critical and may be dictated by other considerations as discussed near the end of Section 10.8.

9. For two-level factors, choice of an optimal single array is based on a modified minimum aberration criterion and the effect ordering principle in parameter design [see (10.16)]. A useful collection of efficient single arrays for 8, 16, 32, and 64 runs is given in Appendix 10A.

10. A compound noise factor should be considered only if its two levels represent the extreme conditions for all settings of the control factors. Unless this assumption can be verified, use of a single array should be considered as an alternative because of its savings in run size and cost.

EXERCISES

1. Suppose that the response y follows the linear model in (10.2). A dispersion measure of y is based on the variance of y with respect to the noise variation in (10.2). Show that this measure is a quadratic function in a control factor. Between the original response y and its dispersion measure, which is easier to model statistically? Find other examples (i.e., models) to support the argument that a response model is easier to fit than a dispersion model.

2. Use the formal tests you learned in Chapter 3 to assess the significance of the effects in the two plots in Figure 10.2. Give p values for the effects that were visually identified as significant in the analysis in (10.7). Comment on your findings.

3. Repeat the previous problem for the two plots in Figure 10.3.

4. (a) Use Lenth's method to determine the p values of the effects chosen for model (10.10). Is it justifiable to include the cluster of four effects (H, M_l, CM_l, AHM_q) in the model?

   (b) Make another half-normal plot after removing D, L, and HL in Figure 10.4. Does the cluster of four effects in Figure 10.4 stand out in the new plot? Relate your finding to the p values for these four effects that were computed in (a). (The analysis here employs the step-down multiple comparison procedure as described in Section 6.7.)

5. By taking the average over the three replicates for each of the control-noise factor combinations in the leaf spring experiment, compute $\bar{y}_i$ and $\ln s_i^2$ for the 16 combinations and model each of them separately in terms of the effects in B, C, D, E, and Q. Use the response modeling approach to determine optimal settings of B, C, D, and E and compare your findings with those in Section 10.5.1.

6. Explain why factor C appears as an important factor in the response model (10.10) for the layer growth experiment while it is not included in the log variance model (10.7) in the location-dispersion approach. (*Hint:* The CM_l term was included as part of a cluster of effects in the half-normal plot in Figure 10.4. If CM_l were to be dropped, then the other three terms should also be dropped. What would happen then?)

7. Based on the plot in Figure 10.6, explain why the $A \times M$ interaction effect and plot will not be significant. If no three-factor interactions were allowed in the fitting of the response model for the layer growth experiment, would it be possible to identify the important role of factor A for variation reduction? What types of three-factor interactions in parameter design experiments should be considered in statistical modeling?

8. One of the objectives of the router bit experiment is to uniformly improve router bit life over the four spindles. With this in mind, reanalyze the data in Chapter 6 by treating spindle as a noise factor. Use the response modeling approach, including the control-by-noise interaction plots. Does it make sense in this case to compute the transmitted variance model $Var(\hat{y})$? Do you find any new information not contained in the analysis of Chapter 6? Explain why.

9. (a) The original layer growth experiment was based on a cross array. Use Theorem 10.1 to analyze its capacity for effect estimation, i.e., what effects are eligible, clear, and strongly clear.

   (b) Suppose a single array format was adopted for the experiment and it was desirable to reduce the total number of runs (i.e., number of wafers) from 128 to 64. Find a single array that allows all the control factor main effects and some of the control-by-noise and control-by-control interaction effects to be clearly estimated. Is this array a minimum aberration 2^{11-5} design?

   (c) Compare the arrays in (a) and (b) in terms of their overall estimation capacities.

   (d) As remarked in Section 10.8, the actual cost of the layer growth experiment is determined by the number of distinct combinations of control factor combinations. Determine this number for each of the two experimental formats. Let c_1 denote the cost of changing a control factor setting and c_2 the cost of changing a noise factor setting (which is the cost of a wafer). Obviously c_1 is much larger than c_2. Discuss the costs of the two experiments in terms of c_1 and c_2. How large would the cost ratio c_1/c_2 have to be for the cross array format to be more economical?

10. For the design in Example 10.2, identify the groups of aliased three-factor interactions for the two remaining degrees of freedom. How useful is the information in these two groups?

11. Suppose that there are four control factors A, B, C, and D and two noise factors r and s, each at two levels.

    (a) It is required that the interactions AB, AC, and AD should be estimable, assuming that the other interactions are negligible. Construct an eight-run control array that satisfies this requirement.

    (b) With the control array in (a) and the 2^2 design for the noise array, describe the corresponding cross array and discuss its estimation capacity based on Theorem 10.1. Show that in the cross array 6 main effects and 12 two-factor interactions can be estimated. Write down the 12 interactions. Is this cross array a good choice in terms of run size economy?

    (c) Assume that only the 6 main effects and the 12 interactions in (b) are of interest. Use a D-optimal design algorithm to find 20-, 22-, and 24-run designs that can estimate these effects optimally according to the D criterion. Compare these designs with the cross array in (b) in terms of estimation capacity, efficiency, and run size economy.

12. Use a D-optimal design algorithm to search for D-optimal designs with 22 or 23 runs for estimating the 5 main effects A, B, C, a, b in Example

10.4 and the 10 linear-by-linear components of their interactions. Comment on the D efficiency increase in terms of the run size and compare the designs you find with the 22- and 23-run central composite designs discussed in Example 10.4.

13. (a) For the array in Table 10A.1 with $k = 5$, $c = 4$, and $n = 1$, show that its defining relation is $\mathbf{I} = ABD = ACa = BCDa$ and derive all the aliases of the main effects and two-factor interactions.

(b) Another array is defined by $\mathbf{I} = aBD = ACa = ABCD$. Derive all the aliases of the main effects and two-factor interactions.

(c) By comparing the aliasing relations in (a) and (b), argue that the array in (a) is preferred based on the effect ordering principle in (10.16).

14. Compare the two arrays in Table 10A.4 with $k = 8$, $c = 6$, and $n = 2$ in terms of the number of clear main effects and two-factor interactions and the effect ordering principle. (*Hint:* Proceed as in Table 10.9 for designs 10.1 and 10.3.)

15. Show that the best 32-run cross array for $k = 7$, $c = 4$, and $n = 3$ has a 2_{IV}^{4-1} design for its control array and a 2_{III}^{3-1} design for its noise array and that it has 4 clear control main effects and 12 clear control-by-noise interactions. By comparison, the corresponding single array in Table 10A.4 has, in addition, 6 clear control-by-control interactions.

16. Show that the best 32-run cross array for $k = 8$, $c = 6$, and $n = 2$ has a 2_{III}^{6-3} design for its control array and a 2^2 design for its noise array and that it has 12 clear control-by-noise interactions and clear noise main effects and interactions. By comparison, the first single array in Table 10A.4 for $k = 8$, $c = 6$, and $n = 2$ has additional 6 clear control main effects.

17. In a robust parameter design experiment, only one control-by-noise interaction effect turns out to be significant. Its corresponding interaction plot is given in Figure 10.9.

(a) From the values in the plot, compute the interaction effect.

(b) Suppose C is a control factor (say, regular or tapered tube) and N is a noise factor (say, tube size, which varies with the incoming parts), and y is the pull-off force. Comment on the choice of the control factor setting if the objective is to maximize pull-off force.

(c) In another scenario, the quality characteristic y is the insert force and the objective is to minimize insert force. Comment on the choice of the setting of C.

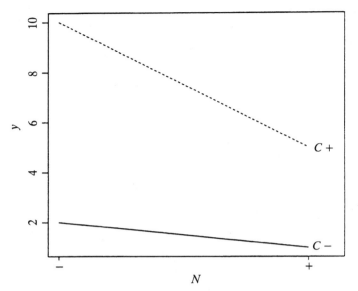

Figure 10.9. A Control-by-Noise Interaction Plot.

18. In a robust parameter design experiment, the control-by-noise inter-
    action effect turns out to be significant. Its corresponding interaction
    plot is given in Figure 10.10.

(a) From the values in the plot, compute the interaction effect.

(b) Suppose that the quality characteristic y has 3 as its nominal value
    (i.e., ideal target) and the specification limits are 3 ± 1. There exists
    another control factor which can be used to make minor adjustments
    to the mean of y without affecting its variance. Comment on the
    choice of the setting of factor C in Figure 10.10.

(c) For the interaction plot in Figure 10.11, the interaction effect has the
    same value as what you calculated in (a). Yet, for the purpose of
    robust design as defined in (b), is there an opportunity from Figure
    10.11? Comment on the difference between the two plots.

19. Suppose a response model consists of a control main effect and a
    control-by-noise interaction but no noise main effect, i.e., the noise main
    effect is zero. Assume that each factor has two levels. Show that

$$\bar{y}(N+|C+) - \bar{y}(N-|C+) = -[\bar{y}(N+|C-) - \bar{y}(N-|C-)].$$

This implies that $\bar{y}$ has the same absolute slope over N for $C-$ and
$C+$, i.e., there is no difference between $C-$ and $C+$ in terms of the
opportunity for robustness. (The plot in Figure 10.11 serves as a good
illustration.) Use what you have just proved to provide an alternative

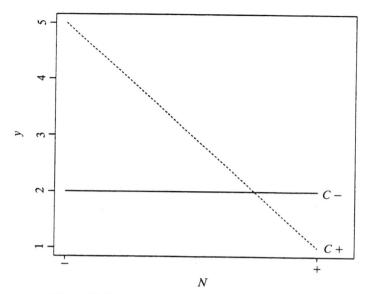

Figure 10.10. A Control-by-Noise Interaction Plot.

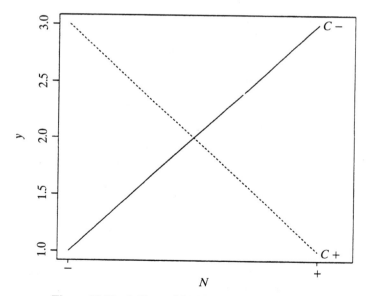

Figure 10.11. A Control-by-Noise Interaction Plot.

Table 10.10 Cross Array and Dishing Data, Gear Experiment

Run	A	B	C	D	E	F=1 G=1 H=1	F=1 G=1 H=-1	F=1 G=-1 H=1	F=1 G=-1 H=-1	F=-1 G=1 H=1	F=-1 G=1 H=-1	F=-1 G=-1 H=1	F=-1 G=-1 H=-1
1	1	1	1	1	1	7.0	12.0	6.5	14.0	3.0	14.0	4.0	16.5
2	1	1	1	-1	-1	13.5	14.5	5.5	17.0	-7.5	15.0	-4.5	12.0
3	1	1	-1	1	-1	3.0	11.0	5.5	18.0	3.0	19.0	1.0	21.0
4	1	1	-1	-1	1	10.5	14.5	6.5	17.5	3.0	14.5	9.0	24.0
5	1	-1	1	1	-1	10.0	23.0	3.5	23.0	4.5	25.5	10.0	21.0
6	1	-1	1	-1	1	6.5	22.0	14.5	23.0	5.5	18.5	8.0	21.5
7	1	-1	-1	1	1	5.5	28.0	7.5	28.0	4.0	27.5	10.5	30.0
8	1	-1	-1	-1	-1	4.0	14.0	6.5	23.0	9.0	25.5	9.0	24.5
9	-1	1	1	1	-1	-4.0	18.5	11.5	26.0	-0.5	13.0	0.0	16.5
10	-1	1	1	-1	1	9.0	19.0	17.5	21.0	0.5	20.0	6.5	18.0
11	-1	1	-1	1	1	17.5	20.0	10.0	23.0	6.5	21.5	0.0	26.0
12	-1	1	-1	-1	-1	7.0	23.5	1.0	20.0	7.0	22.5	4.0	22.5
13	-1	-1	1	1	1	2.5	22.0	12.0	19.5	7.0	27.5	8.5	23.5
14	-1	-1	1	-1	-1	24.0	26.0	14.5	27.5	7.0	22.5	13.0	22.0
15	-1	-1	-1	1	-1	5.5	27.0	2.5	31.0	12.5	27.0	11.5	32.5
16	-1	-1	-1	-1	1	11.0	21.5	12.0	27.0	16.5	29.5	16.0	28.5

explanation to the observed discrepancy in (10.12) that factor A does not appear in the transmitted variance model but the $A \times H \times M_q$ interaction is significant in the response model (10.10).

20. Consider an experiment that studied the geometric distortion of drive gears in a heat treatment process. The design matrix and response data are given in Table 10.10. The response is dishing of the gear. The five control factors are: carbon potential (A), operating mode (B), last zone temperature (C), quench oil temperature (D), and quench oil agitation (E). The three noise factors are furnace track (F), tooth size (G), and part positioning (H).

(a) What fractional factorial design is used for the control array in Table 10.10? With this control array and the 2^3 design for the noise array, describe the cross array in Table 10.10 and discuss its estimation capacity by using Theorem 10.1.

(b) Identify the large number of degrees of freedom in the cross array that are allocated for three-factor and higher interactions. Discuss the cost structure for the control and noise replicates and relate it to the run size economy of the cross array.

(c) Use response modeling to analyze the data in Table 10.10 and identify important effects. Based on the control-by-noise interaction plots and the transmitted variance model, determine control factor settings that are effective for variation reduction.

(d) Based on (b), argue that a much smaller design (in terms of the total number of runs) could have been used for the experiment without sacrificing the estimation capacity. Find a 64-run single array that is a half-fraction of the original cross array and a 32-run single array that is a quarter-fraction of the original cross array. Discuss the estimation capacities of these two single arrays and compare them with that of the cross array.

(e) Repeat the analysis in (c) based on the two single arrays in (d). Compare the identified effects and the recommended settings for the three analyses and argue that the important recommendations from the original experiment could have been obtained by using a much smaller design with half or even a quarter of the original run size.

[*Note:* This real experiment demonstrates the potential savings from using a single array. A more thorough study of this experiment including other quality characteristics and choices of single arrays can be found in Miller et al. (1993).]

21. *Modification of the Two-Step Procedure in* (10.26). Suppose that the expected quadratic loss function $R = E(y - t)^2$ is to be minimized, where t is the target value.

(a) Show that, under (10.24), $R = (\mu(x_1, x_2))^2 P(x_1) + (\mu(x_1, x_2) - t)^2$.

(b) With x_1^* chosen to minimize $P(x_1)$, show that R is minimized by selecting x_2^* to satisfy

$$\mu(x_1^*, x_2^*) = \frac{t}{1 + P(x_1^*)}. \tag{10.29}$$

Argue that (10.29) should replace step (ii) of (10.26) if the expected quadratic loss is to be minimized.

(c) The factor $(1 + P(x_1^*))^{-1}$ in (10.29) is called a *shrinkage factor*. Explain intuitively why a shrinkage factor should be applied in adjusting the location instead of bringing the location to the target value t.

(d) By substituting (10.29) in R, show that

$$R(x_1^*, x_2^*) = \frac{t^2 P(x_1^*)}{1 + P(x_1^*)}, \tag{10.30}$$

which is independent of the adjustment factor x_2. Explain why this justifies step (i) of (10.26). (Further discussion can be found in León, Shoemaker, and Kacker, 1987.)

22. Engel (1992) reported on an experiment to improve an injection molding process. The goal was to determine process parameter settings for which percent shrinkage would be consistently close to a target value. The

Table 10.11   Cross Array and Percent Shrinkage Data, Injection Molding Experiment

								a	−	−	+	+
								b	−	+	−	+
								c	−	+	+	−
Runs	A	B	C	D	E	F	G				
1–4	−	−	−	−	−	−	−	2.2	2.1	2.3	2.3
5–8	−	−	−	+	+	+	+	0.3	2.5	2.7	0.3
9–12	−	+	+	−	−	+	+	0.5	3.1	0.4	2.8
13–16	−	+	+	+	+	−	−	2.0	1.9	1.8	2.0
17–20	+	−	+	−	+	−	+	3.0	3.1	3.0	3.0
21–24	+	−	+	+	−	+	−	2.1	4.2	1.0	3.1
25–28	+	+	−	−	+	+	−	4.0	1.9	4.6	2.2
29–32	+	+	−	+	−	−	+	2.0	1.9	1.9	1.8

control factors are cycle time (A), mold temperature (B), cavity thickness (C), holding pressure (D), injection speed (E), holding time (F), and gate size (G). The noise factors are percentage regrind (a), moisture content (b), and ambient temperature (c). The experiment used a cross array with a 2_{III}^{7-4} design for the control array and a 2_{III}^{3-1} design for the noise array. The design matrix and the percent shrinkage data (in %) are given in Table 10.11.

(a) Use Theorem 10.1 to describe how the 31 degrees of freedom in the the cross array are allocated for effect estimation. Which of them can be clearly estimated?

(b) Analyze the data by using location-dispersion modeling and response modeling. Is there any discrepancy in the analysis results based on the two approaches?

(c) Is any adjustment factor identified?

(d) Determine the control factor settings that are effective for reducing variation.

23. To improve a chemical process that involved two reagents and produced desirable product and some undesirable tars, an experiment (Lawson, 1990) was conducted. One of its goals was to find settings that produced less tars. The control factors are reaction temperature (A), catalyst concentration (B), and excess of reagent 2 (C). The noise factors are purity of reagent 1 (a) and purity of the solvent stream (b). The experiment used a cross array with a Box-Behnken design for the control array and a 2^2 design for the noise array. The design matrix and the observed tar values are given in Table 10.12.

(a) Noting that the control array uses a Box-Behnken design, is it capable of estimating all the first- and second-order effects among the control factors?

Table 10.12   Cross Array and Tar Value Data, Chemical Process Experiment

Runs	A	B	a b C	+ −	− −	+ +	− +
1–4	−	−	0	57.81	37.29	47.07	42.87
5–8	+	−	0	24.89	4.35	14.69	8.23
9–12	−	+	0	13.21	9.51	11.19	10.10
13–16	+	+	0	13.39	9.15	11.23	10.30
17–20	−	0	−	27.71	20.24	24.32	22.28
21–24	+	0	−	11.40	4.48	8.23	5.44
25–28	−	0	+	30.65	18.40	24.45	20.24
29–32	+	0	+	14.94	2.29	8.49	4.30
33–36	0	−	−	42.68	22.42	30.30	21.64
37–40	0	+	−	13.56	10.08	11.38	9.85
41–44	0	−	+	50.60	13.19	30.97	18.84
45–48	0	+	+	15.21	7.44	11.82	9.78
49–52	0	0	0	19.62	12.29	14.54	13.14
53–56	0	0	0	20.60	11.49	13.49	12.06
57–60	0	0	0	20.15	12.20	13.38	14.06

(b) Based on (a) and Theorem 10.1, describe the estimation capacity of the cross array used in the experiment.

(c) The experiment has three center runs for the control factors, each of which has four noise replicates. Explain how you would use the information in the last 12 runs (i.e., runs 49–60) in the analysis. Is it necessary to allocate 12 runs for the center of the control factor region? Is this an economical way to conduct an experiment?

(d) Analyze the data by using location-dispersion modeling and response modeling.

(e) Determine control factor settings that are optimal for reducing tar values and insensitive to noise variation.

24. In an injection molding experiment (Montgomery, 1990), four control factors and three noise factors were chosen to study their effects on the average parts shrinkage and parts shrinkage variation. The four control factors are temperature (A), screw speed (B), holding time (C), and gate size (D), and the three noise factors are cycle time (a), moisture content (b), and holding pressure (c). A single array with the 2_{IV}^{7-3} design with generators $a = ABC$, $b = BCD$, and $c = ACD$ and four center runs was used for the experiment. The run order was randomized. The design matrix, actual run order, and shrinkage data (in %) are given in Table 10.13.

(a) Show that, for the single array used in the experiment, no two-factor interactions are clear.

Table 10.13   Single Array and Response Data, Injection Molding Experiment

Standard Order	Actual Run Order	A	B	C	D	a	b	c	Percent Shrinkage
1	8	−	−	−	−	−	−	−	0.6
2	16	+	−	−	−	+	−	+	1.0
3	18	−	+	−	−	+	+	−	3.2
4	17	+	+	−	−	−	+	+	6.0
5	3	−	−	+	−	+	+	+	4.0
6	5	+	−	+	−	−	+	−	1.5
7	10	−	+	+	−	−	−	+	2.6
8	2	+	+	+	−	+	−	−	6.0
9	9	−	−	−	+	−	+	+	8.0
10	15	+	−	−	+	+	+	−	1.2
11	12	−	+	−	+	+	−	+	3.4
12	6	+	+	−	+	−	−	−	6.0
13	13	−	−	+	+	+	−	−	1.6
14	19	+	−	+	+	−	−	+	5.0
15	11	−	+	+	+	−	+	−	3.7
16	1	+	+	+	+	+	+	+	5.2
17	20	0	0	0	0	0	0	0	2.5
18	4	0	0	0	0	0	0	0	2.9
19	14	0	0	0	0	0	0	0	2.4
20	7	0	0	0	0	0	0	0	2.7

(b) Analyze the data by using the response modeling approach, including the control-by-noise interaction plots and the transmitted variance model. Noting that each control-by-noise interaction effect is aliased with two other two-factor interactions, comment on the reliability of the control-by-noise interaction plots.

(c) What information can the four center runs (in the last four rows of Table 10.13) provide? Does it help in estimating the control-by-noise interaction effects? Comment on the advisability of allocating four runs for the center of the factor region.

(d) In view of (a) and (b), would you consider an alternative single array for the experiment? Make a comparison between the proposed array and the array used in the experiment in terms of estimation capacity and relevance to parameter design. (*Note:* You may consult Appendix 10A to find such an array.)

APPENDIX 10A: TABLES OF SINGLE ARRAYS BASED ON 2^{k-p} DESIGNS

Note: The notation "*I*" in the following tables stands for factor *I*, not the identity in a defining relation.

Table 10A.1 Eight-Run Single Arrays Based on 2^{k-p} Designs ($k-p=3$, $k=4,5$) (k = total number of factors, c = number of control factors, n = number of noise factors, control factors denoted by capital letters A, B, C, etc., noise factors by lowercase letters a, b, c, etc.; the three generating columns denoted by 1, 2, and 3; a cross array is indicated by *.)

k	c	n	Design Generators	Clear Effects
4*	3	1	$A=1$, $B=2$, $C=12$, $a=3$	a, Aa, Ba, Ca
4	3	1	$A=1$, $B=2$, $C=3$, $a=123$	A, B, C, a
4	3	1	$A=1$, $B=2$, $C=3$, $a=12$	C, AC, BC, Ca
4	2	2	$A=1$, $B=2$, $a=3$, $b=13$	B, AB, Ba, Bb
4	2	2	$A=1$, $B=2$, $a=3$, $b=12$	a, Aa, Ba, ab
4	2	2	$A=1$, $B=2$, $a=3$, $b=123$	A, B, a, b
5	4	1	$A=1$, $B=2$, $C=3$, $D=12$, $a=13$	
5	3	2	$A=1$, $B=2$, $C=3$, $a=12$, $b=123$	
5	3	2	$A=1$, $B=2$, $C=3$, $a=12$, $b=13$	

Table 10A.2 Sixteen-Run Single Arrays Based on 2^{k-p} Designs ($k-p=4$, $5 \leq k \leq 8$) (k = total number of factors, c = number of control factors, n = number of noise factors, control factors denoted by capital letters A, B, C, etc., noise factors by lowercase letters a, b, c, etc.; the four generating columns denoted by 1, 2, 3, and 4; a cross array is indicated by *.)

k	c	n	Design Generators	Clear Effects
5	4	1	$A=1$, $B=2$, $C=3$, $D=4$, $a=1234$	all five main effects, all 10 2fi's
5	3	2	$A=1$, $B=2$, $C=3$, $a=4$, $b=1234$	all five main effects, all 10 2fi's
5	2	3	$A=1$, $B=2$, $a=3$, $b=4$, $c=1234$	all five main effects, all 10 2fi's
6	5	1	$A=1$, $B=2$, $C=3$, $D=4$, $E=123$, $a=14$	B, C, E, BD, CD, DE, Ba, Ca, Ea
6	4	2	$A=1$, $B=2$, $C=3$, $D=123$, $a=4$, $b=14$	B, C, D, Ba, Bb, Ca, Cb, Da, Db
6	3	3	$A=1$, $B=2$, $C=3$, $a=4$, $b=123$, $c=1234$	A, B, C, D, Aa, Ac, Ba, Bc, Ca, Cc
6*	3	3	$A=1$, $B=2$, $C=12$, $a=3$, $b=4$, $c=34$	Aa, Ab, Ac, Ba, Bb, Bc, Ca, Cb, Cc
7*	6	1	$A=1$, $B=2$, $C=3$, $D=12$, $E=13$, $F=23$, $a=4$	a, Aa, Ba, Ca, Da, Ea, Fa
7	6	1	$A=1$, $B=2$, $C=3$, $D=4$, $E=234$, $F=134$, $a=124$	A, B, C, D, E, F, a
7	5	2	$A=1$, $B=2$, $C=3$, $D=123$, $E=12$, $a=4$, $b=13$	a, Aa, Ba, Ca, Da, Ea, ab
7	5	2	$A=1$, $B=2$, $C=3$, $D=4$, $E=234$, $a=14$, $b=134$	B, E, Ba, Ea
7	4	3	$A=1$, $B=2$, $C=3$, $D=4$, $a=13$, $b=124$, $c=234$	B, D, Ba, Da
7	4	3	$A=1$, $B=2$, $C=3$, $D=12$, $a=4$, $b=13$, $c=134$	Ba, Bc, Da, Dc

Table 10A.2 (*Continued*)

k	c	n	Design Generators	Clear Effects
8*	7	1	$A = 1, B = 2, C = 3, D = 12, E = 13,$ $F = 23, G = 123, a = 4$	$a, Aa, Ba, Ca, Da, Ea, Fa, Ga$
8	7	1	$A = 1, B = 2, C = 3, D = 4, E = 234,$ $F = 134, G = 123, a = 124$	A, B, C, D, E, F, G, a
8	6	2	$A = 1, B = 2, C = 3, D = 123, E = 12,$ $F = 23, a = 4, b = 13$	$a, Aa, Ba, Ca, Da, Ea, Fa, ab$
8	6	2	$A = 1, B = 2, C = 3, D = 4, E = 123,$ $F = 134, a = 124, b = 234$	A, B, C, D, E, F, a, b
8	5	3	$A = 1, B = 2, C = 3, D = 123, E = 12,$ $a = 4, b = 13, c = 134$	Ea, Ec

Table 10A.3 **Sixteen-Run Single Arrays Based on 2^{k-p} Designs ($k - p = 4$, $9 \le k \le 13$)** (**k = total number of factors, c = number of control factors, n = number of noise factors, control factors denoted by capital letters A, B, C, etc., noise factors by lowercase letters a, b, c, etc.; the four generating columns denoted by 1, 2, 3, and 4.**)

k	c	n	Design Generators
9	8	1	$A = 1, B = 2, C = 3, D = 4, E = 12, F = 13, G = 23, H = 234, a = 14$
9	7	2	$A = 1, B = 2, C = 3, D = 12, E = 13, F = 23, G = 123, a = 4, b = 14$
9	6	3	$A = 1, B = 2, C = 3, D = 4, E = 12, F = 34, a = 13, b = 24, c = 1234$
10	9	1	$A = 1, B = 2, C = 3, D = 4, E = 12, F = 13, G = 14, H = 24, I = 1234,$ $a = 23$
10	8	2	$A = 1, B = 2, C = 3, D = 4, E = 12, F = 13, G = 23, H = 123, a = 24,$ $b = 124$
10	7	3	$A = 1, B = 2, C = 3, D = 4, E = 1234, F = 24, G = 123, a = 134, b = 14,$ $c = 34$
11	10	1	$A = 1, B = 2, C = 3, D = 4, E = 124, F = 12, G = 134, H = 13, I = 1234,$ $J = 123, a = 14$
11	9	2	$A = 1, B = 2, C = 3, D = 4, E = 12, F = 13, G = 23, H = 123, I = 234,$ $a = 24, b = 124$
11	8	3	$A = 1, B = 2, C = 3, D = 4, E = 123, F = 12, G = 24, H = 234,$ $a = 13, b = 134, c = 34$
12	11	1	$A = 1, B = 2, C = 3, D = 4, E = 124, F = 12, G = 134, H = 13, I = 1234,$ $J = 123, K = 234, a = 14$
12	10	2	$A = 1, B = 2, C = 3, D = 4, E = 34, F = 124, G = 1234,$ $H = 24, I = 234, J = 123, a = 14, b = 134$
12	9	3	$A = 1, B = 2, C = 3, D = 4, E = 14, F = 123, G = 12, H = 23, I = 1234,$ $a = 13, b = 134, c = 34$
13	12	1	$A = 1, B = 2, C = 3, D = 4, E = 12, F = 124, G = 24, H = 13, I = 134,$ $J = 34, K = 23, L = 123, a = 14$
13	11	2	$A = 1, B = 2, C = 3, D = 4, E = 34, F = 134, G = 12, H = 123, I = 23,$ $J = 124, K = 24, a = 13, b = 14$
13	10	3	$A = 1, B = 2, C = 3, D = 4, E = 34, F = 12, G = 123, H = 23, I = 124,$ $J = 24, a = 13, b = 14, c = 134$

Note: **No main effect or two-factor interaction is clear** in any of the arrays in this table.

Table 10A.4 Thirty-Two Run Single Arrays Based on 2^{k-p} Designs ($k-p = 5$, $6 \le k \le 13$) (k = total number of factors, c = number of control factors, n = number of noise factors, control factors denoted by capital letters A, B, C, etc., noise factors by lowercase letters a, b, c, etc.; the five generating columns denoted by 1, 2, 3, 4, and 5; a cross array is indicated by *.)

k	c	n	Design Generators	Clear Effects
6	5	1	$A = 1$, $B = 2$, $C = 3$, $D = 4$, $E = 5$, $a = 12345$	all 6 main effects, all 15 2fi's
6	4	2	$A = 1$, $B = 2$, $C = 3$, $D = 4$, $a = 5$, $b = 12345$	all 6 main effects, all 15 2fi's
6	3	3	$A = 1$, $B = 2$, $C = 3$, $a = 4$, $b = 5$, $c = 12345$	all 6 main effects, all 15 2fi's
7	6	1	$A = 1$, $B = 2$, $C = 3$, $D = 4$, $E = 5$, $F = 123$, $a = 1245$	A, B, C, D, E, F, a, AD, AE, BD, BE, CD, CE, DE, DF, EF, Aa, Ba, Ca, Da, Ea, Fa
7	5	2	$A = 1$, $B = 2$, $C = 3$, $D = 4$, $E = 123$, $a = 5$, $b = 1245$	A, B, C, D, E, a, b, AD, BD, CD, DE, Aa, Ab, Ba, Bb, Ca, Cb, Da, Db, Ea, Eb, ab
7	4	3	$A = 1$, $B = 2$, $C = 3$, $D = 4$, $a = 5$, $b = 1234$, $c = 12345$	A, B, C, D, AB, AC, AD, BC, BD, CD, Aa, Ab, Ac, Ba, Bb, Bc, Ca, Cb, Cc, Da, Db, Dc
8	7	1	$A = 1$, $B = 2$, $C = 3$, $D = 4$, $E = 5$, $F = 123$, $G = 124$, $a = 1345$	A, B, C, D, E, F, G, a, AE, BE, CE, DE, EF, EG, Aa, Ba, Ca, Da, Ea, Fa, Ga
8	6	2	$A = 1$, $B = 2$, $C = 3$, $D = 4$, $E = 123$, $F = 124$, $a = 5$, $b = 1345$	A, B, C, D, E, F, a, b, Aa, Ab, Ba, Bb, Ca, Cb, Da, Db, Ea, Eb, Fa, Fb, ab
8	6	2	$A = 1$, $B = 2$, $C = 3$, $D = 4$, $E = 5$, $F = 234$, $a = 124$, $b = 1345$	A, B, C, D, E, F, a, b, AE, BE, CE, DE, EF, Ab, Bb, Cb, Db, Ea, Eb, Fb, ab
8	5	3	$A = 1$, $B = 2$, $C = 3$, $D = 4$, $E = 5$, $a = 124$, $b = 135$, $c = 2345$	A, B, C, D, E, BC, BE, CD, DE, Ac, Bb, Bc, Ca, Cc, Db, Dc, Ea, Ec
8*	5	3	$A = 1$, $B = 2$, $C = 3$, $D = 12$, $E = 13$, $a = 4$, $b = 5$, $c = 45$	Aa, Ab, Ac, Ba, Bb, Bc, Ca, Cb, Cc, Da, Db, Dc, Ea, Eb, Ec
9	8	1	$A = 1$, $B = 2$, $C = 3$, $D = 4$, $E = 5$, $F = 123$, $G = 124$, $H = 134$, $a = 2345$	A, B, C, D, E, F, G, H, a, AE, BE, CE, DE, EF, EG, EH, Aa, Ba, Ca, Da, Ea, Fa, Ga, Ha
9	7	2	$A = 1$, $B = 2$, $C = 3$, $D = 4$, $E = 123$, $F = 124$, $G = 134$, $a = 5$, $b = 2345$	A, B, C, D, E, F, G, a, b, Aa, Ab, Ba, Bb, Ca, Cb, Da, Db, Ea, Eb, Fa, Fb, Ga, Gb, ab
9	7	2	$A = 1$, $B = 2$, $C = 3$, $D = 4$, $E = 5$, $F = 234$, $G = 134$, $a = 124$, $b = 1235$	A, B, C, D, E, F, G, a, b, AE, BE, CE, DE, EF, EG, Ab, Bb, Cb, Db, Eb, Fb, Gb, Ea, ab
9	6	3	$A = 1$, $B = 2$, $C = 3$, $D = 4$, $E = 234$, $F = 134$, $a = 5$, $b = 124$, $c = 1235$	A, B, C, D, E, F, a, b, c, Aa, Ac, Ba, Bc, Ca, Cc, Da, Dc, Ea, Ec, Fa, Fc, ab, ac, bc

Table 10A.4 (*Continued*)

k	c	n	Design Generators	Clear Effects
9	6	3	$A = 1, B = 2, C = 3, D = 12,$ $E = 13, F = 23, a = 4, b = 5,$ $c = 12345$	$a, b, c, Aa, Ab, Ac, Ba, Bb, Bc,$ $Ca, Cb, Cc, Da, Db, Dc, Ea, Eb,$ $Ec, Fa, Fb, Fc, ab, ac, bc$
10	9	1	$A = 1, B = 2, C = 3, D = 4,$ $E = 5, F = 123, G = 124,$ $H = 134, I = 234, a = 15$	$B, C, D, F, G, H, I, BE, CE,$ $DE, EF, EG, EH, EI, Ba, Ca,$ Da, Fa, Ga, Ha, Ia
10	8	2	$A = 1, B = 2, C = 3, D = 4,$ $E = 123, F = 124, G = 134,$ $H = 234, a = 5, b = 15$	$B, C, D, E, F, G, H, Ba, Bb,$ $Ca, Cb, Da, Db, Ea, Eb, Fa, Fb,$ Ga, Gb, Ha, Hb
10	7	3	$A = 1, B = 2, C = 3, D = 4,$ $E = 234, F = 134, G = 123,$ $a = 5, b = 124, c = 1245$	$A, B, C, D, E, F, G, Aa, Ac, Ba,$ $Bc, Ca, Cc, Da, Dc, Ea, Ec, Fa,$ Fc, Ga, Gc
11*	10	1	$A = 1, B = 2, C = 3, D = 4,$ $E = 12, F = 13, G = 23,$ $H = 14, I = 234, J = 1234,$ $a = 5$	$a, Aa, Ba, Ca, Da, Ea, Fa, Ga,$ Ha, Ia, Ja
11	10	1	$A = 1, B = 2, C = 3, D = 4,$ $E = 5, F = 123, G = 124,$ $H = 134, I = 234, J = 125,$ $a = 25$	$C, D, F, G, H, I, Ca, Da,$ Fa, Ga, Ha, Ia
11	9	2	$A = 1, B = 2, C = 3, D = 4,$ $E = 5, F = 123, G = 124,$ $H = 134, I = 234, a = 25,$ $b = 125$	$C, D, F, G, H, I, Ca, Da, Fa,$ Ga, Ha, Ia
11	9	2	$A = 1, B = 2, C = 3, D = 4,$ $E = 12, F = 13, G = 14,$ $H = 234, I = 1234, a = 5,$ $b = 25$	$Ca, Cb, Da, Db, Fa, Fb, Ga, Gb,$ Ha, Hb, Ia, Ib
11	8	3	$A = 1, B = 2, C = 3, D = 4,$ $E = 5, F = 235, G = 135,$ $H = 234, a = 14, b = 125,$ $c = 245$	$B, C, E, F, G, H, Ba, Ca, Ea,$ Fa, Ga, Ha
11	8	3	$A = 1, B = 2, C = 3, D = 4,$ $E = 12, F = 13, G = 23,$ $H = 234, a = 5, b = 14,$ $c = 145$	$Ba, Bc, Ca, Cc, Ea, Ec, Fa, Fc,$ Ga, Gc, Ha, Hc
12*	11	1	$A = 1, B = 2, C = 3, D = 4,$ $E = 12, F = 13, G = 23,$ $H = 14, I = 24, J = 134,$ $K = 234, a = 5$	$a, Aa, Ba, Ca, Da, Ea, Fa, Ga,$ Ha, Ia, Ja, Ka
12	10	2	$A = 1, B = 2, C = 3, D = 4,$ $E = 5, F = 235, G = 123,$ $H = 124, I = 134, J = 234,$ $a = 25, b = 125$	$D, G, H, I, J, Da, Ga, Ha,$ Ia, Ja

Table 10A.4 (*Continued*)

k	c	n	Design Generators	Clear Effects
12	10	2	$A = 1, B = 2, C = 3, D = 4,$ $E = 12, F = 13, G = 23,$ $H = 14, I = 234, J = 1234,$ $a = 5, b = 25$	*Da, Db, Fa, Fb, Ha, Hb, Ia, Ib, Ja, Jb*
12	9	3	$A = 1, B = 2, C = 3, D = 4,$ $E = 5, F = 124, G = 12,$ $H = 123, I = 245, a = 14,$ $b = 135, c = 345$	*E, I, Ca, Gb, Gc, Ha*
12	9	3	$A = 1, B = 2, C = 3, D = 4$ $E = 12, F = 13, G = 23, H = 14,$ $I = 24, a = 5, b = 234, c = 134$	*a, Aa, Ba, Ca, Da, Ea, Fa, Ga, Ha, Ia, ab, ac*
13*	12	1	$A = 1, B = 2, C = 3, D = 4,$ $E = 12, F = 13, G = 23,$ $H = 14, I = 24, J = 134,$ $K = 234, L = 1234, a = 5$	*a, Aa, Ba, Ca, Da, Ea, Fa, Ga, Ha, Ia, Ja, Ka, La*
13	12	1	$A = 1, B = 2, C = 3, D = 4,$ $E = 5, F = 123, G = 124,$ $H = 134, I = 234, J = 125,$ $K = 135, L = 235, a = 145$	*a, A, B, C, D, E, F, G, H, I, J, K, L*
13	11	2	$A = 1, B = 2, C = 3, D = 4,$ $E = 5, F = 235, G = 123,$ $H = 245, I = 124, J = 134,$ $K = 234, a = 25, b = 125$	*G, I, J, K, Ga, Ia, Ja, Ka*
13	11	2	$A = 1, B = 2, C = 3, D = 4,$ $E = 12, F = 13, G = 23,$ $H = 123, I = 14, J = 24,$ $K = 34, a = 5, b = 45$	*Ea, Eb, Fa, Fb, Ga, Gb, Ha, Hb*
13	10	3	$A = 1, B = 2, C = 3, D = 4$ $E = 12, F = 13, G = 23, H = 14,$ $I = 24, J = 134, a = 5,$ $b = 124, c = 234$	*a, Aa, Ba, Ca, Da, Ea, Fa, Ga, Ha, Ia, ab, ac*

Table 10A.5   Sixty-Four Run Single Arrays Based on 2^{k-p} Designs ($k-p=6$, $7 \leq k \leq 13$) ($k =$ total number of factors, $c =$ number of control factors, $n =$ number of noise factors, control factors denoted by capital letters A, B, C, etc., noise factors by lowercase letters a, b, c, etc.; the six generating columns denoted by 1, 2, 3, 4, 5, and 6; a cross array is indicated by *, control-by-noise interactions denoted by CN interactions, control-by-control interactions denoted by CC interactions, and two-factor interactions denoted by 2fi's.)

k	c	n	Design Generators	Clear 2fi's
7	6	1	$A=1$, $B=2$, $C=3$, $D=4$, $E=5$, $F=6$, $a=123456$	all 2fi's
7*	6	1	$A=1$, $B=2$, $C=3$, $D=4$, $E=5$, $F=12345$, $a=6$	all 2fi's
7	5	2	$A=1$, $B=2$, $C=3$, $D=4$, $E=5$, $a=6$, $b=123456$	all 2fi's
7*	5	2	$A=1$, $B=2$, $C=3$, $D=4$, $E=1234$, $a=5$, $b=6$	all 2fi's
7	4	3	$A=1$, $B=2$, $C=3$, $D=4$, $a=5$, $b=6$, $c=123456$	all 2fi's
8	7	1	$A=1$, $B=2$, $C=3$, $D=4$, $E=5$, $F=6$, $G=1234$, $a=1256$	all 2fi's
8	6	2	$A=1$, $B=2$, $C=3$, $D=4$, $E=5$, $F=6$, $a=1234$, $b=1256$	all 2fi's
8	5	3	$A=1$, $B=2$, $C=3$, $D=4$, $E=1234$, $a=5$, $b=6$, $c=1256$	all 2fi's
9	8	1	$A=1$, $B=2$, $C=3$, $D=4$, $E=5$, $F=6$, $G=123$, $H=1345$, $a=1246$	all CN's, all CC's except AB, AC, AG, BC, BG, CG
9	7	2	$A=1$, $B=2$, $C=3$, $D=4$, $E=5$, $F=6$, $G=123$, $a=1245$, $b=1346$	all CN's, all CC's except AB, AC, AG, BC, BG, CG
9	6	3	$A=1$, $B=2$, $C=3$, $D=4$, $E=5$, $F=123$, $a=6$, $b=1345$, $c=1246$	all 2fi's except AB, AC, AF, BC, BF, CF
9	6	3	$A=1$, $B=2$, $C=3$, $D=4$, $E=5$, $F=6$, $a=1245$, $b=1346$, $c=23456$	all CC's, all CN's except Da, Db, Dc
10	9	1	$A=1$, $B=2$, $C=3$, $D=4$, $E=5$, $F=6$, $G=123$, $H=456$, $K=1345$, $a=1246$	all CN's, all $CC's$ except AB, AC, AG, BC, BG, CG, DE, DF, DH, EF, EH, FH
10	8	2	$A=1$, $B=2$, $C=3$, $D=4$, $E=5$, $F=6$, $G=123$, $H=456$, $a=1345$, $b=1246$	all 2fi's except AB, AC, AG, BC, BG, CG, DE, DF, DH, EF, EH, FH
10	7	3	$A=1$, $B=2$, $C=3$, $D=4$, $E=5$, $F=123$, $G=124$, $a=6$, $b=1345$, $c=23456$	all CN's, AE, BE, CE, DE, EF, EG, ab, ac, bc
10	7	3	$A=1$, $B=2$, $C=3$, $D=4$, $E=5$, $F=6$, $G=346$, $a=125$, $b=2356$, $c=1345$	all 2fi's except CD, CF, CG, DF, DG, FG, AB, AE, BE, Aa, Ba, Ea

Table 10A.5 (*Continued*)

k	c	n	Design Generators	Clear 2fi's
11	10	1	$A = 1, B = 2, C = 3, D = 4,$ $E = 5, F = 6, G = 123,$ $H = 124, I = 1345, J = 1346$ $a = 1256$	all CN's, $AE, AF, AI, AJ, BE,$ $BF, BI, BJ, CE, CF, CI, CJ,$ $DE, DF, DI, DJ, EG, EH, FG,$ FH, GI, GJ, HI, HJ
11	9	2	$A = 1, B = 2, C = 3, D = 4,$ $E = 5, F = 123, G = 124,$ $H = 125, I = 1345, a = 6, b = 23456$	all CN's, $AI, BI, CI, DI, EI,$ FI, GI, HI, ab
11	9	2	$A = 1, B = 2, C = 3, D = 4,$ $E = 5, F = 6, G = 123,$ $H = 124, I = 1345, a = 1346,$ $b = 23456$	$AE, BE, CE, DE, EG, EH,$ $AF, BF, CF, DF, FG, FH,$ $AI, BI, CI, DI, GI, HI, ab,$ all CN's except Ea, Fa, Ia
11	8	3	$A = 1, B = 2, C = 3, D = 4,$ $E = 5, F = 6, G = 123,$ $H = 124, a = 1345, b = 1256,$ $c = 23456$	$AE, BE, CE, DE, EG, EH,$ $AF, BF, CF, DF, FG, FH,$ $EF,$ all CN's except Ea, Eb, Ec
11	8	3	$A = 1, B = 2, C = 3, D = 4,$ $E = 5, F = 123, G = 124,$ $H = 125, a = 6, b = 1345, c = 23456$	all CN's, ab, ac, bc
12	11	1	$A = 1, B = 2, C = 3, D = 4,$ $E = 5, F = 6, G = 123,$ $H = 124, I = 134, J = 125, K = 156,$ $a = 2346$	all CN's, $CF, DF, EI, FG, FH,$ $FI, IJ, CK, DK, GK, HK, IK$
12	11	1	$A = 1, B = 2, C = 3, D = 4,$ $E = 5, F = 6, G = 236,$ $H = 1346, I = 1356, J = 2456,$ $K = 12345, a = 126$	$AD, BD, CD, AE, BE, CE,$ $DF, EF, DG, EG, AH, BH,$ $CH, AI, BI, CI, FH, GH, FI,$ $GI, AJ, BJ, CJ, FJ, GJ, AK,$ $BK, CK, FK, GK, Da, Ea, Ha,$ Ia, Ja, Ka
12	10	2	$A = 1, B = 2, C = 3, D = 4,$ $E = 5, F = 123, G = 124,$ $H = 135, I = 145, J = 12345,$ $a = 6, b = 1346$	all CN's, ab
12	10	2	$A = 1, B = 2, C = 3, D = 4,$ $E = 5, F = 236, G = 1346,$ $H = 1356, I = 2456, J = 12345,$ $a = 6, b = 126$	$AB, BD, CD, AE, BE, CE,$ $DF, EF, AG, BG, CG, AH,$ $BH, CH, FG, FH, AI, BI,$ $CI, FI, AJ, BJ, CJ, FJ, Da,$ $Ea, Ga, Ha, Ia, Ja, Db, Eb,$ Gb, Hb, Ib, Jb
12	9	3	$A = 1, B = 2, C = 3, D = 4,$ $E = 5, F = 1346, G = 1356,$ $H = 2456, I = 12345, a = 6,$ $b = 126, c = 236$	$AD, BD, CD, AE, BE, CE,$ $AF, BF, CF, AG, BG, CG,$ $AH, BH, CH, AI, BI, CI,$ $Da, Ea, Fa, Ga, Ha, Ia, Db,$ $Eb, Fb, Gb, Hb, Ib, Dc, Ec,$ Fc, Gc, Hc, Ic

Table 10A.5 (*Continued*)

k	c	n	Design Generators	Clear 2fi's
13	12	1	$A = 1, B = 2, C = 3, D = 4,$ $E = 5, F = 6, G = 123,$ $H = 124, I = 134, J = 125,$ $K = 135, L = 145, a = 2346$	all CN's, $AF, BF, CF, DF, EF,$ FG, FH, FI, FJ, FK, FL
13	11	2	$A = 1, B = 2, C = 3, D = 4,$ $E = 5, F = 6, G = 123,$ $H = 124, I = 134, J = 2345,$ $K = 2346, a = 156, b = 12345$	$BE, BF, BJ, BK, CE, CF,$ $CJ, CK, DE, DF, DJ, DK,$ $EG, EH, EI, FG, FH, FI,$ $GJ, GK, HJ, HK, IJ, IK,$ $Ba, Ca, Da, Ga, Ha, Ia, Bb,$ Cb, Db, Gb, Hb, Ib
13	11	2	$A = 1, B = 2, C = 3, D = 4,$ $E = 5, F = 123, G = 124,$ $H = 134, I = 125, J = 135,$ $K = 145, a = 6, b = 2346$	all CN's, ab
13	10	3	$A = 1, B = 2, C = 3, D = 4,$ $E = 5, F = 6, G = 123,$ $H = 124, I = 134, J = 2345,$ $a = 156, b = 12345, c = 2346$	$BE, BF, BJ, CE, CF, CJ,$ $DE, DF, DJ, EG, EH, EI,$ $FG, FH, FI, GJ, HJ, IJ,$ $Ba, Ca, Da, Ga, Ha, Ia, Bb,$ $Cb, Db, Gb, Hb, Ib, Bc, Cc,$ Dc, Gc, Hc, Ic

Note: **All main effects are clear.**

REFERENCES

Bergman, B. and Hynén, A. (1997), "Testing for Dispersion Effects from Unreplicated Designs," *Technometrics*, 39, 191–198.

Bérubé, J. and Nair, V.N. (1998), "Exploiting the Inherent Structure in Robust Parameter Design Experiments," *Statistica Sinica*, 8, 43–66.

Bérubé, J. and Wu, C.F.J. (1998), "Signal-to-Noise Ratio and Related Measures in Parameter Design Optimization," Technical Report #321, Department of Statistics, University of Michigan .

Bingham, D. and Sitter, R. (1999a), "Minimum-Aberration Two-Level Fractional Factorial Split-Plot Designs," *Technometrics*, 41, 62–70.

Bingham, D. and Sitter, R. (1999b), "Some Theoretical Results for Fractional Factorial Split-Plot Designs," *Annals of Statistics*, 27, 1240–1255.

Borkowski, J.J. and Lucas, J.M. (1997), "Designs of Mixed Resolution for Process Robustness Studies," *Technometrics*, 39, 63–70.

Box, G.E.P. (1988), "Signal-to-Noise Ratios, Performance Criteria, and Transformations," *Technometrics*, 30, 1–17.

Box, G. and Jones, S. (1992), "Split-Plot Designs for Robust Product Experimentation," *Journal of Applied Statistics*, 19, 3–26.

Box, G. and Meyer, R.D. (1986), "Dispersion Effects from Fractional Designs," *Technometrics*, 28, 19–27.

Chan, L.K. and Xiao, P.H. (1995), "Combined Robust Design," *Quality Engineering*, 8, 47–56.

Engel, J. (1992), "Modeling Variation in Industrial Experiments," *Applied Statistics*, 41, 579–593.

Engel, J. and Huele, A.F. (1996), "A Generalized Linear Modeling Approach to Robust Design," *Technometrics*, 38, 365–373.

Fang, K.T. and Wang, Y. (1994), *Number-Theoretic Methods in Statistics*, London: Chapman and Hall.

Freeny, A.E. and Nair, V.N. (1992), "Robust Parameter Design with Uncontrolled Noise Variables," *Statistica Sinica*, 2, 313–334.

Hou, X. and Wu, C.F.J. (1997), "On the Determination of Robust Settings in Parameter Design Experiments," Technical Report #288, Department of Statistics, University of Michigan.

Kackar, R.N. and Shoemaker, A.C. (1986), "Robust Design: A Cost-Effective Method for Improving Manufacturing Processes," *AT&T Technical Journal*, 65, 39–50.

Koehler, J. and Owen, A. (1996), "Computer Experiments," in *Handbook of Statistics: Design and Analysis of Experiments*, (Ed. C.R. Rao), Amsterdam: Elsevier Science B.V., pp. 261–308.

Lawson, J. S. (1990), "Improve a Chemical Process through Use of a Designed Experiment," *Quality Engineering*, 3, 215–235.

León, R.V. and Wu, C.F.J. (1992), "A Theory of Performance Measures in Parameter Design," *Statistica Sinica*, 2, 335–358.

León, R.V., Shoemaker, A.C., and Kacker, R.N. (1987), "Performance Measures Independent of Adjustment," *Technometrics*, 29, 253–285.

Li, W. and Wu, C.F.J. (1999), "An Integrated Method of Parameter Design and Tolerance Design," *Quality Engineering*, 11, 417–425.

MacKay, R.J. and Steiner, S. (1997), "Strategies for Variability Reduction," *Quality Engineering*, 10, 125–136.

Miller, A., Sitter, R., Wu, C.F.J., and Long, D. (1993), "Are Large Taguchi-style Experiments Necessary? A Reanalysis of Gear and Pinion Data," *Quality Engineering*, 6, 21–37.

Montgomery, D.C. (1990), "Using Fractional Factorial Designs for Robust Process Development," *Quality Engineering*, 3, 193–205.

Montgomery, D.C. (1991), *Introduction to Statistical Quality Control*, 2nd ed., New York: John Wiley & Sons.

Moorhead, P.R. and Wu, C.F.J. (1998), "Cost-Driven Parameter Design," *Technometrics*, 40, 111–119.

Nair, V.N. (Ed.) (1992), "Taguchi's Parameter Design: A Panel Discussion," *Technometrics*, 34, 127–161.

Nair, V.N. and Pregibon, D. (1986), "A Data Analysis Strategy for Quality Engineering Experiments," *AT&T Technical Journal*, 65, 73–84.

Nair, V.N. and Pregibon, D. (1988), "Analyzing Dispersion Effects from Replicated Factorial Experiments," *Technometrics*, 30, 247–257.

Rosenbaum, P.R. (1994), "Dispersion Effects from Fractional Factorials in Taguchi's Method of Quality Design," *Journal of the Royal Statistical Society, Series B*, 56, 641–652.

Shoemaker, A.C., Tsui, K.L., and Wu, C.F.J. (1991), "Economical Experimentation Methods for Robust Design," *Technometrics*, 33, 415–427.

Steinberg, D.M. and Bursztyn, D. (1998), "Noise Factors, Dispersion Effects and Robust Design," *Statistica Sinica*, 8, 67–85.

Taguchi, G. (1986), *Introduction to Quality Engineering*, Tokyo, Japan: Asian Productivity Organization.

Taguchi, G. (1987), *System of Experimental Design*, White Plains, NY: Unipub/Kraus International Publications.

Vining, G.G. and Myers, R.H. (1990), "Combining Taguchi and Response Surface Philosophies: A Dual Response Approach," *Journal of Quality Technology*, 22, 38–45.

Welch, W.J., Yu, T.K., Kang, S.M., and Sacks, J. (1990), "Computer Experiments for Quality Control by Parameter Design," *Journal of Quality Technology*, 22, 15–22.

Wolfinger, R.D. and Tobias, R.D. (1998), "Joint Estimation of Location, Dispersion, and Random Effects in Robust Design," *Technometrics*, 40, 62–71.

Wu, C.F.J. and Zhu, Y. (2000), "Optimal Selection of Single Arrays in Parameter Design Experiments," Technical Report, Department of Statistics, University of Michigan.

CHAPTER 11

Robust Parameter Design
for Signal-Response Systems

In some systems, the response takes on different values as a result of changes in a specified factor, called a *signal factor*. This signal-response relationship is of primary importance to the performance of the system. Such a system is called a *signal-response system* (or a system with *dynamic characteristics* in Taguchi's terminology). The nature of the signal-response relationship depends on the control and noise factors of the system. Because the signal-response relationship plays an important role in product/process development, robust parameter design for signal-response systems (also called *dynamic parameter design* in Taguchi's terminology) is an effective and powerful tool for technological development. Three important issues are considered in this chapter: (i) an appropriate performance measure that can be used to evaluate the suitability of a given signal-response relationship for the intended application, (ii) modeling and analysis strategies (performance measure modeling and response function modeling), and (iii) experimentation strategies that extend cross arrays and single arrays by incorporating the signal factor.

11.1 AN INJECTION MOLDING EXPERIMENT

Consider an experiment reported by Miller and Wu (1996) on injection molding. The system (e.g., the injection molding machine) was required to inject different amounts of material for different applications. Thus, a reliable method of controlling the amount of material injected was needed. Part weight was adopted as the response, and high injection pressure was chosen as the signal factor due to its known ability to change the amount of material injected. Seven control factors, each at two levels, were included in the experiment (see Table 11.1). These factors were chosen since they were thought to have the potential of affecting variability in part weight. Four

Table 11.1   Control Factors, Injection Molding Experiment

Factor	Level −	Level +
A. injection speed	0.0	2.0
B. clamp time (seconds)	44	49
C. high injection time (seconds)	6.3	6.8
D. low injection time (seconds)	17	20
E. clamp pressure (psi)	1700	1900
F. water cooling (°F)	70	80
G. low injection pressure (psi)	550	650

Table 11.2   Compound Noise Factor, Injection Molding Experiment

Label	Noise Factor	Level
$X_N = -1$	melt index	18
	percent regrind	5%
	operator	new
	resin moisture	high
$X_N = 1$	melt index	22
	percent regrind	0%
	operator	experienced
	resin moisture	low

noise factors were identified: melt index, percent regrind, operator, and resin moisture. Using a fractional factorial design for these four factors would require at least eight noise replicates (because a four-run design cannot study four factors). Therefore, to reduce the run size and experimental cost, a single compound noise factor at two levels was defined to represent levels of these factors (see Table 11.2). (For the pros and cons of using a compound noise factor, see the discussion at the end of Section 10.8.)

The experiment was run over two days. On the first day, the compound noise factor was set at its low level and the control factors were then varied using a 2^{7-4} design (see Table 11.3). For each control factor combination, the signal factor was varied from 650 to 1000 psi in 50-psi increments. Four parts were made and weighed at each pressure setting. On the second day, the procedure was repeated at the high level of the compound noise factor. The complete data, including the part weight (response) and the corresponding pressure setting (signal factor level), are given in Table 11.4. Analysis of the data will be considered in Section 11.5.

Table 11.3   Control Array, Injection Molding Experiment

Row				Control Factor			
	A	B	C	D	E	F	G
1	+	−	−	−	+	−	+
2	+	−	−	+	−	+	−
3	+	+	+	−	+	+	−
4	+	+	+	+	−	−	+
5	−	−	+	−	−	−	−
6	−	−	+	+	+	+	+
7	−	+	−	−	−	+	+
8	−	+	−	+	+	−	−

Table 11.4   Response Data, Injection Molding Experiment

Row				Signal Factor Level				
	650	700	750	800	850	900	950	1000	X_N
1	639.7	642.3	645.5	653.9	666.6	672.1	692.2	711.6	
	640.5	641.7	644.8	655.1	665.8	670.8	690.6	710.8	−1
	636.2	643.6	646.1	654.7	667.1	673.3	689.7	711.1	
	637.2	644.0	644.3	654.2	665.4	671.1	689.8	710.5	
2	634.4	639.9	642.6	650.2	659.9	666.8	678.4	708.3	
	632.9	640.8	640.4	651.6	660.3	660.3	682.6	710.1	−1
	633.7	641.1	643.1	650.9	657.9	659.8	681.8	707.7	
	635.8	642.4	641.9	653.2	662.1	661.5	683.2	706.6	
3	640.2	646.1	647.2	655.5	666.2	671.0	688.6	708.9	
	638.1	644.4	646.0	654.0	667.3	673.6	687.5	710.0	−1
	637.3	644.4	647.5	653.8	669.1	672.4	691.0	711.3	
	639.1	641.2	644.3	652.8	664.7	672.2	693.1	708.4	
4	641.1	644.5	647.2	652.0	665.3	669.2	688.7	709.8	
	642.1	647.3	644.8	654.6	661.0	671.1	690.4	710.1	−1
	642.0	642.8	646.0	653.8	659.7	670.1	686.3	707.7	
	641.8	643.9	646.3	651.7	662.4	671.1	685.8	706.4	
5	640.8	644.7	647.6	652.3	661.1	673.0	685.7	706.4	
	641.1	645.3	646.8	654.5	662.8	673.2	686.7	707.7	−1
	641.2	644.6	647.3	653.9	659.2	672.5	686.2	706.9	
	641.6	645.0	647.5	653.6	659.9	673.7	686.1	706.3	
6	650.4	655.4	659.7	665.8	671.0	677.7	695.6	716.5	
	650.8	655.0	660.2	665.9	670.8	677.5	696.8	717.0	−1
	651.2	654.6	660.3	665.9	671.2	678.2	694.3	718.3	
	650.7	654.9	659.3	666.4	670.5	677.8	696.1	717.6	
7	639.6	643.8	648.2	655.7	665.2	674.8	691.7	710.1	
	639.4	644.2	647.3	656.0	664.8	675.3	691.4	711.4	−1
	639.9	644.1	647.2	655.5	664.3	675.0	691.8	710.3	
	640.0	644.4	647.8	656.2	663.9	675.1	692.3	711.1	

Table 11.4 (*Continued*)

Row	Signal Factor Level								X_N
	650	700	750	800	850	900	950	1000	
8	636.5	641.8	645.2	653.8	662.8	671.8	689.4	709.7	-1
	636.2	640.6	646.1	653.9	662.3	671.6	689.1	709.6	
	635.7	640.5	645.5	653.9	662.1	671.6	689.6	709.7	
	636.1	640.3	645.0	653.6	662.4	671.6	689.3	709.3	
1	640.1	644.4	647.6	655.2	664.8	674.4	693.2	709.8	1
	641.2	646.2	646.3	657.3	669.7	671.3	689.4	714.2	
	633.6	642.8	647.2	656.4	668.3	676.7	691.1	717.2	
	638.2	643.9	647.8	658.0	669.2	675.1	695.3	704.5	
2	638.6	645.3	645.5	655.1	662.1	670.8	692.3	711.8	1
	636.3	640.2	642.1	654.3	663.6	668.2	691.1	712.3	
	634.4	641.8	642.1	653.3	660.7	672.3	690.5	714.6	
	638.2	641.1	644.3	654.6	667.1	674.3	686.7	710.1	
3	642.6	648.3	650.0	657.3	666.3	675.2	695.2	714.6	1
	640.2	642.9	648.2	659.4	667.3	674.4	691.4	713.7	
	641.6	646.1	647.9	658.1	670.1	676.6	689.9	714.2	
	639.9	645.2	649.9	660.0	671.5	678.2	699.2	709.9	
4	643.8	649.8	650.6	658.3	666.2	673.2	696.6	713.8	1
	641.6	646.3	649.7	657.9	666.8	675.8	691.2	711.7	
	642.2	645.2	648.2	659.1	670.2	675.8	690.2	711.8	
	643.6	647.2	650.1	660.0	671.8	678.2	690.6	712.2	
5	642.6	645.6	647.9	654.6	666.8	672.3	687.9	709.8	1
	641.8	645.8	648.2	655.2	665.7	674.6	688.8	710.2	
	642.0	645.7	648.0	654.7	665.8	673.9	689.3	711.3	
	642.3	646.0	647.8	654.9	669.2	675.4	688.6	710.7	
6	650.6	655.7	660.2	667.8	671.1	678.9	694.7	718.4	1
	650.2	656.2	659.7	666.5	672.0	679.3	693.3	720.2	
	651.3	655.5	659.4	666.7	671.7	679.1	696.8	716.6	
	650.1	656.0	658.9	666.6	671.4	678.6	692.1	717.0	
7	639.9	644.1	647.6	656.3	664.8	675.3	693.1	709.9	1
	640.2	644.6	648.0	656.0	665.2	674.7	692.8	711.4	
	640.3	645.0	648.2	656.4	665.1	674.9	691.9	712.2	
	640.1	644.7	647.8	656.7	665.5	675.2	692.4	711.6	
8	637.7	642.9	647.3	651.1	665.0	673.2	689.6	710.9	1
	638.1	643.4	647.3	655.4	664.7	672.8	689.9	709.3	
	638.2	643.0	646.8	655.4	664.5	673.4	690.7	708.6	
	638.4	642.9	647.0	655.2	664.8	672.8	690.2	709.1	

Note: The run number corresponds to that in Table 11.3.

11.2 SIGNAL-RESPONSE SYSTEMS AND THEIR CLASSIFICATION

In the parameter design problems considered in Chapter 10, the quality characteristic of interest (i.e., the response) is a *single* quantity y, such as layer thickness, voltage, or free height of a spring. It has a specified optimal value, which may be nominal-the-best, smaller-the-better, or larger-the-better. When there is a need for clarification, this type of response will be called a **simple response** and the system a **simple response system**.

For **signal-response systems**, the quality characteristic is the relationship between y and the signal factor M, which can be modeled by

$$y = g(M),$$

and interest lies in the **signal-response relationship** g. For M to be an effective signal factor, it should have a significant effect on y. Typically the relationship g is a monotone function. Because the behavior of g often depends on the control and noise factors of the system, parameter design methodology can be applied.

If the signal-response relationship in a given application is not recognized and is treated as a simple response, severe loss of efficiency may result. Consider, for example, the machining of parts to a specified diameter, whose nominal value is 70 mm. Because the diameter has a fixed target, it could be treated as a nominal-the-best problem. For other applications, however, the required diameter values may change. It is known that cutting depth can be used as a signal factor to effectively change the diameter value. Therefore the "machinability" of the process is characterized by the signal-response relationship between part diameter and cutting depth. A study of this relationship, with the goal of improving it, may lead to the discovery of a more efficient and agile machining process. If the study were conducted with 70 mm as its target diameter value, no such technological advance could be made.

In the terminology of Taguchi (1987, 1991), a simple response is said to have a *static* characteristic and a signal-response relationship is said to have a *dynamic* characteristic. The corresponding parameter design problems are called *static parameter design* and *dynamic parameter design*, respectively. Because the terms "static" and "dynamic" are misleading, they are not adopted as standard terminology in this book. The approach and terminology adopted in this chapter are based on Miller and Wu (1996).

It is useful to classify signal-response systems according to the function of the system, as this affects the manner in which the performance of the system should be evaluated. Two common types of signal-response systems are considered.

The first type is a **multiple target system**, whose function requires that the value of a response be adjusted by changing the level of a signal factor. One example is the injection molding system described in Section 11.1. Another example, which is given in the exercises, involves surface machining. In this case, the quality characteristic of interest is the surface roughness of parts after the surface has been machined using a lathe. Since different applications require different degrees of surface roughness, some method of controlling the surface roughness of machined parts is needed. Experience indicates that the feed rate of the tool bit can effectively be used to alter the roughness of the machined surface, so it is chosen as the signal factor. Factors such as type of lathe, cutting speed, depth of tool cut, type of tool cut, corner radius, cutting edge angle, front escape angle, and side scoop angle may affect the relationship between feed rate and surface roughness. The goal is to select settings of these control factors which will allow surface roughness to be reliably controlled by the feed rate of the tool bit.

A third example involves the braking system of an automobile. Its response is the amount of torque generated while braking and the signal factor is the pedal force. The control factors include pad material, pad taper, pad shape, and rotor material, and the noise factors include road surface conditions, tire conditions, speed, and driver's skills. The goal of parameter design is to find control factor settings (i.e., a new braking design) so that the torque-force relationship satisfies some specified requirement and is robust to noise variation.

The second type of system is a **measurement system**, which is a process used to obtain an estimate of some quantity of interest for a given unit or sample. This may include sampling, sample preparation and calibration, as well as the actual measurement process. The true amount of the quantity present can be considered as an input signal M which the system converts into a measured value or response y. The precision with which M can be estimated based on y is determined by the signal-response relationship $y = g(M)$. Consider an experiment performed at a foundry to optimize an eddy current measurement procedure used to measure part hardness. An eddy current machine uses the intensity of feedback (y) from an electronic probe to estimate hardness (M). Three control factors, frequency, probe temperature, and gain, are identified which can easily be changed and are thought to affect the relationship between hardness and feedback intensity. Two noise factors, part chemistry and part cleanliness, are included in the experiment since these represent conditions that might change from part to part and measurement to measurement and thus introduce variation into the process. The purpose of the experiment is to identify settings of the control factors that will make feedback intensity sensitive to changes in hardness but insensitive to changes in part chemistry and part cleanliness.

Another example arises in the measurement of residual imbalance in manufactured drive shafts. Unbalanced shafts are not acceptable because they cause vibration and noise. The machine reading of imbalance is the response y and the unknown amount of imbalance is the signal factor M. Control factors include master rotor, signal sensitivity, method of rotation at handling time and at measurement, and sequence of corrections of imbal-

Table 11.5   Examples of Signal-Response Systems

I. Multiple Target System	Response (y)	Signal (M)
injection molding	part weight	high injection pressure
injection molding	part size	mold size
machining of parts	part diameter	cutting depth
machining of surface	surface roughness	feed rate
plate coating	coating weight	coated area
automobile braking	torque generated while braking	pedal force
photography	photographic image	true image
paper feeder	paper travel distance	roller rotation

II. Measurement System	Response (y)	Signal (M)
drive shaft imbalance	machine reading	imbalance
eddy current surface hardness	intensity of feedback	surface hardness
engine coolant sensor	output voltage	coolant temperature

III. Control System	Response (y)	Signal (M)
vehicle steerability	vehicle turning radius	steering wheel angle
shower water temperature	water temperature	amount of adjustment
thermostat	on/off	temperature
transducers for adaptive control	force	voltage

ance. The noise factor is represented by drive shafts ranging from moderately unbalanced to severely unbalanced. The experiment and the data are described in the exercises.

A third type of signal-response system involves a control mechanism, which may use feed-forward or feedback control. For example, in adjusting the shower water temperature, the response y is the water temperature and the signal M is the amount of adjustment applied to the tap. There is a time lag in the feed-forward control from M to y. Because more needs to be learned about **control systems** for parameter design applications, it will not be considered in this book.

A variety of signal-response systems is given in Table 11.5, which demonstrates the wide range of applications of this novel approach to quality engineering.

11.2.1   Calibration of Measurement Systems

Before a measurement system can be used to estimate the unknown quantity of interest in a sample, it needs to be calibrated. A linear calibration system can be described by the simple linear regression model

$$y = \alpha + \beta M + \epsilon \qquad (11.1)$$

with $E(\epsilon) = 0$ and $Var(\epsilon) = \sigma^2$. Suppose there are p standards for which the true values of M are known. Let m_j denote their values and y_j the corresponding measured values of y. The least squares estimates for α, β, and σ^2 are

$$\hat{\beta} = \frac{S_{ym}}{S_{mm}}, \qquad \hat{\alpha} = \bar{y} - \hat{\beta}\bar{m},$$

$$s^2 = (p-2)^{-1} \sum_{j=1}^{p} \left[y_j - \bar{y} - \hat{\beta}(m_j - \bar{m}) \right]^2, \tag{11.2}$$

where

$$S_{mm} = \sum_{j=1}^{p} (m_j - \bar{m})^2, \qquad S_{ym} = \sum_{j=1}^{p} (y_j - \bar{y})(m_j - \bar{m}), \qquad \bar{m} = p^{-1} \sum_{j=1}^{p} m_j.$$

These estimates lead to the *calibration line*

$$\hat{\alpha} + \hat{\beta}M, \tag{11.3}$$

which can be used to estimate the unknown value of M for a sample. Suppose that its measured value is y_0. Equating y_0 with (11.3), $y_0 = \hat{\alpha} + \hat{\beta}M$, leads to

$$\hat{M} = \frac{y_0 - \hat{\alpha}}{\hat{\beta}}, \tag{11.4}$$

which estimates the unknown value M for the sample. To obtain a confidence interval for M, we note for a specific value m_0 of M, the $(1 - \alpha) \times 100\%$ prediction interval for y is given by

$$\hat{\alpha} + \hat{\beta}m_0 \pm t_{p-2, \alpha/2} s \sqrt{1 + \frac{1}{p} + \frac{(m_0 - \bar{m})^2}{S_{mm}}}, \tag{11.5}$$

where s^2 is given in (11.2). By inverting this relationship, a $(1 - \alpha) \times 100\%$ confidence interval for m_0, called a *Fieller interval*, can be obtained by using the set of values of M for which the measured value y_0 is in the $(1 - \alpha) \times 100\%$ prediction interval given in (11.5). Therefore the Fieller interval for the unknown value of M consists of all values of m that satisfy

$$\left(y_0 - \hat{\alpha} - \hat{\beta}m \right)^2 \le (t_{p-2, \alpha/2})^2 s^2 \left(1 + \frac{1}{p} + \frac{(m - \bar{m})^2}{S_{mm}} \right). \tag{11.6}$$

For a more in-depth treatment of measurement systems, see Mandel (1964) and Yano (1991). The derivations of (11.5) and (11.6) can be found in Draper and Smith (1998, Chapter 1).

11.3 PERFORMANCE MEASURES FOR PARAMETER DESIGN OPTIMIZATION

An important step in examining a signal-response system is to identify a performance measure which evaluates the suitability of a given signal-response relationship for the intended application. By optimizing the chosen measure, control factor settings which achieve the desired objectives can be identified.

Figure 11.1 illustrates the difference between good and poor signal-response systems. The three systems in Figure 11.1 correspond to different choices of the control factor settings. Each curve represents an observed signal-response relationship for a given set of noise factor conditions. Because the curves in (a) are bundled in a narrower band, the first system is least sensitive (i.e., most robust) to noise variation. The second system in (b) is more sensitive to noise variation but its signal-response relationships are linear while the third system in (c) is nonlinear. Thus the comparison between the first and third systems depends on the desirability of having a linear relationship.

From comparing these three systems, it is clear that **dispersion** and **sensitivity** are the two important aspects of a signal-response system. Assume that the signal-response relationship is linear. Then the system can be

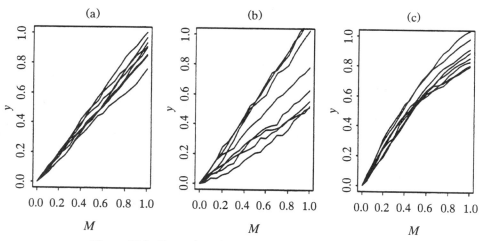

Figure 11.1. Examples of Signal-Response Relationships.

represented by the simple linear regression model in (11.1). In this special case, the dispersion and sensitivity are measured by σ^2 and β, respectively. If the relationship is nonlinear, then β should be replaced by the slope of the best linear approximation to the nonlinear curve. It is always desirable for any signal-response systems to have a small dispersion, i.e., a small σ^2 value. A large sensitivity (i.e., a large β value) is also desirable if it is important to be able to effect the change of the y value through the change of the M value. As will be discussed later, there are situations in which a nominal or small value of β is required. Therefore, an appropriate performance measure should suggest that σ^2 be small, and β be large, close to nominal, or small, depending on the situation.

A performance measure, which converts β and σ^2 into a single measure, is

$$\omega = \ln(\beta^2/\sigma^2). \tag{11.7}$$

Its sample analog

$$\hat{\omega} = \ln(\hat{\beta}^2/s^2), \tag{11.8}$$

where $\hat{\beta}$ and s^2 are given in (11.2), is called the **signal-to-noise (SN) ratio for signal-response systems**, or in Taguchi's terminology the **dynamic SN ratio**. The $\hat{\omega}$ computed for each control factor combination in the experiment can be fitted using a regression model in the control factors. The fitted model is referred to as the *dynamic SN ratio model*. If the experiment also involves noise factors, the $\hat{\beta}$ and s^2 estimates in (11.8) should be based on the observations over all the noise and signal factor levels for the given control factor combination.

As in parameter design for simple response systems (see Section 10.10), the second key element in Taguchi's approach is the notion of an adjustment factor. In the present context, an **adjustment factor** is a control factor that has a significant effect on $\hat{\beta}$ but not on $\hat{\omega}$. Taguchi proposed the following two-step procedure for parameter design optimization:

Taguchi's Two-Step Procedure for Signal-Response Systems

(*i*) *Select the levels of the significant control factors in the dynamical SN ratio model to maximize* $\hat{\omega}$.   (11.9)
(*ii*) *Select the level of an adjustment factor to bring the slope on target*.

In some situations, there may be a need to modify step (ii) to include another adjustment factor to bring the intercept on target.

Another version of the dynamic SN ratio assumes a zero intercept in the model (11.1). The $\hat{\beta}$ and s^2 values in $\hat{\omega}$ are then the least squares estimates based on the simplified model with $\alpha = 0$. Unless the zero intercept assumption is known to hold *a priori* or strongly suggested by the data, the more general version of $\hat{\omega}$ given in (11.8) should be used.

The two-step procedure in (11.9) can be justified if the following relationship holds:

$$Var(y) \propto \beta^2, \tag{11.10}$$

or more precisely, if the signal-response relationship can be represented by the model

$$y = \beta(x_1, x_2)[M + \epsilon(x_1)], \tag{11.11}$$

where $E(\epsilon) = 0$ and $Var(\epsilon) = Q(x_1)$. The model in (11.11) implies that

$$Var(y) = \beta^2(x_1, x_2)Q(x_1)$$

and

$$\frac{\beta^2}{Var(y)} = \frac{1}{Q(x_1)}.$$

Under (11.11), maximizing ω is therefore equivalent to minimizing the variance of ϵ in (11.11). Under model (11.11), Taguchi's two-step procedure can be restated as

(*i*) *Select* x_1^* *to minimize* $Q(x_1)$.
(*ii*) *Select* x_2^* *so that* $\beta(x_1^*, x_2^*) = \beta_T$. $\qquad$ (11.12)

Here β_T is the target slope value.

In this context, x_1 is a vector of the dispersion factors and x_2 is the adjustment factor. Because ω does not depend on the adjustment factor x_2, it is a performance measure independent of adjustment. [For further discussion on performance measures independent of adjustment for signal-response systems, see León, Shoemaker, and Kacker (1987).]

Because model (11.11) is rather specialized, we now consider the development of a two-step procedure for a different class of models. Suppose the signal-response relationship is represented by the model

$$y = \alpha(x_1, x_3) + \beta(x_1, x_2)M + \epsilon(x_1), \tag{11.13}$$

where $E(\epsilon) = 0$ and $Var(\epsilon) = Var(y) = \sigma^2 = Q(x_1)$. It differs from (11.11) in two respects: it has a nonzero intercept α and the variance of y does not depend on β. In model (11.13), x_1 are the dispersion factors, x_2 is an adjustment factor for the slope, and x_3 is an adjustment factor for the intercept. Because $\sigma^2 = Q(x_1)$ does not depend on x_2 or x_3, it is a performance measure independent of adjustment. Suppose that α_T and β_T are the target values for the intercept and slope. We have the following two-step procedure:

Two-Step Procedure for Model (11.13)

(*i*) *Select* $\mathbf{x}_1^*$ *to minimize* $\sigma^2 = Q(\mathbf{x}_1)$.
(*ii*) *Select* x_2^* *and* x_3^* *so that* $\alpha(\mathbf{x}_1^*, x_3^*) = \alpha_T$ *and* $\beta(\mathbf{x}_1^*, x_2^*) = \beta_T$.   (11.14)

In (11.9), (11.12), and (11.14), we assume that the slope and intercept have target values. In many situations, ideal target values are not available or relevant. The slope may be required to be steeper or flatter, depending on the nature of the system or application. Without a target value, the second step in these procedures should be replaced by increasing or decreasing α and β to be larger or smaller as the situation dictates.

The dynamic SN ratio $\hat{\omega}$ in (11.8) has a proper justification for measurement systems. First we consider a heuristic argument. Assume that α and β in model (11.1) are known. Then the estimate $\hat{M}$ in (11.4) becomes

$$(y_0 - \alpha)/\beta$$

and its variance becomes

$$\sigma^2/\beta^2.$$

Then maximizing the dynamic SN ratio is justifiable as it is equivalent to minimizing the variance of $\hat{M}$. A more rigorous justification that does not assume the knowledge of α and β can be made as follows. Recall that the Fieller interval is used as an exact confidence interval for the unknown value of M given the measured value y_0 of y. Therefore the quality of a measurement system can be judged by the length of the interval. A narrower interval suggests a higher precision of the system. It was shown in Miller and Wu (1996) that the length of the Fieller interval given in (11.6) is a decreasing function in $\hat{\beta}^2/s^2$. Therefore *maximizing $\hat{\omega}$ in (11.8) amounts to minimizing the length of the Fieller interval, i.e., maximizing the precision of the measurement system*.

The previous justification has an intuitive explanation. Roughly speaking, maximizing $\hat{\omega}$ in (11.8) has the effect of minimizing s^2 and maximizing $\hat{\beta}^2$. The former is always desirable. The latter is also desirable for measurement systems because the estimate of M involves an inverse regression whose slope is the reciprocal of $\hat{\beta}$. As $\hat{\beta}$ becomes steeper, the error in the inverse regression becomes smaller. This explanation does not hold, however, for many multiple target systems where the inverse regression argument is not applicable. This justification of $\hat{\omega}$ for measurement systems differs from the one in (11.12) in two respects. First, it does not require the existence of an adjustment factor. Second, the error variance in (11.1) is a constant independent of the slope β. Because of the generality of the assumptions underlying this justification, the use of the dynamic SN ratio for measurement systems is on firmer ground.

To summarize, the dynamic SN ratio is a good performance measure in either of the following situations: (i) when the signal-response relationship can be described by model (11.11), which assumes the existence of an adjustment factor and that the response variance is proportional to the square of the slope, and (ii) for measurement systems whose precision can be measured by the length of the Fieller interval.

Situation (i), particularly its assumption that $Var(y)$ is proportional to β^2, is hard to justify. The variance may be proportional to another power of the slope as well as to the intercept. For example, the engine idling experiment in the exercises has a linear signal-response relationship whose response variance is estimated to be proportional to the (-1.5)th power of the slope (see Lunani, Nair, and Wasserman, 1997). A graphical method to determine the power λ in the relation between $Var(y)$ and β^λ is given in the same paper.

Because many multiple target systems cannot be represented by model (11.11), use of the dynamic SN ratio may not be appropriate. Another problem is related to a mathematical property of $\hat{\omega}$. Its maximization has the effect of minimizing s^2 and maximizing $\hat{\beta}^2$. While the former is always desirable, the latter can lead to undesirable results. Given a fixed interval (m_a, m_b) for the signal factor M, a larger $|\hat{\beta}|$ value can give a wider range of y values, which may be outside the specification limits for y. Furthermore, if there is error in the setting of M, this error will be propagated through a larger absolute slope $|\hat{\beta}|$, again resulting in larger variation in y. For example, the analysis of the injection molding data (to be given in Section 11.5) suggests a strong possibility of error in the setting of the signal factor, high injection pressure.

Except in circumstances such as those described in situation (i), a single performance measure is not recommended for multiple target systems. As suggested by model (11.13) and the two-step procedure in (11.14), it is prudent to consider the three quantities σ^2, β, and α. The dispersion σ^2 should be minimized and the slope β and the intercept α should be set to target values or be minimized or maximized (subject to some constraints) as the situation warrants. If the signal-response relationship is described by a more complicated function such as a quadratic function or a spline function, key parameters (e.g., trend, curvature, percentiles) in these functions should be treated in the same way as in the previous discussion of the slope and intercept terms. Another approach to the choice of performance measure for multiple target systems can be found in Miller and Wu (1996).

11.4 MODELING AND ANALYSIS STRATEGIES

Because the purpose of a parameter design experiment is to identify the manner in which control factors affect the performance of the system, the chosen performance measure (PM) should be modeled as a function of the control factors. Two modeling strategies are considered in this section: performance measure modeling and response function modeling.

Performance Measure Modeling (PMM)

(i) *For each control factor combination in the experiment, compute the PM value based on the response values for various combinations of signal and noise factor levels.*

(ii) *Model the PM values in (i) as a function of the control factors. The preferred settings for the control factors are then determined from the fitted model.*   (11.15)

The modeling in step (ii) can employ any analysis technique that was discussed in previous chapters.

Performance measure modeling can be illustrated using the dynamic SN ratio as the PM. Consider a full factorial experiment involving three two-level control factors (C_1, C_2, C_3), two two-level noise factors (N_1, N_2), and a four-level signal factor (M). For each of the eight distinct control factor combinations, there are 16 observations corresponding to the noise and signal factor combinations. These 16 observations are used to fit a regression line [based on model (11.1)], and the parameter estimates are then used to estimate the PM, in this case $\widehat{PM} = \ln(\hat{\beta}^2/s^2)$. The eight $\widehat{PM}$s are then treated as the observations for a 2^3 experiment with C_1, C_2, and C_3 as its factors; standard analysis techniques can be applied.

The performance measure appears as the quantity to be optimized in the first step of the two-step procedures considered in the previous section. In order to complete the data implementation of these two-step procedures, we need to model the quantity in the second step of these procedures. Consider, for example, Taguchi's two-step procedure (11.9). Because its second step involves the adjustment of the slope β, we need to model the slope estimate $\hat{\beta}$, which is computed for each control factor combination in the experiment, as a function of the control factors. Therefore, the PMM approach should involve the modeling of the PM as well as the quantities in the second step of the two-step procedure.

An alternative to Taguchi's two-step procedure is (11.14), which involves σ^2, β, and α in model (11.13). The PMM approach for this procedure requires the modeling of $\ln s^2$, $\hat{\beta}$, and $\hat{\alpha}$, which are computed according to (11.2) for each control factor combination in the experiment. This version of the PMM approach is generally preferred to Taguchi's approach because the latter involves the dynamic SN ratio, which has a limited justification.

The PMM approach can also be viewed as a signal-response analog of the location-dispersion modeling approach for simple response systems (see Section 10.5.1). The former involves the modeling of sensitivity (i.e., $\hat{\beta}$) and dispersion (i.e., $\ln s^2$ or $\hat{\omega}$) and the latter involves the modeling of location (i.e., $\bar{y}$) and dispersion [i.e., $\ln s^2$ or $\hat{\eta}$ in (10.21)].

Response function modeling (RFM) treats the signal-response relationship as the response and models this relationship as a function of the control and noise factors. The specified performance measure is then evaluated with respect to the fitted models in order to select preferred settings of the control factors. It is an extension of the response modeling approach for simple response systems (see Section 10.5.2).

For a linear signal-response relationship, RFM can be stated as the following two-stage procedure:

> (*i*) *For each control/noise factor combination in the experiment, fit a regression line [based on model* (11.1)] *to the response values over the signal factor levels. Denote the estimates for the ith combination by* $\hat{\alpha}_i$, $\hat{\beta}_i$, *and* s_i^2.
>
> (*ii*) *Obtain separate regression models for* $\hat{\alpha}_i$, $\hat{\beta}_i$, *and* $\ln s_i^2$ *as functions of the control and noise factors. The chosen performance measure can then be evaluated with respect to these models and preferred control factor settings can be determined.*   (11.16)

Modeling and analysis in steps (i) and (ii) can employ any technique that was discussed in previous chapters. If the error structure of the problem is more complicated, the modeling in step (ii) will require a more elaborate analysis. One such example is provided by the analysis of the injection molding experiment in the next section. If the signal-response relationship is more complicated, e.g., a quadratic curve or a spline function, (11.16) can be extended in a straightforward manner. For example, for a quadratic relationship, the additional modeling involves the quadratic coefficient in the quadratic regression model.

For an illustration of RFM, consider the example that was used after (11.15) to illustrate PMM. It has 32 combinations of control and noise factor levels. For each of these combinations, there are four observations corresponding to the levels of the signal factor. These four observations are used to fit a regression line [based on model (11.1)] for the signal-response relationship. Then, values of $\hat{\alpha}$, $\hat{\beta}$, and s^2 are obtained for each of the 32 control/noise factor combinations, and regression models for $\hat{\alpha}$, $\hat{\beta}$, and s^2 as functions of the control and noise factors are produced. The chosen PM is then evaluated for different combinations of the control factors using these models.

In general, the PMM procedure is not preferred because it can often obscure useful information present in the data. The PMM approach only provides information on how control factors affect the overall performance of the system. Any information in the data on how specific control factors affect the shape of the signal-response system or interact with specific noise factors is lost. It is this type of information that can be valuable in suggesting directions for future research. The RFM approach does not suffer from this deficiency. The initial modeling of the response will often provide useful insight into the system. The PM is then applied to this model to identify preferred settings of control factors. The advantages of the RFM approach will be clearly demonstrated in the next section using the injection molding experiment.

11.5 ANALYSIS OF THE INJECTION MOLDING EXPERIMENT

In this section, the injection molding experiment is analyzed by the PMM and RFM approaches.

First, a suitable PM must be identified. Suppose that the system must be capable of achieving target values from 650 to 700. Since this range of target values is obtainable for all combinations of control and noise factors used in the experiment, a reasonable PM would be the variation of the response over this range.

Figure 11.2 contains scatter plots of the data (y, M) and the fitted quadratic curves for the signal-response relationship for 4 of the 16 runs, where the response y is the part weight and the signal factor M is the high injection pressure. Obviously, a straight line does not fit the data well. Figure 11.3 contains the residual plots (after the quadratic model fit) against M. Although the quadratic model appears reasonably satisfactory, there appears to be a systematic pattern in the residual plots. In order to investigate this further, the total sum of squares for each control/noise factor combination is decomposed into contrasts using orthogonal polynomials of M. Then a forward selection procedure is used to sequentially test the addition of higher order terms (at the 0.05 significance level) to the models.

In 12 of the 16 cases, the selection procedure indicates that a quadratic polynomial is adequate. In the other 4 cases, the cubic term is also added. An

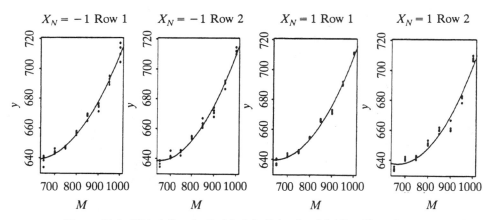

Figure 11.2. Fitted Quadratic Models, Injection Molding Experiment.

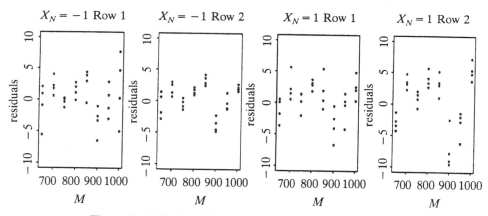

Figure 11.3. Residual Plots, Injection Molding Experiment.

interesting aspect of this analysis is that, in over half the runs, there is an unusually large sum of squares attributed to the sixth degree term. As it is unlikely that the sixth degree term is important, this suggests that at least part of the systematic pattern in the residuals seen in Figure 11.3 may not be due to model inadequacy. For most of the 16 runs, the residuals are mostly negative at $M = 650$, move above or around 0 between 700 and 850, drop suddenly below 0 at $M = 900$, and then start to rise again. If this is a consistent pattern regardless of the control and noise factor combinations, it cannot be explained by these factors. One possible explanation is that a systematic difference exists between the settings of the signal factor and the actual high injection pressure delivered by the system. For example, the residuals corresponding to $M = 900$ are negative in most of the residual plots. This could happen if the system consistently delivers a high injection pressure of less than 900 at a nominal setting of 900, since this would cause the observed part weights to be lower than expected. Similarly one may suspect that the actual high injection pressure is higher than its nominal setting of 850 because the residuals at 850 are mostly positive. Whatever the real explanation for this phenomenon, it is clear that it will not be corrected by using a more complicated model.

Because the quadratic curve adequately captures the essential features of the signal-response relationship, the following model is used to fit the data

$$E(y) = \beta_0 + \beta_1 P_1(M) + \beta_2 P_2(M), \qquad (11.17)$$

where $P_1(M)$ and $P_2(M)$ are orthogonal polynomials of order 1 and 2. Their values for the various signal levels are given in Table 11.6. Orthogonal polynomials are adopted as they make the interpretation of results easier, with β_1 representing the linear component of the signal-response relationship and β_2 the quadratic departure from linearity.

Table 11.6 Levels for Orthogonal Polynomials

	Signal Factor Level							
	650	700	750	800	850	900	950	1000
$P_1(M)$	-7	-5	-3	-1	1	3	5	7
$P_2(M)$	7	1	-3	-5	-5	-3	1	7

11.5.1 PMM Analysis

The first step is to estimate the PM for each control factor combination in Table 11.3. From the residual plots it seems reasonable to assume constant variance across levels of the signal factor (which provides a counterexample to the assumption underlying the use of $\hat{\omega}$ that the response variance is proportional to the square of the slope). For each row in Table 11.3, the least squares estimate of the variance, s^2, for the fitted quadratic model (11.17) is used as the estimated PM. Because there are two noise replicates (at $X_N = \pm 1$) for each control factor combination, s^2 is estimated using the data with $X_N = +1, -1$. The next step is to use $\ln s^2$ as the response in the PMM analysis. (An explanation for analyzing $\ln s^2$ rather than s^2 was given in Section 3.8.) The values of $\ln s^2$ for the eight control factor combinations in Table 11.3 are

$$(2.127, 2.981, 2.204, 2.342, 1.468, 2.277, 0.432, 1.185)^T.$$

The half-normal plot of the factor effects for $\ln s^2$ is given in Figure 11.4. This plot does not clearly indicate that any control factors have an effect on

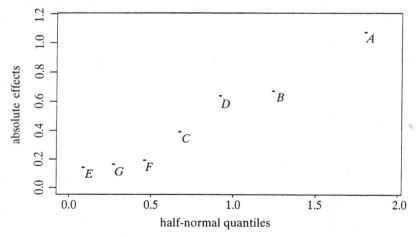

Figure 11.4. Half-Normal Plot for $\ln s^2$, Injection Molding Experiment.

the PM. However, it does appear that A, B, and D warrant further consideration. Because this analysis does not suggest any significant effect for reducing the variance s^2, it makes no sense to implement the second step of the two-step procedure in (11.14), which would require the modeling of the slope parameter. (The modeling of the intercept and linear slope parameters will be given in the following RFM analysis.)

11.5.2 RFM Analysis

The first stage of RFM analysis consists of fitting the quadratic model (11.17). Since there are four replicates at each signal level, it is possible to separate the error variance into a lack-of-fit component and a replicate component (also called a pure error component). Explanation of these two components can be found in a regression text such as Draper and Smith (1998). The replicate component, $\hat{\sigma}_p^2$, reflects part-to-part variation taken over a short time interval because it measures the variation of part weight for the four parts which were molded at each setting of high injection pressure. The lack-of-fit component, $\hat{\sigma}_l^2$, represents longer term variation and, as noted previously, contains an unexplained component which may be due to systematic errors in the recorded values for the signal factor. Again, using the log transformation, $\ln \hat{\sigma}_l^2$ and $\ln \hat{\sigma}_p^2$ are analyzed separately. The estimated values for β_0, β_1, and β_2 in (11.17) and σ_l^2 and σ_p^2 are given in Table 11.7.

At the second stage of the RFM analysis, the effects of the control and noise factors on these parameters are evaluated. Figure 11.5 contains half-normal plots for the β_i parameters, and Figure 11.6 contains half-normal plots for the two components of variation. Because the control/noise factor combinations in Table 11.4 can be viewed as a cross array between the control array in Table 11.3 and the noise array with one compound noise factor, Theorem 10.1 implies that the control and noise main effects are eligible and the seven control-by-noise interactions are clear.

Table 11.7 Estimated Parameters for RFM, Injection Molding Experiment

Run	$X_N = -1$					$X_N = 1$				
-----	$\hat{\beta}_0$	$\hat{\beta}_1$	$\hat{\beta}_2$	$\hat{\sigma}_l^2$	$\hat{\sigma}_p^2$	$\hat{\beta}_0$	$\hat{\beta}_1$	$\hat{\beta}_2$	$\hat{\sigma}_l^2$	$\hat{\sigma}_p^2$
1	665.0	4.98	1.33	27.51	1.20	666.5	5.02	1.16	22.48	7.78
2	660.0	4.69	1.48	107.25	3.20	664.2	5.12	1.44	28.42	4.45
3	665.2	4.86	1.26	25.39	2.70	668.2	4.98	1.22	17.15	4.99
4	664.2	4.55	1.54	14.57	2.64	668.4	4.76	1.25	19.25	3.53
5	664.2	4.46	1.39	10.19	0.56	666.3	4.66	1.35	19.74	0.67
6	674.1	4.33	1.36	53.12	0.30	674.4	4.32	1.32	59.14	1.00
7	666.1	4.91	1.30	7.23	0.18	666.6	4.92	1.31	9.23	0.21
8	663.6	5.02	1.29	15.87	0.12	664.9	4.90	1.25	12.85	0.75

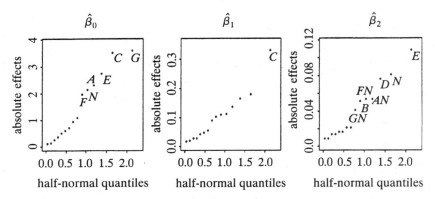

Figure 11.5. Half-Normal Plots for $\hat{\beta}_0$, $\hat{\beta}_1$, and $\hat{\beta}_2$, Injection Molding Experiment.

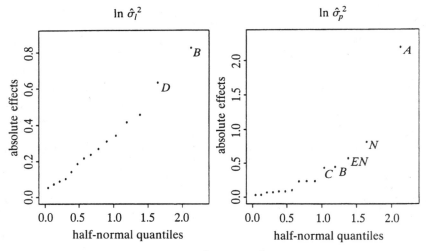

Figure 11.6. Half-Normal Plots for $\ln \hat{\sigma}_l^2$ and $\ln \hat{\sigma}_p^2$, Injection Molding Experiment.

For β_0 there are six effects which appear to be significant: G, C, E, A, N, and F. Only C stands out as being significant for β_1. There is no significant effect for β_2, but E, N, D, B, AN, FN, and GN appear marginally significant. None of the estimated effects are significant for the lack-of-fit variation. For the part-to-part variation A is clearly significant and N, EN, B, and C are all large enough to warrant further attention.

The fitted models for the parameters are

$$\hat{\beta}_0 = 666.4 + 1.2X_A + 1.8X_C$$
$$+ 1.4X_E + 1.0X_F + 1.8X_G + 1.1X_N,$$
$$\hat{\beta}_1 = 4.79 - 0.16X_C,$$
$$\hat{\beta}_2 = 1.33 - 0.03X_B + 0.04X_D - 0.05X_E \qquad (11.18)$$
$$- 0.04X_N + 0.01X_A X_N + 0.03X_F X_N - 0.02X_G X_N,$$
$$\hat{\nu} = 0.12 + 1.10X_A - 0.22X_B + 0.21X_C,$$
$$+ 0.40X_N + 0.28X_E X_N + 0.04X_E,$$

where $\nu = \ln \sigma_p^2$.

To begin, consider the model for $\ln \sigma_p^2$. Part-to-part variation is not the only type of variation which is relevant to the process. However, reducing this type of variation is clearly desirable. In this case, A should be set at the $-$ level. It also appears worthwhile to set B at the $+$ level and C at the $-$ level. The compound noise factor, N, and the EN interaction also affect σ_p^2. Since the EN interaction is significant, E is also included in the model. This can be justified by the strong effect heredity principle. Table 11.8 contains the estimated values of $\ln \sigma_p^2$ for the combinations of levels of E and N assuming $X_A = -1$, $X_B = +1$, and $X_C = -1$. The table indicates that, for E set at $X_E = -1$, the part-to-part variation will be more consistent with respect to changes in the noise factors than for $X_E = 1$.

Next consider the fitted model for β_1. Factor C can be used to adjust the sensitivity of the response to the signal. In this case C would be set at the $-$ level if a wider range of attainable targets is necessary. Otherwise, the level of C should be determined by other considerations. Setting C at the $-$ level would be consistent with the previous recommendation based on $\ln \sigma_p^2$.

Now consider the model for β_0. Due to previous considerations, the levels of A, C, and E have already been determined. This leaves factors F and G, which could be used to make adjustments to β_0 (if necessary). Notice that N does affect β_0, but no interaction has been identified which could offset this.

Finally consider the model for β_2. In this case the levels for B and E have been previously determined. Consider the effect of N. The estimated observed coefficient (given fixed levels of control factors) for N would be

$$(-0.04 + 0.01X_A + 0.03X_F - 0.02X_G)X_N.$$

Table 11.8   Estimated Values for $\ln \sigma_p^2$, Injection Molding Experiment

	$X_E = -1$	$X_E = 1$
$X_N = -1$	-1.57	-2.05
$X_N = 1$	-1.33	-0.69

In order to make the system insensitive to changes in N, we would like to make the absolute value of this coefficient as small as possible. As the setting $X_A = -1$ has already been determined, this suggests setting $X_F = +1$ and $X_G = -1$. The level of D does not have to be chosen because D does not appear with N in (11.18).

Notice that the PMM approach does not clearly indicate that any control factors can be adjusted to improve system performance. Even if certain factors had been identified, it would not have provided insight into how these factors affect the system. On the other hand, RFM not only indicates certain control factors that can be used to improve the system but also provides insight into how these factors affect the system. In particular, a quadratic model is identified as suitably describing the signal-response relationship, and control factors which can be used to alter the parameters of this relationship are found. Further, the residual plots produced from the fitted models for the signal-response relationship indicate the possibility of a systematic error in the signal levels. Finally, the flexibility of the RFM procedure allows the variation to be decomposed into two components, σ_p^2 and σ_l^2, which leads to the conclusion that factor A could be used to reduce part-to-part variation.

11.6 CHOICE OF EXPERIMENTAL PLANS

In signal-response systems, there are three types of factors: control, noise, and signal. Any discussion on the choice of experimental plans should be built on previous work for simple response systems which only involve control and noise factors. Recall from Chapter 10 that there are two types of arrays for simple response systems: cross arrays and single arrays. For notational simplicity, we shall denote them by $C \times N$ and (C, N) respectively, where C and N denote the control and noise factors, "$\times$" denotes the crossing of two arrays, and the parentheses denote the combining of two types of factors in a single array. Because in many signal-response systems the signal factor is relatively easy to adjust in comparison to the control and noise factors, it is common to cross a $C \times N$ or (C, N) array with the signal factor. That is, for each control/noise factor combination in a $C \times N$ or (C, N), the experiment is run at a pre-chosen set of signal factor levels. Using the same notation, these two types of arrays are denoted by

(i) $C \times N \times S$,
(ii) $(C, N) \times S$,

where S stands for the signal factor. The injection molding experiment provides a good example of a type (i) array. The experiment was conducted in two days according to the level of the compound noise factor. Within each day, the eight combinations in the control array were varied, and for each combination the eight signal factor levels were varied. The experiment can be viewed as a split-split-unit experiment, where the noise factor defines the "whole units", the control factors define the "subunits" within the whole units, and the signal factor defines the "sub-subunits" within the subunits.

Two other types of arrays may also be considered:

(iii) $C \times (N, S)$,
(iv) $(C, S) \times N$.

In these arrays the signal factor is combined with either N or C in a single array. The levels of the signal factor change with the factor combinations of N or C, which requires a more elaborate implementation and analysis of the experiment. As a result, these arrays are not as popular as the first two types. The surface machining experiment in the exercises provides a good example for type (iii) arrays. According to the description, it is a split-split-unit experiment. The two replicates define the whole units because they represent the two time periods. Within each period, the apparatus was configured by varying the control factor settings based on a control array. Therefore the control factors define the subunits. For each configuration, three parts were machined using three different cutting depths (i.e., the levels of the signal factor M). Each part was divided into nine portions by the noise factors I and J. To save measurement cost, only three out of the nine portions were measured for each machined part and the $\frac{1}{3}$ fraction was defined by a 3^{3-1} design for the factors M, I, and J. A single array was used for these three factors and the noise/signal factors define the sub-subunits.

In determining what factors should be chosen as whole units, subunits, etc., an important consideration is the difficulty of changing factor levels. To reduce the experimental cost, it is often necessary to *reduce the total number of distinct combinations of the "hard-to-change" factors*. This can often be achieved by maximizing the number of defining relations in the fractional factorial designs that are made entirely of the most hard-to-change factors. Usually such factors are treated as whole-unit factors. For example, furnace temperature may take a long time to stabilize. It may not be feasible to have more than two temperature settings per day, and therefore temperature should be treated as a whole-unit factor. Further discussion on split-unit experiments in the context of parameter design can be found in Box and Jones (1992).

The choice of modeling/analysis strategy depends on the type of the array. Because the PMM approach requires that the PM be evaluated for the *same* set of noise/signal factor combinations, it can only be used for arrays of types (i) and (iii). The RFM approach does not have this restriction and can be used for any of the four types. For type (iv), however, the second stage of the RFM needs more elaborate modeling. The required modification is considered for a special case in the exercises.

Between the arrays of types (i) and (ii), which should be chosen? Because their only difference is in the choice of array for the control and noise

Table 11.9   Comparison of Resolution of Designs

Control/Noise Combinations	Cross Array $C \times N$	Single Array (C, N)
16	(III, II)	III
32	(IV, II)	IV
64	(III, III) or (IV, II)	IV
128	(IV, IV) or (VII, II)	V

Note: The first and second numbers in the parentheses represent the resolution of the control and noise arrays for the given cross array.

factors, their comparison is exactly the same as that between cross arrays and single arrays discussed in Section 10.8 for simple response systems. Generally type (i) is preferred if it is much less costly to run noise replicates than control replicates. Otherwise type (ii) is preferred. The advantage of type (ii) is illustrated by the injection molding experiment, whose cost is mainly determined by the number of distinct control/noise factor combinations. The actual experiment used a cross array with 16 control/noise factor combinations, but to achieve this, it was necessary to use a resolution III design for the control array (seven factors) and what was effectively a resolution II design for the noise array (four factors). Suppose a resolution IV design was desired for both the control and noise arrays. Based on the 2^{k-p} design tables in Chapter 4, the control array must contain at least 16 rows and the noise array must contain at least 8 rows, resulting in a minimum of 128 control/noise combinations. In contrast, one can use a 32-run 2^{11-6}_{IV} to construct a single array for the seven control factors and four noise factors. Table 11.9 compares the maximum resolution of designs which can be constructed using the two approaches for a given number of control/noise combinations. It can be seen that using a single array has the clear advantage of producing higher resolution designs.

   If a type (i) array is too costly to run, an alternative to using a single array is to reduce the size of the noise array in (i) to 2 by using a **compound noise factor** with two levels, which was discussed in Section 10.8. Here we shall use the injection molding experiment to illustrate its potential disadvantages. One key assumption for a compound noise factor to be effective is that the combinations of levels used to define the levels of the compound noise factor should represent the extreme sets of conditions for *all* settings of the control factors. It may be difficult to ascertain the validity of this assumption in a practical situation. For example, a new operator may produce parts which are lighter than those produced by the experienced operator for some control factor settings but produces heavier parts for other settings. Another disadvantage of using a compound noise factor is that the effects of the individual noise factors cannot be separated. This limits the usefulness of the experiment in terms of the information it provides about directions for future research. If, for example, setting factor A (injection speed) to the $+$ level had been found to improve the performance, there is no way of determining the nature of the improvement. Is the process more robust to operators, percent regrind, melt index, resin moisture, or some combination of these? In more technical terms, is the improvement attributed to a significant $A \times N_i$ interaction for a particular noise factor N_i? Answers to questions like this will indicate which sources of variability can be addressed by using factor A and which sources require other solutions. The main problem with the compound noise factor approach stems from the fact that only a few and often only two noise factor combinations are used. In order to have more noise factor combinations in the experiment without increasing the total run size, the only recourse is to use a single array. For the injection molding experiment, it is possible to find a 16-run 2_{III}^{11-7} design as a single array that has all the 16 noise factor combinations. (Details are given in the exercises.)

One remaining issue is the choice of the number of levels for the signal factor. This depends on the nature of the signal-response relationship. If a lower order polynomial can adequately describe the relationship, no more than five levels are required. For the injection molding experiment, eight levels were used. Since the signal-response relationship is quadratic or cubic, hindsight may suggest a smaller number of levels. On the other hand, if the purpose was to map the relationship over a broad range of target part weight, eight levels might be required. If it takes a spline function to describe the relationship, a fine grid of levels for the signal factor will be required. Typically, it will require at least seven levels.

11.7 PRACTICAL SUMMARY

1. There are three broad types of signal-response systems: multiple target systems, measurement systems, and control systems.

2. The dynamic signal-to-noise (SN) ratio in (11.8) is a good performance measure in either of the following situations:

   (i) The signal-response relationship is linear, the response variance is proportional to the square of the slope, and there is an adjustment factor.

   (ii) For measurement systems whose precision can be measured by the length of the Fieller's interval.

   Otherwise, a more prudent approach is to consider several quantities such as the dispersion σ^2, the regression slope (also called sensitivity) β, and the intercept α in the simple linear regression model (11.1) or (11.13).

3. Use the two-step procedures in (11.9) or (11.14) for parameter design optimization. The former is recommended only if the dynamic SN ratio is a good performance measure. The latter is more broadly applicable.

4. There are two strategies for modeling and analysis: performance measure modeling (PMM) and response function modeling (RFM). When both approaches are used in the data analysis and give conflicting results, more information can be obtained from the RFM analysis. Consequently, the RFM is generally preferred.

5. The two types of experimental plans, (i) $C \times N \times S$ and (ii) $(C, N) \times S$, are more commonly used because, in typical applications, it is relatively easy to change the level of the signal factor. Between them, (i) is preferred if it is less costly to run noise replicates than control replicates. More precise conditions for their comparison can be found in Section 10.8 in the discussion of cross arrays and single arrays. Two other types are (iii) $C \times (N, S)$ and (iv) $(C, S) \times N$. The choice of plan is often determined by the split-unit nature of the experiment. The PMM approach can be used only for types (i) and (iii) while the RFM approach can be used for all four types.

EXERCISES

1. As described in Section 11.6, the injection molding experiment can be viewed as a split-split-unit experiment. Reanalyze the data by taking into account its split-unit error structure. Do the conclusions change?

2. For the injection molding experiment, another version of the PMM approach can be considered, that is, to compute the dynamic SN ratio for each row of the control array and model them as a function of the control factors. Compare the results of this analysis with those of the PMM and RFM approaches discussed in Section 11.5.

3. Perform a quadratic model fit and produce a residual plot for each of the 16 runs in the injection molding experiment. Discuss any systematic patterns in the residual plots and explain how they support the claim made in Section 11.5.2 regarding the systematic error in the setting of the high injection pressure.

4. In an experiment on a zinc phosphate coating (Lin and Wen, 1994), low-carbon steel plates are coated with zinc phosphate in an eight-step process involving six different solutions. The experimental factors are amounts of the chemicals used in the solutions and other variables in the coating process, as displayed in Table 11.10. The experimental response is the weight of the phosphate coating. The signal factor is the geometric area of the steel plate. The factor levels are also displayed in Table 11.10. The design matrix for the experiment and the weight response data appear in Table 11.11. The experimental goal is to achieve uniformity by improving linearity, increasing sensitivity, and reducing variability.

Table 11.10 Factors and Levels, Coating Experiment

Control Factor	Level 0	Level 1	Level 2
A. Ni(NO$_3$)$_2$·6H$_2$O (g/L)	2	3	
B. acid-clean time (min)	10	15	20
C. phosphating temperature (°C)	70	80	90
D. phosphating time (min)	5	10	15
E. ZnO (g/L)	0.5	1.0	1.5
F. H$_3$PO$_4$ (g/L)	3	5	7
G. NaNO$_2$ (g/L)	0.15	0.30	0.50
H. NaH$_2$PO$_4$ (g/L)	20	25	30

Signal Factor	Level 1	Level 2	Level 3	Level 4
M. geometric area of low-carbon steel	20	50	100	200

Table 11.11 Design Matrix and Weight Data, Coating Experiment

Run	A	B	C	D	E	F	G	H	M_1	M_2	M_3	M_4
1	0	0	0	0	0	0	0	0	7.6	20.1	35.1	62.2
2	0	0	1	1	1	1	1	1	13.6	34.0	76.9	150.4
3	0	0	2	2	2	2	2	2	12.1	27.5	58.6	110.9
4	0	1	0	0	1	1	2	2	10.0	25.6	53.9	101.2
5	0	1	1	1	2	2	0	0	18.0	47.9	100.2	195.2
6	0	1	2	2	0	0	1	1	19.9	52.9	91.8	224.7
7	0	2	0	1	0	2	1	2	1.5	4.1	8.7	25.0
8	0	2	1	2	1	0	2	0	14.9	36.6	89.7	183.5
9	0	2	2	0	2	1	0	1	15.5	38.8	79.4	161.3
10	1	0	0	2	2	1	1	0	14.8	37.2	81.1	158.9
11	1	0	1	0	0	2	2	1	3.7	11.1	22.6	59.2
12	1	0	2	1	1	0	0	2	17.8	41.4	84.0	145.8
13	1	1	0	1	2	0	2	1	14.1	36.3	66.7	126.0
14	1	1	1	2	0	1	0	2	17.5	40.8	90.9	186.3
15	1	1	2	0	1	2	1	0	10.3	26.5	60.1	136.8
16	1	2	0	2	1	2	0	1	25.6	66.9	129.3	252.5
17	1	2	1	0	2	0	1	2	6.5	12.6	28.6	52.3
18	1	2	2	1	0	1	2	0	13.3	33.2	75.5	136.6

(a) Analyze the data using the RFM approach, assuming a linear signal-response relationship.

(b) Analyze the data using the PMM approach based on the dynamic SN ratio and the slope.

(c) Compare the results in (a) and (b) and comment.

Table 11.12   Control Factors and Levels, Drive Shaft Experiment

Control Factor	Level			
	0	1	2	3
A. testing machine	new	old		
B. master rotor	#1	#2	#3	#4
C. rotations at handling time	current	new		
D. rotations at measurement	current	new		
E. signal sensitivity	10	20	30	40
F. sequence of correction of imbalance	current	reverse	new #1	new #2
G. imbalance correction location	current	new		

5. Unbalanced automobile drive shafts cause vibration and noise. This problem can be corrected by using balance weights, which requires an accurate measurement of the amount of imbalance. An experiment was undertaken with the goal of finding the control factor settings which produce the most precise measurements (Taguchi, 1987). The seven control factors are given in Table 11.12. Three drive shafts (DS1, DS2, DS3) were tested at each combination of control factors. Because these drive shafts were chosen to represent the range of imbalance present in the manufacturing process, they should be treated as a noise factor. DS1 and DS2 were moderately unbalanced and DS3 was severely unbalanced. Four residual imbalance measurements were taken on each drive shaft corresponding to the following conditions:

M_1: Drive shaft measured as is.

M_2: Drive shaft measured with 10 g weight attached to mass-deficient side.

M_3: Drive shaft measured with 20 g weight attached to mass-deficient side.

M_4: Drive shaft measured with 30 g weight attached to mass-deficient side.

Table 11.13   Control Array, Drive Shaft Experiment

Row	Control Factor							Row	Control Factor						
	A	B	C	D	E	F	G		A	B	C	D	E	F	G
1	0	0	0	0	0	0	0	9	1	2	0	0	3	1	0
2	0	1	0	1	1	1	0	10	1	3	0	1	2	0	0
3	0	2	1	0	2	2	0	11	1	0	1	0	1	3	0
4	0	3	1	1	3	3	0	12	1	1	1	1	0	2	0
5	0	1	0	0	2	3	1	13	1	3	0	0	1	2	1
6	0	0	0	1	3	2	1	14	1	2	0	1	0	3	1
7	0	3	1	0	0	1	1	15	1	1	1	0	3	0	1
8	0	2	1	1	1	0	1	16	1	0	1	1	2	1	1

Table 11.14 Response Data, Drive Shaft Experiment

Run	DS1				DS2				DS3			
	M_1	M_2	M_3	M_4	M_1	M_2	M_3	M_4	M_1	M_2	M_3	M_4
1	-4	6	18	27	-4	6	15	25	-20	-10	2	14
2	-7	10	23	42	-3	15	32	46	-34	-20	4	16
3	-4	9	22	34	-7	6	18	30	-26	-15	0	12
4	-2	10	22	36	-4	8	22	34	-30	-18	-7	10
5	-6	6	16	28	-5	6	16	27	-21	-12	3	13
6	-7	13	32	50	-7	14	31	48	-45	-27	-8	13
7	-13	10	30	52	-8	10	30	50	-37	-18	7	26
8	-19	8	27	48	-14	7	29	52	-42	-25	-2	22
9	-10	4	16	29	-8	6	16	26	-29	-20	-14	-14
10	-14	11	32	51	-18	4	25	46	-44	-26	-11	16
11	-3	2	10	16	-4	2	17	13	-13	-8	-5	7
12	-5	5	16	25	-7	3	12	22	-22	-14	-8	12
13	-4	6	18	30	-8	2	16	28	-23	-15	-10	18
14	-6	16	38	62	-16	6	32	55	-44	-25	-6	21
15	-4	2	7	14	-4	2	6	12	-10	-6	-3	7
16	-5	6	16	30	-7	4	16	27	-25	-14	-6	14

An $OA(16, 2^4 4^3)$ was chosen for the control array (see Table 11.13). Table 11.14 displays the response data.

(a) What performance measure is appropriate for the problem?

(b) Analyze the data using the RFM approach and identify significant factors for $\ln s^2$ and $\hat{\beta}$, respectively. Based on the fitted models for $\ln s^2$ and $\hat{\beta}$, identify factor settings that maximize the dynamic SN ratio. Why is the dynamic SN ratio justifiable here?

(c) Analyze the data using the PMM approach and identify significant factors for the dynamic SN ratio $\hat{\omega}$ and factor settings that maximize $\hat{\omega}$.

(d) Compare the results in (a) and (b) and comment on why the RFM approach is preferred for these data.

(e) Based on the experimental description, can it be viewed as a split-split-unit experiment? Which factors define the whole units, sub-units, and sub-subunits?

6. In machining surfaces, a method of controlling the surface roughness is required as different applications require different degrees of surface roughness. Based on previous experience, the feed rate of the tool bit was chosen as the signal factor. The purpose of the experiment (Yano, 1991) was to be able to produce parts with a specified surface roughness by adjusting the feed rate. Eight control factors were selected (see Table 11.15) which were varied according to an 18-run orthogonal array (see Table 11.16). The signal factor (three levels) was included in a

Table 11.15 Control Factors and Levels, Surface Machining Experiment

	Levels		
Control Factor	0	1	2
A. lathe	old	new	—
B. cutting speed (m/min)	150	200	250
C. depth of tool cut (mm)	0.5	1.0	1.5
D. type of tool	P20	cermet	ceramic
E. corner radius (mm)	0.2	0.5	0.8
F. cutting edge angle (deg)	2	6	10
G. front escape angle (deg)	2	4	6
H. side scoop angle (deg)	0	4	8

Table 11.16 Control Array, Surface Machining Experiment

				Control Factor				
Row	A	B	C	D	E	F	G	H
1	0	0	0	0	0	0	0	0
2	0	0	1	1	1	1	1	1
3	0	0	2	2	2	2	2	2
4	0	1	0	0	1	1	2	2
5	0	1	1	1	2	2	0	0
6	0	1	2	2	0	0	1	1
7	0	2	0	1	0	2	1	2
8	0	2	1	2	1	0	2	0
9	0	2	2	0	2	1	0	1
10	1	0	0	2	2	1	1	0
11	1	0	1	0	0	2	2	1
12	1	0	2	1	1	0	0	2
13	1	1	0	1	2	0	2	1
14	1	1	1	2	0	1	0	2
15	1	1	2	0	1	2	1	0
16	1	2	0	2	1	2	0	1
17	1	2	1	0	2	0	1	2
18	1	2	2	1	0	1	2	0

Note: The rows correspond to the 18 rows in Table 11.18.

single array (see Table 11.17) along with replication, R, and two factors, I and J, which represent divisions of the machined part with respect to the circumference and axis of the part, respectively. The runs corresponding to $R = 0$ were all conducted during one time period and the runs corresponding to $R = 1$ were conducted at a later period. Within each time period the apparatus was configured by varying the control factors according to the control array and for each configuration three parts were machined: one at each of the three signal factor levels. Then

Table 11.17 Single Array for Signal and Noise Factors, Surface Machining Experiment

Column	R	M	I	J	Column	R	M	I	J
1	0	0	0	0	10	1	0	0	2
2	0	0	1	1	11	1	0	1	0
3	0	0	2	2	12	1	0	2	1
4	0	1	0	0	13	1	1	0	1
5	0	1	1	1	14	1	1	1	2
6	0	1	2	2	15	1	1	2	0
7	0	2	0	1	16	1	2	0	2
8	0	2	1	2	17	1	2	1	0
9	0	2	2	0	18	1	2	2	1

Note: The column numbers correspond to the 18 columns in Table 11.18.

Table 11.18 Response Data, Surface Machining Experiment

Control Run	1	2	3	4	5	6	7	8	9	10	11	12	13	14	15	16	17	18
1	2.8	2.6	2.4	3.0	2.2	3.0	3.0	3.0	2.2	6.0	4.8	1.9	2.3	2.5	4.0	3.0	4.0	2.0
2	3.2	2.4	3.1	3.8	2.3	3.1	3.0	3.0	2.4	6.0	5.0	2.2	2.0	3.0	3.3	2.9	4.2	2.3
3	3.3	2.7	2.5	3.0	3.0	3.0	3.1	2.9	2.5	6.2	5.0	1.8	1.9	2.0	3.8	3.0	4.0	2.2
4	4.2	3.4	2.9	4.6	3.4	4.3	9.0	3.2	4.1	7.0	5.3	3.0	3.0	4.0	3.8	3.0	4.2	7.4
5	4.1	3.8	3.6	5.2	4.2	4.0	9.6	4.0	4.2	7.2	5.0	2.5	2.6	4.0	4.1	2.9	4.0	7.4
6	4.5	3.8	3.9	4.0	3.3	4.1	9.8	3.9	4.0	7.3	4.3	2.8	2.2	4.0	5.1	2.9	3.8	7.7
7	6.0	8.8	8.6	5.2	10.5	6.5	10.8	5.0	6.1	7.5	10.2	4.2	4.5	9.6	9.0	9.0	6.2	9.0
8	6.0	8.0	8.6	5.2	10.0	6.0	11.0	4.3	5.5	8.0	10.4	4.8	4.0	10.7	10.0	9.0	6.0	8.2
9	7.0	8.2	9.9	6.7	10.0	7.0	10.5	4.7	6.4	8.0	10.0	4.3	4.1	9.8	11.0	9.0	6.0	8.0
10	3.7	2.8	3.5	3.4	2.0	2.7	3.6	2.2	3.0	3.0	3.8	2.1	1.8	3.0	4.0	2.3	4.0	2.4
11	4.0	3.0	3.5	3.2	2.0	3.0	3.6	2.1	3.0	3.0	3.5	2.5	1.5	3.0	4.0	2.7	4.0	2.5
12	3.9	3.0	3.7	4.2	1.7	2.6	3.5	2.0	3.4	3.5	3.5	2.0	1.5	3.0	4.8	2.5	4.0	2.5
13	5.0	4.2	4.0	4.4	3.0	4.2	6.7	2.4	4.1	4.0	4.0	2.8	2.3	5.5	5.4	4.0	4.1	3.0
14	4.8	4.8	4.2	4.2	3.0	4.0	6.0	3.0	4.9	3.9	2.9	2.4	4.0	5.2	5.0	3.5	4.8	3.2
15	4.2	4.5	4.2	4.5	3.2	4.0	5.0	2.5	3.9	4.2	3.0	3.0	2.5	5.2	6.0	3.6	5.0	3.0
16	6.2	11.0	7.0	8.0	6.2	5.0	17.8	4.2	8.5	6.2	9.0	3.0	4.1	11.5	14.2	9.4	6.7	8.5
17	5.0	11.8	6.4	8.4	6.0	4.8	17.0	4.0	10.0	6.0	8.0	3.0	3.2	12.4	14.0	10.0	6.2	8.0
18	6.0	11.0	7.2	8.2	6.0	5.7	17.0	4.0	8.8	6.5	8.3	3.6	4.0	12.4	15.0	10.2	6.5	7.8

three measurements were taken on each part at positions represented by the values of I and J in Table 11.17. Measurements taken at different positions on the same part can be used to evaluate within-part variation. The response data are displayed in Table 11.18.

(a) What performance measure is appropriate for the problem?

(b) Analyze the data by using both the RFM and PMM approaches. Based on the data plot and analysis, is it suitable to assume a straight-line-through-the-origin model? Which approach provides more information about the experiment?

(c) If an array of type (i), (ii), or (iv) is considered for the experiment, can you find a different experimental plan at the same cost (i.e., 108 machined parts and 324 measurements)? Will this plan enjoy other desirable properties than the one actually used?

7. An experiment (Lunani, Nair, and Wasserman, 1997) on engine idling performance investigated the relationship between the fuel flow (signal) and indicated mean effective pressure (response), abbreviated as IMEP, for different engine configurations. Two two-level control factors and six three-level control factors (see Table 11.19) were studied based on an 18-run control array (see Table 11.20). For each of the 18 control factor

Table 11.19   Control Factors, Engine Idling Experiment

A. cylinder head	E. air-fuel ratio
B. fuel injectors	F. spark plug penetration
C. spark plug gap	G. engine speed
D. ignition timing	H. injection timing

Table 11.20   Control Array and Fitted Parameters, Engine Idling Experiment

Control Factor								Parameter	
A	B	C	D	E	F	G	H	$\hat{\beta}$	s
0	0	0	0	0	0	0	0	9.67	3.67
0	0	1	1	1	1	1	1	10.18	3.57
0	0	2	2	2	0	2	2	10.44	3.61
0	1	0	0	1	1	2	2	8.05	4.10
0	1	1	1	2	0	0	0	10.87	3.99
0	1	2	2	0	0	1	1	10.94	2.74
0	2	0	1	0	0	2	1	9.52	3.68
0	2	1	2	1	0	0	2	10.19	3.37
0	2	2	0	2	1	1	0	9.71	4.17
1	0	0	2	2	1	0	1	11.66	2.66
1	0	1	0	0	0	1	2	7.94	3.79
1	0	2	1	1	0	2	0	11.19	3.46
1	1	0	1	2	0	1	2	8.99	3.87
1	1	1	2	0	1	2	0	12.54	2.46
1	1	2	0	1	0	0	1	8.82	3.88
1	2	0	2	1	0	1	0	12.07	3.47
1	2	1	0	2	0	2	1	8.64	4.18
1	2	2	1	0	1	0	2	8.96	3.36

combinations, measurements of fuel flow and IMEP were taken using two test engines: a low-mileage engine and a high-mileage engine. Therefore, the test engine can be treated as a noise factor. Each of the 18 sets of measurements was used to fit the signal-response model $E(y) = \beta M$. The estimated slopes and standard errors for these models are given in Table 11.20.

(a) Analyze $\ln s^2$ and $\hat{\beta}$ as a function of the control factors and identify significant factors in each model.

(b) Assuming that a steeper slope is desired, find optimal factor settings based on the fitted models in (a).

(c) Suppose that the response variance σ^2 is proportional to β^λ. By plotting $\ln s$ against $\ln \hat{\beta}$, determine an empirical estimate of the λ value. Give a heuristic justification of this graphical method.

(d) Based on the estimate of λ in (c) and the discussion in Section 11.3, argue that the use of the dynamic SN ratio is inappropriate and unjustifiable.

8. In addition to the four types of arrays considered in Section 11.6, one may consider the following two types: $(C \times N, S)$ and (C, N, S). Discuss the restrictions that either design may impose and argue why they may not be as popular as types (i)–(iii).

9. In a type (iv) array, a single array (C, S) is used for the control and signal factors. Consider a special case in which there are two different sets of signal factor levels, such as $\{1, 3, 5, 7\}$ and $\{2, 4, 6, 8\}$, depending on the control factor combination that goes with the signal factor. Explain how the second stage of the RFM, as described in (11.16), should be modified and extended to reflect the difference between these two sets of signal factor levels. How would this change affect the analysis? [*Hint:* Each of the three models in (11.16) (ii) should have two versions.]

10. Identify the designs whose resolutions are given in Table 11.9.

11. (a) Find a 16-run 2_{III}^{11-7} design for 7 control factors and 4 noise factors that have all the 16 noise factor combinations. Identify its defining generators. (*Hint:* Is it permissible to have a defining relation that involves only the noise factors?)

(b) Compare the single array in (a) with the cross array in the original experiment in terms of their estimation capacities (which were defined in Chapter 10). Comment on the pros and cons of the two arrays.

REFERENCES

Box, G. and Jones, S. (1992), "Split-Plot Designs for Robust Product Experimentation," *Journal of Applied Statistics*, 19, 3–26.

Draper, N.R. and Smith, H. (1998), *Applied Regression Analysis*, 3rd ed., New York: John Wiley & Sons.

León, R.V., Shoemaker, A.C., and Kacker, R.N. (1987), "Performance Measures Independent of Adjustment: An Explanation and Extension of Taguchi's Signal to Noise Ratio" *Technometrics*, 29, 253–285.

Lin, S.M. and Wen, T.C. (1994), "Experimental Strategy–Application of Taguchi's Quality Engineering Method to Zinc Phosphate Coating Uniformity," *Plating and Surface Finishing*, 81, 59–64.

Lunani, M., Nair, V.N., and Wasserman, G.S. (1997), "Graphical Methods for Robust Design with Dynamic Characteristics," *Journal of Quality Technology*, 29, 327–338.

Mandel, J. (1964), *The Statistical Analysis of Experimental Data*, New York: Interscience.

Miller, A. and Wu, C.F.J. (1996), "Parameter Design for Signal-Response Systems: A Different Look at Taguchi's Dynamic Parameter Design" *Statistical Science*, 11, 122–136.

Taguchi, G. (1987), *System of Experimental Design*, White Plains, NY: Unipub/Kraus International Publications.

Taguchi, G. (1991), *Taguchi Methods, Signal-to-Noise Ratio for Quality Evaluation*, Vol. 3, Dearborn, MI: American Supplier Institute Press.

Yano, H. (1991), *Metrological Control: Industrial Measurement Management*, Tokyo: Asian Productivity Organization.

CHAPTER 12

Experiments for Improving Reliability

Much of the reliability literature is concerned with estimating the reliability of an existing product. In this chapter, we consider how to improve the reliability of products and processes through the use of designed experiments. In such experiments, each unit is tested until it fails or is still functioning when the experiment is terminated. If it fails, the experimental response is its *failure time*. If it is still functioning at the termination time t_0, it is said to be right censored at t_0. There are two major goals in reliability experiments: *reliability improvement* (i.e., to increase the mean failure time) and *robust reliability* (i.e., to reduce the influence of noise variation on reliability). The latter can be achieved using parameter design techniques. Frequentist and Bayesian approaches for analyzing failure time data are considered. For highly reliable products, it may be hard to observe any failure in a reasonable length of time even under accelerated stress conditions. Another source of information is *degradation data*. If the degradation of a product characteristic can be related to reliability, then such data can be used to identify factors that have an impact on reliability. A simple analysis method for degradation data is considered. Design techniques are not specifically considered for reliability and degradation experiments as they are not different from those considered in previous chapters.

12.1 EXPERIMENTS WITH FAILURE TIME DATA

Failure time data were considered in the previous chapters, e.g., the router bit experiment in Chapter 6 and the heat exchanger experiment in the exercises of Chapter 8. In this chapter, three failure time experiments will be analyzed. Others will be given in the exercises.

Table 12.1   Design Matrix and Failure Time Data, Light Experiment

Run	A	B	C	D	E	Failure Time	
1	−	−	−	−	−	(14, 16)	(20, ∞)
2	−	−	+	+	+	(18, 20)	(20, ∞)
3	−	+	−	−	+	(8, 10)	(10, 12)
4	−	+	+	+	−	(18, 20)	(20, ∞)
5	+	−	−	+	−	(20, ∞)	(20, ∞)
6	+	−	+	−	+	(12, 14)	(20, ∞)
7	+	+	−	+	+	(16, 18)	(20, ∞)
8	+	+	+	−	−	(12, 14)	(14, 16)

12.1.1   Light Experiment

Consider an experiment to improve the reliability of fluorescent lights (Taguchi, 1987, p. 930). Five two-level factors denoted by $A-E$ were studied using a 2^{5-2} design with the defining relation $D = -AC$ and $E = -BC$. Each run had two replicates (i.e., two lights were tested) and the experiment was conducted over 20 days with inspections every two days. The design matrix and failure time data appear in Table 12.1. No factor names or levels were given, presumably for proprietary reasons. Besides the main effects, the experimenter also thought that the AB ($= DE$) interaction might be important. If a light had not failed by the 20th day, its failure time is right censored at 20 and is indicated by $(20, \infty)$. There are seven such values in Table 12.1. The rest of the data are interval censored, e.g., $(14, 16)$ means that the particular light failed between the 14th and 16th days. The failure times were short, presumably because the lights were subjected to an accelerating factor which stressed the lights at higher than normal conditions.

Table 12.2   Factors and Levels, Thermostat Experiment

	Level	
Factor	−	+
A. diaphragm plating rinse	clean	contaminated
B. current density (min, amps)	5, 60	10, 15
C. sulfuric acid cleaning (sec)	3	30
D. diaphragm electroclean (min)	2	12
E. beryllium copper grain size (in.)	0.008	0.018
F. stress orientation	perpendicular to seam	in-line with seam
G. humidity	wet	dry
H. heat treatment	45 min, 600°F	4 hr, 600°F
I. brazing machine (sec)	2	3
J. power element electroclean	clean	contaminated
K. power element plating rinse	clean	contaminated

12.1.2 Thermostat Experiment

Bullington et al. (1993) reported the use of a 12-run Plackett-Burman design in an experiment to improve the reliability of industrial thermostats. An increase in early failures had been linked to corrosion-induced pinholes in the diaphragm, a key component of the thermostat. An experiment was needed because of the large number of possible factors affecting the rate of corrosion. Eleven factors, denoted by $A-K$ and given in Table 12.2, were chosen across a 14-stage manufacturing process. Ten thermostats were manufactured at each of the 12 factor settings and tested up to 7342 ($\times 1000$) cycles. The design matrix and failure time data are given in Tables 12.3 and 12.4. If a thermostat was still functioning after 7342 ($\times 1000$) cycles, its right-censored value 7342 is recorded as the response value. There are 22 right-censored data at runs 1, 6, and 11.

Table 12.3 Design Matrix, Thermostat Experiment

Run	A	B	C	D	E	F	G	H	I	J	K
						Factor					
1	−	−	−	−	−	−	−	−	−	−	−
2	−	−	−	−	−	+	+	+	+	+	+
3	−	−	+	+	+	−	−	−	+	+	+
4	−	+	−	+	+	−	+	+	−	−	+
5	−	+	+	−	+	+	−	+	−	+	−
6	−	+	+	+	−	+	+	−	+	−	−
7	+	−	+	+	−	−	+	+	−	+	−
8	+	−	+	−	+	+	+	−	−	−	+
9	+	−	−	+	+	+	−	+	+	−	−
10	+	+	+	−	−	−	−	+	+	−	+
11	+	+	−	+	−	+	−	−	−	+	+
12	+	+	−	−	+	−	+	−	+	+	−

Table 12.4 Failure Time Data (with censoring time of 7342), Thermostat Experiment

Run	Failure Time									
1	957	2846	7342	7342	7342	7342	7342	7342	7342	7342
2	206	284	296	305	313	343	364	420	422	543
3	63	113	129	138	149	153	217	272	311	402
4	75	104	113	234	270	364	398	481	517	611
5	97	126	245	250	390	390	479	487	533	573
6	490	971	1615	6768	7342	7342	7342	7342	7342	7342
7	232	326	326	351	372	446	459	590	597	732
8	56	71	92	104	126	156	161	167	216	263
9	142	142	238	247	310	318	420	482	663	672
10	259	266	306	337	347	368	372	426	451	510
11	381	420	7342	7342	7342	7342	7342	7342	7342	7342
12	56	62	92	104	113	121	164	232	258	731

In the light experiment and the thermostat experiment, no explicit noise factors were used. Noise variation was represented by the use of replicates for each control run. In the next experiment, five noise factors were identified and systematically varied in the experiment.

12.1.3   Drill Bit Experiment

Consider an experiment for improving the reliability of drill bits used in fabricating multilayer printed circuit boards. A drill bit's failure time is defined by the number of holes drilled before it breaks. In designing multilayer circuit boards, small-diameter holes are desired because they allow more room for the circuitry. Small-diameter drill bits are weaker and more prone to break, however. This is a serious problem because broken bits cannot be removed from the boards so that the boards have to be scrapped. There were 11 control factors (A at four levels and $B-J$ and L at two levels) dealing with selected drill bit material composition and geometric characteristics of the drill bits and five noise factors ($M-Q$ at two levels) dealing with different types of multilayer circuit boards to be drilled and the feed rate. The factors and levels are given in Table 12.5. The experiment employed a cross array consisting of a 16-run control array [using an $OA(16, 2^{10}4^1)$ with

Table 12.5   Factors and Levels, Drill Bit Experiment

Control Factor	Level			
	0	1	2	3
A. carbide cobalt (%)	A1	A2	A3	A4
	Level			
	−		+	
B. body length (in.)	minimum		minimum + 30%	
C. web thickness (% diameter)	C1		C2	
D. web taper	D1		D2	
E. moment of inertia (in.^{4})	standard		standard + x	
F. radial rake	F1		F2	
G. helix angle	G1		G2	
H. axial rake	H1		H2	
I. flute length (in.)	minimum		minimum + 50%	
J. point angle	J1		J2	
L. point style	standard		strong	
Noise Factor	Level			
	−		+	
M. feed rate (in./min)	10		20	
N. backup material	hard board		phenolic	
O. pcb material	epoxy		polyamide	
P. number of layers	8		12	
Q. four 2-oz layers	no		yes	

Table 12.6 Cross Array and Failure Time Data (with censoring time of 3000), Drill Bit Experiment

											Noise Factor							
										M	−	−	−	−	+	+	+	+
										N	−	−	+	+	−	−	+	+
										O	−	−	+	+	+	+	−	−
										P	−	+	−	+	−	+	−	+
										Q	−	+	+	−	+	−	−	+
			Control Factor															
A	D	B	C	F	G	H	I	E	J	L								
0	−	−	−	−	−	−	−	−	−	−	1280	44	150	20	60	2	65	25
0	−	−	−	−	+	+	+	+	+	+	2680	125	120	2	165	100	795	307
0	+	+	+	+	−	−	−	−	+	+	2670	480	762	130	1422	280	670	130
0	+	+	+	+	+	+	+	+	−	−	2655	90	7	27	3	15	90	480
1	−	−	+	+	−	−	+	+	+	−	3000	440	480	10	1260	5	1720	3000
1	−	+	+	+	+	−	+	−	−	+	2586	6	370	45	2190	36	1030	16
1	+	+	−	−	−	−	+	+	+	−	3000	2580	20	320	425	85	950	3000
1	+	+	−	−	+	+	−	−	−	+	800	45	260	250	1650	470	1250	70
2	−	+	−	+	−	+	−	+	−	−	3000	190	140	2	100	3	450	840
2	−	+	−	+	+	−	+	−	+	+	3000	638	440	145	690	140	1180	1080
2	+	−	+	−	−	+	−	+	+	+	3000	970	180	220	415	70	2630	3000
2	+	−	+	−	+	−	+	−	−	−	3000	180	870	310	2820	240	2190	1100
3	−	+	+	−	−	+	+	−	−	+	3000	612	1611	625	1720	195	1881	2780
3	−	+	+	−	+	−	−	+	+	−	3000	1145	1060	198	1340	95	2509	345
3	+	−	−	+	−	+	+	−	+	−	3000	3000	794	40	160	50	495	3000
3	+	−	−	+	+	−	−	+	−	+	680	140	809	275	1130	145	2025	125

$A = (1, 2, 12)$, $D = 3$, $B = -13$, $C = -23$, $F = 123$, $G = 4$, $H = -14$, $I = -24$, $E = 124$, $J = -34$, $L = 234$] and an 8-run noise array (using a 2_{III}^{5-2} design with $M = 1$, $N = 2$, $O = -12$, $P = 3$, $Q = 123$). The cross array and the data (Hamada, 1993) are given in Table 12.6, where the four levels of A are denoted by 0, 1, 2, and 3 and the two levels of the other factors are denoted by − and +. Note that the noise array in Table 12.6 is displayed in a sideways manner. Its eight columns represent the eight factor combinations and the five rows represent the five noise factors. In the experiment, 16 different drill bit designs specified by the 16 runs in the control array were used in the eight different production conditions specified by the noise array. The testing was stopped after 3000 holes were drilled. If a drill bit did not break before 3000 holes, its failure time was right censored at 3000. About 11% of the drill bits were right censored. The experimental goal was to identify control factor settings whose resulting reliability was high and robust to noise variation.

12.2 REGRESSION MODEL FOR FAILURE TIME DATA

Suppose t denotes the failure time (or lifetime) of a unit. Because t only takes nonnegative values, a standard linear regression modeling of t is not appropriate. It should be transformed to take values from $-\infty$ to ∞. The most obvious choice is the log transformation $t \to y = \ln t$. The following two failure time distributions are commonly used for reliability studies:

lognormal with density

$$f(t) = \frac{1}{\sqrt{2\pi\sigma^2}} t^{-1} \exp\left(\frac{-(\ln t - \mu)^2}{2\sigma^2}\right), \qquad t > 0, \quad (12.1)$$

and *Weibull* with density

$$f(t) = (\psi/\lambda)(t/\lambda)^{\psi-1} \exp\left[-(t/\lambda)^\psi\right], \qquad t > 0. \quad (12.2)$$

The exponential density $f(t) = (1/\lambda)\exp(-t/\lambda)$ is a special case of the Weibull density with $\psi = 1$.

If t has a lognormal distribution with density (12.1), it is easy to show that $y = \ln t$ has a normal distribution $N(\mu, \sigma^2)$. By using a standard result on the transformation of random variables, it can be shown that, for the Weibull distribution, $y = \ln t$ has the *extreme-value* distribution whose density is

$$f(y) = \frac{1}{\sigma} \exp\left[\frac{y-\mu}{\sigma} - \exp\left(\frac{y-\mu}{\sigma}\right)\right], \qquad (12.3)$$

with $\mu = -\ln \lambda$ and $\sigma = \psi^{-1}$. For $\mu = 0$ and $\sigma = 1$, it is called the standard extreme-value distribution. In the Weibull distribution, λ is the scale parameter and ψ is the shape parameter. On the log-transformed scale, $\mu = -\ln \lambda$ and $\sigma = \psi^{-1}$ become respectively the location and scale parameters of the extreme-value distribution.

A linear regression model can now be applied to the y data as follows:

$$
\begin{aligned}
y_i = \ln t_i &= \beta_0 + \beta_1 x_{i1} + \cdots + \beta_p x_{ip} + \sigma\epsilon_i \\
&= \mathbf{x}_i^T \boldsymbol{\beta} + \sigma\epsilon_i, \qquad i = 1, \ldots, N,
\end{aligned}
\qquad (12.4)
$$

where the $\{t_i\}$ are the failure times, $\mathbf{x}_i^T = (1, x_{i1}, \ldots, x_{ip})$ are the corresponding vectors of covariate values, $\boldsymbol{\beta} = (\beta_0, \beta_1, \ldots, \beta_p)^T$ is the vector of regression parameters, and σ is the scale parameter. The errors $\{\epsilon_i\}$ have independent standard extreme-value distributions if the failure times follow a Weibull distribution and have independent $N(0, 1)$ distributions if the failure times follow a lognormal distribution. Note that the lognormal regression model is the same as the general linear model of (1.1) for the log failure times $y_i = \ln t_i$.

For reliability improvement experiments such as those given in Sections 12.1.1 and 12.1.2, the covariates correspond to an intercept, the factor main effects, and possibly some interactions. For robust reliability experiments such as the drill bit experiment in Section 12.1.3, the covariates correspond to an intercept, the control main effects, possibly some control-by-control interactions, the control-by-noise interactions, the noise main effects, and possibly some noise-by-noise interactions, where control and noise refer to control factor and noise factor, respectively.

Because the sample size used in reliability experiments is typically small, there may be little qualitative difference between the use of either model. The choice of either model therefore needs to be based on other considerations such as interpretability of the distribution or sophistication of the computing environment.

12.3 A LIKELIHOOD APPROACH FOR HANDLING FAILURE TIME DATA WITH CENSORING

If the experimental data are lognormally distributed and none is censored, i.e., all failure times are observed, standard regression techniques applied to the data $\ln t_i$ would be appropriate. For the Weibull regression model or when some of the data are censored for both regression models in Section 12.2, this approach is not adequate. In this section a likelihood-based approach is considered.

First we need to define different types of censoring. If the true response, which can be the failure time or other characteristics, is known to be above a certain value t_o, it is said to be *right censored* at t_o. This happens if the unit is still functioning at time t_o when the study is terminated. (In clinical trials, this would mean that the patient is still alive when the trial is terminated.) Similarly, if the true response is known to be below a certain value t_o, i.e., it is in the interval $[0, t_o)$ or $(-\infty, t_o)$, it is said to be *left censored* at t_o. A good example of left censoring is given in the exercises on a window forming experiment, in which the true response is the window size. If the true window size is smaller than a certain value (which acts as the left-censored value), the window will not open and the only recorded value is the left-censored value. A third type is *interval censoring*, where the true response is known to lie within an interval. The light experiment in Section 12.1.1 provides a good example. Interval censoring occurs commonly in reliability studies when a unit is inspected at specified times until it fails so that its failure time is between consecutive inspection times, i.e., an interval-censored response.

One analysis method *treats the right-censored times as actual failure times* and then analyzes them by using standard methods such as failure time regression modeling. Ignoring the censoring information, however, can lead to wrong decisions because the unobserved failure times and right-censored times may differ greatly, depending on the particular factor level combination. A simulation study in Hamada and Wu (1991) showed that this method can perform poorly by missing some important effects and misidentifying spurious effects. Because of its simplicity, this method continues to be used in practice, especially when the investigators are not familiar with the more appropriate techniques discussed in this chapter or do not have access to the software supporting these techniques. Its practice may still lead to valid results and recommendations if the proportion of censored observations is small and if the difference between censoring times and the unobserved failure times is not great. For example, the cast fatigue experiment in Section 7.1 has one right-censored value at 7.0 (run 5 in Table 7.2). The analysis in Chapter 8 treated 7.0 as the actual logged lifetime. A valid method should treat 7.0 as a right-censored value and use the likelihood or Bayesian method for estimation and variable selection. It turns out that the analysis results and recommended factor settings are the same as those given in Chapter 8. This can be explained by the fact that only 1 out of 12 observations in the experiment is censored and the difference between 7.0 and the conditional expected lifetime over $(7.0, \infty)$ for run 5 is not large. Another example is the router bit experiment reported in Chapter 6. The data are actually interval censored and right censored but the analysis in Chapter 6 treated them as actual lifetimes. Results from this analysis are not qualitatively different from those based on the methods considered in this chapter. A comparison is left as an exercise.

The likelihood approach can easily handle both failure and censored data. Assume that the failure times t_i follow the regression model in (12.4), where $y_i = \ln t_i$. If a unit is right censored at y_i, its contribution to the likelihood is the probability that the true failure time exceeds y_i. For the lognormal distribution, it is

$$1 - \Phi\big((y_i - \mathbf{x}_i^T \boldsymbol{\beta})/\sigma\big), \tag{12.5}$$

where Φ is the standard normal cumulative distribution function, and for the Weibull distribution, it is

$$\exp\big\{-\exp\big[(y_i - \mathbf{x}_i^T \boldsymbol{\beta})/\sigma\big]\big\}, \tag{12.6}$$

which is obtained by integrating the extreme-value density in (12.3) (with $\mu = \mathbf{x}_i^T \boldsymbol{\beta}$) from y_i to ∞. For left censoring, the corresponding contribution to the likelihood is one minus the term given in (12.5) and (12.6). For interval censoring with the interval (a, b), the corresponding contribution is

$$\Phi\big[(\ln b - \mathbf{x}_i^T \boldsymbol{\beta})/\sigma\big] - \Phi\big[(\ln a - \mathbf{x}_i^T \boldsymbol{\beta})/\sigma\big] \tag{12.7}$$

for the lognormal distribution, and

$$\exp\big\{-\exp\big[(\ln a - \mathbf{x}_i^T \boldsymbol{\beta})/\sigma\big]\big\} - \exp\big\{-\exp\big[(\ln b - \mathbf{x}_i^T \boldsymbol{\beta})/\sigma\big]\big\} \tag{12.8}$$

for the Weibull distribution.

For clarity of presentation, we will derive the likelihood only for the situation in which failure and right-censored data are observed. By collecting the contributions from the failure and right-censored data, the likelihood can be shown to be

$$\mathscr{L}(\boldsymbol{\beta}, \sigma^2) = \prod_{i \in FAIL} (2\pi\sigma^2)^{-1/2} \exp\left\{(-\tfrac{1}{2})\left[(y_i - \mathbf{x}_i^T\boldsymbol{\beta})/\sigma\right]^2\right\}$$

$$\times \prod_{i \in CEN} \left\{1 - \Phi\left[(y_i - \mathbf{x}_i^T\boldsymbol{\beta})/\sigma\right]\right\}, \tag{12.9}$$

for the lognormal regression model and

$$\mathscr{L}(\boldsymbol{\beta}, \sigma^2) = \prod_{i \in FAIL} (1/\sigma)\exp\left\{\left[(y_i - \mathbf{x}_i^T\boldsymbol{\beta})/\sigma\right] - \exp\left[(y_i - \mathbf{x}_i^T\boldsymbol{\beta})/\sigma\right]\right\}$$

$$\times \prod_{i \in CEN} \exp\left\{-\exp\left[(y_i - \mathbf{x}_i^T\boldsymbol{\beta})/\sigma\right]\right\} \tag{12.10}$$

for the Weibull regression model, where *FAIL* and *CEN* denote the sets of observed failure times and censoring times, respectively. For left-censored or interval-censored data, the likelihood can be easily derived by incorporating the corresponding terms as discussed in the previous paragraph.

Once the likelihood $\mathscr{L}(\boldsymbol{\beta}, \sigma^2)$ is obtained, *maximum likelihood estimates* (MLEs) $(\hat{\boldsymbol{\beta}}, \hat{\sigma}^2)$ can be obtained by maximizing $\mathscr{L}(\boldsymbol{\beta}, \sigma^2)$ with respect to $\boldsymbol{\beta}$ and σ^2. To test the significance of the estimates, they need to be compared with their standard errors. Let $\boldsymbol{\theta} = (\boldsymbol{\beta}, \sigma^2)$. Standard errors can be obtained from the Fisher information matrix I or the observed information matrix I_O defined respectively as

$$I(\hat{\boldsymbol{\theta}}) = E\left(\frac{-\partial^2 \mathscr{L}(\boldsymbol{\theta})}{\partial \theta_i \, \partial \theta_j}\right)\Bigg|_{\boldsymbol{\theta} = \hat{\boldsymbol{\theta}}} \tag{12.11}$$

and

$$I_O(\hat{\boldsymbol{\theta}}) = \frac{-\partial^2 \mathscr{L}(\boldsymbol{\theta})}{\partial \theta_i \, \partial \theta_j}\Bigg|_{\boldsymbol{\theta} = \hat{\boldsymbol{\theta}}}, \tag{12.12}$$

which are evaluated at the MLE $\hat{\boldsymbol{\theta}}$. For large samples, it is known that $\hat{\boldsymbol{\theta}} - \boldsymbol{\theta}$ is approximately distributed as $N(0, \mathbf{I}^{-1})$ or $N(0, \mathbf{I}_O^{-1})$. Let $\hat{\sigma}_i^2$ denote the ith diagonal element of $\mathbf{I}^{-1}$ or $\mathbf{I}_O^{-1}$. Then the asymptotic normality result justifies the use of

$$\left(\hat{\theta}_i - z_{\alpha/2}\,\hat{\sigma}_i, \hat{\theta}_i + z_{\alpha/2}\,\hat{\sigma}_i\right) \tag{12.13}$$

as a confidence interval with confidence coefficient $1 - \alpha$. If this interval does not contain zero, the effect associated with θ_i is declared to be significant at level α. If $(p + 1)/N \geq 0.1$, where $p + 1$ is the dimension of $\boldsymbol{\beta}$ and N is the total sample size, $\hat{\sigma}_i^2$ should be replaced by $\hat{\sigma}_i^2[N/(N - p - 1)]$ to reduce the bias of $\hat{\sigma}_i^2$.

Likelihood ratio tests provide an alternative method for assessing significance of the ith parameter in $\boldsymbol{\theta}$:

$$\Gamma = -2\ln\frac{\mathscr{L}\left(\hat{\boldsymbol{\theta}}_{(-i)}\right)}{\mathscr{L}(\hat{\boldsymbol{\theta}})}, \tag{12.14}$$

where the ith covariate is dropped from the model and $\boldsymbol{\theta}_{(-i)}$ are the corresponding parameters. Under the null hypothesis $\theta_i = 0$, Γ is distributed asymptotically as χ^2 with one degree of freedom. Therefore, $\theta_i = 0$ is rejected at level α if $\Gamma > \chi_{1,\alpha}^2$. See Meeker and Escobar (1998) for details on methods of estimation and testing and the Splus functions for these computations. Commercial software is also available, such as the RELIABILITY procedure in SAS (SAS Institute, 1989).

12.3.1 Estimability Problem with MLEs

One problem with the likelihood approach for censored data is that the MLEs may be *infinite* so that testing for important effects cannot be done by comparing the MLEs with their standard errors. Consider the simplest case of a one-factor experiment with two levels. Suppose that all the observations at the high level are right censored while those at the low level are not, i.e., the right-censored data and the failure time data are *separated* according to the level of a covariate. Then the likelihood function for any reasonable regression model increases to the maximum value as the regression parameter (i.e., the factor main effect) tends toward infinity. Therefore, the MLE for the main effect tends to be infinite when the true factor effect is large and the opportunity for improvement is great. Similarly for multi-factor experiments, if right-censored data and failure time data are separated according to the level of a covariate, then the MLE for the effect of the covariate will tend to infinity. The covariate can correspond to a factor main effect or an interaction involving several factors. The exact conditions for the existence of MLEs in model (12.4) can be found in Silvapulle and Burridge (1986). In the reliability context, Hamada and Tse (1992) concluded that estimability problems tend to occur often for 2^{k-p} designs in which the fitted model has nearly the same number of parameters as the number of observations and especially when there is a single replicate. The commercially available software mentioned above do not check for infinite MLEs. Numerical instability in the MLE computation and large standard error relative to the MLE are two indications of a serious estimability problem. Nonzero first derivatives or a nonpositive definite observed information matrix are two other warning signs. While the MLEs may be infinite, the likelihood is well defined so that the likelihood ratio tests can still be used. See Clarkson and Jennrich (1991) for an algorithm to compute the maximized likelihood. This algorithm is implemented in an IMSL survival analysis subroutine.

Because a separation pattern between censored data and failure time data would strongly suggest a significant effect of the underlying covariate and a great opportunity for improvement, the estimability problem for the MLEs needs to be addressed. A natural way to circumvent the estimability problem is to use a Bayesian approach, which will be considered in Section 12.5.

12.4 DESIGN-DEPENDENT MODEL SELECTION STRATEGIES

The likelihood-based estimation and testing considered in the previous section assume that a fixed model [i.e., the $\mathbf{X}$ matrix consisting of the $\mathbf{x}_i$ in (12.4) as its rows] has been chosen. The immediate question is how to choose a model that is compatible with the data. The model selection strategy depends on the nature of the design. Recall from Chapter 7 that factorial designs can be classified into two broad types: regular and nonregular.

Model selection for regular designs is fairly simple. For a regular design, each degree of freedom is associated with a string of aliased effects. Apart from the usual problem of de-aliasing the aliased effects (see Section 4.4), one can use the saturated model consisting of all the degrees of freedom in the design for estimation and testing. Consider the 2_V^{5-1} design with $\mathbf{I} = ABCDE$ for illustration. Its 15 degrees of freedom consist of the 5 main effects and 10 two-factor interactions (2fi's). Because it has resolution V, all the main effects are strongly clear and the 2fi's are clear. Therefore their respective aliases can be assumed to be negligible according to the hierarchical ordering principle. This leads to the choice of a 15-term comprehensive model. Consider next a more complicated situation given by the 2_{IV}^{8-4} design with $5 = 123, 6 = 124, 7 = 134$, and $8 = 234$. Because it has resolution IV, its eight main effects are clear. Each of the remaining seven degrees is associated with a string of four aliased 2fi's. Except for the usual problem of de-aliasing the four 2fi's, the comprehensive model for the design consists of the eight main effects and seven 2fi's selected from the strings of aliases. Because the effects in the comprehensive model can be separately tested, little or no effort is required to search for submodels.

Model selection for nonregular designs requires more work. Due to the problems associated with complex aliasing (see Chapter 8), there is no comprehensive model that contains all other models as submodels. An iterative model selection strategy such as those discussed in Sections 8.4 and 8.5 is thus required. Consider, for example, the 12-run Plackett-Burman design with 8 factors. Its 28 2fi's have very complex aliasing relations with the main effects and among themselves. Assuming that the 8 main effects are retained in the model, there are still $\binom{28}{3} = 3276$ choices of the 2fi's for the remaining 3 terms in the model, resulting in a total of 3276 models with 11 terms. With a large number of candidate models, it is necessary to use a fast and efficient model selection algorithm such as the stochastic search algorithm considered in Section 8.5. An extension of this algorithm for censored data will be considered in the next section.

12.5 A BAYESIAN APPROACH TO ESTIMATION AND MODEL SELECTION FOR FAILURE TIME DATA

A Bayesian approach is natural for handling the estimability problem of MLEs because the important (but unknown) factor effects might be large but not infinite. By using proper prior distributions, well-behaved posterior distributions can be calculated and their finite modes or medians used to estimate the effects. Posterior distributions also allow the importance of factorial effects to be assessed without using asymptotic approximations which are required in the likelihood approach. A simple way to judge whether a factor effect is significant is to see if zero is in the upper or lower $\frac{1}{2}\alpha$ tail of the marginal posterior distribution for the effect. Here we shall focus on the lognormal regression model (12.4). A similar development for the Weibull regression can be done but will be left as an exercise.

The discussion here is very similar to the Bayesian formulation given in Section 8.5. Recall that the posterior distribution $\pi(\theta \mid y)$ is proportional to $f(y \mid \theta)\pi(\theta)$, where $f(y \mid \theta)$ is the probability density function given in (12.9), $\pi(\theta)$ is the prior distribution, and $\theta = (\beta, \sigma^2)$. For $\pi(\theta)$, we use

$$\pi(\theta) \propto \sigma^{-(p+1)}\exp\left[-(\beta - \beta_0)^T A_0(\beta - \beta_0)/2\sigma^2\right]$$
$$\times \sigma^{-(\nu_0+1)}\exp(-\nu_0 s_0^2/2\sigma^2), \tag{12.15}$$

where $p + 1$ is the number of regression parameters in β. Straightforward calculations show that $\pi(\theta)$ in (12.15) can be factored into the product of (i) the marginal prior for σ^2, which has an inverted gamma distribution [see (8.27)] with parameters $(\nu_0/2, \nu_0 s_0^2/2)$ and (ii) the conditional prior of β given σ^2, which has a multivariate normal distribution with mean β_0 and covariance matrix $\sigma^2 A_0^{-1}$.

If there are no censored values, the density [which is the same as the likelihood in (12.9)] takes the simpler form

$$f(y \mid \theta) = \prod_i (2\pi\sigma^2)^{-1/2}\exp\left\{(-1/2)\left[(y_i - x_i^T\beta)/\sigma\right]^2\right\}. \tag{12.16}$$

By comparing (12.15) and (12.16), $\pi(\theta)$ and $f(y \mid \theta)$ have the same functional form in β and σ^2. Therefore, $\pi(\theta)$ is called a *natural conjugate prior* (Berger, 1985) for $f(y \mid \theta)$ in (12.16). The posterior distribution $\pi(\theta \mid y)$ can be shown to be the product of (i) the marginal distribution of σ^2, which has an inverted gamma distribution with parameters $(\nu_1/2, \nu_1 s_1^2/2)$,

$$\sigma^2 \sim IG(\nu_1/2, \nu_1 s_1^2/2), \qquad (12.17)$$

and (ii) the conditional distribution of $\boldsymbol{\beta}$ given σ^2, which has a multivariate normal distribution with mean $\tilde{\boldsymbol{\beta}}$ and covariance matrix $\sigma^2 \mathbf{M}^{-1}$,

$$\boldsymbol{\beta} | \sigma^2 \sim MN(\tilde{\boldsymbol{\beta}}, \sigma^2 \mathbf{M}^{-1}), \qquad (12.18)$$

where

$$\nu_1 = N + \nu_0, \qquad \mathbf{M} = \mathbf{X}^T \mathbf{X} + \mathbf{A}_0,$$

$$\tilde{\boldsymbol{\beta}} = \mathbf{M}^{-1}(\mathbf{X}^T \mathbf{X} \hat{\boldsymbol{\beta}} + \mathbf{A}_0 \boldsymbol{\beta}_0),$$

$$\nu_1 s_1^2 = \nu_0 s_0^2 + \nu s^2 + (\tilde{\boldsymbol{\beta}} - \boldsymbol{\beta}_0)^T \mathbf{A}_0 (\tilde{\boldsymbol{\beta}} - \boldsymbol{\beta}_0) \qquad (12.19)$$

$$+ (\tilde{\boldsymbol{\beta}} - \hat{\boldsymbol{\beta}})^T \mathbf{X}^T \mathbf{X} (\tilde{\boldsymbol{\beta}} - \hat{\boldsymbol{\beta}}).$$

The rows of the matrix X are the $\mathbf{x}_i^T$, and $\hat{\boldsymbol{\beta}}$ and νs^2 are the least squares estimates and residual sum of squares, respectively, based on $\mathbf{y}$.

In the presence of censored data, the prior in (12.15) is no longer a conjugate prior for the likelihood given in (12.9), and the posterior distribution does not have a simple form as given above. In order to take advantage of the fast computation provided by formulas (12.17)–(12.19), we need to *impute* the censored data based on the current values of the unknown parameters. (An imputed value may be obtained by randomly sampling the conditional distribution of the true failure time or using the conditional mean or median.) Then we treat the imputed values as pseudo failure time data and apply (12.17)–(12.19) to the combined failure time and pseudo failure time data. The following algorithm for implementing this approach can be viewed as a Gibbs sampling algorithm because it treats the unobserved failure times $\tilde{\mathbf{y}}$ as additional parameters in the model and samples successively from the conditional distribution of $\tilde{\mathbf{y}}$ given $(\boldsymbol{\beta}, \sigma^2)$ and the conditional distribution of $(\boldsymbol{\beta}, \sigma^2)$ given the failure time data and the imputed values of $\tilde{\mathbf{y}}$. The specific steps are given as follows:

Gibbs Sampling Algorithm for Censored and Failure Time Data   (12.20)

(i) Sample from $f(\tilde{\mathbf{y}} | \boldsymbol{\beta}, \sigma^2)$ for the current values of $(\boldsymbol{\beta}, \sigma^2)$. For each right-censored observation y_i, $\tilde{y}_i$ is sampled from the conditional normal density

$$\sigma^{-1}\phi(\{\tilde{y}_i - \mathbf{x}_i^T \boldsymbol{\beta}\}\sigma^{-1})[1 - \Phi(\{y_i - \mathbf{x}_i^T \boldsymbol{\beta}\}\sigma^{-1})]^{-1} \quad \text{for} \quad \tilde{y}_i > y_i,$$

where ϕ and Φ are the standard normal probability density and cumulative distribution functions. More generally, for an interval-censored observation (a, b), $\tilde{y}_i$ is sampled from the conditional normal density

$$\sigma^{-1}\phi\left(\{\tilde{y}_i - \mathbf{x}_i^T\boldsymbol{\beta}\}\sigma^{-1}\right)/\left[\Phi\left(\{\ln b - \mathbf{x}_i^T\boldsymbol{\beta}\}\sigma^{-1}\right) - \Phi\left(\{\ln a - \mathbf{x}_i^T\boldsymbol{\beta}\}\sigma^{-1}\right)\right]$$

for $\ln a < \tilde{y}_i < \ln b$.

(ii) Sample from $f(\boldsymbol{\beta}, \sigma \mid \mathbf{y}^*)$, where $\mathbf{y}^* = (\mathbf{y}, \tilde{\mathbf{y}})$, $\mathbf{y}$ denotes the failure time data, and $\tilde{\mathbf{y}} = (\tilde{y}_i)$ are the values obtained in (i). The sampled values of $(\boldsymbol{\beta}, \sigma^2)$ become the current values to be used in (i) for the next iteration. Note that $f(\boldsymbol{\beta}, \sigma \mid \mathbf{y}^*)$ is the product of the two densities given in (12.17) and (12.18) with $\mathbf{y}$ replaced by $\mathbf{y}^*$.

After sufficient iteration between (i) and (ii), the Gibbs sampler will converge so that samples drawn in (ii) are from the posterior distribution of $(\boldsymbol{\beta}, \sigma^2)$. See the discussion on the convergence of the Gibbs sampler in Section 8.5.

How should the parameters in the prior distribution be chosen? Noting that the estimability problem with the MLEs only occurs for some covariates, the prior parameters should be chosen so that the Bayesian estimates are close to the MLEs when the latter exist and are finite. The fact that the MLEs are the same as the Bayesian estimates for the non-informative prior having density $\pi(\boldsymbol{\beta}, \sigma) = \sigma^{-1}$ suggests the use of a relatively diffuse prior as follows:

1. For $\boldsymbol{\beta}$, use $\boldsymbol{\beta}_0$ equal to the vector of zeros and the entries of $\mathbf{A}_0$ small, say 0.01 or 0.1, which correspond to prior variances for the $\boldsymbol{\beta}$ components of $100\sigma^2$ or $10\sigma^2$, respectively.
2. For ν_0 and s_0^2, $(1, 0.001)$ gives a prior distribution for σ^2 with central 99% covering $(0.0001, 22.6805)$ and $(1, 0.01)$ gives a prior distribution for σ^2 with central 99% covering $(0.0013, 226.8023)$.

Given the samples from the joint posterior of $(\boldsymbol{\beta}, \sigma)$, assessment of the importance of an effect can be done by using its corresponding marginal posterior to get a 95 or 99% credible interval. A $(1 - \alpha) \times 100\%$ *credible interval* is the central $(1 - \alpha) \times 100\%$ interval of the marginal posterior with both tail probabilities equal to $\alpha/2$. An effect is significant if the corresponding credible interval does not contain zero.

The Bayesian approach developed thus far has focused on estimation under a *fixed* model. How should a model be chosen that is compatible with the data? As pointed out in Section 12.4, there are two model selection strategies. For regular designs, we can work with a comprehensive model and apply the Bayesian estimation method just described to the parameters in the model. All the insignificant effects (as judged by the marginal posterior distributions) in this model will be dropped and the reduced model is the model of choice. For nonregular designs, we need to develop an efficient model selection strategy by modifying the stochastic search algorithm given in Section 8.5.2. For censored data, the Gibbs sampling algorithm in (8.17)–(8.18) needs an additional step for imputing the censored data. Using the notation and definitions from Section 8.5 and (12.20), the modified algorithm is stated as follows. Recall from Section 8.5 that δ is a vector of zeros and ones that indicate the inclusion or exclusion of effects in the model.

Bayesian Variable Selection for Censored and Failure Time Data   (12.21)

(i) Sample from the density $f(\tilde{y} \mid \beta, \sigma^2, \delta)$ for the current values of $(\beta, \sigma^2, \delta)$. Since δ is fixed, this density has the same formula as given in Step (i) of (12.20).

(ii) Define y^* as in step (ii) of (12.20). Perform Gibbs sampling according to (8.17)–(8.18) (and the formulas in Appendix 8.A) with y replaced by y^* and $f(y^* \mid \beta, \sigma^2, \delta)$ following the lognormal regression model (12.4). The sampled values of $(\beta, \sigma^2, \delta)$ become the current values to be used in (i) for the next iteration.

Steps (i) and (ii) of (12.21) are repeated until a certain convergence criterion (see Section 8.5) is satisfied. Model posterior probabilities are then computed based on the Markov Chain Monte Carlo samples obtained in (12.21). Regarding the choice of the prior tuning constants for (8.17)–(8.18), follow the recommendations given in Section 8.5.3 with the following adaptation: for the set of complete responses, use the mid-points for interval-censored data and the censoring times for left- and right-censored data.

12.6 ANALYSIS OF RELIABILITY EXPERIMENTS WITH FAILURE TIME DATA

In this section, the three experiments presented in Section 12.1 are analyzed using the likelihood and Bayesian approaches.

12.6.1 Analysis of Light Experiment

For the fluorescent light experiment, first consider a likelihood based analysis using the lognormal regression model. Table 12.7 presents the MLEs, likelihood ratio statistics and p values for the five main effects (A–E) and the AB interaction which the experimenters thought might be important. Note that the likelihood ratio statistic for each effect is based on Γ in (12.14) and its p value equals $Prob(\chi_1^2 \geq \Gamma)$. Based on these results, the main effects D, B, and E are significant in descending order.

Table 12.7   MLEs, Likelihood Ratio (LR) Statistics, and p Values for Lognormal Regression Model, Light Experiment

Effect	MLE	LR	p Value
intercept	2.939	2089.0	0.00
A	0.117	3.6	0.06
B	−0.201	11.1	0.00
AB	−0.049	0.6	0.43
C	−0.051	0.7	0.41
D	0.273	19.3	0.00
E	−0.153	6.0	0.02
σ	1.590		

Table 12.8   Posterior Quantiles, Light Experiment

Effect	Quantile				
	0.005	0.025	0.500	0.975	0.995
intercept	2.784	2.822	3.007	3.497	3.734
A	−0.064	−0.016	0.167	0.621	0.848
B	−0.942	−0.728	−0.257	−0.074	−0.029
C	−0.779	−0.562	−0.101	0.081	0.127
D	0.106	0.147	0.331	0.801	1.026
E	−0.878	−0.666	−0.204	−0.021	0.026
AB	−0.780	−0.559	−0.099	0.083	0.132
BD	−0.714	−0.498	−0.047	0.135	0.188
σ	0.109	0.122	0.191	0.341	0.430

An additional effect BD ($=AE$) can be entertained if a saturated model is fitted, but the MLEs do not exist for this model since both replicates at the fifth run are right censored. [This nonexistence result follows from Hamada and Tse (1992).] The Bayesian approach of Section 12.5 should be taken. A relatively diffuse prior for (β, σ^2) is used whose parameters are: the vector β_0 is zero everywhere; $A_0 = diag(0.01, 0.1, 0.1, 0.1, 0.1, 0.1, 0.1, 0.1)$ (i.e., a prior variance for the intercept of $100\sigma^2$ and for the factorial effects of $10\sigma^2$), and $\nu_0 = 1$ and $s_0^2 = 0.001$ so that $(0.0001, 22.6805)$ is the central 99% of the prior distribution for σ^2. The Gibbs sampler was run for a burn-in period of 10,000 iterations and then run for another 50,000 iterations. The quantiles corresponding to central 0.95 and 0.99 intervals as well as the median of the posterior distribution are displayed in Table 12.8. These results are based on the last 50,000 iterations of the Gibbs sampler. Nearly the same results were obtained by taking every 10th or 50th sample. These results show that BD is not significant and confirm the importance of the main effects D, B, and E. (Note that the posterior distribution for σ is smaller than the MLE for σ but recall that the BD effect was not included in the model analyzed by MLE.)

The sign of these effects suggests that reliability will be improved at the recommended setting $B_-D_+E_-$, where the subscript indicates the recommended level.

12.6.2 Analysis of Thermostat Experiment

Taking the likelihood approach for the thermostat experiment, a lognormal regression model with 11 factor main effects $(A-K)$ is fitted. The results, including the MLEs, the likelihood ratio statistics, and p values, are given in Table 12.9 under Model 1. Effects E and H are the most significant. All other factor effects have magnitudes in a narrow range $(0.22, 0.39)$. From the partial aliasing relations of a 12-run Plackett-Burman design, it is known that the other factor main effects (not E nor H) have a $\frac{1}{3}$ or $-\frac{1}{3}$ partial aliasing coefficient with EH. Therefore, the narrow range of the other factor main effects strongly suggests that they are due to the presence of an EH interaction. (The same inference was used and explained in Sections 8.3 and 8.4.)

In order to add EH to the model, we must drop a main effect from the model. Consequently, a model (Model 2) is fitted in which the B main effect (the least significant from Model 1) is dropped and replaced by the EH interaction. The results in Table 12.9 (Model 2) indicate that E, H, and EH emerge as the only significant effects. The other factor main effects are no longer significant even at the 0.1 level. This confirms the previous suggestion that EH was responsible for the observed significance of the other factor main effects (i.e., other than E and H). Because the signs of E and H are negative and the sign of EH is positive, the mean logged failure time is maximized by setting E and H at their $-$ levels, i.e., a grain size (E) of 0.008 inches and heat treatment (H) of 45 minutes at 600°F.

Table 12.9 **MLEs, Likelihood Ratio (LR) Statistics, and p Values for Lognormal Regression Models, Thermostat Experiment**

	Model 1				Model 2		
Effect	MLE	LR	p Value	Effect	MLE	LR	p Value
A	-0.312	18.3	0.0000	A	-0.091	0.7	0.3890
B	0.221	9.2	0.0024	EH	0.663	9.2	0.0024
C	-0.319	19.1	0.0000	C	-0.098	0.9	0.3474
D	0.285	15.2	0.0000	D	0.064	0.4	0.5174
E	-1.023	192.3	0.0000	E	-1.023	192.3	0.0000
F	0.231	10.0	0.0016	F	0.010	0.0	0.9219
G	-0.390	28.4	0.0000	G	-0.169	2.6	0.1075
H	-0.557	57.0	0.0000	H	-0.557	57.0	0.0000
I	-0.332	20.7	0.0000	I	-0.112	1.1	0.2872
J	-0.277	14.4	0.0000	J	-0.056	0.3	0.5958
K	-0.352	23.3	0.0000	K	-0.131	1.5	0.2149

Table 12.10   Posterior Model Probabilities, Thermostat Experiment

Model	Prob
E, H, EH	0.251
E, G, H, EH	0.013
E, H, I, EH	0.012
E, H, CH, EH	0.011
E, H, EH, HI	0.011

The results of applying the Bayesian variable selection procedure in (12.21) are displayed in Table 12.10 and confirm the findings above; that is, (E, H, EH) is the most probable model.

12.6.3   Analysis of Drill Bit Experiment

For the drill bit experiment, the lognormal model and the Weibull model lead to the same conclusions. For the purpose of illustration, we use a Weibull regression model to analyze the data. Recall that the experiment used a cross array consisting of a 16-run control array for 11 control factors ($A-J$ and L) and an 8-run noise array for 5 noise factors ($M-Q$). For A, which has four levels denoted by 0, 1, 2, and 3, the contrasts A_1, A_2, and A_3 from (6.5) are used, which can also be interpreted as linear, quadratic, and cubic effects. In addition to the other control main effects ($B-L$), BE and A_1J can be estimated. These effects exhaust all the 15 degrees of freedom for the control array. For the noise array, only the 5 noise main effects ($M-Q$) are considered. According to Theorem 10.1, all the 65 ($= 13 \times 5$) control-by-noise interactions can also be estimated. Table 12.11 presents only the MLEs and the p values of their corresponding likelihood ratio statistics [based on (12.14)] for the significant effects.

Once the data have been fitted, recommendations for the important control factor settings need to be made. For a simple model with few noise factors, they may be apparent from examining the signs and magnitudes of the significant effects and the control-by-noise interaction plots. For complicated models, this approach is tedious and does not address the dual goals of reliability improvement and robust reliability.

An alternative is to specify a meaningful performance measure and evaluate its distribution for a suitable noise factor distribution. In the reliability context, one such measure is the *probability of exceeding a warranty period w*,

$$Prob(T > w | \mathbf{x}_C, \mathbf{x}_N), \tag{12.22}$$

where T is the failure time random variable and $[0, w]$ is the warranty period. The probability depends on the control factors $\mathbf{x}_C$ and noise factors $\mathbf{x}_N$ through the fitted model. For fixed $\mathbf{x}_C$, $Prob(T > w | \mathbf{x}_C, \mathbf{x}_N)$ is a random

Table 12.11 MLEs and p Values of Weibull Regression Model, Drill Bit Experiment

Effect	MLE	p Value	Effect	MLE	p Value
intercept	6.147	0.000	A_2M	−0.182	0.012
A_1	0.458	0.000	GM	0.173	0.022
A_2	−0.165	0.009	LM	0.187	0.016
A_3	0.335	0.000	A_3O	0.164	0.004
B	0.130	0.034	EO	−0.427	0.063
C	0.186	0.004	FO	−0.138	0.024
D	0.221	0.001	GO	0.351	0.000
F	−0.245	0.000	HO	−0.128	0.041
G	−0.131	0.038	IO	−0.353	0.000
H	−0.195	0.002	LO	0.249	0.000
I	0.316	0.000	BP	0.176	0.024
J	0.292	0.000	DP	0.319	0.000
L	0.247	0.000	IP	0.283	0.000
O	−0.872	0.000	GQ	−0.336	0.000
P	−0.751	0.000	σ	0.513	
A_1J	−0.136	0.021			

Table 12.12 Best Factor Settings, Drill Bit Experiment

			Setting								Reliability			
A	B	C	D	E	F	G	H	I	J	L	Mean	1st Quartile	Median	3rd Quartile
3	+	+	+	−	−	−	−	+	+	+	0.65035	0.35810	0.72769	0.84833
3	+	+	+	−	−	+	−	+	+	+	0.55407	0.34545	0.56422	0.75080
3	−	+	+	−	−	−	−	+	+	+	0.53549	0.17120	0.58133	0.77319
3	+	+	+	+	−	+	−	+	+	+	0.52547	0.10351	0.54249	0.81814
3	+	+	+	+	−	−	−	+	+	+	0.51859	0.00435	0.23076	0.96942
3	+	+	+	−	−	−	−	+	−	+	0.51629	0.15138	0.55744	0.73905
0	+	+	+	−	−	−	−	+	+	+	0.51498	0.16519	0.53859	0.74946
3	−	+	+	+	−	−	−	+	+	+	0.49256	0.00009	0.25611	0.94802
3	+	−	+	−	−	−	−	+	+	+	0.48980	0.11996	0.51872	0.71202
2	+	+	+	−	−	−	−	+	+	+	0.48793	0.02777	0.38044	0.85262

variable induced by the variation in the noise factors $\mathbf{x}_N$. Its distribution can be summarized by the mean, median, and first and third quartiles.

For the drill bit experiment, the five noise factors are assumed to take on the two levels $(-, +)$ independently and equally often. That is, 32 $(= 2^5)$ noise factor settings are assumed to occur equally often. The distribution for (12.22) was evaluated for each of the 4096 $(= 4 \times 2^{10})$ control factor settings and ranked according to the mean using the Weibull regression model identified in Table 12.11 and $w = 3000$. The choice of 3000 is arbitrary, but coincides with the right-censoring time in the experiment. The best 10 results are presented in Table 12.12. Among the 10 settings, all have $D_+F_-H_-I_+L_+$ and the majority (i.e., at least 8 out of 10) have $A_3B_+C_+G_-J_+$, where A_3 denotes level 3 (i.e., the last level) of factor A. The level for factor E can be chosen at $-$ or be based on other considerations.

12.7   RELIABILITY EXPERIMENTS WITH DEGRADATION DATA

In today's economy most products are expected to function properly for a long period (or at least beyond the warranty period from the point of view of the manufacturer). Under normal conditions they rarely fail during a reliability study whose period tends to be short for obvious practical reasons. If very few or no units fail at the termination of the study, most data are censored and the few failure times will provide little information about the long-term reliability of the product. A common practice in engineering is to accelerate failures by subjecting the units to *stress* conditions which are more severe than *normal* or *use* conditions. The method is referred to as *accelerated life testing*. Examples of stress variables include elevated temperature (i.e., much higher than room temperature) and voltage for electronic components and increased loads on mechanical parts. After observing (presumably more) failures under stress conditions, inference is to be made about the reliability under use conditions. Because the use conditions and the stress conditions are in most cases far apart, it requires extrapolation well beyond the experimental region. A physical or statistical model has to be employed that relates the product reliability under the two sets of conditions. This model cannot, however, be validated by data because no (or few) failure time data are observed at use conditions. Unless the model can be justified on physical grounds or from prior experience, reliability prediction based on accelerated life testing is on shaky ground. Another problem is that a good accelerating factor may not easily be found, e.g., a factor for accelerating the failure of hybrid electronic components has not been found.

As products become increasingly reliable, failures can only be observed at higher stress levels. For example, the reliability requirements for undersea cables are so demanding that it would take a very high load or extremely harsh environmental conditions to induce any failure. With ever increasing stress conditions, the uncertainty of reliability prediction becomes even more acute. An alternative to accelerated life testing is to select some product characteristics that can be related to the product reliability and observe its degradation over time. In order to observe a detectable degree of degradation over a reasonably short period of time, the units may still be subject to stress acceleration. A *degradation* characteristic is appropriate for reliability study if it is a *good* predictor of failure and only requires *mild* stress conditions to induce a detectable degree of degradation within the study period. Examples include crack growth of metals, deterioration of electronic performance of semiconductors, and breakdown of insulation. For crack growth, failure occurs when the crack reaches a given size. Details on modeling, estimation, and prediction from accelerated life test data and degradation data can be found in Nelson (1990) and Meeker and Escobar (1998).

In this and the next sections we shall focus on the planning and analysis of degradation experiments. An experiment on fluorescent lamps (Tseng, Hamada, and Chiao, 1995) will be used to illustrate the methodology.

Because degradation was observed under normal conditions, no extrapolation was required. A key quality characteristic of a fluorescent lamp is its luminosity, measured in lumens. Because it degrades over time as the fluorescent material darkens, failure is defined in terms of the amount of degradation in luminosity. More specifically, the industry's standard definition for failure time is the time t when a lamp's luminosity $\Lambda(t)$ falls below 60% of its luminosity after 100 hours of use or aging, i.e., $0.6\Lambda(100)$. Because fluorescent lamps are highly reliable products, collecting degradation data is a natural approach to take.

Based on Lin (1976), a model for relative luminosity $\ln[\Lambda(t)/\Lambda(100)]$ at time t is

$$\ln[\Lambda(t)/\Lambda(100)] = -\lambda(t - 100). \tag{12.23}$$

The parameter λ is the rate of degradation. Using (12.23), the failure time t is:

$$t = -\ln 0.6 /\lambda + 100. \tag{12.24}$$

Figure 12.1(a) depicts the luminosity degradation path for a single lamp as given by the dotted line. The lamp fails when the relative luminosity crosses 0.6, which is at time 18,651, as indicated by the dashed vertical line.

Since all of the lamps do not fail at the same time, the failure times in (12.24) follow some distribution. Thus, (12.24) implies that the rate of degradation λ must be a random variable. Again, from (12.24), $\ln(t - 100)$ and $\ln \lambda$ are linearly related. Therefore, if λ has a lognormal distribution, then the failure time t has a lognormal distribution. Figure 12.1(b) shows the simulated degradation paths as defined by (12.23) of 100 lamps with lognormally distributed λ. The degradation paths are plotted until failure. A histogram of the corresponding failure times is presented in Figure 12.1(c), which exhibits the characteristic skewness of the lognormal distribution.

The darkening of the fluorescent material, which is associated with the luminosity degradation, has three possible explanations: (i) the electric discharge material wears out, (ii) oxidation of the electric discharge material, and (iii) diffusion of impurities arising in the manufacturing process. From the seven-step manufacturing process of fluorescent lamps, three factors related to these explanations were studied:

A—the amount of electric current in the exhaustive process,

B—the concentration of the mercury dispenser in the mercury dispenser coating process, and

C—the concentration of argon in the argon filling process.

The three factors were studied using a 2^{3-1} design with the defining relation $\mathbf{I} = ABC$. The four runs are indicated by

$1\ (-,-,-),\quad 2\ (-,+,+),\quad 3\ (+,-,+),\quad \text{and}\quad 4\ (+,+,-).$

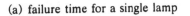

(a) failure time for a single lamp

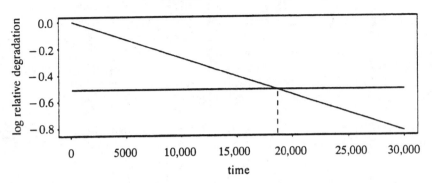

(b) failure times for 100 lamp

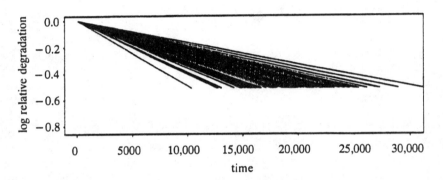

(c) histogram of 100 lamp failure times

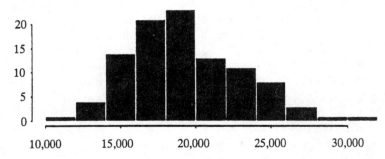

Figure 12.1. The Relationship Between Degradation and Failure Time.

Table 12.13 Relative Luminosity Data, Fluorescent Lamp Experiment

Run 1

Lamp	500	1000	2000	3000	4000	5000	6000	7000	8000	9000	10,000	11,000	12,000
1	-0.0822	-0.0903	-0.1112	-0.1225	-0.1958	-0.2187	-0.2285	-0.1905	-0.2109	-0.2312	-0.2414	-0.2516	-0.2567
2	-0.0817	-0.0999	-0.1322	-0.1444	-0.2186	-0.2136	-0.2237	-0.2278	-0.2480	-0.2649	-0.2751	-0.2850	-0.2902
3	-0.0702	-0.0898	-0.1209	-0.1564	-0.2054	-0.2279	-0.2522	-0.2278	-0.2480	-0.2649	-0.2751	-0.2850	-0.2902
4	-0.0719	-0.1094	-0.1417	-0.1785	-0.2282	-0.2374	-0.2761	-0.2436	-0.2640	-0.2843	-0.2944	-0.3046	-0.3095
5	-0.0912	-0.0983	-0.1172	-0.1634	-0.2244	-0.2468	-0.2712	-0.2360	-0.2559	-0.2764	-0.2865	-0.2967	-0.3016

Run 2

Lamp	500	1000	2000	3000	4000	5000	6000
1	-0.0302	-0.0575	-0.1152	-0.1362	-0.1475	-0.1585	-0.1705
2	-0.0556	-0.0829	-0.1297	-0.1403	-0.1730	-0.1955	-0.2076
3	-0.0556	-0.1031	-0.1407	-0.1621	-0.1840	-0.2067	-0.2076
4	-0.0486	-0.0762	-0.1346	-0.1562	-0.1809	-0.1900	-0.2231
5	-0.0473	-0.0654	-0.1131	-0.1345	-0.1572	-0.1783	-0.2159

Run 3

Lamp	500	1000	2000	3000	4000	5000	6000
1	-0.0205	-0.0304	-0.0968	-0.1257	-0.1663	-0.1667	-0.2099
2	-0.0215	-0.0442	-0.1263	-0.1293	-0.1555	-0.1560	-0.2302
3	-0.0315	-0.0660	-0.1078	-0.1511	-0.1773	-0.1927	-0.2519
4	-0.0205	-0.0550	-0.1103	-0.1257	-0.1379	-0.1667	-0.2099
5	-0.0203	-0.0414	-0.1213	-0.1367	-0.1489	-0.1777	-0.1913

Run 4

Lamp	500	1000	2000	3000	4000	5000	6000	7000	8000	9000	10,000	11,000	12,000
1	-0.0496	-0.0938	-0.1721	-0.1504	-0.1692	-0.1796	-0.2105	-0.2712	-0.2460	-0.2648	-0.2827	-0.3481	-0.3320
2	-0.0319	-0.0621	-0.1486	-0.1333	-0.1817	-0.2104	-0.2325	-0.3185	-0.2881	-0.3231	-0.3385	-0.4149	-0.4038
3	-0.0261	-0.0676	-0.1429	-0.1098	-0.1480	-0.1775	-0.2041	-0.2653	-0.2358	-0.2751	-0.3032	-0.3540	-0.3710
4	-0.0229	-0.0470	-0.1320	-0.0963	-0.1804	-0.1666	-0.2062	-0.2727	-0.2313	-0.2540	-0.2781	-0.3254	-0.3235
5	-0.0264	-0.0528	-0.1396	-0.1120	-0.1527	-0.1680	-0.1856	-0.2562	-0.2364	-0.2526	-0.2909	-0.3417	-0.3403

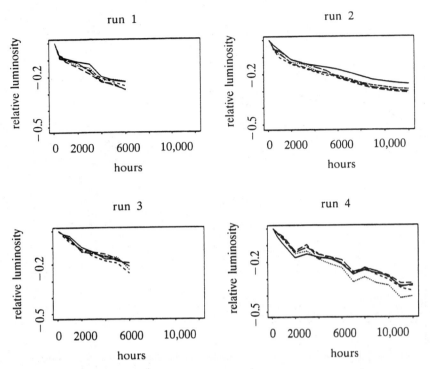

Figure 12.2. Degradation Paths, Fluorescent Lamp Experiment.

The two settings $(-, +)$ represent the range in which the electric current (A) in amperes is typically controlled, the region in which the concentration of the mercury dispenser (B) in milligrams is usually set, and the range in which the ideal concentration of argon (C) in torrs is thought to be. From a production run of each of the four lamp types (as specified by the four runs), five lamps were randomly selected for testing. A spectrophotometer was used to measure luminosity of the lamps at the following inspection times: 100, 500, 1000, 2000, 3000, 4000, 5000, and 6000 hours for all the runs and 7000–12,000 in increments of 1000 hours for runs 2 and 4. The relative luminosities $\ln[\Lambda(t)/\Lambda(100)]$ for all the inspections after 100 hours are displayed in Table 12.13 and plotted in Figure 12.2 for the four runs.

A simple analysis of the fluorescent lamp data will be given in the next section.

12.8   A SIMPLE ANALYSIS FOR DEGRADATION DATA

Here we assume that (12.23) holds with λ lognormally distributed so that failure times are lognormally distributed. Consider the following simple two-stage analysis.

Two-Stage Analysis of Degradation Data

(i) For each unit (e.g., a lamp), fit model (12.23) to its degradation path and predict its failure time.

(ii) Analyze the predicted failure times from step (i) in terms of the control factors using standard methods. (12.25)

Using the degradation paths in Figure 12.2 of the relative luminosity $L(t) = \ln[\Lambda(t)/\Lambda(100)]$, the model

$$L(t) = -\lambda(t - 100) + \varepsilon \tag{12.26}$$

based on (12.23) is fitted to the degradation path for each lamp. The error ε is assumed to follow a normal distribution $N(0, \sigma_\varepsilon^2)$. The least squares estimates $\hat{\lambda}$ are presented in Table 12.14. Residual plots (not shown here) show that the model fits well and the normality assumption is reasonable. The root mean square errors are small and do not exceed 0.007.

Next we use (12.24) to predict the failure time of each lamp based on $\hat{\lambda}$:

$$\hat{t}_{ij} = -\ln 0.6 / \hat{\lambda}_{ij} + 100, \tag{12.27}$$

where i indicates the run number and j the lamp number. The predicted failure times for the 20 lamps are given in Table 12.14.

Table 12.14 Estimates for Degradation Model, Predicted Failure Times and Location-Scale Parameters of the Lognormal Distribution, Fluorescent Lamp Experiment

Run	Lamp	$\hat{\lambda}_{ij}$ (10^{-5})	$\hat{t}_{ij}$	$\hat{\mu}_i$	$\hat{\sigma}_i$
	1	4.44	11603.91		
	2	4.61	11181.04		
1	3	4.82	10689.06	9.27	0.066
	4	5.28	9771.20		
	5	5.17	9964.97		
	1	2.56	20038.17		
	2	2.96	17363.95		
2	3	2.99	17181.96	9.77	0.079
	4	3.15	16339.77		
	5	3.03	16974.84		
	1	3.78	13620.22		
	2	3.90	13203.61		
3	3	4.41	11685.35	9.49	0.064
	4	3.71	13884.22		
	5	3.74	13767.50		
	1	3.21	16020.39		
	2	3.76	13678.42		
4	3	3.27	15740.21	9.66	0.073
	4	3.06	16779.90		
	5	3.10	16557.92		

In step (ii), we use the predicted failure times $\hat{t}_{ij}$ to identify the important factors that affect lamp reliability. The normal probability plot of $\ln \hat{t}_{ij}$ in Figure 12.3 shows that the lognormal distribution reasonably fits the predicted failure times, i.e., the logged failure times follow a normal distribution. Let (μ_i, σ_i) denote the normal location and scale parameters of the logged failure times at run i. Their least squares estimates $(\hat{\mu}_i, \hat{\sigma}_i)$ are given in Table 12.14. Using the test in (3.39) or Bartlett's test, the hypothesis of equal σ_i's (H_0: $\sigma_1 = \sigma_2 = \sigma_3 = \sigma_4$) is not rejected at the 0.05 level. Consequently, standard methods for analyzing replicated fractional factorial designs can be applied to identify the important factors.

Fitting a main effects model to $\ln \hat{t}_{ij}$, the logged predicted failure times, gives the results in Table 12.15, which indicate that only factors B and C are significant. Thus, the fitted model for the log predicted failure time is

$$\ln \hat{y} = 9.55 + 0.17x_B + 0.08x_C. \tag{12.28}$$

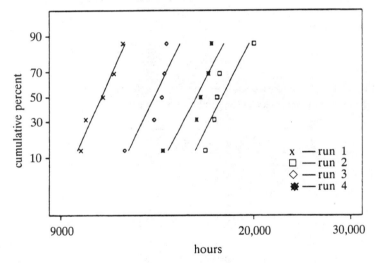

Figure 12.3. Lognormal Plot of Predicted Failure Times, Fluorescent Lamp Experiment.

Table 12.15   Least Squares Estimates, Standard Errors, t Statistics, and p Values from Predicted Failure Times, Fluorescent Lamp Experiment

Effect	Estimate	Standard Error	t	p Value
intercept	9.55	0.017	560.95	0.00
A	0.03	0.017	1.59	0.13
B	0.17	0.017	9.92	0.00
C	0.08	0.017	4.81	0.00

Based on (12.28), reliability can be improved by setting factors B and C at their high levels $(+, +)$. The estimated failure time at the recommended levels $(x_B = +1, x_C = +1)$ from (12.28) is

$$\hat{y} = \exp[9.55 + 0.17(+1) + 0.08(+1)] = 18{,}034 \text{ hours},$$

which is 65% higher than 10,938 hours, the estimated failure time at the original process settings (i.e., all three factors at their $-$ levels, which leads to $\hat{y} = \exp[9.55 + 0.17(-1) + 0.08(-1)] = 10{,}938$).

The simple approach described in this section will not be appropriate if the error variance σ_ϵ^2 is substantial. An alternative to the least squares method employed is to maximize the likelihood with respect to both λ and σ_ϵ^2. More complicated models can be considered. For example, we may let both μ and σ depend on the control factors. If noise factors are included in a degradation experiment, the data can be analyzed using parameter design methodology. An example can be found in Chiao and Hamada (1996), who presented a light-emitting diode (LED) experiment for achieving robust reliability. Degradation data can also be modeled as nonlinear random effect models (Davidian and Giltinan, 1995) and stochastic processes (Doksum and Hoyland, 1992).

12.9   PRACTICAL SUMMARY

1. The response of a reliability experiment is the failure time of the unit under test. Because of the need to terminate the study in a reasonable period of time, often some of the units do not fail at the end of the study, resulting in the failure time being censored. There are three types of censoring: right censoring, interval censoring, and left censoring. Two common failure time distributions are the lognormal and Weibull given in (12.1) and (12.2).

2. Because of the presence of censored observations, standard regression techniques cannot be applied in a straightforward manner. It is common to use the maximum likelihood method for point estimation, confidence intervals [see (12.13)], and likelihood ratio tests for assessing effect significance [see (12.14)].

3. The maximum likelihood estimate (MLE) is infinite when the failure time data and censored data are separated according to the level of a covariate. This happens when the effect of the covariate is large and the opportunity for improvement is great. A natural way to circumvent this estimability problem is to use a Bayesian approach based on the Gibbs sampling algorithm in (12.20).

4. The model selection strategy depends on the nature of the design used in the experiment.

(i) For a regular design, an estimation method (MLE or Bayesian) can be applied to a comprehensive model which corresponds to all the degrees of freedom in the design. Apart from the usual problem of de-aliasing the aliased effects, each effect in the model can be tested separately.

(ii) For a nonregular design, there is no comprehensive model, and an iterative model search strategy like the Bayesian variable selection in (12.21) is required.

5. For highly reliable products it may be hard to observe any failures in a reasonable period of time even under severe stress conditions. If the degradation of a product characteristic is related to reliability, an alternative is to collect degradation data under a variety of experimental conditions. A simple two-stage modeling and analysis strategy is given in Section 12.8.

EXERCISES

1. The lifetime data in the router bit experiment in Chapter 6 are either interval censored or right censored. The analysis in Chapter 6 used the midpoint of the interval and right-censored value as the response values. Reanalyze the data by using the likelihood or Bayesian approach. Do the conclusions change? Explain your findings.

2. In the cast fatigue experiment in Section 7.1, the value 7.0 was actually right censored. Reanalyze the data using the likelihood approach. For the final model and the parameter estimates identified in your analysis, impute the censored value for run 5 by taking the conditional mean of the log failure time distribution over $(7, \infty)$. How far is this value from 7? Explain why the reanalysis does not reveal any new information beyond what was found in Chapter 8.

3. The lifetime data in the heat exchanger experiment in the exercises of Chapter 8 are either interval censored or right censored. Reanalyze the data by using the likelihood or Bayesian approach. Do the conclusions differ from those in Chapter 8? Explain your findings.

(To save computational effort in searching for models for Exercises 1–3, you may start with the model identified in previous chapters.)

4. Reanalyze the drill bit experiment using the lognormal regression model. Do the conclusions change?

5. Repeat the computations in Table 12.12 by ranking according to the median of the probability distribution in (12.22). Do the recommended factor settings change substantially?

6. The drill bit experiment used a cross array with 128 runs. Because each control/noise factor combination corresponds to a new drilling process, the experimental cost is proportional to the total number of runs. To save cost and time, an alternative to the cross array format is to use a single array as was explained in Chapter 10. Find 32-run and 64-run single arrays to accommodate the 11 control and 5 noise factors. Study their estimation capacities and compare them with that of the cross array. (*Hint*: Use the tables in the appendices of Chapter 6.)

7. For Weibull regression develop a Bayesian estimation and model selection procedure similar to the one in Section 12.6 for lognormal regression.

8. Suppose that the failure times for the fluorescent lamp degradation experiment follow a Weibull distribution. Describe how the analysis should be done. Explain why the results will not be qualitatively different and the lognormal assumption is preferred.

9. Give detailed derivations for the posterior distribution $\pi(\theta \,|\, y)$ given in (12.17)–(12.19).

10. In an experiment to improve ball bearing reliability (Hellstrand, 1989), three factors involving the ball bearing design were studied: (A) inner ring heat treatment, (B) outer ring osculation, and (C) cage design. The experiment employed a 2^3 design, given in Table 12.16, where the two levels are standard and modified values, respectively. The response y is the lifetime in hours. Analyze the data by using the lognormal failure time regression model in (12.4). Show that the analysis for these data is the same as standard regression analysis as applied to the log failure times. Identify significant effects and optimal factor settings.

Table 12.16 Design Matrix and Failure Time Data, Ball Bearing Experiment

Run	Factor A	B	C	Failure Time
1	−	−	−	17
2	+	−	−	26
3	−	+	−	25
4	+	+	−	85
5	−	−	+	19
6	+	−	+	16
7	−	+	+	21
8	+	+	+	128

Table 12.17 Design Matrix and Lifetime Data ($\times 10^3$ cycles, censored at 2000), Transmission Shaft Experiment

Run	\multicolumn{7}{c}{Factor}							\multicolumn{2}{c}{Replication}	
	A	B	C	D	E	F	G	1	2
1	1	1	2	2	2	2	1	322	2000
2	1	2	2	1	1	1	2	95	95.4
3	1	3	2	3	3	3	1	2000	125
4	1	1	1	2	1	3	1	747	414
5	1	2	1	1	3	2	1	821	192
6	1	3	1	3	2	1	2	2000	2000
7	1	1	3	1	2	1	1	972	2000
8	1	2	3	3	1	3	1	2000	1920
9	1	3	3	2	3	2	2	2000	2000
10	2	1	2	3	3	1	1	739	285
11	2	2	2	2	2	3	2	1080	634
12	2	3	2	1	1	2	1	2000	1940
13	2	1	1	1	3	3	2	2000	1790
14	2	2	1	3	2	2	1	2000	617
15	2	3	1	2	1	1	1	2000	2000
16	2	1	3	3	1	2	2	1380	1110
17	2	2	3	2	3	1	1	2000	2000
18	2	3	3	1	2	3	1	2000	2000

11. Davis (1995) described a fatigue testing experiment on center shafts of an automatic transmission. The experiment studied seven factors, two at two levels and five at three levels: (A) spline end profile (spherical, grooved); (B) annealing amount (nil, 1 h, 4 h); (C) shaft diameter (16.1, 17.7, and 18.8 mm); (D) shot intensity (3 Almen, 6 Almen, 9 Almen); (E) shot coverage (200, 400, 600%); (F) tempering temperature (140, 160, 180°C); (G) shot blasting (without, with). The experimental plan was obtained from an $OA(18, 2^1 3^6)$ by collapsing factor G (i.e., level 3 in the original array was set to level 1). The response is time until a crack in the center shaft is detected. A shaft was monitored for 2000 ($\times 10^3$) cycles. The experiment was replicated twice. The design matrix and the lifetime data are given in Table 12.17. A lifetime of 2000 indicates the shaft had not cracked when the monitoring of the shaft was terminated.

(a) Analyze the experimental data using a lognormal regression model. Note that the levels for factors B and C are not equally spaced.

(b) Reanalyze the experimental data using a Weibull regression model. Do the results change?

Table 12.18 Original Window Size Data (WNO indicates window not opened), Window Forming Experiment

Run	Window Size									
1	WNO	WNO	WNO	WNO	WNO	WNO	WNO	WNO	WNO	WNO
2	2.32	2.23	2.30	2.56	2.51	2.22	2.33	2.34	2.15	2.35
3	2.98	3.14	3.02	2.89	3.16	3.15	3.08	2.78	WNO	2.86
4	WNO	WNO	WNO	WNO	WNO	WNO	WNO	WNO	WNO	WNO
5	WNO	WNO	WNO	WNO	WNO					
6	2.45	2.19	2.14	2.32	2.12	WNO	WNO	WNO	WNO	WNO
7	WNO	WNO	WNO	WNO	WNO	WNO	WNO	WNO	WNO	WNO
8	WNO	WNO	WNO	WNO	WNO	2.89	2.97	3.13	3.25	3.19
9	3.16	2.91	3.12	3.18	3.11	2.43	2.35	2.14	2.40	2.28
10	2.0	1.75	1.97	1.91	1.72	WNO	2.7	WNO	2.61	2.73
11	2.76	3.09	3.22	3.05	3.04	3.12	3.21	WNO	2.71	2.27
12	3.24	3.08	WNO	2.89	2.72	3.5	3.71	3.52	3.53	3.71
13	2.54	2.63	2.88	2.31	2.71	WNO	WNO	WNO	WNO	WNO
14	WNO	1.74	2.24	2.07	2.38	WNO	WNO	WNO	WNO	WNO
15	WNO	WNO	WNO	WNO	WNO					
16	WNO	WNO	WNO	WNO	WNO	WNO	WNO	WNO	WNO	WNO
17	3.09	2.91	3.06	3.09	3.29	3.39	2.5	2.57	2.62	2.35
18	3.39	3.34	3.45	3.44	3.33					

12. In the window forming experiment (see Table 8.12 in the exercises of Chapter 8), another response is the window size, whose data are given in Table 12.18. A WNO in Table 12.18 indicates that the window was not open. When the window was open, its size (in micrometers) was measured and given in the table. The target window size is 3 μm.

(a) Analyze the data by using a lognormal regression model and treating a WNO as an interval-censored observation in $[0, 1.7]$.

(b) Based on the fitted model in (a), identify factor settings that give expected window size close to the target.

13. Watkins, Bergman, and Horton (1994) discussed an experiment to reduce the tooling costs of cutting threads on pipes to make electrical conduit. The cutting tool is called a chaser and the angle at which it approaches the pipe is called the rake angle. The experimental factors are (A) chaser type, (B) coolant type, (C) speed, (D) conduit manufacturer, and (E) rake angle. Two levels for each factor were investigated using a 2^{5-1} design with $E = ABCD$. The response is the time until poor thread quality is observed or the product has to be scrapped. The threading machine had two spindles (left and right denoted by L and R); each

Table 12.19 Design Matrix and Lifetime Data, Threading Machine Experiment

Run	A	B	C	D	E	F	1	2	3
1	−	−	−	−	−	L	137	24	58
2	−	−	−	+	+	R	89	41	26
3	−	−	+	−	+	L	56	34	199
4	−	−	+	+	−	L	545	105	106
4	−	−	+	+	−	L	122	132	168
4	−	−	+	+	−	R	74	66	49
5	−	+	−	−	+	L	428	157	188
6	−	+	−	+	−	R	352	68	97
7	−	+	+	−	−	L	320	750	988
8	−	+	+	+	+	R	632	73	529
9	+	−	−	−	+	R	·44	41	7
10	+	−	−	+	−	R	2	3	4
11	+	−	+	−	−	R	112	64	59
12	+	−	+	+	+	R	59	45	9
13	+	+	−	−	−	R	40	7	32
14	+	+	−	+	+	R	21	42	120
15	+	+	+	−	+	R	14	19	28
15	+	+	+	−	+	R	168	34	58
16	+	+	+	+	−	R	70	57	18

Column headers: Factor (A, B, C, D, E, F); Replication (1, 2, 3).

spindle can handle a pipe. The design matrix and lifetime data are given in Table 12.19. Note that no observation was censored, setting number 4 has three runs, and setting number 15 has two runs.

(a) Analyze the failure time data by using the lognormal and Weibull regression models. Do they lead to similar conclusions?

(b) Based on the fitted model in (a), identify factor settings that maximize the lifetime.

14. Consider an experiment to improve a window wiper switch which used an $OA(8, 4^1 2^4)$ to study one four-level factor (A) and four two-level factors ($B–E$). Four switches are available for each of the eight runs. For each switch, the initial voltage drop across multiple contacts is recorded (i.e, first inspection) and then recorded every 20,000 cycles thereafter up to 180,000 cycles, resulting in 10 inspections. The design matrix and voltage drop data are given in Table 12.20.

(a) Analyze the data using a two-stage analysis for degradation data.

(b) Suppose that a voltage drop of 120 is unacceptable and the corresponding time (in cycles) is defined as the failure time. What factors impact the failure times of the switch? What factor levels are recommended? Assume that the four-level factor is quantitative with evenly spaced levels.

Table 12.20 Design Matrix and Voltage Drop Data, Wiper Switch Experiment

Factor					Inspection									
A	B	C	D	E	1	2	3	4	5	6	7	8	9	10
0	−	−	−	−	24	37	40	65	72	77	90	101	117	128
					22	36	47	64	71	86	99	118	127	136
					17	34	40	52	66	79	91	98	115	119
					24	30	38	46	57	71	73	91	98	104
0	+	+	+	+	45	60	79	90	113	124	141	153	176	188
					51	68	84	104	122	136	148	166	191	197
					42	58	70	82	103	119	128	143	160	175
					41	56	56	70	81	89	98	108	113	128
1	−	−	+	+	28	40	56	69	87	86	110	121	132	146
					46	50	81	95	114	130	145	161	185	202
					45	54	79	90	111	132	143	168	185	202
					37	58	81	99	123	143	166	191	202	231
1	+	+	−	−	54	51	64	66	78	84	90	93	106	109
					47	45	50	53	58	57	61	55	61	66
					47	54	63	68	70	77	88	86	91	102
					53	55	66	68	91	90	98	104	118	120
2	−	+	−	+	18	35	48	56	65	81	89	98	117	124
					20	37	52	53	67	75	85	95	112	122
					32	54	76	98	119	143	158	181	205	231
					28	39	54	73	89	98	117	127	138	157
2	+	−	+	−	44	50	48	46	55	63	65	71	68	76
					43	44	55	56	58	62	66	66	72	72
					40	46	45	49	55	62	61	61	64	66
					55	67	73	75	91	88	102	111	115	119
3	−	+	+	−	39	47	58	72	84	104	109	129	143	154
					29	42	55	67	82	91	104	117	130	136
					36	45	56	80	93	101	121	138	154	170
					31	40	60	72	82	98	103	117	130	146
3	+	−	−	+	61	67	69	86	86	88	95	103	107	118
					68	75	82	90	95	109	107	118	120	133
					60	72	85	84	87	98	99	111	113	125
					65	68	69	75	79	84	95	96	101	100

REFERENCES

Berger, J.O. (1985), *Statistical Decision Theory and Bayesian Analysis*, 2nd ed., New York: Springer-Verlag.

Bullington, R.G., Lovin, S.G., Miller, D.M., and Woodall, W.H. (1993), "Improvement of an Industrial Thermostat Using Designed Experiments," *Journal of Quality Technology*, 25, 262–270.

Chiao, C.H. and Hamada, M. (1996), "Using Degradation Data from an Experiment to Achieve Robust Reliability for Light Emitting Diodes," *Quality and Reliability Engineering International*, 12, 89–94.

Clarkson, D.B. and Jennrich, R.I. (1991), "Computing Extended Maximum Likelihood Estimates for Linear Parameter Models," *Journal of the Royal Statistical Society, Series B*, 53, 417–426.

Davidian, M. and Giltinan, D. (1995), *Nonlinear Models for Repeated Measurement Data*, London: Chapman and Hall.

Davis, T.P. (1995), "Analysis of an Experiment Aimed at Improving the Reliability of Transmission Centre Shafts," *Lifetime Data Analysis*, 1, 275–306.

Doksum, K.A. and Hoyland, A. (1992), "Models for Variable-Accelerated Life Testing Experiments Based on Wiener Processes and the Inverse Gaussian Distribution," *Technometrics*, 34, 74–82.

Hamada, M. (1993), "Reliability Improvement via Taguchi's Robust Design," *Quality and Reliability Engineering International*, 9, 7–13.

Hamada, M. and Tse, S.K. (1992), "On Estimability Problems in Industrial Experiments with Censored Data," *Statistica Sinica*, 2, 381–391.

Hamada, M. and Wu, C.F.J. (1991), "Analysis of Censored Data from Highly Fractionated Experiments," *Technometrics*, 33, 25–38.

Hellstrand, C. (1989), "The Necessity of Modern Quality Improvement and Some Experience with Its Implementation in the Manufacture of Rolling Bearings," *Philosophical Transactions of the Royal Society of London A*, 327, 529–537.

Lin, W.Z. (1976), *Fundamentals of Fluorescent Lamp*, Taiwan: Taiwan Fluorescent Lamp Co.

Meeker, W.Q. and Escobar, L.A. (1998), *Statistical Methods for Reliability Data*, New York: John Wiley & Sons.

Nelson, W. (1990), *Accelerated Testing–Statistical Models, Test Plans, and Data Analysis*, New York: John Wiley & Sons.

SAS Institute (1989), *SAS Institute Inc. SAS/QC Software: Reference, Version 6, First Edition*, Cary, NC: SAS Institute.

Silvapulle, M.J. and Burridge, J. (1986), "Existence of Maximum Likelihood Estimates in Regression Models for Grouped and Ungrouped Data," *Journal of the Royal Statistical Society, Series B*, 48, 100–106.

Taguchi, G. (1987), *System of Experimental Design*, White Plains, NY: Unipub/Kraus International Publications.

Tseng, S.T., Hamada, M., and Chiao, C.H. (1995), "Using Degradation Data to Improve Fluorescent Lamp Reliability," *Journal of Quality Technology*, 27, 363–369.

Watkins, D., Bergman, A., and Horton, D. (1994), "Optimization of Tool Life on the Shop Floor Using Design of Experiments," *Quality Engineering*, 6, 609–620.

CHAPTER 13

Experiments with Nonnormal Data

With the exception of the Weibull failure time distribution, the experimental response has been assumed to be normally or approximately normally distributed in previous chapters. On the log scale, the lognormally distributed response in Chapter 12 is also normally distributed. In many practical applications, however, the experimental data are not normally distributed. In this chapter we consider other data types and distributions. First we consider the class of *generalized linear models* (GLMs), which is a generalization of the normal linear model and includes discrete distributions like Poisson and binomial and continuous distributions like gamma and inverse Gaussian. The analysis strategy is based on standard methods for GLMs. Ordinal (also called ordered categorical) data are also considered. Both the likelihood-based and Bayesian methods of analysis are presented.

13.1 A WAVE SOLDERING EXPERIMENT WITH COUNT DATA

Solder-joint defects, a quality problem in electronics manufacturing, arise when components are attached to electronic circuits card assemblies in a wave soldering process. The soldering process involves baking and preheating the circuit card and then passing it through a solder wave by conveyer. An experiment was performed to reduce solder defects (Condra, 1993, Chapter 7) that investigated seven two-level factors: prebake temperature (A), flux density (B), conveyer speed (C), preheat condition (D), cooling time (E), ultrasonic solder agitator (F), and solder temperature (G).

A 2_{IV}^{7-3} design with the defining relation $E = ABD$, $F = ACD$, and $G = BCD$ was used to study the seven factors each at two levels. This design was chosen to study seven main effects and the six interactions AB, AC, AD, BC, BD, and CD, which the experimenters thought might be potentially important. The response was the number of defects on a board. Three boards

563

Table 13.1 Design Matrix and Defect Counts, Wave Soldering Experiment

| Run | \multicolumn{7}{c}{Factor} | | | | | | | \multicolumn{3}{c}{Board} | | |
	A	B	C	D	E	F	G	1	2	3
1	−	−	−	−	−	−	−	13	30	26
2	−	−	−	+	+	+	+	4	16	11
3	−	−	+	−	−	+	+	20	15	20
4	−	−	+	+	+	−	−	42	43	64
5	−	+	−	−	+	−	+	14	15	17
6	−	+	−	+	−	+	−	10	17	16
7	−	+	+	−	+	+	−	36	29	53
8	−	+	+	+	−	−	+	5	9	16
9	+	−	−	−	+	+	−	29	0	14
10	+	−	−	+	−	−	+	10	26	9
11	+	−	+	−	+	−	+	28	173	19
12	+	−	+	+	−	+	−	100	129	151
13	+	+	−	−	−	+	+	11	15	11
14	+	+	−	+	+	−	−	17	2	17
15	+	+	+	−	−	−	−	53	70	89
16	+	+	+	+	+	+	+	23	22	7

from each run were assessed for defects. The design matrix and response data are given in Table 13.1.

This type of data is commonly modeled by the Poisson distribution, whose density is

$$f(y) = \mu^y \exp(-\mu)/y!, \qquad y = 0, 1, 2, \ldots. \qquad (13.1)$$

For a Poisson distributed response y with parameter μ, its mean and variance are both μ. Then y is approximately distributed as $N(\mu, \mu)$ for large μ and its square root transformation $\sqrt{y}$ is approximately distributed as $N(\sqrt{\mu}, 1/4)$ (see Section 2.5 for derivations). Standard regression techniques, which are based on the normality assumption, may be applied to the square root of the count data. This approach is appealing for its simplicity but has some disadvantages. First, the normal approximation is not accurate for small μ. In the experiment some boards have low defect counts, and the experimental goal was to minimize the number of defects. Second, the equal mean and variance relationship for the Poisson distribution may not hold for some count data. Using generalized linear models would avoid these two problems.

13.2 GENERALIZED LINEAR MODELS

In this and the next section, we will give a brief review of **generalized linear models** (GLMs), and skip most of the derivations and justifications. For a comprehensive treatment, see McCullagh and Nelder (1989). First, recall the normal linear model from (1.1):

$$y = \beta_0 + \beta_1 x_1 + \beta_2 x_2 + \cdots + \beta_p x_p + \epsilon, \tag{13.2}$$

where y is the response, $\beta_0 + \beta_1 x_1 + \beta_2 x_2 + \cdots + \beta_p x_p$ is the *systematic* component involving p covariates $x_1, \ldots, x_p$, and ϵ, the *random* component, is independent and distributed as $N(0, \sigma^2)$.

To see how GLMs generalize the normal linear model, consider the following equivalent description of (13.2):

$$\begin{aligned}
y &\sim N(\mu, \sigma^2), \\
E(y) = \mu &= \beta_0 + \beta_1 x_1 + \beta_2 x_2 + \cdots + \beta_p x_p.
\end{aligned} \tag{13.3}$$

GLMs extend the normal linear model in two directions: the assumption about the distribution of the response y and the form of the systematic component. A GLM can be specified as follows:

Specification of a GLM

(*i*) *The distribution of the response y.*
(*ii*) *The link function defining the scale on which the effects of*   (13.4)
     the explanatory variables are assumed to combine additively.

The random and systematic components of a GLM will be discussed in the next two subsections.

13.2.1 The Distribution of the Response

The Poisson distribution for defect counts in the wave soldering experiment is a member of the class of distributions called the **one-parameter exponential family**, whose log-likelihood (i.e., log of the likelihood function denoted by l) takes the following general form:

$$l(\theta; y) = [y\theta - b(\theta)]/\phi + c(y, \phi). \tag{13.5}$$

In this formulation, θ is called the *canonical parameter* and ϕ is the *dispersion parameter*. Note that the dispersion parameter ϕ is defined in the context of a GLM and should not be confused with the dispersion effects in variation reduction studies even though they are identical or closely related for some distributions.

The Poisson distribution with mean μ is a member of this class because its log-likelihood [i.e., the log of (13.1)] takes the form

$$l(\mu; y) = -\mu + y \ln(\mu) - \ln(y!).$$

To match this up with (13.5), let $\theta = \ln(\mu)$, so that $\mu = \exp(\theta)$ and $\phi = 1$. This gives the form

$$l(\mu; y) = y\theta - \exp(\theta) - \ln(y!),$$

which shows that $b(\theta) = \exp(\theta)$ and $c(y, \phi) = -\ln(y!)$. Thus, $\ln(\mu)$ is the canonical parameter and 1 is the dispersion parameter.

The normal distribution $N(\mu, \sigma^2)$ is also a member of this class since its log-likelihood,

$$l(\mu; y) = -\tfrac{1}{2}\ln(2\pi\sigma^2) - (y - \mu)^2/(2\sigma^2),$$

can be rewritten to look like (13.5) by expanding the term $(y - \mu)^2$ to obtain

$$l(\mu; y) = (y\mu - \mu^2/2)/\sigma^2 - \left[y^2/\sigma^2 + \ln(2\pi\sigma^2)\right]/2.$$

Letting $\theta = \mu$, $b(\theta) = \theta^2/2$, $\phi = \sigma^2$ and $c(y, \phi) = -[y^2/\sigma^2 + \ln(2\pi\sigma^2)]/2$ give the expressions in the form of (13.5). Thus for the normal distribution, the canonical parameter θ is equal to the mean μ, and the dispersion parameter ϕ is equal to the variance σ^2.

It turns out that the binomial, gamma, and inverse Gaussian distributions are also members of this class. The binomial density is

$$f(y) = \binom{m}{y} \mu^y (1 - \mu)^{m-y}, \qquad y = 0, \ldots, m, \qquad 0 \le \mu \le 1; \quad (13.6)$$

the gamma $G(\mu, \nu)$ density is

$$f(y) = \frac{1}{y\Gamma(\nu)} \left(\frac{\nu y}{\mu}\right)^\nu \exp\left(-\frac{\nu y}{\mu}\right), \qquad y > 0, \qquad \nu > 0, \qquad \mu > 0; \quad (13.7)$$

and the inverse Gaussian $IG(\mu, \sigma^2)$ density is

$$f(y) = \frac{1}{\sqrt{2\pi\sigma^2 y^3}} \exp\left(-\frac{1}{2\sigma^2\mu^2 y}(y - \mu)^2\right),$$

$$y > 0, \qquad \sigma^2 > 0, \qquad \mu > 0. \qquad (13.8)$$

For the distributions in (13.5), it can be shown that the variance of the response y takes the form

$$Var(y) = \phi V(\mu), \qquad (13.9)$$

where ϕ is the dispersion parameter given in (13.5) and $V(\mu)$ is called the *variance function*. The variance function $V(\mu)$ thus represents the part of the

Table 13.2 Form of the Variance for Some GLMs

Distribution	Dispersion Parameter	Variance Function
normal	σ^2	1
Poisson	1	μ
binomial	$1/m$	$\mu(1-\mu)$
gamma	$1/\nu$	μ^2
inverse Gaussian	σ^2	μ^3

variance of the response y that depends on the mean μ and the dispersion parameter ϕ the part that does not. For the normal distribution, $\phi = \sigma^2$ and $V(\mu) = 1$. Table 13.2 presents the components of the response variance for some GLMs. For the gamma distribution, its variance is μ^2/ν so that the part of the variance not depending on the mean is $1/\nu$, where ν is the shape parameter for the gamma distribution.

13.2.2 The Form of the Systematic Effects

In the normal linear model, the systematic effects are assumed to combine additively; that is, the systematic part of the model $E(y)$ takes the form $\mu = \beta_0 + \beta_1 x_1 + \beta_2 x_2 + \cdots + \beta_p x_p$ for p covariates $x_1, x_2, \ldots, x_p$. In GLMs, this linear combination also forms a *linear predictor* η, which in turn is assumed to be some monotonic transformation of the mean μ. Notationally, this can be expressed as:

$$\eta = \beta_0 + \beta_1 x_1 + \beta_2 x_2 + \cdots + \beta_p x_p = \mathbf{x}^T \boldsymbol{\beta} = g(\mu), \qquad (13.10)$$

where $g(\cdot)$ is the **link function**, $\mathbf{x}^T = (1, x_1, \ldots, x_p)$ and $\boldsymbol{\beta}^T = (\beta_0, \beta_1, \ldots, \beta_p)$. The link function thus defines the scale on which the systematic effects are assumed to be additive. For example, with Poisson regression models in which the effects are assumed multiplicative on the μ scale, the link function is the log, i.e., $\eta = \ln(\mu)$ produces effects that combine additively on the η scale. Table 13.3 shows some well-known models described in GLM terminology.

One view of the model for Bernoulli data (i.e., binomial with $m = 1$) is that they are generated by an underlying continuous distribution. There is a cutpoint on the continuous scale that defines the probability of observing a zero, i.e., $\mu = Prob(y = 0)$ equals the probability that the underlying random variable is less than the cutpoint. For GLMs, the intercept in the linear predictor is the cutpoint and the link function implies the underlying distribution. For example, the logistic link corresponds to the logistic distribution, the probit link corresponds to the standard normal distribution, and the complementary-log-log link corresponds to the standard extreme-value distribution. A detailed discussion of the models in Table 13.3 can be found in McCullagh and Nelder (1989).

Table 13.3 Some GLMs and Their Constituents

Model	Distribution	Link $\eta = g(\mu)$
normal linear	normal	identity $(\eta = \mu)$
Poisson regression	Poisson	log $(\eta = \ln \mu)$
loglinear model	Poisson	log $(\eta = \ln \mu)$
logistic regression	binomial	logit $(\eta = \ln[\,\mu/(1 - \mu)])$
probit regression	binomial	probit $[\eta = \Phi^{-1}(\mu)]$
binary regression (underlying exponential)	binomial	complementary-log-log $(\eta = \ln[-\ln(1 - \mu)])$
gamma regression	gamma	reciprocal $(\eta = \mu^{-1})$
inverse Gaussian regression	inverse Gaussian	$\eta = \mu^{-2}$

13.2.3 GLM versus Transforming the Response

In Section 2.5, the transformed responses $g(y)$ were assumed to follow a normal linear model. That is, the transformed responses are normally distributed and have constant variances independent of the mean and the mean consists of the systematic effects that combine additively. Thus, a single transformation is asked to produce three things simultaneously: *normality*, *constant variance*, and *linearity*. For many distributions it is known that no single transformation exists that can achieve all the three goals. In the case of the Poisson distribution, for approximate normality, the transformation $y \longrightarrow y^{2/3}$ is required, for approximately constant variance $y \longrightarrow y^{1/2}$, and for multiplicative effects (i.e., additivity and linearity on the log scale) $y \longrightarrow \ln y$. In practice the square-root transformation ends up being used; in other words constant variance is chosen over normality and linearity. It might be argued that the linearity of effects should take precedence and the log transformation should be used for Poisson counts. Another problem arises with zero counts because zeros cannot be logged. The transformation can be adjusted by using $\ln(y + \frac{1}{2})$, but now the variances will not be constant and the errors will not be approximately normal. Furthermore, adding a half will distort the estimates of the systematic effects.

These problems disappear when GLMs are used to fit the data. First, a natural distribution for the data can be adopted, e.g., Poisson for count data. Second, the mean μ instead of the response y is transformed to achieve linearity. This makes sense because linearity is a function of μ rather than y. Third, the choice of variance function to model the random component is entirely separate from the choice of link function to achieve linearity of the systematic effects. Thus, a single transformation is no longer trying to do several jobs.

Despite the advantages of GLMs, transformation may continue to play a practical role. Situations may arise when analysis in the original metric is clearly not appropriate, but it is not obvious what member of the exponential family should be used to model the response distribution. Yet some transformation yields a simple normal model that fits the data well. A GLM, likely to give similar results, could then be identified by finding the implied variance on the original scale as a function of the mean. Also, an important requirement of a good model is that it be interpretable so that only transformations that make sense in a particular context need to be considered. A GLM using the transformed response could be considered; this extension provides an even wider range of applications.

13.3 ANALYSIS OF GENERALIZED LINEAR MODELS

As with the Weibull regression model in Chapter 12, a GLM is fitted by maximum likelihood. The maximum likelihood estimates of the regression parameters $\hat{\boldsymbol{\beta}}$ can be obtained by the following *iteratively reweighted least squares algorithm*. Noting that $\eta = g(\mu)$ is assumed to have a linear structure, standard regression methodology should be applied to η, not μ. First we consider the following *adjusted dependent variable z*, which is a linearized form of the link function g applied to the data. Let t index the current iteration and $\hat{\mu}^{(t)}$ the current estimate of μ. Then $z^{(t)}$ is defined as follows:

$$z^{(t)} = \hat{\eta}^{(t)} + (y - \hat{\mu}^{(t)}) \frac{\partial \eta}{\partial \mu}\bigg|_{\hat{\mu}^{(t)}}, \tag{13.11}$$

which is computed for each of the N observations in the data. The next estimate $\hat{\boldsymbol{\beta}}^{(t+1)}$ is obtained by weighted least squares for the response vector $\mathbf{z}^{(t)} = (z_1^{(t)}, \ldots, z_N^{(t)})$, the covariates $x_1, \ldots, x_p$, and the current weights $W^{(t)}$ as specified by

$$W^{(t)^{-1}} = \left(\frac{\partial \eta}{\partial \mu}\bigg|_{\hat{\mu}^{(t)}}\right)^2 V(\hat{\mu}^{(t)}), \tag{13.12}$$

where $V(\cdot)$ is the variance function. That is,

$$\hat{\boldsymbol{\beta}}^{(t+1)} = (\mathbf{X}^T \mathbf{W}^{(t)} \mathbf{X})^{-1} \mathbf{X}^T \mathbf{W}^{(t)} \mathbf{z}^{(t)}, \tag{13.13}$$

where $\mathbf{X}$ is the model matrix whose rows are $\mathbf{x}_i^T$ and $\mathbf{W}^{(t)}$ is the diagonal matrix of weights $W^{(t)}$. From $\hat{\boldsymbol{\beta}}^{(t+1)}$, one obtains

$$\hat{\boldsymbol{\eta}}^{(t+1)} = \mathbf{X}\hat{\boldsymbol{\beta}}^{(t+1)} \quad \text{and} \quad \hat{\boldsymbol{\mu}}^{(t+1)} = g^{-1}(\hat{\boldsymbol{\eta}}^{(t+1)}). \tag{13.14}$$

Iteration then continues until convergence.

One suggestion for starting values is to use $\hat{\boldsymbol{\mu}}_0 = \mathbf{y}$ or $\hat{\boldsymbol{\mu}}_0 = \mathbf{y}/m$ (for m Bernoulli observations per run). A problem occurs for the Poisson distribution and log link with zero counts or for the binomial distribution with various links and observed proportions being zero or one. For these cases, use $\frac{1}{6}$ for a zero Poisson count and $(y + 1/2)/(m + 1)$ for a zero or m binomial count. Several software packages support the fitting of GLMs such as GLIM (Nelder, 1974), GENSTAT (Payne et al., 1987), and the GENMOD procedure in SAS (SAS Institute, 1996). See Hilbe (1994) for a comparison of current software.

Given the MLEs for the regression parameters, the significance of β_i can be tested by comparing its maximum likelihood estimate $\hat{\beta}_i$ with its standard error $\hat{\sigma}_i$ based on the Fisher information or observed information matrix. See (12.11)–(12.13). Alternatively, the likelihood ratio test in (12.14) can be used.

For the normal linear model, the *residual sum of squares* (*RSS*) with its corresponding *degrees of freedom* (*df*) measures the discrepancy between the data and the fitted values generated by a model. In GLMs, the *RSS* is generalized to the **deviance**, which is defined as

$$D(\mathbf{y}; \hat{\boldsymbol{\mu}}) = \left[-2l(\hat{\boldsymbol{\mu}}, \mathbf{y}) + 2l(\mathbf{y}, \mathbf{y}) \right] \phi, \tag{13.15}$$

where $l(\hat{\boldsymbol{\mu}}, \mathbf{y}) = \sum_i l(\hat{\mu}_i; y_i)$ is the sum of the log-likelihood values over all the observations evaluated at the maximum likelihood estimates $\hat{\boldsymbol{\mu}}$ and $l(\mathbf{y}, \mathbf{y}) = \sum_i l(y_i; y_i)$ is the corresponding quantity when the complete model is fitted, i.e., each fitted value equals the observation. The *degrees of freedom* for the deviance in (13.15) equal the total run size minus the number of parameters in $\boldsymbol{\beta}$ for the model used in obtaining the $\hat{\boldsymbol{\mu}}$ values. Application of (13.15) to the normal distribution yields

$$-2l(\hat{\mu}; y) = \ln(2\pi\sigma^2) + (y - \hat{\mu})^2/\sigma^2 \quad \text{and} \quad D(\mathbf{y}; \hat{\boldsymbol{\mu}}) = \sum_i (y_i - \hat{\mu}_i)^2.$$

That is, the deviance is the *RSS*. The deviance is a function of the data only because ϕ in (13.15) cancels the same term in the denominator of $l(\mu, y)$ in (13.5). For the Poisson distribution,

$$D(\mathbf{y}; \hat{\boldsymbol{\mu}}) = 2\sum_i \left[y_i \ln(y_i/\hat{\mu}_i) - (y_i - \hat{\mu}_i) \right],$$

which is sometimes called G^2. Deviances can be used for significance testing and model selection. For models M_1 and M_2 with M_1 nested within M_2, let D_1 and D_2 denote the corresponding deviances with df_1 and df_2 degrees of freedom. Then it can be shown that, under the null hypothesis that model M_1 is adequate (i.e., the additional effects in model M_2 that are not in M_1 are zero),

$$F = \frac{(D_1 - D_2)/(df_1 - df_2)}{D_2/df_2} \tag{13.16}$$

has approximately an F distribution with parameters $(df_1 - df_2, df_2)$. If the F value in (13.16) exceeds $F_{df_1 - df_2, df_2, \alpha}$, the upper α percentile of $F_{df_1 - df_2, df_2}$, the additional effects in model M_2 that are not in M_1 are declared significant at level α.

Finally, *residuals*, a vital component of graphical model-checking techniques, also generalize in a straightforward way to what are called *deviance residuals*. The deviance consists of a nonnegative component d_i for each observation summed over all observations, where

$$d_i = \left[-2l(\hat{\mu}_i, y_i) + 2l(y_i, y_i) \right] \phi.$$

The signed square roots of the deviance components

$$\text{sign}(y_i - \hat{\mu}_i)\sqrt{d_i} \tag{13.17}$$

are called deviance residuals and generalize the normal residuals, $y - \hat{\mu}$, to the wider GLM class of distributions.

For the normal linear model, the *residual mean-squared error* (*RMSE*),

$$RSS/df,$$

is a measure of the variation unaccounted for by the model. It is used to estimate the variance σ^2 and appears as a scaling factor for the sampling variances and covariances of the parameter estimates. Usually there is no prior knowledge of σ^2, so that the *RMSE* is used as an estimate, assuming that the fitted model has adequately removed all the systematic variation. In GLMs, *RMSE* is replaced by the *mean deviance*

$$\hat{\phi} = D(\mathbf{y}; \hat{\mathbf{\mu}})/df, \tag{13.18}$$

where df is the degrees of freedom for the deviance. For the normal distribution, the *RMSE* and mean deviance are the same. The mean deviance can be used as an estimate of ϕ, the dispersion parameter, which is itself a generalization of σ^2. For the Poisson and Bernoulli distributions, ϕ has an *a priori* value of 1, so that one expects to find a mean deviance close to 1 if the model fits well. In practice, discrete data often give values of the mean deviance considerably greater than 1. This is called *over-dispersion* and may be explained by the presence of unknown systematic effects or between-run random variations which are not incorporated in the model. The corresponding GLM models can be adapted to cope with the presence of over-dispersion (and less frequently under-dispersion) as follows. The mean deviance is used to estimate ϕ and becomes a multiplier in estimating variances and covariances of the parameter estimates, as in the normal case. The estimates of the systematic effects are, however, unaffected by this modification. Justification for this adaptation requires the idea of *quasi-likelihood*. See McCullagh and Nelder (1989, Chapter 9) for details.

13.4 ANALYSIS OF THE WAVE SOLDERING EXPERIMENT

It is reasonable to use the Poisson distribution to model the defect counts in the wave soldering experiment. This choice is supported by the observation that the range of the defect counts in Table 13.1 increases with the mean. Thus, we consider using a Poisson regression model to analyze the data. In GLM terminology, a Poisson distribution for the response and a log link are adopted. There is also a clear outlier in the data; for the second board in run 11, a count of 173 defects is not in line with the two other values of 19 and 28. This outlier is excluded in the analysis that follows.

As described in Section 13.1, the experiment employed a 2_{IV}^{7-3} design. Straightforward calculations show that the design has 7 clear main effects and 21 two-factor interactions in seven groups of aliased effects. The six interactions AB, AC, AD, BC, BD, and CD, which were thought to be potentially important by the investigator, take up six of the seven degrees of freedom, with CE and its aliases being the seventh. The remaining degree of freedom is for the effect ABC and its aliases. This consideration suggests that a saturated model be fitted with terms in the linear predictor consisting of all main effects $(A-G)$, the 6 two-factor interactions (AB, AC, AD, BC, BD, CD) and 2 additional effects CE and ABC. The mean deviance of 4.5 ($= 139.51/31$) is much higher than 1, which suggests a strong possibility of over-dispersion. Consequently, appropriate F tests need to be used to determine the significance of the effects. The MLEs, the F statistics for the effects (based on (13.16) with M_2 being the saturated model and M_1 the model obtained by dropping the effect under test from M_2), and their corresponding p values are given in Table 13.4. Examination of these

Table 13.4   Estimates, F Statistics, and p Values, Wave Soldering Experiment

Effect	Estimate	F	p Value
intercept	3.071		
A	0.114	2.41	0.13
B	-0.136	3.45	0.07
C	0.417	32.43	0.00
D	-0.059	0.65	0.43
E	-0.096	1.69	0.20
F	-0.015	0.04	0.84
G	-0.382	27.73	0.00
AB	-0.021	0.09	0.77
AC	0.176	5.76	0.02
AD	0.096	1.73	0.20
BC	-0.080	1.18	0.29
BD	-0.300	16.66	0.00
CD	0.047	0.41	0.52
CE	0.005	0.00	0.94
ABC	0.015	0.04	0.83

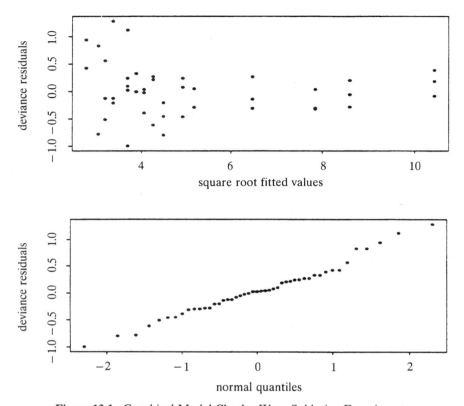

Figure 13.1. Graphical Model Checks, Wave Soldering Experiment.

results suggests that C, G, AC, and BD are significant. Since BD is aliased with CG, which effect should be included in the final model? Noting that the parent factors C and G are also significant, the strong effect heredity principle (see Section 8.5.1) would suggest that the effect significance be attributed to CG. Based on the signs of the estimates, recommended factor settings are $A_+C_-G_+$, where the subscript denotes the recommended level. Since factors B, D, E, and F are not important, their recommended levels can be set based on other considerations like cost.

Next, we consider graphical displays for model checking. Figure 13.1 shows two graphs which are useful in assessing the consistency of the fit. The top graph shows the deviance residuals plotted against the fitted values on the square-root scale. [For a justification of the square-root scale, see Nelder (1990).] If the model is adequate, the residuals should be evenly distributed about zero with no systematic pattern. The lower graph gives the normal probability plot of the deviance residuals. If the fit is good, the plot should follow approximately a straight line (Pierce and Schafer, 1986). The reasonably even distribution of residuals in the top graph and the approximate linearity of the lower graph confirm the adequacy of the chosen model.

13.5 OTHER USES AND EXTENSIONS OF GENERALIZED LINEAR MODELS

GLMs have been used in many applications, most of which are not in engineering or industry. See McCullagh and Nelder (1989), Aitkin et al. (1989), *The GLIM System Release* 4 *Manual* (Francis, Green, and Payne, 1993, Chapter 12), GLIM newsletters (Numerical Algorithms Group), and the GLIM conference proceedings (e.g., *GLIM* 82, 1982). Hamada and Nelder (1997) showed how GLMs can be used for industrial problems. Here we briefly outline some of its applications.

GLMs can be used for analyzing reliability experiments. For example, the gamma and inverse Gaussian failure time distributions are members of the GLM distributions. GLM software can also handle the Weibull distribution and right-censored data (Aitkin and Clayton, 1980). GLMs can also be used to model Bernoulli response data. For example, in manufacturing windshield moldings, debris in the stamping process is carried into the die and appears as dents in the product (called slugs). In a windshield molding experiment to improve the slugging condition (see Table 13.8 in the exercises), the response is the number of good parts without slugs out of 1000.

A table of counts classified by two or more variables is called a contingency table. For example, a sample of defective parts might be taken with each sampled part being classified according to what shift (a.m./p.m.) and what day (Monday–Friday) of the week it was produced. Such data can be modeled by a log-linear model (i.e., Poisson response and log link) to see whether there are shift or day effects.

Parameter design experiments with nonnormal responses can be analyzed by GLMs. The experiment in Section 13.6 with ordinal data is a good example. Another application is to data with replicates whose mean and variance depend on the experimental factors. Grego (1993) showed how the sample mean and sample standard deviation for each run could be analyzed using two different GLMs. Nelder and Lee (1991) showed how an extension of GLM methodology, called *extended quasi-likelihood*, could model the mean and variance simultaneously.

Estimability problems also arise in fitting GLMs. For example, if there are too many zero Poisson counts or if too many runs have binomial zero or m counts, the MLEs will likely be infinite. As in Section 12.5 for analyzing reliability experiments, a Bayesian approach for analyzing GLM experiments is natural for handling estimability problems. Priors for the regression parameters β and the dispersion parameter ϕ (if it is not equal to 1) need to be specified. Then Gibbs sampling can be used. With GLMs, it is not as easy to sample from the full conditionals. Dellaportas and Smith (1993) showed how adaptive rejection sampling can be used to draw samples from the full conditionals. Some restrictions on the choice of link function are required, however. This method is implemented in BUGS (Bayesian inference using Gibbs sampling) software (MRC Biostatistics Unit, 1995).

13.6 A FOAM MOLDING EXPERIMENT WITH ORDINAL DATA

An experiment was performed to reduce voids in a molded urethane-foam product. A product is rated in terms of the extent of the voids as very good, acceptable, or needs repair. Thus, the response is one of three ordered categories, which we will refer to later as "good," "ok," and "poor." Such data are called **ordered categorical** or **ordinal data**. The experiment used a cross array to study seven control factors (with a 2^{7-4} design) and two noise factors (with a 2^2 design). The factors and levels are given in Table 13.5. The experimental goal was to make a product with as few voids as possible that is robust to the noise factors. The cross array and response data (Chipman and

Table 13.5 Factors and Levels, Foam Molding Experiment

Control Factor	Level −	+
A. shot weight	185	250
B. mold temperature (°F)	70–80	120–130
C. foam block	use	do not use
D. RTV insert	use	do not use
E. vent shell	vented	unvented
F. spray wax ratio	2:1	4:1
G. tool elevation	level	elevated

Noise Factor	Level −	+
H. shift	second	third
I. shell quality	good	bad

Table 13.6 Cross Array and Frequencies (good (I), ok (II), poor (III)), Foam Molding Experiment

							Noise Factor											
							H −	+		−		+						
		Control Factor					*I* −			−		+			+			
A	*B*	*C*	*D*	*E*	*F*	*G*	I	II	III	I	II	III	I	II	III	I	II	III
−	−	−	−	−	−	−	3	6	1	6	4	0	1	4	5	0	10	0
−	−	−	+	+	+	+	0	3	7	3	4	3	0	6	4	0	7	3
−	+	+	−	−	+	+	0	0	10	0	1	9	0	0	10	0	0	10
−	+	+	+	+	−	−	0	0	10	0	10	0	0	3	7	0	9	1
+	−	+	−	+	−	+	3	5	2	3	7	0	3	5	2	1	6	3
+	−	+	+	−	+	−	2	8	0	4	5	1	0	5	5	1	5	4
+	+	−	−	+	+	−	2	7	1	2	5	3	2	7	1	1	6	3
+	+	−	+	−	−	+	0	4	6	1	7	2	0	4	6	0	3	7

Hamada, 1996) are given in Table 13.6. The two rows in the noise array represent the two noise factors H and I. At each combination of the control and noise factors specified by the cross array, 10 parts were classified into one of the three categories. Table 13.6 gives the frequencies of the three categories denoted by I, II, and III.

13.7   MODEL AND ANALYSIS FOR ORDINAL DATA

In this section, we consider regression models for ordinal data which express the probability of an observation y_i being in category j as a function of the covariates. McCullagh (1980) suggested models of the form

$$g\left(Prob\left(y_i \leq j\right)\right) = \gamma_j - \mathbf{x}_i^T\boldsymbol{\beta} \quad \text{for} \quad j = 1, \ldots, J - 1 \text{ and } i = 1, \ldots, N,$$

$$(13.19)$$

where the ordered categories are labeled $1, \ldots, J$, the link $g(\cdot)$ as in GLMs is a monotone increasing function that maps the interval $(0, 1)$ onto the real line $(-\infty, \infty)$, the γ_j are *cutpoints* with $\gamma_1 < \gamma_2 < \cdots < \gamma_{J-1}$, $\mathbf{x}_i^T$ is a vector of covariates, and $\boldsymbol{\beta}$ is a vector of regression parameters. The probit and logit functions are two common choices for g. The monotonicity of γ_j and the constancy of $\boldsymbol{\beta}$ across categories ensure that $Prob(y \leq j)$ is an increasing function in j. A natural ordering of categories is obtained by modeling cumulative category probabilities rather than individual probabilities. Although each response category has a corresponding cutpoint, the regression parameters $\boldsymbol{\beta}$ are constant across categories. If $\boldsymbol{\beta}$ could vary with j, $Prob(y \leq j)$ would have the undesirable property of not being an increasing function in j over all $\mathbf{x}_i$'s. The γ_j's may be thought of as intercepts, since there is no intercept term in the vector $\boldsymbol{\beta}$.

Once a link function is specified, the parameters $(\boldsymbol{\gamma}, \boldsymbol{\beta})$ [with $\boldsymbol{\gamma} = (\gamma_j)$] can be estimated by maximum likelihood, where the likelihood is based on (13.19). The significance of β_i can be tested by comparing its maximum likelihood estimate $\hat{\beta}_i$ with its standard error $\hat{\sigma}_i$ based on the Fisher information or observed information matrix. See (12.11)–(12.13). Alternatively, the likelihood ratio test in (12.14) can be used. The maximum likelihood estimation is implemented in the LOGISTIC procedure in SAS (SAS Institute, 1989).

One problem with the maximum-likelihood approach is that estimates for certain regression parameters or cutpoints may be infinite. For example, consider a binary response and a factor A with two levels. If all of the observations at the low level of A are in the first category and all of the observations at the high level of A are in the second category, then the estimated cell probabilities are 0 and 1. This corresponds to an estimate of $+\infty$ for the effect of A. In multifactor experiments, this problem is likely to occur, especially when the number of effects in the model is about equal to the number of runs. This estimability problem is similar to the one discussed in Section 12.3.1.

A Bayesian approach that assumes some prior knowledge of the regression parameters is a natural way to handle this estimability problem. Even when knowledge about the regression parameters is minimal, this approach can still be justified; one suspects that the parameters are large but the data are insufficient to distinguish between large and infinite values. On the probability scale such assumptions imply that probabilities of exactly 0 or 1 in (13.19) are unlikely, but values close to 0 and 1 are plausible. As in Sections 8.5 and 12.5, we will use a Gibbs sampler to obtain the posterior of $\boldsymbol{\beta}$.

13.7.1 The Gibbs Sampler for Ordinal Data

To implement the Gibbs sampler, we use model (13.19) with the probit link, i.e., $g = \Phi^{-1}$, the normal inverse cumulative distribution function, a multivariate normal prior for the regression parameters $\boldsymbol{\beta}$, and an ordered multivariate normal prior for the cutpoints $\boldsymbol{\gamma}$. The form of the full conditional distributions is simplified by the assumption of an underlying (but unobserved) continuous variable associated with the observed categorical response. This continuous variable is referred to as a *latent variable*. In many applications, a latent variable exists, but an ordered categorical version of it is observed due to cost, technical, or time constraints. Consider, for example, the window forming experiment in the exercises of Chapter 12. If there is no high-precision measuring device, the size of the window may not be measured precisely but be visually rated as "not open," "small," "adequate," or "large." The latent variable is then the unknown diameter of the window.

Inclusion of a latent variable in the model greatly simplifies the Gibbs sampling procedure, to be discussed below. Because the latent variable z has a normal regression relationship with the covariates $\mathbf{x}$, the procedure is very similar to that developed in Section 8.5 with the additional steps of draws for the latent variable and the cutpoints.

It is assumed that for each response y_i, there are covariates $\mathbf{x}_i$ and an unobserved variable z_i on the continuous scale. The correspondence between z_i and the J categories of the response is via the cutpoints $\gamma_0, \gamma_1, \ldots, \gamma_{J-1}, \gamma_J$, where $-\infty = \gamma_0 < \gamma_1 < \cdots < \gamma_{J-1} < \gamma_J = \infty$, that is,

$$\gamma_{j-1} \le z_i < \gamma_j \quad \text{if and only if} \quad y_i = j. \tag{13.20}$$

This means that z_i has the distribution function $g^{-1}(z_i - \mathbf{x}_i^T\boldsymbol{\beta})$, since, from (13.19),

$$Prob(z_i < \gamma_j) = Prob(y_i \le j) = g^{-1}(\gamma_j - \mathbf{x}_i^T\boldsymbol{\beta}). \tag{13.21}$$

For a probit link,

$$g = \Phi^{-1} \quad \text{and} \quad z_i \sim N(\mathbf{x}_i^T\boldsymbol{\beta}, 1), \tag{13.22}$$

where the variance is chosen to be 1 so that (13.21) is satisfied. In what follows, vector notation will be used, with $\mathbf{y}, \mathbf{z}, \boldsymbol{\gamma}$ as vectors of lengths N, N, and $J - 1$, respectively, and $\mathbf{X}$ (whose rows are $\mathbf{x}_i^T$) denoting the model matrix. From (13.22),

$$\mathbf{z} \sim MN(\mathbf{X}\boldsymbol{\beta}, \mathbf{I}), \qquad (13.23)$$

where MN stands for multivariate normal.

Conditional on the observed $\mathbf{y}$, full conditional distributions are now derived for arbitrary independent priors on $\boldsymbol{\beta}$ and $\boldsymbol{\gamma}$ as well as an arbitrary link function. First, consider the joint distribution of $\mathbf{y}, \boldsymbol{\gamma}, \boldsymbol{\beta}, \mathbf{z}$. This distribution is degenerate since knowledge of $(\mathbf{z}, \boldsymbol{\gamma})$ determines $\mathbf{y}$ exactly [see (13.20)]. Its density can be written as

$$f(\boldsymbol{\beta}, \boldsymbol{\gamma}, \mathbf{z}, \mathbf{y}) = f(\boldsymbol{\beta}, \boldsymbol{\gamma}, \mathbf{z}) \mathscr{I}(\mathbf{y}; \mathbf{z}, \boldsymbol{\gamma}),$$

where the function $\mathscr{I}$ is 1 when $\gamma_{y_i - 1} \leq z_i < \gamma_{y_i}$ for $i = 1, \ldots, N$, and 0 otherwise.

Using conditioning and assuming independence of priors for $\boldsymbol{\gamma}$ and $\boldsymbol{\beta}$ yields

$$\begin{aligned} f(\boldsymbol{\beta}, \boldsymbol{\gamma}, \mathbf{z}, \mathbf{y}) &= f(\mathbf{z} | \boldsymbol{\beta}, \boldsymbol{\gamma}) \pi(\boldsymbol{\beta}, \boldsymbol{\gamma}) \mathscr{I}(\mathbf{y}; \mathbf{z}, \boldsymbol{\gamma}) \\ &= f(\mathbf{z} | \boldsymbol{\beta}, \boldsymbol{\gamma}) \pi(\boldsymbol{\beta}) \pi(\boldsymbol{\gamma}) \mathscr{I}(\mathbf{y}; \mathbf{z}, \boldsymbol{\gamma}). \end{aligned}$$

Note that the distribution of $\mathbf{z}$ given $(\boldsymbol{\beta}, \boldsymbol{\gamma})$ does not depend on $\boldsymbol{\gamma}$. This follows, since the only dependence of $\mathbf{z}$ on $\boldsymbol{\gamma}$ is through $\mathbf{y}$, which is not being conditioned upon in the first term of the expression above. Thus,

$$f(\boldsymbol{\beta}, \boldsymbol{\gamma}, \mathbf{z}, \mathbf{y}) = f(\mathbf{z} | \boldsymbol{\beta}) \pi(\boldsymbol{\beta}) \pi(\boldsymbol{\gamma}) \mathscr{I}(\mathbf{y}; \mathbf{z}, \boldsymbol{\gamma})$$

for arbitrary densities $\pi(\boldsymbol{\beta}), \pi(\boldsymbol{\gamma}), f(\mathbf{z} | \boldsymbol{\beta})$. As mentioned in (13.22) and (13.23), the link determines $f(\mathbf{z} | \boldsymbol{\beta})$. The priors used here are

$$\boldsymbol{\beta} \sim MN(\mathbf{0}, \boldsymbol{\Sigma}_{\boldsymbol{\beta}}) \quad \text{and} \quad \boldsymbol{\gamma} \propto MN(\mathbf{0}, \mathbf{D}),$$

where $\mathbf{D}$ is a diagonal matrix and $\boldsymbol{\gamma}$ is restricted by $\gamma_1 < \gamma_2 < \cdots < \gamma_{J-1}$, so that the density of $\boldsymbol{\gamma}$ is that of an ordered multivariate normal, i.e., the conditional density of $MN(\mathbf{0}, \mathbf{D})$ constrained to the subspace $\gamma_1 < \gamma_2 < \cdots < \gamma_{J-1}$. Typically $\boldsymbol{\Sigma}_{\boldsymbol{\beta}}$ is also diagonal, with $\boldsymbol{\Sigma}_{\boldsymbol{\beta}} = \sigma_{\beta}^2 \mathbf{I}$ for covariates with comparable scales.

The practical reason for these choices is that all the full conditional distributions have forms that make it easier to perform Gibbs sampling. The normal priors allow considerable flexibility and will help illustrate the application of these techniques without diverting too much attention to computational issues. Because the experiments considered are primarily screening experiments, a zero prior mean for $\boldsymbol{\beta}$ seems reasonable. As mentioned previously, $\boldsymbol{\beta}$ does not contain an intercept term, meaning that a zero mean for all elements is justifiable. The priors for $\boldsymbol{\gamma}$ are also justifiable, because they allow considerable flexibility and include two important special cases: a uniform prior on the probabilities of the ordered categories (by taking the probit link and $\sigma_{\gamma_j} = 1, j = 1, \ldots, J-1$) and an uninformative prior on the continuous scale (by taking $\sigma_{\gamma_j} \to \infty, j = 1, \ldots, J-1$). By using a parameterization which puts all the intercepts in the same vector $\boldsymbol{\gamma}$ rather than having one in $\boldsymbol{\beta}$, independent priors on $\boldsymbol{\beta}$ and $\boldsymbol{\gamma}$ seem reasonable.

With these priors, the full conditional distributions can readily be expressed as follows:

$$f(\boldsymbol{\beta}|\boldsymbol{\gamma},\mathbf{z},\mathbf{y}) \propto f(\mathbf{z}|\boldsymbol{\beta})\pi(\boldsymbol{\beta})$$

$$= MN\Big(\big(\boldsymbol{\Sigma}_{\boldsymbol{\beta}}^{-1} + \mathbf{X}^T\mathbf{X}\big)^{-1}\mathbf{X}^T\mathbf{z}, \big(\boldsymbol{\Sigma}_{\boldsymbol{\beta}}^{-1} + \mathbf{X}^T\mathbf{X}\big)^{-1}\Big), \qquad (13.24)$$

$$f(\boldsymbol{\gamma}|\boldsymbol{\beta},\mathbf{z},\mathbf{y}) \propto \pi(\boldsymbol{\gamma})\mathscr{F}(\mathbf{y};\mathbf{z},\boldsymbol{\gamma})$$

$$= I(\gamma_1 < \cdots < \gamma_{J-1})\prod_{j=1}^{J-1} N\big(0,\sigma_{\gamma_j}^2\big)\times\mathscr{F}(\mathbf{y};\mathbf{z},\boldsymbol{\gamma})$$

$$= \prod_{j=1}^{J-1} N\big(0,\sigma_{\gamma_j}^2\big)\times I\Big(\max_i\{z_i|y_i=j\} < \gamma_j \le \min_i\{z_i|y_i=j+1\}\Big),$$

$$(13.25)$$

$$f(\mathbf{z}|\boldsymbol{\beta},\boldsymbol{\gamma},\mathbf{y}) \propto f(\mathbf{z}|\boldsymbol{\beta})\mathscr{F}(\mathbf{y};\mathbf{z},\boldsymbol{\gamma})$$

$$= N(\mathbf{X}^T\boldsymbol{\beta},\mathbf{I})\mathscr{F}(\mathbf{y};\mathbf{z},\boldsymbol{\gamma})$$

$$= \prod_{i=1}^{N} N(\mathbf{x}_i^T\boldsymbol{\beta},1)I(\gamma_{y_i-1} \le z_i < \gamma_{y_i}). \qquad (13.26)$$

The generic indicator function I is 1 whenever its argument is true and 0 otherwise. The identity in (13.24) follows from a standard formula in Bayesian statistics (Broemeling, 1985, p. 6) with $\mathbf{z}|\boldsymbol{\beta} \sim MN(\mathbf{X}\boldsymbol{\beta},\mathbf{I})$ and $\boldsymbol{\beta} \sim MN(\mathbf{0},\boldsymbol{\Sigma}_{\boldsymbol{\beta}})$. The indicator functions in (13.25) may be combined into a single indicator, which is 1 only when γ_j lies between z_i elements corresponding to $y_i=j$ and $y_i=j+1$. This is based on the assumption that there is at least one observation in each outcome category; otherwise neighboring intervals are merged until all outcome categories contain at least one observation. The truncated distribution for $\mathbf{z}$ in (13.26) is similar with each element bounded by a pair of cutpoints. Notice that in both (13.25) and (13.26), the elements of $\mathbf{z}$ and $\boldsymbol{\gamma}$ have independent full conditional distributions, implying that the elements of the vectors may be sampled one at a time.

Thus the Gibbs sampler [see (8.17)] will sample $\mathbf{z}$ and $\boldsymbol{\gamma}$ one element at a time and $\boldsymbol{\beta}$ all at once. The order of sampling is $\mathbf{z},\boldsymbol{\gamma},\boldsymbol{\beta}$, and all the distributions are easy to sample. As in the discussions in Chapters 8 and 12, at convergence the Gibbs sampler is sampling the desired posterior distribution.

13.8   ANALYSIS OF FOAM MOLDING EXPERIMENT

The control array in Table 13.6 used a saturated 2^{7-4} design, which allows only the control main effects to be estimated. Because the noise array is a 2^2 design, H, I, and HI are estimable. According to Theorem 10.1, all the control-by-noise interactions are also estimable. This consideration suggests an initial model consisting of the main effects for factors $A-I$, HI and the control-by-noise interactions between factors $A-G$ and $H-I$. In total, there are 24 effects plus two cutpoints which define the three ordered categories. The model was fitted with diffuse priors for γ ($\sigma_{\gamma_j} = \infty$) and a $N(0, I)$ prior for β (i.e., $\sigma_\beta = 1$).

The first analysis task is to identify the factors with the largest effects using the posteriors generated by the Bayesian analysis. The 0.005, 0.025, 0.5, 0.975, and 0.995 posterior quantiles of the regression parameters as well as the cutpoints γ_1 and γ_2 are given in Table 13.7. These posteriors can be used to identify terms that can be dropped from the model. In this case, D, BH, CH, DH, EH, FH, GH, AI, BI, CI, DI, FI, and GI can be dropped, because zero is between the 0.005 quantile and the 0.995 quantile for each of these terms. This leads to the final model with terms A, B, C, E, F, G, H, I, HI, AH, and EI. Because the posteriors for the final model are qualitatively very similar to those given in Table 13.7, the results will not be given here.

Using the important effects identified above, optimization can now be considered where the goal is to find settings which yield a high probability content in certain outcome categories. The vector of the probabilities of the ordered categories or outcome proportions

$$\mathbf{p} = (p_1, \ldots, p_J),$$

where $p_j = Prob(y = j)$, can also be thought of as a parameter with a posterior distribution since it is a function of β, γ, and $\mathbf{X}$. To emphasize the distinction between the parameter $\mathbf{p}$ and its posterior, the terminology *proportion* $\mathbf{p}$ of outcomes and the posterior distribution of this proportion will be used. Because the posterior of $\mathbf{p}$ depends on the covariates $\mathbf{x}$ (e.g., main effects and interactions), optimal factor settings that have a "good" posterior for the proportions can be identified. "Goodness" is defined to be one with the most posterior distribution in the ideal category and as little variation as possible.

The robust design aspect of this experiment adds a further challenge in the optimization process. First, the response is modeled as a function of both

Table 13.7 Marginal Posterior Quantiles, Foam Molding Experiment

Effect	0.005	0.025	Quantiles 0.500	0.925	0.995
A	−1.103	−0.995	−0.636	−0.397	−0.334
B	0.477	0.545	0.795	1.152	1.245
C	0.191	0.256	0.505	0.865	0.964
D	−0.602	−0.491	−0.132	0.134	0.191
E	0.132	0.198	0.458	0.810	0.927
F	−0.932	−0.828	−0.470	−0.213	−0.148
G	0.304	0.376	0.640	1.007	1.108
H	−0.824	−0.704	−0.367	−0.147	−0.080
I	0.052	0.114	0.329	0.651	0.729
HI	0.004	0.048	0.190	0.335	0.384
AH	0.068	0.128	0.349	0.695	0.807
BH	−0.562	−0.458	−0.116	0.100	0.165
CH	−0.600	−0.495	−0.157	0.063	0.119
DH	−0.391	−0.328	−0.106	0.232	0.348
EH	−0.219	−0.163	0.066	0.403	0.519
FH	−0.397	−0.285	0.054	0.276	0.334
GH	−0.455	−0.350	−0.014	0.210	0.259
AI	−0.418	−0.322	0.007	0.218	0.277
BI	−0.331	−0.282	−0.070	0.256	0.348
CI	−0.201	−0.138	0.069	0.395	0.489
DI	−0.540	−0.461	−0.140	0.073	0.124
EI	−0.728	−0.646	−0.316	−0.110	−0.060
FI	−0.190	−0.126	0.078	0.408	0.502
GI	−0.224	−0.164	0.045	0.375	0.465
γ_1	−2.415	−2.363	−1.956	−1.649	−1.568
γ_2	−0.341	−0.213	0.094	0.331	0.442

the control and noise factors. Then the noise factors are assumed to follow a specific distribution. This allows control settings that desensitize the response to the noise variation to be identified. For the Bayesian model, the distribution of $\mathbf{p}$ has two different components—the posterior for $(\boldsymbol{\beta}, \boldsymbol{\gamma})$ and the distribution of the noise factors. The calculation of the distribution of proportions is simple, and the additional distribution associated with the noise factors is easily integrated into this calculation. First, values of the control factors are set at fixed levels indexed by l. For each $(\boldsymbol{\beta}_l, \boldsymbol{\gamma}_l)$ in the posterior sample, the noise factors are drawn from their distribution, and $\mathbf{p}$ is calculated using (13.19). Repeating this for the entire sample yields a sample from the distribution of proportions for a given setting of the control factors.

In the foam experiment, H and I are the noise factors, so that settings of the control factors A, B, C, E, F, G which maximize *Prob*(good) need to be

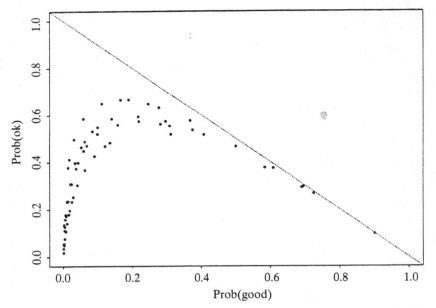

Figure 13.2. Posterior Proportion of "good" and "ok" Parts for Control Factor Settings, Foam Molding Experiment.

found. For illustrative purposes, $(H, I) \sim N(0, (1/3)^2 I)$ is assumed, i.e., the "+" and "−" levels used in the experiment represent plus and minus three standard deviations of the noise distributions. At each $(\boldsymbol{\beta}_l, \boldsymbol{\gamma}_l)$, H_l and I_l are drawn and *Prob*(good) is calculated. The optimization is straightforward by considering all combinations of the two levels for the six control factors. Using $E(\mathbf{p})$, promising settings can be identified, where the expectation is taken over $\boldsymbol{\beta}$, $\boldsymbol{\gamma}$, H, and I. Note that for given $(\boldsymbol{\beta}_l, \boldsymbol{\gamma}_l)$, the expectation of *Prob*(good) over H and I is the *Prob*(good) for a given control factor setting. Then the expectation over $(\boldsymbol{\beta}_l, \boldsymbol{\gamma}_l)$ yields the posterior *Prob*(good). The expected proportions for the $2^6 = 64$ different control settings are plotted in Figure 13.2. Only two components of $E(\mathbf{p})$ are plotted, as the three probabilities sum to 1. Thus, settings which have the most mass on "good" and most of the remaining mass on "ok" are desirable—namely those near the lower right of the plot and near the diagonal line *Prob*(good) + *Prob*(ok) = 1. In this case, there appears to be a clear winner corresponding to the lower rightmost point in Figure 13.2; this setting has levels $(A, B, C, E, F, G) = (+, -, -, +, -, -)$ with posterior proportions of 90.1% good parts and 9.8% ok parts. Of course the promise of this setting needs to be confirmed by a follow-up experiment.

13.9 SCORING: A SIMPLE METHOD FOR ANALYZING ORDINAL DATA

A simple method for analyzing ordered categorical data is to score the categories. Scores can be assigned or based on data. Assigned scores can reflect differences between categories through the distances between the assigned scores. For example, assigning the score j to the jth category implies equal distance between neighboring categories. An example of data-based scores are mid-ranks. Let f_j, $j = 1, \ldots, J$, be the observed frequencies of the categories over the entire experiment. Then, the *mid-rank score* m_j for the jth category is defined as

$$m_j = \sum_{i=1}^{j-1} f_i + f_j/2. \qquad (13.27)$$

Suppose that t_j is the score assigned to the jth category. Naturally t_j should be *monotonic* in j. For the ith run, compute the *scored response* R_i:

$$R_i = \sum_{j=1}^{J} t_j f_{ij}, \qquad (13.28)$$

where f_{ij} is the frequency of the jth category for the ith run. Then, the scored responses can be analyzed as if they were normally distributed from an unreplicated experiment. The scores should be meaningful and need to be strictly increasing or strictly decreasing. There are, however, problems with this method when a middle category is the ideal and scores are chosen to reflect a cost which increases in both directions as the category moves away from the ideal.

Consider a score-based analysis of the foam molding experiment using mid-ranks and equally spaced scores. The mid-rank scores for the three ordered categories are $(19, 116, 257)$ and the equally spaced scores are chosen to be $(1, 2, 3)$. The half-normal plots of factorial effects for these two sets of scores appear in Figures 13.3(a) and (b), respectively. Note that none of the unsigned estimated effects falls off the line in either plot so that no effect would be declared significant. Compare these results with the Bayesian analysis in the preceding section. There, 11 effects were identified as significant. In terms of the relative rankings of the effects, the top 11 effects in both scoring analyses are the same and differ by one effect with the Bayesian analysis. The scoring analyses identify B, G, and A as the top 3 effects, as does the Bayesian analysis, but then there are differences in the ordering of the next 8 effects.

The scoring method is attractively simple but does not have the same inferential validity as the MLE or Bayesian methods based on the logit model. The MLE or Bayesian methods are preferred when they are readily available.

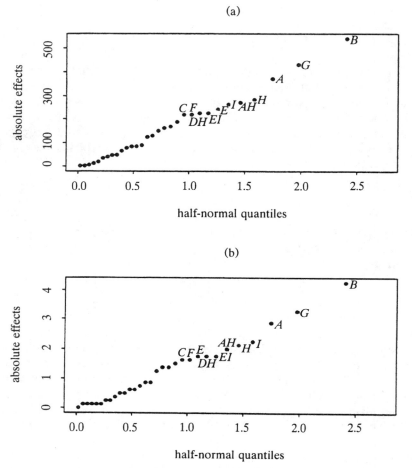

Figure 13.3. Half-Normal Plots with Scores (a) Mid-Ranks and (b) Equally Spaced, Foam Molding Experiment.

13.10 PRACTICAL SUMMARY

1. Regression data with nonnormal response like binomial, Poisson, gamma, and inverse Gaussian can be described by a generalized linear model (GLM). Advantages of using a GLM over the method of transforming the response are discussed in Section 13.2.3.

2. A GLM has two components: a random component for the distribution of the response and a systematic component based on a link function that defines the scale on which the systematic effects are assumed to be additive. A summary of distributions and link functions is given in Table 13.3.

3. The regression parameters in a GLM can be estimated by the maximum likelihood estimate (MLE) via the iteratively reweighted least squares algorithm (13.11)–(13.14). The deviance in (13.15) plays the role of the residual sum of squares in regression. Effect significance can be tested by the F statistic in (13.16). The deviance residuals in (13.17) are extensions of residuals in regression and the mean deviance in (13.18) can be used to test for over-dispersion.

4. Ordinal (also called ordered categorical) data can be described by the logit model in (13.19). The model parameters can be estimated by the MLE.

5. As with censored data in Chapter 12, the MLE for a GLM or a logit model may be infinite when the effect for a covariate is large and the opportunity for improvement is great. A natural way to circumvent this estimability problem is to use the Bayesian approach in Section 13.7.1, which is based on Gibbs sampling.

6. An alternative method for analyzing ordered categorical data is to assign scores to the response categories, compute a scored response for each control factor setting, and analyze the scored responses as if they were normally distributed. The scoring method is attractively simple but does not have the same inferential validity as the MLE or Bayesian methods.

EXERCISES

1. (a) For the wave soldering experiment, identify all the aliased two-factor interactions, which take up seven degrees of freedom, as well as the aliases of *ABC*.

   (b) In the analysis of Section 13.4, *AC* was identified as significant. Because it is aliased with *DF* and *EG*, it may be argued that, based on the weak effect heredity principle, the effect significance could be attributed to either *AC* or *EG*. If *EG* instead of *AC* were included in the final model, how would it affect the choice of recommended factor settings?

   (c) If additional runs need to be conducted to resolve the ambiguity between *AC* and *EG*, discuss different options available based on the techniques in Section 4.4.

2. During the stamping process in manufacturing windshield moldings, debris carried into the die appears as dents in the product (referred to as slugs) and results in imperfect parts. An experiment was conducted to improve the slugging condition. Four factors, each at two levels, were studied: (*A*) poly-film thickness (0.0025, 0.0017), (*B*) oil mixture (1:30, 1:10), (*C*) gloves (cotton, nylon), and (*D*) metal blanks (dry, oily),

Table 13.8 Design Matrix and Response Data (Number of Good Parts Out of 1000), Windshield Molding Experiment

Run	Factor				Good Parts
	A	B	C	D	
1	−	−	−	−	341
2	−	−	+	+	828
3	−	+	−	−	345
4	−	+	+	+	642
5	+	−	−	+	908
6	+	−	+	−	970
7	+	+	−	+	961
8	+	+	+	−	969

where the values of the two levels are given in parentheses. The experiment used a 2^{4-1} design with $\mathbf{I} = -ACD$ as given in Table 13.8. The response y is the number of parts (out of 1000) classified as good.

(a) Could a better design have been chosen?

(b) Identify the distribution of the response as a member of GLMs. Which link function would you choose?

(c) Use a GLM to analyze the data and identify significant effects and optimal factor settings. Your analysis should include MLEs, analysis of deviance, examination of residuals, and study of over-dispersion.

(d) Since the total number of parts per run in the experiment is very large, the normal approximation to binomial is accurate, i.e., $\hat{p} \sim N(p, p(1-p)/n)$, where $\hat{p}$ is the percent of good parts and n is the total number of parts. Show that $\arcsin(2\hat{p} - 1) \sim N(\arcsin(2p - 1), n^{-1})$, that is, the arcsine transformation is a variance stabilizing transformation.

(e) Because $\arcsin(2\hat{p} - 1)$ has a constant variance, it may be analyzed by standard methods based on the normal assumption. Explain why it should not be used as a routine practice and particularly for the given data. (*Hint:* What are the range of the arcsine transformation and its lack of sensitivity to $\hat{p}$ values near 0 or 1?)

3. In the ball bearing experiment in the exercises of Chapter 12, an analysis using a lognormal regression model was suggested.

(a) Reanalyze the experiment with GLM by treating the failure times as coming from a gamma distribution and using the reciprocal link.

(b) Compare the findings in (a) and those based on lognormal regression in the exercises of Chapter 12. Is there any real advantage in using the GLM approach for failure time data?

(c) Another option is to analyze the failure time data using standard regression techniques. Compare the model chosen by this method and the model identified in (a). Which one is more parsimonious?

[*Note*: An analysis of the windshield molding experiment and the ball bearing experiment can be found in Hamada and Nelder (1997).]

4. For the wave soldering experiment, compute the mean deviance for each run and analyze it as a function of the control factors. Do any factor effects appear to be significant? How would this affect the choice of optimal factor settings in terms of minimizing the defect counts and its variation?

5. (a) If the $-$ and $+$ levels of H and I in the foam molding experiment are two standard deviations from the center of the noise distribution, how would this information change the choice of the variance in the noise variation?

   (b) Repeat the work that led to Figure 13.2 by using $\frac{1}{4}$ for the variance of H and I. Are the conclusions different?

6. Instead of using the plot in Figure 13.2 to judge the goodness of factor settings, the following two quantitative measures may be considered.

   (a) Compute the Euclidean distance between $\mathbf{p}$ and $\mathbf{t} = (1, 0, 0)$. Rank the control factor settings according to the distribution of this distance over the posterior and the noise distribution.

   (b) Score the categories and compute an overall score. Rank the control factor settings according to the distribution of the overall score over the posterior and the noise distribution. Discuss the choice of scores for the foam molding experiment, including those in Section 13.9.

   (c) Compare the results from (a) and (b) with those in Figure 13.2. Is there any substantial difference?

7. Prove the identity in (13.24).

8. In an experiment on an arc welding process performed by the National Railway Corporation of Japan and reported in Taguchi and Wu (1980), welding quality was assessed in terms of the workability of a welded section between two steel plates and by x-ray. Workability is the degree of difficulty in welding the two steel plates together, which is classified as easy, normal, or difficult (denoted by E, N, and D). The x-ray response is classified as good, normal, or bad. The experimenters were initially interested in nine factors (A–H, J) and four two-factor interactions (AG, AH, AC, GH). The experiment used a 2^{9-5} design in which one workability and four x-ray observations were taken as displayed in Table 13.9.

Table 13.9   Design Matrix and Response Data, Arc Welding Experiment

				Factor							X-Ray		
Run	A	B	C	D	E	F	G	H	J	Workability	Good	Normal	Bad
1	−	−	−	−	−	−	−	−	−	N	3	1	0
2	−	−	+	+	+	+	−	−	+	N	3	1	0
3	−	+	+	−	−	−	−	+	−	N	4	0	0
4	−	+	−	+	+	+	−	+	+	E	2	2	0
5	−	+	+	−	−	+	+	−	+	E	2	2	0
6	−	+	−	+	+	−	+	−	−	D	4	0	0
7	−	−	−	−	−	+	+	+	+	E	2	2	0
8	−	−	+	+	+	−	+	+	−	D	3	1	0
9	+	+	+	−	+	−	−	−	+	N	2	1	1
10	+	+	−	+	−	+	−	−	−	N	3	1	0
11	+	−	−	−	+	−	−	+	+	N	3	1	0
12	+	−	+	+	−	+	−	+	−	N	4	0	0
13	+	−	−	−	+	+	+	−	−	E	0	3	1
14	+	−	+	+	−	−	+	−	+	D	3	0	1
15	+	+	+	−	+	+	+	+	−	N	1	3	0
16	+	+	−	+	−	−	+	+	+	N	4	0	0

(a) Analyze the x-ray data by scoring using the assigned scores $(1, 2, 3)$ and the mid-ranks $(21.5, 52.0, 62.5)$ (based on the frequencies $43, 18, 3$). Do the two scoring methods lead to different results? Are any effects identified as significant by either scoring method?

(b) Consider the use of the more extreme assigned scores $(1, 2, 6)$ and $(1, 2, 10)$, which give far more weight to the "bad" category. Do they lead to different results? Comment on the sensitivity of results to the choice of scores.

(c) Use the Bayesian method in Section 13.7.1 to analyze the x-ray data. Are the results different from those based on the method of scoring in (a) and (b)?

(d) Explain why the scoring method is not appropriate for the workability data. Analyze the workability data by using the Bayesian method in Section 13.7.1.

[*Note:* For various analyses of these data, see Hamada and Wu (1990) and its discussion.]

9. Consider an experiment (Anderson and Wu, 1996) involving the balancing of engine flywheels. A rotating part like a flywheel needs to be precisely balanced to prevent excessive vibration. The response was the location on the circumference of the part where a corrective adjustment would be required to balance the part. A 2^4 design was run with 10 observations at each set of factor combinations. The four factors thought

to influence the dispersion of the imbalance are as follows:

(A)—The location of a buttweld to the flywheel, either fixed (F) or random (R). In current production, the selection of the location for joining two pieces was determined randomly.

(B)—Flywheel radius grade, either low (L) or high (H). Currently low-grade flywheels were used.

(C)—Flywheel thickness grade, either L or H. Currently, low-grade thickness was used.

(D)—Size of counterweight attached at $0°$, either L or H. The current setting was at the low level.

In addition, a fifth factor, ring gear imbalance (E), which is very difficult and expensive to control, was observed at three levels (L, M, H). The data are provided in angular form (measured in degrees) in Table 13.10 with the level of factor E in parentheses below. A dispersion measure for directional data can be defined as follows (for more details see Anderson and Wu, 1996). For a sample of n directions denoted by $(\theta_1, \ldots, \theta_n)$, the n vectors for the directions are $(\cos \theta_i, \sin \theta_i)$, $i = 1, \ldots, n$. The sum of these n vectors is called the resultant vector, which has length $R = (c^2 + s^2)^{1/2}$, with $c = \Sigma_i \cos \theta_i$ and $s = \Sigma_i \sin \theta_i$. Define the *circular variance* as $S_0 = 1 - R/n$. The values of S_0 range from 0 to 1. For $S_0 = 0$, all vectors are concentrated on one direction, i.e., there is no variation. For $S_0 = 1$, the vectors are uniformly distributed on the circumference of the circle. It is common to use $\ln S_0$ (or other power transformation of S_0) as the response for regression modeling.

(a) For the flywheel data, compute the S_0 value for each of the 16 control settings in Table 13.10. (There are 10 replicates per setting.) Analyze the 16 $\ln S_0$ values as a function of the four control factors.

(b) Based on the fitted model found in (a), find optimal factor settings in terms of minimizing the dispersion of the balancing directions.

10. Reanalyze the foam molding experiment by using the maximum likelihood-method. Is the chosen model different from the one given in Section 13.8?

11. Grove and Davis (1992, Section 3.3) described an experiment on an injection molding process that made plastic lids for automotive glove compartments. A quality problem is flow marks which mar the surface of the plastic. The experiment studied eight factors with two levels: (A) injection stroke (40 mm, 70 mm); (B) melt temperature (200°C, 250°C); (C) mold temperature (40°C, 80°C); (D) change-over point (13 mm, 23 mm); (E) nozzle aperture (4 mm, 6 mm); (F) mold clamping force (480 tons, 630 tons); (G) first-stage injection speed (15 mm/s, 25 mm/s); (H) second-stage injection speed (20 mm/s, 30 mm/s). The experiment used

Table 13.10 Design Matrix and Response Data, Flywheel Balancing Experiment

Run	A	B	C	D	Response in degrees (with levels of E)									
1	R	L	L	L	133 (L)	175 (M)	178 (M)	178 (M)	153 (M)	190 (M)	221 (M)	177 (M)	281 (H)	190 (H)
2	R	L	L	H	139 (L)	61 (L)	109 (M)	187 (M)	74 (M)	351 (M)	309 (M)	236 (M)	69 (H)	320 (H)
3	R	L	H	L	111 (L)	122 (L)	105 (M)	49 (M)	189 (M)	188 (M)	177 (M)	151 (M)	62 (H)	329 (H)
4	R	L	H	H	170 (L)	19 (M)	337 (M)	171 (M)	114 (M)	341 (M)	10 (M)	266 (H)	201 (H)	162 (H)
5	R	H	L	L	127 (L)	215 (L)	125 (M)	188 (M)	187 (M)	175 (M)	162 (M)	172 (M)	169 (H)	82 (H)
6	R	H	L	H	150 (L)	84 (L)	113 (M)	318 (M)	84 (M)	353 (M)	301 (M)	12 (M)	82 (H)	351 (H)
7	R	H	H	L	152 (L)	164 (L)	180 (M)	187 (M)	159 (M)	149 (M)	127 (M)	148 (M)	175 (H)	201 (H)
8	R	H	H	H	184 (L)	128 (L)	177 (L)	186 (M)	163 (M)	178 (M)	196 (M)	155 (M)	150 (H)	120 (H)
9	F	L	L	L	154 (L)	200 (L)	147 (M)	133 (M)	171 (M)	318 (M)	100 (M)	108 (M)	86 (H)	73 (H)
10	F	L	L	H	165 (L)	31 (M)	51 (M)	314 (M)	84 (M)	267 (M)	135 (M)	318 (H)	14 (H)	198 (H)
11	F	L	H	L	345 (L)	43 (L)	4 (M)	295 (M)	75 (M)	138 (M)	149 (M)	141 (M)	198 (H)	175 (H)
12	F	L	H	H	153 (L)	194 (L)	136 (M)	144 (M)	206 (M)	151 (M)	202 (M)	104 (H)	188 (H)	207 (H)
13	F	H	L	L	140 (L)	170 (M)	62 (M)	109 (M)	127 (M)	132 (M)	116 (M)	94 (M)	183 (M)	134 (H)
14	F	H	L	H	340 (L)	111 (L)	128 (M)	327 (M)	81 (M)	301 (M)	3 (M)	335 (M)	215 (H)	334 (H)
15	F	H	H	L	160 (L)	152 (M)	187 (M)	158 (M)	143 (M)	91 (M)	200 (M)	143 (M)	84 (H)	191 (H)
16	F	H	H	H	171 (L)	156 (L)	171 (M)	195 (M)	159 (M)	153 (M)	188 (M)	125 (M)	107 (H)	98 (H)

a 2^{8-4} design. The response is the count of flow marks on 50 lids. The design matrix and the count data are given in Table 13.11.

(a) How was the design constructed?

(b) Analyze the experimental data using a Poisson regression model.

(c) Compute the average flow mark count by dividing by 50. Do the results change by analyzing the average data by a normal regression model.

Table 13.11 Design Matrix and Flow Mark Count Data, Glove Compartment Experiment

Run	A	B	C	D	E	F	G	H	Flow Mark Count
1	−	−	−	−	−	−	−	−	310
2	+	−	−	+	−	+	+	−	260
3	−	+	−	+	−	+	−	+	215
4	+	+	−	−	−	−	+	+	150
5	−	−	+	+	−	−	+	+	265
6	+	−	+	−	−	+	−	+	200
7	−	+	+	−	−	+	+	−	0
8	+	+	+	+	−	−	−	−	95
9	−	−	−	−	+	+	+	+	315
10	+	−	−	+	+	−	−	+	290
11	−	+	−	+	+	−	+	−	300
12	+	+	−	−	+	+	−	−	150
13	−	−	+	+	+	+	−	−	165
14	+	−	+	−	+	−	+	−	290
15	−	+	+	−	+	−	−	+	0
16	+	+	+	+	+	+	+	+	0

12. Taguchi (1986, Chapter 5) recounted the now famous Ina tile experiment to reduce the number of off-grade tiles produced. The experiment studied seven factors with two levels: (A) lime additive content (5%, 1%); (B) additive granularity (coarse, fine); (C) agalmatolite content (43%, 53%); (D) agalmatolite type (current, cheaper); (E) charge quantity (1300 kg, 1200 kg); (F) waste return quantity (0%, 4%); (G) felspar content (0%, 5%); The response was the number of off-grade tiles out of a sample of 100. The 2^{7-4} design matrix and response data are given in Table 13.12.

   (a) Analyze the experimental data using a logistic regression model.

   (b) Since none of the off-grade counts are close to zero or 100, a simpler analysis method would be to treat the data as nearly normal and use standard regression analysis. Compare the results with those in (a).

13. Six weed control treatments were studied to control poppies, a weed in the cereal oats (Bartlett, 1936). The experiment used a randomized block design with the response being the number of poppy plants in a $3\frac{3}{4}$ ft^2 area as provided in Table 13.13. The treatment is given in the parentheses. Treatment 1 was the control treatment.

   (a) Analyze the experiment using a Poisson regression model. Identify any pairs of treatments that are significantly different.

**Table 13.12   Design Matrix and Off-Grade Tile Count (Out of 100),
Tile Experiment**

Run	A	B	C	D	E	F	G	Off-Grade Tiles
				Factor				
1	−	−	−	−	−	−	−	16
2	−	−	−	+	+	+	+	17
3	−	+	+	−	−	+	+	12
4	−	+	+	+	+	−	−	6
5	+	−	+	−	+	−	+	6
6	+	−	+	+	−	+	−	68
7	+	+	−	−	+	+	−	42
8	+	+	−	+	−	−	+	26

**Table 13.13   Poppy Counts (plants per $3\frac{3}{4}$ ft^2, treatment in parentheses),
Weed Infestation Experiment**

Block	Counts					
1	438(1)	17(4)	538(2)	18(5)	77(3)	115(6)
2	61(3)	422(2)	57(6)	442(1)	26(5)	31(4)
3	77(5)	157(3)	87(4)	100(6)	377(2)	319(1)
4	315(2)	380(1)	20(5)	52(3)	16(4)	45(6)

    (b) An alternative is to take the square root of each count and analyze
the transformed data using standard regression analysis. Identify any
pairs of treatments that are significantly different.

    (c) Are the results in (a) and (b) different?

14. Five treatments for cockchafer larvae were studied using a randomized
block design (Bartlett, 1936) as given in Table 13.14. The response is the
number of larvae classified into two age classes for the five treatments.
There were eight blocks. In the first two blocks, counts were carried out
over a whole plot. In the remaining blocks, counts were carried out over
one-quarter of a plot. Perform an analysis of the experiment using a
Poisson regression model for each of the two age classes. Identify any
pairs of treatments that are significantly different. Care must be taken to
address the difference in the count data between the first two blocks and
the remaining blocks.

15. An experiment was conducted to study the germination of wheat (Bartlett, 1936). Six treatments were studied in four randomized blocks. The
response is the number of seeds failing to germinate out of 50 seeds, as
given in Table 13.15.

Table 13.14 Larvae Counts, Larvae Control Experiment

| | Age 1 | | | | | Age 2 | | | | |
| | Treatment | | | | | Treatment | | | | |
Block	1	2	3	4	5	1	2	3	4	5
1	13	16	13	20	16	28	12	40	31	22
2	29	12	23	15	17	61	49	48	44	45
3	5	4	4	1	2	7	2	4	5	2
4	5	12	1	5	3	14	5	14	9	8
5	0	2	2	2	0	3	3	2	7	0
6	1	1	1	3	5	7	6	7	7	4
7	1	3	1	0	1	10	5	8	3	6
8	4	4	7	3	1	13	11	10	12	8

Table 13.15 Unsuccessful Germination Counts (Out of 50 Seeds), Wheat Experiment

| | Treatment | | | | | | |
Block	1	2	3	4	5	6	7
1	10	11	8	9	7	6	9
2	8	10	3	7	9	3	11
3	5	11	2	8	10	7	11
4	1	6	4	13	7	10	10

(a) Analyze the experiment using a logistic regression model. Identify any pairs of treatments that are significantly different.

(b) Would it be appropriate to use standard regression analysis by treating the data as nearly normal?

16. The window size data in Table 13.16 are classified into five ordered categories and scored 1–5; the third category is the target. In the original window size data (see Table 12.18 of the exercises in Chapter 12) the window was either open or not and, if the window was open, the size of the window was measured. The five categories correspond to "window not open," $(0, 2.25)$, $[2.25, 2.75)$, $[2.75, 3.25]$, $(3.25, \infty)$.

(a) Analyze the ordinal data in Table 13.16 by using the Bayesian method and the scoring method with scores $(1, 2, 3, 4, 5)$. How different are the results based on the two methods?

(b) Compare the findings in (a) with those based on the censored data for the same experiment in Table 12.18. Noting that the ordinal data are obtained from the censored data in Table 12.18, argue that the ordinal data analysis here is less informative than the censored data analysis in the exercises of Chapter 12.

(c) Repeat (a) with the maximum likelihood method. Does its computation encounter any numerical problem?

Table 13.16 Window Size Data, Window Forming Experiment

Run	Window Size									
1	1	1	1	1	1	1	1	1	1	1
2	2	4	4	5	5	2	2	3	3	3
3	4	4	4	4	4	1	4	4	4	4
4	1	1	1	1	1	1	1	1	1	1
5	1	1	1	1	1					
6	2	2	2	3	3	1	1	1	1	1
7	1	1	1	1	1	1	1	1	1	1
8	1	1	1	1	1	4	4	4	4	4
9	4	4	4	4	4	2	3	3	3	3
10	2	2	2	2	2	1	1	3	3	3
11	4	4	4	4	4	1	2	3	3	4
12	1	3	4	4	4	5	5	5	5	5
13	3	3	3	4	4	1	1	1	1	1
14	1	2	2	2	3	1	1	1	1	1
15	1	1	1	1	1					
16	1	1	1	1	1	1	1	1	1	1
17	4	4	4	5	5	3	3	3	3	5
18	5	5	5	5	5					

REFERENCES

Aitkin, M. and Clayton, D.G. (1980), "The Fitting of Exponential, Weibull and Extreme Value Distributions to Complex Censored Survival Data Using GLIM," *Applied Statistics*, 29, 156–163.

Aitkin, M., Anderson, D., Francis, B., and Hinde, J. (1989), *Statistical Modelling in GLIM*, London: Oxford University Press.

Anderson, C.M. and Wu, C.F.J. (1996), "Dispersion Measures and Analysis for Factorial Directional Data with Replicates," *Applied Statistics*, 45, 47–61.

Bartlett, M.S. (1936), "The Square Root Transformation in Analysis of Variance," *Journal of the Royal Statistical Society, Series B*, 1, 68–78.

Broemeling, L.D. (1985), *Bayesian Analysis of Linear Models*, New York: Marcel Dekker.

Chipman, H. and Hamada, M. (1996), "A Bayesian Approach for Analyzing Ordinal Data from Industrial Experiments," *Technometrics*, 38, 1–10.

Condra, L.W. (1993), *Reliability Improvement with Design of Experiments*, New York: Marcel Dekker.

Dellaportas, P. and Smith, A.F.M. (1993), "Bayesian Inference for Generalized Linear and Proportional Hazards Models via Gibbs Sampling," *Applied Statistics*, 42, 443–459.

Francis, B., Green, M., and Payne, C. (Eds.) (1993), *The GLIM System Release 4 Manual*, Oxford: Clarendon Press.

GENSTAT 5 Committee (1987), *GENSTAT 5 Reference Manual*, London: Oxford University Press.

GLIM 82: *Proceedings of the Conference on Generalized Linear Models* (1982), New York: Springer-Verlag.

Grego, J.M. (1993), "Generalized Linear Models and Process Variation," *Journal of Quality Technology*, 25, 288–295.

Grove, D.M. and Davis, T.P. (1992), *Engineering Quality and Experimental Design*, Essex: Longman Scientific and Technical.

Hamada, M. and Nelder, J.A. (1997), "Generalized Linear Models for Quality-Improvement Experiments," *Journal of Quality Technology*, 29, 292–304.

Hamada, M. and Wu, C. F. J. (1990), "A Critical Look at Accumulation Analysis and Related Methods (with discussion)," *Technometrics*, 32, 119–162.

Hilbe, J.M. (1994), "Generalized Linear Models," *American Statistician*, 48, 255–265.

McCullagh, P. (1980), "Regression Models for Ordinal Data," *Journal of the Royal Statistical Society, Series B*, 42, 109–142.

McCullagh, P. and Nelder, J.A. (1989), *Generalized Linear Models*, 2nd ed., London: Chapman and Hall.

MRC Biostatistics Unit (1995), BUGS 0.5 Bayesian Inference Using Gibbs Sampling Manual.

Nelder, J.A. (1974), *The GLIM Manual*, Oxford: Numerical Algorithms Group.

Nelder, J.A. (1990), "Nearly Parallel Lines in Residual Plots," *American Statistician*, 44, 221–222.

Nelder, J.A. and Lee, Y. (1991), "Generalized Linear Models for the Analysis of Taguchi-Type Experiments," *Applied Stochastic Models and Data Analysis*, 7, 107–120.

Numerical Algorithms Group, *The GLIM Newsletter*, Oxford: Numerical Algorithms Group.

Payne, R.W. et al. (1987), *Genstat 5 Reference Manual*, London: Oxford University Press.

Pierce, D.A. and Schafer, D.W. (1986), "Residuals in Generalized Linear Models," *Journal of the American Statistical Association*, 81, 977–986.

SAS Institute (1989), *SAS/STAT User's Guide, Version 6 Fourth Edition, Volume 2*, Cary: SAS Institute.

SAS Institute (1996), *SAS/STAT Software: Changes and Enhancements through Release 6.11*, Cary: SAS Institute.

Taguchi, G. (1986), *Introduction to Quality Engineering*, Tokyo: Asian Productivity Organization.

Taguchi, G. and Wu, Y. (1980), *Introduction to Off-Line Quality Control*, Nagoya, Japan: Central Japan Quality Control Association.

Upper Tail Probabilities of the Standard Normal Distribution, $\int_{z}^{\infty} \frac{1}{\sqrt{2\pi}} e^{-u^2/2} \, du$

z	0.00	0.01	0.02	0.03	0.04	0.05	0.06	0.07	0.08	0.09
0.0	0.5000	0.4960	0.4920	0.4880	0.4840	0.4801	0.4761	0.4721	0.4681	0.4641
0.1	0.4602	0.4562	0.4522	0.4483	0.4443	0.4404	0.4364	0.4325	0.4286	0.4247
0.2	0.4207	0.4168	0.4129	0.4090	0.4052	0.4013	0.3974	0.3936	0.3897	0.3859
0.3	0.3821	0.3783	0.3745	0.3707	0.3669	0.3632	0.3594	0.3557	0.3520	0.3483
0.4	0.3446	0.3409	0.3372	0.3336	0.3300	0.3264	0.3228	0.3192	0.3156	0.3121
0.5	0.3085	0.3050	0.3015	0.2981	0.2946	0.2912	0.2877	0.2843	0.2810	0.2776
0.6	0.2743	0.2709	0.2676	0.2643	0.2611	0.2578	0.2546	0.2514	0.2483	0.2451
0.7	0.2420	0.2389	0.2358	0.2327	0.2296	0.2266	0.2236	0.2206	0.2177	0.2148
0.8	0.2119	0.2090	0.2061	0.2033	0.2005	0.1977	0.1949	0.1922	0.1894	0.1867
0.9	0.1841	0.1814	0.1788	0.1762	0.1736	0.1711	0.1685	0.1660	0.1635	0.1611
1.0	0.1587	0.1562	0.1539	0.1515	0.1492	0.1469	0.1446	0.1423	0.1401	0.1379
1.1	0.1357	0.1335	0.1314	0.1292	0.1271	0.1251	0.1230	0.1210	0.1190	0.1170
1.2	0.1151	0.1131	0.1112	0.1093	0.1075	0.1056	0.1038	0.1020	0.1003	0.0985
1.3	0.0968	0.0951	0.0934	0.0918	0.0901	0.0885	0.0869	0.0853	0.0838	0.0823
1.4	0.0808	0.0793	0.0778	0.0764	0.0749	0.0735	0.0721	0.0708	0.0694	0.0681
1.5	0.0668	0.0655	0.0643	0.0630	0.0618	0.0606	0.0594	0.0582	0.0571	0.0559
1.6	0.0548	0.0537	0.0526	0.0516	0.0505	0.0495	0.0485	0.0475	0.0465	0.0455
1.7	0.0446	0.0436	0.0427	0.0418	0.0409	0.0401	0.0392	0.0384	0.0375	0.0367
1.8	0.0359	0.0351	0.0344	0.0336	0.0329	0.0322	0.0314	0.0307	0.0301	0.0294
1.9	0.0287	0.0281	0.0274	0.0268	0.0262	0.0256	0.0250	0.0244	0.0239	0.0233
2.0	0.0228	0.0222	0.0217	0.0212	0.0207	0.0202	0.0197	0.0192	0.0188	0.0183
2.1	0.0179	0.0174	0.0170	0.0166	0.0162	0.0158	0.0154	0.0150	0.0146	0.0143
2.2	0.0139	0.0136	0.0132	0.0129	0.0125	0.0122	0.0119	0.0116	0.0113	0.0110
2.3	0.0107	0.0104	0.0102	0.0099	0.0096	0.0094	0.0091	0.0089	0.0087	0.0084
2.4	0.0082	0.0080	0.0078	0.0075	0.0073	0.0071	0.0069	0.0068	0.0066	0.0064
2.5	0.0062	0.0060	0.0059	0.0057	0.0055	0.0054	0.0052	0.0051	0.0049	0.0048
2.6	0.0047	0.0045	0.0044	0.0043	0.0041	0.0040	0.0039	0.0038	0.0037	0.0036
2.7	0.0035	0.0034	0.0033	0.0032	0.0031	0.0030	0.0029	0.0028	0.0027	0.0026
2.8	0.0026	0.0025	0.0024	0.0023	0.0023	0.0022	0.0021	0.0021	0.0020	0.0019
2.9	0.0019	0.0018	0.0018	0.0017	0.0016	0.0016	0.0015	0.0015	0.0014	0.0014
3.0	0.0013	0.0013	0.0013	0.0012	0.0012	0.0011	0.0011	0.0011	0.0010	0.0010
3.1	0.0010	0.0009	0.0009	0.0009	0.0008	0.0008	0.0008	0.0008	0.0007	0.0007
3.2	0.0007	0.0007	0.0006	0.0006	0.0006	0.0006	0.0006	0.0005	0.0005	0.0005
3.3	0.0005	0.0005	0.0005	0.0004	0.0004	0.0004	0.0004	0.0004	0.0004	0.0003
3.4	0.0003	0.0003	0.0003	0.0003	0.0003	0.0003	0.0003	0.0003	0.0003	0.0002
3.5	0.0002	0.0002	0.0002	0.0002	0.0002	0.0002	0.0002	0.0002	0.0002	0.0002

Upper Percentiles of the *t* Distribution

ν	0.25	0.1	0.075	0.05	0.025	α 0.01	0.005	0.0025	0.001	0.0005
1	1.000	3.078	4.165	6.314	12.706	31.821	63.657	127.321	318.309	636.619
2	0.816	1.886	2.282	2.920	4.303	6.965	9.925	14.089	22.327	31.599
3	0.765	1.638	1.924	2.353	3.182	4.541	5.841	7.453	10.215	12.924
4	0.741	1.533	1.778	2.132	2.776	3.747	4.604	5.598	7.173	8.610
5	0.727	1.476	1.699	2.015	2.571	3.365	4.032	4.773	5.893	6.869
6	0.718	1.440	1.650	1.943	2.447	3.143	3.707	4.317	5.208	5.959
7	0.711	1.415	1.617	1.895	2.365	2.998	3.499	4.029	4.785	5.408
8	0.706	1.397	1.592	1.860	2.306	2.896	3.355	3.833	4.501	5.041
9	0.703	1.383	1.574	1.833	2.262	2.821	3.250	3.690	4.297	4.781
10	0.700	1.372	1.559	1.812	2.228	2.764	3.169	3.581	4.144	4.587
11	0.697	1.363	1.548	1.796	2.201	2.718	3.106	3.497	4.025	4.437
12	0.695	1.356	1.538	1.782	2.179	2.681	3.055	3.428	3.930	4.318
13	0.694	1.350	1.530	1.771	2.160	2.650	3.012	3.372	3.852	4.221
14	0.692	1.345	1.523	1.761	2.145	2.624	2.977	3.326	3.787	4.140
15	0.691	1.341	1.517	1.753	2.131	2.602	2.947	3.286	3.733	4.073
16	0.690	1.337	1.512	1.746	2.120	2.583	2.921	3.252	3.686	4.015
17	0.689	1.333	1.508	1.740	2.110	2.567	2.898	3.222	3.646	3.965
18	0.688	1.330	1.504	1.734	2.101	2.552	2.878	3.197	3.610	3.922
19	0.688	1.328	1.500	1.729	2.093	2.539	2.861	3.174	3.579	3.883
20	0.687	1.325	1.497	1.725	2.086	2.528	2.845	3.153	3.552	3.850
21	0.686	1.323	1.494	1.721	2.080	2.518	2.831	3.135	3.527	3.819
22	0.686	1.321	1.492	1.717	2.074	2.508	2.819	3.119	3.505	3.792
23	0.685	1.319	1.489	1.714	2.069	2.500	2.807	3.104	3.485	3.768
24	0.685	1.318	1.487	1.711	2.064	2.492	2.797	3.091	3.467	3.745
25	0.684	1.316	1.485	1.708	2.060	2.485	2.787	3.078	3.450	3.725
26	0.684	1.315	1.483	1.706	2.056	2.479	2.779	3.067	3.435	3.707
27	0.684	1.314	1.482	1.703	2.052	2.473	2.771	3.057	3.421	3.690
28	0.683	1.313	1.480	1.701	2.048	2.467	2.763	3.047	3.408	3.674
29	0.683	1.311	1.479	1.699	2.045	2.462	2.756	3.038	3.396	3.659
30	0.683	1.310	1.477	1.697	2.042	2.457	2.750	3.030	3.385	3.646
40	0.681	1.303	1.468	1.684	2.021	2.423	2.704	2.971	3.307	3.551
60	0.679	1.296	1.458	1.671	2.000	2.390	2.660	2.915	3.232	3.460
120	0.677	1.289	1.449	1.658	1.980	2.358	2.617	2.860	3.160	3.373
∞	0.674	1.282	1.440	1.645	1.960	2.326	2.576	2.807	3.090	3.291

α = upper tail probability, ν = degrees of freedom.

Upper Percentiles of the χ^2 Distribution

ν	0.25	0.1	0.075	0.05	α 0.025	0.01	0.005	0.0025	0.001
1	1.32	2.71	3.17	3.84	5.02	6.63	7.88	9.14	10.83
2	2.77	4.61	5.18	5.99	7.38	9.21	10.60	11.98	13.82
3	4.11	6.25	6.90	7.81	9.35	11.34	12.84	14.32	16.27
4	5.39	7.78	8.50	9.49	11.14	13.28	14.86	16.42	18.47
5	6.63	9.24	10.01	11.07	12.83	15.09	16.75	18.39	20.52
6	7.84	10.64	11.47	12.59	14.45	16.81	18.55	20.25	22.46
7	9.04	12.02	12.88	14.07	16.01	18.48	20.28	22.04	24.32
8	10.22	13.36	14.27	15.51	17.53	20.09	21.95	23.77	26.12
9	11.39	14.68	15.63	16.92	19.02	21.67	23.59	25.46	27.88
10	12.55	15.99	16.97	18.31	20.48	23.21	25.19	27.11	29.59
11	13.70	17.28	18.29	19.68	21.92	24.72	26.76	28.73	31.26
12	14.85	18.55	19.60	21.03	23.34	26.22	28.30	30.32	32.91
13	15.98	19.81	20.90	22.36	24.74	27.69	29.82	31.88	34.53
14	17.12	21.06	22.18	23.68	26.12	29.14	31.32	33.43	36.12
15	18.25	22.31	23.45	25.00	27.49	30.58	32.80	34.95	37.70
16	19.37	23.54	24.72	26.30	28.85	32.00	34.27	36.46	39.25
17	20.49	24.77	25.97	27.59	30.19	33.41	35.72	37.95	40.79
18	21.60	25.99	27.22	28.87	31.53	34.81	37.16	39.42	42.31
19	22.72	27.20	28.46	30.14	32.85	36.19	38.58	40.88	43.82
20	23.83	28.41	29.69	31.41	34.17	37.57	40.00	42.34	45.31
21	24.93	29.62	30.92	32.67	35.48	38.93	41.40	43.78	46.80
22	26.04	30.81	32.14	33.92	36.78	40.29	42.80	45.20	48.27
23	27.14	32.01	33.36	35.17	38.08	41.64	44.18	46.62	49.73
24	28.24	33.20	34.57	36.42	39.36	42.98	45.56	48.03	51.18
25	29.34	34.38	35.78	37.65	40.65	44.31	46.93	49.44	52.62
26	30.43	35.56	36.98	38.89	41.92	45.64	48.29	50.83	54.05
27	31.53	36.74	38.18	40.11	43.19	46.96	49.64	52.22	55.48
28	32.62	37.92	39.38	41.34	44.46	48.28	50.99	53.59	56.89
29	33.71	39.09	40.57	42.56	45.72	49.59	52.34	54.97	58.30
30	34.80	40.26	41.76	43.77	46.98	50.89	53.67	56.33	59.70
40	45.62	51.81	53.50	55.76	59.34	63.69	66.77	69.70	73.40
50	56.33	63.17	65.03	67.50	71.42	76.15	79.49	82.66	86.66
60	66.98	74.40	76.41	79.08	83.30	88.38	91.95	95.34	99.61

α = upper tail probability, ν = degrees of freedom.

Upper Percentiles of the *F* Distribution

$\alpha = 0.10$

								Degrees of freedom for the numerator (v_1)											
v_2	1	2	3	4	5	6	7	8	9	10	12	15	20	24	30	40	60	120	∞
1	39.86	49.50	53.59	55.83	57.24	58.20	58.91	59.44	59.86	60.19	60.71	61.22	61.74	62.00	62.26	62.53	62.79	63.06	63.33
2	8.53	9.00	9.16	9.24	9.29	9.33	9.35	9.37	9.38	9.39	9.41	9.42	9.44	9.45	9.46	9.47	9.47	9.48	9.49
3	5.54	5.46	5.39	5.34	5.31	5.28	5.27	5.25	5.24	5.23	5.22	5.20	5.18	5.18	5.17	5.16	5.15	5.14	5.13
4	4.54	4.32	4.19	4.11	4.05	4.01	3.98	3.95	3.94	3.92	3.90	3.87	3.84	3.83	3.82	3.80	3.79	3.78	3.76
5	4.06	3.78	3.62	3.52	3.45	3.40	3.37	3.34	3.32	3.30	3.27	3.24	3.21	3.19	3.17	3.16	3.14	3.12	3.10
6	3.78	3.46	3.29	3.18	3.11	3.05	3.01	2.98	2.96	2.94	2.90	2.87	2.84	2.82	2.80	2.78	2.76	2.74	2.72
7	3.59	3.26	3.07	2.96	2.88	2.83	2.78	2.75	2.72	2.70	2.67	2.63	2.59	2.58	2.56	2.54	2.51	2.49	2.47
8	3.46	3.11	2.92	2.81	2.73	2.67	2.62	2.59	2.56	2.54	2.50	2.46	2.42	2.40	2.38	2.36	2.34	2.32	2.29
9	3.36	3.01	2.81	2.69	2.61	2.55	2.51	2.47	2.44	2.42	2.38	2.34	2.30	2.28	2.25	2.23	2.21	2.18	2.16
10	3.29	2.92	2.73	2.61	2.52	2.46	2.41	2.38	2.35	2.32	2.28	2.24	2.20	2.18	2.16	2.13	2.11	2.08	2.06
11	3.23	2.86	2.66	2.54	2.45	2.39	2.34	2.30	2.27	2.25	2.21	2.17	2.12	2.10	2.08	2.05	2.03	2.00	1.97
12	3.18	2.81	2.61	2.48	2.39	2.33	2.28	2.24	2.21	2.19	2.15	2.10	2.06	2.04	2.01	1.99	1.96	1.93	1.90
13	3.14	2.76	2.56	2.43	2.35	2.28	2.23	2.20	2.16	2.14	2.10	2.05	2.01	1.98	1.96	1.93	1.90	1.88	1.85
14	3.10	2.73	2.52	2.39	2.31	2.24	2.19	2.15	2.12	2.10	2.05	2.01	1.96	1.94	1.91	1.89	1.86	1.83	1.80
15	3.07	2.70	2.49	2.36	2.27	2.21	2.16	2.12	2.09	2.06	2.02	1.97	1.92	1.90	1.87	1.85	1.82	1.79	1.76
16	3.05	2.67	2.46	2.33	2.24	2.18	2.13	2.09	2.06	2.03	1.99	1.94	1.89	1.87	1.84	1.81	1.78	1.75	1.72
17	3.03	2.64	2.44	2.31	2.22	2.15	2.10	2.06	2.03	2.00	1.96	1.91	1.86	1.84	1.81	1.78	1.75	1.72	1.69
18	3.01	2.62	2.42	2.29	2.20	2.13	2.08	2.04	2.00	1.98	1.93	1.89	1.84	1.81	1.78	1.75	1.72	1.69	1.66
19	2.99	2.61	2.40	2.27	2.18	2.11	2.06	2.02	1.98	1.96	1.91	1.86	1.81	1.79	1.76	1.73	1.70	1.67	1.63
20	2.97	2.59	2.38	2.25	2.16	2.09	2.04	2.00	1.96	1.94	1.89	1.84	1.79	1.77	1.74	1.71	1.68	1.64	1.61
21	2.96	2.57	2.36	2.23	2.14	2.08	2.02	1.98	1.95	1.92	1.87	1.83	1.78	1.75	1.72	1.69	1.66	1.62	1.59
22	2.95	2.56	2.35	2.22	2.13	2.06	2.01	1.97	1.93	1.90	1.86	1.81	1.76	1.73	1.70	1.67	1.64	1.60	1.57
23	2.94	2.55	2.34	2.21	2.11	2.05	1.99	1.95	1.92	1.89	1.84	1.80	1.74	1.72	1.69	1.66	1.62	1.59	1.55
24	2.93	2.54	2.33	2.19	2.10	2.04	1.98	1.94	1.91	1.88	1.83	1.78	1.73	1.70	1.67	1.64	1.61	1.57	1.53
25	2.92	2.53	2.32	2.18	2.09	2.02	1.97	1.93	1.89	1.87	1.82	1.77	1.72	1.69	1.66	1.63	1.59	1.56	1.52
26	2.91	2.52	2.31	2.17	2.08	2.01	1.96	1.92	1.88	1.86	1.81	1.76	1.71	1.68	1.65	1.61	1.58	1.54	1.50
27	2.90	2.51	2.30	2.17	2.07	2.00	1.95	1.91	1.87	1.85	1.80	1.75	1.70	1.67	1.64	1.60	1.57	1.53	1.49
28	2.89	2.50	2.29	2.16	2.06	2.00	1.94	1.90	1.87	1.84	1.79	1.74	1.69	1.66	1.63	1.59	1.56	1.52	1.48
29	2.89	2.50	2.28	2.15	2.06	1.99	1.93	1.89	1.86	1.83	1.78	1.73	1.68	1.65	1.62	1.58	1.55	1.51	1.47
30	2.88	2.49	2.28	2.14	2.05	1.98	1.93	1.88	1.85	1.82	1.77	1.72	1.67	1.64	1.61	1.57	1.54	1.50	1.46
40	2.84	2.44	2.23	2.09	2.00	1.93	1.87	1.83	1.79	1.76	1.71	1.66	1.61	1.57	1.54	1.51	1.47	1.42	1.38
60	2.79	2.39	2.18	2.04	1.95	1.87	1.82	1.77	1.74	1.71	1.66	1.60	1.54	1.51	1.48	1.44	1.40	1.35	1.29
120	2.75	2.35	2.13	1.99	1.90	1.82	1.77	1.72	1.68	1.65	1.60	1.55	1.48	1.45	1.41	1.37	1.32	1.26	1.19
∞	2.71	2.30	2.08	1.94	1.85	1.77	1.72	1.67	1.63	1.60	1.55	1.49	1.42	1.38	1.34	1.30	1.24	1.17	1.00

603

$\alpha = 0.05$

Degrees of freedom for the numerator (ν_1)

ν_2	1	2	3	4	5	6	7	8	9	10	12	15	20	24	30	40	60	120	∞
1	161.45	199.50	215.71	224.58	230.16	233.99	236.77	238.88	240.54	241.88	243.91	245.95	248.01	249.05	250.10	251.14	252.20	253.25	254.31
2	18.51	19.00	19.16	19.25	19.30	19.33	19.35	19.37	19.38	19.40	19.41	19.43	19.45	19.45	19.46	19.47	19.48	19.49	19.50
3	10.13	9.55	9.28	9.12	9.01	8.94	8.89	8.85	8.81	8.79	8.74	8.70	8.66	8.64	8.62	8.59	8.57	8.55	8.53
4	7.71	6.94	6.59	6.39	6.26	6.16	6.09	6.04	6.00	5.96	5.91	5.86	5.80	5.77	5.75	5.72	5.69	5.66	5.63
5	6.61	5.79	5.41	5.19	5.05	4.95	4.88	4.82	4.77	4.74	4.68	4.62	4.56	4.53	4.50	4.46	4.43	4.40	4.36
6	5.99	5.14	4.76	4.53	4.39	4.28	4.21	4.15	4.10	4.06	4.00	3.94	3.87	3.84	3.81	3.77	3.74	3.70	3.67
7	5.59	4.74	4.35	4.12	3.97	3.87	3.79	3.73	3.68	3.64	3.57	3.51	3.44	3.41	3.38	3.34	3.30	3.27	3.23
8	5.32	4.46	4.07	3.84	3.69	3.58	3.50	3.44	3.39	3.35	3.28	3.22	3.15	3.12	3.08	3.04	3.01	2.97	2.93
9	5.12	4.26	3.86	3.63	3.48	3.37	3.29	3.23	3.18	3.14	3.07	3.01	2.94	2.90	2.86	2.83	2.79	2.75	2.71
10	4.96	4.10	3.71	3.48	3.33	3.22	3.14	3.07	3.02	2.98	2.91	2.85	2.77	2.74	2.70	2.66	2.62	2.58	2.54
11	4.84	3.98	3.59	3.36	3.20	3.09	3.01	2.95	2.90	2.85	2.79	2.72	2.65	2.61	2.57	2.53	2.49	2.45	2.40
12	4.75	3.89	3.49	3.26	3.11	3.00	2.91	2.85	2.80	2.75	2.69	2.62	2.54	2.51	2.47	2.43	2.38	2.34	2.30
13	4.67	3.81	3.41	3.18	3.03	2.92	2.83	2.77	2.71	2.67	2.60	2.53	2.46	2.42	2.38	2.34	2.30	2.25	2.21
14	4.60	3.74	3.34	3.11	2.96	2.85	2.76	2.70	2.65	2.60	2.53	2.46	2.39	2.35	2.31	2.27	2.22	2.18	2.13
15	4.54	3.68	3.29	3.06	2.90	2.79	2.71	2.64	2.59	2.54	2.48	2.40	2.33	2.29	2.25	2.20	2.16	2.11	2.07
16	4.49	3.63	3.24	3.01	2.85	2.74	2.66	2.59	2.54	2.49	2.42	2.35	2.28	2.24	2.19	2.15	2.11	2.06	2.01
17	4.45	3.59	3.20	2.96	2.81	2.70	2.61	2.55	2.49	2.45	2.38	2.31	2.23	2.19	2.15	2.10	2.06	2.01	1.96
18	4.41	3.55	3.16	2.93	2.77	2.66	2.58	2.51	2.46	2.41	2.34	2.27	2.19	2.15	2.11	2.06	2.02	1.97	1.92
19	4.38	3.52	3.13	2.90	2.74	2.63	2.54	2.48	2.42	2.38	2.31	2.23	2.16	2.11	2.07	2.03	1.98	1.93	1.88
20	4.35	3.49	3.10	2.87	2.71	2.60	2.51	2.45	2.39	2.35	2.28	2.20	2.12	2.08	2.04	1.99	1.95	1.90	1.84
21	4.32	3.47	3.07	2.84	2.68	2.57	2.49	2.42	2.37	2.32	2.25	2.18	2.10	2.05	2.01	1.96	1.92	1.87	1.81
22	4.30	3.44	3.05	2.82	2.66	2.55	2.46	2.40	2.34	2.30	2.23	2.15	2.07	2.03	1.98	1.94	1.89	1.84	1.78
23	4.28	3.42	3.03	2.80	2.64	2.53	2.44	2.37	2.32	2.27	2.20	2.13	2.05	2.01	1.96	1.91	1.86	1.81	1.76
24	4.26	3.40	3.01	2.78	2.62	2.51	2.42	2.36	2.30	2.25	2.18	2.11	2.03	1.98	1.94	1.89	1.84	1.79	1.73
25	4.24	3.39	2.99	2.76	2.60	2.49	2.40	2.34	2.28	2.24	2.16	2.09	2.01	1.96	1.92	1.87	1.82	1.77	1.71
26	4.23	3.37	2.98	2.74	2.59	2.47	2.39	2.32	2.27	2.22	2.15	2.07	1.99	1.95	1.90	1.85	1.80	1.75	1.69
27	4.21	3.35	2.96	2.73	2.57	2.46	2.37	2.31	2.25	2.20	2.13	2.06	1.97	1.93	1.88	1.84	1.79	1.73	1.67
28	4.20	3.34	2.95	2.71	2.56	2.45	2.36	2.29	2.24	2.19	2.12	2.04	1.96	1.91	1.87	1.82	1.77	1.71	1.65
29	4.18	3.33	2.93	2.70	2.55	2.43	2.35	2.28	2.22	2.18	2.10	2.03	1.94	1.90	1.85	1.81	1.75	1.70	1.64
30	4.17	3.32	2.92	2.69	2.53	2.42	2.33	2.27	2.21	2.16	2.09	2.01	1.93	1.89	1.84	1.79	1.74	1.68	1.62
40	4.08	3.23	2.84	2.61	2.45	2.34	2.25	2.18	2.12	2.08	2.00	1.92	1.84	1.79	1.74	1.69	1.64	1.58	1.51
60	4.00	3.15	2.76	2.53	2.37	2.25	2.17	2.10	2.04	1.99	1.92	1.84	1.75	1.70	1.65	1.59	1.53	1.47	1.39
120	3.92	3.07	2.68	2.45	2.29	2.18	2.09	2.02	1.96	1.91	1.83	1.75	1.66	1.61	1.55	1.50	1.43	1.35	1.25
∞	3.84	3.00	2.60	2.37	2.21	2.10	2.01	1.94	1.88	1.83	1.75	1.67	1.57	1.52	1.46	1.39	1.32	1.22	1.00

$\alpha = 0.01$

| | | | | | | | | Degrees of freedom for the numerator (ν_1) | | | | | | | | | | | |
ν_2	1	2	3	4	5	6	7	8	9	10	12	15	20	24	30	40	60	120	∞
1	4052	5000	5403	5625	5764	5859	5928	5981	6022	6056	6106	6157	6209	6235	6261	6287	6313	6339	6366
2	98.5000	99.00	99.17	99.25	99.30	99.33	99.36	99.37	99.39	99.40	99.42	99.43	99.45	99.46	99.47	99.47	99.48	99.49	99.50
3	34.12	30.82	29.46	28.71	28.24	27.91	27.67	27.49	27.35	27.23	27.05	26.87	26.69	26.60	26.50	26.41	26.32	26.22	26.13
4	21.20	18.00	16.69	15.98	15.52	15.21	14.98	14.80	14.66	14.55	14.37	14.20	14.02	13.93	13.84	13.75	13.65	13.56	13.46
5	16.26	13.27	12.06	11.39	10.97	10.67	10.46	10.29	10.16	10.05	9.89	9.72	9.55	9.47	9.38	9.29	9.20	9.11	9.02
6	13.75	10.92	9.78	9.15	8.75	8.47	8.26	8.10	7.98	7.87	7.72	7.56	7.40	7.31	7.23	7.14	7.06	6.97	6.88
7	12.25	9.55	8.45	7.85	7.46	7.19	6.99	6.84	6.72	6.62	6.47	6.31	6.16	6.07	5.99	5.91	5.82	5.74	5.65
8	11.26	8.65	7.59	7.01	6.63	6.37	6.18	6.03	5.91	5.81	5.67	5.52	5.36	5.28	5.20	5.12	5.03	4.95	4.86
9	10.56	8.02	6.99	6.42	6.06	5.80	5.61	5.47	5.35	5.26	5.11	4.96	4.81	4.73	4.65	4.57	4.48	4.40	4.31
10	10.04	7.56	6.55	5.99	5.64	5.39	5.20	5.06	4.94	4.85	4.71	4.56	4.41	4.33	4.25	4.17	4.08	4.00	3.91
11	9.65	7.21	6.22	5.67	5.32	5.07	4.89	4.74	4.63	4.54	4.40	4.25	4.10	4.02	3.94	3.86	3.78	3.69	3.60
12	9.33	6.93	5.95	5.41	5.06	4.82	4.64	4.50	4.39	4.30	4.16	4.01	3.86	3.78	3.70	3.62	3.54	3.45	3.36
13	9.07	6.70	5.74	5.21	4.86	4.62	4.44	4.30	4.19	4.10	3.96	3.82	3.66	3.59	3.51	3.43	3.34	3.25	3.17
14	8.86	6.51	5.56	5.04	4.69	4.46	4.28	4.14	4.03	3.94	3.80	3.66	3.51	3.43	3.35	3.27	3.18	3.09	3.00
15	8.68	6.36	5.42	4.89	4.56	4.32	4.14	4.00	3.89	3.80	3.67	3.52	3.37	3.29	3.21	3.13	3.05	2.96	2.87
16	8.53	6.23	5.29	4.77	4.44	4.20	4.03	3.89	3.78	3.69	3.55	3.41	3.26	3.18	3.10	3.02	2.93	2.84	2.75
17	8.40	6.11	5.18	4.67	4.34	4.10	3.93	3.79	3.68	3.59	3.46	3.31	3.16	3.08	3.00	2.92	2.83	2.75	2.65
18	8.29	6.01	5.09	4.58	4.25	4.01	3.84	3.71	3.60	3.51	3.37	3.23	3.08	3.00	2.92	2.84	2.75	2.66	2.57
19	8.18	5.93	5.01	4.50	4.17	3.94	3.77	3.63	3.52	3.43	3.30	3.15	3.00	2.92	2.84	2.76	2.67	2.58	2.49
20	8.10	5.85	4.94	4.43	4.10	3.87	3.70	3.56	3.46	3.37	3.23	3.09	2.94	2.86	2.78	2.69	2.61	2.52	2.42
21	8.02	5.78	4.87	4.37	4.04	3.81	3.64	3.51	3.40	3.31	3.17	3.03	2.88	2.80	2.72	2.64	2.55	2.46	2.36
22	7.95	5.72	4.82	4.31	3.99	3.76	3.59	3.45	3.35	3.26	3.12	2.98	2.83	2.75	2.67	2.58	2.50	2.40	2.31
23	7.88	5.66	4.76	4.26	3.94	3.71	3.54	3.41	3.30	3.21	3.07	2.93	2.78	2.70	2.62	2.54	2.45	2.35	2.26
24	7.82	5.61	4.72	4.22	3.90	3.67	3.50	3.36	3.26	3.17	3.03	2.89	2.74	2.66	2.58	2.49	2.40	2.31	2.21
25	7.77	5.57	4.68	4.18	3.85	3.63	3.46	3.32	3.22	3.13	2.99	2.85	2.70	2.62	2.54	2.45	2.36	2.27	2.17
26	7.72	5.53	4.64	4.14	3.82	3.59	3.42	3.29	3.18	3.09	2.96	2.81	2.66	2.58	2.50	2.42	2.33	2.23	2.13
27	7.68	5.49	4.60	4.11	3.78	3.56	3.39	3.26	3.15	3.06	2.93	2.78	2.63	2.55	2.47	2.38	2.29	2.20	2.10
28	7.64	5.45	4.57	4.07	3.75	3.53	3.36	3.23	3.12	3.03	2.90	2.75	2.60	2.52	2.44	2.35	2.26	2.17	2.06
29	7.60	5.42	4.54	4.04	3.73	3.50	3.33	3.20	3.09	3.00	2.87	2.73	2.57	2.49	2.41	2.33	2.23	2.14	2.03
30	7.56	5.39	4.51	4.02	3.70	3.47	3.30	3.17	3.07	2.98	2.84	2.70	2.55	2.47	2.39	2.30	2.21	2.11	2.01
40	7.31	5.18	4.31	3.83	3.51	3.29	3.12	2.99	2.89	2.80	2.66	2.52	2.37	2.29	2.20	2.11	2.02	1.92	1.80
60	7.08	4.98	4.13	3.65	3.34	3.12	2.95	2.82	2.72	2.63	2.50	2.35	2.20	2.12	2.03	1.94	1.84	1.73	1.60
120	6.85	4.79	3.95	3.48	3.17	2.96	2.79	2.66	2.56	2.47	2.34	2.19	2.03	1.95	1.86	1.76	1.66	1.53	1.38
∞	6.63	4.61	3.78	3.32	3.02	2.80	2.64	2.51	2.41	2.32	2.18	2.04	1.88	1.79	1.70	1.59	1.47	1.32	1.00

α = upper tail probability, ν_2 = degrees of freedom for the denominator.

Upper Percentiles of the Studentized Range Distribution

$$\alpha = 0.10$$

ν	k																		
	2	3	4	5	6	7	8	9	10	11	12	13	14	15	16	17	18	19	20
1	8.93	13.44	16.36	18.49	20.15	21.51	22.64	23.62	24.48	25.24	25.92	26.54	27.10	27.62	28.10	28.54	28.96	29.35	29.71
2	4.13	5.73	6.77	7.54	8.14	8.63	9.05	9.41	9.72	10.01	10.26	10.49	10.70	10.89	11.07	11.24	11.39	11.54	11.68
3	3.33	4.47	5.20	5.74	6.16	6.51	6.81	7.06	7.29	7.49	7.67	7.83	7.98	8.12	8.25	8.37	8.48	8.58	8.68
4	3.01	3.98	4.59	5.03	5.39	5.68	5.93	6.14	6.33	6.49	6.65	6.78	6.91	7.02	7.13	7.23	7.33	7.41	7.50
5	2.85	3.72	4.26	4.66	4.98	5.24	5.46	5.65	5.82	5.97	6.10	6.22	6.34	6.44	6.54	6.63	6.71	6.79	6.86
6	2.75	3.56	4.07	4.44	4.73	4.97	5.17	5.34	5.50	5.64	5.76	5.87	5.98	6.07	6.16	6.25	6.32	6.40	6.47
7	2.68	3.45	3.93	4.28	4.55	4.78	4.97	5.14	5.28	5.41	5.53	5.64	5.74	5.83	5.91	5.99	6.06	6.13	6.19
8	2.63	3.37	3.83	4.17	4.43	4.65	4.83	4.99	5.13	5.25	5.36	5.46	5.56	5.64	5.72	5.80	5.87	5.93	6.00
9	2.59	3.32	3.76	4.08	4.34	4.54	4.72	4.87	5.01	5.13	5.23	5.33	5.42	5.51	5.58	5.65	5.72	5.79	5.85
10	2.56	3.27	3.70	4.02	4.26	4.47	4.64	4.78	4.91	5.03	5.13	5.23	5.32	5.40	5.47	5.54	5.61	5.67	5.73
11	2.54	3.23	3.66	3.96	4.20	4.40	4.57	4.71	4.84	4.95	5.05	5.15	5.23	5.31	5.38	5.45	5.51	5.57	5.63
12	2.52	3.20	3.62	3.92	4.16	4.35	4.51	4.65	4.78	4.89	4.99	5.08	5.16	5.24	5.31	5.37	5.44	5.49	5.55
13	2.50	3.18	3.59	3.88	4.12	4.30	4.46	4.60	4.72	4.83	4.93	5.02	5.10	5.18	5.25	5.31	5.37	5.43	5.48
14	2.49	3.16	3.56	3.85	4.08	4.27	4.42	4.56	4.68	4.79	4.88	4.97	5.05	5.12	5.19	5.26	5.32	5.37	5.43
15	2.48	3.14	3.54	3.83	4.05	4.23	4.39	4.52	4.64	4.75	4.84	4.93	5.01	5.08	5.15	5.21	5.27	5.32	5.38
16	2.47	3.12	3.52	3.80	4.03	4.21	4.36	4.49	4.61	4.71	4.80	4.89	4.97	5.04	5.11	5.17	5.23	5.28	5.33
17	2.46	3.11	3.50	3.78	4.00	4.18	4.33	4.46	4.58	4.68	4.77	4.86	4.93	5.01	5.07	5.13	5.19	5.24	5.30
18	2.45	3.10	3.49	3.77	3.98	4.16	4.31	4.44	4.55	4.65	4.75	4.83	4.90	4.97	5.04	5.10	5.16	5.21	5.26
19	2.45	3.09	3.47	3.75	3.97	4.14	4.29	4.42	4.53	4.63	4.72	4.80	4.88	4.95	5.01	5.07	5.13	5.18	5.23
20	2.44	3.08	3.46	3.74	3.95	4.12	4.27	4.40	4.51	4.61	4.70	4.78	4.85	4.92	4.99	5.05	5.10	5.16	5.20
24	2.42	3.05	3.42	3.69	3.90	4.07	4.21	4.34	4.44	4.54	4.63	4.71	4.78	4.85	4.91	4.97	5.02	5.07	5.12
30	2.40	3.02	3.39	3.65	3.85	4.02	4.16	4.28	4.38	4.47	4.56	4.64	4.71	4.77	4.83	4.89	4.94	4.99	5.03
40	2.38	2.99	3.35	3.60	3.80	3.96	4.10	4.21	4.32	4.41	4.49	4.56	4.63	4.69	4.75	4.81	4.86	4.90	4.95
60	2.36	2.96	3.31	3.56	3.75	3.91	4.04	4.16	4.25	4.34	4.42	4.49	4.56	4.62	4.67	4.73	4.78	4.82	4.86
120	2.34	2.93	3.28	3.52	3.71	3.86	3.99	4.10	4.19	4.28	4.35	4.42	4.48	4.54	4.60	4.65	4.69	4.74	4.78
∞	2.33	2.90	3.24	3.48	3.66	3.81	3.93	4.04	4.13	4.21	4.28	4.35	4.41	4.47	4.52	4.57	4.61	4.65	4.69

$\alpha = 0.05$

ν									k										
	2	3	4	5	6	7	8	9	10	11	12	13	14	15	16	17	18	19	20
1	17.97	26.98	32.82	37.08	40.41	43.12	45.40	47.36	49.07	50.59	51.96	53.20	54.33	55.36	56.32	57.22	58.04	58.83	59.56
2	6.08	8.33	9.80	10.88	11.74	12.44	13.03	13.54	13.99	14.39	14.75	15.08	15.38	15.65	15.91	16.14	16.37	16.57	16.77
3	4.50	5.91	6.82	7.50	8.04	8.48	8.85	9.18	9.46	9.72	9.95	10.15	10.35	10.52	10.69	10.84	10.98	11.11	11.24
4	3.93	5.04	5.76	6.29	6.71	7.05	7.35	7.60	7.83	8.03	8.21	8.37	8.52	8.66	8.79	8.91	9.03	9.13	9.23
5	3.64	4.60	5.22	5.67	6.03	6.33	6.58	6.80	6.99	7.17	7.32	7.47	7.60	7.72	7.83	7.93	8.03	8.12	8.21
6	3.46	4.34	4.90	5.30	5.63	5.90	6.12	6.32	6.49	6.65	6.79	6.92	7.03	7.14	7.24	7.34	7.43	7.51	7.59
7	3.34	4.16	4.68	5.06	5.36	5.61	5.82	6.00	6.16	6.30	6.43	6.55	6.66	6.76	6.85	6.94	7.02	7.10	7.17
8	3.26	4.04	4.53	4.89	5.17	5.40	5.60	5.77	5.92	6.05	6.18	6.29	6.39	6.48	6.57	6.65	6.73	6.80	6.87
9	3.20	3.95	4.41	4.76	5.02	5.24	5.43	5.59	5.74	5.87	5.98	6.09	6.19	6.28	6.36	6.44	6.51	6.58	6.64
10	3.15	3.88	4.33	4.65	4.91	5.12	5.30	5.46	5.60	5.72	5.83	5.93	6.03	6.11	6.19	6.27	6.34	6.40	6.47
11	3.11	3.82	4.26	4.57	4.82	5.03	5.20	5.35	5.49	5.61	5.71	5.81	5.90	5.98	6.06	6.13	6.20	6.27	6.33
12	3.08	3.77	4.20	4.51	4.75	4.95	5.12	5.27	5.39	5.51	5.61	5.71	5.80	5.88	5.95	6.02	6.09	6.15	6.21
13	3.06	3.73	4.15	4.45	4.69	4.88	5.05	5.19	5.32	5.43	5.53	5.63	5.71	5.79	5.86	5.93	5.99	6.05	6.11
14	3.03	3.70	4.11	4.41	4.64	4.83	4.99	5.13	5.25	5.36	5.46	5.55	5.64	5.71	5.79	5.85	5.91	5.97	6.03
15	3.01	3.67	4.08	4.37	4.59	4.78	4.94	5.08	5.20	5.31	5.40	5.49	5.57	5.65	5.72	5.78	5.85	5.90	5.96
16	3.00	3.65	4.05	4.33	4.56	4.74	4.90	5.03	5.15	5.26	5.35	5.44	5.52	5.59	5.66	5.73	5.79	5.84	5.90
17	2.98	3.63	4.02	4.30	4.52	4.70	4.86	4.99	5.11	5.21	5.31	5.39	5.47	5.54	5.61	5.67	5.73	5.79	5.84
18	2.97	3.61	4.00	4.28	4.49	4.67	4.82	4.96	5.07	5.17	5.27	5.35	5.43	5.50	5.57	5.63	5.69	5.74	5.79
19	2.96	3.59	3.98	4.25	4.47	4.65	4.79	4.92	5.04	5.14	5.23	5.31	5.39	5.46	5.53	5.59	5.65	5.70	5.75
20	2.95	3.58	3.96	4.23	4.45	4.62	4.77	4.90	5.01	5.11	5.20	5.28	5.36	5.43	5.49	5.55	5.61	5.66	5.71
24	2.92	3.53	3.90	4.17	4.37	4.54	4.68	4.81	4.92	5.01	5.10	5.18	5.25	5.32	5.38	5.44	5.49	5.55	5.59
30	2.89	3.49	3.85	4.10	4.30	4.46	4.60	4.72	4.82	4.92	5.00	5.08	5.15	5.21	5.27	5.33	5.38	5.43	5.47
40	2.86	3.44	3.79	4.04	4.23	4.39	4.52	4.63	4.73	4.82	4.90	4.98	5.04	5.11	5.16	5.22	5.27	5.31	5.36
60	2.83	3.40	3.74	3.98	4.16	4.31	4.44	4.55	4.65	4.73	4.81	4.88	4.94	5.00	5.06	5.11	5.15	5.20	5.24
120	2.80	3.36	3.68	3.92	4.10	4.24	4.36	4.47	4.56	4.64	4.71	4.78	4.84	4.90	4.95	5.00	5.04	5.09	5.13
∞	2.77	3.31	3.63	3.86	4.03	4.17	4.29	4.39	4.47	4.55	4.62	4.68	4.74	4.80	4.85	4.89	4.93	4.97	5.01

$\alpha = 0.01$

ν	2	3	4	5	6	7	8	9	10	11	12	13	14	15	16	17	18	19	20
									k										
1	90.03	135.0	164.3	185.6	202.2	215.8	227.2	237.0	245.6	253.2	260.0	266.2	271.8	277.0	281.8	286.3	290.4	294.3	298.0
2	14.04	19.02	22.29	24.72	26.63	28.20	29.53	30.68	31.69	32.59	33.40	34.13	34.81	35.43	36.00	36.53	37.03	37.50	37.95
3	8.26	10.62	12.17	13.33	14.24	15.00	15.64	16.20	16.69	17.13	17.53	17.89	18.22	18.52	18.81	19.07	19.32	19.55	19.77
4	6.51	8.12	9.17	9.96	10.58	11.10	11.55	11.93	12.27	12.57	12.84	13.09	13.32	13.53	13.73	13.91	14.08	14.24	14.40
5	5.70	6.98	7.80	8.42	8.91	9.32	9.67	9.97	10.24	10.48	10.70	10.89	11.08	11.24	11.40	11.55	11.68	11.81	11.93
6	5.24	6.33	7.03	7.56	7.97	8.32	8.61	8.87	9.10	9.30	9.48	9.65	9.81	9.95	10.08	10.21	10.32	10.43	10.54
7	4.95	5.92	6.54	7.00	7.37	7.68	7.94	8.17	8.37	8.55	8.71	8.86	9.00	9.12	9.24	9.35	9.46	9.55	9.65
8	4.75	5.64	6.20	6.62	6.96	7.24	7.47	7.68	7.86	8.03	8.18	8.31	8.44	8.55	8.66	8.76	8.85	8.94	9.03
9	4.60	5.43	5.96	6.35	6.66	6.91	7.13	7.33	7.49	7.65	7.78	7.91	8.03	8.13	8.23	8.33	8.41	8.49	8.57
10	4.48	5.27	5.77	6.14	6.43	6.67	6.87	7.05	7.21	7.36	7.49	7.60	7.71	7.81	7.91	7.99	8.08	8.15	8.23
11	4.39	5.15	5.62	5.97	6.25	6.48	6.67	6.84	6.99	7.13	7.25	7.36	7.46	7.56	7.65	7.73	7.81	7.88	7.95
12	4.32	5.05	5.50	5.84	6.10	6.32	6.51	6.67	6.81	6.94	7.06	7.17	7.26	7.36	7.44	7.52	7.59	7.66	7.73
13	4.26	4.96	5.40	5.73	5.98	6.19	6.37	6.53	6.67	6.79	6.90	7.01	7.10	7.19	7.27	7.35	7.42	7.48	7.55
14	4.21	4.89	5.32	5.63	5.88	6.08	6.26	6.41	6.54	6.66	6.77	6.87	6.96	7.05	7.13	7.20	7.27	7.33	7.39
15	4.17	4.84	5.25	5.56	5.80	5.99	6.16	6.31	6.44	6.55	6.66	6.76	6.84	6.93	7.00	7.07	7.14	7.20	7.26
16	4.13	4.79	5.19	5.49	5.72	5.92	6.08	6.22	6.35	6.46	6.56	6.66	6.74	6.82	6.90	6.97	7.03	7.09	7.15
17	4.10	4.74	5.14	5.43	5.66	5.85	6.01	6.15	6.27	6.38	6.48	6.57	6.66	6.73	6.81	6.87	6.94	7.00	7.05
18	4.07	4.70	5.09	5.38	5.60	5.79	5.94	6.08	6.20	6.31	6.41	6.50	6.58	6.65	6.73	6.79	6.85	6.91	6.97
19	4.05	4.67	5.05	5.33	5.55	5.73	5.89	6.02	6.14	6.25	6.34	6.43	6.51	6.58	6.65	6.72	6.78	6.84	6.89
20	4.02	4.64	5.02	5.29	5.51	5.69	5.84	5.97	6.09	6.19	6.28	6.37	6.45	6.52	6.59	6.65	6.71	6.77	6.82
24	3.96	4.55	4.91	5.17	5.37	5.54	5.69	5.81	5.92	6.02	6.11	6.19	6.26	6.33	6.39	6.45	6.51	6.56	6.61
30	3.89	4.45	4.80	5.05	5.24	5.40	5.54	5.65	5.76	5.85	5.93	6.01	6.08	6.14	6.20	6.26	6.31	6.36	6.41
40	3.82	4.37	4.70	4.93	5.11	5.26	5.39	5.50	5.60	5.69	5.76	5.83	5.90	5.96	6.02	6.07	6.12	6.16	6.21
60	3.76	4.28	4.59	4.82	4.99	5.13	5.25	5.36	5.45	5.53	5.60	5.67	5.73	5.78	5.84	5.89	5.93	5.97	6.01
120	3.70	4.20	4.50	4.71	4.87	5.01	5.12	5.21	5.30	5.37	5.44	5.50	5.56	5.61	5.66	5.71	5.75	5.79	5.83
∞	3.64	4.12	4.40	4.60	4.76	4.88	4.99	5.08	5.16	5.23	5.29	5.35	5.40	5.45	5.49	5.54	5.57	5.61	5.65

α = upper tail probability, ν = degrees of freedom, k = number of treatments.

Upper Percentiles of the Studentized Maximum Modulus Distribution

$$\alpha = 0.10$$

ν	1	2	3	4	5	6	7	8	9	10	11	12	13	14	15	16	17	18
2	2.92	3.83	4.38	4.77	5.06	5.30	5.50	5.67	5.82	5.96	6.08	6.18	6.28	6.37	6.45	6.53	6.60	6.67
3	2.35	2.99	3.37	3.64	3.84	4.01	4.15	4.27	4.38	4.47	4.55	4.63	4.70	4.76	4.82	4.88	4.93	4.98
4	2.13	2.66	2.98	3.20	3.37	3.51	3.62	3.72	3.81	3.89	3.96	4.02	4.08	4.13	4.18	4.23	4.27	4.31
5	2.02	2.49	2.77	2.96	3.12	3.24	3.34	3.43	3.51	3.58	3.64	3.69	3.75	3.79	3.84	3.88	3.92	3.95
6	1.94	2.38	2.64	2.82	2.96	3.07	3.17	3.25	3.32	3.38	3.44	3.49	3.54	3.58	3.62	3.66	3.70	3.73
7	1.89	2.31	2.56	2.73	2.86	2.96	3.05	3.13	3.19	3.25	3.31	3.35	3.40	3.44	3.48	3.51	3.55	3.58
8	1.86	2.26	2.49	2.66	2.78	2.88	2.96	3.04	3.10	3.16	3.21	3.26	3.30	3.34	3.37	3.41	3.44	3.47
9	1.83	2.22	2.45	2.60	2.72	2.82	2.90	2.97	3.03	3.09	3.13	3.18	3.22	3.26	3.29	3.32	3.35	3.38
10	1.81	2.19	2.41	2.56	2.68	2.77	2.85	2.92	2.98	3.03	3.08	3.12	3.16	3.20	3.23	3.26	3.29	3.32
11	1.80	2.17	2.38	2.53	2.64	2.73	2.81	2.88	2.93	2.98	3.03	3.07	3.11	3.15	3.18	3.21	3.24	3.26
12	1.78	2.15	2.36	2.50	2.61	2.70	2.78	2.84	2.90	2.95	2.99	3.03	3.07	3.10	3.14	3.17	3.19	3.22
13	1.77	2.13	2.34	2.48	2.59	2.67	2.75	2.81	2.87	2.91	2.96	3.00	3.04	3.07	3.10	3.13	3.16	3.18
14	1.76	2.12	2.32	2.46	2.57	2.65	2.72	2.79	2.84	2.89	2.93	2.97	3.01	3.04	3.07	3.10	3.13	3.15
15	1.75	2.11	2.31	2.44	2.55	2.63	2.70	2.76	2.82	2.87	2.91	2.95	2.98	3.01	3.04	3.07	3.10	3.12
16	1.75	2.10	2.29	2.43	2.53	2.62	2.69	2.75	2.80	2.85	2.89	2.93	2.96	2.99	3.02	3.05	3.08	3.10
17	1.74	2.09	2.28	2.42	2.52	2.60	2.67	2.73	2.78	2.83	2.87	2.91	2.94	2.97	3.00	3.03	3.06	3.08
18	1.73	2.08	2.27	2.41	2.51	2.59	2.66	2.72	2.77	2.81	2.85	2.89	2.92	2.96	2.99	3.01	3.04	3.06
19	1.73	2.07	2.26	2.40	2.50	2.58	2.64	2.70	2.75	2.80	2.84	2.88	2.91	2.94	2.97	3.00	3.02	3.04
20	1.72	2.07	2.26	2.39	2.49	2.57	2.63	2.69	2.74	2.79	2.83	2.86	2.90	2.93	2.96	2.98	3.01	3.03
21	1.72	2.06	2.25	2.38	2.48	2.56	2.62	2.68	2.73	2.78	2.81	2.85	2.88	2.91	2.94	2.97	2.99	3.02
22	1.72	2.05	2.24	2.37	2.47	2.55	2.61	2.67	2.72	2.77	2.80	2.84	2.87	2.90	2.93	2.96	2.98	3.00
23	1.71	2.05	2.24	2.36	2.46	2.54	2.61	2.66	2.71	2.76	2.80	2.83	2.86	2.89	2.92	2.95	2.97	2.99
24	1.71	2.05	2.23	2.36	2.46	2.53	2.60	2.66	2.70	2.75	2.79	2.82	2.85	2.88	2.91	2.94	2.96	2.98
25	1.71	2.04	2.23	2.35	2.45	2.53	2.59	2.65	2.70	2.74	2.78	2.81	2.85	2.88	2.90	2.93	2.95	2.97
26	1.71	2.04	2.22	2.35	2.44	2.52	2.59	2.64	2.69	2.73	2.77	2.81	2.84	2.87	2.89	2.92	2.94	2.97
27	1.70	2.03	2.22	2.34	2.44	2.52	2.58	2.64	2.68	2.73	2.76	2.80	2.83	2.86	2.89	2.91	2.94	2.96
28	1.70	2.03	2.21	2.34	2.43	2.51	2.57	2.63	2.68	2.72	2.76	2.79	2.82	2.85	2.88	2.91	2.93	2.95
29	1.70	2.03	2.21	2.34	2.43	2.51	2.57	2.62	2.67	2.71	2.75	2.79	2.82	2.85	2.87	2.90	2.92	2.94
30	1.70	2.03	2.21	2.33	2.43	2.50	2.57	2.62	2.67	2.71	2.75	2.78	2.81	2.84	2.87	2.89	2.92	2.94
35	1.69	2.01	2.19	2.32	2.41	2.48	2.55	2.60	2.65	2.69	2.72	2.76	2.79	2.82	2.84	2.87	2.89	2.91
40	1.68	2.01	2.18	2.30	2.40	2.47	2.53	2.58	2.63	2.67	2.71	2.74	2.77	2.80	2.82	2.85	2.87	2.89
45	1.68	2.00	2.18	2.30	2.39	2.46	2.52	2.57	2.62	2.66	2.69	2.73	2.76	2.78	2.81	2.83	2.86	2.88
50	1.68	1.99	2.17	2.29	2.38	2.45	2.51	2.56	2.61	2.65	2.68	2.72	2.75	2.77	2.80	2.82	2.84	2.86
60	1.67	1.99	2.16	2.28	2.37	2.44	2.50	2.55	2.59	2.63	2.67	2.70	2.73	2.76	2.78	2.80	2.83	2.85
80	1.66	1.98	2.15	2.26	2.35	2.42	2.48	2.53	2.58	2.61	2.65	2.68	2.71	2.74	2.76	2.78	2.80	2.82
100	1.66	1.97	2.14	2.26	2.34	2.41	2.47	2.52	2.57	2.60	2.64	2.67	2.70	2.72	2.75	2.77	2.79	2.81
200	1.65	1.96	2.13	2.24	2.33	2.40	2.45	2.50	2.54	2.58	2.62	2.65	2.67	2.70	2.72	2.74	2.76	2.78
∞	1.64	1.95	2.11	2.23	2.31	2.38	2.43	2.48	2.52	2.56	2.59	2.62	2.65	2.67	2.70	2.72	2.74	2.76

$$\alpha = 0.10$$

ν	19	20	21	22	23	24	25	26	27	28	29	30	31	32	33	34	35
2	6.74	6.80	6.85	6.91	6.96	7.01	7.05	7.10	7.14	7.18	7.22	7.26	7.30	7.33	7.37	7.40	7.43
3	5.02	5.07	5.11	5.15	5.18	5.22	5.25	5.28	5.31	5.34	5.37	5.40	5.42	5.45	5.47	5.50	5.52
4	4.35	4.38	4.42	4.45	4.48	4.51	4.54	4.56	4.59	4.61	4.64	4.66	4.68	4.70	4.72	4.74	4.76
5	3.99	4.02	4.05	4.08	4.10	4.13	4.16	4.18	4.20	4.22	4.25	4.27	4.29	4.30	4.32	4.34	4.36
6	3.76	3.79	3.82	3.84	3.87	3.89	3.92	3.94	3.96	3.98	4.00	4.02	4.04	4.06	4.07	4.09	4.10
7	3.61	3.63	3.66	3.69	3.71	3.73	3.75	3.78	3.80	3.81	3.83	3.85	3.87	3.88	3.90	3.92	3.93
8	3.50	3.52	3.55	3.57	3.59	3.61	3.64	3.66	3.67	3.69	3.71	3.73	3.74	3.76	3.77	3.79	3.80
9	3.41	3.44	3.46	3.48	3.50	3.53	3.55	3.56	3.58	3.60	3.62	3.63	3.65	3.66	3.68	3.69	3.71
10	3.34	3.37	3.39	3.41	3.43	3.45	3.47	3.49	3.51	3.53	3.54	3.56	3.57	3.59	3.60	3.62	3.63
11	3.29	3.31	3.34	3.36	3.38	3.40	3.42	3.43	3.45	3.47	3.48	3.50	3.51	3.53	3.54	3.56	3.57
12	3.24	3.27	3.29	3.31	3.33	3.35	3.37	3.39	3.40	3.42	3.43	3.45	3.46	3.48	3.49	3.50	3.52
13	3.21	3.23	3.25	3.27	3.29	3.31	3.33	3.35	3.36	3.38	3.39	3.41	3.42	3.44	3.45	3.46	3.47
14	3.17	3.20	3.22	3.24	3.26	3.28	3.29	3.31	3.33	3.34	3.36	3.37	3.39	3.40	3.41	3.43	3.44
15	3.15	3.17	3.19	3.21	3.23	3.25	3.26	3.28	3.30	3.31	3.33	3.34	3.36	3.37	3.38	3.39	3.41
16	3.12	3.15	3.17	3.19	3.20	3.22	3.24	3.26	3.27	3.29	3.30	3.31	3.33	3.34	3.35	3.37	3.38
17	3.10	3.12	3.14	3.16	3.18	3.20	3.22	3.23	3.25	3.26	3.28	3.29	3.30	3.32	3.33	3.34	3.35
18	3.08	3.10	3.13	3.14	3.16	3.18	3.20	3.21	3.23	3.24	3.26	3.27	3.28	3.30	3.31	3.32	3.33
19	3.07	3.09	3.11	3.13	3.14	3.16	3.18	3.19	3.21	3.22	3.24	3.25	3.26	3.28	3.29	3.30	3.31
20	3.05	3.07	3.09	3.11	3.13	3.15	3.16	3.18	3.19	3.21	3.22	3.23	3.25	3.26	3.27	3.28	3.30
21	3.04	3.06	3.08	3.10	3.12	3.13	3.15	3.16	3.18	3.19	3.21	3.22	3.23	3.24	3.26	3.27	3.28
22	3.03	3.05	3.07	3.08	3.10	3.12	3.13	3.15	3.16	3.18	3.19	3.21	3.22	3.23	3.24	3.25	3.27
23	3.02	3.04	3.05	3.07	3.09	3.11	3.12	3.14	3.15	3.17	3.18	3.19	3.21	3.22	3.23	3.24	3.25
24	3.01	3.03	3.04	3.06	3.08	3.10	3.11	3.13	3.14	3.16	3.17	3.18	3.19	3.21	3.22	3.23	3.24
25	3.00	3.02	3.03	3.05	3.07	3.09	3.10	3.12	3.13	3.15	3.16	3.17	3.18	3.20	3.21	3.22	3.23
26	2.99	3.01	3.03	3.04	3.06	3.08	3.09	3.11	3.12	3.14	3.15	3.16	3.17	3.19	3.20	3.21	3.22
27	2.98	3.00	3.02	3.04	3.05	3.07	3.08	3.10	3.11	3.13	3.14	3.15	3.16	3.18	3.19	3.20	3.21
28	2.97	2.99	3.01	3.03	3.04	3.06	3.08	3.09	3.10	3.12	3.13	3.14	3.16	3.17	3.18	3.19	3.20
29	2.96	2.98	3.00	3.02	3.04	3.05	3.07	3.08	3.10	3.11	3.12	3.14	3.15	3.16	3.17	3.18	3.19
30	2.96	2.98	3.00	3.01	3.03	3.05	3.06	3.08	3.09	3.10	3.12	3.13	3.14	3.15	3.16	3.18	3.19
35	2.93	2.95	2.97	2.99	3.00	3.02	3.03	3.05	3.06	3.07	3.09	3.10	3.11	3.12	3.13	3.14	3.15
40	2.91	2.93	2.95	2.97	2.98	3.00	3.01	3.03	3.04	3.05	3.06	3.08	3.09	3.10	3.11	3.12	3.13
45	2.90	2.91	2.93	2.95	2.97	2.98	2.99	3.01	3.02	3.03	3.05	3.06	3.07	3.08	3.09	3.10	3.11
50	2.88	2.90	2.92	2.94	2.95	2.97	2.98	3.00	3.01	3.02	3.03	3.05	3.06	3.07	3.08	3.09	3.10
60	2.87	2.88	2.90	2.92	2.93	2.95	2.96	2.98	2.99	3.00	3.01	3.02	3.04	3.05	3.06	3.07	3.08
80	2.84	2.86	2.88	2.89	2.91	2.92	2.94	2.95	2.96	2.97	2.99	3.00	3.01	3.02	3.03	3.04	3.05
100	2.83	2.85	2.86	2.88	2.89	2.91	2.92	2.93	2.95	2.96	2.97	2.98	2.99	3.00	3.01	3.02	3.03
200	2.80	2.82	2.83	2.85	2.86	2.88	2.89	2.90	2.92	2.93	2.94	2.95	2.96	2.97	2.98	2.99	3.00
∞	2.77	2.79	2.81	2.82	2.84	2.85	2.86	2.87	2.89	2.90	2.91	2.92	2.93	2.94	2.95	2.96	2.97

$$\alpha = 0.05$$

ν	1	2	3	4	5	6	7	8	9	10	11	12	13	14	15	16	17	18
2	4.30	5.57	6.34	6.89	7.31	7.65	7.93	8.17	8.38	8.57	8.74	8.89	9.03	9.16	9.28	9.39	9.49	9.59
3	3.18	3.96	4.43	4.76	5.02	5.23	5.41	5.56	5.69	5.81	5.92	6.01	6.10	6.18	6.26	6.33	6.39	6.45
4	2.78	3.38	3.74	4.00	4.20	4.37	4.50	4.62	4.72	4.82	4.90	4.97	5.04	5.11	5.17	5.22	5.27	5.32
5	2.57	3.09	3.40	3.62	3.79	3.93	4.04	4.14	4.23	4.31	4.38	4.45	4.51	4.56	4.61	4.66	4.70	4.74
6	2.45	2.92	3.19	3.39	3.54	3.66	3.77	3.86	3.94	4.01	4.07	4.13	4.18	4.23	4.28	4.32	4.36	4.39
7	2.36	2.80	3.06	3.24	3.38	3.49	3.59	3.67	3.74	3.80	3.86	3.92	3.96	4.01	4.05	4.09	4.13	4.16
8	2.31	2.72	2.96	3.13	3.26	3.36	3.45	3.53	3.60	3.66	3.71	3.76	3.81	3.85	3.89	3.93	3.96	3.99
9	2.26	2.66	2.89	3.05	3.17	3.27	3.36	3.43	3.49	3.55	3.60	3.65	3.69	3.73	3.77	3.80	3.84	3.87
10	2.23	2.61	2.83	2.98	3.10	3.20	3.28	3.35	3.41	3.47	3.52	3.56	3.60	3.64	3.68	3.71	3.74	3.77
11	2.20	2.57	2.78	2.93	3.05	3.14	3.22	3.29	3.35	3.40	3.45	3.49	3.53	3.57	3.60	3.63	3.66	3.69
12	2.18	2.54	2.75	2.89	3.00	3.09	3.17	3.24	3.29	3.35	3.39	3.43	3.47	3.51	3.54	3.57	3.60	3.63
13	2.16	2.51	2.72	2.86	2.97	3.06	3.13	3.19	3.25	3.30	3.34	3.39	3.42	3.46	3.49	3.52	3.55	3.57
14	2.14	2.49	2.69	2.83	2.94	3.02	3.09	3.16	3.21	3.26	3.30	3.34	3.38	3.41	3.45	3.48	3.50	3.53
15	2.13	2.47	2.67	2.81	2.91	2.99	3.06	3.13	3.18	3.23	3.27	3.31	3.35	3.38	3.41	3.44	3.46	3.49
16	2.12	2.46	2.65	2.78	2.89	2.97	3.04	3.10	3.15	3.20	3.24	3.28	3.31	3.35	3.38	3.40	3.43	3.46
17	2.11	2.44	2.63	2.77	2.87	2.95	3.02	3.08	3.13	3.17	3.21	3.25	3.29	3.32	3.35	3.38	3.40	3.43
18	2.10	2.43	2.62	2.75	2.85	2.93	3.00	3.05	3.11	3.15	3.19	3.23	3.26	3.29	3.32	3.35	3.38	3.40
19	2.09	2.42	2.61	2.73	2.83	2.91	2.98	3.04	3.09	3.13	3.17	3.21	3.24	3.27	3.30	3.33	3.35	3.38
20	2.09	2.41	2.59	2.72	2.82	2.90	2.96	3.02	3.07	3.11	3.15	3.19	3.22	3.25	3.28	3.31	3.33	3.36
21	2.08	2.40	2.58	2.71	2.81	2.88	2.95	3.01	3.05	3.10	3.14	3.17	3.21	3.24	3.26	3.29	3.31	3.34
22	2.07	2.39	2.57	2.70	2.79	2.87	2.94	2.99	3.04	3.08	3.12	3.16	3.19	3.22	3.25	3.27	3.30	3.32
23	2.07	2.39	2.57	2.69	2.78	2.86	2.92	2.98	3.03	3.07	3.11	3.14	3.18	3.21	3.23	3.26	3.28	3.31
24	2.06	2.38	2.56	2.68	2.77	2.85	2.91	2.97	3.02	3.06	3.10	3.13	3.16	3.19	3.22	3.25	3.27	3.29
25	2.06	2.37	2.55	2.67	2.77	2.84	2.90	2.96	3.01	3.05	3.09	3.12	3.15	3.18	3.21	3.23	3.26	3.28
26	2.06	2.37	2.54	2.67	2.76	2.83	2.90	2.95	3.00	3.04	3.08	3.11	3.14	3.17	3.20	3.22	3.25	3.27
27	2.05	2.36	2.54	2.66	2.75	2.83	2.89	2.94	2.99	3.03	3.07	3.10	3.13	3.16	3.19	3.21	3.23	3.26
28	2.05	2.36	2.53	2.65	2.74	2.82	2.88	2.93	2.98	3.02	3.06	3.09	3.12	3.15	3.18	3.20	3.22	3.25
29	2.05	2.35	2.53	2.65	2.74	2.81	2.87	2.93	2.97	3.01	3.05	3.08	3.11	3.14	3.17	3.19	3.22	3.24
30	2.04	2.35	2.52	2.64	2.73	2.80	2.87	2.92	2.96	3.00	3.04	3.07	3.11	3.13	3.16	3.18	3.21	3.23
35	2.03	2.33	2.50	2.62	2.71	2.78	2.84	2.89	2.93	2.97	3.01	3.04	3.07	3.10	3.13	3.15	3.17	3.19
40	2.02	2.32	2.49	2.60	2.69	2.76	2.82	2.87	2.91	2.95	2.99	3.02	3.05	3.08	3.10	3.12	3.14	3.17
45	2.01	2.31	2.48	2.59	2.68	2.75	2.80	2.85	2.90	2.93	2.97	3.00	3.03	3.06	3.08	3.10	3.12	3.14
50	2.01	2.30	2.47	2.58	2.66	2.73	2.79	2.84	2.88	2.92	2.95	2.99	3.01	3.04	3.06	3.09	3.11	3.13
60	2.00	2.29	2.45	2.56	2.65	2.72	2.77	2.82	2.86	2.90	2.93	2.96	2.99	3.02	3.04	3.06	3.08	3.10
80	1.99	2.28	2.44	2.55	2.63	2.69	2.75	2.80	2.84	2.87	2.91	2.94	2.96	2.99	3.01	3.03	3.05	3.07
100	1.98	2.27	2.43	2.53	2.62	2.68	2.74	2.78	2.82	2.86	2.89	2.92	2.95	2.97	3.00	3.02	3.04	3.06
200	1.97	2.25	2.41	2.51	2.59	2.66	2.71	2.75	2.79	2.83	2.86	2.89	2.92	2.94	2.96	2.98	3.00	3.02
∞	1.96	2.24	2.39	2.49	2.57	2.63	2.68	2.73	2.77	2.80	2.83	2.86	2.88	2.91	2.93	2.95	2.97	2.98

$\alpha = 0.05$

ν	I																
	19	20	21	22	23	24	25	26	27	28	29	30	31	32	33	34	35
2	9.68	9.77	9.85	9.92	10.00	10.07	10.13	10.20	10.26	10.32	10.37	10.43	10.48	10.53	10.58	10.63	10.67
3	6.51	6.57	6.62	6.67	6.71	6.76	6.80	6.84	6.88	6.92	6.95	6.99	7.02	7.06	7.09	7.12	7.15
4	5.37	5.41	5.45	5.49	5.52	5.56	5.59	5.63	5.66	5.69	5.71	5.74	5.77	5.79	5.82	5.84	5.87
5	4.78	4.82	4.85	4.89	4.92	4.95	4.98	5.00	5.03	5.06	5.08	5.11	5.13	5.15	5.17	5.19	5.21
6	4.43	4.46	4.49	4.52	4.55	4.58	4.60	4.63	4.65	4.68	4.70	4.72	4.74	4.76	4.78	4.80	4.82
7	4.19	4.22	4.25	4.28	4.31	4.33	4.35	4.38	4.40	4.42	4.44	4.46	4.48	4.50	4.52	4.53	4.55
8	4.02	4.05	4.08	4.10	4.13	4.15	4.18	4.20	4.22	4.24	4.26	4.28	4.29	4.31	4.33	4.34	4.36
9	3.90	3.92	3.95	3.97	4.00	4.02	4.04	4.06	4.08	4.10	4.12	4.14	4.15	4.17	4.18	4.20	4.21
10	3.80	3.82	3.85	3.87	3.89	3.91	3.94	3.95	3.97	3.99	4.01	4.03	4.04	4.06	4.07	4.09	4.10
11	3.72	3.74	3.77	3.79	3.81	3.83	3.85	3.87	3.89	3.91	3.92	3.94	3.95	3.97	3.98	4.00	4.01
12	3.65	3.68	3.70	3.72	3.74	3.76	3.78	3.80	3.82	3.83	3.85	3.87	3.88	3.90	3.91	3.92	3.94
13	3.60	3.62	3.64	3.67	3.69	3.71	3.72	3.74	3.76	3.78	3.79	3.81	3.82	3.84	3.85	3.86	3.88
14	3.55	3.58	3.60	3.62	3.64	3.66	3.68	3.69	3.71	3.73	3.74	3.76	3.77	3.78	3.80	3.81	3.82
15	3.51	3.54	3.56	3.58	3.60	3.62	3.63	3.65	3.67	3.68	3.70	3.71	3.73	3.74	3.75	3.77	3.78
16	3.48	3.50	3.52	3.54	3.56	3.58	3.60	3.61	3.63	3.64	3.66	3.67	3.69	3.70	3.71	3.73	3.74
17	3.45	3.47	3.49	3.51	3.53	3.55	3.56	3.58	3.60	3.61	3.63	3.64	3.65	3.67	3.68	3.69	3.70
18	3.42	3.44	3.46	3.48	3.50	3.52	3.54	3.55	3.57	3.58	3.60	3.61	3.62	3.64	3.65	3.66	3.67
19	3.40	3.42	3.44	3.46	3.48	3.49	3.51	3.53	3.54	3.56	3.57	3.59	3.60	3.61	3.62	3.64	3.65
20	3.38	3.40	3.42	3.44	3.46	3.47	3.49	3.50	3.52	3.53	3.55	3.56	3.57	3.59	3.60	3.61	3.62
21	3.36	3.38	3.40	3.42	3.44	3.45	3.47	3.48	3.50	3.51	3.53	3.54	3.55	3.57	3.58	3.59	3.60
22	3.34	3.36	3.38	3.40	3.42	3.43	3.45	3.47	3.48	3.49	3.51	3.52	3.53	3.55	3.56	3.57	3.58
23	3.33	3.35	3.37	3.38	3.40	3.42	3.43	3.45	3.46	3.48	3.49	3.50	3.52	3.53	3.54	3.55	3.56
24	3.31	3.33	3.35	3.37	3.39	3.40	3.42	3.43	3.45	3.46	3.48	3.49	3.50	3.51	3.52	3.54	3.55
25	3.30	3.32	3.34	3.36	3.37	3.39	3.40	3.42	3.43	3.45	3.46	3.47	3.49	3.50	3.51	3.52	3.53
26	3.29	3.31	3.33	3.34	3.36	3.38	3.39	3.41	3.42	3.43	3.45	3.46	3.47	3.48	3.50	3.51	3.52
27	3.28	3.30	3.31	3.33	3.35	3.36	3.38	3.39	3.41	3.42	3.44	3.45	3.46	3.47	3.48	3.49	3.50
28	3.27	3.29	3.30	3.32	3.34	3.35	3.37	3.38	3.40	3.41	3.42	3.44	3.45	3.46	3.47	3.48	3.49
29	3.26	3.28	3.29	3.31	3.33	3.34	3.36	3.37	3.39	3.40	3.41	3.43	3.44	3.45	3.46	3.47	3.48
30	3.25	3.27	3.29	3.30	3.32	3.33	3.35	3.36	3.38	3.39	3.40	3.42	3.43	3.44	3.45	3.46	3.47
35	3.21	3.23	3.25	3.26	3.28	3.30	3.31	3.32	3.34	3.35	3.36	3.37	3.39	3.40	3.41	3.42	3.43
40	3.18	3.20	3.22	3.24	3.25	3.27	3.28	3.29	3.31	3.32	3.33	3.34	3.36	3.37	3.38	3.39	3.40
45	3.16	3.18	3.20	3.21	3.23	3.24	3.26	3.27	3.28	3.30	3.31	3.32	3.33	3.34	3.35	3.36	3.37
50	3.15	3.16	3.18	3.20	3.21	3.23	3.24	3.25	3.27	3.28	3.29	3.30	3.31	3.32	3.33	3.34	3.35
60	3.12	3.14	3.16	3.17	3.19	3.20	3.21	3.23	3.24	3.25	3.26	3.27	3.28	3.30	3.31	3.31	3.32
80	3.09	3.11	3.12	3.14	3.15	3.17	3.18	3.19	3.21	3.22	3.23	3.24	3.25	3.26	3.27	3.28	3.29
100	3.07	3.09	3.11	3.12	3.13	3.15	3.16	3.17	3.19	3.20	3.21	3.22	3.23	3.24	3.25	3.26	3.27
200	3.04	3.05	3.07	3.08	3.10	3.11	3.12	3.13	3.15	3.16	3.17	3.18	3.19	3.20	3.21	3.22	3.22
∞	3.00	3.02	3.03	3.04	3.06	3.07	3.08	3.09	3.11	3.12	3.13	3.14	3.15	3.16	3.16	3.17	3.18

$$\alpha = 0.01$$

ν	1	2	3	4	5	6	7	8	9	10	11	12	13	14	15	16	17	18
2	9.93	12.73	14.44	15.65	16.59	17.35	17.99	18.53	19.01	19.43	19.81	20.15	20.46	20.75	21.02	21.26	21.49	21.71
3	5.84	7.13	7.91	8.48	8.92	9.28	9.58	9.84	10.06	10.27	10.45	10.61	10.76	10.90	11.03	11.15	11.27	11.37
4	4.60	5.46	5.99	6.36	6.66	6.90	7.10	7.27	7.43	7.57	7.69	7.80	7.91	8.00	8.09	8.17	8.25	8.32
5	4.03	4.70	5.11	5.40	5.63	5.81	5.97	6.11	6.23	6.33	6.43	6.52	6.60	6.67	6.74	6.81	6.87	6.93
6	3.71	4.27	4.61	4.86	5.05	5.20	5.33	5.45	5.55	5.64	5.72	5.80	5.86	5.93	5.99	6.04	6.09	6.14
7	3.50	4.00	4.30	4.51	4.68	4.81	4.93	5.03	5.12	5.20	5.27	5.33	5.39	5.45	5.50	5.55	5.59	5.64
8	3.36	3.81	4.08	4.27	4.42	4.55	4.65	4.74	4.82	4.89	4.96	5.02	5.07	5.12	5.17	5.21	5.25	5.29
9	3.25	3.67	3.92	4.10	4.24	4.35	4.45	4.53	4.61	4.67	4.73	4.79	4.84	4.88	4.92	4.96	5.00	5.04
10	3.17	3.57	3.80	3.97	4.10	4.20	4.29	4.37	4.44	4.50	4.56	4.61	4.66	4.70	4.74	4.78	4.81	4.84
11	3.11	3.48	3.71	3.87	3.99	4.09	4.17	4.25	4.31	4.37	4.42	4.47	4.51	4.55	4.59	4.63	4.66	4.69
12	3.05	3.42	3.63	3.78	3.90	4.00	4.08	4.15	4.21	4.26	4.31	4.36	4.40	4.44	4.48	4.51	4.54	4.57
13	3.01	3.36	3.57	3.71	3.83	3.92	4.00	4.06	4.12	4.18	4.22	4.27	4.31	4.34	4.38	4.41	4.44	4.47
14	2.98	3.32	3.52	3.66	3.77	3.85	3.93	3.99	4.05	4.10	4.15	4.19	4.23	4.26	4.30	4.33	4.36	4.39
15	2.95	3.28	3.47	3.61	3.71	3.80	3.87	3.93	3.99	4.04	4.08	4.12	4.16	4.20	4.23	4.26	4.29	4.31
16	2.92	3.25	3.43	3.57	3.67	3.75	3.82	3.88	3.94	3.99	4.03	4.07	4.11	4.14	4.17	4.20	4.23	4.25
17	2.90	3.22	3.40	3.53	3.63	3.71	3.78	3.84	3.89	3.94	3.98	4.02	4.06	4.09	4.12	4.15	4.17	4.20
18	2.88	3.19	3.37	3.50	3.60	3.68	3.74	3.80	3.85	3.90	3.94	3.98	4.01	4.04	4.07	4.10	4.13	4.15
19	2.86	3.17	3.35	3.47	3.57	3.65	3.71	3.77	3.82	3.86	3.90	3.94	3.97	4.01	4.03	4.06	4.09	4.11
20	2.85	3.15	3.32	3.45	3.54	3.62	3.68	3.74	3.79	3.83	3.87	3.91	3.94	3.97	4.00	4.03	4.05	4.07
21	2.83	3.13	3.30	3.42	3.52	3.59	3.66	3.71	3.76	3.80	3.84	3.88	3.91	3.94	3.97	3.99	4.02	4.04
22	2.82	3.11	3.28	3.40	3.50	3.57	3.63	3.69	3.73	3.78	3.81	3.85	3.88	3.91	3.94	3.96	3.99	4.01
23	2.81	3.10	3.27	3.39	3.48	3.55	3.61	3.66	3.71	3.75	3.79	3.83	3.86	3.89	3.91	3.94	3.96	3.98
24	2.80	3.09	3.25	3.37	3.46	3.53	3.59	3.64	3.69	3.73	3.77	3.80	3.83	3.86	3.89	3.91	3.94	3.96
25	2.79	3.07	3.24	3.35	3.44	3.51	3.57	3.63	3.67	3.71	3.75	3.78	3.81	3.84	3.87	3.89	3.92	3.94
26	2.78	3.06	3.23	3.34	3.43	3.50	3.56	3.61	3.65	3.70	3.73	3.76	3.79	3.82	3.85	3.87	3.90	3.92
27	2.77	3.05	3.21	3.33	3.41	3.48	3.54	3.59	3.64	3.68	3.71	3.75	3.78	3.81	3.83	3.86	3.88	3.90
28	2.76	3.04	3.20	3.32	3.40	3.47	3.53	3.58	3.62	3.66	3.70	3.73	3.76	3.79	3.81	3.84	3.86	3.88
29	2.76	3.04	3.19	3.30	3.39	3.46	3.52	3.57	3.61	3.65	3.68	3.72	3.75	3.77	3.80	3.82	3.84	3.87
30	2.75	3.03	3.18	3.29	3.38	3.45	3.50	3.55	3.60	3.64	3.67	3.70	3.73	3.76	3.78	3.81	3.83	3.85
35	2.72	2.99	3.15	3.25	3.33	3.40	3.46	3.50	3.55	3.58	3.62	3.65	3.68	3.70	3.73	3.75	3.77	3.79
40	2.70	2.97	3.12	3.22	3.30	3.37	3.42	3.47	3.51	3.54	3.58	3.61	3.63	3.66	3.68	3.71	3.73	3.74
45	2.69	2.95	3.10	3.20	3.28	3.34	3.39	3.44	3.48	3.51	3.55	3.58	3.60	3.63	3.65	3.67	3.69	3.71
50	2.68	2.94	3.08	3.18	3.26	3.32	3.37	3.42	3.46	3.49	3.52	3.55	3.58	3.60	3.62	3.65	3.66	3.68
60	2.66	2.91	3.06	3.15	3.23	3.29	3.34	3.38	3.42	3.46	3.49	3.51	3.54	3.56	3.59	3.61	3.63	3.64
80	2.64	2.89	3.02	3.12	3.19	3.25	3.30	3.34	3.38	3.41	3.44	3.47	3.49	3.52	3.54	3.56	3.58	3.59
100	2.63	2.87	3.01	3.10	3.17	3.23	3.28	3.32	3.36	3.39	3.42	3.44	3.47	3.49	3.51	3.53	3.55	3.56
200	2.60	2.84	2.97	3.06	3.13	3.19	3.23	3.27	3.31	3.34	3.37	3.39	3.41	3.44	3.46	3.47	3.49	3.51
∞	2.58	2.81	2.93	3.02	3.09	3.14	3.19	3.23	3.26	3.29	3.32	3.34	3.36	3.38	3.40	3.42	3.44	3.45

α = upper tail probability, ν = degrees of freedom, I = number of compared effects.

$$\alpha = 0.01$$

ν	19	20	21	22	23	24	25	26	27	28	29	30	31	32	33	34	35
2	21.91	22.11	22.29	22.46	22.63	22.78	22.93	23.08	23.21	23.35	23.47	23.59	23.71	23.83	23.94	24.04	24.14
3	11.47	11.56	11.65	11.74	11.82	11.90	11.97	12.04	12.11	12.17	12.23	12.29	12.35	12.41	12.46	12.51	12.56
4	8.39	8.45	8.51	8.57	8.63	8.68	8.73	8.78	8.83	8.87	8.92	8.96	9.00	9.04	9.07	9.11	9.15
5	6.98	7.03	7.08	7.13	7.17	7.21	7.25	7.29	7.33	7.36	7.40	7.43	7.46	7.49	7.52	7.55	7.58
6	6.18	6.23	6.27	6.31	6.34	6.38	6.41	6.45	6.48	6.51	6.54	6.57	6.59	6.62	6.64	6.67	6.69
7	5.68	5.72	5.75	5.79	5.82	5.85	5.88	5.91	5.94	5.96	5.99	6.01	6.04	6.06	6.08	6.10	6.13
8	5.33	5.36	5.39	5.43	5.45	5.48	5.51	5.54	5.56	5.59	5.61	5.63	5.65	5.67	5.69	5.71	5.73
9	5.07	5.10	5.13	5.16	5.19	5.21	5.24	5.26	5.29	5.31	5.33	5.35	5.37	5.39	5.41	5.43	5.45
10	4.88	4.91	4.93	4.96	4.99	5.01	5.03	5.06	5.08	5.10	5.12	5.14	5.16	5.18	5.19	5.21	5.23
11	4.72	4.75	4.78	4.80	4.83	4.85	4.87	4.89	4.91	4.93	4.95	4.97	4.99	5.01	5.02	5.04	5.05
12	4.60	4.62	4.65	4.67	4.70	4.72	4.74	4.76	4.78	4.80	4.82	4.83	4.85	4.87	4.88	4.90	4.91
13	4.50	4.52	4.55	4.57	4.59	4.61	4.63	4.65	4.67	4.69	4.71	4.72	4.74	4.75	4.77	4.78	4.80
14	4.41	4.44	4.46	4.48	4.50	4.52	4.54	4.56	4.58	4.60	4.61	4.63	4.65	4.66	4.67	4.69	4.70
15	4.34	4.36	4.39	4.41	4.43	4.45	4.47	4.48	4.50	4.52	4.53	4.55	4.56	4.58	4.59	4.61	4.62
16	4.28	4.30	4.32	4.34	4.36	4.38	4.40	4.42	4.43	4.45	4.47	4.48	4.50	4.51	4.52	4.54	4.55
17	4.22	4.25	4.27	4.29	4.31	4.32	4.34	4.36	4.38	4.39	4.41	4.42	4.44	4.45	4.46	4.48	4.49
18	4.18	4.20	4.22	4.24	4.26	4.28	4.29	4.31	4.33	4.34	4.36	4.37	4.38	4.40	4.41	4.42	4.44
19	4.13	4.15	4.18	4.19	4.21	4.23	4.25	4.26	4.28	4.30	4.31	4.32	4.34	4.35	4.36	4.38	4.39
20	4.10	4.12	4.14	4.16	4.17	4.19	4.21	4.22	4.24	4.25	4.27	4.28	4.30	4.31	4.32	4.33	4.35
21	4.06	4.08	4.10	4.12	4.14	4.16	4.17	4.19	4.20	4.22	4.23	4.25	4.26	4.27	4.28	4.30	4.31
22	4.03	4.05	4.07	4.09	4.11	4.13	4.14	4.16	4.17	4.19	4.20	4.21	4.23	4.24	4.25	4.26	4.27
23	4.01	4.03	4.04	4.06	4.08	4.10	4.11	4.13	4.14	4.16	4.17	4.18	4.20	4.21	4.22	4.23	4.24
24	3.98	4.00	4.02	4.04	4.05	4.07	4.09	4.10	4.12	4.13	4.14	4.16	4.17	4.18	4.19	4.20	4.21
25	3.96	3.98	4.00	4.01	4.03	4.05	4.06	4.08	4.09	4.11	4.12	4.13	4.14	4.16	4.17	4.18	4.19
26	3.94	3.96	3.98	3.99	4.01	4.03	4.04	4.06	4.07	4.08	4.10	4.11	4.12	4.13	4.14	4.15	4.17
27	3.92	3.94	3.96	3.97	3.99	4.01	4.02	4.03	4.05	4.06	4.07	4.09	4.10	4.11	4.12	4.13	4.14
28	3.90	3.92	3.94	3.96	3.97	3.99	4.00	4.02	4.03	4.04	4.06	4.07	4.08	4.09	4.10	4.11	4.12
29	3.89	3.90	3.92	3.94	3.95	3.97	3.98	4.00	4.01	4.03	4.04	4.05	4.06	4.07	4.08	4.09	4.11
30	3.87	3.89	3.91	3.92	3.94	3.95	3.97	3.98	4.00	4.01	4.02	4.03	4.04	4.06	4.07	4.08	4.09
35	3.81	3.83	3.84	3.86	3.87	3.89	3.90	3.92	3.93	3.94	3.95	3.97	3.98	3.99	4.00	4.01	4.02
40	3.76	3.78	3.80	3.81	3.83	3.84	3.85	3.87	3.88	3.89	3.90	3.91	3.93	3.94	3.95	3.96	3.97
45	3.73	3.74	3.76	3.78	3.79	3.80	3.82	3.83	3.84	3.85	3.87	3.88	3.89	3.90	3.91	3.92	3.93
50	3.70	3.72	3.73	3.75	3.76	3.78	3.79	3.80	3.81	3.82	3.84	3.85	3.86	3.87	3.88	3.89	3.89
60	3.66	3.68	3.69	3.71	3.72	3.73	3.75	3.76	3.77	3.78	3.79	3.80	3.81	3.82	3.83	3.84	3.85
80	3.61	3.63	3.64	3.65	3.67	3.68	3.69	3.70	3.71	3.73	3.74	3.75	3.76	3.76	3.77	3.78	3.79
100	3.58	3.60	3.61	3.62	3.64	3.65	3.66	3.67	3.68	3.69	3.70	3.71	3.72	3.73	3.74	3.75	3.76
200	3.52	3.54	3.55	3.56	3.58	3.59	3.60	3.61	3.62	3.63	3.64	3.65	3.66	3.67	3.68	3.68	3.69
∞	3.47	3.48	3.49	3.50	3.52	3.53	3.54	3.55	3.56	3.57	3.58	3.59	3.60	3.60	3.61	3.62	3.63

Coefficients of Orthogonal Contrast Vectors

Level	$k=3$ P_1	P_2	$k=4$ P_1	P_2	P_3	$k=5$ P_1	P_2	P_3	P_4
1	-1	1	-3	1	-1	-2	2	-1	1
2	0	-2	-1	-1	3	-1	-1	2	-4
3	1	1	1	-1	-3	0	-2	0	6
4			3	1	1	1	-1	-2	-4
5						2	2	1	1
C	2	6	20	4	20	10	14	10	70
λ	1	3	2	1	$\frac{10}{3}$	1	1	$\frac{5}{6}$	$\frac{35}{12}$

Level	$k=6$ P_1	P_2	P_3	P_4	P_5	$k=7$ P_1	P_2	P_3	P_4	P_5	P_6
1	-5	5	-5	1	-1	-3	5	-1	3	-1	1
2	-3	-1	7	-3	5	-2	0	1	-7	4	-6
3	-1	-4	4	2	-10	-1	-3	1	1	-5	15
4	1	-4	-4	2	10	0	-4	0	6	0	-20
5	3	-1	-7	-3	-5	1	-3	-1	1	5	15
6	5	5	5	1	1	2	0	-1	-7	-4	-6
7						3	5	1	3	1	1
C	70	84	180	28	252	28	84	6	154	84	924
λ	2	$\frac{3}{2}$	$\frac{5}{3}$	$\frac{7}{12}$	$\frac{21}{10}$	1	1	$\frac{1}{6}$	$\frac{7}{12}$	$\frac{7}{20}$	$\frac{77}{60}$

k = number of levels, λ = constants, P_i = contrast vector for ith degree orthogonal polynomial, C = squared length for P_i.

Critical Values for Lenth's Method

IER Critical Values (IER$_\alpha$)

I	α 0.4	0.3	0.2	0.1	0.09	0.08	0.07	0.06	0.05	0.04	0.03	0.02	0.01	0.009	0.008	0.007	0.006	0.005	0.004	0.003	0.002	0.001
7	0.74	0.93	1.20	1.71	1.79	1.89	2.00	2.13	2.30	2.53	3.09	3.76	5.07	5.30	5.58	5.91	6.32	6.84	7.52	8.50	10.08	13.39
8	0.78	0.94	1.20	1.67	1.75	1.83	1.93	2.05	2.20	2.39	2.85	3.53	4.70	4.89	5.11	5.37	5.69	6.07	6.56	7.28	8.35	10.60
11	0.78	0.97	1.24	1.71	1.78	1.87	1.96	2.08	2.21	2.38	2.72	3.22	4.08	4.22	4.39	4.58	4.82	5.12	5.50	6.04	6.86	8.52
15	0.79	0.99	1.25	1.70	1.77	1.84	1.93	2.03	2.16	2.31	2.52	2.95	3.63	3.74	3.86	4.00	4.16	4.37	4.63	4.99	5.53	6.54
17	0.80	1.00	1.26	1.70	1.76	1.84	1.92	2.02	2.14	2.28	2.47	2.87	3.48	3.58	3.69	3.82	3.97	4.16	4.39	4.69	5.16	6.03
19	0.80	1.00	1.26	1.70	1.76	1.83	1.91	2.01	2.12	2.26	2.44	2.81	3.38	3.47	3.57	3.68	3.82	3.98	4.19	4.46	4.86	5.62
23	0.81	1.01	1.27	1.69	1.75	1.82	1.90	1.99	2.10	2.23	2.40	2.71	3.23	3.30	3.39	3.49	3.60	3.74	3.91	4.14	4.47	5.09
26	0.81	1.01	1.27	1.68	1.75	1.81	1.89	1.98	2.08	2.21	2.37	2.66	3.15	3.22	3.30	3.39	3.49	3.62	3.77	3.97	4.27	4.82
27	0.81	1.01	1.27	1.68	1.74	1.81	1.89	1.98	2.08	2.20	2.36	2.64	3.12	3.19	3.26	3.35	3.45	3.58	3.73	3.93	4.22	4.75
31	0.82	1.02	1.27	1.68	1.74	1.80	1.88	1.96	2.06	2.19	2.34	2.59	3.04	3.11	3.18	3.26	3.36	3.47	3.61	3.79	4.05	4.52
35	0.82	1.02	1.27	1.68	1.74	1.80	1.87	1.96	2.05	2.17	2.32	2.56	2.99	3.05	3.12	3.19	3.28	3.39	3.52	3.69	3.93	4.36
53	0.83	1.03	1.28	1.67	1.73	1.79	1.86	1.93	2.02	2.13	2.27	2.47	2.85	2.90	2.96	3.02	3.09	3.18	3.29	3.42	3.61	3.95
63	0.83	1.03	1.28	1.66	1.72	1.78	1.85	1.92	2.01	2.12	2.26	2.44	2.80	2.85	2.90	2.96	3.03	3.11	3.21	3.33	3.51	3.82
80	0.83	1.03	1.28	1.66	1.72	1.78	1.84	1.92	2.00	2.11	2.24	2.42	2.75	2.80	2.85	2.90	2.97	3.04	3.14	3.26	3.42	3.71
127	0.84	1.03	1.28	1.66	1.71	1.77	1.83	1.90	1.99	2.09	2.21	2.38	2.68	2.72	2.77	2.82	2.88	2.95	3.03	3.14	3.28	3.53

EER Critical Values (EER$_\alpha$)

I	α 0.4	0.3	0.2	0.1	0.09	0.08	0.07	0.06	0.05	0.04	0.03	0.02	0.01	0.009	0.008	0.007	0.006	0.005	0.004	0.003	0.002	0.001
7	1.80	2.06	2.42	3.69	3.85	4.04	4.26	4.52	4.87	5.33	6.03	7.17	9.72	10.14	10.66	11.32	12.08	13.02	14.31	16.01	18.78	24.32
8	1.84	2.06	2.39	3.65	3.83	4.03	4.26	4.53	4.87	5.30	5.90	6.80	8.63	8.95	9.32	9.76	10.29	10.95	11.82	12.99	14.90	18.88
11	2.02	2.25	2.74	3.56	3.69	3.83	3.99	4.19	4.44	4.76	5.22	5.95	7.41	7.65	7.95	8.28	8.71	9.23	9.88	10.77	12.23	14.97
15	2.14	2.36	2.84	3.51	3.61	3.73	3.87	4.04	4.24	4.50	4.86	5.40	6.45	6.62	6.82	7.05	7.33	7.66	8.08	8.65	9.51	11.23
17	2.20	2.41	2.87	3.49	3.58	3.69	3.82	3.98	4.16	4.40	4.72	5.21	6.12	6.28	6.45	6.64	6.87	7.15	7.53	8.02	8.75	10.25
19	2.24	2.45	2.90	3.48	3.57	3.68	3.80	3.94	4.12	4.34	4.63	5.07	5.88	6.02	6.16	6.34	6.55	6.82	7.14	7.57	8.20	9.43
23	2.31	2.55	2.93	3.46	3.54	3.64	3.74	3.87	4.02	4.20	4.46	4.82	5.51	5.61	5.73	5.87	6.03	6.22	6.48	6.83	7.32	8.22
26	2.35	2.62	2.96	3.47	3.55	3.64	3.73	3.85	3.98	4.16	4.38	4.70	5.30	5.40	5.50	5.63	5.78	5.94	6.18	6.46	6.89	7.62
27	2.37	2.64	2.96	3.45	3.53	3.62	3.72	3.83	3.96	4.13	4.35	4.68	5.27	5.36	5.46	5.58	5.73	5.90	6.13	6.44	6.87	7.65
31	2.42	2.70	2.98	3.45	3.52	3.60	3.69	3.80	3.92	4.08	4.28	4.57	5.10	5.18	5.27	5.38	5.50	5.66	5.84	6.09	6.42	7.08
35	2.46	2.73	3.00	3.45	3.51	3.59	3.68	3.77	3.89	4.03	4.22	4.50	4.97	5.05	5.13	5.23	5.34	5.48	5.64	5.88	6.19	6.80
53	2.66	2.84	3.08	3.46	3.52	3.58	3.65	3.73	3.82	3.94	4.08	4.29	4.64	4.70	4.76	4.83	4.91	5.00	5.12	5.27	5.50	5.88
63	2.72	2.89	3.12	3.47	3.52	3.58	3.64	3.72	3.80	3.91	4.04	4.23	4.56	4.61	4.67	4.73	4.80	4.89	5.00	5.14	5.33	5.67
80	2.79	2.96	3.16	3.49	3.54	3.59	3.65	3.72	3.79	3.89	4.01	4.18	4.46	4.51	4.56	4.61	4.67	4.75	4.84	4.96	5.13	5.43
127	2.93	3.07	3.26	3.55	3.59	3.63	3.68	3.74	3.81	3.89	3.99	4.13	4.37	4.41	4.45	4.49	4.54	4.61	4.68	4.77	4.90	5.13

α = upper tail probability, I = number of compared effects.

Author Index

Subject Index

624

WILEY SERIES IN PROBABILITY AND STATISTICS
ESTABLISHED BY WALTER A. SHEWHART AND SAMUEL S. WILKS

Editors
Vic Barnett, Noel A. C. Cressie, Nicholas I. Fisher,
Iain M. Johnstone, J. B. Kadane, David G. Kendall, David W. Scott,
Bernard W. Silverman, Adrian F. M. Smith, Jozef L. Teugels;
Ralph A. Bradley, Emeritus, J. Stuart Hunter, Emeritus

Probability and Statistics Section

*Now available in a lower priced paperback edition in the Wiley Classics Library.

*Now available in a lower priced paperback edition in the Wiley Classics Library.

*Now available in a lower priced paperback edition in the Wiley Classics Library.

*Now available in a lower priced paperback edition in the Wiley Classics Library.

*Now available in a lower priced paperback edition in the Wiley Classics Library.

Texts and References Section

*Now available in a lower priced paperback edition in the Wiley Classics Library.

Texts and References (Continued)

FREEMAN and SMITH · Aspects of Uncertainty: A Tribute to D. V. Lindley

GROSS and HARRIS · Fundamentals of Queueing Theory, *Third Edition*

HALD · A History of Probability and Statistics and their Applications Before 1750

HALD · A History of Mathematical Statistics from 1750 to 1930

HELLER · MACSYMA for Statisticians

HOEL · Introduction to Mathematical Statistics, *Fifth Edition*

HOLLANDER and WOLFE · Nonparametric Statistical Methods, *Second Edition*

HOSMER and LEMESHOW · Applied Survival Analysis: Regression Modeling of Time to Event Data

JOHNSON and BALAKRISHNAN · Advances in the Theory and Practice of Statistics: A Volume in Honor of Samuel Kotz

JOHNSON and KOTZ (editors) · Leading Personalities in Statistical Sciences: From the Seventeenth Century to the Present

JUDGE, GRIFFITHS, HILL, LÜTKEPOHL, and LEE · The Theory and Practice of Econometrics, *Second Edition*

KHURI · Advanced Calculus with Applications in Statistics

KOTZ and JOHNSON (editors) · Encyclopedia of Statistical Sciences: Volumes 1 to 9 wtih Index

KOTZ and JOHNSON (editors) · Encyclopedia of Statistical Sciences: Supplement Volume

KOTZ, REED, and BANKS (editors) · Encyclopedia of Statistical Sciences: Update Volume 1

KOTZ, REED, and BANKS (editors) · Encyclopedia of Statistical Sciences: Update Volume 2

LAMPERTI · Probability: A Survey of the Mathematical Theory, *Second Edition*

LARSON · Introduction to Probability Theory and Statistical Inference, *Third Edition*

LE · Applied Categorical Data Analysis

LE · Applied Survival Analysis

MALLOWS · Design, Data, and Analysis by Some Friends of Cuthbert Daniel

MARDIA · The Art of Statistical Science: A Tribute to G. S. Watson

MASON, GUNST, and HESS · Statistical Design and Analysis of Experiments with Applications to Engineering and Science

MURRAY · X-STAT 2.0 Statistical Experimentation, Design Data Analysis, and Nonlinear Optimization

PURI, VILAPLANA, and WERTZ · New Perspectives in Theoretical and Applied Statistics

RENCHER · Linear Models in Statistics

RENCHER · Methods of Multivariate Analysis

RENCHER · Multivariate Statistical Inference with Applications

ROSS · Introduction to Probability and Statistics for Engineers and Scientists

ROHATGI · An Introduction to Probability Theory and Mathematical Statistics

RYAN · Modern Regression Methods

SCHOTT · Matrix Analysis for Statistics

SEARLE · Matrix Algebra Useful for Statistics

STYAN · The Collected Papers of T. W. Anderson: 1943–1985

TIERNEY · LISP-STAT: An Object-Oriented Environment for Statistical Computing and Dynamic Graphics

WONNACOTT and WONNACOTT · Econometrics, *Second Edition*

WU and HAMADA · Experiments: Planning, Analysis, and Parameter Design Optimization

*Now available in a lower priced paperback edition in the Wiley Classics Library.

WILEY SERIES IN PROBABILITY AND STATISTICS
ESTABLISHED BY WALTER A. SHEWHART AND SAMUEL S. WILKS

Editors
Robert M. Groves, Graham Kalton, J. N. K. Rao, Norbert Schwarz,
Christopher Skinner

Survey Methodology Section

BIEMER, GROVES, LYBERG, MATHIOWETZ, and SUDMAN · Measurement
Errors in Surveys
COCHRAN · Sampling Techniques, *Third Edition*
COUPER, BAKER, BETHLEHEM, CLARK, MARTIN, NICHOLLS, and O'REILLY
(editors) · Computer Assisted Survey Information Collection
COX, BINDER, CHINNAPPA, CHRISTIANSON, COLLEDGE, and KOTT (editors) ·
Business Survey Methods
*DEMING · Sample Design in Business Research
DILLMAN · Mail and Telephone Surveys: The Total Design Method
GROVES and COUPER · Nonresponse in Household Interview Surveys
GROVES · Survey Errors and Survey Costs
GROVES, BIEMER, LYBERG, MASSEY, NICHOLLS, and WAKSBERG ·
Telephone Survey Methodology
*HANSEN, HURWITZ, and MADOW · Sample Survey Methods and Theory,
Volume 1: Methods and Applications
*HANSEN, HURWITZ, and MADOW · Sample Survey Methods and Theory,
Volume II: Theory
KISH · Statistical Design for Research
*KISH · Survey Sampling
KORN and GRAUBARD · Analysis of Health Surveys
LESSLER and KALSBEEK · Nonsampling Error in Surveys
LEVY and LEMESHOW · Sampling of Populations: Methods and Applications,
Third Edition
LYBERG, BIEMER, COLLINS, de LEEUW, DIPPO, SCHWARZ, TREWIN (editors) ·
Survey Measurement and Process Quality
SIRKEN, HERRMANN, SCHECHTER, SCHWARZ, TANUR, and TOURANGEAU
(editors) · Cognition and Survey Research

*Now available in a lower priced paperback edition in the Wiley Classics Library.